3 In the Brazilian Amazon, World Bank supported projects are flooding thousands [of] square kilometers of rainforest and [killing] indigenous people.

4 In tropical forests, agricul[tural clear]ing can cause irreversibl[e damage. Afte]r just a few years of farming, [the nutri]ents are depleted and the la[nd can no lon]ger support crops. Grasses c[an be grow]n to feed cattle, but onl[y for a short ti]me.

ENVIRONMENTAL SCIENCE
Action for a Sustainable Future

Also of interest from the Benjamin/Cummings Series in the Life Sciences

General Biology

N. A. Campbell
Biology, second edition (1990)

Evolution, Ecology, and Behavior

F. J. Ayala and J. W. Valentine
Evolving: The Theory and Process of Organic Evolution (1979)

R. J. Lederer
Ecology and Field Biology (1984)

M. Lerman
Marine Biology: Environment, Diversity, and Ecology (1986)

D. McFarland
Animal Behavior (1985)

E. Minkoff
Evolutionary Biology (1983)

R. Trivers
Social Evolution (1985)

Plant Biology

M. G. Barbour, J. H. Burk, and W. D. Pitts
Terrestrial Plant Ecology, second edition (1987)

J. Mauseth
Plant Anatomy (1988)

E. Zeiger and L. Taiz
Plant Physiology (available 1991)

Animal Biology

R. Chase
Bassett Atlas of Human Anatomy (1989)

H. E. Evans
Insect Biology: A Textbook of Entomology (1984)

P. E. Lutz
Invertebrate Zoology (1986)

E. N. Marieb
Human Anatomy and Physiology (1989)

L. G. Mitchell, J. A. Mutchmor, W. D. Dolphin
General Zoology (1988)

A. P. Spence
Basic Human Anatomy, second edition (1986)

A. P. Spence and E. B. Mason
Human Anatomy and Physiology, third edition (1987)

Genetics

F. J. Ayala
Population and Evolutionary Genetics: A Primer (1982)

F. J. Ayala and J. A. Kiger, Jr.
Modern Genetics, second edition (1984)

J. B. Jenkins
Human Genetics (1983)

R. Schleif
Genetics and Molecular Biology (1986)

J. Watson, et al.
Biology of the Gene, fourth edition (1987)

Microbiology

I. E. Alcamo
Fundamentals of Microbiology, third edition (available 1991)

R. M. Atlas and R. Bartha
Microbial Ecology: Fundamentals and Applications, second edition (1987)

M. Dworkin
Developmental Biology of the Bacteria (1985)

G. J. Tortora, B. R. Funke, and C. L. Case
Microbiology: An Introduction, third edition (1989)

P. J. VanDemark and B. L. Batzing
The Microbes (1987)

ENVIRONMENTAL SCIENCE
Action for a Sustainable Future

THIRD EDITION

DANIEL D. CHIRAS

THE BENJAMIN/CUMMINGS PUBLISHING COMPANY, INC.

Redwood City, California • Fort Collins, Colorado • Menlo Park, California
Reading, Massachusetts • New York • Don Mills, Ontario • Wokingham, U.K.
Amsterdam • Bonn • Sydney • Singapore • Tokyo • Madrid • San Juan

For Kathleen in whom all hope is born

Sponsoring Editor: Melinda Adams

Associate Editors: Lisa Donohoe, Laura Bonazzoli

Developmental Editor: Valerie Kuletz

Production Editor: Eleanor Renner Brown

Designer: Gary Head

Cover Designer: Victoria Philp

Photo Research: Mark Childs

Cover photo of Cloud Forest in Costa Rica is by Randy Hayes of the Rainforest Action Network

Photo and text credits appear after the Glossary

Library of Congress Cataloging in Publication Data

Chiras, Daniel D.
    Environmental science: action for a sustainable future /
    Daniel D. Chiras.—3rd ed.
        p.   cm.

    ISBN 0-8053-1031-2

    1. Environmental policy. 2. Environmental protection.
I. Title.
HC79.E5C485   1990
363.7—dc20   90-507
                                                                CIP

2 3 4 5 6 7 8 9 10   MU   95 94 93 92 91

The Benjamin/Cummings Publishing Company, Inc.
390 Bridge Parkway
Redwood City, California 94065

Preface

The first and second editions of *Environmental Science: A Framework for Decision Making* reached a large audience. Feedback from users and reviewers helped me prepare this third edition. My goal for the new edition was to add new material, update statistics, and polish the writing—in short, to make this the most interesting and readable book on the market.

As in the first two editions, I wanted the book to be user friendly, not laden with irrelevant statistics. I wanted to continue to present important facts and concepts in a clear and exciting way and to minimize bias by presenting both sides of issues and by including Point/Counterpoints on important, controversial issues. My objective was to write a book that helps students learn the facts behind environmental issues and solutions so that they can make up their own minds about what should be done.

This book provides a broad overview of the many environmental problems facing humanity and describes a wide range of solutions. The chapters contain important information on ecology, anthropology, evolution, earth science, biology, ethics, economics, and other areas to enable students to understand more fully the sometimes overwhelming assortment of environmental problems facing the world. The melding of these disciplines results in new ways of looking at our environmental problems and opens up many avenues for solving them.

Themes

The central theme of this book is that time for action is running short; overpopulation, resource depletion, pollution, and indifference are rapidly catching up with us.

The second major theme is that the long–term well being of this planet and its inhabitants requires the development of a sustainable society—one that conserves natural resources, recycles, relies on renewable resources whenever possible, reduces pollution, and controls population growth. Such a society, based on the lessons from ecology, may seem foreign or even unattainable, but it remains our only realistic hope for prosperity in the long term. Careful planning and implementation will usher in a sustainable future.

Finally, this book stresses that we are all part of the problem and must therefore be part of the solution. Air pollution is not just a problem of inadequate laws or corporate neglect, but also the result of our own wasteful practices. Individual action is essential. Solving these problems need not mean reverting to old-fashioned ways or even making tremendous sacrifices. It does mean using energy and other resources much more wisely, conserving all resources, recycling all that we can, using renewable resources, and limiting our family size. Numerous suggestions are given in each chapter and in the Environmental Action Guide that accompanies this text for such personal solutions.

Organization

This book is divided into five parts and organized around three central issues—population, resources, and pollution. Part 1 provides a base of knowledge in ecology, earth science, chemistry, biology, evolution, and human social development. Part 2 covers population growth, the impact of population, and population control. Part 3 deals with a variety of resource issues and outlines a plan for developing a sustainable society. Part 4 discusses pollution and the legal, technical, and personal solutions for it. Part 5, the capstone of the book, places the population, resource, and pollution crisis against a social backdrop by looking at ethics, economics, and politics. It suggests ways to make the transition to a sustainable society.

Special Features

The following special features from the first and second editions have been retained to keep this text informative and useful and to increase student interest and involvement:

Models

One of the key features of this book is the use of conceptual models, which in this edition have been integrated in appropriate chapters. These models are easy to understand and are designed to encourage holistic thinking, emphasizing the systems approach to environmental problems. Below is a brief description of each model:

- *Population, Resource, and Pollution Model:* presents a fuller view of the human niche, and helps students see the way we affect our environment and vice versa.
- *Multiple Cause and Effect Model:* helps students analyze the causes of many of our current environmental dilemmas by exhibiting the web of cause and effect.
- *Impact Analysis Model:* shows the various impacts that we have on the environment and the ways in which we are affected by our own actions.
- *Risk Analysis Model:* examines the risks and benefits associated with today's new and existing hazards.

Chapter Supplements

Chapter supplements, found at the end of some chapters, provide more detailed coverage of important topics and provide an added degree of flexibility. Such topics of current interest include acid rain, indoor air pollution, stratospheric ozone depletion, radiation pollution, nuclear war, and environmental law.

Point/Counterpoints and Viewpoints

As might be expected, complex environmental issues often result in hotly contested debates:

- Is outer space the answer to our population and resource problems?
- Are we responsible to future generations?
- Is population growth good or bad for us?
- Does environmental protection cost us jobs?
- Are we losing the war against cancer?
- Are we playing God with nature?

These and many other important and timely issues are debated in Point/Counterpoint or discussed in Viewpoint by such luminaries as Norman Myers, Ben Bova, Garrett Hardin, Julian Simon, Amory and Hunter Lovins, Frederic Krupp, and others. These editorials can stimulate individual thinking as well as classroom discussion on many complex problems.

Color Galleries

Four color galleries are included in this book to emphasize some of the key concepts and issues. They are: the earth, the biomes, endangered species, and resource misuse.

Case Studies

To give students insight to the timely issues of our day, this third edition includes new case studies written by leaders in Environmental Science. Examples of topics discussed include tropical deforestation, genetic engineering, Yellowstone's controversy over fire control, the protection of Antarctica, and solid waste control.

Chapter Summaries

Each chapter is followed by a succinct summary of the important concepts and terms, designed to reinforce the key points. These summaries may also be valuable study tools.

Coverage of the Basic Sciences

I've added more information from the basic sciences to help students better understand environmental issues. The formation of the earth, the evolution of life, geological processes, chemistry, and other fundamental topics are covered in Chapter 2 and integrated in other chapters.

New in the Third Edition

Updated Coverage

The third edition has been thoroughly updated with new discoveries, new concepts, new environmental laws, the most recent statistics on resources, population, and pollution, and new suggested readings. New essays, viewpoints, photographs, tables, and line drawings have been added as well.

Global Orientation

To help make the book even more global, numerous examples have been added from both third world and developed countries. Examples include preservation of the rainforests in Belize and Brazil, depletion of the ozone layer in the Arctic, sustainable agricultural practices in third world countries, family planning efforts in India and the impact of whale hunting in Japan. In addition, I have incorporated more examples of Canadian environmental issues. These include government support for logging, acidification of lakes and rivers, and preservation of Canada's fishing industry, forests and fuel reserves.

Critical Thinking

New to this edition is a section on Critical Thinking skills. Critical Thinking enables students to discern fact from conjecture and will help them analyze complex issues and make decisions. Beginning with an introductory exercise on the scientific method, this feature is carried throughout many chapters. Students will be asked to exercise critical thinking skills after reading case studies, viewpoints, and point/counterpoints.

Supplements

Environmental Action Guide

A concern of environmental science instructors is that their students leave this course with a sense of what the individual can do to effect change. To address this need, this edition is published with a new and unique manual, *The Environmental Action Guide*. Written by Ann S. Causey of Auburn University, this resource provides information on environmentally sound products, investments, careers, community action groups, letter writing campaigns, and an overall low–impact lifestyle.

Instructor's Guide

Ann S. Causey and I have completely updated and revised the instructor's guide. It includes chapter outlines and test questions. (Black–line transparency masters will be packaged separately.) In addition, we include more case studies and critical thinking problems for further class discussion.

Laboratory Manual

The third edition laboratory manual is by Dr. Merle Alexander, Director of Environmental Studies, Baylor University. This manual includes 14 lab exercises, each designed to be conducted in a single class section. Students learn to apply textbook theory to practical application in the areas of community composition and species, population, resources and pollution, among others.

Acknowledgments

This book is the offspring of a great many people, for whom a mere thanks seems terribly inadequate. First and foremost are the thousands of scholars in anthropology, biology, chemistry, demography, natural resources, political science, economics, ecology, and dozens of other disciplines. Their ideas, their research, indeed their lives, form the foundation on which this book rests. To them a world of thanks and an enormous debt of gratitude.

A warm and very special thanks to my friend and colleague, Teresa Audesirk, who reviewed the second edition and helped make it more accurate and more readable and who helped supply me with mounds of reference material needed in the update.

A genuine thanks to the staff at Benjamin/Cummings who labored over this book as if it were their own. A special thanks to Melinda Adams, my editor; to Developmental Editor, Valerie Kuletz, who coordinated much of the long process, who read and commented on the manuscript throughout, who infused the project with enthusiasm and creativity, and whose high standards are evident everywhere; to Associate Editor, Lisa Donohoe, who took over for Valerie at the beginning of the production phase, reading and commenting on the text; to Eleanor Renner Brown, Production Coordinator, who guided this book through the tedious production stages; to Mark Thomas Childs, who located the new photographs for this text; to Associate Editor, Laura Bonazzoli, who developed and coordinated the supplements to this third edition; to first edition research assistants Dave Shugarts, Cynthia Stuart, Diane Short, and Ann Beckenhauer and second edition research assistants Rae Nelson, Carmen Bal, and Elizabeth Yerkes, who helped update the book.

Finally, an extra special word of thanks to my wife, Kathleen, who helped update statistics and proofread chapters, and who endured complaints of tired eyes and sore back from long hours bent over my word processor. To her, much more than thanks.

Many manuscript reviewers provided helpful and constructive criticism on both the first and second editions of *Environmental Science: A Framework for Decision Making*:

David M. Armstrong, *University of Colorado, Boulder*
Robert Auckerman, *Colorado State University*
Terry Audesirk, *University of Colorado, Denver*
Michael Bass, *Mary Washington College*
Bayard H. Brattstrom, *California State University, Fullerton*
Lester Brown, *Worldwatch Institute*
Ann S. Causey, *Auburn University*
Donald Collins, *Montana State University*

Sally DeGroot, *St. Petersburg Junior College*

Joseph Farynairz, *Mattatuck Community College*

Ted Georgian, *St. Bonaventure University*

James H. Grosklags, *Northern Illinois University*

Richard Haas, *California State University, Fresno*

William S. Hardenbergh, *Southern Illinois University*

John P. Harley, *Eastern Kentucky University*

John N. Hoefer, *University of Wisconsin, La Crosse*

Gary James, *Orange Coast College*

John Jones, *Miami Dade Community College*

Alan R. P. Journet, *Southeast Missouri State University*

Thomas L. Keefe, *Eastern Kentucky University*

Suzanne Kelly, *Scottsdale Community College*

Thomas G. Kinsey, *State University College at Buffalo*

Kip Kruse, *Eastern Illinois University*

David Lovejoy, *Westfield State College*

Timothy F. Lyon, *Ball State University*

Glenn P. Moffat, *Foothill College*

Charles Mohler, *Cornell University*

Joseph Moore, *California State University, Northridge*

Bryan C. Myres, *Cypress College*

John H. Peck, *St. Cloud State University*

Michael Picker, *Western Director for the National Toxics Campaign*

David Pimental, *Cornell University*

Michael Priano, *Westchester Community College*

Joseph Priest, *Miami University*

Martha W. Rees, *Baylor University*

Robert J. Robel, *Kansas State University*

Jack Schlein, *York College, City University of New York*

Michael P. Shields, *Southern Illinois University*

Rocky Smith, *Colorado Environmental Coalition*

Laura Tamber, *Nassau Community College*

Roger E. Thibault, *Bowling Green State University*

Leland Van Fossen, *DeAnza College*

Bruce Webb, *Animal Protection Institute*

Ross W. Westover, *Cañada College*

Jeffrey R. White, *Indiana University*

Ray E. Williams, *Rio Hondo College*

Larry Wilson, *Miami Dade Community College*

Stephen W. Wilson, *Central Missouri State University*

Susan Wilson, *Miami Dade Community College*

Robert Wiseman, *Lakewood Community College*

Richard J. Wright, *Valencia Community College*

Paul A. Yambert, *Southern Illinois University*

I am very thankful for their helpful comments.

DANIEL D. CHIRAS

Evergreen, Colorado

Brief Contents

Detailed Contents

Prologue

In an outlying village in Ethiopia, two children are lowered into a communal grave that houses the bodies of others who have died in recent days. Villagers stare vacantly at the men who cover the bodies with dirt; to the friends and relatives of these children who watch, death has lost much of its significance. Against the constant hunger and death, few mourn another child's passing.

Worldwide, 700 people will die from starvation, extreme malnutrition, or infectious disease stemming from food shortages in the half hour it takes you to watch the evening news. This year alone, the death toll from hunger and associated diseases is estimated to be 40 million people. This is the equivalent of 300 jumbojets, each carrying 400 passengers, crashing with no survivors every day of the year. Almost half of the victims are children. Despite an outpouring of aid from the rich nations, hundreds of millions more will die in years to come.

A False Sense of Security?

For Africans of the southern Sahara, the future looks bleak. Long-term drought, overpopulation, continued misuse of the land, and political struggles all create spreading deserts that swallow farmland at an alarming rate. In this dilemma, nature dictates an extreme solution: people must die to reestablish the balance.

But what about those of us in the wealthy nations of the world? Need we worry? To many people, the answer is no. Resource shortages are a thing of the past. Newspaper headlines assure us of an "oil glut" that has forced the oil-producing countries to slash prices, a move that has helped ensure economic stability in many countries. Some critics believe that our sense of security is illusory. But why not feel secure; with an ally as powerful as technology, how could we not prosper?

Part of the answer may lie in the way we mistreat our soil, perhaps our greatest resource of all. In the United States, for example, farmers currently cultivate 170 million hectares (421 million acres) of land. According to estimates by the Department of Agriculture, nearly one half of the United States' farmland is eroding faster than it can be replaced by natural processes. Making matters worse, there is very little land in reserve to replace the prime land now eroding away. Some experts believe that crop production could fall by 10% to 30% in the United States in the next 50 years if soil erosion continues unchecked. Costs of food will rise as good farmland is destroyed. The United States may lose its position as a leading food exporter. Grain shipments to hungry nations may be reduced as well, unless something is done . . . quickly.

Consider also one of our most valuable resources, oil, thought by many to be the lifeblood of industrial societies. Oil's economic importance to developed nations became clear in the 1970s when per-barrel prices jumped from $3.00 to over $35. A whirlwind of inflation began, perilously gripping the industrial world, nearly halting industrial production. The American economy was driven to its knees. Millions of workers were laid off as inflation brought industrial production to a near standstill.

Despite current, short-term gluts and falling prices, the long-range future of oil is dim. Estimated worldwide oil supplies will last only 65 more years at current consumption. Should consumption rise, as expected, even fewer oil years await us. Clearly, time is running out for oil.

Long before our wells run dry, however, the rich, oil-dependent nations could begin to flounder. By some estimates, somewhere around 2000 or 2010 global oil production will fall short of demand, sending prices sharply upward. The inflation of the 1970s will seem like warm spring breezes compared to the hurricane winds of global inflation.

You and I, and millions of people like us, will very likely see the end of oil within our lifetimes. The time is ripe for charting new paths, but this nation and others are sitting back, doing very little to develop alternative fuels and cut existing waste.

Declining resources are only part of the threat to modern society. Pollution and development also threaten to destroy the delicate web of life. Foremost on the list of pollutants is acid rain and snow.

Today, over 245 ponds and lakes in the Adirondacks have lost their aquatic life because of acids from industry and transportation. Deposited by rain and snow, these acids kill fish, algae, and aquatic plants. In southern Sweden 20,000 lakes are without or soon to be without fish because of widespread acid deposition. In Canada, 100 lakes have met a similar fate. But the effect of acid rain is felt much wider. For instance, much of the once-rich Black Forest in Germany has been poisoned by this toxic rain.

As these examples suggest, the environment is in trouble—and so are we. Despite more than 20 years of effort and significant gains in environmental legislation, most of our environmental problems are growing worse. Consider some examples:

- Since 1970, world population has increased by 1.6 billion people, climbing from 3.7 billion to 5.3 billion. Today, we're adding nearly 90 million people to the world population each year.
- Since 1970, the number of species on the official list of endangered and threatened species has increased from 92 to 539 (in 1989).
- Since 1970, annual global carbon dioxide emissions have increased from 3.9 billion metric tons to over 5.2 billion tons.
- Since 1970, the number of African elephants has declined from 4.5 million to only about 500,000.

The past 20 years has seen America grow to be a world leader in waste production. Today, Americans throw away 160 million tons of municipal garbage each year. That's enough to fill the superdome two times a day, 365 days a year—the equivalent of about 1300 pounds of trash for every man, woman and child each year.

Each year, American industries produce an estimated 250–280 million tons of hazardous wastes (over 2000 pounds of hazardous waste for every man, woman, and child in this country).

Pollution is choking our cities. According to the Environmental Protection Agency, 110 million Americans live in air considered hazardous to their health. An estimated 50,000 Americans die prematurely each year as a result of air pollution.

The long-term future of the world is in jeopardy. It is not just the poor of Ethiopia or Chad or Sudan who stand to lose, but also the wealthy residents who make up one-fourth of the world's population but consume 80% of its resources. The rich and the poor are locked in a crisis created by overpopulation, vanishing resources, and excessive pollution.

Tragedy of Our Times

Paul Valery once noted that the tragedy of our times is that the future is not what it used to be. In reality, though, the future is rarely what we think it will be. The tragedy of our times is that few people realize that the future has changed. We are, as a whole, going about our daily lives as if nothing has happened, lulled into complacency by old and fairly unrealistic dreams. Oil gluts, falling gasoline prices, and economic stability have given us a false sense of security at a time when we need, more than anything, three key ingredients: foresight, planning, and action—both individually and collectively.

This book examines the crisis of population, resources, and pollution that engulfs humankind. You will find it a hopeful book, filled with solutions. It views our dilemma in much the same way that the Chinese view crises. Their word for crisis is *wei-chi*. The first part means "beware of danger." The second part means "opportunity for change."

In this spirit, I invite you to look at the critical paths we are now on. You will see that the human race can survive the human race and prosper. But changes must be made—big changes in the way we think and the ways we act.

The Secrets of Nature

What alterations in our course are necessary? Experts disagree, but many believe that the key to our long-term survival lies in the widely ignored lessons of nature. Consider these facts: undisturbed ecosystems persist for decades, centuries, even millions of years. The rate of extinction in such ecosystems is low. Human society, on the other hand, now wipes out a vertebrate (backboned) species every nine months and itself faces global extinction after only a relatively short stay on earth. Why is it that nature persists while we deplete and destroy? The secret of nature is that survival hinges on a sustainable system—a system that perpetuates itself without destroying the very things that permit life to continue.

Nature capitalizes on four major strategies to meet this end. The first is recycling. The global ecosystem is a consummate recycler. Water, carbon, oxygen, nitrogen, and all other substances are used over and over. As a result, new generations are built from the old. The long-term future of humankind depends on following a similar direction.

Nature's second secret is the use of renewable resources—resources that renew themselves through natural biological or physical and chemical processes.

Wood, water, and wind are examples. For millennia, humankind heated its homes with wood, reaped the riches of the biological world for food, and fashioned its goods from flax and other plant products. Only in the past 200 years has our allegiance to renewable resources wavered. Today, we depend heavily on a variety of non-renewable substances: fuel, plastics, and synthetic fabrics made from oil; metals; and so on. Our new dependency, many think, is a dangerous trap. It cannot be sustained indefinitely. Our long-term future requires a greater dependency on resources that can renew themselves. Protecting these resources is a form of self care.

Nature's third secret is conservation. A fat wildebeest or an obese ostrich do not exist in nature. For the most part, organisms use what they need—no more, no less. Modern industrial societies, on the other hand, are often gluttonous, overeating, wastefully consuming, and recklessly depleting. Ecologists warn us that we cannot do so forever with impunity.

The fourth secret of nature is population control. Through a variety of ways, populations of living things are kept from living beyond their means. Predators trim the prey populations. Diseases eliminate the weak and aged. Environmental conditions keep populations from exploding. For humans, technological advances, medicine, and sanitation have removed many of the natural barriers to human population explosion. The upshot of the rapid human population expansion is often foul-smelling skies, filthy water, and landscapes devoid of vegetation and animal life. The ecosystem is sacrificed to continue population growth. Most ecologists agree that we must learn to control our numbers to preserve the global ecosystem.

Such are the secrets of nature: recycling, renewable resources, conservation, and population control. It is ironic that today we must go back to nature to relearn these forgotten lessons. If we are to survive for thousands of years to come, we must build a sustainable society, a society that lives in harmony with nature, not a society that seeks complete domination over all living things or destroys its renewable resource base. Building a sustainable society does not mean reverting to a primitive existence, it means using resources in a pattern laid down by nature.

Frontiers

A great frontier lies ahead of us. It is not the great expanse of space or the oceanic depths that we must conquer, but rather it is ourselves. Ahead of us lies the greatest and sometimes most inaccessible frontier—that of self-understanding and self-control. Before we race further into space to satisfy our needs, we must learn to look deeper within ourselves and find ways to build a sustainable society.

We can achieve such a society within our lifetimes, but each of us must help. Individuals must do more than pay lip service to recycling, conservation, renewable resources, and population control, and they must take action now. This book looks at the problems and suggests ways to build a sustainable society. It concerns itself more with the long-term future of humankind, recognizing fully that we must make changes now to transition smoothly into sustainability. Some of you may wonder why we should worry about future generations. Shouldn't we let them fend for themselves? And why should we change our ways now? Part of the answer is that we hold the future in our hands. At no time in history has the present generation had such potential to shape the future. The decisions we make on nuclear energy, acid rain, and tropical rainforests will affect our sons and daughters and theirs more profoundly than they will affect us. It is for this reason alone that we must rethink the past and redefine the future.

A sustainable society will not be a radical departure from our current way. In fact, many examples of sustainability are now commonplace, like bottle bills, battery recycling, water conservation, and wilderness preservation. It takes only a small effort and a little wisdom to get back on track. Abraham Lincoln said it best, "As the times are new, so we must think and act anew." Let this be our challenge: to see that the future is no longer what it used to be and to build an enduring future.

ENVIRONMENTAL SCIENCE
Action for a Sustainable Future

PRINCIPLES OF ENVIRONMENTAL SCIENCE

1

Environmental Science: Meeting the Challenge

The ability of our minds to imagine, coupled with the ability of our hands to devise our images, brings us a power almost beyond our control.

JOAN MCINTYRE

"It was the best of times, it was the worst of times," wrote Charles Dickens in his classic novel *A Tale of Two Cities*. Were he alive today, Dickens might look upon the present with an equal mixture of feelings. At no time have humanity's prospects seemed so bright yet so gloomy. Modern medicine, computers, and lightning-fast communications systems are three of our marvelous accomplishments. Despite our technological prowess, we seem besieged by an insurmountable array of problems: war and social unrest; devastating hunger in the Third World at the same time that US farmers are falling into bankruptcy; water shortages caused by farmers who pump groundwater faster than it can be replaced; plant and animal extinction on an unprecedented scale; soil erosion of epic proportions; widespread destruction of the rich tropical rain forests; acid rain and snow that threaten sensitive ecosystems; air pollution that endangers the global climate; and toxic chemical spills and abandoned waste dumps that pollute our precious groundwater.

A Modern Response to the Environmental Crisis

This book focuses on the problems we now face and offers—with enthusiasm and optimism—many solutions to them. The solutions spring from a new kind of science called **environmental science**. Environmental science is really just a new name for an activity that humans have been engaged in throughout time: learning how to live on this planet without damaging it unnecessarily or threatening our existence in the process.

Today, environmental science is practiced by many thousands of specialists, ranging from researchers testing new metals to naturalists searching for undiscovered plants to scientists learning how to better monitor the earth's activities from outer space (Figure 1-1).

What Is Environmental Science?

Just what is environmental science? To understand the term, take each word separately. The word *environmental* refers broadly to everything around us: the air, the water, and the land as well as the plants, animals, and microorganisms that inhabit them. *Science*, of course, refers to a body of knowledge about the world and all its parts. It is also a method for finding new information. Science seeks exactness through measurement, insight through close observation, and foresight through its theories. How scientists go about their work is discussed in Chapter Supplement 1-1.

Environmental science came into existence as a recognized discipline to cope with the vast problems spawned by overpopulation, resource depletion, and pollution. It has become a key tool in our survival.

Modern environmental science is aimed at helping us control our own actions in the natural world to avoid irreparable damage. In this sense, environmental science means learning to master ourselves.

Welcome to a New Kind of Science

To solve the highly complex problems of overpopulation, resource depletion, and pollution requires a knowledge of many scientific fields. Environmental science calls on chemistry, biology, geology, and a great many other disciplines, including sociology, climatology, anthropology, forestry, and agriculture. Spanning this wide range of knowledge, environmental science offers an integrated view of the world and our part in it.

Environmental science takes on the colossal task of understanding complex issues. Because of this it is an often awkward melding of science, engineering, and liberal arts that requires broadly educated men and women in an age that leans heavily toward specialization. More and more, though, human survival depends on the lessons we can learn from this broad, interdisciplinary science.

Environmental science differs from the traditional "pure," or objective, science, which seeks knowledge for

Figure 1-1 The earth from outer space. This finite planet is a closed system, powered by the sun. The materials all living things need to survive must be recycled over and over.

its own sake. Instead, it offers a great deal of urgent advice and reaches many conclusions that challenge cherished beliefs and practices. You may find this true as you read this book. In contrast to the astronomer in a mountaintop observatory or the cell biologist in a laboratory, environmental scientists are often in the thick of things, at the heart of today's hottest debates.

Environmental science is the study of the environment, its living and nonliving components, and the interactions of these components. By choice it focuses on the many ways that humans affect the environment and the ways our actions come back to haunt us. Crossing many traditional boundaries, it attempts to find answers to complex, interrelated problems of population, resources, and pollution, problems that threaten the welfare and long-term survival of humanity.

Outlines of a Crisis

The environmental crisis can be divided into three main categories: overpopulation, depletion, and pollution. Entwined throughout these three is a fourth, less tangible problem involving values and feelings; it will be discussed under the heading of the human failing.

This book takes its structure from these four problems. Part 1 describes the principles and practices of environmental science; Part 2 discusses population; Part 3 deals with resources; and Part 4 covers pollution. Part 5, in many ways the capstone of this book, looks into questions of the values, economics, and politics of the present time and those needed to build a sustainable future.

Overpopulation: Too Many People

In 1990 the world's population exceeded 5.3 billion, and it was swelling at a rate of 1.7% a year. This rate may not seem very high but it would mean a doubling of our numbers in only 41 years. By the time you retire, 10 billion people could inhabit the earth. Despite several decades of agricultural research aimed at increasing the food supply, one of every three persons living in the poor, developing nations is unable to find enough to eat. Worldwide, twelve million people die of starvation each year, and thirty million more perish from diseases made worse by hunger. Most startling, in the areas hardest hit by hunger, population is doubling every 17 to 30 years!

This explosion of world population is an outgrowth of the industrial age. Three key factors are to blame: increased food production, disease control, and better sanitation. These advances greatly increased the survival rate of newborns, but this great change occurred in many countries without any decrease in the number of births. As a result, populations worldwide have exploded. Most experts agree that until people start having fewer children, especially in the poorer nations, population will continue to soar, and problems of depletion and pollution will worsen.

Depletion: Eroding the Basis of Life

As world population continues to increase, many resources necessary for human survival—and the survival of the millions of species that share this planet with us—are falling into short supply. Consider firewood: A rural Indian peasant needs only a few pieces of wood each day for cooking. But throughout India millions of poor people venture from their homes in search of the tiny ration of wood that will get them through the day; such actions add up to an alarming total. Forests around many villages have been stripped of dead limbs. When dead wood is gone, villagers turn to live trees. As live trees fall, peasants are forced to forage farther and farther from their home villages. The circle of destruction widens.

In India and other poor countries of the world, many people spend the majority of their waking hours in search of fuel. In an attempt to compensate for the shortages, many villagers now burn dried cow dung. Although this practice satisfies the short-term demand for fuel, it robs the farmland on which they depend of an important source of fertilizer.

The earth's supply of **nonrenewable resources**—those that cannot be regenerated, such as oil, silver, and coal—

is finite, or limited. As populations swell and demand increases, nonrenewable resources fall more quickly into short supply. As you learned in the prologue, time is running out for oil, the lifeblood of our modern industrial society. Many of us will see the end of the oil supply within our lifetimes. The time is ripe for charting new paths, converting to renewable fuels, and practicing conservation, but little is being done to make the needed changes on the scale required.

Consider another example of resource depletion. Based on existing estimates of world reserves and projections of consumption, at least 18 economically essential minerals are bound to fall into short supply—some within a decade or two—even if nations greatly increase recovery and recycling. Gold, silver, mercury, lead, sulfur, tin, tungsten, and zinc are members of this endangered group. Even if new discoveries and new technologies make it possible to mine five times the currently known reserves of these materials, this group will be 80% depleted by 2040—in many of your lifetimes. Declining mineral supplies combined with falling oil supplies could cripple the world economy unless we make radical changes, and soon.

Even **renewable resources**, which have the capacity to regenerate, can be depleted if demand exceeds replenishment. Countless examples exist. Stands of ancient trees, the **oldgrowth forests**, which support such magnificent trees as the giant redwood of California and the sitka spruce of Oregon and Washington, are falling faster than they can regrow. The giant redwoods are turned into picnic tables, decks, lawn furniture, and siding for houses. Ancient trees from Oregon and Washington are being exported to Japan at an alarming rate and at ridiculously low prices. Numerous species of plants and animals are also being depleted because of overharvesting. (For more on the oldgrowth controversy see Chapter 9.)

In Haiti, overpopulation, hunger, and poverty have spawned similar devastation. The tiny island of Haiti is the poorest country in the Western hemisphere. Its population of over 6 million people is expected to double in the next 30 years if the current rate of growth continues. Today, one-third of Haiti's land is seriously eroded, and one of every three people is malnourished.

Northwest Haiti was once a heavily vegetated region, but today the soil is barren. Only the stumps of trees stick up through the baked ground. Villagers have exhausted the soil and stripped the trees to make charcoal to sell to city residents. Widespread deforestation has brought severe erosion. Because of this widespread deforestation in the northwestern part of Haiti, foresters turned to the last remnants of forest in the southwestern part of the country. Rich and lush just two years ago, this area is beginning to show the signs of ruin. Deep gullies cut through the country and entire hillsides have washed away. This island nation is a frightening example of what happens when a natural system collapses due to overpopulation and resource abuse.

In the United States, excessive demand and poor land management practices are currently destroying our land at a surprising rate, depleting a crucial renewable resource. Two of the major problems are erosion and **desertification**, the formation of desert in once productive and healthy land. We currently lose .5 million hectares (1.25 million acres) of farmland each year to erosion. Should this continue our farmland will disappear in just 300 years!

But we are not alone. Worldwide, 6 million hectares (15 million acres) of cropland, pasture, and rangeland become desert each year. One-third of the world's cropland is currently eroding faster than it can be replaced.

Nowhere is the vicious cycle of depletion more evident than in the tropical rain forests. Once covering an area the size of the United States, tropical rain forests have been reduced by at least one-third, perhaps as much as one-half. At least 11 million hectares (27 million acres) of tropical rain forest fall each year. At the current rate of destruction, half of the forests of the Ivory Coast of Africa will be destroyed in the next 10 years. Two-thirds of the world's species inhabit the tropics. Rampant deforestation will inevitably drive many of these species to extinction.

Today, one vertebrate species disappears every nine months. Add to that the plants, microorganisms, and invertebrates, and the rate of extinction climbs to an alarming one species every day! With continued population growth and economic development, especially in the biologically rich tropics, biologists fear that the rate of extinction could climb to one species every hour by the end of this decade. A million species could succumb to humankind's massive global colonization between 1980 and 2000.

Tropical deforestation is a symptom of a pervasive problem—widespread habitat destruction—that is destroying the homes of countless species. Pure population pressure and excess consumption lie behind resource depletion. But wasteful industrial processes that squander resources and produce excessive pollution are also at the root of our environmental crisis. Despite the many steps taken worldwide to defuse the crisis, problems continue and grow worse.

Pollution: Fouling the Land, Water, and Air

In the spring of 1983 atmospheric scientists flying over the Arctic made the first measurements of a thick orange-brown layer of industrial air pollution covering an area the size of North America and extending more than six miles into the atmosphere. In 1989 US astronauts return-ing on a space shuttle flight deplored the filthy blanket of air encircling the earth.

In 1988, the United States Environmental Protection Agency (EPA) announced that more than 60 American cities fail to comply with clean air standards, even though the standards have been eased and the deadlines for meeting federal standards have been extended repeatedly since the Clean Air Act was first passed in 1970. As a result, tens of millions of Americans breathe unhealthy air.

Air pollution in our cities and surrounding suburbs is the most visible sign of the pollution crisis. Far more widespread, and perhaps more dangerous, are the problems of acid deposition, global warming, and stratospheric ozone depletion.

Today, over 230 lakes in the Adirondack Mountains of New York State have become critically acidified by sulfuric and nitric acids produced from sulfur dioxide and nitrogen oxides generated during the combustion of fossil fuels. In these lakes fish, algae, and aquatic plants have perished. In 1988 the National Wildlife Federation published a list of US lakes that either have high acid levels or are sensitive to acid precipitation. The report showed that eastern lakes have been particularly hard hit by acid precipitation. One-third of the lakes in Florida, for instance, are acidic enough to be harmful to aquatic life. One-fifth of the lakes in Massachusetts, New Hampshire, New York, and Rhode Island are in the same condition.

Throughout much of the developed world, the story is the same: In southern Sweden 20,000 lakes are without (or soon to be without) fish because of widespread acid-ification. In Canada, the prospects for lakes and rivers are quickly dimming as acids fall from the sky. Nine of Nova Scotia's famous salmon-fishing rivers have already lost their fish populations, and the populations of eleven more rivers are teetering on the brink of destruction. In southern Ontario and Quebec, acid deposition has destroyed at least 100 lakes. Scientists predict that by the year 2000 nearly half of Quebec's 48,000 lakes will be unable to support life.

According to a 1988 report by the Environmental Defense Fund (EDF), acid deposition contributes at least 25% of the nitrogen entering Chesapeake Bay each year from human sources. Acid deposition, therefore, ranks second only to fertilizer runoff as a source of nitrogen. In Chesapeake Bay, the excess nitrogen stimulates an overgrowth of algae—*algal blooms*. Prolific algae consume dissolved oxygen, suffocating fish and other aquatic organisms. Algal blooms also reduce sunlight penetration to rooted plants on the floor of the bay, thus killing them.

EDF scientists warn that acid deposition may be a significant problem along the entire eastern seaboard, adding nitrogen to already polluted waters and negating efforts to control water pollution coming from farms, cities, and sewage treatment plants.

Acids also affect cropland and forests and are implicated in the widespread destruction of statuary and historic buildings in Europe and North America. Much of the once-lush Black Forest in Germany has died as a result of acids deposited from the sky.

Continued dependence on fossil fuels is poisoning our planet with acids and may be setting the stage for a disastrous global warming trend. Five of the hottest years in over a century have occurred between 1980 and 1989. If global temperature continues to rise as it has in the past 20 years, the earth's climate will soon be warmer than it has been in at least 100,000 years. Predictions based on models of the greenhouse effect indicate that global temperatures will accelerate during the next decade.

Several pollutants produced by modern industrial societies are believed responsible for global warming (see Chapter 15). They absorb heat leaving the earth and radiate it back, thus warming the earth. Many scientists predict that global carbon dioxide levels could double in the next four decades, raising average global temperature by two to five degrees Celsius.

This increase in global temperature would dramatically alter our planet's climate. Much of the United States and Canada, computer models suggest, would be warmer and drier. The productive agricultural plains states might become too dry to support dry-land farming. The United States and Canada, both major exporters of food, could find themselves short of food for their own populations. A global warming trend would melt polar ice caps and glaciers, raising the sea level 200 to 300 feet, thus flooding up to 20% of the world's land mass. Florida and most of the US coastal cities would be under water. New York and Los Angeles could be completely flooded. Farmland, already suffering because of lower rainfall, would shrink as the US population moved inland.

Modern society is as careless with its land and water as it is with its air. Each year, American factories produce an estimated 250 to 270 million metric tons of waste considered hazardous by state and federal standards— over a ton of hazardous waste for every man, woman, and child. Until quite recently, 90% of that waste ended up in our waterways, landfills, and groundwater. Ill-conceived and irresponsible waste disposal practices have left a legacy of polluted groundwater and contaminated land. Cleaning them up could cost as much as $100 billion.

Today, because of stricter controls on domestic hazardous waste disposal, some American companies are exporting their wastes to Third-World nations taking advantage of their cheap disposal rates and cash-hungry economies. But in these countries waste is more often than not dumped carelessly. The United States is not alone. Many other industrial nations see the poor, developing nations as a cheap place to dump their toxic garbage.

Another pollution problem is the millions of leaking underground storage tanks that are contaminating the nation's precious groundwater. Containing petroleum by-products, toxic chemicals, and hazardous wastes, these tanks deteriorate over time. A tiny leak can contaminate millions of liters of groundwater in a single day! No one knows how much groundwater is currently contaminated in the United States, but a 1981 EPA study showed that 28% of 950 cities with populations over 10,000 had some contamination.

Many lakes and rivers have been cleaned up in the last two decades in the United States and abroad by placing controls on the most visible pollutants, such as raw sewage and industrial effluents. However, many invisible, cancer-causing substances still pollute these waterways. In some areas, they are causing an epidemic of cancer in fish and threatening the health of people who regularly consume fish. Truly, we are poisoning ourselves and leaving a legacy of ill health for our children.

One of the decade's most dramatic pollution news stories occurred in Bhopal, India. On December 3, 1984, while many people were asleep, a deadly cloud of methyl isocyanate, used to make pesticides, leaked from a storage tank at a Union Carbide Corporation plant. Within days the death toll had risen to 2500. The number injured was estimated in the tens of thousands. According to one estimate, as many as 100,000 survivors may suffer permanent injury, including sterility, blindness, and brain damage. The Bhopal disaster was considered the worst industrial accident in history.

Then, in 1986 the Chernobyl nuclear reactor in the Soviet Union exploded. Tons of radioactive debris escaped from the reactor for many days, spreading throughout much of the western USSR and Europe. Thirty-one people died within six months of the accident. No one knows the toll in human cancer, but some experts believe that 5000 to 50,000 additional cancer fatalities will result from that accident. In 1989, an Exxon supertanker ran aground off the coast of Alaska, spilling over 40 million liters (11 million gallons) of crude oil into the pristine, biologically rich waters of Prince William Sound. These dramatic examples are only part of the problem, however. Less visible pollutants, such as acid precipitation and carbon dioxide, spawn a host of problems far more dangerous to the planet.

The Human Failing: A Crisis of Spirit

The ecological crisis is the sum of the interconnected problems of overpopulation, resource depletion, and pollution. Underlying it is a more subtle problem, a crisis of the human spirit, something awry in the way we perceive the world and our place in it.

Dedicate the '90s to the Environment: International Commitment Could Be Our Gift to a New Century

James Gustave Speth

Dr. James Gustave Speth is President of World Resources Institute, and past Chairman of the Council on Environmental Quality.

Today's environmental problems are closely interlinked, planetary in scale and deadly serious. They cannot be addressed issue-by-issue or by one nation or even by a small group of nations acting alone. They will not yield to modest commitments of resources in the face of a doubling of world population and a quintupling of world economic activity in the lifetimes of today's children.

The nations of the world must come together in recognition of that grave challenge. They should declare that a priority mission of international cooperation and diplomacy in the 1990s will be to deliver a gift to the new century of a planet sustained and whole. To give long-term structure to this commitment they should establish a 10-year program, the International Environmental Decade.

This program could be inaugurated with the high-level international conference on the environment promised by President Bush during the campaign—a pledge that he has since reiterated. Born in politics, this conference in fact presents an opportunity to do something of genuine historical importance.

The timing could not be more propitious. The world's environment is in trouble. The buildup of carbon dioxide and other gases in the atmosphere threatens far-reaching climate change, and one class of these gases—chlorofluorocarbons— is also depleting the Earth's ozone layer, which shields us from the sun's ultraviolet radiation. Over large areas of the globe, air pollutants are escaping urban-industrial areas and invading the countryside, seriously damaging aquatic life, forests and crops.

In the developing world, pressures on natural resources intensify daily. The deserts expand while the forests, with their

immense wealth of life forms, retreat. An acre of tropical forests disappears every second. Hundreds of millions of people live in absolute poverty, destroying the resources on which their future depends because no alternative is open to them.

Although the seriousness of these challenges is increasingly acknowledged by political leaders as diverse as Margaret Thatcher, Mikhail Gorbachev and Rajiv Gandhi, only modest efforts have been launched to deal with them. As yet there is no concerted international response on a scale equal to the challenges. The international conference on the environment would be a major step in closing the gap between the increasing realization that grave problems exist and the paucity of the political response to date.

An International Environmental Decade could both stimulate and coordinate action on three fronts. The first of these is providing the poor of the world with environmentally non-destructive livelihoods; sustainable development is critical both to meet basic human needs and to take pressure off a deteriorating resource base. The second front is transforming industrial systems away from the pollution-prone technologies of the 20th Century to the environmentally benign technologies that will be essential if economic growth is to continue. And the third is stabilizing populations both in nations where growth is explosive and on a global basis before the world's population doubles again.

Most basically, world political leaders must devise a new system of international responsibility—burden-sharing to sustain the Earth and its people—that does not exist today. Unless such a system is worked out, the next great international crisis is likely to be about the environment.

Nowhere is such a system needed more than in slowing down the global warming brought on by the greenhouse effect. To be successful, we will need a series of international conventions that not only respond to the complexity of that issue but also provide complementary solutions to other environmental problems.

We need to secure swift international approval of the ozone-layer protection protocol signed in Montreal last year and have the nations of the world come back to negotiate a complete, swift phaseout of chlorofluorocarbons. We need an overall global climate-protection convention—a prime goal of which should be to reduce carbon-dioxide emissions through greatly improved energy efficiency and other measures.

We need an international agreement to protect the world's tropical forests and to reforest the spreading wasteland areas in many developing countries. Relief for Third World countries from the stifling burden of international debt will be required for progress on this front.

Bush's sponsorship of an environmental conference of world leaders can put the world on the road to this new system of international responsibility. The 1990s are all the time that we have left to prepare what can be our most important gift to the new century.

The Aswan Dam: Ecological Backlash from Blind Cost-Benefit Analysis

The Nile River flows from its headwaters in Sudan and Ethiopia through an arid region of Egypt and eventually spills into the Mediterranean Sea. For centuries this great river carried 50 million to 100 million tons of silt each year to the land stretching along its banks and to the Mediterranean Sea. In the sea this nutrient-rich silt nourished a variety of microorganisms that, in turn, were food for thriving fish populations. On the land the silty water flooded nutrient-impoverished soil as the river spilled over its banks, enriching the soil. The nutrients robbed by agriculture were replaced by the natural floods that occurred each year in late summer and early fall, floods caused by the monsoon rains along the river's headwaters.

In the early 1960s, however, Egypt built the Aswan Dam along the Nile to provide electricity for the growing city of Cairo and to provide irrigation water for the lower Nile basin. The government, of course, had studied the proposal to build the dam, but it had looked primarily at the benefits that would result from the dam versus the economic cost of construction. Little attention was given to possible ecological backlashes.

Not long after the dam was completed and Lake Nassar began to fill, the people of Egypt and the world grew alarmed. Numerous problems began to make themselves painfully evident. First, the periodic flooding that had provided an annual fertilization of the land ceased. As a result, farmers along the Nile had to import fertilizer for their land at an exorbitant cost. Second, the sardine fishery in the eastern Mediterranean collapsed. Nutrient-rich silt that had once poured into the sea almost stopped. The sardine catch plummeted from 18,000 to 500 tons per year in only a few years. Third, the rising waters of Lake Nassar threatened the Ramses Temple at Abu Simbel, built over 3000 years ago. Engineers and construction workers sponsored by the United States, Egypt, and the United Nations dismantled the huge temple piece by piece and moved it to a site 60 meters (200 feet) above its original level, where it would be safe from the rising waters. The cost of this project was astronomical. Fourth, the incidence of schistosomiasis (a debilitating, sometimes fatal disease) in humans increased in Egypt as a result of the dam. The organism that causes this disease is carried by snails. Snails require a constant supply of water, which the lake and the irrigation channels provided. The spread of this disease, for which there is no known cure, is almost certain.

If a cost-benefit analysis had been done before construction of the dam began, chances are that it might not have been built. See Chapter 14 for an example of an approach you might take to making a cost-benefit analysis of the Aswan Dam.

The crisis of the human spirit is manifest in our escape into materialistic life-styles and our view of humans as apart from nature and immune to its laws. It is found in our single-minded pursuit of economic wealth. It is the foundation of our general lack of concern for future generations and for the long-term well-being of our planet.

This crisis of the spirit can be blamed for much of the damage that occurs in the name of progress. Therefore, solutions to the environmental crisis lie not only in new scientific discoveries and new technologies but also in finding answers to the internal crisis. Part 5 of this book, Environment and Society, will examine this topic of values and ethics and show how we can change our ways through fundamental changes in our beliefs.

Beyond Despair

Many experts agree that the problems of overpopulation, depletion, pollution, and the human failing have created an ecological crisis that threatens the natural systems of which humans are a part. This threat to the **biosphere**, the living skin of planet earth, is worsened by a recurring failure to consider the full impacts of our actions. In failing to do so, we often create additional problems whose solutions may be beyond our grasp. Consider what is happening in many African nations now actively engaged in eliminating **infectious diseases**, those caused by bacteria and viruses that are transmitted from person to person. Pesticides to eradicate vectors (animals and insects that carry diseases) and new drugs to treat infectious disease have successfully lowered the death rate in many countries, but in many cases reductions in birth rate have not accompanied the rapid decline in deaths. As a result the populations of many countries have exploded, placing enormous demands on farmland, wildlife populations, and other resources. Such unanticipated, adverse effects are called **ecological backlashes**.

Consider another example: Pesticides and fertilizers are used to increase agricultural output to feed the 83 million new world residents each year. These chemicals are released into the environment, where they may poison wildlife and pollute drinking water. In some cases they have been implicated in birth defects in babies, miscarriages in pregnant women, and cancer in adults.

Many observers believe that our daunting list of problems can be solved without devastating ecological backlashes. These people see the dangers posed by overpopulation, pollution, and resource depletion as an opportunity to strike out in new directions. Such an optimistic view is not meant to minimize the severity of our predicament but, rather, to give us a refreshingly new outlook.

In his studies of societal attitudes toward the future, the Dutch futurist Fred Polak showed that people's image of what is to come has the power to shape that future. This is a form of self-fulfilling prophecy. Our view of the future is much like a barometer that foretells the success or failure of a society. Polak concluded that "bold visionary thinking is in itself the prerequisite for effective social change." When pessimism abounds, the future will probably be a gloomy one, but a positive outlook increases the possibility of a bright future.

The challenge before us is immense: to create a society that can thrive within the limits of a finite planet. To do so we must redirect technology, economics, industry, and government with a common goal in mind: preserving the foundation on which life depends—a healthy biosphere. Our goal is to build a sustainable future that robs neither the earth of its beauty and strength nor its inhabitants of their place for survival (Figure 1-2).

Figure 1-2 Solar housing in Colorado. They have cut their energy bills substantially by superinsulating their buildings and tapping the generous power of the sun.

The Population, Resource, and Pollution Model: A New Perspective

Many people find environmental issues complex and confusing. Even experts disagree on the causes of our problems and the ways to solve them. Others find environmental problems clear-cut. "Politics are to blame," one person proclaims. "It's our attitudes toward nature," another argues. "Technology and energy waste are the roots of our troubles," a third declares. Such people often support narrow solutions, which fall short of the mark and may even create ecological backlashes and additional financial burdens on society.

This section presents a "conceptual model" to satisfy the needs of these three different groups—a need for order, on the one hand, and a need to expand understanding, on the other. Careful study of the model shown in Figure 1-3 will reward you many times over.

Models are one of the most powerful tools of the working environmental scientist. They can be adapted to computers and used to study complex issues such as world food supply, pollution, or even plant and animal extinction. The study of complex systems with computers is called **system dynamics**.

System dynamics computer models in environmental science are computer programs that deal with many environmental problems on many different levels. The first computerized global model was introduced in 1970 by J. W. Forrester. He used it to study the future of human civilization by looking at some of the most pressing issues, including population growth, pollution, and the depletion of natural resources. Figure 1-4 depicts just one part (population) of an updated version of the model, showing how complex it is. Not all models need be this detailed.

Perhaps you have worked with a computer to manage your budget. This type of program, like all computer programs, is developed from conceptual models much like the one presented in Figure 1-4. On a home computer, budget programs can forecast the effects of variations in living expenses on your savings or spending money. In a similar manner computer models can predict the effects of pollution on human health, the effects of oil scarcity on economics and resource demand, and much more.

The Population, Resource, and Pollution (PRP) model (Figure 1-3) seems general on first glance. It shows in the most basic terms how humans interact with their environment. Don't be deceived by its simplicity. A more careful look will show that it contains a wealth of useful information.

Vital Links: Humans and the Environment

The PRP model in Figure 1-3 says that human populations acquire resources from the environment. In that respect, humans are no different from any other organism. The acquisition and use of these resources produces pollu-

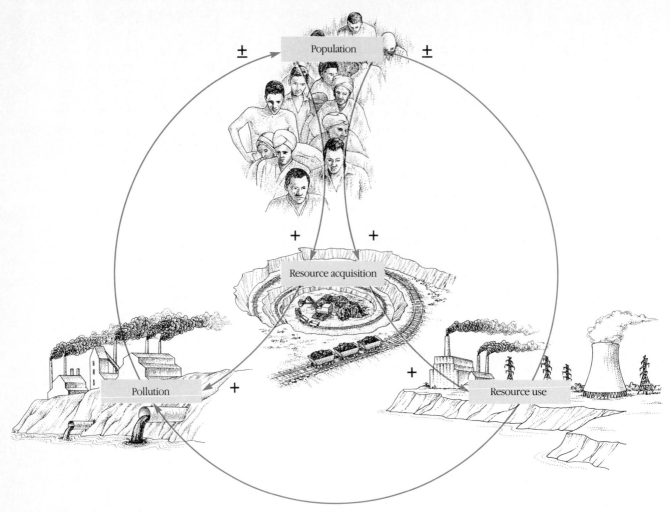

Figure 1-3 Population, Resource, and Pollution (PRP) model. Plus marks indicate positive-feedback loops, and minus signs indicate negative-feedback loops. Note that some activities, such as resource use, can have positive and negative influences on others, such as population.

tion. If you were to take a walk through a forest being clear-cut for timber in the Pacific Northwest, you would see how resource acquisition produces extraordinary amounts of pollution. Improper cutting—say, on steep slopes—can result in the erosion of tons of sediment into nearby streams.

We use a great many other resources as well. Oil and gasoline made from oil, for example, power much of modern society. Drilling for oil and burning gasoline in our automobiles both produce pollution. **Pollution**, by definition, is a change in the chemical composition of air, water, and soil that reduces their ability to support life.

Some pollution is caused by natural events; this is called natural pollution. The rest comes from human activities and is called **anthropogenic** pollution. Pollutants may be **biodegradable**, which means that they can be broken down by living organisms, usually single-celled bacteria

and fungi. Sunlight and temperature, however, may also degrade some pollutants. **Nonbiodegradable** pollutants, those that are not readily broken down, can persist in the environment. Both degradable and biodegradable pollutants can cause extensive harm.

The effects of pollution vary considerably, depending on a number of factors. For example, some water pollutants kill fish and the birds that feed on them. Pollution can make water unsuitable for drinking or swimming and, in some cases, can even cause serious human illness and death.

Scientists tend to classify pollutants by the medium they contaminate, for example, air pollution, water pollution, and land pollution. In recent years, scientists are finding that pollution readily crosses the boundaries between these media, a phenomenon called **cross-media contamination**. Air pollution, for example, washes from the sky and is deposited in lakes and forests. Some water pollutants evaporate from lakes and streams, entering the atmosphere, only to be deposited downwind in rain and snow. Hazardous wastes in the land drip into groundwater.

In sum, the PRP model says that human populations acquire and use resources from the environment. While resource acquisition and use can degrade the environment and harm our health, they also greatly enhance our survival. Food and fiber make it possible for humanity to survive on earth. This relationship is indicated in the model by the plus signs on the arrows leading from resources to population. This response is an example of a **positive-feedback loop**, where one factor leads to the growth of a second factor, which in turn stimulates the first one in a repetitive cycle. Increasing the use of the earth's resources invariably increases population size. The spiraling effect is the result of increased resource demand by a growing population. As a rule, positive-feedback mechanisms are dangerous, creating a devastating cycle of depletion and pollution.

Another type of feedback is the **negative-feedback loop**, where one factor shuts off another. The furnace in your

home is an excellent example. As you know, the furnace switches on when the temperature drops below the setting on the thermostat. Heat from the furnace then warms the room air until the air temperature reaches the desired level. When it does, the furnace shuts off.

Negative-feedback loops are the chief mechanism for controlling biological organisms on many different levels. For instance, the levels of glucose in your blood are maintained by a negative-feedback mechanism.

How do negative-feedback loops figure into the model? As shown in Figure 1-3, pollution may actually decrease population size. High levels of pollution have been known to kill people. Resource depletion may have similar effects. The current spread of deserts in Africa, resulting in large part from poor land management, is leaving millions without a way to feed their families, and millions are dying each year. Pollution and resource depletion are negative-feedback loops. (Chapter 6 covers controls on population size in more detail.)

The PRP model represents a fundamental ecological relationship true to all living organisms. It's a guide to animal activity. It's as relevant to humans as it is to black bears. What is more, it provides some visual insight into

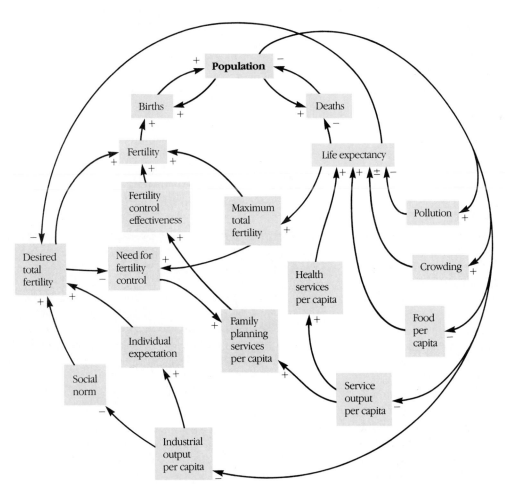

Figure 1-4 Don't be dismayed. This segment of a world computer model is shown here to illustrate the model's complexity, as evident in the number of elements and connections. Each factor is assigned a numerical value, and the computer calculates how changes in one or more factors affect others. Pluses indicate positive feedback, in which one factor causes an increase in another. Minuses indicate negative feedback, in which one factor diminishes another.

Table 1-1   Breakdown of Factors of the Population, Resource, and Pollution Model

Population	Resources	Pollution
Size	Acquisition	Water
Distribution	Use	Land
Density	Supply	Air
Growth rate	Demand	
	Character	

the human-environment interaction. It allows us to predict the impacts of human actions. Change one variable in the model and all the rest change as well. Add more people, for example, and resource use is bound to rise. Acquire more resources and pollution is likely to climb.

The model also provides insight into solutions. To solve rising pollution levels, resource demand might be cut. To control resource depletion, controls on population growth might be desirable.

The PRP model takes on a greater significance when we expand its principal elements. For example, population may be expanded to include factors such as growth rate and distribution of people within a country. Both have important implications for pollution levels. Table 1-1 lists some additional factors discussed in more detail throughout this book. You might want to study it before reading the next section. Chapter Supplement 1-1 discusses additional models that will prove useful in your study of environmental problems and your quest to find solutions.

Studying the Interactions: Cross-Impact Analysis

The three elements of the PRP model interact with one another in so many ways that only a computer could keep track of them. We can get a glimpse of how they affect one another by using a simple technique called **cross-impact analysis**. This technique helps break down complex interactions into simpler ones that are more easily understood.

Figure 1-5 is a simplified cross-impact analysis chart. In this chart the three elements of the PRP model are lined up in two perpendicular columns. In box 1 you can describe how population factors affect resource factors, such as supply, demand, acquisition, and use. In box 2 you can describe how population factors affect pollution, and so on.

Take some time to fill out the chart. You will find that you already know a great deal about the ways populations, resources, and pollution are linked. At the end of the course you may want to repeat the exercise; if you have

been conscientious in your study, you will find that your knowledge has grown tremendously and that you have a deeper appreciation of the complexity of modern society. Table 1-2 is an example of a completed cross-impact analysis.

A Glimpse of What Is to Come

Today, modern society and its underlying beliefs are under attack from all sides. One thing that this book will make clear is that modern industrial society as we know it cannot continue. Part 2 will show that the curve of population growth, growing steeper every year, must eventually stop rising. Part 3 will document how the resources we rely on—oil, metals, land, and water—are reaching or, in some cases, have already reached their limits. Part 4 will describe the facts about pollution and its effects. Part 5 will outline the attitudes and practices of both economics and politics that are fast becoming outdated, even dangerous, in a world of limits.

Building a Sustainable Society

Many experts agree that if human society is to endure, not just for another century but for thousands and thousands of years, we need to build a **sustainable society** that (1) controls population size, (2) uses resources wisely, (3) recycles all nonrenewable resources, and (4) relies on renewable resources wherever possible. A growing number of scientists and world leaders have taken an active role in developing such a world. A sustainable society is so profoundly different from the way we live that it cannot be achieved without considerable effort. You will learn how to build a sustainable society as you read this book, and you will also learn where you can help. You may want to jot down these ideas in a notebook for later reference.

Column A			
Population	X	1	2
Resources	3	X	4
Pollution	5	6	X
Column B	Population	Resources	Pollution

Figure 1-5   Cross-impact analysis chart, used to compare the interactions between the three factors in the Population, Resource, and Pollution model. By comparing factors in Column A with those in Column B, you can gain a deeper understanding of environmental problems. Why not take a few minutes to test your understanding? Start by comparing ways that population affects resources, then look at ways that population affects pollution, and so on.

Table 1-2 Completed Cross-Impact Analysis

Box	Summary of Interactions

1   Effects of Population on Resources

The size of a population determines, in part, the resource demand, how resources are acquired, and how much is used. Social, economic, and technological development of a country (all considered population factors) affect the demand for resources as well as the manner of acquiring them and how they are used. More developed countries tend to have more complex resource needs and tend to use resources that are nonrecyclable and nonrenewable. Growth rate affects resource allocation and use. Resource demand in rapidly growing populations may result in less concern for the consequences of resource allocation and use and, thus, more damage than in populations with slower growth. Population distribution affects resource supply, allocation, and use.

2   Effects of Population on Pollution

Populations create pollution through resource allocation and use. Pollution may result from using a resource as a depository for human and industrial waste. In addition, allocation of resources (coal, oil, gas) may result in environmental degradation. The amount of resources and manner in which these resources are acquired and used determine the amount of pollution.

3   Effects of Resources on Population

Positive effect. Discovery and use of new resources (oil, coal) can increase the population size, growth rate, and distribution, as well as social, economic, and technological development. Resources allow humans to move to new habitats and extract and use resources not previously available. In addition, resource development allows habitation in inimical environments.

Negative effect. Depletion of resources can limit population growth, size, and distribution, as well as social, economic, and technological development. Degradation of resources (air pollution) by misuse can theoretically reduce population size or eliminate populations.

4   Effects of Resources on Pollution

The amount and manner of resource allocation and use can affect pollution. The more resources are allocated and used, the more pollution, although methods of use and allocation greatly affect pollution. Resource depletion can reduce pollution.

5   Effects of Pollution on Population

Pollution can limit population size, growth rate, and distribution as well as social, economic, and technological development. Pollution can increase mortality and morbidity, thus having a social and economic impact. Pollution has an aesthetic impact. Pollution can change attitudes, which serve to change laws and the ways resources are allocated and used.

6   Effects of Pollution on Resources

Pollution of one medium (air) can contribute to the destruction of another. New laws designed to reduce pollution could shift resource demand, supply, acquisition, and use.

Changing Our Ways

Perhaps the first order of business is to convince governments to look beyond immediate needs and to learn how to cooperate better internationally to control population, pollution, and resource depletion. Environmental science will continue to prove helpful in achieving these goals. This new science has already helped us recognize overpopulation as one of the root causes of our environmental crisis. It has already done much to expand our understanding of global resource depletion and our knowledge of ways to treat and prevent global pollution. From the knowledge of human impact may come the necessary transformations in values and institutions needed to build a sustainable society.

New values are essential to building a sustainable society. Some of the new attitudes we need are clear. The attitude that "there is always more"—labeled the **frontier mentality** in this book—must be replaced with an attitude that this world is limited: "there is not always more." Limits are all around us, set by the availability of fossil fuel energy, minerals, and land. **Technological fixes** (new technologies to solve our problems) may not be possible, given the finite resource base on which our society is built. Clearly, we must learn to live within the limits set by nature. We must also learn to be more careful with natural processes. As you will learn in the chapters on ecology, the earth is a **closed system**. All of the elements needed to maintain life come from within and are recycled over and over. We humans interrupt these natural recycling systems at our own expense.

The Role of Environmental Science

Changing our ways will be a colossal task, a process that will take generations to complete. It will involve arduous work in many fields. The moon landing was a weekend home-improvement chore compared with the job ahead. The study of environmental science is a cornerstone of change.

But what has science to say to a philosopher, a public policymaker, a social leader, an economist, or a religious thinker? Aren't these people outside the sphere of science? As we will see, the boundaries between the world of science and the rest of society are being eliminated today by the problems we face. Scientific understanding is essential to all who ponder the fate of humanity and actively seek to make changes. To voters, scientific knowledge of the environment can become a tool for wise decision making.

Some pessimists believe that the human species is already doomed: that resource depletion, poisonous pollutants, radiation, and changes in global climate, alone or together, will slowly eradicate humanity. At the least, they say, things will get worse: we could lose centuries of technological and economic progress in the next few decades if trends continue. Our wonderfully diverse biological world, the product of millions of years of evolution, could be wiped out in a fraction of the history of living things.

Many other people see a glimmer of hope. They hold that the technology that got us into this mess can also get us out of it. Technological fixes for current problems are their answer. But such "technological optimists" often paint an unrealistic picture of the future and do not always distinguish what is technologically possible from what is feasible and affordable. Carried away with technological optimism, they often see outer space as a source of new minerals, free energy, and more living space for the crowded citizenry of the world. They propose new technologies such as nuclear fusion to solve the energy crisis. They view new pollution control technologies as the answer to our polluted waters and skies. Little emphasis is given, however, to changing life-styles, cutting back on resource demand, or stopping population growth.

Beyond the incompatible visions of these two groups, a few things are certain. The first is that time is short. Indeed, shortage of time may be our greatest shortage. Another certainty is that during your lifetime vast changes will take place; some will be bad, others good. What we do today will determine which predominate.

Humans have grown into a force rivaling nature itself; as Walt Whitman wrote, "Here at last is something in the doings of men that corresponds with the broadcast doings of the day and night." To build a sustainable society we must learn ways to treat our earth better, not to rival nature but to cooperate with it to live in harmony. For that great task, environmental science is our single most important tool.

Life can only be understood backwards; but it must be lived forwards.

KIERKEGAARD

Summary

Environmental science is the study of the environment, its non-living and living components, and the interactions between them. The importance of environmental science is that it helps us understand and solve a growing number of problems revolving around three central issues: pollution, resources, and overpopulation. Environmental science is a new synthesis; unlike more traditional sciences it calls on dozens of disciplines—from psychology to atmospheric science—in an effort to understand the interaction between humans and the environment.

Environmental science deals with population, resource, and pollution issues, each of which has profound effects on the others. In 1989 the world's population was 5.3 billion. If the current growth rate continues, the world population will double in 41 years. Many people—one of every three in the developing nations—do not get enough to eat now. What is the prospect for future generations?

Such unprecedented growth in human numbers springs primarily from medical and scientific developments, which have drastically cut the death rate but have not been accompanied by a fall in the birth rate. If populations continue to swell, resource shortages are bound to get worse. They may even limit future growth or cause a reduction in human numbers.

Overpopulation, excessive consumption, and wasteful practices create another problem for all nations of the world—pollution. Signs of global pollution are everywhere.

Underlying the environmental crisis is a more subtle problem: a distorted view of our place in the natural world, which leads to reckless exploitation of the earth's resources. Solving our problems will require a change in our attitudes. Knowledge of our impact on the environment may help us adopt holistic values that help us change population growth, resource acquisition and use, pollution, consumption, and technology. Such an approach will go a long way toward solving the ecological crisis and help us prevent unfortunate **ecological backlashes**.

This book views crisis as an opportunity for change and views societal attitudes toward the future as a barometer of the ultimate fate of that society. An optimistic outlook can lead to a brighter future, but it must be realistic and accompanied by action.

The **PRP model** can help us formulate a new direction by helping us understand where we are now. Illustrating the fundamental ties between populations, resources, and pollution, it can help us predict the impacts of future actions. A better understanding of these interrelationships can be gained by using **cross-impact analysis**.

Building a **sustainable society** requires four achievements: control of population growth, wise use of resources (conservation), widespread reliance on recycling, and use of renewable resources wherever possible. Environmental science can help us attain these goals by continuing to provide insight and information.

Discussion Questions

In a study group or by yourself, answer each of the following:

1. Name one of the key differences between environmental science and chemistry or another traditional science.

2. The environmental crisis is composed of four principal elements. Name them and describe each one.

3. Describe the Population, Resource, and Pollution model, defining the three factors that make up the model and illustrating their interaction.

4. Describe how the factors in Column A might affect those in Column B. Give examples.

A	B
Population	Resources
Resource allocation	Pollution
Resource use	Pollution
Pollution	Population

5. Describe the foundations of a sustainable society.

6. Give examples in the United States of activities consistent with a sustainable society. What areas need improvement?

7. Make a list of things you do that are consistent with a sustainable society. What area could use some improvement?

8. In your opinion, should we be optimistic about the future or pessimistic? Why?

9. How can optimism, or pessimism for that matter, contribute to the future of a society? Can you think of some examples?

Suggested Readings

Adams, J. H., et al. (1985). *An Environmental Agenda for the Future.* Washington, DC: Island Press. Synopsis of major environmental problems facing the United States and the world with recommendations for action.

Borelli, P., (ed). (1988). *Crossroads: Environmental Priorities for the Future.* Washington, DC: Island Press. Wonderful collection of essays on environmental problems continuing to affect the world and on needs for change in the environmental movement.

Brown, L. R., et al. (1990). *State of the World.* New York: Norton. Survey of problems and progress toward building a sustainable society.

Capra, F. (1982). *The Turning Point: Science, Society, and the Rising Culture.* New York: Bantam. Insightful book.

Chiras, D. D. (1990). *Beyond the Fray: Reshaping America's Environmental Response.* Boulder: Johnson Books. Prescription for building a stronger environmental movement.

Hardin, G. (1980). An Ecolate View of the Human Predicament. In *Global Resources: Perspectives and Alternatives*, ed. C. N. McRostie. Baltimore: University Park Press. Brilliant essay on global sharing.

Hardin, G. (1985). *Filters against Folly.* New York: Viking. Excellent treatise on ecology, economics, and science.

Hillary, E., ed. (1984). *Ecology 2000: The Changing Face of Earth.* New York: Beaufort Books. Graphically illustrated account of world environmental problems.

Myers, N., ed. (1984). *Gaia: An Atlas of Planet Management.* Garden City, N.Y.: Anchor Books. Comprehensive, graphic presentation of global population, resource, and pollution problems.

Sancton, T. A. (1989). What on Earth are We Doing? *Time* 133 (1): 24–30. Passionate plea for global environmental action. See this entire issue for other articles outlining the state of the environment.

Tobias, M., ed. (1985). *Deep Ecology.* San Diego: Avant Books. Collection of writings discussing world environmental problems and solutions.

Western, D. (1985). The Last Stand. *Discover* 6 (10): 102–107. An excellent case study of the interactions between population and resources.

Science and Society: Ways of Understanding Our World

To many outsiders, science is an arcane body of knowledge that has little relevance to day-to-day living. And many outsiders think that the scientist's work sounds dull and uncreative. Still others see science as a mysterious realm not easily understood by the general public.

The truth of the matter, however, is that science is an exciting field that requires tremendous creativity and insight. It is also essential to our understanding of the modern world and the many issues before us. And it is not that hard to master, once we become acquainted with its special vocabulary.

You don't need to become a scientist to understand what science is all about, but the time you spend improving your understanding of this fascinating field will pay lifetime dividends. Science may help you decide what type of house to build or what type of energy to heat your water with. It may even help you vote more intelligently on important issues. Almost without exception, the more you know about science, the better off you will be. Society as a whole benefits from science, too, as it charts a safe course into the future. Without it we could bumble from one mistake to another.

Throughout this book you will learn both some fundamental principles of science and many of the scientific facts behind the headlines. The laws of ecology, the laws of thermodynamics, and the principles of toxicology—to name a few—will make you a more informed citizen.

Values and Science

Your study of environmental science will yield many insights into human society and nature. What you do with that information, however, depends in large part on your values. Values are subjective. They dictate what is right and wrong. Science, on the other hand, is supposedly objective. It does not tell us what to do, but tells us only what the impacts of our actions will be. Environmental science helps society see what is not obvious, for instance, the importance of forests in maintaining carbon dioxide levels. Science can give us information on which we can plot action, but that action is the result of our values.

Our values forge political and economic decisions that determine the course of the world. Unfortunately, the link between science and values is not always so clear-cut. For example, Thomas Aquinas's view of nature reflected the feudal hierarchy that prevailed during the 13th century. Organisms, he said, are diverse but unequal. Each has a distinct obligation and role that benefits others. Furthermore, he believed that change was unlikely, just as it was unlikely for a serf to become a knight. Nature was viewed as rigidly structured and independent, unchanging. The scientific interpretation of nature was heavily tainted by the social conditions of the time.

Just as social conditions may have molded scientific thinking, so did science mold or reinforce social values. In 18th-century England, for instance, the theory of evolution was seen as a process of change that led to superior forms of life, a view that scientists now see as erroneous. To the 18th-century English, however, evolution meant that some life forms were better than others. This view of science was used to justify the prosperity of a small elite class of business owners while workers were suffering in dangerous factories. This disparity was justified as a form of the survival of the fittest. The application of the theory of evolution to social ethics is called **social Darwinism**.

This misapplication of Darwinian evolution suggests that scientific information must be carefully understood and not loosely used to justify social practices. Many scientists would agree. As this book points out, however, societal values can and should be based on laws of science, especially the practical rules of ecology. Values that do not take into account the scientific principles that govern life can be destructive and may limit the future of humanity.

The Methods of Science

The study of science is often orderly and precise. The process begins with observations and measurements of the world—the rate of oxygen consumption by cells, the growth of tree rings, or the rate of soil erosion (Figure S1-1). From observations, scientists often formulate generalizations or hypotheses. A **hypothesis** is a tentative explanation of what scientists observe, for example, an explanation of why a rattlesnake shakes its rattles or why a desert plant sheds its leaves in dry spells. Hypotheses based on observation and measurement are derived by a type of thinking called **inductive reasoning**. Inductive reasoning includes any thought process in which generalities, such as hypotheses, are derived from specific facts and observations.

Even though the term *inductive reasoning* may be foreign to you, you have probably been using the process for many years. It may have allowed you to solve the mystery of an anonymous love note in high school, for example, or perhaps to reason why a good friend was angry with you.

Let's look at a simple example. Suppose that you were driving your car at night. Each time you hit a bump in the road, one of your headlights flickered. But when you were on newly paved blacktop, the lights worked fine. Without being aware that you were engaged in the scientific method, your mind quickly searched for answers. Within seconds you arrived at a hypothesis: perhaps bumps in the road were jiggling the filament in the headlight, causing it to flicker. This simple exercise is inductive reasoning.

Once a hypothesis is made, it is up to the scientist—in this case, you—to determine how valid it is. To test the validity of their hypotheses, researchers perform experiments. You would no doubt do the same. For example, you might experiment by replacing the headlamp and then by taking your car out on the same roads to see if the problem had been corrected. If the headlight continued to flicker, you would conclude that your hypothesis was invalid.

In a similar fashion the results of scientific experiments either support or refute initial hypotheses. If a hypothesis is refuted, a new one is generally substituted. In your case you might conjecture that the electrical wiring was faulty. By simply wiggling the electrical wires connected to the headlight, you could test your new hypothesis. If the lights flickered, you would know that your new hypothesis was correct (and that you had wasted your money buying a new headlamp).

Scientific study is a lot more involved than finding reasons why your car stalls or your garbage disposal leaks. But the methodology is still the same. Observations lead to hypotheses, which are tested by experiments.

In many scientific experiments it is necessary to set up experimental and control groups of animals or people to test the effects of various treatments. The **experimental group** is the one that you "experiment" on, perhaps giving a new drug treatment. The second group is treated identically, except that it is not given the drug. This group is the **control** group. By setting up their experiments in this manner, scientists test the effect of a single variable, the drug. Any differences in disease rates should be the result of the experimental treatment.

Through careful study, scientists check out the validity of their own hypotheses. Other researchers might also test them by setting up the same experiment. Scientific knowledge grows little by little. As the facts accumulate, they often begin to create a larger picture of reality—a unifying concept that ties the facts together. This is a **theory**. A theory, therefore, is an explanation that accounts for the assembled facts. Atomic theory, for example, is the current explanation of atomic structure, knitting together many observations about the atom. Theories are not immutable, as you shall soon see.

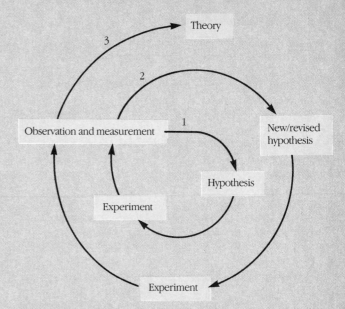

Figure S1-1 The scientific method starts with observation and measurement, from which scientists develop hypotheses, or tentative explanations. The scientist next tests the hypothesis by experimentation. If the results do not support the hypothesis, a new or modified hypothesis is created, which can then be retested and refined until a theory is formed.

The Rise and Fall of Theories

Scientists must be open-minded about theories and must be willing to replace their most cherished ones with new ones should new information render them obsolete. As new results are published, scientists often find that different interpretations of natural events and biological processes emerge. These interpretations may replace explanations that have prevailed for decades. As a result, it is often necessary to alter or even abandon theories that have enjoyed a faithful following for many years.

New research techniques cause much of the turmoil in the scientific community. The findings from such techniques shine light on old theories and shaky hypotheses, which sometimes crumble as a result.

Perhaps the best known example of changing theory is the Copernican revolution in astronomy. The Greek astronomer Ptolemy hypothesized in AD 140 that the earth was the center of the solar system (the geocentric view). The moon, the stars, and the sun, he argued, all revolve around the earth. This notion held sway for hundreds of years. In 1580, however, Nicolaus Copernicus showed that the observations were better explained by assuming that the sun was the center of the solar system. Copernicus was not the first to suggest this heliocentric view. Early Greek astronomers had proposed the idea, but it gained little attention until Copernicus's time.

The new view of the universe proposed by Copernicus was condemned as heretical by the Catholic Church. When he published his theory, it was placed on the papal index of forbidden books.

Supertheories, or Paradigms

The dominant set of assumptions that underlies any branch of science is called a **paradigm**, a word coined by the philosopher and historian of science Thomas Kuhn. A paradigm is likened to a "supertheory." It is the basic model or reality in any science. Evolution is an example from the life sciences.

Paradigms govern the way scientists think, form theories, and interpret the results of experiments. They govern the way non-scientists think, too.

Once a paradigm is accepted, it is rarely questioned. New observations are interpreted according to the paradigm; those that are inconsistent with it are often ignored. However, phenomena that fail to fit conventional wisdom may amass to a point at which they can no longer be ignored, causing scientists to rethink their most cherished beliefs and, sometimes, toss them aside. This unsettling event is called a **paradigm shift**.

The central ideas of biology, ecology, chemistry, philosophy, health, and education are all paradigms; sacred as they may seem, they are not immutable. As scientists make new observations, new theories emerge that may shake apart the foundations of a society. This book, and especially Part 5, examines the fundamental attitudes we hold toward nature (philosophical paradigms) and ways in which a shift in these attitudes could reshape human civilization.

Critical Thinking Skills

Science provides a great deal of information to modern society. One of the chief tools of a good scientist (and of a great many other professions) is **critical thinking**: thought based on logic and a careful analysis of factors that get in the way of logic. There's no single formula for critical thinking, but most critical thinkers would agree that several key steps are required for this important process.

The first requirement of critical thinking is a clear understanding of terms. Define all terms. Make sure you understand them and demand the same of others.

The second requirement of critical thinking is that individuals question the methods by which facts are derived. Were the facts derived from experimentation? Can they be verified? Was the experiment correctly run? Or is a generalization derived from a single observation? How easy is it for a single event to make a lasting impression and taint our thinking? A newscast showing an angry mob in New York, for example, may give the impression that the entire country is in turmoil.

A third rule of critical thinking requires us to question the source of the facts, that is, who is telling them. When the American Tobacco Institute argues that the link between cigarette smoking and lung cancer hasn't been proven, a critical thinker would be skeptical. When a business says that their pollution isn't causing any harm, one might again question the assertion. Even environmentalists are prone to exaggeration. They are fair game for your critical thinking skills. Sadly, bias taints our views and creates a distorted view of reality.

The fourth requirement of critical thinking is to question the conclusions derived from facts. Do the facts support the conclusions? There are a surprising number of examples in which conclusions drawn from research simply are not supported by the facts. For example, one of the earliest studies on lung cancer showed a correlation between lung cancer and sugar consumption. A careful re-examination of the patients showed that the wrong conclusion had been drawn. It turned out that cigarette smokers actually consumed more sugar than nonsmokers. Thus, the link between sugar and lung cancer was incorrect. The real link lay between smoking and lung cancer.

This example illustrates a key principle of science worth remembering: correlation doesn't necessarily mean causation. Just because the economy improves when a certain politician is in office doesn't necessarily mean that the politician and his or her policies had anything to do with the improvement. Bias and assumptions taint how we interpret our observations. They are impediments to critical thinking.

The fifth rule of critical thinking requires us to examine the big picture. In 1988, researchers at Monsanto announced that they had discovered a way to genetically alter wheat, making it resistant to a fungus that causes enormous crop damage each year. To control the fungus now, farmers often rotate wheat from year to year with crops that the fungus does not infect. Crop rotation prevents the fungus from proliferating, keeping the pest in check. With the new genetically altered strain, researchers say, farmers will not have to rotate their crops. They can plant wheat in the same field year after year and can even plant larger crops. This may sound good at first, but when one considers the bigger picture, it is clear that the solution is an invitation to disaster. Why?

Crop rotation helps build soil fertility. Rotating beans, clover, and alfalfa with wheat, for instance, adds nitrogen to the soil. This helps maintain soil fertility. Crop rotation also prevents insect pest populations from getting out of hand, because their food supply is not constant. Eliminating crop rotation could result in an outbreak of insect pests, requiring more pesticide application.

In solving the fungus problem, then, science may contribute to several more problems. A careful examination of the bigger picture—the ecological relationships—often throws into question the apparent wisdom of new actions.

The Population, Resource, and Pollution model presented in Chapter 1 provides some insight into the big picture. To expand our understanding of relationships, additional models can be useful.

Multiple Cause and Effect Model

Human thinking is often narrow and simplistic, especially when it comes to understanding problems and finding solutions to them. Critical thinking demands a broader view of causes and effects, an approach that will increase our insight into environmental problems and increase the likelihood of solving them. Consider an example: In the late 1960s and early 1970s, when environmental issues came to the forefront of attention, two noted scientists stimulated a debate over the underlying causes. Paul Ehrlich, a Stanford University biologist, argued that the root of all our pollution and resource problems was overpopulation—too many people. His argument was hotly contested by another biologist, Barry Commoner, who argued that technology was primarily to blame. Their debate engaged but misled the scientific community and the public for years. On closer examination it becomes clear that both scientists were

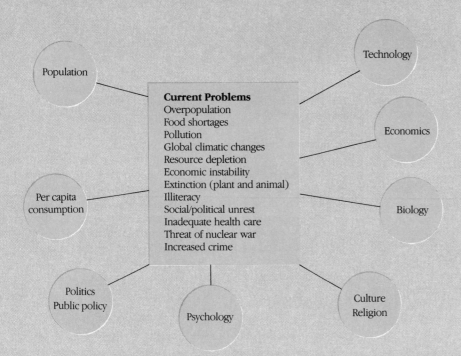

Figure S1-2 Multiple Cause and Effect Model for analyzing specific environmental issues. The circled factors contribute in many ways to the problems facing the world. Understanding the many causes of a problem helps us to find better solutions.

wrong. The complex problems of poverty, resource depletion, pollution, wildlife extinction, and food shortages are not the result of technology or overpopulation alone but rather are caused by a variety of factors, as shown in Figure S1-2.

To illustrate the multiple etiology of environmental problems further, we will use the Multiple Cause and Effect model (Figure S1-2) to look at wildlife extinction, for which commercial and sport hunting and habitat destruction are often blamed. As we will see, this view—like Ehrlich's and Commoner's—gives us a simplified picture of a problem that has many hidden causes. To see why this is so, and to see how the model is used, let us examine each element of the model individually.

Population

The human population destroys wildlife habitat in many ways. We build roads through forests once occupied by bears; we build airports and housing tracts on land that once provided food and shelter for birds and other animals; we pollute streams once the home of fish and otters; we strip-mine for coal and gravel, in the process destroying vegetation that was once consumed by deer. These activities enhance human survival and prosperity but decrease the amount of wildlife habitat. The extent of environmental damage depends on numerous population factors such as population size, growth rate, and geographical distribution. For example, the larger our population, the greater its impact. Likewise, the faster our population grows, the more impact we have. But population is only one of many underlying causes of extinction.

Per Capita Consumption

The amount of resources used by each member of society— per capita consumption—also plays a key role in the destruction of wildlife habitat. For example, in the United States per capita

consumption of water has increased 300% in the past 80 years. Electrical energy consumption has more than doubled in the past 20 years. This rapid increase in resource consumption strains the earth's available land and water resources. It also pollutes the planet. The cars we drive pollute the air animals breathe; the oil we use to make the gasoline that powers our automobiles can, if accidentally released, pollute the ocean and rivers. Thus, how much a society consumes also plays a significant role in wildlife extinction.

Politics and Public Policy

The legal system also has an important role in wildlife extinction. Laws can affect how much habitat is destroyed, how much hunting and poaching occur, and which species will or will not be hunted. In addition, laws influence population growth, which, in turn, affects wildlife habitat. (For instance, tax laws give families deductions for each child, reducing the financial burden of rearing children, and thus may promote population growth.) Laws also affect resource acquisition and use, for example, how much coal is mined, how it is mined, and how reclamation is carried out. Before 1977 there were few legal controls on surface mining. At that time more than 1 million acres of land had been surface-mined in the United States, and about two-thirds of this land (mostly in the East) had been left barren. New laws now require the revegetation of surface-mined land and create a fund for reclaiming abandoned, unreclaimed lands. Thus, laws must also be taken into account when discussing animal extinction.

Economics

Economics is a key element in animal extinction. It plays an important role in determining how resources are acquired and used. For example, surface mining of coal is preferred to underground mining worldwide because it is much cheaper. Because surface mining is more environmentally destructive, the decision to acquire coal this way, dictated by economics, affects wildlife habitat.

Economics affects many of our daily business decisions. On the surface, there is nothing wrong with this. However, economics often fails to account for external costs: pollution, habitat destruction, and other environmental impacts that are not entered into the cost of producing goods. Air pollution from factories, cars, and power plants, for example, may cause up to $16 billion worth of damage to crops, health, forests, and buildings in the United States each year. These external costs are borne by members of the public who suffer ill health, pay a higher price for food, or pay for repairing buildings. If companies invested in pollution control devices, the external costs would be greatly reduced, but consumers would pay a higher price for energy and goods. Thus, a system of economics in which the consumer pays to reduce the external cost would diminish environmental pollution and its threat to wildlife habitat. But because economic thinking often ignores economic externalities, it, too, is a prime causative agent in the reduction of the world's wildlife.

Psychology, Culture, and Religion

How we behave toward the environment is a function of our attitudes. Underlying all of our behavior is a psychology that tends to put immediate needs before the long-term good of the environment and future generations. Rather than thinking about the future, we tend to be concerned with things that affect us in the short term and ways we can make the present as comfortable as possible. Although this type of thinking is shortsighted, it is very much a biological characteristic. Survival in the animal world is generally the result of animals' satisfying immediate needs by using environmental resources. Being animals ourselves, we tend to think and act with an eye to the immediate. Because our population is so large and our technologies are so well developed, however, short-term thinking leads to many problems, such as wildlife extinction.

Another crucial psychological element is the way we perceive our place in the environment. Do we see ourselves as superior to and at odds with the natural world? Or are we a part of nature, willing to live in accordance with ecological laws? An attitude of being superior and separated from nature predominates today. It gives many license to run roughshod over wildlife and plants, contributing to the extinction of many species.

Social, cultural, and religious factors to a large extent determine our psychological makeup. Cultural and religious attitudes in which humans are viewed as supreme and apart from nature, as we have seen, can have devastating effects on wildlife. If animals are revered for cultural or religious reasons, however, extinction from hunting and habitat destruction may be curbed.

Technology

Technology, like population, is only one of a complex of factors contributing to wildlife extinction. We typically think of technology as an instrument or tool to make better use of the world's resources in meeting our basic survival needs. Technology offers us many advantages in the struggle for survival. As a result

of technological developments we travel freely around the world and through space, inhabit new regions, and acquire resources from distant sites. Technology allows us to prosper in environments where survival might otherwise be impossible.

Our travels, new settlements, and resource acquisitions all contribute to the shrinking supply of wildlife habitat. This happens partly because technology enhances population growth. The more people there are, the more resources are needed and the greater the conflict over living space.

Technological advances have spelled trouble for wildlife for other reasons, too. For instance, improvements in guns used for whaling have allowed humans to severely deplete the populations of many whales. New drugs have eradicated diseases, reducing infant mortality; in the process, though, they have contributed to rapid population growth. New chemical pesticides have increased agricultural productivity but have also devastated populations of birds and other animals. Although technology has been a boon to society, it has exacted a price, too.

Biology

Numerous biological features also play a role in wildlife extinction: adaptability, number of offspring, and sensitivity to environmental pollutants are a few. Highly specialized species, like the California condor, are generally unable to adapt to changes brought about by human beings and are more vulnerable than less specialized animals like the coyote, which seems immune to human presence. The number of offspring a species produces also affects the resistance of the species to human pressures such as hunting, habitat alteration, and pesticide use. Susceptibility to environmental pollutants also varies considerably among plants and animals.

The Multiple Cause and Effect model enlarges our focus and contributes to a more organized, scientific view of the world around us. It will help us avoid the oversimplified diagnosis and treatment of contemporary problems.

Suggested Readings

Cole, K. C. (1985). Is There Such a Thing as Scientific Objectivity? *Discover* 6 (9): 98–99. Insightful look at science and the scientific method.

Hardin, G. (1985). *Filters against Folly: How to Survive Despite Economists, Ecologists, and the Merely Eloquent.* New York: Viking. Eloquent writing on scientific bias. See Chapters 1–7.

Kelley, D. (1988). *The Art of Reasoning.* New York: W. W. Norton and Co. Especially good is Part 4, "Inductive Reasoning."

Klemke, E. D., Hollinger, R., Kline, A., eds. (1988). *Introductory Readings in the Philosophy of Science.* New York: Prometheus Books. An excellent introduction to critical thinking and its application in science; covers science-vs-nonscience, explanation and law, theory and observation, confirmation and acceptance, science and values, and science and culture.

Kuhn, T. (1970). *The Structure of Scientific Revolutions.* Chicago: University of Chicago Press. The original description of paradigms.

Milbrath, L. W. (1989). *Envisioning A Sustainable Society: Learning Our Way Out.* Albany: State University of New York Press. Asks important questions about underlying thrust and philosophy of some fields in modern science.

Rifkin, J. (1985). *Declaration of a Heretic.* Boston: Routledge and Kegan Paul. Part 1 of this book examines our modern scientific view.

2

New Visions of Life:
Evolution of a Living Planet

Nobody knows the age of the human race, but everyone agrees that it is old enough to know better.

ANONYMOUS

In 1948 the noted British astronomer Sir Frederick Hoyle predicted that "once a photograph of the earth, taken from the outside, is available . . . a new idea as powerful as any in history will let loose." It was not too many years later that the first photograph of earth from outer space came to us, and Hoyle's prediction bore fruit. The earth, a sparkling red, white, and blue planet circling around the sun, came into sharp view in the minds of many people throughout the world. The vision was both breathtaking and disturbing, for it showed our home, which we had always seen as inexhaustible, as a tiny sphere alone in a mighty universe that extends beyond our imaginations. Sparkling in the sun's rays, the earth seemed exquisite, fragile, and, knowing how we have treated it, vulnerable.

The late Buckminster Fuller, well known for his contributions to environmental science, dubbed this tiny planet **Spaceship Earth**, likening it to a self-contained spacecraft whose life-support systems recycle all matter necessary for astronauts to survive. This mechanistic analogy stuck. It drove home one of the most important lessons of ecology: that all life on earth is part of an enormous recycling system. Nutrients are a common thread from the nonliving environment to the plants to animals and back again, binding together the web of life on the earth.

Subsequent journeys into space have taught us important new lessons. Lewis Thomas, a biomedical researcher and philosopher, captured one of the newest and most popular lessons when he wrote, "The astonishing thing

21

about the Earth . . . is that it is alive." New evidence from space suggests that the biosphere regulates its physical and chemical environment in much the same way that organisms control their own internal environment.

The intricate control of life's essential processes is now threatened by human civilization. Through our use of fossil fuels such as coal and oil, some noted scientists say, the planet's temperature balance is being thrown out of kilter, causing a rise in global temperature that changes the earth's climate and raises the sea level. Even the powerful ozone layer that screens out harmful ultraviolet rays is being slowly eroded by air pollutants. Widespread skin cancer and serious plant damage may result.

How did we reach such a precarious place? To answer this question we will journey back in time 15 billion to 20 billion years to the beginning of the universe to trace the origin of our solar system and the emergence of life. Next we will look at the beginnings of humankind and the ways in which human society has developed.

Origin of the Earth

Gaze out at the stars at night as far as you can see with the naked eye. Billions upon billions of stars float overhead in the dark void of space. Your eyes behold only a tiny fraction of the universe, for all stars you see are part of the Milky Way galaxy, a grouping of perhaps a trillion stars along with gas and dust. The Milky Way galaxy also houses our solar system—the sun and its nine orbiting planets. Beyond the Milky Way are countless other galaxies, a great, immeasurable sea of space and matter.

Unfathomable though they may be, the dimensions of space can be better understood by considering these facts: Light travels at a rate of 300,000 kilometers per second (186,000 miles per second). At this speed a beam of light can travel 7.5 times around the earth's equator in a second. By comparison, a supersonic jet traveling at the speed of sound (340 meters per second) would cover the same distance in 1.5 days. Now suppose you could board a spaceship traveling at the speed of light to travel across the Milky Way galaxy. This modest journey into space would take 100,000 years. Since a **light year** is defined as the distance that light travels in a year, the Milky Way galaxy is about 100,000 light years across.

The Milky Way galaxy is only one of 20 galaxies that belong to a small galaxy cluster called the **Local Group**. And the Local Group is only one of many clusters of galaxies in the universe. For your final journey suppose you decided to travel from earth to the Andromeda galaxy, located on the periphery of the Local Group. This journey to the edge of the Local Group would take 2.2 million years at the speed of light.

Formation of the Universe

Astronomers believe that the universe began 15 billion to 20 billion years ago. All matter making up the universe today, they think, was compressed into an infinitely small and infinitely dense volume that exploded with a fury unknown to humankind. This explosion is appropriately called the **big bang**. Physicists estimate that the temperature after the explosion was an incredible 100 billion degrees celsius—far hotter than the centers of the hottest stars. As in a nuclear blast, energy and matter catapulted from the center of the explosion and streamed out into space, where it would give rise to the universe over many millions of years.

For about a million years, in fact, so much energy existed that stable atoms could not form. **Atoms** are the fundamental units of all matter. They consist of three smaller particles called **subatomic particles**—protons, neutrons, and electrons. **Electrons** are negatively charged particles that spin furiously around the **nucleus**, as shown in Figure 2-1. The nucleus is the dense central region of an atom consisting of **protons** and **neutrons**. These particles are much larger and heavier than electrons and constitute the bulk of the atom.

When the universe had cooled, after a million years of continuous expansion, protons, neutrons, and electrons combined to form the first stable atoms, hydrogen and helium. These materials, over time, gave rise to the remaining elements, which ultimately formed the galaxies and solar systems, and you and me as well.

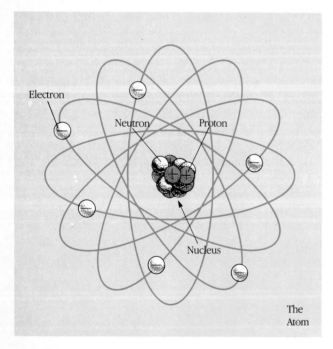

Figure 2-1 Atoms consist of a dense central region, the nucleus, containing protons and neutrons. Electrons orbit around the nucleus in the electron cloud.

Contrary to popular misconception, the scientific evidence supporting the big bang is surprisingly complete. The big bang theory is based on a rather surprising amount of observation and calculation. Careful measurements, for example, show that the galaxies, huge clusters of stars described below, are still rushing apart at very near the speed of light—propelled by this ancient explosion. Extrapolating back in time, scientists conclude that all galaxies must have been much closer at one time, so close in fact that neither galaxies, nor stars, nor even atoms or atomic nuclei could have had a separate existence.

Formation of Galaxies and Stars

The universe was slow to cool after the fiery big bang. A million years afterward the universe was still 3000° C (5400° F). All matter and energy were racing from the point of explosion in a uniform flood of energy and mass. Within the expanding universe at this time, however, huge clouds of matter began forming. These clouds, called **protogalaxies**, were only slightly denser than the surrounding material. They were to give rise to the star-studded galaxies.

The fate of protogalaxies was rich and varied. Some merged with others to form supergalaxies; others shrank and formed dwarf galaxies. All had one thing in common: the cloud of cosmic dust condensed into a huge, swirling mass. Condensation was brought about by gravitation, the mutual attraction of matter within the cloud, causing the cloud to contract. Over time the protogalaxies continued to contract, and as they did, they began to spin faster and faster. Spherical clouds compressed into the flattened shapes typical of many known galaxies (Figure 2-2).

Billions of years later, out of the protogalaxies stars were born. The swirling clouds of matter fragmented into smaller clumps, like lumps in your Thanksgiving gravy, which scientists call **protostars**. Collapsing under gravitation, the protostars grew denser and hotter. When a critical density was reached in the core of the protostar, the temperature became so high that nuclear fusion reactions began. Each **nuclear fusion** reaction in stars fuses four protons (hydrogen nuclei) to form a helium nucleus. In the process, energy is given off in the form of heat and light. Many nuclear reactions every second in stars like our sun allow them to shine brightly for billions of years. Once the nuclear reactions begin, a protostar's collapse ends, and a star is born. Billions of stars formed within the swirling galactic dust, and even today new stars are constantly being formed in the galaxies.

Formation of the Solar System

Within the galaxies solar systems also formed. A **solar system** is a group of planets revolving around a star. Our

Figure 2-2   Side view of a typical spiral galaxy.

solar system, consisting of the sun (a medium-sized star) and nine planets in elliptical trajectories around it, formed about 4.6 billion years ago. Astronomers now know much about the formation of our solar system because of evidence of other solar systems in formation in the Milky Way. This process began with a spherical cloud of dust and gases in the galaxy. Many astronomers believe that an exploding star, or **supernova**, near this cloud of gas triggered the contraction and subsequent condensation of the gas and dust. In the center a star formed—the sun. Leftover materials on the periphery formed the planets by accretion of cosmic dust, much as a snowball grows if you roll it through wet snow. The planets remain in orbit around the sun, drawn to it by gravitation. Their surface temperature is primarily determined by their distance from the sun. Were it a little closer to the sun, the earth would be too hot to support life; were it a little farther away, it would be too cold.

The Evolution of Life

The primeval earth was at first a spherical mass of rock and metal. The condensation of matter, radioactive decay, and solar heat melted this material, turning it into a molten mass. Over time the earth cooled, radiating its intense heat into space. Heavy molten metals, nickel and iron, sank inward and formed the **core** (Figure 2-3). The least dense materials remained on the surface and cooled to form a thin outer **crust**. Sandwiched between the two was the **mantle**, made up of materials of intermediate density. Geologists estimate the age of the earth to be 4.6 billion years.

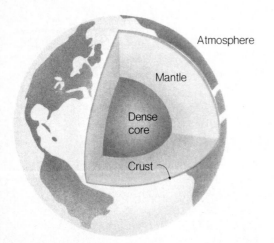

Figure 2-3 Cutaway model of the earth showing the crust, mantle, and core. Each of the inner planets of the solar system (Mercury, Venus, Earth, and Mars) has a differentiated structure, in which a dense, iron-rich core rests beneath a less dense mantle made primarily of silicon and oxygen. The outermost few kilometers form the rocky crust of the planet.

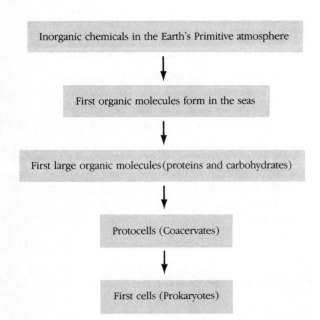

Figure 2-4 The Russian scientist Oparin's hypothesis on the formation of life.

How do scientists make such determinations? The age of fossils and rocks can be determined by complex scientific procedures that measure their levels of radioactive material, then use knowledge of the decay rate, that is, the rate of radioactive emission, to calculate the age of rocks. The technique used to determine the age of the earth relies on measurements of a radioactive material called rubidium. It gives off radiation, but very slowly. The time required for a radioactive material to emit half of its radiation is called a **half life**. Rubidium has a half life of 49 million years. Through complex mathematical equations, scientists can determine the age of ancient rocks. These have helped them determine the age of the earth as well.

Because of the earth's own heat and its proximity to the sun, lighter elements such as hydrogen and helium gas were driven off. An atmosphere could not form until the sun and the earth had both cooled.

Chemical Evolution of Life's Molecules

Over time an atmosphere emerged around the infant planet earth. It was first made of small inorganic molecules—water vapor, carbon dioxide, ammonia, hydrogen—and one organic compound—methane. Many of these came from the gases emitted by volcanoes. Biologists believe that these simple molecules gave rise to life (Figure 2-4). But how did such wondrous living things as saguaro cacti, redwoods, and blue whales emerge from such an inauspicious beginning?

For millennia as the earth and its atmosphere cooled, water fell from the skies and filled the oceans. Dissolving inorganic and organic molecules of the primitive atmosphere as it fell, rain formed the oceans and lakes. Ultraviolet light from the sun, heat from volcanoes, and lightning energized the molecules dissolved in the waters, causing them to react and form organic molecules that would give rise to the first living things. Thus, the dilute inorganic soup was turned into an organic broth containing many of the key elements of life.

In the shallow pools along the margins of the seas, these organic compounds united to form simple proteins and other essential organic molecules (Figure 2-4). Piece by piece the organic chemicals needed for life began to appear.

The idea that organic compounds formed from the chemicals found in the earth's primitive atmosphere is called the **theory of chemical evolution**. A Russian scientist by the name of Oparin first proposed the theory, but an American chemist, Stanley Miller, set out to test this unusual notion while still a graduate student. Miller devised an apparatus to test the hypothesis (shown in Figure 2-5). This glass structure was first sterilized to avoid contamination. It was then filled with methane, water, ammonia, and hydrogen gas. Electrical sparks and heat provided energy, in much the same way that volcanoes, lightning, and sunlight energized the early chemical reactions on earth. After one week the clear mixture of gases and water turned brown. Chemical analyses revealed the startling truth for which Miller would receive a Nobel Prize: simple organic molecules necessary for life could be formed in conditions similar to those found many millions of years ago on earth. Miller's experimental apparatus yielded several amino acids, some simple carbohydrates, urea, and a host of small organic compounds. All of a sudden, Oparin's far-fetched hypothesis seemed plausible. But, scientists wondered, could larger molecules, such as proteins, be formed under similar conditions? Subsequent research showed that they could. Amino acids heated in a sterilized apparatus united to form small proteins. A whole host of larger organic chemicals were also produced under similar conditions. Ultimately, researchers found that virtually all of the essential organic building blocks of life could be formed in conditions thought to mimic those found on earth some four billion years ago.

How cells arose from the organic stew, no one knows. But biologists believe that proteins and other polymers (long chains of molecules, also including the carbohydrates) combined to form **protocells**, small globules that looked and even acted like living organisms (Figure 2-6). Laboratory studies show that such amalgamations can grow, divide, and selectively take up materials from the environment, much like living organisms. Over millions of years, scientists believe, these simple aggregates of organic molecules took on a life of their own. They developed ways to capture energy from the sun, ways to reproduce themselves, and a way of storing genetic information: DNA.

The First Cells

The Emergence of Microbes
The first cells on earth were probably bacteria that arose at least 3.7 billion years ago, according to the latest findings. They may have lived off the organic molecules that formed earlier. An astonishing development of this exciting but little understood period of evolutionary history was the emergence of photosynthesis. **Photosynthesis** is a series of reactions by which organic molecules, such as sugar, are made from carbon dioxide, water, and energy from the sun. Plants and algae are the principal photosynthetic organisms alive today. During the early stages of evolution photosynthesis solved an important problem facing the newly forming biological world: how to replenish the rapidly declining supplies of organic materials in the oceans, the only foodstuffs available to the early cells. Photosynthesis provided a way for organisms to make organic materials from atmospheric carbon dioxide, which was abundant but was not being converted fast enough abiotically (through chem-

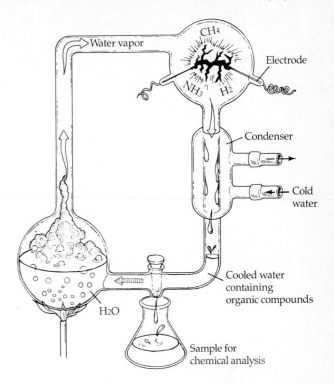

Figure 2-5 Stanley Miller's experimental apparatus in which he demonstrated that organic molecules necessary for life could form from inorganic molecules in the earth's primitive atmosphere.

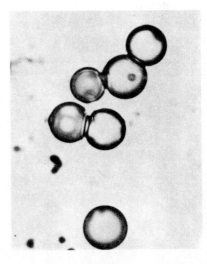

Figure 2-6 Proteinoids. These protein globules were made in the laboratory under conditions that simulated the earth's primitive environment. It is thought that proteinoids or other similar globules gave rise to the first cells, and for this reason they are called protocells.

ical evolution) to keep up with the demands of the billions of new cells inhabiting the oceans. Without photosynthesis life today would be drab indeed, most likely limited to giant bacterial colonies living along the coastlines, feeding off one another as they perished from natural causes. The land surface would be barren rock without plants or animals.

At first photosynthesis in bacteria was a relatively simple process that captured sunlight and generated organic matter. Unlike photosynthesis in plants today, it generated no oxygen. Over millions of years, however, photosynthesis changed in ways that would once again revolutionize life: it began to produce oxygen by splitting water molecules. Thus arose the cyanobacteria (once called blue-green algae). Before them, little free oxygen could be found in the atmosphere because it quickly reacted with chemicals. With the advent of this remarkable new process, oxygen slowly began to accumulate in the atmosphere. The first appreciable amounts were found 2.2 billion years ago.

These accumulating levels of oxygen precipitated a crisis, the first global pollution. Oxygen was probably harmful to the single-celled organisms that lived in the shallow seas. It combined with organic molecules and destroyed them. Biologists believe that many species that did not require oxygen perished over the ensuing millennia. Others sought refuge deep in muds. Still others adapted, putting this former toxin to good use and forming a whole host of new species.

The Emergence of Eukaryotic Cells   Undoubtedly, one of the most important results of the global oxygen crisis was the evolution of aerobic (oxygen-requiring) cells, complete with nuclei and energy-releasing organelles. These nucleated cells are called **eukaryotes**; the first fossil remains of these cells appear in rock estimated to be 1.5 billion years old. Amoebas and paramecia are two modern forms that most closely resemble the first eukaryotic cells (Figure 2-7).

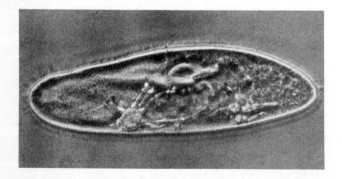

Figure 2-7   Early eukaryotic organisms no doubt resembled this paramecium commonly found in fresh water today. Magnified 400 times its original size.

The most widely accepted explanation for the origin of eukaryotes is called the **theory of endosymbiotic evolution**. It holds that the many cellular organelles in eukaryotic cells—the tiny inclusions such as mitochondria and chloroplasts, where specific functions are carried out apart from the rest of the cell's metabolism—had once been free-living bacterial cells but were engulfed by other cells. This endosymbiotic union is thought to have enhanced the survival of both cells. In their new home the internal symbionts flourished, becoming the cell's organelles. Biologists believe that eventually a permanent relationship beneficial to both developed and persisted so that organelles could be passed from one generation to the next, *ad infinitum*. (For more on cells see Chapter Supplement 2-1.)

This odd relationship, like photosynthesis, was a major leap in evolution. Many new single-celled organisms arose. New and markedly different strategies for survival evolved, and out of this flood of single-celled life emerged a world of many-celled plants, animals, and fungi (Figure 2-8).

After the formation of eukaryotes the pace of evolution greatly accelerated. More than three billion years had been required to make the first nucleated cells, the eukaryotes. Five hundred million years more would pass before the sponges emerged. Four hundred million years would lumber by before mollusks (such as snails) appeared. But in a flash of geologic time plants emerged from the waters and colonized the land. In even less time animals came ashore to make the earth their home. From the land animals arose the ancestors of human beings, who in a mere four to six million years have advanced from gathering seeds and fruits to space exploration and genetic engineering, a technology so powerful it could alter the course of evolution.

The Process of Evolution

To understand how we got here and the profound effects our species can have on the course of evolution, we must understand a little about the process of evolution. Evolution explains how life formed on the earth. Evolution also explains how life has changed over many millions of years and why it is so diverse today.

Most of what we know about the early history of life comes from fossils—preserved bones, imprints of organisms captured in rock, or even footprints of early animals. By examining these fossils embedded in rock strata whose age can be determined through various means, scientists have been able to construct a biological history of the earth.

The **theory of evolution** is often attributed to Charles Darwin, a 19th-century British naturalist. Darwin, however important, did not come up with the idea; evolution was already being widely discussed in his time, and the

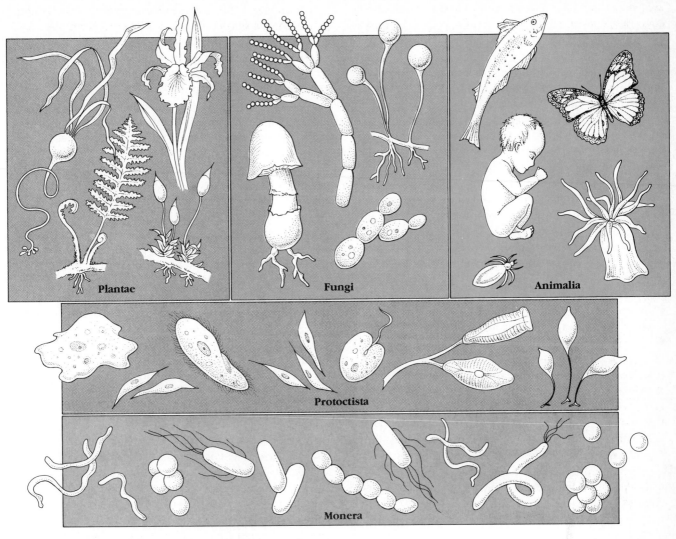

Figure 2-8 The world's organisms are divided into five major kingdoms. The first to arrive were the monerans, single-celled organisms represented today by bacteria. These gave rise to the protists, the single-celled organisms with distinct nuclei and cellular organelles. These, in turn, gave rise to plants, animals, and fungi during millions of years of evolution.

idea dates back to ancient Greece. What Darwin did was to propose a believable mechanism by which evolutionary changes could come about. This was his theory of **natural selection**, the idea that organisms best adapted to their environment are likely to leave more surviving offspring than other individuals, so that future generations tend to resemble the successful reproducers, as discussed in more detail below.

Darwin's ideas on natural selection were controversial at the time, for they opposed a prevalent religious belief that God gave us the diverse array of living creatures. Consequently, Darwin delayed publishing his theory until 1859, more than 20 years after he had first proposed the idea of natural selection, and then went ahead—as legend

has it—only because Alfred Russell Wallace was about to publish a virtually identical theory.

Like many other great ideas, natural selection took many years to be understood and appreciated. Not until the 1940s did Darwin's and Wallace's ideas on natural selection become widely accepted, nearly a hundred years after they developed this revolutionary concept. Even today the ideas of evolution and natural selection have their critics.

Mutation and Variation   Evolution is a process in which species change over time, making them better adapted to their environment. In extreme cases new species may emerge. In sexually reproducing species the process begins by mutations in the genetic material, or DNA, of the germ cells (those that produce ova in females and sperm in males). DNA is a storehouse of information needed by the cells of the body to carry out their many functions. Sections of the DNA molecule, called **genes**, regulate cell structure and function. Mutations in the genes may be caused by ultraviolet light or other high-energy radiation, chemicals in the environment, and cosmic rays from the sun. (For more on mutation see Chapter 14.) Many mutations are quickly repaired by the cells. Some mutations may be harmful and can lead to the cell's death. Harmful mutations passed to offspring can result in birth defects or fetal death. Other germ cell mutations are neutral; that is, they have no effect at all. Still others, which provide the basis for evolutionary change, increase survivability of the offspring, for example, by improving an organism's ability to escape predators, tolerate cold temperatures, or find food. Genetically based characteristics such as these, which increase an organism's chances of passing on its genes, are called **adaptations**. Genetic changes may also arise during the production of ova or sperm cells, known as gametes. During the production of germ cells, genetic material may be shifted from one chromosome (long strand of DNA with associated protein) to another. This process is called **crossing over**. The result is new genetic combinations, some of which may provide a selective advantage to the offspring. Another source of new genetic combinations is sexual reproduction. **Sexual reproduction** occurs when offspring are produced by a union of sperm from males and ova from females. Each offspring contains half of each parent's genetic information. Offspring, therefore, represent a new genetic combination that may provide unforeseen benefits.

Sexual reproduction, beneficial genetic mutations, and crossing over lead to genetic **variation** in individuals within a population, creating a broader genetic base that gives rise to structural, functional, and behavioral variation among individuals. This variation provides more leeway to cope with changing environmental conditions, resulting in some members that are better adapted to environmental conditions than others and hence more likely to reproduce and leave offspring behind. For these reasons biologists refer to variation as the "raw material of evolution." Without variation, evolutionary change would be impossible.

Favorable mutations and crossing over may give an organism an advantage over other members of the same species, called a **selective advantage**. This translates into more successful reproduction. The offspring of the more successful organisms gradually increase in number within a population and, over time, come to predominate. In time the genetic composition of the population, or the **gene pool**, will shift. If it changes significantly, a new species may form.

Adaptation   The molding or driving force of evolution is natural selection. Darwin described natural selection as the process in which slight variations, if useful, are preserved. It is important to note that the environment does not in any way *direct* these changes in the genetic material; the mutations are spontaneous. The environment only preserves ("selects") those organisms in a population with new traits that confer some advantage over the rest. Natural selection is the mechanism that results in organisms better adapted to fit their environment. Some think of natural selection as survival of the fittest.

Fitness is commonly thought of as a measure of strength or survivability. Strictly speaking, this is not right. To a biologist, **fitness** is really just a measure of reproductive success and thus the genetic influence an individual has on future generations. By definition, the fittest individuals leave the largest number of descendants in subsequent generations. They make up a higher proportion of future generations. Their influence on the genetic makeup of those generations is, therefore, greater than less fit individuals. Natural selection favors the fittest individuals. These individuals are said to be well adapted, but by no means are they perfectly adapted. Evolutionary theory, therefore, tells us that over time organisms will come to match their environment closely but probably not perfectly. This process of continuous matching is called **adaptation**.

Speciation   Genetic changes in organisms can be quite drastic. In some cases these changes can result in the formation of new species, a process called **speciation**. A **species** is a group of organisms that is similar in form, function, and behavior and that generally cannot interbreed with other species, or, if it does, produces infertile offspring. The rhinoceros and the water buffalo, for example, are different species. They look different, they behave differently, and they cannot interbreed.

How does speciation occur? Perhaps the most common mechanism of speciation occurs when members of a species are separated (Figure 2-9). In scientific language this is called **geographic isolation**. Geographic isolation can result from new mountain ranges or rivers forming in an organism's natural range. Separated by physical barriers and exposed to different environmental conditions, the two populations, over time, can evolve in quite different directions. If the populations are separated long enough, their members may lose their ability to interbreed. This is called **reproductive isolation**. When geographic isolation leads to reproductive isolation, new species are always the result. When the new species emerge in different regions, the process is appropriately called **allopatric speciation** (*allopatric* is derived from the Greek for "other" and "fatherland"). New species may also form without geographical isolation. This is common in plants and is termed **sympatric speciation** ("same fatherland").

Figure 2-10 illustrates an important evolutionary process leading to the development of new species. As shown in the figure, the ancient reptiles gave rise to terrestrial, aquatic, and flying species. This process, in which one life form gives rise to many others that occupy different environments, is called **adaptive radiation**. It has also occurred in mammals. This diversification process may occur on a large scale, as in the case of the radiation of reptiles; a medium scale; or a much smaller scale. Darwin saw evidence of adaptive radiation on a small scale when his travels took him to the Galápagos Islands, off the coast of Ecuador. Here he found 14 species of finches that probably arose from a single species of mainland finch (Figure 2-11). Occupying different islands, the mainland finch evolved into new species, diverging to make best use of the varying food sources.

Unrelated species adapt to similar environments in similar ways—for example, bats and flying dinosaurs, lions and carnivorous tyrannosauruses, as in Figure 2-10. Evolutionary biologists call this tendency for organisms to develop the same types of adaptations in response to similar environmental conditions **convergent evolution**.

How rapidly a species evolves is the subject of much debate. For many years evolutionists believed that such changes occurred gradually over many millions of years. They coined the term **gradualism** as a result. If this were true, paleontologists reasoned, the fossil record should contain many intermediate forms of organisms, a sort of geological recording of the gradual transformation. With few exceptions, however, the fossil record shows few intermediate stages. This gap led two noted paleontologists, Stephen Gould of Harvard University and Niles Eldredge of the American Museum of Natural History, to propose a slightly different theory: that the history of life has been characterized by long, fairly quiet periods punctuated by times of fairly rapid change. This theory is called **punctuated equilibrium**. It accounts for the lack of geological evidence to support gradualism.

Coevolution When most people think about natural selection they think of environmental conditions as the chief agents of evolutionary change. However, they are only part of the picture. Organisms themselves are also selective agents. Owls, for example, hunt voles (small mouselike rodents) and mice at night. Changes in the owls' ability to find prey could profoundly affect the prey population, in much the same way that a change in weather would weed out the less fit. But the relationship does not end here. Changes in the voles' ability to elude the barn owl can act as a selective force on the predator as well.

When members of two species interact, changes in one species can result in changes in the other. Thus, each species generates selective forces that can affect the evolution of the other. This process is called **coevolution**.

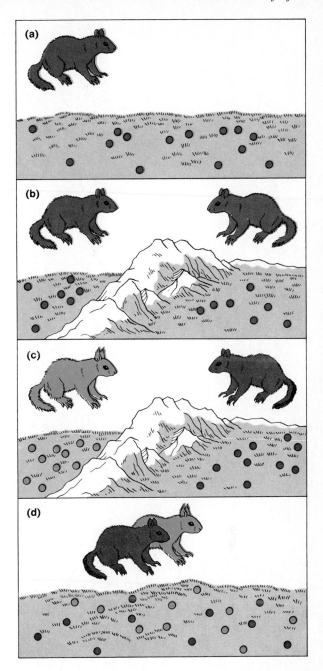

Figure 2-9 Geographical isolation. When a population is split because of a new geographic barrier, new populations may arise over time because the two groups are subjected to different selective forces. This illustration shows how the Grand Canyon split the population of long-eared squirrels, over time creating two new species.

Coevolution has been likened to an arms race between predator and prey. Thus, each improvement in predatory ability is followed by an improvement in the prey's ability to avoid or resist attack. Coevolution may also occur between plants and the animals that feed on them. Hard evidence for coevolution, say some scientists, is still lacking.

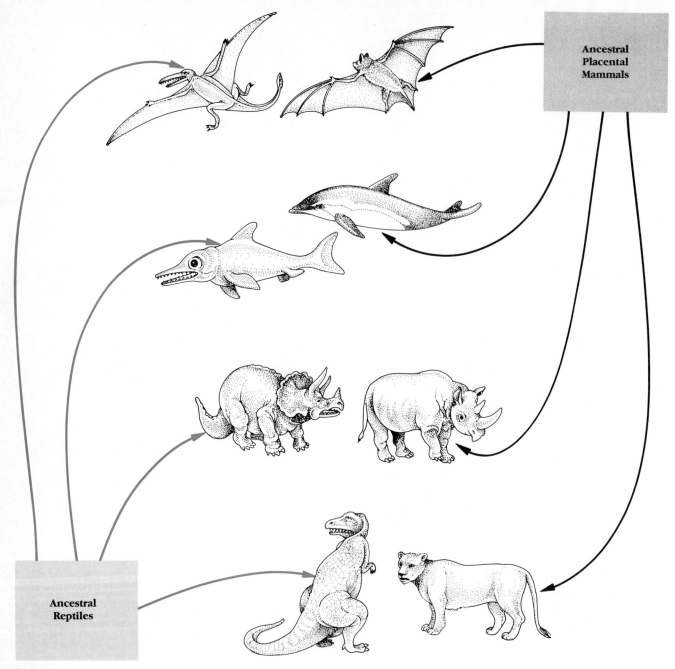

Figure 2-10 Adaptive radiation occurs as organisms evolve to fill ecological niches, making best use of available resources. The drawing shows the niches first filled by dinosaurs and later in evolution by placental mammals.

Applied Evolution Why study evolution in a course on environmental science? For one, it helps us understand just how our rich natural resources—trees, wildlife, fish, and the like—got here and how long it would take nature to evolve replacements. Second, it helps us understand the importance of our impacts on other living things. Consider, for example, the evolution of penicillin-resistant bacteria.

Penicillin was the wonder drug of the 1940s and 1950s, introduced to treat bacterial infections. Today, however, 90% of the staphylococci, common and troublesome bacteria, are resistant to this drug. What happened? Over the years, continued use of penicillin created superstrains of penicillin-resistant staphylococci, just as a natural environmental factor selects favorable adaptations. Penicillin kills off sensitive bacteria, leaving behind those that are

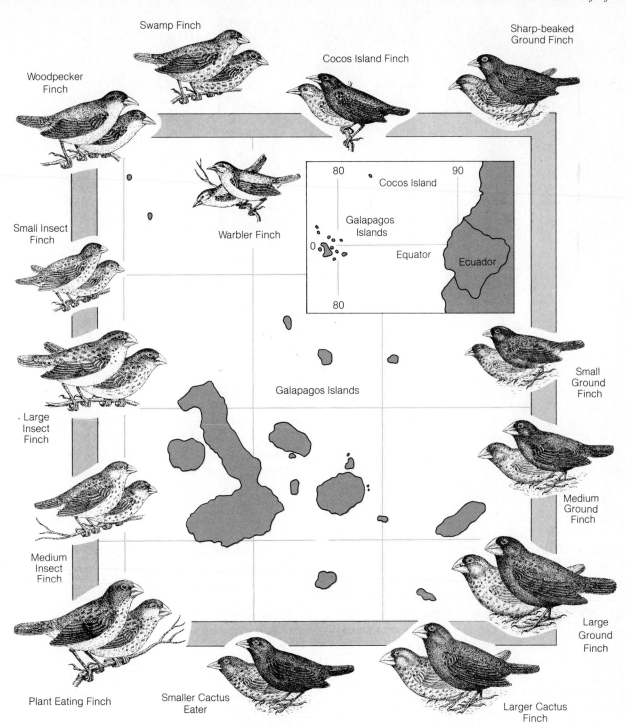

Figure 2-11 The finches of the Galápagos Islands are found nowhere else on earth, although other finch species do inhabit the mainland of South America. The Galápagos finches are not uniformly distributed over all the islands. Only the warbler finch is found on all the islands, while the medium insect finch is found only on one island. Adaptation has given rise to different species of Galápagos finches that fill the ecological niches occupied by hummingbirds, flycatchers, and woodpeckers elsewhere in the world.

Figure 2-12 (*a*) Tarsier, (*b*) lemur, and (*c*) monkey. Notice the front-facing eyes characteristic of all primates.

resistant. Over time the resistant forms predominate. Higher doses might work, but genetically resistant forms persist, eventually producing superstrains that are resistant to penicillin even at high doses. More than a hundred new antibiotics have been developed in recent years to combat this problem, but each of these faces the same problem: growing genetic resistance. Resistant strains develop almost as fast as the pharmaceutical companies turn out new antibiotics.

Another example with equally disturbing implications is genetic resistance to pesticides, among them DDT. DDT was used widely in the years after World War II to combat malaria-carrying mosquitoes. In Sri Lanka, for example, over a million cases of malaria were reported each year in the 1940s. Widespread spraying of homes to control mosquitoes and the use of several drugs to treat malaria victims drastically reduced the incidence of the disease. In recent years, however, several epidemics have broken out, with as many as a half a million cases. Today the incidence of malaria is on the rise. The reason is twofold. First, DDT-resistant mosquitoes have emerged. Second, the malarial parasite (plasmodium) carried by the mosquito has also developed a resistance to the drugs used to treat victims. (Additional examples of insect resistance are discussed in Chapter 17.)

Penicillin and DDT resistance are examples of human-directed selection, or **artificial selection** to distinguish it from natural selection, which does not result from human activities. Both DDT and penicillin resistance were unintentional byproducts of human intervention. In some instances, though, humans have taken an active, intentional role in evolution. Consider the changes that have

taken place in dogs, whose predecessor the wolf is a far cry from the pet poodle. Cattle breeding and plant breeding are additional examples of artificial selection. Artificial selection has helped humans improve crop and meat production.

Humans manipulate many environments through deforestation or elimination of certain species. These actions disturb communities with long evolutionary histories and can also disrupt many coevolved relationships. By altering the environment humans act as powerful selective agents. The consequences of these actions to the human population are often difficult to predict. For the species involved in such disruptions, however, the results are often catastrophic. Change may occur so rapidly that species cannot adapt in sufficient time to survive. Destroying the grassland habitat for mice in Missouri, for example, greatly reduces the number of barn owls. If damage occurs quickly and widely, the barn owl is likely to become extinct in much of its habitat. Destroying the tropical rain forests of Brazil, Central America, Africa, and Polynesia will likely eliminate many thousands of species because they cannot adapt to changing conditions.

Human Evolution

Our Biological Roots

The humorist Willy Cuppy wrote, "All modern men are descended from a wormlike creature, but it shows more in some people." Actually, scientists believe that humans evolved from a tree shrew that lived 80 million years ago in Africa. With handlike paws the shrew moved about the trees at night, feeding on insects. Over 50 million years the shrews evolved into tree-dwelling primates similar to modern-day tarsiers, lemurs, and monkeys, with further refinements in their hands and front-facing eyes to give them binocular vision and better depth perception (Figure 2-12).

From the monkey line evolved the dryopithecines, the ancestors of the great apes and hominids—humans and their fossil relatives. Their emergence is estimated at roughly 20 million to 30 million years ago. Dryopithecines walked on all four limbs, much like modern apes. From them sprang the first hominids, **australopithecines** (Figure 2-13). Living in southern Africa nearly 4 million years ago, australopithecines roamed the grasslands, walking upright in search of food and shelter. Roughly 2 million years ago *Homo habilis* emerged. The first tool and weapon makers, they spread from Africa to Europe and Asia.

About 500,000 years later a new form arose, *Homo erectus*. With a brain slightly smaller than ours, *Homo erectus* made more sophisticated tools and weapons such as hand axes and spears. Anthropologists have found evi-

dence in China that *Homo erectus* used fire to cook, to warm their caves, and to frighten away predators.

Evidence of *Homo sapiens'* emergence is scanty. Paleontologists believe that somewhere around 400,000 years ago our species came into being. One of the best-known examples is the European inhabitant, the Neanderthal. Like their predecessors, Neanderthals lived in caves, cooked their food on fires, and hunted animals with tools. But Neanderthals stood fully erect, had slightly larger brains than we do today, and even buried their dead. Paleontologists have found bear skulls in burial sites along with food and flowers, which they believe are a form of offering left in a ritual performed by clans that roamed the land in search of food.

Modern humans, or Cro-Magnons, emerged 30,000 years ago, perhaps in Africa. They rapidly replaced the Asiatic and European Neanderthals, for reasons still unknown. Characterized by domed heads, smooth eyebrows, and prominent chins, they perfected the stone and bone tools of their ancestors. Anthropologists believe that Cro-Magnons had a fully developed language.

At the end of the last great glacial age, Cro-Magnons spread across Siberia to the New World. Sweeping across

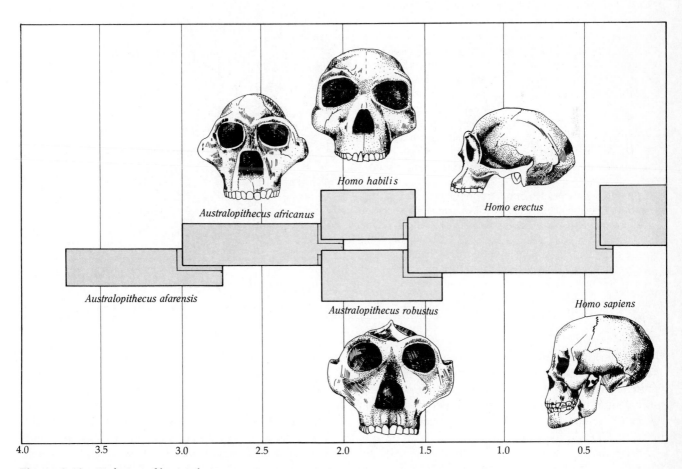

Homo habilis

Homo erectus

Australopithecus africanus

Australopithecus afarensis

Australopithecus robustus

Homo sapiens

| 4.0 | 3.5 | 3.0 | 2.5 | 2.0 | 1.5 | 1.0 | 0.5 |

Figure 2-13 Evolution of human beings.

The Forgotten People

The term *endangered species* suggests the problem of plants and animals destroyed or threatened by habitat alteration, hunting, and pollution. But certain human populations—tribal peoples—are also threatened with extinction. About one out of every 25 humans alive today is an Eskimo, Pygmy, Bushman, Indian, aborigine, or some other tribal member. These people live the way their ancestors did thousands of years ago, as hunters and gatherers or subsistence-level farmers.

Tribal peoples have been uprooted in the name of progress to develop farms, mineral deposits, timber, dams, reservoirs, and wildlife parks. Often driven from their homeland, they are forced into an area unlike that in which they have lived for centuries. In Paraguay, for example, the remnants of the Toba-Maskoy tribe have been moved to an arid region where their survival is in doubt.

Those who are allowed to stay are susceptible to new diseases brought by the developer. Brazil's Indian population has shrunk from 6 million to 200,000 since the first Portuguese explorers arrived in the early 1500s. Although war was responsible for some of the deaths, diseases brought from foreign lands were, and still are, the greatest killers. Barbara Bentley, the director of Survival International (an organization dedicated to the protection of tribal people), says, "The easiest way to dispose of these isolated tribal people is by sneezing."

Some tribes have been "assimilated" into the invading culture with disastrous results. Having suddenly been catapulted two or three centuries in time, they become lost, frightened, and confused by modern technology. Often, they return to their homeland only to find it destroyed. Loss of homeland and traditional values can lead to fatal psychic trauma: from Brazil to Australia, alcoholism, severe depression, and destitution take their toll. Once-skilled hunters are often reduced to begging.

The stories go on: copper mining in Panama threatens thousands of Guaymi Indians; Kalinga tribes in the Philippines fight the construction of hydroelectric dams that would flood their rice terraces; 300,000 Chilean Mapuche Indians have recently been told that their land will be opened for timber cutting; in Peru, the long-isolated Amuesha Indians are threatened by a new highway that would link them to civilization.

The elimination of these cultures will mean not only the end of age-old languages, myths, and social customs but also the irreversible loss of knowledge, including information on medicinal plants, dyes, and diet. Tribal peoples are responsible for discovering more than 3000 plant species with antifertility properties, a boon for birth control research. Some of their plant materials give promising clues to cancer prevention and cure.

The question of whether we can afford to allow these tribal people a continued existence is changing. Now the question is, can we afford to live without them?

Bushwoman from the Kalahari Desert, in Namibia, resting on her digging stick.

North America, they may have been responsible for the extinction of many animal species, starting a trend that continues today. Ten thousand years ago, anthropologists estimate, there were about 10 million people living on the earth. Today our numbers have swollen to over 5 billion, a 500-fold increase.

Homo habilis, present some 2 million years ago, had roughly the same form we have today. Although we know nothing of their skin, hair, or facial expression, these hominids were probably not much different from you or me.

The most drastic changes in our evolution occurred millions of years ago as we evolved from shrews to dryopithecines; the physical refinements since the emergence of australopithecines have been relatively slight. Thus, the changes that brought us to the industrial age

Table 2-1 Classification of Human Social Systems

Social System	Features
Hunting and gathering	1. The people were nomadic or semipermanent. 2. They benefited from their intelligence and ability to manipulate tools and weapons. 3. They were knowledgeable about the environment, and skilled at finding food and water. 4. On the whole, they were generally exploitive of their resources. 5. The environmental impact was generally small because of low population density and lack of advanced technology. 6. They lived healthy lives, were well fed and experienced low disease rates. 7. Their widespread use of fire may have caused significant environmental damage in some locations.
Agriculture	1. Farmers were generally either subsistence level or urban based. 2. They benefited from new technologies to enhance crops and resource acquisition needed for their survival. 3. They were knowledgeable about domestic crops and animals. 4. They were highly exploitive of their resources. 5. The impact of subsistence-level farming was significant, but because population size was small, damage was minimized. The impact of urban-based agriculture was much larger because of new technologies, trade in food products, increasing population, and lack of good land-management practices. 6. Disease was more common among city dwellers because of increased population density. 7. Poor agriculture, overgrazing, and excessive timber cutting caused widespread environmental damage.
Industry	1. Industry includes early and advanced forms. 2. It relies on new technologies, energy, energy-intensive forms of transportation, tremendous input of materials, reduced number of workers, and, recently, biotechnology. 3. Mass production and modern technology are transferred to the farm. 4. Industry is highly exploitive, more so than earlier societies; devoted to maximum material output and consumption. 5. Impact is enormous and includes pollution, species extinction, waste production, dehumanization. 6. Humans become subject to infectious disease and new industrial-age diseases including ulcers, heart disease, and mental illness. 7. Widespread environmental damage results from industry, agriculture, and population growth.

came through cultural evolution—through learned behavior and changes in the way we tap the earth's renewable and nonrenewable resources. A closer look at these changes will help us understand where we have gone wrong and where we need to go if we are to build a sustainable society.

Human Society and Nature: The Changing Relationship

Anthropologists recognize three major social groupings of human societies: hunting and gathering, agricultural, and industrial (Table 2-1). Regardless of their differences, all interact with the environment in fundamentally the same way. As illustrated in the Population, Resource, and Pollution Model in Chapter 1, human societies, regardless of their form, all require resources. The acquisition and

use of these resources can result in pollution and depletion. The level of pollution and the extent of depletion depend on how big the population is, the population's demand for resources, and the type of resources it relies on. Discrepancies in the three societies' levels of pollution and depletion, combined with fundamental attitudinal differences, set them apart from one another in ways worth studying, ways that will give us a better understanding of the environmental crisis we now face.

Hunting and Gathering Societies Hunting and gathering societies were the dominant form of social organization throughout most of human history. It is only in the last few fleeting moments of geological time that we have taken up agriculture and then industry as ways of life.

(a)

(b)

Figure 2-14 Hunting and gathering societies wander the land in search of food and water. (*a*) The men hunt. (*b*) The women are the primary food and fuel gatherers, foraging for wood, roots, berries, bark, and other plant products.

Hunting and gathering societies made a living doing just that: hunting animals and gathering numerous varieties of seeds, fruits, vegetables, and other plant matter, depending on the season (Figure 2-14). New anthropological evidence suggests that these people may have been much less skilled at hunting large animals than we commonly think. Pat Shipman, an anthropologist at Johns Hopkins University, asserts that hunters and gatherers probably gained a substantial amount of their meat by scavenging—picking the bones of animals that had been killed by others or had died from natural causes.

In any event, members of these societies had a profound knowledge of the environment. Experts in survival, they knew where to find water, edible plants, and animals; how to predict the weather; and what plants had medicinal properties. This profound ecological knowledge is evident today in surviving hunting and gathering socie-

ties. The Bushmen of southern Africa, for example, can find water in the desert where most others would fail. The Australian aborigines can locate and catch a variety of lizards, insects, and grubs far better than trained field biologists.

Studies of present-day hunters and gatherers also suggest that, contrary to popular conception, our early ancestors did not live with the constant threat of starvation and did not spend the greater part of their lives in search of food. Studies also suggest that they were healthy and well nourished and suffered from few diseases.

Many hunters and gatherers were nomads, wanderers who foraged for plants and captured a variety of animals using only primitive weapons. Because their technology did not give them a great advantage over other species, their populations never grew very large.

Judging from existing hunting and gathering societies,

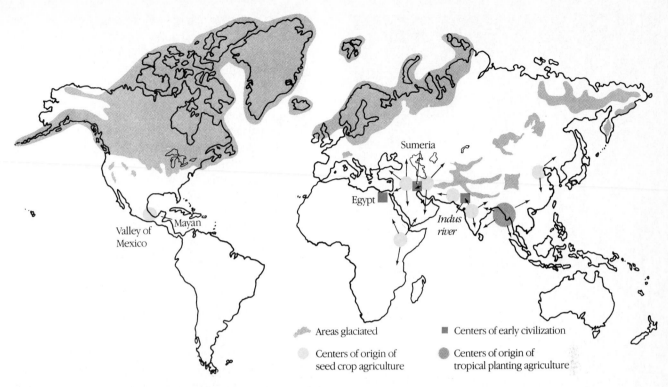

Figure 2-15 Roots of agriculture. The areas where tropical planting and seed crop agriculture originated.

anthropologists argue that our ancient ancestors probably had a deep reverence for the environment and the plants and animals they ate. But they were by no means environmentally benign. Hunters and gatherers, for example, may have ignited grass fires to drive animal herds over cliffs for slaughter, killing many more than they needed. Such a wasteful action does not fit the image of a wise steward of the land. The plains tribes of Canada were sophisticated users of fire. They burned clearings regularly to keep them open to maintain the preferred habitat of deer. Today, Canada is dotted with large open meadows; ecologists believe that without natural fires and the fires of native tribes it would be covered with unbroken coniferous forests.

Hunters and gatherers fashioned tools from sticks, stones, and animal bones to enhance their survival. On the whole these humans presented little danger to wild species. Even so, many scientists believe that they were responsible for killing off many species of large animals in the years following the last ice age. They may have decimated the cave bear, giant sloth, mammoth, giant bison, mastodon, giant beaver, saber-toothed tiger, and Irish elk. They killed these animals directly, drove them out of their preferred habitats, and may have wiped them out by killing their prey. Some hunters and gatherers developed semipermanent life-styles, setting up homes near rich hunting or fishing grounds that could provide a year-round supply of food. These groups were more likely to cause noticeable damage.

New research on hunting and gathering societies suggests that, for thousands of years and well before the advent of the agricultural age, many groups grew food and raised animals to feed themselves and to trade with other groups on a limited scale. Trade with neighboring tribes established the first system of commerce. Cave dwellers in Europe 28,000 to 10,000 years ago, for example, probably participated in extensive networks set up to trade food and other valuables.

On balance, the hunting and gathering societies had little impact on the environment. Their small numbers, their generally nomadic life-style, their inefficient technology, and very possibly their reverence for the life-giving earth kept them from depleting the renewable resource base on which they depended.

Agricultural Societies Anthropologists believe that **agricultural societies** emerged between 10,000 and 6000 BC. The roots of agriculture can be traced to Southeast Asia (Figure 2-15). Here in the moist rain forests, early humans practiced **slash-and-burn farming**. In slash-and-burn agriculture farmers cleared small jungle plots to plant their crops. They harvested and planted the same plot for several years; because the jungle soils were poor in nutrients, however, crops failed after several years, and

Figure 2-16 This ancient city was once surrounded by rich forests and grasslands. Deforestation, overgrazing, and poor agricultural practices denuded the land, turning it into desert.

plots were abandoned for new clearings. Native species invaded, returning the land to its original state. Damage to the jungle was negligible.

The early agricultural societies of Southeast Asia also domesticated many animals, such as the pig and fowl. These became vital food sources, greatly supplementing their crops.

Seed crops originated in a wide region extending from India and China to eastern Africa (Figure 2-15). The first farmers cleared woodlands, known for their rich soils. But a new development, the plow, opened up the fertile grassland soil, which had previously been too difficult to cultivate because of the heavy sod and thick roots of native grasses. The first plow was nothing more than a tree limb with a branch that cut through the topsoil. Archeological evidence indicates that the plow came into use in the Middle East around 3000 BC. In ensuing years, more elaborate plows were developed and pulled by oxen. This enabled farmers to cultivate grassland soils, dramatically increasing crop production. A variety of domesticated animals, such as goats, sheep, and cows, supplemented the human diet.

The plow gave agricultural societies the means to greatly increase the productivity of the land. As a result, they achieved a greater degree of control over their lives, making an important shift in the human–environment interaction. For the first time human populations could, by manipulating their environment, expand beyond the limits previously set by the natural food supply.

At least two major changes followed this shift. First, the human population began to swell. Fewer people were needed to provide food, so many left the farms and moved to villages and cities, where they took up crafts and small-scale manufacturing. Second, cities grew and eventually became centers of trade, commerce, government, and religion. The face of civilization changed for good.

The plow marks a pivotal point in our cultural evolution. It is the beginning of modern technology. The growth of population and small-scale industry, in many ways caused by the plow's success, placed greater demands on the environment for wood, metals, and stone. Heightened exploitation accompanied by poor land management resulted in widespread destruction of the natural environment. Many fertile areas were destroyed by overgrazing, excessive timber cutting, and poor agricultural practices.

The shift to mass-produced food had a potentially more harmful effect: it severed, in large part, the link to nature. Farmers concentrated on a small number of plants and animals to provide food. The profound knowledge of the environment characteristic of the hunters and gatherers all but vanished. Agriculture became a tool to dominate nature, replacing the cooperation and understanding that had marked earlier cultures.

The environmental impact of agricultural societies and their large urban centers was enormous. Archaeological and historical records show us that overgrazing, widespread destruction of forests, and poor farming practices changed many fruitful regions into barren landscapes (Figure 2-16). Ancient civilizations perished as a result, either directly as their crops failed or indirectly as other displaced peoples invaded.

This decline was especially evident in the Middle East, North Africa, and the Mediterranean from 5000 BC to AD 200. As an example, the Babylonian Empire once occupied most of what is now Iran and Iraq. At the outset this land was covered with productive forests and grasslands. Huge herds of cattle, goats, and sheep overgrazed the grasslands, however, and eventually destroyed the natural vegetation. Forests were cut to provide timber and create more pasture. The loss of grassland and forest vegetation decreased rainfall and eventually parched the land. Sediment washed from the barren soils, robbing them of nutrients and filling irrigation canals. These changes and a succession of invading armies eventually destroyed this once-great empire.

This story has been repeated all around the Mediterranean Sea. For example, throughout Saharan Africa in what are now dry and uninhabitable regions, remnants of once wealthy cities can be found buried in the sand.

Agricultural societies were far more damaging than hunting and gathering societies. Agriculturalists, for example, developed new tools to increase the productivity—and destruction—of the land. Rising production

Figure 2-17 The Industrial Revolution marked a new day for human societies. Populations grew, resource demands skyrocketed, and pollution became widespread.

resulted in an upsurge in population and, in turn, the emergence of towns and cities, both of which placed incredible demands on the surrounding countryside. Agricultural societies ushered in the age of commerce. As economics emerged, humans began to see the natural world in a much different way—as a source of wealth. As a result of these and other changes, agriculturalists lost the connection with nature felt by their hunting and gathering predecessors.

The Industrial Society The **industrial society** is a recent phenomenon in human history. It was brought on by the **Industrial Revolution**, a drastic change in manufacturing marked by a shift from small-scale production by hand to large-scale production by machine. Starting in England in the 1700s and in the United States in the 1800s, the Industrial Revolution brought about further changes in society, which continued to strain the once-close ties between humans and nature.

With the advent of coal-powered machines, manufacturing became more capital- and energy-intensive and less labor-intensive (Figure 2-17). People became less vital to industry than machines and production. With the increase in machine labor came a dramatic escalation in energy demand and a need for new means of transportation to move goods to and from the city. As industries

grew, the influx of materials—fuel, food, minerals, and timber—into the city rose sharply.

The shift to machine production changed the working environment, the city, and the surrounding countryside that supplied the resources. The new manufacturing technologies were the fruit of scientific and engineering advances. They were complex, often made work meaningless and boring, and produced large quantities of smoke, ash, and other wastes.

Mechanization also swept our farms (Figure 2-18). Technological advances such as Jethro Wood's cast-iron plow with interchangeable parts and Cyrus McCormick's reaper brought on a rapid increase in agricultural production. Perhaps one of the most significant advances was the invention of the internal-combustion engine, which made horse-drawn implements obsolete. The motor-powered tractor alone could plow as much land in a week as one of our forebears could work in a lifetime using hand tools. Because of more efficient farming methods, fewer farm workers were needed. Unemployed workers migrated to cities, swelling their populations.

Other significant advances included the development of fertilizers, which allowed an increase in agricultural productivity without increasing the amount of land cultivated, and plant breeding, which produced higher-yield crops.

Figure 2-18 Modern agriculture depends heavily on machinery, energy, and additional resources. Large fields are worked to achieve maximum output.

New medicines and better control of infectious disease through insecticides and improved sanitation also grew out of the Industrial Revolution. These important developments enhanced human survival. People began to live longer, and the human population began its rapid increase.

Population growth, the agricultural transformation, the rise of industry, the growing demand for resources, and the further isolation from nature had tremendous environmental repercussions. Pollution became more widespread. Increased agricultural output destroyed wildlife habitat, depleted soils, and caused severe soil erosion, which polluted waterways. Dredging new harbors destroyed productive fishing grounds and shellfish beds. Mine waste, city sewage, and industrial discharges polluted waters and wiped out native fish populations such as the Atlantic salmon.

The shift to an industrial society further distorted the human view of nature. Industrial societies sought more control over the environment to ensure their own survival. Industrial people came to view themselves as more and more apart from nature and superior to it. The prevailing attitude today can be traced back to the 17th-century English philosopher John Locke, who argued that the purpose of government was to allow people the freedom to exercise their power over nature to produce wealth. "The negation of nature," Locke argued, "is the way toward happiness." People must become "emancipated from the bonds of nature." These notions of disassociation and superiority have been passed from generation to generation throughout the industrial age. Locke also preached unlimited economic growth and expansion, with the belief that individual wealth was

socially important for a harmonious society, an idea that has also persisted.

Industrial societies were swept into a battle with nature. New medicines to combat disease, an arsenal of chemicals to fight pests, and new technologies to extract resources more efficiently were the key weapons. On the whole, people ate better, lived better, and began to live longer. But there was a price to be paid in environmental deterioration.

Advanced industrial societies arose in the period after World War II. Several major features distinguished this new form of society: (1) a marked rise in production and consumption, (2) a shift toward synthetics such as plastics and nonrenewable resources such as oil and metals, and (3) huge increases in energy demand for farming, industry, and day-to-day living.

The advanced industrial society, caught in an ever-escalating production–consumption cycle, has only begun to awaken to the costs of environmentally irresponsible behavior. Domination of nature continues as the central theme of modern industrial societies, economic growth retains its commanding allure, despite evidence that both features threaten the long-term future of our planet.

One threat of considerable importance is synthetic substances so common in advanced industrial societies. Among them are plastics, nylon, and chemical pesticides, all derived from petroleum. Synthetics can create problems in nature, because naturally occurring bacteria in soil and water, which decompose natural materials, are frequently unable to break synthetics down. Synthetics may therefore persist in the environment for decades. The persistent insecticide DDT, for example, can accumulate in the fatty tissues of birds and disrupt reproduction. (See Chapters 14 and 17 for a more thorough discussion of this phenomenon.) Plastic pollution in water has proven to be a major problem for aquatic organisms (see Chapter 16). Synthetics are generally derived from nonrenewable petroleum resources; plastics, for instance, come from nonrenewable petroleum, and metals are made from nonrenewable ores.

Another threat is carbon dioxide, released from combustion sources such as cars, jet aircraft, factories, and power plants. As discussed further in Chapter 15, carbon dioxide released in large quantities now threatens to change the global climate in ways that will disrupt agriculture and radically change where we live.

The evolution of human society through agricultural and industrial eras has been marked by increasing specialization. Farmers in agricultural societies took on the role of feeding people. Craftsmen and craftswomen assumed the role of making clothing, pottery, and other necessities. Traders saw to it that goods moved freely from one region to another, while bankers managed the supply of money. In industrial societies specialization has continued at an alarming pace. Today, energy experts provide

Understanding the Earth

The late Buckminster Fuller coined the term "Spaceship Earth" to describe our watery planet. The earth, like a spaceship, is a closed system capable of recycling the materials necessary for life. Similar to a spaceship, the earth is vulnerable to disruptions of its life-giving systems.

In the 1970s, the British scientist J. E. Lovelock began to consider the earth as a living entity rather than a spaceship, for the earth in many ways behaves like an organism. It maintains a constant temperature, monitors the chemical composition of its lakes and oceans, and regulates its atmospheric gases in much the same way that an organism maintains its internal constancy.

Although this notion, known as the Gaia Hypothesis, has drawn much criticism, it nonetheless is an elegant metaphor that underscores a key principle of ecology: that all living things operate together. The sequoia and the gazelle, the lion and the rosebush—all are integral parts of a system as interdependent as the cells in your brain and kidneys.

This gallery takes a glimpse at the processes of earth. It shows the active building and tearing down of land, the delicate energy balance, and the oceanic and wind currents that affect life in innumerable ways.

1 The view of earth from the moon initially led to the notion of the earth as a spaceship, capable of recycling its life-sustaining materials. This metaphor has now largely been replaced by the Gaia Hypothesis of the earth as similar to an organism.

2 The earth's water participates in an enormous recycling system—the hydrological cycle. Heated by the sun, water evaporates from the earth's surface into the skies where it forms clouds. Falling back to earth, much of the water evaporates again while the rest flows back to the oceans along rivers and lakes. A small fraction seeps into the earth's surface and migrates slowly as groundwater.

Rain wears at the earth's surface, removing soil and washing it into the sea and lakes. People worsen erosion by stripping the land of its vegetation. Some 20 billion tons of farmland topsoil is washed away each year by wind and rain, often because of poor land management.

The earth's energy balance is as important as the hydrological cycle.

About 30% of the sunlight reaching the earth is reflected by the atmosphere, while approximately 70% is absorbed by the earth's surface and atmosphere and radiated back into space as heat. Water vapor and gases such as carbon dioxide in the atmosphere block much of the outgoing heat and radiate it back to earth, keeping the earth habitable. Unfortunately, modern civilization has added billions of tons of carbon dioxide and other gases to this thermal blanket, causing a gradual rise in global temperature. Some scientists predict that as pollution continues this warming of the earth could eventually melt enough ice to flood nearly 20% of the earth's land surface, with disastrous results for agriculture and our food supply.

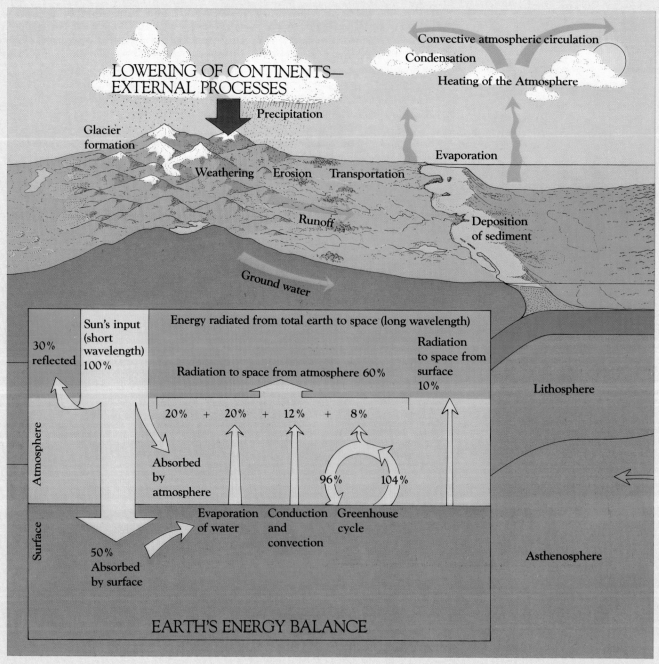

Erosion is to some extent counterbalanced by internal geological processes that build up the land. The action of tectonic plates in this respect is explained on the next page. Lava from volcanoes also adds to the buildup.

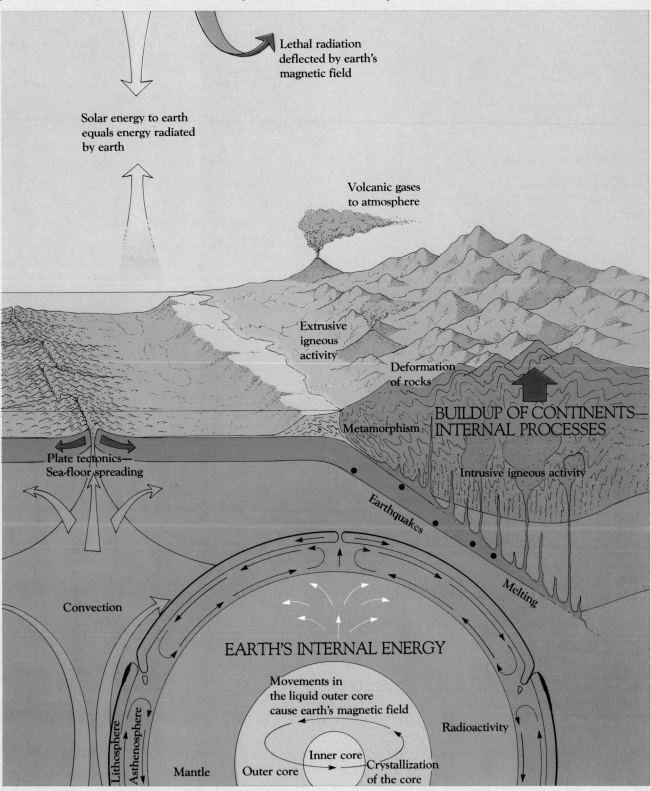

Lethal radiation deflected by earth's magnetic field

Solar energy to earth equals energy radiated by earth

Volcanic gases to atmosphere

Extrusive igneous activity

Deformation of rocks

Metamorphism

BUILDUP OF CONTINENTS—INTERNAL PROCESSES

Plate tectonics—Sea-floor spreading

Intrusive igneous activity

Earthquakes

Melting

Convection

EARTH'S INTERNAL ENERGY

Movements in the liquid outer core cause earth's magnetic field

Lithosphere

Asthenosphere

Radioactivity

Mantle

Outer core

Inner core

Crystallization of the core

3 The earth's crust consists of about ten huge movable plates, called tectonic plates, eight of which are shown here. The plates are propelled by convection of the earth's molten interior. Where they pull apart, molten rock can flow upwards and form new ocean floor. Plates that slip under others cause the buildup of continents through mountain formation. Although new crust and new minerals are formed as the earth's crust regenerates itself, the minerals are largely out of reach of modern civilization.

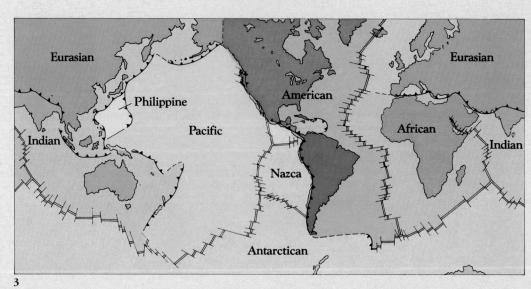

3

4 As the earth orbits the sun, the planet spins on its axis, causing large swirling currents in the earth's atmosphere. These prevailing winds are also generated by the uneven heating of the earth's surface, which leads to differences in atmospheric pressure. Wind occurs when air moves from regions of high pressure to low-pressure zones. For example, hot air rising at the equator causes the low-pressure region known as the doldrums. The northeast and southeast trade winds blow to the equator from the colder high-pressure regions at 30 north and south latitude, respectively. The westerlies blow strongly from the subtropical highs to the so-called subpolar lows at 60 north and south latitude, cooling the climate in those regions.

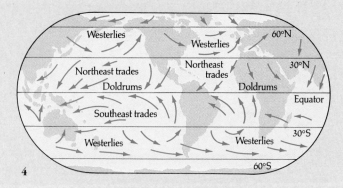

4

5 The major ocean currents are determined by the earth's rotation, by the force of the prevailing winds, and by the location of landmasses that block the currents. The flow of the warm (red) and cold (blue) currents influences regional climates. Because of the presence of the continents the currents move in loops known as gyres. The gyres flow clockwise in the northern hemisphere and counterclockwise in the southern hemisphere, due to the earth's rotation. The Gulf Stream, for example, carries warm water from the equator to the British Isles, keeping them warmer than expected at that latitude. The Humboldt Current, in contrast, brings cold water from the Antarctic to the equator, cooling the western coast of South America.

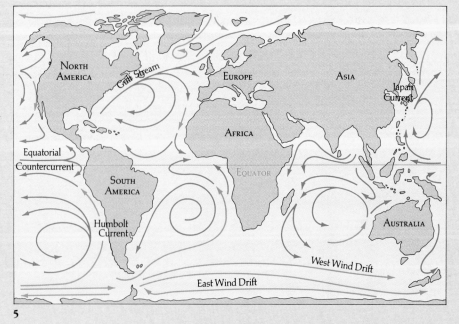

5

fuel for our automobiles, manufactured in turn by an army of specialists, each trained at individual tasks. Builders construct our homes, farmers continue to supply our food, teachers educate our children, and government specialists pass laws and regulations to control many aspects of our lives. Specialization requires that we give up control over our lives. The average citizen is left with only the responsibilities of making money and finding entertainment, says author and philosopher Wendell Berry.

Although specialization has bettered our lives in many ways, it also creates incredible dependency and anxiety, says Berry. Moreover, our specialization leads us to think narrowly. We are caught up in a system that lacks a sense of wholeness and cooperation. Interests have become too narrowly defined. Ford and General Motors, for example, fight improvements in automobile efficiency while the entire planet is in peril from pollutants coming in large part from our cars. Such narrow thinking evolved as human society grew apart from nature. It is a big part of the environmental crisis.

Another significant change from the days of hunting and gathering is a dramatic increase in energy use, a trend with serious consequences. In hunting and gathering societies, for instance, individuals required only about 2000 to 5000 kilocalories of energy per day. A **kilocalorie** is a measure of energy. One kilocalorie is equal to one calorie, a measure used by dieters. The 2000 to 5000 kilocalories used by individuals provided metabolic energy and additional heat to cook meat and light a cave. Early agricultural societies, however, required twice that amount per person to grow crops. Advanced agricultural societies required twice as much again, or about 20,000 kilocalories per person per day. Increased energy use resulted from improvements in agriculture and the growth of crafts, commerce, and cities. In industrial societies per capita energy use climbed again, increasing to 60,000 kilocalories per day per person. But modern industrial society with its mechanized farming, energy-intensive industries, and affluent life-styles doubled that amount, raising energy consumption to about 120,000 kilocalories per person per day. Especially wasteful nations like the United States, Canada, and the Soviet Union consume, on average, 250,000 kilocalories per person per day.

What's the significance of this? Energy use is a major source of environmental decay. It's the source of most of our air pollution. As the giant oil spill off the coast of Valdez, Alaska reminds us, it can be a major source of water pollution as well.

Twentieth-century technology and industry and the underlying belief that human beings have dominion over the world and are apart from nature have brought us to the present environmental crisis. Advances in technology, increasing control over our environment, rising population, and our growing isolation from nature have all contributed greatly to the present environmental crisis. One thing that this book will make clear is that modern society as we know it cannot continue. Part 2 will show that the population growth curve must eventually stop rising and that we can halt the upward trend in population if only we want to. Part 3 will document how the resources we rely on are reaching or, in some cases, have already reached their limits. It will point out ways in which we must change. Part 4 will describe the facts about pollution and its effects and will show how we can keep from poisoning our planet. Finally, Part 5 will outline attitudes and practices of both economics and politics that are outdated, even dangerous, in a finite world and will suggest alternatives that may allow us to achieve a more peaceful coexistence with our living and giving planet.

New Visions: A Final View from Outer Space

We started this overview in outer space with a cataclysmic explosion that gave birth to the universe. It is only fitting that we end in space, too, hoping that another explosion, this one of our own making, will not steal from us the time needed to mend the damage we have wrought on the tiny, fragile planet we call home.

Traveling through space at 107,000 kilometers per hour (66,000 miles per hour), this watery planet is more than a spaceship, supporting life through its vast recycling networks. The earth is, as alluded to earlier, like a living thing. Just as you and I do, the earth regulates its temperature within a fairly narrow range and has, geologists believe, done so for several million years. The earth creates and maintains a unique chemical atmosphere conducive to life and entirely different from what scientists would predict based on the earth's chemistry.

The British scientist James Lovelock coined the term **Gaia** (pronounced GAY-a) **hypothesis**, after the Greek goddess of the earth, to describe this ability of the planet to maintain constant physical and chemical conditions necessary for life. As a metaphor it goes beyond the concept of spaceship earth, for it emphasizes a lifelike unity of all things that make up the biosphere. This key principle of ecology may humble our self-centered view of human existence and help us build a sustainable society that protects the living fabric of which we are a part.

Measure your health by your sympathy with morning and spring.

HENRY DAVID THOREAU

Summary

Pictures taken during our voyages into space have helped us reshape how we think about the planet and have given rise to the concept of the **spaceship earth**, a vessel carrying millions of passengers and sustained by intricate recycling mechanisms. More recent voyages have given rise to a new metaphor, that of the earth as a living thing. This is called the **Gaia hypothesis**.

Our beginnings can be traced back to the big bang 15 to 20 billion years ago. At that time, scientists speculate, all matter that makes up the universe was compressed in an infinitely small and dense volume that then exploded. The universe, made up of matter and energy, rapidly expanded. It took a million years for the universe to cool enough for stable atoms to form. **Atoms**, a basic unit of much matter, formed as **electrons**, **protons**, and **neutrons** united. Atoms consist of a central, dense **nucleus** containing protons and neutrons. Circling around the nucleus are the negatively charged electrons. They are found in the **electron cloud**. Atoms unite by various chemical bonds to form **molecules**.

Within the universe of atoms and subatomic particles slightly more dense clouds of immense size formed; these **protogalaxies** contracted under gravitation—the mutual attraction of all matter. Swirling in space the protogalaxies formed smaller clouds of matter that eventually gave rise to stars and planets.

The primeval earth was a spherical mass of rock and metal made molten by the condensation of matter, radioactive decay, and solar heat. Over time the earth cooled, forming a thin **crust** of lighter elements. The heaviest elements sank to the **core**, and intermediate-weight elements remained in the intervening **mantle**.

An atmosphere of carbon dioxide, methane, ammonia, hydrogen, and water vapor formed as the earth continued to cool. Water vapor condensed during this cooling and filled the oceans, dissolving inorganic and organic molecules from the atmosphere. Ultraviolet light from the sun energized these molecules in the dilute oceans and caused them to react with one another. Gradually, organic molecules necessary for life accumulated. These simple sugars and amino acids combined to form proteins, the genetic material DNA, and many other compounds. Larger-molecular-weight compounds, scientists think, united in small globules known as **protocells**, which looked and acted like living organisms. Over millions of years these simple aggregates took on a life of their own as they developed metabolic pathways and ways to reproduce.

The first real cells to emerge from the protocells were probably bacteria that lived by absorbing nutrients from the waters. Soon they were joined by photosynthesizing cells capable of making their own organic molecules from carbon dioxide and sunlight energy. Photosynthetic bacteria evolved a more complex biochemistry that began releasing oxygen.

As oxygen accumulated in the atmosphere, the **eukaryotes** evolved. These were nucleated cells; that is, they contained genetic material in membrane-bound nuclei. Eukaryotes contained cellular organelles, distinct structures specialized to carry out many cell functions, such as energy production. The **theory of endosymbiotic evolution** describes how these organelles arose. It states that organelles were once free-living organisms that became incorporated in other cells and there set up a permanent relationship, beneficial to both host and internal symbiont. At this point in evolution the pace of change quickened. From the unicellular eukaryotes evolved the plants, fungi, and animals.

Evolution describes how life formed here on earth and how it has changed over many millions of years. It explains why life is so diverse today. But evolution also tells of a common ancestry and a common strategy for life.

Evolution results in genetic changes within a species that may lead to the formation of new species. A **species** is a group of organisms that are different anatomically, physiologically, and behaviorally from all others and generally cannot interbreed with other such groups. Evolution can be summarized as follows: There is variation in all traits, or **adaptations** to the environment, in populations of organisms. New adaptations arise from random **mutations**, changes in the genetic material, **crossing over**, the exchange of DNA during gamete formation, and new genetic combinations resulting from **sexual reproduction**. Whether an organism survives in a changing environment depends on its inherited adaptations. **Natural selection** is the process by which the environment determines which organisms survive and reproduce. Biologists call reproductive success **fitness**. The fittest individuals leave the largest number of descendants in subsequent generations. The genetic makeup of a species may change over time because of natural selection, creating a new species. This process is called **speciation**.

Organisms may be selective agents in evolution. When members of two different species interact, changes in one species may lead to changes in the other, a process called **coevolution**.

Humans can also alter evolution by changing environmental conditions or by selective breeding, which brings out desired traits. These are called **artificial selections**.

The origin of the human race can be traced back to tree shrews that lived in Africa 80 million years ago. They gave rise to monkeylike creatures over the next 50 million years, which in turn evolved into the first animals with a distinctly human form, the **australopithecines**, arising about 4 million years ago. From this point humans changed very little anatomically. Throughout most of our 4-million-year history we have made a living by hunting and gathering. It was not until 10,000 BC that humans turned to agriculture, a way of life that persisted until the 18th and 19th centuries. Agricultural societies were quickly replaced by industrial societies.

The shift from hunting and gathering societies to industrial societies involved several major changes. First, our control over the environment increased, gradually at first and then, once industry started, at rocket speed. Knowledge of science and technology brought about this revolution in our control over the environment. Second, we became less attached philosophically and morally to the earth. We became powerful, almost godlike, in our manipulation of the world around us. More and more we saw ourselves as apart from, rather than a part of, nature. Third, our control and our emotional severance led to widespread depletion and pollution.

A new vision of the earth may help us return to a position more likely to be sustained. That vision is called the Gaia hypothesis. It holds that the earth is like a living thing, capable of maintaining its own environment, in much the same way as your body controls blood pressure, body temperature, and dozens of other functions. As a metaphor it emphasizes the unity of the biosphere, a key philosophical realization that may guide us on our way to building a sustainable society.

Discussion Questions

1. Summarize the formation of the universe, galaxies, stars, and our solar system. In what ways are the formations of galaxies, stars, and solar systems similar?

2. How would you set up an experiment to test the theory of chemical evolution?

3. What features did protocells have to acquire to become bacteria?

4. Define the theory of endosymbiotic evolution.

5. Describe the theory of evolution using the following terms in your discussion: Charles Darwin, natural selection, species, mutation, variation, gene pool, adaptation, speciation, geographic isolation, allopatric speciation, adaptive radiation, and convergent evolution.

6. Describe each of the major forms of society—hunting and gathering, agricultural, and industrial. How were they similar, and how were they different? How did the human–environment interaction shift over time?

7. List things that your body maintains in a relatively constant state, for example, body temperature. Make a parallel list for the earth. From your experience is it safe to conclude that the earth behaves like an organism?

Suggested Readings

Berry, W. (1977). *The Unsettling of America: Culture and Agriculture.* San Francisco: Sierra Club Books. Extraordinary collection of essays on the environment.

Carter, V. G. and Dale, T. (1974). *Topsoil and Civilization.* Norman: University of Oklahoma Press. An in-depth account of the impacts of early societies on soils.

Clapham, W. B. (1981). *Human Ecosystems.* New York: Macmillan. Lengthy account of the stages of cultural development and the impact of societies on the environment.

Commoner, B. (1976). *The Poverty of Power: Energy and the Economic Crisis.* New York: Bantam. Insightful coverage of the economic and energy problems of industrial societies.

Conservation Foundation. (1987). *State of the Environment. A View Toward the Nineties.* Valuable resource on global environmental problems.

Foster, R. J. (1982). *Earth Science.* Menlo Park, CA: Benjamin/Cummings. A good beginning book on the earth sciences.

Goldsmith, D. (1985). *The Evolving Universe.* Menlo Park, CA: Benjamin/Cummings. Eloquently written treatise on the evolution of the universe, galaxies, and solar systems.

Isaacs, J. (1980). *Australian Dreaming: 40,000 Years of Aboriginal History.* Sydney, Australia: Lansdown Press. A wonderfully illustrated book on the hunters and gatherers of Australia.

McPhee, J. (1989). *The Control of Nature.* New York: Farrar Straus Giroux. Three case studies showing the extent to which modern society will go to attempt to control natural forces.

Rifkin, J. (1985). *Declaration of a Heretic.* Boston: Routledge and Kegan Paul. Important, controversial reading on the advanced industrial society.

Russell, P. (1983). *The Global Brain.* Los Angeles: Tarcher. Superbly written view of possible evolutionary changes that could radically reshape our society. A must.

Shipman, P. (1985). Silent Bones, Broken Stones. *Discover* 6 (8): 66–69. Insightful look at how anthropologists study earlier societies, bias, and some new insights into human evolution.

Van Matre, S. and Weiler, B. (1983). *The Earth Speaks.* Warrenville, IL: Institute for Earth Education. Good collection of writing on the earth by poets and scientists. See especially the essay by Loren Eiseley for a beautiful description of evolution.

World Commission on Environment and Development. (1987). *Our Common Future.* Oxford: Oxford University Press. Chilling survey of global environmental problems.

World Resources Institute and International Institute for Environment and Development. (1988). *World Resources 1988/1989.* New York: Basic Books.

Principles of Ecology: Ecosystem Structure and Function

Never does nature say one thing, and wisdom another.

JUVENAL

An urban dweller awakens to the buzz of his alarm clock, eats a hurried breakfast, and heads to the concrete, steel, and glass towers of Downtown, USA. Bumper to bumper on the ten-lane highway, encased in his shiny automobile, he listens to the radio to quell his mounting tension. After half an hour on the freeway he parks his car and rides the elevator to his 15th-floor office.

Sitting back in his chair, looking out at the skyscrapers and the paved arteries bringing more office workers like him to the city, he entertains only vague notions of nature. To him, ecology may mean recycling aluminum cans or turning off the lights to save energy, activities he hasn't the time or the inclination to do. His connection with the environment is obscured by business reports to prepare, bills to pay, college educations to plan for, and a myriad of annoying chores around his suburban home. He does not think of himself as a part of the natural environment, subject to the rules and constraints that govern all living organisms. Why should he? Everything around him—his air-conditioned office, the city skyline, his home, even the carefully tended park across the street—bespeaks human mastery over nature.

Albert Camus wrote, "Man is the only creature that refuses to be what he is." It is too tempting to think of ourselves as apart from, and even above, nature. It is too easy to think that technology has made us immune to the rules that govern all living organisms. In the words of the ecologist Raymond Dasmann, however, "A human apart from environment is an abstraction—in reality no such thing exists." We humans are very much a part of the environment. Our lives are rooted in the soil and dependent on the air, water, plants, and algae. They are subject to the ecological rules that govern all the earth's living things.

This book will show how dependent we are on the earth and on the myriad forms of life that share the planet with us. This chapter and the next, which examine ecology, can help you understand the danger of breaking the rules of nature.

Strictly speaking, **ecology** is the study of living organisms and their relationship to one another and to the environment. Ecology takes the entire living world as its domain in an attempt to understand all organism–environment interactions. Given the vast number of organisms in the world, the realm of ecology is immense.

How Is the Living World Organized?

Our study of ecology will begin with a look at the way life is organized on earth, starting on a grand scale and successively focusing on the smaller organizational patterns.

The Biosphere

The part of the earth that supports life is called the **biosphere**, or **ecosphere**. As shown in Figure 3-1, the biosphere extends from the floor of the ocean, approximately 11,000 meters (36,000 feet) below the surface, to the tops of the highest mountains, about 9000 meters (30,000 feet) above sea level. If the earth were the size of an apple, the biosphere would be a thin layer only about as thick as its skin.

Life forms are scarce at the far extremes of the biosphere: on the highest mountaintops only inert spores of bacteria and fungi can be found, and on the deep ocean floor few organisms can survive. Life evolved under more moderate conditions than the biospheric extremes, and it is in these conditions that most species thrive. The zone of abundance is a narrow band extending from less than 200 meters (660 feet) below the surface of the ocean to about 6000 meters (20,000 feet) above sea level.

Life exists mostly at the intersection of land (lithosphere), air (atmosphere), and water (hydrosphere), as shown in Figure 3-2. And from these vast domains come the ingredients that make life possible—minerals from the soil, oxygen and carbon dioxide from the air, and water from oceans and lakes. Put these ingredients in a test tube, however, and you have simply a mixture of air, water, and soil. But when these molecules are uniquely organized in living things they produce a fascinating array of shapes and forms that, among other things, grow, reproduce, and respond to various stimuli.

The biosphere—this living skin of planet earth—is, in many ways, like a sealed terrarium. If carefully set up with soil, water, plants, and a snail or two, a sealed terrarium operates without interference; water is reused over and over, as are minerals. Plant growth is trimmed back by hungry snails, who also consume the oxygen the plants release. Carbon dioxide from the snails is taken up by plants and used in photosynthesis to make more plant life. In short, the terrarium is a sustainable system, less complicated but much like the biosphere, which recycles water, oxygen, carbon, and other substances over and over in a perpetual cycle. Only one thing must come from the outside for life to continue in the terrarium and in the biosphere: sunlight. Unlike the materials that make life possible, sunlight energy cannot be recycled.

Because the biosphere recycles all matter, it is called a **closed system**. The closed nature of the biosphere makes all species throughout all time cousins of sorts; the muscle tissue in your arm may contain carbon atoms that were part of the muscle of Julius Caesar or of a dinosaur. The tear in your eye may have been a tear in Napoleon's eye at Waterloo or a raindrop that fell on ancient hunters and gatherers. The atoms in your body may some day be part of the great leaders or the citizens of a society living centuries in the future.

The first law we glean from our study of ecology is that life on earth is possible only because of the recycling of matter. It is a rule, many ecologists believe, that we humans break at our peril and the peril of those yet to come.

Biomes

The terrestrial portion of the biosphere is divided into **biomes**, regions with a particular climate and thus a particular type of vegetation and animal life. A dozen biomes spread over millions of square kilometers and span entire continents like pieces of a giant puzzle (Figure 3.3).

Climate, the average weather conditions in a given region, determines the boundaries of a biome and the abundance of plants and animals found in it. Climatic conditions also determine the adaptations found in all living things within a biome. The most important climatic factors are precipitation and temperature (Figure 3-4, page 49). (For more details on biomes see Chapter Supplement 3-1 and Gallery 2.)

Aquatic Life Zones

The sea is also divided into more or less distinct zones, including coral reefs, estuaries, the deep ocean, and continental shelves. These are regions of different plant and animal life. The major differences among them can be

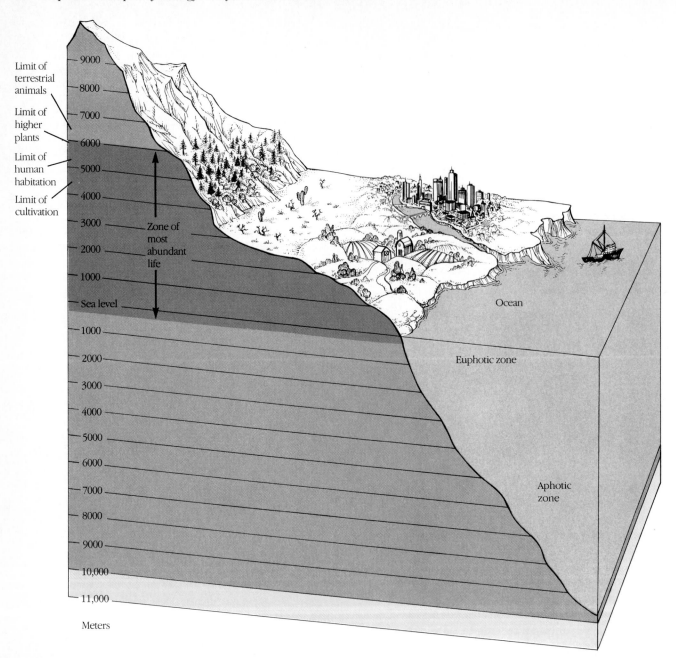

Limit of terrestrial animals

Limit of higher plants

Limit of human habitation

Limit of cultivation

Zone of most abundant life

Sea level

Ocean

Euphotic zone

Aphotic zone

Meters

Figure 3-1.   Vertical dimensions of the biosphere. Life exists in a broad band extending from the highest mountain peaks to the depths of the ocean. However, life at the extremes is rare, and most organisms are restricted to the narrow zone shown here.

traced back to levels of dissolved nutrients, water temperature, depth of sunlight penetration, and other environmental factors.

The aquatic life zones vary widely in their abundance and diversity of life. The richest areas in the ocean are generally those where land and sea meet, the **estuarine zone**. This zone includes **estuaries**, the mouths of rivers or inlets where salt and fresh water mix, and **coastal wetlands**, including mangrove swamps, salt marshes, and lagoons. In the estuarine zone nutrients from upstream soil erosion support a rich and diverse group of microorganisms, insects and fish. So rich is this region that it is often likened to the tropical rain forest. (For more information on estuaries and wetlands see Chapter Supplement 10-1.) The open ocean, in contrast, is relatively barren and is sometimes likened to the deserts.

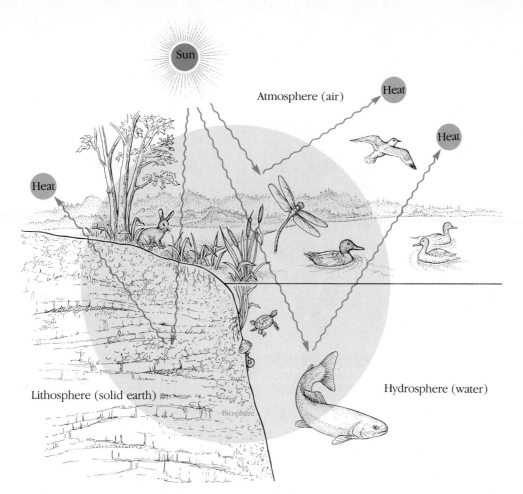

Sun

Atmosphere (air)

Heat

Heat

Heat

Lithosphere (solid earth)

Biosphere

Hydrosphere (water)

Figure 3-2. Life exists primarily at the intersection of land (lithosphere), air (atmosphere), and water (hydrosphere). As shown here, the biosphere is energized by sunlight.

Ecosystems

Ecologists invented the word **ecosystem**, an abbreviated form of **ecological system**, to describe a network consisting of organisms, their environment, and all of the interactions that exist in a particular place. In short, an ecosystem is an interdependent and dynamic (ever-changing) biological, physical, and chemical system.

The biosphere is a global ecosystem. Because it is too complex to study, ecologists generally limit their view to smaller regions, setting up more manageable boundaries. For the sake of simplicity an ecosystem might be a pond, a cornfield, a river, a field, a terrarium, or a small clearing in the forest. Accordingly, ecosystems vary considerably in complexity, too. Some may be quite simple—for example, a rock with lichens growing on it. Others, like the tropical rain forests, are quite complex. They contain an abundance of living organisms and a wide variety of species as well.

On your walks through forests and fields or even your drives in the country, you have probably noticed that adjacent ecosystems merge with one another, often without clear boundaries. For example, the edge of many mountain meadows is a zone that is half forest and half meadow, creating a unique mixture of plant and animal species. These transition zones, found between virtually all adjoining ecosystems, are called **ecotones**. Animals and plants of adjacent ecosystems intermingle in the ecotone. In addition, the ecotone supports many species not found in either of the adjacent ecosystems. These species are uniquely adapted to the conditions found there. As a result, ecotones generally have a greater number of species than surrounding areas and deserve special protection.

To gain a better understanding of the ecosystem, let us look at the two major components of all ecosystems: abiotic and biotic.

Abiotic Factors The **abiotic**, or nonliving, **factors** of an ecosystem are its physical and chemical components, for example, rainfall, temperature, sunlight, and nutrient supplies. Each of the earth's many organisms is finely tuned (through adaptations) to its environment and operates within a range of chemical and physical conditions, the **range of tolerance** (Figure 3-5, page 50). Although the range of tolerance is wide, most organisms thrive

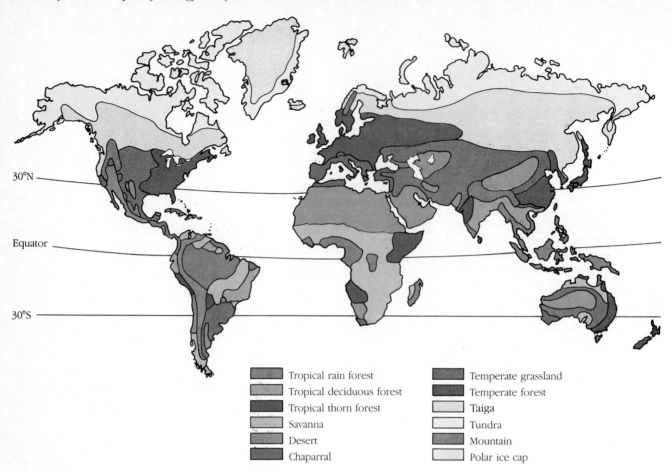

30°N

Equator

30°S

Tropical rain forest
Tropical deciduous forest
Tropical thorn forest
Savanna
Desert
Chaparral

Temperate grassland
Temperate forest
Taiga
Tundra
Mountain
Polar ice cap

Figure 3-3. The earth's major biomes.

within a narrower range of conditions. This is called the **optimum range**. Outside this is the **zone of physiological stress**, where survival is possible, but difficult. Further outside the optimum range is the **zone of intolerance**, where an organism cannot survive. Fish, for example, generally tolerate a narrow range of water temperature. If the water cools below the lower limits of their range of tolerance, they will die or escape to warmer water. Water temperatures exceeding their upper limits of tolerance may also cause death or flight.

One of the problems with modern society is that it shifts environmental conditions, making regions hotter and drier, for example. Such changes can make life more difficult, if not impossible, for other organisms. Changes in water temperature or the chemical composition of lakes, for instance, have dramatic impacts on fish and other aquatic organisms, as you will see in Chapter 16.

Organisms are affected by all of the chemical and physical factors of their surroundings. Nonetheless, one factor usually outweighs the others in determining growth. It is, therefore, called a **limiting factor**. The concept of limiting factors was introduced in 1840 by the German scientist Justus von Liebig, who was studying the effects of

chemical nutrients on plant growth. The limiting factor is the primary determinant of growth in an ecosystem, because it easily falls below the range of tolerance of key organisms. In certain aquatic ecosystems, for instance, phosphorus is a limiting factor. It is the first nutrient to be used up. When phosphorus is reduced, the growth of algae is impaired. Nitrate can also be a limiting factor in water or soil. A clean body of water can turn green overnight with algae, for instance, if phosphate-containing detergents from a sewage treatment plant flow into it.

The limiting factor is analogous to the slowest elephant in an African expedition. The entire party's pace through the jungle is set by this slowest elephant. In the same fashion, the entire structure of an ecosystem is determined by the limiting factor. Should the slowest elephant die, a new pace would be set by the second-slowest one. The same is true in ecosystems. An increase in the availability of a limiting factor allows accelerated growth, but new limits are invariably set by another factor.

For most terrestrial ecosystems rainfall is *the* limiting factor. As we see in Figure 3-4, rainfall determines whether land is covered by forest, grassland, or desert. If annual rainfall exceeds 70 to 80 centimeters (27 to 31

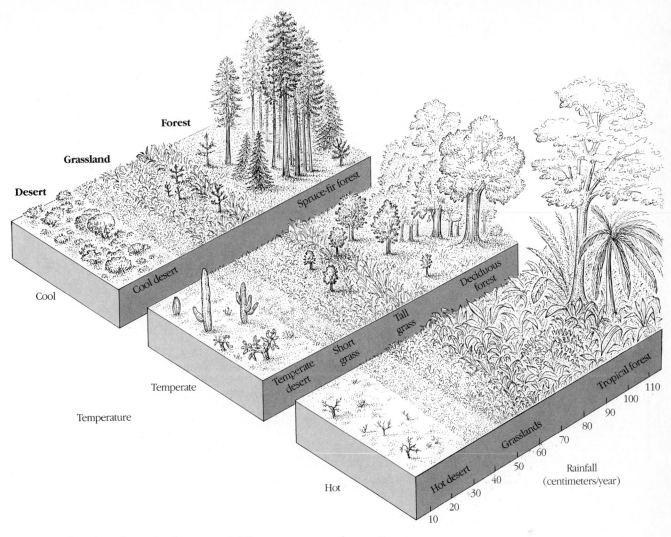

Figure 3-4. The relationship between rainfall, temperature, and vegetation. Rainfall determines the basic type of vegetation, and temperature is responsible for alterations in this basic type.

inches), for example, forests develop. Slightly drier climates support grasslands, and the driest regions are always desert.

Differences in temperature are responsible for variations on these three basic themes (Figure 3-4). Take the wettest ecosystems as an example. Warm, wet climates support tropical rain forests with huge trees reaching 60 meters (200 feet) into the air. Slightly cooler regions are characterized by deciduous forests. In areas that are even cooler, coniferous forests grow.

Biotic Factors The **biotic components** of an ecosystem are its living things—fungi, plants, animals, and microorganisms (Figure 3-6). Organisms live in **populations**, groups of the same species occupying a given region. Populations are dynamic groups, changing in size, age, structure, and genetic composition in correspon-

dence to changes in the environment. Populations never live alone in an ecosystem. They always share resources with others, forming a community. A **community** is defined as the organisms living within a given area. In Yellowstone National Park in Wyoming and Montana is a community familiar to many Americans. It consists of grizzly bears, elk, mule deer, coyotes, ravens, and many other organisms, including dozens of plants, all forming a complex community of living things.

Within communities are many different kinds of relationships—some cooperative, some not. One of the most familiar relationships is the predator–prey interaction. **Predation** occurs when one organism kills another for food, for instance, when a robin catches a worm or a grizzly catches a ground squirrel. Biologists even consider grazing a type of predation. Elk, for example, prey on grass.

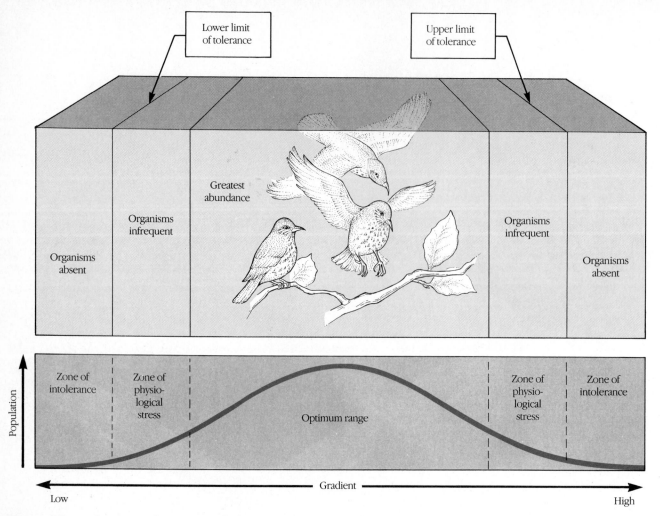

Figure 3-5. Each organism has a range of tolerance for abiotic factors. It survives within this range and can be killed by shifts below or above the range.

The interaction between predator and prey has obvious benefits to predators, because it is often their chief means of acquiring food. Predation also benefits prey populations, because predators often kill the weakened, sick, and aged members of a population. This helps reduce prey population size and ensures that the remaining members have a larger share of the available food supply—a benefit especially evident in winter months, when food is in naturally short supply. (For an example of the impacts of destroying predator–prey relationships see Case Study 3-2.)

Another type of interaction found in living systems is **commensalism**, a relationship between two species that is beneficial to one, but neutral (neither good nor bad) to the other. Humans, for instance, form bacteria in their intestines. For the most part, these organisms provide little or no benefit (they produce a small amount of vitamin K and a few other vitamins). The bacteria, however,

benefit enormously by this association, receiving an abundant food supply.

Mutualism is a relationship that is beneficial to both organisms. A classic example is lichen, an organism that clings to barren rock, capturing water from rainfall and gathering nutrients from the rock by eating away at its surface. Lichens are fungi that house algae inside them. The algae are photosynthetic. They capture sunlight and produce a whole host of organic molecules used by the lichen for energy and growth. The fungi, in turn, provide shelter and moisture. Both organisms benefit from this relationship.

Not all species interact in an ecosystem. This "unrelationship" is called **neutralism**. Seabirds on the coast of Louisiana and Texas, for instance, do not interact with songbirds living in nearby trees.

Some relationships can be harmful to one member. Parasitism is an example. **Parasitism** occurs when one

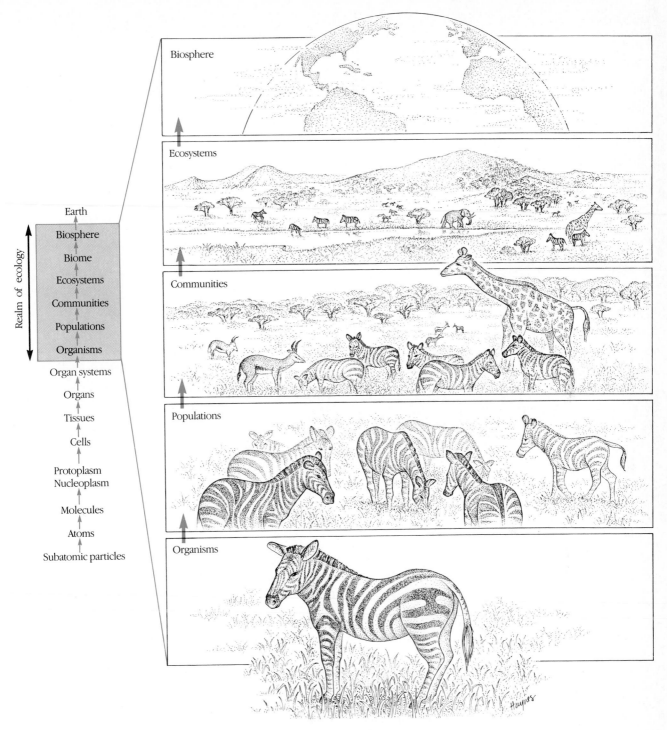

Figure 3-6. The organization of living matter. Ecologists concern themselves with the study of individual organisms, populations, and communities as well as ecosystems and the biosphere.

species lives on or even inside another, called the host. In this relationship the parasite obtains food by slowly eating the host or by dining on fluids inside the host. A parasite may be a temporary resident, such as a wood tick, or it may set up a long-term relationship, as tape-worms do. When the host is healthy, parasites simply dine on the surplus and have little effect. However, parasitism can kill a debilitated host that is suffering from disease, is aged and weak, or is under considerable stress. In Third World countries, humans who do not get enough to eat

Case Study 3-1

Benign Neglect in Texas: An Ecological Solution to a Perennial Problem

Texas highway officials and taxpayers have found that "benign neglect" is both economically and ecologically the best way to conduct roadside maintenance along the state's 71,000 miles of highways.

Fifty years ago the Lone Star state began sowing wildflower seeds along its nearly 800,000 acres of roadside, encouraging the brightly colored blossoms of the state's 5000 native wildflowers.

Texas is the leader in this ecological solution to roadside maintenance, which elsewhere has typically involved the planting of delicate grasses, regular mowing, irrigation, fertilization, and spraying with insecticides and herbicides—all for roadside lawns as well-groomed as a golf course. In Texas road crews mow a few times a year to help spread the flower seeds, but that's about the extent of roadside maintenance.

Besides beautifying the landscape, the hardy flowers reduce erosion and provide habitat for native species. More importantly, native wildflowers require only the water that falls naturally from the sky, and they need little if any fertilizer and

care. They resist drought, insects, and freezing weather, because they inhabit the land they have evolved over millions of years to live in.

Benign neglect makes good sense economically and environmentally. In 1988, the Federal Highway Commission took a step that delighted conservationists throughout the nation. It voted to spend 25 cents of every federal dollar invested in roadside landscaping on planting native wildflowers. The wildflower rule was spearheaded in Congress by Senator Lloyd Bentsen of Texas, who says that his state's lengthy experience with planting native wildflowers shows that they save taxpayers' money as well as beautify highways. Wildflowers seem to deter people from littering as well.

Montaigne, the French essayist and philosopher, once wrote, "Let nature have her way, she understands her business better than we do." Clearly, this is one case in point. The ecological wisdom of roadside flowers also turns out to make good sense economically. Can you think of other examples that might have the same effect?

often suffer enormously from parasitic diseases.

The final relationship worth noting is **competition**, a vying for resources between members of the same species (**intraspecific**) or between members of different species (**interspecific**). Competition is an extremely important force, for it can become a major component of natural selection (see Chapter 2). When interspecific competition is intense, one species may displace another.

Habitat and Niche

If you were to describe yourself to a foreigner, you would probably begin with where you live. Then you might tell what you do, who your friends (and enemies) are, how you interact with others, and what you do in your spare time. If you asked an ecologist studying armadillos to describe them, she would probably follow a similar approach, first telling you where these armored minitanks live and then what they do. Where an organism lives is its **habitat**. Where it lives and how it fits into the ecosystem are its **ecological niche**, or simply its **niche**. To describe an organism's niche requires that we tell what it eats, what eats it, where it lives, and how it interacts with other living and nonliving components of the ecosystem. This is no easy task.

No two species occupy the same niche, but similar species living in the same habitat often have similar lifestyles. Consider the red-tailed hawk and the great horned owl. On the surface these species may seem quite similar, maybe identical. Hawks and owls both prey on mice and other rodents. They live in trees. But owls generally live in dense forests and come out at night to feed, whereas hawks prefer daytime hunting. Despite their apparent similarities, therefore, these species occupy slightly different niches. The evolution of species to fit different niches has two beneficial effects: First, it reduces competition for limited resources by members of different species (interspecific competition). Second, it makes full use of environmental resources.

Because no two species occupy the same niche, two similar species can coexist in a given habitat without threatening each other's survival. For example, elk and mule deer live in the same habitat in the Rocky Mountains. But elk feed primarily on grasses and herbaceous plants, whereas deer feed primarily on shrubs.

The practical lesson of the **exclusion principle**, as it is called, is that two species cannot occupy the same niche without drastic consequences. This is another important law of ecology with important implications for wildlife officials, who in seeking to increase hunting opportunities

(a)   *(b)*

Figure 3-7. (a) The koala is a specialist, because it eats only the leaves of eucalyptus trees. A loss of this food source would result in its extinction. (b) The coyote is a generalist. Like humans, it is an opportunist, capable of eating a wide variety of foods. Its existence is not so delicately balanced.

may introduce species into an environment that come in direct conflict with existing species because of identical or extensively overlapping niches. Such species are called **ecological equivalents.**

An organism's niche may be generalized or specialized. Take the koala of Australia and the coyote of North America as examples. The koala is a specialist that feeds exclusively on eucalyptus leaves. Without them it would surely perish (Figure 3-7a). The coyote, on the other hand, feeds on a wide variety of animals—rodents, rabbits, ground-nesting birds, snakes, and an occasional sheep (Figure 3-7b). The coyote occupies a wide range of habitats, too, unlike the koala. Because of its versatility the coyote is considered a generalist. Human beings, like coyotes, are generalists in many ways. We live in diverse climates and eat a wide variety of plants and animals.

Generalists tend to be less subject to changes in the abiotic and biotic environment and, therefore, are less apt to become extinct. Should a food source disappear, a generalist will not suffer. Specialists, on the other hand, walk a tightrope. Their existence may easily be threatened by the loss of a single food supply. Today the giant panda of China is threatened by the loss of its sole food source, the bamboo.

Although humans are generalists, our recent evolutionary history has been marked by increasing specialization. Today, for example, we depend on a handful of metals, a relatively small number of crops and livestock species, and a few energy sources. If supplies of any of these should be cut off, catastrophe could follow. Avoiding such hardship, many experts believe, requires us to build a future less dependent on nonrenewable (depletable)

resources. Almost certainly, renewable substitutes must be found to provide energy, building materials, and raw materials for the many chemicals we use. To avoid hardship we must also protect our vast biological resources, for example, the genetic stock of the tropics, from which many of our important cultivated crops once came (Chapters 7 and 8). We must manage our forests and grasslands well, and protect commercial fish species (Chapter 9). More than ever we are learning that protecting these biological resources means reducing air and water pollution (Chapters 14–18).

Studying the niches of organisms is of great practical importance today. It can help science predict the impacts of human action on species and can be used to help save organisms from extinction as well. Knowing the habitat requirements, important food sources, and other important facts about an organism's niche will help society learn to accommodate its fellow inhabitants. Scientists are developing a habitat database that may help them understand the niche of the California condor, a species on the brink of extinction. All condors are in captivity now. Researchers hope to breed them successfully and one day be able to release them into the wild. The database will help scientists and conservationists find a suitable new habitat that meets their requirements.

How Do Ecosystems Work?

In this section we turn our attention to how ecosystems function. We will see how producers and consumers are related in an ecosystem and how energy and chemical nutrients flow through the biosphere.

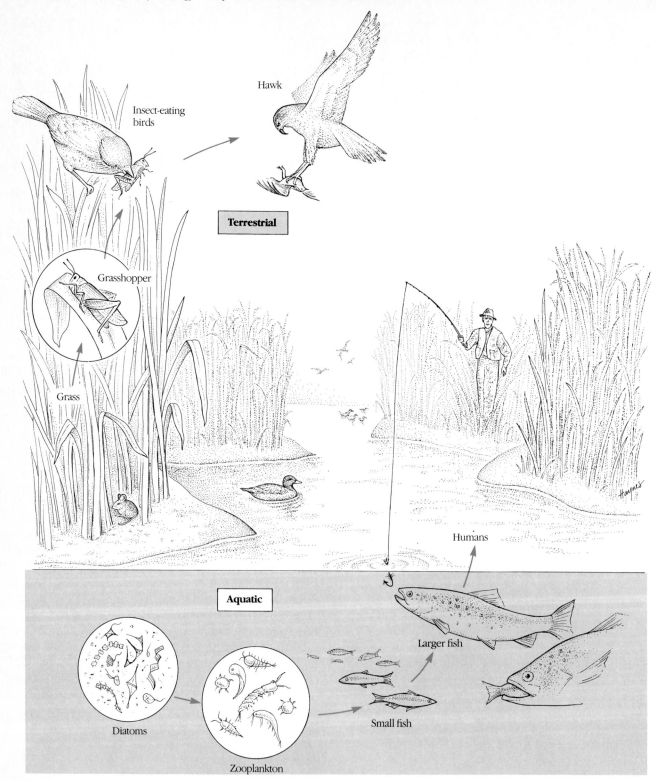

Figure 3-8. Examples of grazer food chains occurring on land and in water.

Food Chains and Food Webs

In the biological world you are one of two things, either a producer or a consumer. (Only rarely can you be both.) **Producers** are the organisms that support the entire living world through photosynthesis. Plants, algae, and cyanobacteria are the key producers of energy-rich organic materials. They are also called **autotrophs** (from the Greek root "troph"—to feed, nourish), because they literally nourish themselves photosynthetically, that is, by using sunlight and atmospheric carbon dioxide to make the food materials they need to survive. **Consumers** feed on plants and other organisms and are called **heterotrophs**, because they are nourished by consuming other organisms.

Consumer organisms that feed exclusively on plants are called **herbivores**. Cattle, deer, elk, and tomato hornworms are examples. Those consumers that feed exclusively on other animals, such as the mountain lion, are **carnivores**. Those consumers that feed on both plants and animals, such as humans, bears, and raccoons, are **omnivores**.

The interconnections among producers and consumers are visible all around us. Mice living in and around our homes, for example, eat the seeds of domestic and wild plants and, in turn, are preyed on by cats and hawks.

A series of organisms, each feeding on the preceding one, forms a **food chain** (Figure 3-8). Two basic types of food chains exist in nature: grazer and decomposer. **Grazer food chains**, like the one discussed above, are so named because they start with plants and with grazers, organisms that feed on plants. Figure 3-8 illustrates familiar terrestrial and aquatic grazer food chains.

In the second type—the **decomposer**, or **detritus**, **food chain**—organic waste material is the major food source (Figure 3-9). **Detritus** is organic waste which comes from plants and animals and is consumed on two levels. In the grasslands of Africa, for instance, a wildebeest that dies of old age is consumed by vultures, hyenas, and the larvae of various flies. These are called **detritus feeders**, or **macroconsumers**. The actual process of decomposition is carried out primarily by microscopic bacteria and fungi, known as **microconsumers**. All of these organisms, microscopic or not, derive energy and essential organic building blocks from detritus. In the process they liberate carbon dioxide, water, and other nutrients needed by plants to make more plant material and maintain the perpetual cycle.

Food chains are conduits for the flow of energy and nutrients through ecosystems. The sun's energy is first captured by plants and stored in organic molecules, which then pass through the grazer and decomposer food chains. In addition, plants incorporate a variety of inorganic materials such as nitrogen, phosphorus, and magnesium from the soil. These **chemical nutrients** become part of the plant's living matter. When the green plant is consumed, these nutrients enter the food chain. They are eventually returned to the environment by the decomposer food chain.

Classifying consumers Biologists are avid namers, happiest when things are categorized, tagged, and dissected. The study of ecology has not been spared this obsession. For instance, ecologists categorize consumers by their position in the food chain (Figure 3-10). In a grazer food chain, for example, herbivores are called **primary consumers**, since they are the first organisms to consume the plants. Organisms that feed on primary consumers are **secondary consumers**, and so on.

The feeding level an organism occupies in a food chain is called the **trophic level**. The **first trophic level** marks the beginning of the food chain and is made up of the producers or autotrophs (self-feeders). Primary consumers occupy the **second trophic level**, and secondary, tertiary, and quaternary consumers occupy the third, fourth, and fifth trophic levels, respectively. All consumers are heterotrophs. Figure 3-10 shows an example of a food chain broken down into trophic levels.

Food chains exist only on the pages of ecology texts; in reality, virtually all food chains are woven into a much more complex network called a **food web**. The food web gives a complete picture of who eats whom (Figure 3-11, page 58).

Trophic levels can be assigned in food webs just as in food chains; in a food web, however, many species occupy more than one trophic level. As illustrated in Figure 3-12 on page 59, a grizzly bear feeding on berries and roots is acting as a primary consumer; it occupies the second trophic level. When feeding on marmots, an animal similar to the woodchuck, however, the grizzly is considered a secondary consumer and occupies the third trophic level. In other instances a grizzly may feed on

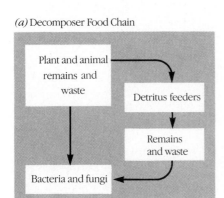

(a) Decomposer Food Chain

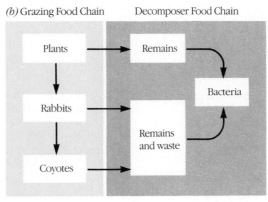

(b) Grazing Food Chain   Decomposer Food Chain

Figure 3-9. (a) Decomposer food chain. Bacteria and other organisms feed on plant and animal remains. (b) Link between the grazer and the decomposer food chains.

Ecology or Egology? The Role of the Individual in the Environmental Crisis

Peter Russell

The author has been closely involved with the development of the Learning Methods Group, an international organization that helps people make fuller use of their mental potential. His books include The TM Technique, The Brain Book, Meditation, *and* The Global Brain, *which presents an optimistic vision of our long-term future.*

The environmental sciences usually focus on understanding the many intricate relationships and interdependencies that have evolved between the millions of living systems inhabiting the earth. Rather than looking further into the many facets of these sciences, I wish to explore this question: "Why are such studies necessary in the first place? Why is it that one species out of millions can disrupt the natural balance in so many ways and with such dire consequences?"

The rapid and liberal development of technology is clearly part of the problem. *Homo sapiens* has always been a manipulative species. A million years ago, long before civilization appeared, our opposable thumbs and large brain singled us out as the creature most capable of modifying its environment. In more recent history technology has amplified this capacity, and to such an extent that today one person's decision may have global repercussions.

Yet it is also clear that technological might is not the sole cause of environmental mismanagement. It is ultimately human beings who choose how to use technology, and they initiate the actions that result in ecological disturbances. In some cases the disturbance is totally unforeseen. In other cases the impacts are foreseeable but those concerned ignore the warnings or, worse, seek evidence to the contrary, displaying an apparent lack of care for other species, the biosystem as a whole, and paradoxically, their own long-term welfare. How is it that people can adhere to policies that by nearly all projections look suicidal?

Such behavior stems from individuals' not seeing beyond their short-term welfare and from their perceiving their interests to be different from the interests of humanity as a whole. For most people an immediate personal fulfillment is more attractive than some distant long-term benefit, and so they naturally go for the former. Education and social controls may help curb the pull of immediate gratification, but the pull is so strong that these are unlikely to be sufficient. But what is it that makes us want to satisfy our short-term needs to the detriment of our long-term welfare? The answer appears to lie deep within our psyche.

In the more developed nations—those responsible for the major environmental problems today—the basic needs for food, clothing, shelter, and health care are fairly well attended

to. What emerges then is the need for psychological welfare, in particular, the need to be liked. We need people to recognize and reaffirm our worth, and we spend considerable time and effort fulfilling this need. Much of our activity is really a search for personal reinforcement (what are often termed "positive psychological strokes"). Some psychologists estimate that as much as 80% of all our actions may be motivated by this search. And when people stand back and listen to themselves, they often find that as much as 90% of their casual conversation is prompted by this need for approval. The need to be liked has become one of our most powerful drives.

One of the most common ways we try to win approval and prestige is through material possessions. We collect many of the various accoutrements of modern living—new cars, fashionable clothes, expensive furniture—not because we need them physically but because we need them psychologically.

This might not be so bad if our various possessions satiated our psychological needs; but they don't. Our insecurity is rooted far deeper in our psyche than that. Instead of spotting the obvious flaw, we search for yet more things in the outer world that we hope might fill the inner gap.

Some implications of this for the way we treat the environment are obvious. We gobble up irreplaceable resources, with little regard for the long-term future, partly because the various products they are transformed into may briefly satisfy our search for identity. Thus, the consumers are as much exploiters of the environment as the corporations.

Yet there is more to it than that. We do have choices in how we satisfy our misguided search for material well-being. We might even be able to do so in ecologically sane ways. Ultimately there are people somewhere making decisions to exploit this or that particular resource, initiate industrial processes with various environmental effects, and set in motion other activities that in one way or another upset the ecological balance. And such people are usually motivated by similar needs for approval and recognition from their peers and by the mistaken belief that financial security brings inner security. They are caught in the same trap as everyone else. The only difference is that they are more visible.

Another profound consequence of our inner insecurity is a lack of true caring and compassion, either for others or for the environment. Each of us, at our core, is a compassionate being capable of deep empathy and caring. If we can get back in contact with this deeper self, we can begin to experience compassion not only for other people but also for the rest of the world.

Letting go of our self-gratifying patterns of behavior, we find not only compassion but humility. We rediscover respect for the unfathomable complexity of life on earth and for all its many species.

Our efforts to solve pressing environmental problems will remain incomplete until they take account of the human psychology that gives rise to their necessity. The only approach that will be successful in the long term is one that attends both to the external environment and to the internal environment, the system we call "I."

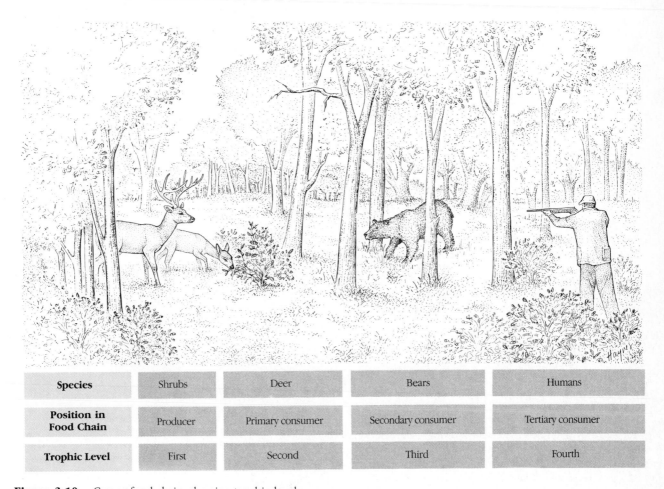

Species	Shrubs	Deer	Bears	Humans
Position in Food Chain	Producer	Primary consumer	Secondary consumer	Tertiary consumer
Trophic Level	First	Second	Third	Fourth

Figure 3-10.   Grazer food chain, showing trophic levels.

insect-eating chipmunks. It thus occupies the fourth trophic level. The grizzly also feeds on **carrion**, animal carcasses it finds. In such instances, the grizzly becomes a member of the decomposer food chain.

The Flow of Energy and Matter Through Ecosystems

In the previous section we saw that photosynthetic organisms such as plants are the starting point of all food chains,

because they alone are capable of tapping the sun, the ultimate source of energy for all ecosystems. But plants capture only 1% or 2% of the energy that the sun transmits to earth (Figure 3-13, page 61). Still, on this small fraction of the sunlight is built the entire living world. In this section we study the flow of energy and matter through ecosystems. But first let's take a brief look at energy.

What Is Energy?   Energy and love. We are all familiar with the terms, but we still stumble when trying to define

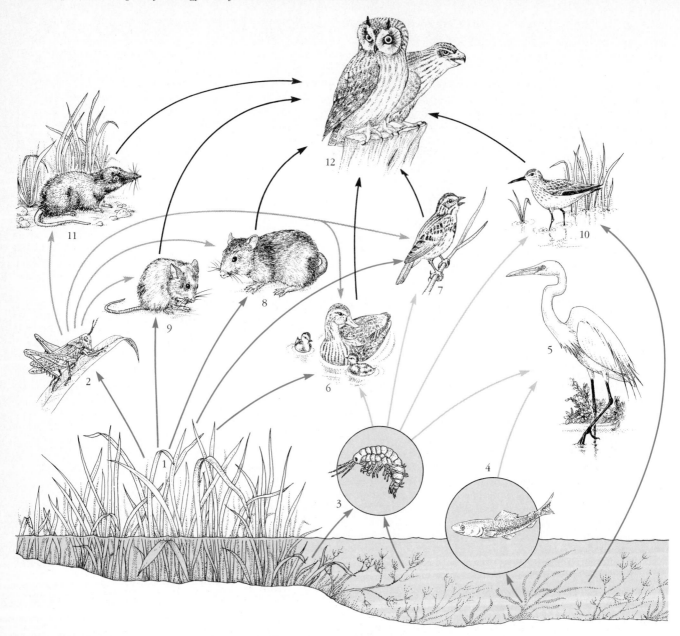

Figure 3-11. A simplified food web, showing the relationships between producers and consumers. In a salt marsh (San Francisco Bay area), producer organisms, terrestrial and salt marsh plants (1), are consumed by herbivorous invertebrates, represented by the grasshopper and the snail (2). The marine plants are consumed by herbivorous marine and intertidal invertebrates (3). Fish, represented by smelt and anchovy (4), feed on vegetative matter from both terrestrial and marine environments. The fish in turn are eaten by first-level carnivores, represented by the great blue heron and the common egret (5). Continuing through the food web, we have the following omnivores: clapper rail and mallard duck (6), savannah and song sparrows (7), Norway rat (8), California vole and salt marsh harvest mouse (9), and eastern and western sandpipers (10). The vagrant shrew (11) is a first-level carnivore, while the top carnivores (second level) are the marsh hawk and the short-eared owl (12).

Figure 3-12. The grizzly eats widely from the food web. When individual food chains are drawn separately, this becomes evident.

(a)

Trophic level	Grazer Food Chains			Decomposer Food Chain
Fourth			Grizzly	
Third		Grizzly	Chipmunk	
Second	Grizzly	Elk, Marmot	Insects	Grizzly
First	Berries, Roots, and Grasses	Grasses	Grasses	Carrion

(b)

them. One of the reasons is that, like *amour*, energy comes in many forms. Heat, light, sound, electricity, coal, oil, natural gas, mechanical energy, geothermal energy, wind, nuclear power, hydroelectric power, and magnetic energy are some of the more familiar ones. As different as these forms of energy are, however, they have one thing in common: the capacity to do work. To a physicist, then, **energy** is defined as the capacity to do work. **Work** is performed on an object—be it a mountain or a mole hill—when it is moved over some distance.

All forms of energy fall into two groups: potential and kinetic. **Potential energy** is stored energy—energy stored in coal, oil, or even the food you eat. When released, it can perform work. **Kinetic energy**, in contrast, is the energy possessed by objects in motion. A falling rock and a swinging hammer can do work and, therefore, are said to possess kinetic energy.

All forms of energy follow basic laws, called the **laws of thermodynamics**. Understanding these laws will help

you better understand ecology and many current environmental issues.

The First Law The **first law of thermodynamics** is often called the law of **conservation of energy**. It states that energy can be neither created nor destroyed but can only be transformed from one form to another.

Let's look at a familiar example. The gasoline in your car contains an enormous amount of potential energy that, when released in combustion, speeds your car down the highway. As you drive along, the gas gauge shows how much gasoline you have burned. Contrary to what you may think, however, you have not destroyed that energy. Instead, you have converted it into other forms—electricity to run your radio and windshield wipers, heat to keep you warm and defrost your windows, light to show you where you are going, and, of course, mechanical movements that propel your car along the highway. Careful measurement of the amount of energy your car is

Ecosystem Imbalance in the Great Barrier Reef: Human Impact on a Delicate and Important Ecosystem

Australia's Great Barrier Reef has been described by some as the eighth great natural wonder of the world. However, scientists today are wondering whether this magnificent reef will survive to the year 2000.

For many years, Dr. Robert Endean of the University of Queensland has argued that the Great Barrier Reef is in trouble. Many marine biologists agree. They think that the delicate life of the coral reef is threatened by a once rare species of starfish, known as the Crown of Thorns. This starfish feeds on stationary marine animals called coral polyps, which gave rise to the coral reef over many years. Until recently the starfish was so rare that it had little effect on coral reefs.

In the 1960s, however, the population of the Crown of Thorns starfish rose dramatically in the Great Barrier Reef, killing off large sections of the reef. The central third of the Great Barrier Reef has been infested with Crown of Thorns starfish, and scientists now report outbreaks on the northern and southern ends as well.

The cause of this outbreak is unknown, but many scientists think that it is the unintentional handiwork of humans. By destroying the starfish's natural predators, humans have unleashed a destructive force that could wipe out the Great Barrier Reef in a decade.

One of those predators is a mollusk, the giant triton. Over the years, it has been collected in great numbers and sold for food. This mollusk, which once helped keep starfish populations in check, is becoming rare on the reefs. In addition, predatory fish, such as cod and grouper, have been severely overfished in the waters surrounding the reef, contributing to the starfish's sudden increase in numbers.

Natural predators have controlled starfish populations for many hundreds of years. Human intervention has severed an important relationship that is needed to keep the coral reef alive. To counteract the dramatic increase in starfish, the Queensland government has banned the capture of starfish predators. However, the Great Barrier Reef is 2000 kilometers long (1200 miles)—roughly the distance from San Francisco to Denver. It lies up to 160 kilometers (100 miles) offshore, making it difficult to patrol.

Should the reef be destroyed, Queensland fishing companies could face great economic hardship, because many commercially important species live in and among the reefs. Tour operations could also be greatly affected, because the Great Barrier Reef is a natural attraction for scuba divers and sightseers the world over. Losing the coral reef would be an aesthetic loss as well. For the species that live among the coral, it would be a tragedy of epic proportion.

---

consuming and the amount of energy being produced in these various forms shows that the two are equal. In simple terms, energy input is equal to energy output.

Many energy conversions take place in biological systems. Sunlight, for instance, is trapped by plants and stored in organic food materials the plant makes during photosynthesis. These molecules are ingested by herbivores and broken down to provide energy needed to perform a variety of cellular functions.

Modern society also relies on energy conversions to perform millions of activities every day. Coal, for example, is burned in power plants to generate electricity, which is used, in turn, to power light bulbs, neon signs, and electric motors. Surely, evidence of the first law is everywhere, but how important is it?

The first law helps us understand that energy exists in many forms. Coal, for example, is a type of potential energy. It can be converted into heat and light and electricity—all forms of kinetic energy. Of these forms, the most widely used is electricity. Electricity is a useful form

compared with heat, much of which escapes from the boiler. Factories and power plants that efficiently convert coal—or any other fuel—to electricity are obviously the most desirable.

The first law is also popularly referred to as the no-free-lunch law. What that means is that in all systems, energy output can never be greater than energy input. Even today, people claim to invent machines that create more energy than they use. Beware of such promises. Years of research show them to be false. Careful calculations always show that energy input and energy output are equal. The first law of thermodynamics is often said to mean that you can't get something for nothing.

The first law reminds us to balance our energy calculations. When developing new energy sources, look carefully at energy put into the system to get the resource out. Suppose you had a seam of coal on your property and wanted to burn it to heat your house. To get the coal out, you'd have to dig it up. Then you'd have to crush it and transport it to your furnace. When calculating the

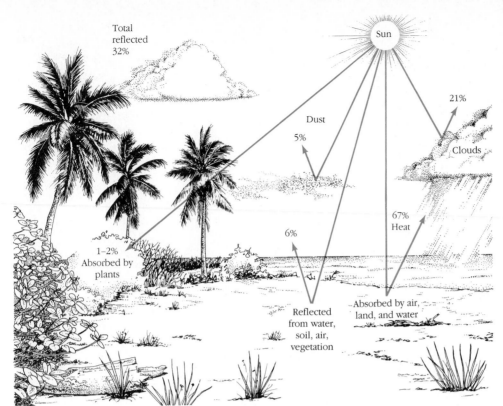

Total
reflected
32%

Sun

21%

Dust
5%

Clouds

1–2%
Absorbed by
plants

6%

67%
Heat

Reflected
from water,
soil, air,
vegetation

Absorbed by air,
land, and water

Figure 3-13. Distribution of incoming solar radiation. Note that plants absorb only a small fraction of the sun's energy.

amount of energy you get from coal—to see if it is worth your effort—you must factor in all of the energy it took to mine, transport, and process the coal. The energy output of your system is quite low considering the energy investment to obtain the coal in the first place.

The Second Law   The first law deals with energy conversion and involves quantities—energy inputs and outputs. The second law also deals with energy conversion, but it involves a different aspect of energy—its quality.

More specifically, the **second law of thermodynamics** explains what happens to energy quality when energy changes from one form to another. The law simply states that energy is "degraded" during such conversions (Figure 3-14). Another way of saying this is that energy goes from a concentrated to a less concentrated form during a transformation. For example, when gasoline is burned in an automobile, it is converted from a very concentrated form to much less concentrated heat, which is no longer available for useful work. Concentrated energy forms are said to have a great deal of **available work**. The less concentrated forms have a lower capacity and are said to have less available work. Consider some examples. Oil and natural gas are high-quality energy sources. They are concentrated fuels that, when burned, can be put to good use. Most heat, however, is a low-quality energy source. It is quickly diluted or dispersed and is a less useful form

of energy to power machinery. All heat produced on earth eventually dissipates into space and is lost to us.

The second law of thermodynamics has many implications for our lives. It tells us that when we burn fossil fuels, our supply of highly concentrated energy (our finite fossil fuel reserve) shrinks. It tells us also that we cannot recycle high-quality energy, because when it is burned it is dissipated into heat and lost into space. And it warns us not to waste this precious resource; the source we now tap is all that we have.

Entropy   The universe is a system in decay. Why? Because all of the high-quality energy available to it is being converted to heat. With this conversion, matter also undergoes a significant change. A chunk of coal with its highly ordered molecules, for example, is broken apart when it burns, producing many molecules of carbon dioxide gas, which are dispersed randomly in the atmosphere. When coal is burned, then, disorder is created.

Randomness, or disorder, is called **entropy**. Your college dorm room is a perfect example. It's a system in constant decay. Entropy always increases. The only way you can stop it is to invest energy in keeping things straight.

In the universe, entropy prevails. Randomness increases as time passes. But within the chaos are pockets of order. One of the most important is life itself. Living

Figure 3-14. Second law of thermodynamics. During energy transformations the amount of useful work diminishes as we go from high-quality (concentrated) energy forms to low-quality (less concentrated) energy forms.

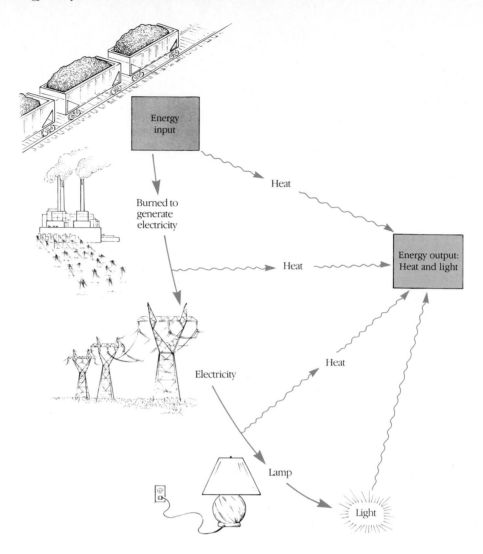

systems create order from disorder. Plants, for example, remove carbon dioxide molecules from the atmosphere, uniting them to form organic food molecules. Thus, entropy can be reversed. But it can only be reversed or held at bay by expending energy. In plants, that energy comes from sunlight.

Sunlight, which plants use to create order here on earth, is produced by the gradual decay of the sun. The sun is slowly being dissipated to produce energy. Overall, the order created on earth by living systems is much smaller than the disorder created by the sun's decay. On balance, then, disorder prevails.

The second law of thermodynamics says that energy is degraded when it is converted from one form to another. Another way of putting this is to say that the universe proceeds towards disorder, or entropy.

Biomass and Ecological Pyramids The laws of thermodynamics go much further. They rule the living

world, from the tiniest bacterium to the largest whale. They limit ecosystems, as you will soon see, and can be used to guide us along a sustainable pathway. They can bring common sense to our politics and help us avoid foolish economic mistakes. Let us look at the implications of these laws from an ecological perspective. We will begin with the concept of **biomass**.

Biomass is organic matter created by plants and other photosynthetic organisms and passed up the food chain. Because organisms vary considerably in their water content, water must be excluded when determining biomass. Biomass is, therefore, the dry weight of living things and can be measured for each trophic level. To sample the biomass of plants, for example, a single square meter of vegetation would be removed, roots and all, then dried and weighed. The result is the plant's biomass. The biomass at the first trophic level in almost all ecosystems represents a large amount of potential (chemical) energy and tissue-building materials for the second trophic level. Studies of ecosystems show, however, that not all of the

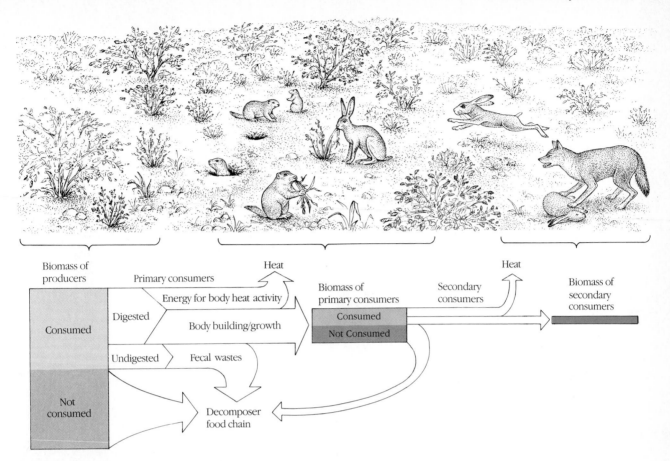

Figure 3-15. Energy and nutrient flow through the ecosystem. Note that a large fraction of the biomass of the first trophic level is not consumed. Also note the losses from one trophic level to the next.

biomass of the first trophic level is converted into biomass in the second trophic level (Figure 3-15). In other words, not all plant matter becomes animal matter. Several reasons can be given for this fact. The first is that only a small part of the plant matter in any given ecosystem is eaten by organisms of the next higher level, as shown in Figure 3-15. Second, not all of the biomass eaten by the herbivores is digested; some passes through the gastrointestinal tract unchanged and is excreted. Third, most of what is digested is broken down into carbon dioxide, water, and the energy used to move about, to breathe, and to maintain body temperature. The second law of thermodynamics tells us that this energy is converted into heat, which is dissipated into the environment and is eventually lost into space.

Because of these factors the biomass of the second trophic level is greatly reduced. Ecologists once thought that 10% of the biomass at one trophic level was transferred to the next. They called this the **ten percent rule**. Further research showed, however, that the amount of

biomass transferred from one level to the next varies from 5% to 20% in different food chains. Graphically represented, biomass at the different trophic levels forms the **pyramid of biomass** (Figure 3-16).

Biomass is the substance that makes up living things. The chemical bonds that hold the organic compounds of biomass together contain enormous amounts of stored, or potential, energy. This energy can be released when organic matter burns or when it is broken down by cells in plants, animals, and microorganisms. If the energy content of biomass is graphed as the biomass was, it, too, forms a pyramid, called the **pyramid of energy** (Figure 3-17).

What are the implications of the ecological pyramids? As you learned earlier, biomass decreases in the upper trophic levels of a food chain. Thus, fewer organisms can be supported at these higher levels. Graphically represented this forms a **pyramid of numbers**. As illustrated in Figure 3-18, over twice as many herbivores (trophic level two) can be supported in a grassland biome as carnivores

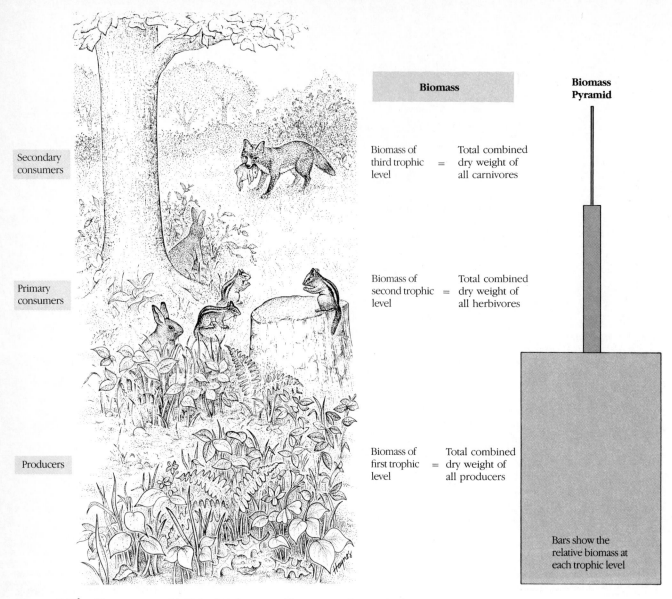

Figure 3-16.   Biomass pyramid. Note that the pyramid corresponds to a single food chain at successive trophic levels.

(trophic levels three and four). Carnivores, in fact, always have the smallest populations in ecosystems. Thus, they are often the first endangered by human activities that disrupt the food chain. A reduction in biomass produced at the first trophic level, for instance, can have profound impacts on carnivores because it decreases prey populations. Take away its prey and the carnivore is left to starve or enter into fierce competition with its own kind for food (intraspecific competition).

Another implication of ecological pyramids is that more organisms can be supported in an ecosystem if they can feed at lower trophic levels. In human terms, then, consuming meat is much more wasteful of solar energy than

eating plants, calorie for calorie. If 20,000 kilocalories worth of corn were fed to a steer, for example, 2000 kilocalories of beef would be produced (using a 10% conversion). This would feed only one person (assuming that a person can survive on 2000 kilocalories per day). If the 20,000 calories of corn were eaten directly, however, it would feed ten people for a day. In improving food supply, then, it makes more sense to increase supplies of grain rather than meat, which is generally the approach most countries take. Encouraging people to eat lower on the food chain could also help increase available food supplies. (Chapter 7 offers additional solutions to world hunger.)

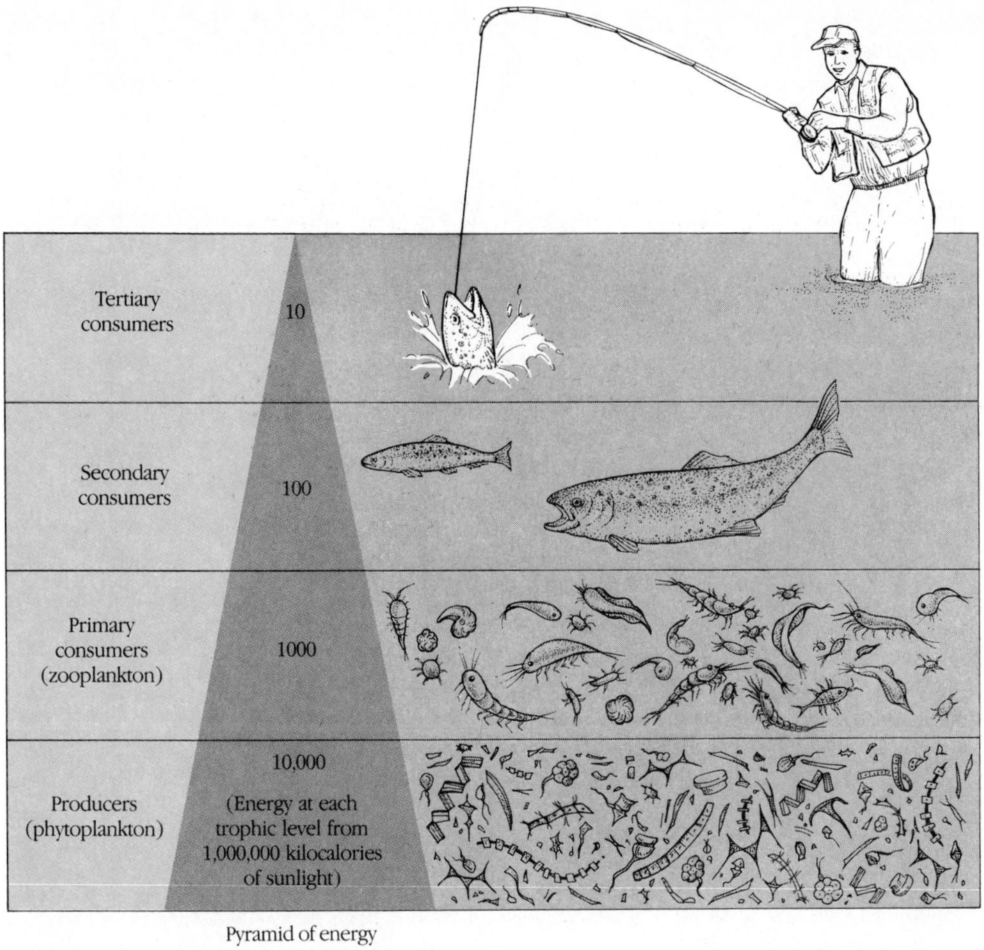

Figure 3-17. Pyramid of energy. Note the rapid decrease in potential energy as we ascend the food chain.

Tertiary consumers	10
Secondary consumers	100
Primary consumers (zooplankton)	1000
Producers (phytoplankton)	10,000 (Energy at each trophic level from 1,000,000 kilocalories of sunlight)

Pyramid of energy

The third implication of ecological pyramids is that the loss of biomass from one trophic level to the next sets limits on the length of the food chain. Food chains usually have no more than four trophic levels, because the amount of biomass at the top of the trophic structure is not sufficient to support another level.

Productivity We measure output in factories and mines in terms of productivity: how much steel is produced and how much coal is mined per hour of labor. In ecosystems, ecologists measure productivity in a similar way. The most common measure is kilocalories of biomass produced in a year, usually per square meter of surface area (kcal/m²/year). In an ecosystem, then, productivity is the rate at which sunlight energy is converted into the potential energy of biomass. The overall rate of biomass production is called the **gross primary productivity** (GPP). Like a worker's gross pay, the GPP is subject to some deductions. For plants the chief deduction comes in the form of energy used to meet their own needs. Therefore, by subtracting the biomass broken down to release energy, which is called cellular respiration (R), ecologists arrive at the **net primary productivity** (NPP). Much like your net pay, the NPP is what's left over after

deductions. The simple mathematical equation for net primary productivity is NPP = GPP − R.

Studies of biomass production in the earth's biomes help us understand the importance of various regions to the earth's carbon and oxygen cycles, since productivity (photosynthetic output) is directly related to oxygen consumption and carbon production by plants. Table 3-1 reveals some important results: Estuaries and coral reefs

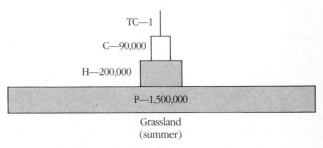

TC—1
C—90,000
H—200,000
P—1,500,000

Grassland (summer)

Figure 3-18. Pyramid of numbers for a grassland community in the summer. The numbers represent individuals per 1000 square meters. P = producers, H = herbivores, C = carnivores, TC = top carnivores (highest consumer in a food chain).

Table 3-1 Estimated Annual Gross Primary Productivity of the Biosphere and Major Ecosystems

Ecosystem	Area (10^6 km^2)	Gross Primary Productivity (kcal/m^2/year)	Worldwide Annual Gross Primary Production (10^{16} kcal)*
Marine			
Open ocean	326.0	1,000	32.6
Coastal zones	34.0	2,000	6.8
Upwelling zones	0.4	6,000	0.2
Estuaries and reefs	2.0	20,000	4.0
Marine total	**362.4**		**43.6**
Terrestrial			
Deserts and tundras	40.0	200	0.8
Grasslands and pastures	42.0	2,500	10.5
Dry forests	9.4	2,500	2.4
Boreal coniferous forests	10.0	3,000	3.0
Cultivated lands with little or no energy subsidy	10.0	3,000	3.0
Moist temperate forests	4.9	8,000	3.9
Fuel-subsidized (mechanized) agriculture	4.0	12,000	4.8
Wet tropical and subtropical (broadleaved evergreen) forests	14.7	20,000	29.0
Terrestrial total	**135.0**		**57.4**
Biosphere total (round figures, not including ice caps)	**500.0**	**2,000**	**100.0**

*This column is calculated by multiplying the area by the productivity of each region. It tells the relative importance of each zone to total biospheric productivity. *Source:* Odum, E. (1971). *Fundamentals of Ecology* (3rd ed.). Philadelphia: Saunders.

are the richest areas of the ocean. On land, moist temperate forests, agricultural land, and tropical forests have the highest gross primary productivity. Protecting these areas is important for the well-being of our planet and ourselves, reaffirming a fundamental law of ecology—that planet care is self-care.

How important each biome is to total global biomass production, however, depends on the total area it occupies. Taking into account both productivity and surface area, the largest producers of biomass are the open oceans, grasslands and pastures, and tropical forests (Table 3-1).

Nutrient Cycles

The economy of a country is based, in large part, on the flow of goods from agricultural and industrial producers to the people, the consumers. The "economy" of the living world is very similar. For instance, biomass and inorganic matter move from plants, the producers, to animals, the consumers. In nature these nutrients flow in a cyclic fash-

ion so materials can be reused. Modern industrial societies flagrantly violate this fundamental rule by discarding materials, taking them out of the natural cycles so they cannot be reused easily.

Nutrients flow in cycles, but energy is an entirely different matter. It flows through the biosphere from producer to consumer but cannot be recycled, as implied by the second law of thermodynamics. Thus, as we have seen, all of the energy that enters a food chain is eventually lost as heat. This low-quality energy is dissipated into space and cannot be reused.

The cycles that move nutrients through the biosphere are called **biogeochemical cycles**, or **nutrient cycles**. As shown in Figure 3-19, the nutrient cycles involve two general phases: the **environmental phase**, in which the chemical nutrient is in the soil, water, or air, and the **organismic phase**, in which the nutrient becomes part of the living tissue of organisms.

Of the 92 naturally occurring elements, about 40 are essential to life. Six of these elements—carbon, oxygen, hydrogen, nitrogen, phosphorus, and sulfur—form 95%

of the mass of all plants, animals, and microorganisms. These elements, and a few others needed in relatively large quantities, are called **macronutrients**. Others such as iron, copper, zinc, and iodine are required in only small amounts and are called **micronutrients**.

This chapter examines three nutrient cycles: carbon, nitrogen, and phosphorus. The water cycle is discussed in Chapter 10.

The Carbon Cycle   One of the most important nutrient cycles is the carbon cycle. It is responsible for the recycling of carbon dioxide given off by all living things. The carbon cycle consists of two halves. The first half is **photosynthesis**. This is the phase during which carbon dioxide is taken up by plants and algae and converted to food molecules with the aid of sunlight. The chief products of photosynthesis are oxygen and organic molecules. Half the oxygen in the atmosphere is replenished each year by plants and algae. Both oxygen and organic molecules are essential nutrients for nonphotosynthetic organisms, which constitute the second half of the cycle. The nonphotosynthetic organisms consume the oxygen and organic food molecules, giving off carbon dioxide to complete this ever-continuing cycle of mutual dependence. Thus, virtually all animal life exists on earth because of plants. Plants, in turn, survive thanks to carbon dioxide released by animals.

A simplified version of the carbon cycle is shown in Figure 3-20. Atmospheric carbon dioxide enters terrestrial and aquatic ecosystems. Within plants, carbon dioxide molecules react to form organic molecules such as glucose. Energy for these reactions comes from sunlight through the process of photosynthesis. Photosynthesis can be written as follows:

carbon dioxide + water + sunlight ⟶ glucose + oxygen

$$6\ CO_2\ +\ 6\ H_2O\ +\ E\ \longrightarrow\ C_6H_{12}O_6\ +\ 6\ O_2$$

As illustrated in Figure 3-20, the organic molecules produced during photosynthesis are passed through the food web. Thus, carbon flows from the atmosphere into and through the organismic phase of the cycle, travelling through grazing and decomposer food chains.

Carbon returns to the atmosphere in several ways. In plants and animals, for instance, some of the organic molecules are broken down to generate cellular energy. This process, called **cellular respiration**, results in the production of usable energy, heat, carbon dioxide, and water. It can be written as follows:

glucose + oxygen ⟶ carbon dioxide + water + energy

$$C_6H_{12}O_6\ +\ 6\ O_2\ \longrightarrow\ 6\ CO_2\ +\ 6\ H_2O\ +\ E$$

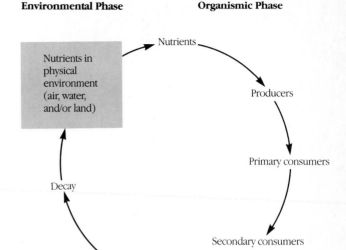

Figure 3-19. Nutrient cycles have two basic phases: environmental and organismic.

This reaction is the reverse of photosynthesis. The carbon dioxide gas released during cellular respiration reenters the environmental phase of the cycle for reuse.

Carbon can also return to the atmosphere through the decomposer food chain. Here, decomposers consume organic wastes from plants and animals and convert them into living tissue and energy, while liberating carbon dioxide through cellular respiration. Carbon also returns when plant materials are burned by natural causes such as lightning and forest fires or as a result of human activities such as combustion of wood and coal or deliberately set fires.

Humans intervene in the global carbon cycle in two major ways: (1) by removing forests and vegetation that use atmospheric carbon dioxide to make organic molecules, and (2) by liberating carbon dioxide during the combustion of coal, oil, and natural gas—carbon sources that were once isolated deep beneath the earth's surface. Combined, such activities have increased the global carbon dioxide concentrations by over 19% since 1870. Further increases could have a devastating effect on global climate and life. (For more details see Chapter 15.)

The Nitrogen Cycle   Nitrogen forms part of many essential organic molecules, notably amino acids (the building blocks of proteins) and the genetic materials RNA and DNA. Fortunately for plants and animals, nitrogen is an abundant element; approximately 79% of the air is nitrogen gas (N_2). However, plants and animals cannot use nitrogen in this form. To be usable it must first be converted into ammonia (NH_3) or nitrate (NO_3) (Figure 3-21).

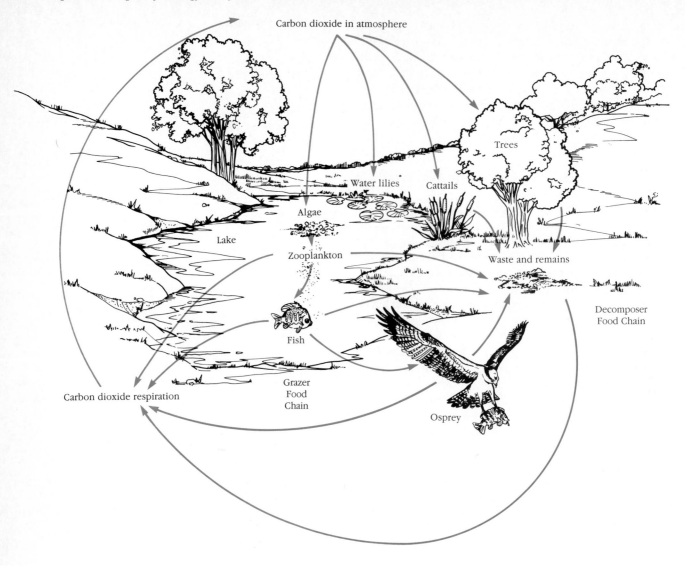

Figure 3-20. The carbon cycle. Carbon cycles back and forth between the environment (air and water) and organisms. It's trapped by plants and passed to herbivores, carnivores, and decomposers. Carbon dioxide is released by the breakdown of organic materials for energy in these organisms, thus continuing the cycle.

The conversion of atmospheric nitrogen into nitrate and ammonia is called **nitrogen fixation**, and it occurs mainly in certain bacteria in the soil and water. Without these organisms life as we know it could not exist. One nitrogen-fixing bacterium, called *Rhizobium*, invades the roots of a group of plants called **legumes**, which include beans, peas, alfalfa, clover, and others. The roots respond by forming tiny nodules that serve as sites for nitrogen fixation. Formed either in the soil or in the root nodules, the nitrogen compounds are taken into plants and there used to synthesize amino acids, proteins, DNA, and RNA.

Animals, in turn, receive the nitrogen they need by eating plants (and other animals).

Nitrogen is returned to the soil by the decay of detritus (Figure 3-21). Within the soil certain species of bacteria and fungi decompose the nitrogen-rich wastes from plants and animals. Nitrogen is released in the form of ammonia and ammonium salts. Ammonia is further converted into nitrites and then into nitrates.

The ammonium salts, nitrates, and nitrites may all be incorporated by the roots of plants and reused. Some nitrite is converted into a gas, nitrous oxide (N_2O), and

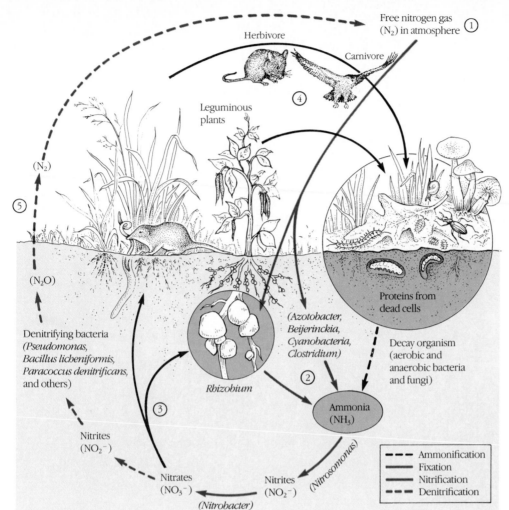

Figure 3-21. The nitrogen cycle. (1) Nitrogen gas is abundant in the atmosphere, but it can't be used by plants or animals in this form. (2) It must first be converted to ammonia and nitrates by bacteria in the soil and in the roots of certain plants. (3) It can then be incorporated by plants and used to make amino acids. Herbivores acquire the amino acids they need from plants and carnivores get them from herbivores that they feed on. (4) Nitrogen is returned to the soil in the urine and feces of animals or when dead plant and animal material decays. (5) Some nitrate is converted back into atmospheric nitrogen by denitrifying bacteria.

released into the atmosphere.

In various forms nitrogen travels from air to soil to plant to animal and then back to soil and atmosphere in a never-ending cycle. Human farming practices, discussed in Chapter 7, greatly affect the nitrogen cycle. For example, some crops such as corn absorb large amounts of nitrogen from the soil; if nitrogen is not replaced in alternate years, the soil may become unproductive. Farmers may replace soil nitrogen by applying artificial fertilizers that contain nitrates. Frequently used in excess, these nitrates may wash into streams, causing serious water pollution (Chapter 16).

The Phosphorus Cycle Phosphorus, found in living organisms as phosphate (PO_4), is an important part of RNA and DNA. Phosphorus is also found in fats (phospholipids) in cell membranes.

Phosphate is slowly dissolved (leached) from rocks by rain and melting snow and is carried to waterways (Figure 3-22). Dissolved phosphates are incorporated by plants and then passed to animals in the food web. Phosphorus reenters the environment in at least two ways: (1) some is excreted directly by animals, and (2) some is returned when detritus decays.

Each year large quantities of phosphate are washed into the oceans, where much of it settles to the bottom and is incorporated into marine sediments. Sediments may release some of the phosphate needed by aquatic organisms. The rest may become buried and taken out of circulation.

Phosphate is a major component of fertilizer. By applying excess fertilizer farmers may alter the phosphorus cycle. Since phosphorus is a limiting factor in aquatic ecosystems, excesses entering from farmland runoff may cause rapid growth in algae and other aquatic plants. This phenomenon is discussed in more detail in Chapter 7.

Figure 3-22. The phosphorus cycle has two interconnected parts, a terrestrial portion and an aquatic portion.

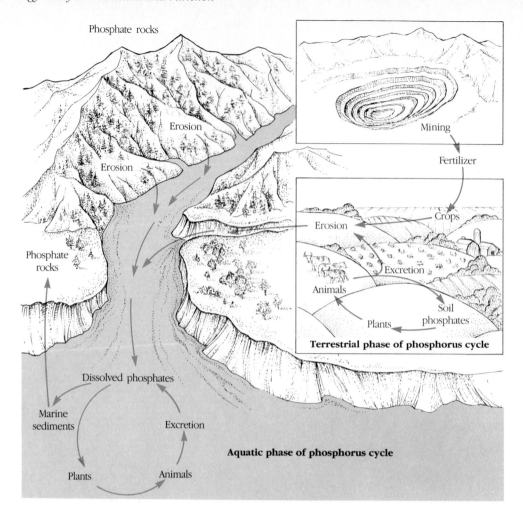

This chapter has discussed the structure of the living world, starting with the largest of all biological subdivisions, the biomes, and focusing on smaller and smaller units, including ecosystems, communities, and populations. We've seen how ecosystems are organized and have briefly examined the way ecosystems work, gaining a better understanding of how human populations alter them. With this background let us turn our attention to ecosystem balance, paying particular attention to the forces that help achieve balance. From the discussion of balance, you will see how ecosystems are altered by natural events and human intervention.

Nothing can survive on the planet unless it is a cooperative part of larger global life.

BARRY COMMONER

Summary

Ecology is the study of the interactions between organisms and their environment. It takes the entire living world as its domain in an attempt to understand the complex web of life in the **biosphere**. The biosphere, like a sealed terrarium, is a **closed system**. All of the materials essential to life come from within it and must be recycled over and over for life to continue.

The biosphere is divided into regions, called **biomes**, each with unique plant and animal life. The boundaries of a biome are determined by the **climate**, the average of the weather conditions over long periods of time. Precipitation and temperature are the most important climatic factors.

Ecological systems, or **ecosystems**, are networks consisting of organisms, their environment, and all the interactions that exist between them. Ecosystems consist of **biotic** (living) and

abiotic (nonliving) components. The abiotic components are physical and chemical factors needed for life.

Each organism operates within a range of chemical and physical factors called its **range of tolerance**. When the upper or lower limits of the range are exceeded, survival is threatened. In all ecosystems one abiotic factor usually limits growth and is therefore called a **limiting factor**.

The organisms are the **biotic** components of the ecosystem. Groups of organisms of the same species occupying a specific region form a **population**. The populations of plants, animals, and microorganisms in an ecosystem form a **community**. Each organism within a community occupies a specific region, its **habitat**. Its **niche** includes this habitat and all its relationships with abiotic and biotic components of the environment. No two species occupy the same niche in the same ecosystem.

Organisms may be specialists or generalists. **Generalists** occupy many different habitats and eat a wide variety of foods. **Specialists** generally live in one habitat and consume one or only a few organisms, making them more susceptible to changes in the habitat and more prone to extinction.

The living web is held together by **food chains**, a series of organisms each feeding on the preceding one. The **grazer food chain** is the nutrient and energy pathway starting with plants that are consumed by grazers. The **decomposer food chain** is a nutrient and energy pathway that starts with **detritus**. Individual food chains are part of more complex **food webs**.

The terms *primary*, *secondary*, *tertiary*, and *quaternary* are applied to consumers in the food chain. **Primary consumers** are the first level of consumers; they feed on plants in the grazer food chain and on detritus in the decomposer food chain. **Secondary consumers** feed on primary consumers (herbivores).

The feeding level an organism occupies in a food chain is called the **trophic level**. Producers form the first trophic level. Primary consumers form the second trophic level, and secondary, tertiary, and quaternary consumers occupy the third, fourth, and fifth levels, respectively.

Energy is defined as the capacity to do work. Organic matter in living organisms contains **potential energy**. Energy of motion is called **kinetic energy**. Energy is governed by the **laws of thermodynamics**. The **first law** states that energy is neither created nor destroyed but only converted from one form to another. The **second law** states that when energy is converted from one form to another, there is a loss of available work; thus, energy has been degraded or converted from a more concentrated to a less concentrated form. The laws of thermodynamics have many implications in ecology and everyday life.

Organic matter produced by living things is called **biomass**. It is a form of potential energy passed from one trophic level to the next in an ecosystem. As biomass "flows" through a food chain from one trophic level to the next, it decreases. Graphically represented, biomass at the different trophic levels forms a **pyramid of biomass**. Since biomass contains energy, the amount of potential energy at higher trophic levels also decreases and forms a **pyramid of energy**. Finally, the number of organisms decreases from one trophic level to the next higher one; this forms a **pyramid of numbers**.

Productivity is the measure of biomass production. Regions with high productivity, such as tropical rain forests, are important to humans because they supply many valuable resources.

Organic and inorganic matter move within ecosystems in **nutrient cycles**. Examples are the carbon, nitrogen, and phosphorus cycles. These cycles are essential for life and can be altered by human activities.

Discussion Questions

1. What is ecology?

2. Define the term *biosphere*. Why is the biosphere considered a closed system?

3. Define the term *biome*. What determines the type of vegetation in a biome? After studying the biome map in this chapter, name the biomes that you have visited in your lifetime.

4. Define the term *ecosystem*.

5. Discuss the concept *range of tolerance*. Draw a graph showing the optimum range, zones of physiological stress, and zones of intolerance. Explain each one.

6. What is a limiting factor? What is the most important limiting factor in terrestrial ecosystems?

7. How are a *niche* and a *habitat* different?

8. In what ways are humans specialists? In what ways are we generalists?

9. What is a food chain? What are the two major types? How are they different? How are they similar?

10. Sketch several simple food chains and indicate all producers and consumers. Also indicate the trophic level of each organism. Can one organism occupy several different trophic levels? Give an example.

11. Why does biomass decrease as we ascend the food chain?

12. What are the implications of decreasing biomass in the food chain?

13. Draw a detailed picture of the carbon cycle, and describe what happens during the various parts of the cycle.

14. Why are nitrogen-fixing bacteria vital to life on earth?

Suggested Readings

Attenborough, D. (1979). *Life on Earth*. Boston: Little, Brown. Superb study of life.

Bergen, M., Harper, J. L., and Townsend, C. R. (1986). *Ecology: Individuals, Populations, and Communities*. Sunderland, Massachusetts: Sinauer. For students wanting more detailed information on ecology.

Clapham, W. B. (1983). *Natural Ecosystems* (2nd ed.). New York: Macmillan. Advanced treatise on ecology.

Colinvaux, P. (1986). *Ecology*. New York: Wiley. An excellent textbook of basic ecology.

Committee on the Application of Ecological Theory to Environmental Problems (1986). *Ecological Knowledge and Environmental Problem-Solving. Concepts and Case Studies*. Washington, D.C.: National Academy Press. Important reference for students who are interested in a career in applied ecology.

Ehrlich, P. R. and Roughgarden, J. (1987). *The Science of Ecology*. New York: Macmillan. Higher-level coverage of ecology.

Smith, R. L. (1985). *Elements of Ecology* (2nd ed.). New York: Harper and Row. Advanced readings on ecology and environmental problems.

The Biomes

Look outside the window. You probably see tree-lined streets, homes or campus buildings, neatly tended lawns, or possibly a parking lot. The vegetation growing around you is probably not "natural" at all. Grasses may have been imported from Kentucky. The trees may come from Norway or the Soviet Union. Even some of the birds, such as starlings and house sparrows, are aliens, brought in from England. To get a glimpse of the natural vegetation, you may have to journey outside your town. There you may find grassland, forest, or desert that resembles the environment before humanity began to reshape it.

Grasslands, deserts, and forests are three of the relatively homogeneous zones called biomes. Biomes, as we have seen, are large regions, each with its own distinctive climate, plant and animal life, and soil type. A dozen biomes spread over millions of square kilometers, spanning entire continents. This supplement looks at some of the major world biomes and the way humans have altered them.

Tundra

The **tundra** is a vast, virtually treeless plain on the far northern borders of North America, Europe, and Asia (Gallery 2, Figure 11). It lies between a region of perpetual ice and snow to the north and a band of coniferous forests (the taiga) to the south. It is one of the largest biomes, covering about one-tenth of the earth's surface.

The tundra receives very little precipitation (less than 25 centimeters, or 10 inches, per year), most of it during the summer. Contrary to popular belief, very little snow falls during the long, cold winter.

The gently rolling tundra is dominated by herbaceous plants (grasses, sedges, rushes, and heather), mosses, lichens, and dwarf willows. Deep-rooted plants such as trees cannot grow there, because much of the subsoil remains frozen year-round, hence the name **permafrost**. Only about 10 centimeters (6 inches) of soil thaw in the summer months, creating a very shallow root zone for plants. Most tundra plants are stunted because of the short growing season, the annual freeze–thaw cycle, which tears and crushes roots and impairs growth, and windblown ice and snow, which abrade vegetation.

The summer days are long and warm, and the land becomes dotted with thousands of shallow lakes, ponds, and bogs because of the permafrost, which hinders drainage. Millions of birds come north to nest and feed on the tundra's abundant insects.

In North America musk ox, caribou, and a variety of small rodents and mammals inhabit the tundra year-round. They share their habitat with snowy owls and ptarmigan (ground-dwelling birds similar to grouse).

The tundra is an extremely fragile biome, containing few species, a condition believed to make it more vulnerable to change (Chapter 4). Vegetation takes decades to grow back after it has been destroyed by vehicles. Large-tired "rolligons" and new vehicles that ride on a layer of air are now being used experimentally and may help us reduce our impact on the tundra (Figure S3-1).

Figure S3-1. "Rolligons" are sometimes used in Alaska to carry heavy equipment across the tundra with minimal damage. Specially designed tires spread the weight over a large surface. Where normal trucks are used, the tundra can become badly damaged.

Taiga

South of the tundra, extending across North America, Europe, and Asia, is a broad band of coniferous (evergreen) forests. This biome is called the **taiga** (Gallery 2, Figures 6 and 7). The average annual precipitation in the taiga is higher than in the tundra, as is the average daily temperature.

With a growing season of 150 days and a complete thawing of the subsoil, the taiga supports an abundance of life forms. Conifers are the dominant form of plant life, for they are well adapted to the long, cold, and snowy winters. Water loss is minimized by waxy coatings on their needles. Reduced evaporation is critical in the winter, when water transport from the root systems halts. The flexible limbs are an adaptation that allows conifers to bend without breaking when covered with snow. Some deciduous trees (ones that shed their leaves each year) inhabit the taiga in localized regions, particularly in areas that are recovering after fires or heavy timber cutting.

Animal life is abundant and diverse. Bears, moose, wolves, lynx, wolverines, martens, porcupines, and numerous small rodents inhabit the forests, along with a variety of birds such as grouse, ravens, hairy woodpeckers, and great horned owls. Insects, such as mosquitoes, also thrive during the warm summer months. Over 50 species of insects that feed on conifers live in the taiga, including the spruce budworm, pine beetle, tussock moth, and pine sawfly. These can cause widespread damage to trees during droughts, which weaken trees' resistance.

Numerous lakes, ponds, and bogs spot the landscape. In North America trappers once traveled by boat into the interior of the taiga on the interconnected waterways in quest of beaver and other fur bearers. The taiga has long been of interest to the logging industry. Through past practices loggers have stripped large sections of land by **clear-cutting** forests. Until recently little or no effort was made to replant denuded areas, resulting in severe soil erosion, destruction of wildlife habitat, and pollution of streams and lakes with sediment.

Temperate Deciduous Forest

The **temperate deciduous forest** biome is located in the eastern United States, Europe, and northeastern China. A region with a warm, mild growing season and abundant rainfall, the temperate deciduous forest supports a wide variety of plants and animals (Gallery 2, Figure 8). The dominant vegetation consists of broadleafed deciduous trees such as maple, oak, and hickory. However, the dominant plant species can vary from region to region, depending on the amount of precipitation.

Temperate deciduous forests support a variety of organisms, including white-tailed deer, opossums, raccoons, squirrels, chipmunks, foxes, rabbits, black bears, mice, shrews, wrens, downy woodpeckers, and owls. During the summer, warblers and other bird species migrate into and nest in the rich green forests.

Deciduous trees act as "nutrient pumps," drawing inorganic chemical nutrients from the subsoil up through the roots to the leaves. When the leaves fall and decay, they add valuable organic and inorganic nutrients to the superficial layers of soil. These nutrients are important for plant growth.

The temperate deciduous forest biome has been extensively exploited by humans. In North America this biome once extended from the Mississippi River to the Atlantic Coast. Now, however, only about .1% of the original forest remains; most has been cleared away for farms, orchards, and cities. Early settlers often failed to practice good soil conservation methods on their farms, letting the land erode. Farmers then moved westward into virgin territory, where they often continued the same practices. Soil erosion continues today on America's farmland (Chapter 7).

Grassland

Grasslands exist in both temperate and tropical regions where rainfall is relatively low or uneven. The temperate grasslands in North America occupy a region known as the Great Plains, which extends from the Rocky Mountains to the Mississippi River. Moist regions of the Great Basin, lying between the Sierra Nevada and the Rockies, also support grasslands. Temperate grasslands are found in South America, Australia, Europe, and Asia; tropical grasslands exist in Africa and South America.

Temperate and tropical grasslands are remarkably similar in appearance (Gallery 1, Figures 12 and 13). Both experience periodic drought and are characterized by flat or slightly rolling terrain. Large grazers such as the bison in North America and the zebra in Africa feed off the lush grasses in this biome. The native grasses are well adapted to drought, as their roots penetrate deep into the subsoil where they can always find water.

Because their soils are rich in inorganic and organic nutrients, grasslands have been widely used for agriculture. As discussed in Chapter 7, wind and rain take their toll on grassland soils that are farmed improperly.

Desert

Deserts are found throughout the world. The Sahara, the largest desert in the world, stretches across the African continent and is nearly as large as the United States. In North America deserts lie primarily on the downwind side of mountain ranges. The reason is that warm, moist air flowing toward the mountains is propelled upward as it comes in contact with the mountain range; as the moist air ascends, it is cooled, and water vapor in the air condenses, forming droplets too large to remain in suspension. These droplets form rain, which falls mostly on the windward side, leaving the downwind side dry. When rain does fall in a desert, it is often intense and frequently results in flash floods and severe soil erosion. Water evaporates quickly from the soil because the vegetation is sparse and because of the high temperatures reached on the desert floor in the summer (Gallery 2, Figure 2).

Contrary to the view held by many people, most deserts are not lifeless lands. A surprising variety of plants and animals have adapted to the desert's unique environmental conditions. Plants include cacti, other succulents, and shrubs (mesquite, acacia, greasewood, and creosote bush). Small, fast-growing annual herbs are also found. Many of these are wildflowers that bloom only in the spring or after drenching rains. Such sudden bursts of growth can turn the desert into a colorful landscape almost overnight (Gallery 2, Figure 3).

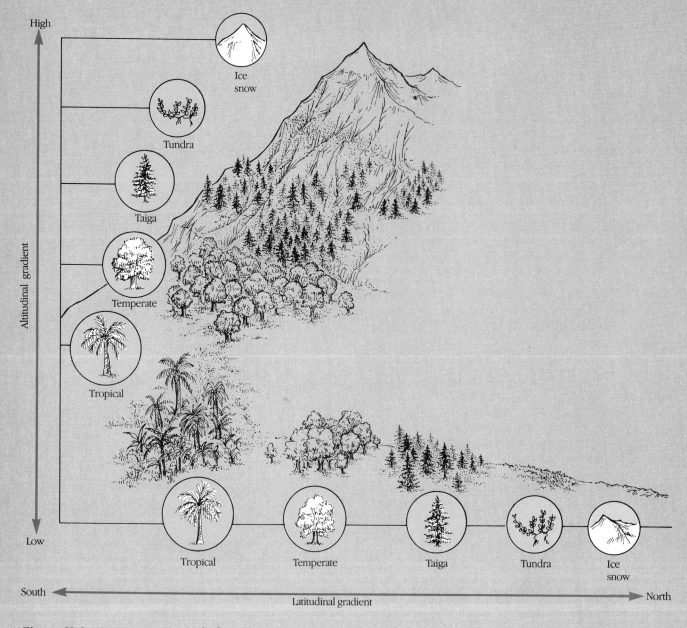

Figure S3-2. Vegetation varies with altitude in any given region. Altitudinal gradients mirror latitudinal gradients.

Plants have developed a number of adaptations to cope with the dry conditions of the desert. These include (1) thick, water-proof outer layers, which reduce water loss; (2) an absence of leaves or a reduction in leaf size, both ways to reduce the surface area from which water can evaporate; (3) hairs and thorns, which reflect sunlight and shade the plants; (4) the ability to drop leaves (a strategy used by the ocotillo) when moisture levels are low; (5) extensive, shallow root systems to absorb as much water as possible during cloudbursts; (6) deep taproots, which penetrate into groundwater; (7) succulent water-retaining tissues to store water; (8) recessed pores (stomata), which reduce water loss; (9) wide spacing between plants to reduce competition for available water; and (10) short life spans, which allow plants to develop to maturity quickly following rainstorms.

A surprising number of animals also live in the desert. Like the plants, they have evolved a number of strategies to survive in the harsh climate. Many animals, such as the ring-tailed cat, are active only at night or during the early and late hours of the day when the heat is less intense. Snakes have thick scales that prevent water loss. The kangaroo rat conserves body moisture by excreting a solid urine containing nitrogen wastes (in the form of uric acid). Some species obtain water from vegetation, and others from the blood and tissue fluids of their prey. Gila monsters combat the heat by remaining underground during the hottest daylight hours.

Large cities have sprung up in the American desert. Farmers have plowed the poor soil to plant their crops, but the desert soil contains little of the organic matter needed to retain moisture and contains only small amounts of the nitrogen required for plant growth. Consequently, crops are successful only if nitrogen and water are provided. City dwellers and farmers now compete for the limited water in the American West. With rising population and years of irrigation, water supplies have begun to decline, forcing many farms to shut down.

Human populations have expanded the borders of the world's deserts by allowing livestock to overgraze, deforesting lands, and following poor agricultural practices. The Sahara's southern boundaries are spreading into the Sahel region because of intense grazing in bordering grassland and a shift of rainfall northward, possibly due to global warming. Each year millions of acres of land are engulfed by spreading deserts.

Tropical Rain Forest

Tropical rain forests are located near the equator on several continents and islands. With an average annual temperature of approximately 18° C (64° F) and rainfall of 200 to 400 centimeters (80 to 160 inches) per year, the tropical rain forest is one of the richest and most complex biomes.

The dominant vegetation consists of trees that tower as high as 50 to 60 meters (165 to 200 feet) above the forest floor (Gallery 2, Figure 4). The tops of these trees interlace and form a dense canopy which blocks out much of the sunlight. Smaller trees further reduce light penetration. Only 1% of the sunlight reaches the forest floor, and therefore only a few plants adapted to low light can grow on the ground.

The tropical rain forest is well known for its species diversity. As many as 100 tree species can be found, whereas a northern coniferous forest would have only 2 or 3. The rain forest supports a diverse array of insects, birds, and other animals, too. There are as many species of butterflies (500 to 600) on a single tropical island as there are in the entire United States.

Since the ground is bare, many of the animals and insects live in the treetops, where the food is. The treetops are without doubt the most heavily and diversely populated regions of the tropical rain forest.

Woody vines grow in the rain forest, and a large number of plants, called **epiphytes**, live dangling among the branches of the taller trees where sunlight is available. Epiphytes, like Spanish moss, which grows in the southern United States in temperate deciduous forests, gather moisture and nutrients from the humid air. They have no roots and need no soil. Researchers recently discovered twelve species of trees in the Amazon rain forest whose roots originate in the soil but turn upward to climb up the trunks of other trees. There it is believed they scavenge nutrients from rain since the soil is so poor.

Trees of the tropical rain forest have shallow root systems because most of the nutrients are near the surface and because rainfall is abundant, which makes deep root systems unnecessary. Since their root systems are so shallow, many trees develop wide bases (known as buttresses) to prevent them from toppling over.

The soils of the tropical rain forests are thin and extremely poor in nutrients. In fact, almost all of the inorganic and organic nutrients found in this biome are tied up in the vegetation. Plant and animal wastes are rapidly decomposed on the forest floor in the decomposer food chain; organic matter does not build up, as it does in deciduous forests and grasslands. Minerals returned during decomposition are rapidly absorbed by the roots of trees or are leached from the soil by rain and carried to the groundwater, where they are often unavailable to plants.

Because they are nutrient-poor, rain forest soils are not suitable for conventional agriculture. In addition, many of the soils in tropical rain forests are composed of a red clay known as **laterite** (from the Latin *later*, "brick"), which forms an impenetrable crust one or two years after the forest is cleared to make farmland. A third problem is that once trees are removed to make room for farms, soils are easily washed away by the frequent rains. Despite these major drawbacks, people throughout the world still clear tropical rain forests for farming, grazing, and timber production. An area the size of Georgia is deforested each year.

Destruction of tropical rain forests is so extensive that ecologists project that this biome, which constitutes about half the world's forest, will be gone or at least severely damaged by the year 2000. With it, they assert, thousands of animal species will vanish. (For more on tropical deforestation, see Chapters 8 and 9.)

Altitudinal Biomes

Hiking or driving through the Rocky Mountains, the Sierra Nevada, or the Cascade Mountains reveals an interesting shift in plant communities similar to that seen when travelling north from the grasslands through Canada's coniferous forests to the tundra (Figure S3-2). Why does this shift occur? Climbing up a mountain creates the same effect as moving northward; in fact, each 120-meter (400-foot) gain in altitude is like a trip one degree north in latitude. Rainfall and temperature vary with altitude, and thus so does the plant and animal life. In the high Rockies climatic conditions are similar to those of the arctic tundra; so are the plant and animal life. Because of these similarities high mountain regions are called **alpine tundra**. As in the arctic tundra, the growing seasons are short, and the winters are cold. Annual precipitation is fairly low. Vegetation and animal life are similar to those of the far northern tundra (Gallery 2, Figures 10 and 11). Areas below the tundra resemble the taiga. The forests in these regions are dominated by spruce and fir.

4

Principles of Ecology: Ecosystem Balance and Imbalance

And this our life, exempt from public haunt, finds tongues in trees, books in running brooks, sermons in stones, and good in everything.

SHAKESPEARE

In 1884 the water hyacinth was introduced into Florida from South America. The hyacinth remained, for a while, simply a beautiful ornamental flowering plant in a private pond. Unfortunately, the plant accidentally entered the waterways of Florida. In these nutrient-rich waters it spread like cancer throughout the canals and rivers that lace the state. Aided by a remarkable ability to reproduce—10 plants can multiply to 600,000 in just eight months—the hyacinth now clogs waterways throughout Florida, choking out native species and making navigation impossible in some areas (Figure 4-1).

From Florida the plant has spread throughout much of the southern United States. Today nearly 800,000 hectares (2 million acres) of rivers and lakes from Florida to California are choked with thick surface mats of hyacinths. Florida, Louisiana, and Texas, where the problem is most severe, spend nearly $11 million a year to relieve the stranglehold these plants have on their waterways. (For a practical application of the water hyacinth, see Case Study 4-1.)

For most of its existence humankind has worried about the ways in which nature affects people's lives. As this example points out, however, the danger may not be what nature will do to us but what we will do to nature (and ourselves) by unleashing some of its forces.

The story of the water hyacinth is a lesson in ecological balance or, more correctly, imbalance, a lesson in how an ecosystem can be thrown out of kilter by humankind's zeal to fashion the environment to its liking. Chapter 3

Figure 4-1. The water hyacinth was introduced accidentally into Florida waters. It has proliferated at a tremendous rate, choking canals and streams and outcompeting many native species.

looked at the structure of ecosystems and the way energy and matter flow within them. A complete picture of ecosystems is not possible, though, until we look at ecosystem balance and imbalance to answer some important questions: How can ecosystems be altered, and in what ways can they recover from damage? How far can we push a natural community before permanently changing it?

Ecosystem Stability Defined

Before the introduction of the water hyacinth Florida's waterways were considered fairly stable ecosystems. In other words, things may have changed from day to day, but on the whole they remained more or less the same. Ecologists describe this condition as a state of **dynamic equilibrium**, or a steady state. In other words, it is a state of **stability**. To understand this term suppose that you studied a mature forest ecosystem near your home each spring for an extended period, say, 20 years. If the system were stable, you would find that (1) the total number of species was fairly constant from year to year, (2) the same species were present each year, and (3) the population size of each species was approximately the same from year to year. This system is stable.

Stability doesn't mean that all of the parts of an ecosystem operate in perfect harmony. Not at all. Ecosystem stability or balance, in fact, is often achieved through competition and apparent conflict: animals competing for a limited food supply, disease organisms killing off the weak, and predators feeding on prey. The net result, or the more or less constant condition, is what ecologists refer to as stability.

Ecosystems, then, are complex self-regulating systems. As in all such systems, a change in one variable results in a corresponding change in another, bringing the system back into alignment. The chief form of regulation is negative feedback, discussed in Chapter 1. The ability to

bounce back from small changes is sometimes called **resilience**.

Perhaps the key to understanding stability is to realize that it does not mean that conditions are absolutely constant. Just as your blood sugar levels vary slightly from hour to hour, so do conditions within an ecosystem. The number of plants growing in a field in Florida or Alabama, for instance, may increase or decrease in concert with natural variations in climate—with wet spells or drought. A stable ecosystem, however, can weather such variation. That's because species operate within a range of conditions—the range of tolerance described in Chapter 3. When conditions are shifted out of the range of tolerance, an ecosystem may begin to deteriorate, losing individual species that are vital links in food webs.

What Keeps Ecosystems Stable?

Life is a continuous balancing act. Good things and bad things, income and debt, happiness and sadness, friends and enemies all make your life what it is. Too much of the good life, and life may become meaningless. Too much sadness, and despair creeps in. The secret of life is achieving a balance between the good and the bad.

Nature, too, is a balancing act of growth and decline, predator and prey, sickness and health; the secret of living systems is achieving a balance and maintaining it. Ecosystem health, like your own health, is dependent on this precarious balance.

Population Growth and Environmental Resistance

Balance, or stability, in an ecosystem is the result of opposing forces that constantly work to regulate the size of populations. These forces can be broken down into two

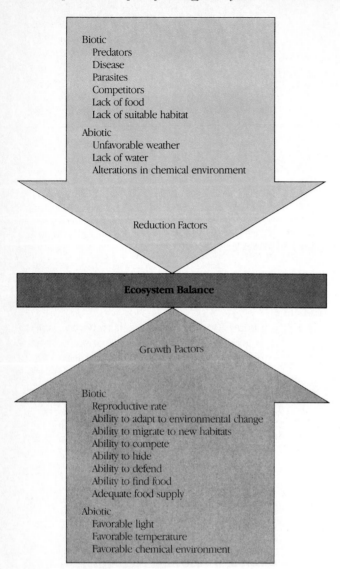

Figure 4-2. Ecosystem balance is affected by forces that tend to increase population size and forces that tend to decrease it. Growth and reduction factors consist of abiotic and biotic components.

groups: factors that tend to increase population size, or **growth factors**, and those that tend to decrease it, called **reduction factors**, terms coined by the author to describe positive and negative feedback mechanisms that help regulate populations. As illustrated in Figure 4-2, growth and reduction factors can be biotic or abiotic.

At any given moment population size is determined by the interplay of these factors. Since ecosystems contain many species, the entire ecosystem balance can be crudely related to the sum of the individual population balances.

The biotic factors that stimulate population growth include the ability to produce many offspring, to adapt to new or changing environments, to migrate into new territories, to compete with other species, to blend into the environment, to defend against enemies, and to find food (Figure 4-2). Certain favorable abiotic conditions also tend to increase population size. Favorable light, temperature, and rainfall, for example, all promote maximum plant growth and, because animals depend on plants, often promote increases in animal populations. The success of the water hyacinth in southern waterways is attributed, in large part, to its prolific breeding, its ability to use nutrients dissolved in surface waters, its ability to spread into new territories, and, of course, the favorable climate of the South.

Opposing the positive influence on growth are a host of abiotic and biotic factors. Ecologists describe these factors collectively as **environmental resistance**. Predators, disease, parasites, and competition by other species all effectively reduce population size, as do unfavorable weather and lack of food and water. The water hyacinth faces little environmental resistance. Even plant-eating native fish are no match for its reproductive success. Florida's waterways are out of balance; growth far exceeds environmental resistance.

Before the invention of the plow unbalanced ecosystems were a rare thing. Expansion of agriculture, urban growth, and industrial development, discussed in Chapter 2, changed the picture entirely, making balanced ecosystems increasingly difficult to find in many parts of the world.

On the rolling hills of eastern Kansas, however, is a reminder of the days past: an ecosystem operating as it has for tens of thousands of years. We will focus on a tiny portion of that ecosystem to illustrate how nature ensures its harmony.

Living in the grass-covered Flint Hills is a mouselike rodent called the prairie vole. Its population size depends on many factors such as light, rainfall, available food supply, predation by coyotes and hawks, temperature, and disease.

The prairie vole has a high reproductive capacity. When raised in the laboratory under *optimum* conditions, a pair of voles becomes remarkably prolific, producing a litter of seven pups every three weeks, month after month, for several years. The optimum conditions in a laboratory are low temperature (slightly above freezing), long days (14 hours of light a day), and plenty of water and food. Under these conditions a female will give birth to a litter and mate a few hours later. Her second litter is born about the time she weans her first. This can go on, in the laboratory, for several years. For the captive-raised vole, motherhood is no picnic.

Fortunately, in the wild, optimum conditions never completely coincide. In the summer, for instance, when food, water, and day length are optimal, the temperature is too warm. Reproduction occurs at a much more reasonable pace. In the winter, when the temperature is just

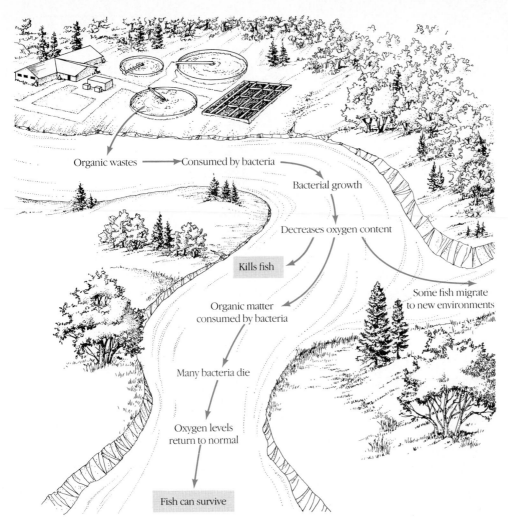

Figure 4-3. The events that follow the dumping of organic wastes into a stream.

Organic wastes → Consumed by bacteria

Bacterial growth

Decreases oxygen content

Kills fish

Some fish migrate to new environments

Organic matter consumed by bacteria

Many bacteria die

Oxygen levels return to normal

Fish can survive

right for breeding, the days are short, and the food supply is reduced. As a result, the population balance is maintained.

Resisting Change

The key word in ecosystems is stability or balance. As all of us can attest, to preserve balance it is easiest to resist change in the first place. In an ecosystem this is called **inertia**. Small changes in water chemistry, for example, may have no effect on aquatic organisms. As a result, the aquatic system resists change. If change does occur, ecosystems may "bounce back" or recover rapidly; this is called **resilience**. This section discusses the phenomenon of resilience.

In the living world change comes from shifts in growth and reduction factors. The introduction of new predators, a shortage of food, low rainfall, or unfavorable temperature, for example, all tend to decrease population size. Other factors, such as an abundance of food, may cause explosive growth in populations.

Changes in abiotic and biotic conditions occur with great regularity in ecosystems. Minor fluctuations are of little consequence, however, for species have evolved numerous mechanisms either to resist change or to recover quickly. Prairie voles may increase their reproductive rate when a cold winter kills off a larger than normal number of their population; tropical monkeys may exploit new food sources when a rainy season fails to materialize and traditional food supplies are inadequate; wolves may migrate into new territory if crowding occurs.

In many ways nature is a series of checks and balances that preserve the integrity of the whole. These checks and balances help minimize human impact, too. Sewage dumped into a stream, for example, adds organic and inorganic chemicals to the water (Figure 4-3). The organic molecules are consumed by naturally occurring bacteria whose population is normally low. The number of bacteria increases. Since bacteria use up oxygen when they consume organic materials, the level of dissolved oxygen in the stream usually drops. This decline kills fish and

Figure 4-4. The number of mammals (shown here) and most other species varies considerably with latitude. The highest diversity is found in the tropics; the lowest is found in the tundra. Species diversity may help create ecosystem stability, but some ecologists think that diversity is a product of climatic stability. (After Simpson, 1964.)

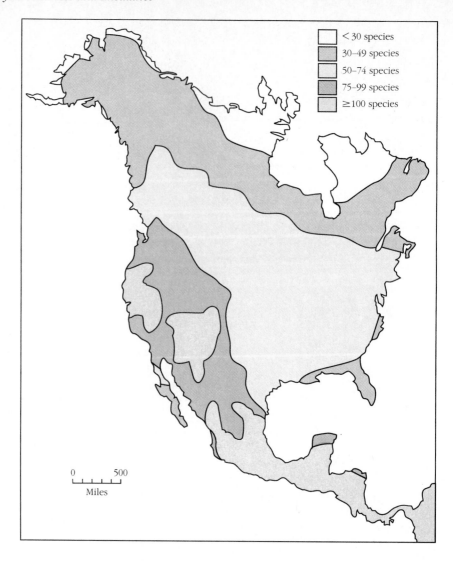

other organisms or forces them to migrate to new areas. In time, though, the stream will return to normal if further spills are prevented. The bacteria that proliferated after the spill will perish as the level of organic pollutants falls off. The levels of dissolved oxygen will also return to normal. Fish will return. This is an example of resilience. Chapter 16 treats this topic in detail.

Species Diversity and Stability

Ecosystem stability is significantly affected by **species diversity**, which, roughly speaking, is a measure of the number of species in a community. Figure 4-4 shows how species diversity among mammals varies with latitude in North and Central America. In the frozen northern regions of Canada and Alaska, for example, species diversity is low. Heading south, diversity increases until one reaches the tropics of Central America, where the highest diversity can be found. The relationship between species

diversity and latitude is found in virtually all groups. Latitude, therefore, is an important factor affecting species diversity, but it is not the only one.

Some ecologists believe that ecosystem stability is largely the result of species diversity. The higher the diversity, they say, the greater the stability. In support of this idea are observations that extremely complex ecosystems, such as rain forests, remain unchanged almost indefinitely if undisturbed. Simple ecosystems such as the tundra are more volatile. They can experience sudden, drastic shifts in population size. Other simplified ecosystems such as fields of wheat and corn also show extreme vulnerability to change, and they collapse if abiotic or biotic factors shift.

To see how ecologists explain this phenomenon, look at the differences in food webs in simple and complex ecosystems. As illustrated in Figure 4-5, the number of species in a food web in a mature ecosystem is large. So is the number of interactions among these organisms. In

Case Study 4-1

Putting a Pest to Work: An Ecological Solution to Water Pollution

It's not often a pest makes good, but such is the case of the water hyacinth. Long infamous for clogging canals, rivers, and ponds and costing state governments millions for cleanup, the water hyacinth has been put to work as a water purifier in sewage treatment plants.

Unlike many other plants, water hyacinths need no soil. Their roots simply dangle below the surface and absorb the nutrients the plant needs to grow. The nutrients these plants need are, coincidentally, also major water pollutants—nitrates, phosphates, and potassium. Thus, the water hyacinth acts as a biological filter, removing these harmful pollutants from the water. Grown in special ponds in which wastes from sewage plants are dumped, the hyacinths provide an inexpensive alternative to conventional wastewater treatment (see Chapter 16 for more detail on this subject). As an added benefit, the water hyacinth absorbs certain toxic wastes, pesticides, and heavy metals.

At Disney World near Orlando, Florida, officials have constructed five ponds, each roughly the size of a football field. Wastewater from the park is pumped into these ponds, purified by the water hyacinths, and then returned for reuse. San Diego has a similar treatment facility.

The advantages of this approach are several. Conventional wastewater treatment is costly, requiring large amounts of electricity and numerous maintenance workers. Hyacinth treatment, in contrast, may cost about half as much because of the low energy requirements and the lack of personnel. It also allows for recycling of water, a boon to water-short regions. Also, periodic harvesting of the hyacinths yields food for livestock or fertilizer for fields. The water hyacinth provides an important ecological solution to help us build a sustainable society.

a complex ecosystem the elimination of one species would probably have little effect on the ecosystem balance. In sharp contrast, the number of species in the food web of a simple ecosystem is small. The elimination of one species could have repercussions on all other species.

Some ecologists believe that the link between stability and species diversity has not been adequately tested. They argue that the reason tropical rain forests are stable is that the climate is relatively uniform throughout the year. On the tundra, a relatively simple and somewhat unstable ecosystem, the climate shifts dramatically from season to season. Thus, the stability of the tropical rain forest may result not from species diversity but from its constant climate.

What is the upshot of this debate? Although it is not known for certain whether diversity creates stability, we can say that simplifying ecosystems by reducing species diversity can have deleterious effects. Several examples are discussed in this chapter.

Correcting Imbalance in Ecosystems: Succession

Small shifts in the growth and reduction factors in an ecosystem, whether brought about by natural causes or

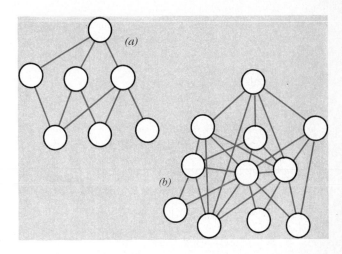

Figure 4-5. (a) Food web in a simple ecosystem. Circles represent organisms. Note the lack of links in the simplified web. (b) Food web in a complex ecosystem. Many ecologists believe that complex ecosystems are more stable because of the increased number of links, but there is no general agreement on this reason for stability in ecosystems.

by humans, are fairly common. Ecosystems respond to these changes in a way that ensures the survival of the community. As this section points out, however, drastic shifts can seriously, sometimes irreversibly, upset ecosys-

▶◀ *Point/Counterpoint*

Environmentalism: On Trial
Ecology Is a Subversive Science

William Tucker

The author, a journalist and critic of the environmental movement, has written numerous articles on environmental issues. His latest book, Progress and Privilege, *is a thought-provoking and controversial discussion of environmentalism.*

One of the key realizations of ecology is that the earth is a kind of living system governed by many self-regulating (homeostatic) mechanisms. The earth is in a state of equilibrium. If pushed too far in any one direction, the self-regulating mechanisms can become overloaded and break down, resulting in radical changes.

In its scientific aspects ecology seems to offer an extraordinary broadening of our understanding of life on the planet. Yet with its transfer into the public domain, it has become little more than a sophisticated way of saying "we don't want any more progress." Somehow this exciting discipline has been translated into a very conservative social doctrine. People have often waved the flag of "ecology" as a new way of saying that nature must be preserved and human activity minimized. Ecology is sometimes viewed as "subversive" to technological progress. It supposedly tells us that our ignorance of natural systems is too great for us to proceed any further with human enterprise. Just as nationalistic conservatives always try to throw a veil of reverence around such concepts as "patriotism" and "national tradition," so environmentalists try to maintain the same indefinable quality around ecosystems.

The lesson environmentalists drive home is that since we do not understand ecosystems in their entirety, and never will, we dare not touch them. Our knowledge is too limited, and nothing should be done until we understand more fully the implications of our actions.

To say that ecology is the science that does not yet grasp the complex interrelationships of organisms is like trying to define medicine as the science that does not yet know how to cure cancer. Environmentalists emphasize the negative parts of the discipline because it fits their concept that we have already had enough technological progress.

The lessons of ecology tell us many things. They tell us that organisms cannot go on reproducing uncontrollably. But they also tell us that many organisms have developed behavioral systems that keep their populations from exploding. The laws of ecology tell us that we cannot throw things away into the environment without having them come back to haunt us. But they also tell us that nature evolved intricate ways of recycling wastes long before human beings appeared, and ecosystems are not as fragile as they seem.

In fact, the whole notion of "fragile ecosystem" is somewhat contradictory. If these systems are so fragile, how could they

have survived this long? If ecology teaches us anything, it enhances our appreciation of how resilient nature is, and how tenaciously creatures cling to life in the most severe circumstances. This, of course, should not serve as an invitation for us to see how efficiently we can wipe them out. But it does suggest that the rumors of our powers for destruction may be exaggerated.

The environmentalist interpretation of ecology has been that ecosystems have somehow perfectly evolved, and that human intervention always leads to degradation. It should be clear that even if a particular ecosystem did represent biological perfection, that is not reason in and of itself to preserve it at the expense of human utility. Our ethical position cannot be one of completely detached aesthetic appreciation. We must first be human beings in making our ethical judgments. We cannot be completely on the side of nature.

We are not a group of imbeciles aimlessly poking into the backs of watches or tossing rocks into the gears of Creation. There is purpose to what we do, and it is essentially the same as nature's. We are trying to rearrange the elements of nature for our own survival, comfort, and welfare. We can certainly act stupidly, but we can also act out of wisdom. It is foolish to argue that everything is already perfect and must be left alone. To portray humans as meddling outsiders in an already perfected world is nonsense. In going to this extreme to reaffirm nature, we only deny that we are a part of it.

Environmental writers suggest that we practice an "ecological ethic," extending our moral concerns to other animals, plants, ecosystems, and the entire biosphere. I would accept this proposal, with one important qualification: that is, that our ethical concerns still retain a hierarchy of interest. We should extend our moral concerns to plants, trees, and animals, but not at the expense of human beings. Our first obligation is to humanity. We should avoid actions that are destructive to the biosphere, but we must recognize that at some point our interests are going to impinge upon other living things.

Critical Thinking

1. Review the critical thinking skills presented in Chapter Supplement 1-1.

2. Summarize William Tucker's major points.

3. Use your critical thinking skills to analyze and comment on his views. Does he violate any rules?

Ecology Is a Neutral Science

Daniel D. Chiras

In January 1987, crossing between the South and North islands of New Zealand, our ship overtook a Greenpeace vessel returning from Antarctica. Greenpeace is an international environmental organization renowned for its stands against nuclear weapons, dumping of hazardous wastes, whaling, and nuclear power.

When the captain of our ferry announced the presence of the Greenpeace ship, supporters flooded the deck to wave. The enthusiasm was not universal, however; a few passengers—presumably those who thought that Greenpeace goes too far—grumbled; a few others simply shrugged their shoulders and slouched back in their seats. The response of the crowd on board points out the wide range of public sentiment toward environmentalism.

The environmental movement, like the public's opinion of it, is varied. Founded largely on lessons gleaned from ecology, it attracts a following generally interested in preserving the environment. But that means many things to many people. There is, in fact, no "typical" environmentalist. People sympathetic with the plight of the natural world have not adopted uniform values. Instead, environmentalists' values lie on a broad continuum, from the "complete preservation" philosophy, which tolerates little human intervention, to the "preserve-it-in-parks" philosophy, which seeks to protect small pieces of the environment for future generations to enjoy, while developing most of the rest for human use.

Nowhere are environmentalists' beliefs more fractured than on the notion of progress. William Tucker's preceding essay in this Point/Counterpoint casts environmentalists as narrow-minded obstructionists who condemn all human progress and technological development. No doubt, many business people feel the same—and for good reason. Environmentalists have threatened many a proposed housing project, dam, highway, and mine. In this narrow vein, few could deny that environmentalism is subversive to progress.

The truth, however, is not so simple. Progress is not one thing to all people. The debate over protection versus progress hinges, quite precariously, on this one fundamental statement.

To the environmentalist, progress has a meaning much broader than economic growth, faster jets, growing markets, and new gadgets to make life a little easier. Progress means prospering within the limits of nature, living on earth without destroying the air, water, and soil on which our lives depend. But environmental protection ultimately means protecting people. Planet care is self-care.

Environmentalism looks forward to a better tomorrow, to progress, but with a large measure of realism. It sees that humanity cannot prosper without fundamental changes in the way we go about our day-to-day business.

Environmentalism seeks to restrain unprincipled avarice to prevent society from destroying the earth's generous and renewable gifts. Ecology has taught environmentalists to look to the future with their eyes wide open, rather than narrowly focused on material welfare, economic wealth, and new conveniences. Environmentalists tell us that we must be on the side of nature and that in attempting to rearrange the elements of nature to satisfy our needs, we must tread carefully.

Ecology teaches us that human activities often have profound influences on the natural world and that what we do to the earth and water and sky we do to ourselves. The dust bowl days, the great deserts spreading through Africa, and destruction of the once-rich Fertile Crescent are blatant reminders of our impact, as are the threats of acid precipitation, the greenhouse effect, and stratospheric ozone depletion. Human intervention doesn't always lead to degradation, as Tucker implies, but it is also not a force to be trivialized.

Few people claim the natural world is perfect. Most of those who argue for protection of natural diversity do so because they have strong feelings about saving other species. The crux of the disagreement between environmentalists and "developmentalists" lies in each group's view of progress and of the relative importance of humankind in the natural world. Environmentalists recognize the interdependence of all living things. But the reasons why environmentalists seek to protect other species vary. Some see environmentalism as a way to protect human life, knowing how much we depend on other species for food, fiber, and enjoyment. Others assert that non-human species have a right to exist so we must protect them.

Ecology, transferred to the public domain, is a neutral science. It provides knowledge on which we can make decisions. Our decisions, however, are strongly influenced by our values. How we define progress and the relative importance of humans in the scheme of events is the issue of greatest concern to environmentalists. Ultimately, our fate hinges on new definitions of progress that fit more closely with the scientific realities that we learn from the study of ecology.

Critical Thinking

1. Summarize the author's main points.

2. Use your critical thinking skills to analyze and comment on his views. Does he violate any rules?

3. Whose views most closely represent yours? Why?

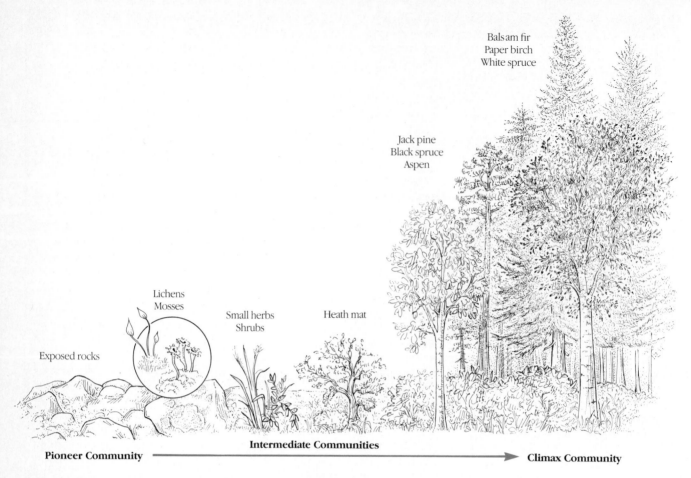

Balsam fir
Paper birch
White spruce

Jack pine
Black spruce
Aspen

Lichens
Mosses

Small herbs
Shrubs

Heath mat

Exposed rocks

Intermediate Communities

Pioneer Community → **Climax Community**

Figure 4-6. A representation of primary succession on Isle Royale, located in Lake Superior. Rock exposed by the retreat of glaciers is colonized by lichens, then mosses, followed by other plant communities leading to a climax community. One biotic community replaces another until a mature community is formed. During succession the plants of each community alter their habitat so drastically that conditions become more suitable for other species.

tem balance. Nevertheless, nature has its way of healing.

Walk along a forest path in the aftermath of a volcanic eruption or a forest fire sparked by lightning. Or travel the dusty path over abandoned surface mines or carelessly logged forests. Life sprouts out of the ashes or in the unstable soils; but for these places to return to their predisaster stages, years or decades must pass. If the soil washes away before nature can apply her healing craft, recovery may be impossible.

A biotic community destroyed by natural or human causes often recovers in a series of changes in which one community is gradually replaced by another until the mature, or climax, community is reached. This process is called **succession**, but some biologists use the term **biotic development**. Biotic communities can also form where none previously existed; they start out slowly but can eventually form mature communities. This is also called succession. We will take a look at both kinds of

succession, focusing on a classical view that scientists are finding simplifies a highly variable process.

Primary Succession

Primary succession is the sequential development of a biotic community where none had existed. For instance, when the great glaciers began to retreat 15,000 years ago, large areas of barren land and rock were exposed. The exposed rock became populated with lichens (Figure 4-6). The lichens thrived for a while but were gradually replaced by mosses. The mosses were eventually replaced by ferns, grasses, and larger plants, which eventually gave way to shrubs and trees.

Understanding the process of succession requires a closer look at each plant community in this procession from barren rock to the climax community. Lichens, the first inhabitants, spread over a rock's surface. Clinging to the rock, living off moisture from the rain and organic nutrients from photosynthesis, these organisms secrete a weak acid, called carbonic acid. Carbonic acid dissolves rock, liberating nutrients and helping to make soil. Carbonic acid in normal rainfall also helps wear down the rock. Tiny insects may join the lichens, forming a **pioneer community**, the first community to become established in a once-barren environment.

The lichens and insects gradually change their environment; the consequences of such changes are dire for them. Capturing windblown dirt particles, for example, the lichens promote further soil development. Dead lichens crumble and become part of the soil, along with the remains of insects, fungi, and bacteria. Over time, enough soil develops for mosses to take root. The mosses shade the lichens and eventually kill them. Mosses, fungi, bacteria, and insects form a new community. This community, however, also brings about changes, which will eventually usher in still another community. The rise and fall of communities proceeds until the climax community forms.

An additional example of primary succession is the establishment of plants on newly formed volcanic islands. The Hawaiian Islands are a good example. These islands arose from lava deep from within the earth. When the lava cooled, plant life began to take root. The plants came from two principal sources: Many came from seeds deposited in the feces of sea birds that happened upon the islands. Other plants are thought to have come from neighboring continents and islands. Uprooted vegetation, say biologists, may have drifted to the newly formed islands over many thousands of years, and some took root there, turning the once-barren mass of rock into a rich tropical garden.

Another example of primary succession is the transformation of wetland to forest, shown in Figure 4-7. Sediment can fill in ponds and marshes, allowing trees to take root in the soil. Eventually, a pond may become forest. Wetland ecologists are quick to point out, and rightly so, that wetlands become forests very rarely. But for years, ecologists thought that the process was much more common. Long-term studies have helped clear up the controversy. Researchers have found that, over many years, wetlands go through cycles of high and low water. Trees may advance in dry years, but be driven back in wet years. The wetland, therefore, is a dynamic balance of forest and pond. Only rarely does a forest entirely replace this ever-changing system.

Secondary Succession

Secondary succession is the sequential development of biotic communities after the complete or partial destruction of an existing community. Usually the long, slow development of soil that takes place in primary succession

is unnecessary. A climax community or intermediate community may be destroyed by natural events such as volcanic eruptions, floods, droughts, fires, and storms. The eruption of Mount Saint Helens in 1980 and the devastating fires in Yellowstone in 1988 are modern examples. Established communities are also commonly destroyed by human intervention such as agriculture, intentional flooding, fire, or mining.

Abandoned farm fields provide an excellent opportunity to observe secondary succession (Figure 4-8). Former farmland is first invaded by hardy species, such as crabgrass or broom sedge, depending on the area. These plants are well adapted to survive in bare, sun-baked soil. In the eastern United States crabgrass, insects, and mice invade abandoned fields, forming pioneer communities. But crabgrass is soon joined by tall grasses and other herbaceous plants. The newcomers' shade eventually eliminates the sun-loving crabgrass. Tall grasses and other herbaceous plants dominate the ecosystem for a few years along with a variety of animals, such as mice, woodchucks, rabbits, insects, and seed-eating birds.

In time, pine seeds settle in the area, and pine seedlings begin to spring up in the open field. Like crabgrass the pine trees flourish in the sunny fields. Over the next three decades pines shade out the grasses and herbs. Animals that feed on grasses, such as woodchucks, move on to more hospitable environments. Squirrels and chipmunks, which prefer a wooded habitat, invade the new ecosystem.

Shade from the pines gradually creates an inhospitable environment for pine seedlings and a favorable environment for the growth of shade-tolerant hardwood trees such as maple and oak. As a result, hardwoods begin to take root and eventually tower over the pines; their shade gradually kills the pines, which had invaded 60 years earlier.

Succession is a race to catch the sun. Each plant stretches out its leaves to monopolize the sunlight. Succession is a process of change that is internally driven. Plants shape the environment first by making soil. The success of a plant community, however, always leads to its destruction, until the final stage is reached and a climax community is formed. At this point, the struggle to monopolize sunlight is over.

During succession, animal populations shift with the changing plant communities. The early stages in succession are characterized by a low species diversity, but the plant and animal populations rapidly expand to exploit the untapped resources of the ecosystem (Table 4-1, page 88). Because there are fewer species in the early stages of succession, food webs tend to be simple, and populations tend to be volatile. Mature biotic communities, on the other hand, have a high species diversity and stable populations. Food webs are intricate, and stability is therefore assured if changes are not too drastic.

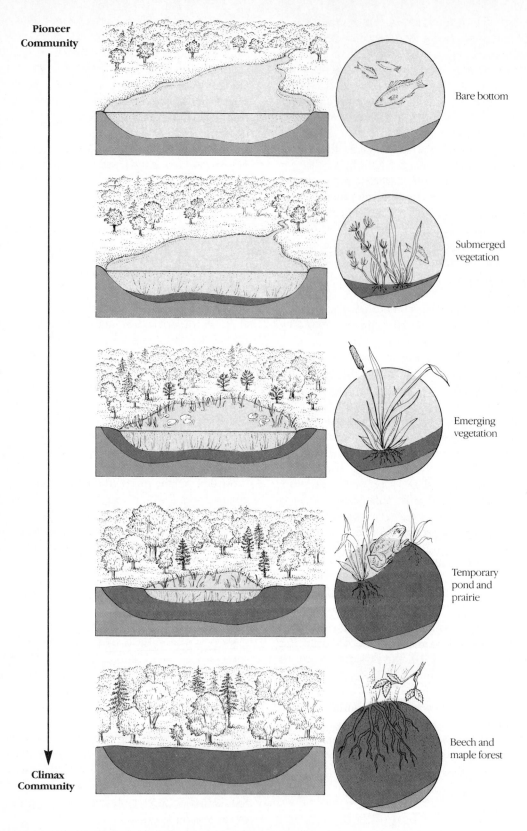

Pioneer
Community

Bare bottom

Submerged
vegetation

Emerging
vegetation

Temporary
pond and
prairie

Beech and
maple forest

Climax
Community

Figure 4-7. Primary succession. In this example a pond in New England is gradually converted into a forest ecosystem. Sediments accumulate on the bottom. Bottom-dwelling vegetation takes root. A series of biotic communities develop until the mature ecosystem is reached.

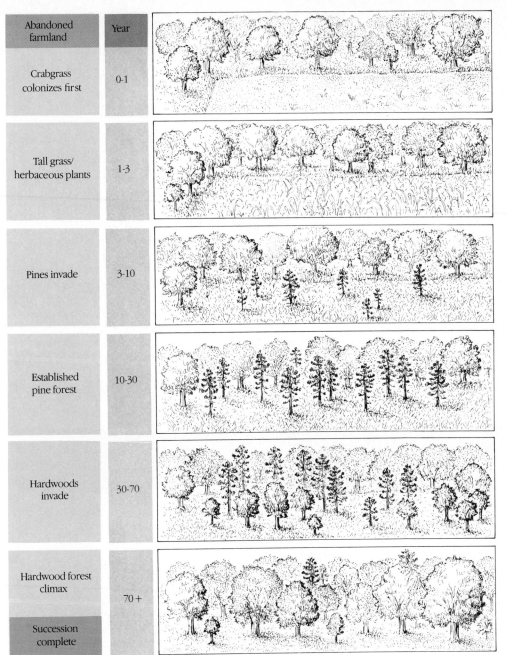

Abandoned farmland	Year	
Crabgrass colonizes first	0-1	
Tall grass/ herbaceous plants	1-3	
Pines invade	3-10	
Established pine forest	10-30	
Hardwoods invade	30-70	
Hardwood forest climax	70+	
Succession complete		

Figure 4-8. Secondary succession. Here, abandoned eastern US farmland is gradually replaced by crab-grass, which in turn gives way to other herbaceous plants. Trees move in, and over time a mature hardwood ecosystem is formed.

As ecosystems become increasingly complex, the food chains become woven into more complex food webs. The grazer food chain accounts for the bulk of the bio-mass flow in immature or developing ecosystems. In mature ecosystems, however, most of the biomass flows through the detritus food chain. In fact, in a mature forest less than 10% of the net primary productivity is consumed by grazers.

In mature ecosystems the nutrients are cycled more efficiently with less loss. Whereas immature, or devel-oping, ecosystems tend to lose a considerable amount of their nutrients because of erosion and other factors, mature systems are better able to entrap and hold nutrients.

Pioneer and intermediate communities are in a state of imbalance. The imbalance exists because the abiotic and biotic factors that regulate population size are con-stantly changing. For instance, there is initially little envi-ronmental resistance to crabgrass or lichens, so there are few limitations to growth. Thus, the growth factors such as the availability of food, favorable light, and optimum temperature stimulate a population expansion much like that seen in the hyacinth-choked waters of Florida.

During succession each community goes through a

Case Study 4-2:

Upsetting the Balance: The Accidental Parasite

One of the benefits of studying environmental science is that it helps us become aware of hidden relationships, especially ecological backlashes that can cause incredible harm. Consider an example: Ten years ago a seemingly innocuous event, a shipment of oysters from California to the north coast of France, began the steady decline of a once-lucrative oyster-growing industry.

The California relative of the European oyster brought with it a protozoan parasite. In the ten years that followed, the parasite invaded oyster beds throughout the coast. Today, in regions where 50,000 oyster farmers once made a living, only 10,000 manage a meager income, and then only by growing Japanese oysters, which don't do well in the colder waters. Today, European oysters account for just 10% of Brittany's oyster harvest. The parasite has all but wiped them out.

Henri Grizel, an internationally known mollusk pathologist, warned Brittany's oyster growers early in the infestation not to transfer oysters from bay to bay. He argued that this would probably spread the disease. Oyster growers typically move their oysters between different waters as many as five times before selling them, which apparently speeds up growth and improves the flavor. Unfortunately, the farmers insisted on moving the diseased oysters to areas free of the disease, helping spread the protozoan and ultimately nearly destroying the oyster industry they relied on.

Researchers are now trying to reestablish a healthy European oyster industry by taking three- and four-year-old European oysters that are apparently immune to the protozoan, and transferring them to a region free of the disease. Slowly, they may be able to reestablish healthy breeding colonies.

With careful work and time, the rich oyster beds may someday return to the north of France. Had the oyster growers heeded the advice of scientists early on, this disaster might never have taken place.

Table 4.1 Characteristics of Mature and Immature Ecosystems

Characteristic	Immature Ecosystem	Mature Ecosystem
Food chains	Linear, predominantly grazer	Weblike, predominantly detritus
Net productivity	High	Low
Species diversity	Low	High
Niche specialization	Broad	Narrow
Nutrient cycles	Open	Closed
Nutrient conservation	Poor	Good
Stability	Low	Higher

Source: Modified from Odum, E. (1969). The Strategy of Ecosystem Development. *Science* 164: 262–270. Copyright 1969 by the American Association for the Advancement of Science.

phase of increasing environmental resistance as competitors rise in number and as biotic conditions shift (Figure 4-9). When environmental resistance accelerates, the population declines, and a new community establishes itself.

Those of you who will go on to study ecology will learn that the view of succession presented here is the classical interpretation. New research has thrown into question some of the key tenets of this theory. Research is showing that the process of succession is more complex than once thought, varying considerably in different regions and with different types of vegetation.

Here's a case in point: In the Rocky Mountains in New Mexico, Colorado, Wyoming, and Montana, avalanches can destroy stands of conifers (subalpine fir and Engleman spruce). The opening in the forest created by an avalanche is first filled by grasses, wildflowers, and aspen trees. Over the years, dense stands of aspen grow. Next, shade-tolerant fir and spruce grow up in the forests, eventually reestablishing the climax community. In this system only a few stages are required to reestablish the climax community. What's more important, though, is that species diversity in the intermediate community exceeds that

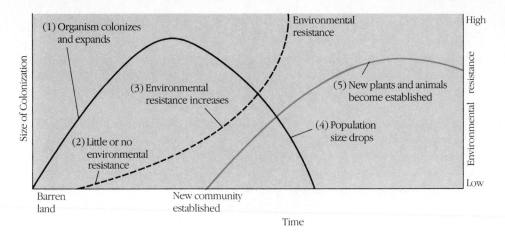

Figure 4-9. A simplified graphic representation of changing population size and environmental resistance as a community develops and is eventually replaced during succession. Note that environmental resistance increases as a new community develops. Resistance is caused by unfavorable conditions created by the community itself.

found in the mature system. Aspen forests support grasses, herbs, wildflowers, and a great many birds and mammals, such as deer and elk. As the climax community becomes established, species diversity actually falls because the dense forests do not permit grasses and other sun-loving plants to grow. The needle-covered floors of a fir and spruce forest support relatively few species.

New research is also showing that the relay sequence of succession, the sequential replacement of one community by another, is not always as clear-cut as once thought. In the succession of an eastern farmfield to forest, for example, trees are often present early in the vegetational development. Their presence is more noticeable later in the process, however, because trees grow more slowly and are often browsed, which suppresses their growth.

This discussion of natural succession may give the impression that all damaged ecosystems are returned to the mature climax stage. That's not always true. When damage is severe, natural revegetation may be impossible. In a huge region where Tennessee, North Carolina, and Georgia meet, for instance, pollution from a copper smelter, deforestation, and erosion have created a bleak landscape, so badly damaged that plants cannot take root. It is a scar visible from outer space. A journalist described it as a "vast, raw plain, cooking in the summer sun." Left on its own, it might never recover. But thanks to the diligent work of government officials, the land is being replanted and coaxed back to life. Similar devastation is occurring in the tropical rain forests. Each year millions of hectares are cleared and abandoned. Soils are so badly eroded that forest may take centuries to regrow.

Human Impact on Ecosystems

Throughout human history we have used our knowledge to gain control over the environment—to shape it to our

liking and enhance our survival. But our understandable search for control and security has not always produced the desired outcome. Water hyacinth–choked rivers, polluted streams, acidic rainfall, radioactive spills, energy shortages, runaway population growth, and a host of other common maladies are often the unanticipated results of our actions.

This section examines two general ways in which humans alter the biosphere: they tamper with either the biotic or the abiotic factors. This material is worth close study. What we do to the biosphere, we also do to ourselves.

Tampering with Biotic Factors

Introducing Competitors Early in the morning on June 9, 1985, Bill Wilson, who operates a front-end loader for the Chevron Corporation in southern California, watched in amazement as a swarm of bees attacked and killed a rabbit, stinging it to death. To his further surprise, when they had finished the kill, the swarm of several thousand bees came after him. Fortunately, he was protected from the onslaught by a glass enclosure on his vehicle. The bees that had attacked Wilson were so-called killer bees, newly arrived from South America.

The killer bee is actually an African honeybee which lives in hives with 30,000 to 80,000 ill-tempered cohorts. It was brought to the New World quite intentionally. In 1956 a geneticist, Warwick Kerr, imported some African honeybees to Venezuela in an attempt to develop a successful stock of honey producers. The docile European bees used previously had fared poorly in the tropical climate. Kerr thought that interbreeding the two might yield a more successful tropical strain.

Knowing their aggressiveness, the researcher isolated the bees in screened-in hives. But in 1957 a visitor unwittingly lifted the screen, allowing 26 queens to escape with their entourage. Trouble soon began. The killer bees

Figure 4-10. (a) Rabbits introduced in small numbers to Australia increased dramatically in a few years, posing a major environmental problem. They ate vegetation intended for sheep and cost the government and ranchers millions of dollars. (b) Widespread destruction caused by rabbits in Australia. Range on the right was protected by rabbit-proof fencing.

(a)

(b)

quickly spread, moving a remarkable 350 to 500 kilometers (200 to 300 miles) a year, interbreeding with honeybees, and destroying the honey industry in Venezuela and other countries. And that's not all. Bees assaulted people, horses, and livestock that crossed their paths.

Killer bees arrived in the United States by ship, many experts think, accompanying a load of pipe from South America. So far, California officials have located at least four colonies and have destroyed hundreds of commercial beehives to destroy any killer bees that may have mixed with the colonies.

Experts worry about the impact of the bee on the $140-million US honey industry. Should it spread into northern climates and breed with its tamer cousin, the killer bee could put an end to honey production. Hives would have to be destroyed to prevent further spread and attacks. The loss of honey production would, however, be minor compared with the indirect effect of losing America's honeybees. Each year honeybees pollinate 90 major crops, worth $19 billion.

Many biologists hoped that the northward spread of the killer bee would be halted by the colder climates. A study published in 1988, however, showed that killer bees could survive at a 0° C temperature for six months. It is

feared that the bees could migrate as far north as Canada, causing widespread damage to the North American honey industry and to crops pollinated by native bees.

The saga of the African honeybee is one of a number of biological nightmares created by the introduction of a foreign species into a new region. But not all such introductions have adverse effects. The ring-necked pheasant and chukar partridge, both aliens in this country, have done well in some areas. In other cases alien species have perished without a trace. Hardy species such as the killer bee, however, are the ones that demand our attention and remind us of the folly of careless introductions.

Consider what happened when rabbits were intentionally introduced into Australia. Twelve pairs of European rabbits were brought into the country in 1859 and released on a private ranch. In a few years the rabbits had proliferated wildly and begun eating grass intended for sheep, creating a major national environmental problem (Figure 4-10). Five rabbits eat as much grass as one sheep. Despite a major campaign to remove rabbits from Australia, by 1953 over a billion were inhabiting 3 million square kilometers (1.2 million square miles).

Plants such as the prickly pear cactus or water hyacinth can also reproduce uncontrollably in foreign environments. Taking over a new territory, they can wipe out native populations that compete for the same habitat. The results can be ecologically and economically disastrous.

Eliminating or Introducing Predators Predators have never fared well in human societies. Early hunters and gatherers killed them for food and because they viewed them as competition for prey. Modern societies have carried on this dangerous tradition, killing bears, eagles, hawks, wolves, coyotes, and mountain lions with a vengeance, often with serious ecological consequences.

One such example took place on the Kaibab Plateau, on the north rim of the Grand Canyon. In the early 1900s the state of Arizona put a bounty on wolves, coyotes, and mountain lions, which triggered their wholesale slaughter. Within 15 years virtually all of these predators had been eliminated. This intervention spawned an ecological catastrophe. Without its natural control, the deer population soared from 4000 to about 100,000 by 1924. The deer overgrazed the plateau; approximately 60,000 died from starvation in the following winters. The vegetation has still not completely recovered.

Not all scientists agree that eliminating predators in the Kaibab Plateau caused the dramatic population explosion. Some say that populations of grazing and browsing animals fluctuate naturally with changes in food supply brought on by changes in climate, especially rainfall. The rapid increase in deer populations in the Kaibab Plateau, they argue, may be linked more to natural change in food supply than to the elimination of predators. No one knows for certain. Predators are one of many factors that regulate prey population, helping to hold ecosystems in balance. Case Study 3-2 on the Great Barrier Reef illustrates the profound impact of eliminating predators on a rich aquatic life zone off the coast of Australia.

Occasionally, however, predators are introduced into areas with adverse effects. The mosquito fish, a native of the southeastern United States, has been introduced into many subtropical regions throughout the world because it eats the larvae of mosquitoes and therefore helps to control malaria, a mosquito-borne disease. Unfortunately, the mosquito fish also feeds heavily on zooplankton, single-celled organisms that consume algae. By depleting zooplankton populations, the mosquito fish removes environmental resistance that curbs algal growth. This causes algae to proliferate and form thick mats that reduce light penetration and plant growth in aquatic ecosystems.

These examples illustrate that altering the trophic structure by introducing or eliminating predators can drastically affect ecosystems and human populations as well.

Introducing Disease Organisms Pathogenic (disease-producing) organisms are a natural part of ecosystems. Humans have unwittingly introduced pathogens into new environments where there are no natural controls. There, they have reproduced at a high rate and caused serious damage.

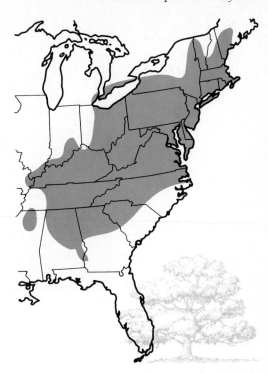

Figure 4-11. Former range of the American chestnut, a species nearly wiped out by the accidental introduction of a harmful fungus.

In the late 1800s, for example, a fungus that infects Chinese chestnuts was introduced accidentally into the United States. It had been carried in with several Chinese chestnut trees brought to the New York Zoological Park. The Chinese chestnut has evolved mechanisms to combat the fungus and is immune to it. But the American chestnut, once a valuable commercial tree found in much of the eastern United States, had no resistance at all and was virtually eliminated from this country between 1910 and 1940 (Figure 4-11). Now chestnut wood is found only in antiques, and the once-abundant American chestnut, a tasty seed commonly roasted in embers, is only a memory of our grandparents. (See Case Study 4-2 for a description of the accidental transplantation of a parasite that has devastated the oyster industry of northern France.)

Tampering with Abiotic Factors

Humans also tamper with many abiotic factors by polluting the air, water, and land and by depleting resources, such as water. The consequences can be as significant as those from our manipulation of the biotic world.

Pollution Water pollution and air pollution create an unfavorable environment for many living organisms. Chlorinated compounds released into rivers from wastewater treatment plants, for example, can virtually elimi-

Figure 4-12. Parts of Miami are built on coastal areas that were once marshland.

nate native fish. Oil spills on lakes, rivers, and oceans destroy fish, reptiles, and birds. Toxic pesticides eliminate birds that feed on contaminated insects and fish. Thermal pollution from power plants kills fish and many aquatic organisms on which fish feed. Possible global changes in temperature brought about by increasing atmospheric carbon dioxide could alter climate in many regions of the world, possibly eliminating thousands of species of plants and animals. In these and many other cases, human activities create an unfavorable abiotic environment that can reduce or eliminate species and upset the ecological balance. (See Chapter 16 for a description of the impact of oil from the Exxon supertanker *Valdez* on the Prince William Sound in Alaska.)

Resource Depletion Human populations deplete or destroy resources used by other species, too. The diversion of mountain streams to supply growing cities, for instance, has left many waterways dry. Housing developments along coastlines are often made possible by filling in estuaries and marshes with dirt (Figure 4-12). Full-scale oil shale development may destroy the valuable wintering grounds of thousands of mule deer in Colorado, possibly resulting in the decimation of deer populations in the western part of the state.

Simplifying Ecosystems

Tampering with abiotic and biotic factors tends to simplify an ecosystem by reducing species diversity. A reduction in species diversity may cause ecosystem imbalance and eventual collapse.

Ecosystem simplification is best seen when natural ecosystems are converted into farmland. Grassland contains many species of plants and animals. When plowed under and planted in one crop, called a **monoculture**, the field becomes simplified and vulnerable to insects, disease, drought, wind, and adverse weather.

The reasons for this susceptibility are many. Perhaps one of the most important is that monocultures provide a virtually unlimited food source for insects and plant pathogens, especially viruses and fungi. As crops grow, food supplies increase dramatically, favoring massive growth of pest populations. Viruses, fungi, and insects become major pests. Monocultures provide little or no environmental resistance.

The need to protect monocultures leads to many far-reaching problems. For instance, farmers have long used chemical pesticides to control outbreaks of pests. Chemical pesticides used to control fungi, viruses, and insects may be carried in the air or water to natural ecosystems, where they may poison beneficial organisms such as honeybees. The widespread use of DDT (dichlorodiphenyltrichloroethane) in the United States contaminated many terrestrial and aquatic ecosystems. Passed through the food chain, DDT and its chief breakdown product DDE reached high levels in top-level consumers, including ospreys, brown pelicans, and peregrine falcons (Figure 4-13). This poison did not kill the birds outright; rather, it interfered with the deposition of calcium in their eggshells, resulting in thinning of the shells. The fragile eggs were easily broken. Few embryos survived, and populations fell sharply, reminding us of an important biological principle: an organism lives as it breeds.

The DDT incident also illustrates an important biological phenomenon called **biological magnification** or simply **biomagnification**, the accumulation of certain substances in food chains, with the highest concentrations found in the highest trophic levels. Because of this phenomenon, fairly low levels of DDT in the environment

Figure 4-13. Peregrine falcon and her chick. The peregrine, once a nearly extinct species in the United States, has made a remarkable comeback with the assistance of researchers who have incubated eggs in captivity and released birds in the wild. DDT used to protect human crops caused eggshell thinning in peregrines and other species, nearly eliminating the natural populations of these birds.

can result in dangerously high levels in organisms. (For more on biomagnification see Chapter 17.)

The peregrine falcon, which nests on rocky ledges throughout the United States, was nearly destroyed by DDT (Figure 4-13). By the time scientists had determined that the decline in reproduction was the result of DDT and DDE, none of the 200 known pairs east of the Mississippi River was successfully producing young. Fortunately for the peregrine falcon, DDT was banned in the United States (but still manufactured here until 1984), and a determined program was mounted that may well save these birds. By 1982 over 1000 peregrines had been raised in captivity and released into the wild by ornithologists from Cornell and Colorado State universities.

The story of the peregrine falcon, like so many other species that have been brought back from the brink of extinction, is one of personal dedication and triumph. By using ecological caution in releasing a powerful toxin into the environment, however, we could have avoided a great deal of the cost and effort.

Impact Analysis Model

In Chapter 2 we saw how human life-styles and needs have changed over the millennia. In particular, we looked at the changing impact of human populations—how humans have become a major molding force in the environment and how increasing technological development and increasing population size have affected the environment.

To develop a deeper understanding of human impact, let us briefly examine the model illustrated in Figure 4-14. This simplified version of the **Impact Analysis Model** shows that humans have an impact on the environment— the air, water, and land—as well as the biota—the plant, animal, and microbial inhabitants of the environment. Impact analysis is far more than an intellectual exercise; it is the basis of environmental impact statements, which must be prepared for almost all major construction projects (see Chapter Supplement 21-1). It is also a vital tool in developing critical thinking skills.

The Impact of Coal Use

To illustrate the Impact Analysis Model, let us take a brief look at the mining and combustion of coal in the United States. Environmental impacts include air, water, and land pollution (Figure 4-14). Air pollution may result from mining or related activities, such as trucks travelling on dirt roads near mines. Combustion of coal to generate electricity creates a much larger air pollution problem. A single 1000-megawatt coal-fired power plant, which serves roughly one million people, emits approximately 1500 to 30,000 metric tons (1650 to 33,000 tons) of particulates (smoke and ash) and 11,000 to 110,000 metric tons (12,000 to 121,000 tons) of sulfur dioxide gas each year.

Water pollution can also be caused by coal mining. In abandoned underground coal mines in the eastern United States, for example, certain sulfur-bearing minerals (iron pyrite) react with oxygen and water to produce

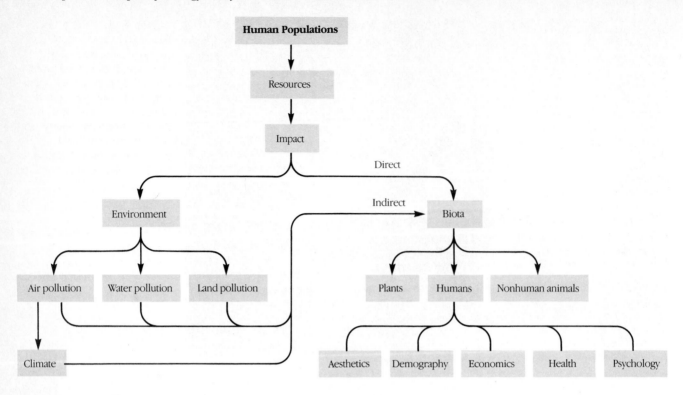

Figure 4-14. Impact Analysis Model, showing the range of impacts caused by human activities.

highly caustic sulfuric acid. This acid drains from the mine and pollutes rivers and lakes downstream (see Chapter 11).

Land pollution occurs from the actual mining of coal. In the West huge surface mines are used to reach the thick beds of coal. Large draglines rip up the surface to dig down to the coal, in the process destroying the vegetation.

Indirect Effects of Human Activities All of the forms of pollution described above affect organisms. For example, the sulfur dioxide produced by coal-fired power plants is converted to sulfuric acid in the atmosphere; carried as small droplets in the air, this acid may be transported many miles from the source and then washed from the sky by rain or snow. This phenomenon, acid precipitation, occurs widely in the eastern half of the United States and in Canada. In particularly susceptible regions such as southern Canada and the Adirondack Mountains of New York, acids have destroyed several hundred lakes. Lakes once known for sport fishing have turned acidic, and virtually all life has been destroyed in them. (For more on acid precipitation see Chapter Supplement 15-2.)

Water pollutants affect a variety of organisms. Acid draining from coal mines in the East kills fish and aquatic plants. It corrodes bridges, locks, barges, pumps, and

other metallic structures, costing millions of dollars a year for repair and replacement.

These examples illustrate a few of the many impacts of coal mining and combustion on the environment (for additional examples, see Chapter 11), showing us the indirect ways in which environmental pollutants affect biological organisms, including humans.

Direct Effects of Human Activities Human activities may also have direct impacts on the biota. Surface mining in the West, for example, destroys the habitat of elk, deer, prong-horn antelope, and prairie dogs. In the Yellowstone River basin, an area richly endowed with thick coal seams, about 36,440 hectares (90,000 acres) of land will be mined between 1977 and 2000. The impacts on native wildlife populations will be severe. Disturbance of land will also affect underground water supplies, requiring landowners to redrill wells that supply water for livestock and human uses.

A large increase in coal mining may result in changes in populations in nearby towns. These are called **demographic changes**. The town of Craig in northwestern Colorado, for example, experienced a rapid influx of miners and their families between 1970 and 1980. With this came a tripling in the crime rate. Many other towns in the United States, like Craig, were hit hard by the sudden upturn in coal production in the late 1970s. In most cases

Impacts		High Probability	Medium Probability	Low Probability	Unknown Probability
Air pollution	+				
	−				
Water pollution	+				
	−				
Land pollution	+				
	−				
Plants	+				
	−				
Humans	+				
	−				
Nonhuman animals	+				
	−				

Figure 4-15. Impact analysis chart is used for summarizing the positive and negative effects of a given technology or potential hazard. As an aid in decision making, impacts are listed according to the degree of probability of their occurrence.

they did not have enough housing for newcomers. Schools, hospitals, sewage systems, water supply systems, stores, and recreational facilities were inadequate for their swelling populations and had to be upgraded at considerable expense. As a consequence of the influx of miners, the whole social fabric of communities has changed almost overnight. Quaint agricultural towns have become noisy and crowded.

Why Study Impacts?

The Impact Analysis Model is helpful when studying the effects of a particular technology, such as coal or oil shale development. It shows the areas that should be examined when making an impact analysis and helps us review new technologies more critically.

We study impact before we build roads, dams, power plants, gravel pits, and other projects, to better understand how we will affect the environment. Commonly, too little is known about the impact of human activities to make predictions concerning overall damage. As a result, we must often rely on educated guesswork or speculation.

One of the dangers of speculation is that it is imprecise. Estimates may be off by 100% or more; people forget that estimates are based on many assumptions that may not be accurate. Even though the uncertainties and assumptions are disclosed in scientific studies, people have a tendency to omit the qualifiers. Thus, what is a tentative judgment becomes an ironclad fact. The statement "Under

certain weather conditions, the proposed coal-fired power plant may cause damage to neighboring crops" may be transformed into "Coal plant will damage crops." The dangers of this distortion are obvious.

Assessing the Probability of Impacts

The critical thinking skills you learned in Chapter Supplement 1-1 suggest that a broader look at issues helps us better understand our problems and ways to solve them. The discussion on science and scientific method in that section noted that much of what we know in science is uncertain. The degree of uncertainty of any given impact is important to know. Critical thinkers need to ask about the degree of probability of a certain impact, as well as the likelihood that a given solution will actually solve a problem. Figure 4-15 illustrates a simple chart that can help you become a more critical thinker. It breaks down the probability of an event into four categories: high, medium, low, and unknown. High probability is indicated when the data supporting a given conclusion are strong and numerous. Medium probability is indicated when the impact is suspected but there is some uncertainty. Low probability is assigned when the impact might theoretically happen but the data do not support the contention. The fourth level of uncertainty—unknown probability—might also be added when there are no data at all showing either an impact or the absence of one.

Figure 4-15 also takes into account a much overlooked variable: the fact that not all impacts are deleterious. As mentioned earlier, removing trees from forests in large numbers opens up meadows, which are used by elk for grazing. But not all clearings are equal. It appears that 3-hectare (8-acre) cuts are optimum. Larger cuts may not prove so beneficial.

Restoration Ecology: Reestablishing the Balance

In the United States, ecological destruction began in large part with the early colonists. As they cut down forests and plowed the land, our ancestors began to change the face of this great continent, a process that continues today at a much faster rate. With industrialization, the damage increased dramatically (see Chapter 2).

Consider what happened to the fresh water marshes of New Jersey: Majestic white cedars once grew around these biologically rich wetlands. But early settlers cut the trees down to build homes and factories. Dam-builders cut off the flow of fresh water to coastal freshwater swamps, letting in the sea and destroying the habitat for fish and other animals. Some of these devastated lands are now on the mend.

The Birth of a New Science

In 1985, a man by the name of Ed Garbisch arrived in New Jersey to begin the long and costly process of restoring an ancient freshwater wetland. In a $4-million project, Garbisch and his workers seeded lands with marsh grass, cut channels to restore water flow, and built knolls for ducks to construct their nests. Despite initial skepticism, Garbisch and his coworkers restored the swamp to its previous condition.

Garbisch and others like him have given rise to a new field of ecology, called **restoration ecology**. Scientists often prefer to call this new science **conservation biology**, the study of how ecosystem recovery occurs and how it can be facilitated. In this relatively new field of study, the principles of science are used to repair land and waters damaged by decades of misuse. In many respects conservation biology is the rehabilitative medicine of the ecological sciences. Some of the earliest restoration projects were designed to reclaim badly damaged land that had been surface-mined for coal and other minerals. Today, however, ecologists and others are working to restore marshes, tropical rain forests, streams, and prairies throughout the world.

John Berger, who holds a PhD in ecology, has done as much to stimulate restoration as any person. His book *Restoring the Earth* describes what people are doing in the United States to correct past mistakes. Berger also founded a nonprofit environmental group called Restoring the Earth, whose purpose is to repair the damaged American landscape. Because of their work and the work of Ed Garbisch, hundreds of marshes have been restored along the east coast from Maine to Virginia, reestablishing native plants and habitats for many species.

One of the most ambitious projects on the books today is that of Daniel Janzen, a University of Pennsylvania biologist. With the help of others, Janzen hopes to reforest 3900 hectares (9600 acres) of dry tropical forest in northwest Costa Rica. Janzen's project will take 100 years or so to complete, one fifth of the time it would take for nature to reestablish forest that has been cut and burned over the years.

Controversy over Restoration

There are few natural systems that haven't been affected by humans. The potential for restoration is enormous. But restoration ecology has its critics. The Sierra Club, for instance, is concerned that developers who want to build on desirable existing wetlands and other natural ecosystems will offer to replace them with wetlands "created" elsewhere. Supporters of restoration ecology, however, insist that they are advocating restoration to repair previous damage, not to legitimize further destruction.

In Washington and other states, developers are already looking at restoration as a new tool to build on wetlands. If they can make a wetland elsewhere, they wonder, why shouldn't they be allowed to destroy an existing one?

Critical thinking suggests several reasons. For one, an existing wetland is a complex ecosystem. A flooded field planted with some swamp vegetation is a far cry from the previous system. It may, over a period of a decade, come to resemble the lost swamp, but it doesn't replace the wetland that's lost. Rebuilding wetlands or other ecosystems is a complex and costly task. It requires expert attention and follow-up.

Benefits of Restoration

Restoring marshes has economic as well as aesthetic benefits. To protect an eroding shoreline with concrete or rocks, for example, can cost $500 per meter. Planting a 7-meter- (21-foot) wide strip of salt marsh to protect the shoreline may cost only $50–$80 per meter. This can result in a savings of $100,000 or more to owners of large shoreline property who are trying to safeguard residential or commercial property. Plants can also turn a barren, desolate beachfront into dense, lush, emerald marshes.

Each year, the US Army Corps of Engineers dredges shipping channels to keep them open. The sediment from these operations has routinely been dumped along coastal waters, where it kills aquatic life. The sediment also tends to migrate back into the channel from which it was dredged. Consequently, the Corps of Engineers now dumps its dredged material on land, which is very costly. Garbisch, however, encourages the Corps to deposit the sediment in regions needing wildlife improvement. For instance, the Corps dredged a channel near Barren Island near Chesapeake Bay in 1981. The sediment was deposited in two-hectare (five-acre) islands barely above the water's surface. Garbisch planted these with native grasses, establishing a habitat for, at the very least, terns, a bird threatened by extinction.

Creating a marsh costs $5,000 to $25,000 per hectare ($2,000–10,000/acre in 1983 dollars). A marsh's value cannot easily be calculated in dollars and cents. Biologists in Louisiana estimated some years ago that an acre of salt marsh was worth $82,000 solely for its ability to reduce pollutants such as sediment.

Marshes are one of the world's most productive ecosystems. A single hectare of marsh produces more than 25 tons of organic matter per year, which is more than twice the yield of a corn field and 10 times the yield of coastal waters. (Further benefits of marshland are discussed in Chapter Supplement 10-1.)

Many people see marshes as muddy, uninhabitable places that should be drained or filled to build houses, airports, industrial parks, or garbage dumps. Today, approximately one half of this country's salt marshes have already been destroyed. When a salt marsh disappears, shorelines erode at an accelerated rate, fish populations collapse, birds vanish, wildlife retreats, and some of nature's remarkable plant communities are destroyed.

For the first time in the whole history of evolution, responsibility for the continued unfolding of evolution has been placed upon the evolutionary material itself. . . . Whether we like it or not, we are now the custodians of the evolutionary process on earth. Within our own hands—or rather, within our own minds— lies the evolutionary future of the planet.

PETER RUSSELL

Summary

Mature ecosystems change with time but remain more or less the same from one year to the next. They are said to be in a state of **dynamic equilibrium**, or a **steady state**. Ecosystem **stability** is the ability of an ecosystem to resist change (**inertia**) or return to its original condition after it is disturbed (**resilience**).

Biotic and abiotic components of an ecosystem cause the increase and decrease of populations and ultimately determine ecosystem balance. Those that increase population size are called **growth factors**; those that depress it are called **reduction factors**, which collectively produce **environmental resistance**.

Species diversity is considered by many ecologists to be one of the more important factors determining ecosystem stability, although direct evidence supporting this hypothesis is rare. We do know that simplifying ecosystems by reducing species diversity tends to make them more vulnerable to insects, adverse weather, and disease-producing organisms.

Small shifts in the biotic or abiotic growth or reduction factors may temporarily tip ecosystem balance, but appropriate responses within the biological community can return the system to normal. Larger shifts may result in a dramatic destabilization of the ecosystem, resulting in its collapse. A biological community destroyed by such large shifts may recover during a process known as **secondary succession**, in which new communities develop sequentially on the remains of the old until a mature, or climax, community is formed. **Primary succession**, on the other hand, is the sequential development of communities where none previously existed.

Humans cause imbalance in an ecosystem by altering its biotic or abiotic components. Introducing or eliminating competitors, predators, and pathogenic organisms, for example, can permanently disrupt ecosystem balance. Pollution and resource depletion may have the same effect. Tampering with abiotic or biotic factors in the ecosystem can reduce species diversity and simplify ecosystems. **Monocultures**, huge single-crop plantings, represent extreme ecosystem simplifications that are vulnerable to insects, disease, and adverse weather.

Ecosystem damage continues today, but a growing number of people are finding ways to repair the damage. **Conservation biology** is a new field aimed at studying recovery and particular ways that humans can facilitate the process.

Discussion Questions

1. Describe ecosystem stability. Give an example of an ecosystem you are familiar with. Is it stable? Why or why not?

2. If you were to examine a mature ecosystem over the course of 30 years at the same time each year, would you expect the number of species and the population size of each species to be the same from year to year? Why or why not?

3. What is environmental resistance? What role does it play in population balance? What role does it play in ecosystem balance?

4. Give evidence that species diversity affects ecosystem stability. Is there any evidence to contradict this idea? What is it?

5. What is a mature ecosystem? What are its major features?

6. Describe temporary imbalances caused in ecosystems you are familiar with and how the ecosystems return to normal.

7. Why is one biotic community eventually replaced by another during succession?

8. What is the difference between primary and secondary succession? Give examples of both.

9. What is a pioneer community?

10. Describe how introducing competitors into an ecosystem can affect ecosystem stability. Give some examples.

11. How do humans tamper with abiotic ecosystem components? Give some examples, and trace the effect on plants and animals.

12. "Humans are like all other organisms in many respects, except they greatly simplify their ecosystem." Is this statement true? Why is it necessary for humans to simplify ecosystems? How can it be avoided? Give some examples.

13. From an evolutionary standpoint, discuss why simplified ecosystems (monocultures) are highly susceptible to fungi, viruses, and insects.

14. In what ways has the study of ecology broadened your view of life? Has it made you reconsider any of your views?

Suggested Readings

Attenborough, D. (1979). *Life on Earth*. Boston: Little, Brown. Superb study of life.

Barnhardt, W. (1987). The Death of Ducktown. *Discover* 8 (10): 34–43. Excellent reading, showing the devastation of human activities and the efforts to restore a ruined landscape.

Berger, J. J. (1985). *Restoring the Earth. How Americans are Working to Renew our Damaged Environment*. New York: Knopf. A delightful book that's a must.

Bergon, M., Harper, J. L., and Townsend, C. R. (1986). *Ecology: Individuals, Populations, and Communities*. Sunderland, Massachusetts: Sinauer. For students wanting more detailed information on ecology.

Clapham, W. B. (1983). *Natural Ecosystems* (2nd ed.). New York: Macmillan. Advanced treatise on ecology.

Colinvaux, P. (1986). *Ecology*. New York: Wiley. An excellent textbook of basic ecology.

Committee on the Application of Ecological Theory to Environmental Problems (1986). *Ecological Knowledge and Environmental Problem-Solving. Concepts and Case Studies*. Washington, D.C.: National Academy Press. Important reference for students interested in a career in applied ecology.

Ehrlich, P. R. and Roughgarden, J. (1987). *The Science of Ecology*. New York: Macmillan. Higher-level coverage of ecology.

Rodgers, C. L. and Kerstetter, R. E. (1974). *The Ecosphere: Organisms, Habitats, and Disturbances*. New York: Harper and Row. A good account on biomes and human disturbance.

Smith, R. L. (1985). *Elements of Ecology* (2nd ed.). New York: Harper and Row. Advanced readings on ecology and environmental problems.

Nuclear War: Pathway to Environmental Catastrophe

Air pollution, water pollution, and resource depletion—all of these are major environmental problems facing modern civilization. Their effects, however, pale by comparison with the environmental imbalance that would follow a nuclear war. A glimpse into the consequences of such an event provides a sometimes shocking view of the potential that humanity has for disrupting the biosphere.

The Nuclear Detonation

A nuclear explosion produces (1) enormous amounts of heat and light, (2) an explosive blast and high winds, (3) intense radiation (radiation is described in Chapter Supplement 12-1), (4) a pulse of electromagnetic energy, and (5) radioactive dust called **fallout** (Figure S4-1).

Heat and Light

About a third of the energy from a nuclear explosion is released as heat and light. The light flash may produce temporary blindness in victims looking at the explosion. This is called **flash blindness** and lasts only a few minutes in most people. However, individuals gazing directly at the fireball may never completely recover their eyesight; some may be permanently blinded.

The most common injury among survivors of atomic explosions is burns resulting from the intense heat flash. Even people standing 11 kilometers (7 miles) from a 1-megaton explosion, for example, would suffer first-degree burns (similar to sunburn) on their exposed skin; 1.6 kilometers (1 mile) closer, and their exposed skin would blister from the heat. Another 1.6 kilometers closer, and their exposed skin would be charred by the heat from the blast.

Severe burns are one of the greatest threats to human life in an atomic explosion. In a major city many burn victims would die from shock soon after an explosion. The heat flash would also ignite any combustible material on its path, turning the city into an inferno and killing many other people.

Explosive Blast and Winds

Several seconds after the heat and light flashes sweep through the area surrounding a bomb detonation site, air pressure rises. This increase is called **static overpressure** and is measured in pounds per square inch (Table S4-1). Although the pressure increase is rather small, it is forceful enough to topple houses and office buildings and rupture eardrums. Tens of thousands of people would be crushed by collapsing buildings if a bomb exploded near a major city during working hours.

Accompanying the static overpressure are intense winds (Table S4-1). A bomb with the explosive power of 1 million tons of TNT (1 megaton) produces winds of 290 kilometers per hour (180 miles per hour) 6.5 kilometers (4 miles) from the detonation site. These would fan fires and damage power lines, homes, and trees. Flying debris would injure any people and animals within range.

Figure S4-1. Fireball and mushroom cloud produced by a thermonuclear blast.

Table S4-1 Blast Effects of Nuclear Explosions[1]

Distance from Explosion	1-Megaton Weapon	10-Megaton Weapon
1 mile	Overpressure: 43 psi Winds: 1700 mph Many humans killed	Above 200 psi Above 2000 mph Buried 20-centimeter-thick concrete arch destroyed
2 miles	Overpressure: 17 psi Winds: 400 mph Humans battered to death; lung hemorrhage; eardrums ruptured; heavy machinery dragged severely	50 psi 1800 mph Humans fatally crushed; severe damage to buried light corrugated steel arch
5 miles	Overpressure: 4 psi Winds: 130 mph Bones fractured; 90% of trees down; many buildings flattened	14 psi 330 mph Eardrums ruptured; lung hemorrhage; reinforced concrete building severely damaged
10 miles	Overpressure: 4.4 psi Winds: 150 mph Bones fractured; 90% of trees down; many buildings flattened	
20 miles	Overpressure: Below 1 psi Winds: Below 35 mph Many broken windows	1.5 psi 55 mph Cuts and blows from flying debris; many buildings moderately damaged

[1]The range of effects required dictates the optimum height at which the weapon would be detonated.
Source: Goodwin, P. (1981). *Nuclear War: The Facts on Our Survival.* New York: Rutledge Press, pp. 28–29.

Direct Nuclear Radiation

A highly intense blast of radiation also spreads out in all directions from a nuclear explosion (Table S4-2). For large weapons the irradiated area is much smaller than the area affected by the heat or static overpressure; therefore, radiation would not be the principal cause of death. For smaller weapons, however, the range of intense radiation is greater than the range of heat and static overpressure; in such cases radiation would be the major cause of death. Some of the effects of radiation on human health are summarized in Table S4-3 and are discussed in Chapter Supplement 12-1.

Electromagnetic Pulse

Nuclear detonations create a single momentary pulse of electromagnetic radiation that spreads in all directions. This energy is similar to the electrical signal given off by lightning but much more powerful. The electromagnetic pulse would short-circuit unshielded radios, computers, and other electrical equipment, including telephone systems. The resulting blackout would occur at a time when communications would be badly needed to help coordinate rescue and health care. A single large bomb detonated in space 400 kilometers (250 miles) above Omaha,

for example, could effectively cripple communications in the entire United States.

Fallout

The heat and force of a nuclear explosion thrusts thousands of tons of dust into the atmosphere. In the process much of the dirt becomes radioactive. Radioactive dust particles return to the earth as fallout. Some of the dust particles may fall back immediately, landing near the site of the explosion and making the area intensely radioactive. Some fallout is carried higher into the atmosphere, only to fall back to earth many days later. Some fallout may be transported so high that it circulates in the upper atmosphere for decades, falling from the sky very gradually but exposing large areas. Radioactivity from fallout poses a threat to human health and the welfare of many other animal species.

Combined Injuries

One consequence of nuclear war that is sometimes overlooked is the effect of combined injuries. For example, victims of a nuclear exchange might be burned as well as exposed to non-

Table S4-2 Effects of Radiation Exposure

Main Organ Involved	Distinguishing Signs	Convalescence Period	Incidence of Death	Death Occurs Within
	Above 50 rads: slight changes in blood cells			
Blood-forming systems: bone marrow, lymph glands	Moderate fall in white blood cell count	Several weeks		
Blood-forming systems: bone marrow, lymph glands	Severe loss of white blood cells; bleeding; infection; anemia (loss of red blood cells); loss of hair (above 300 rads)	1–12 months	0–90%	2–12 weeks
Blood-forming systems: bone marrow, lymph glands		Long	90–100%	2–12 weeks
Gastrointestinal tract	Diarrhea; fever; shock symptoms		100%	2–14 days
Central nervous system	Convulsions; tremor; involuntary movements; lethargy		100%	days

Source: Goodwin, P. (1981). *Nuclear War: The Facts on Our Survival.* New York: Rutledge Press, pp. 44–45.

lethal doses of radiation. Broken bones, cuts, and abrasions might add to their injuries. Studies on laboratory animals suggest that nonfatal injuries can add up, producing death.

Nuclear Winter

Imagine a heavy nuclear exchange in the Northern Hemisphere. Some people could survive the initial shock wave, the intense heat, the radiation, and the resulting fallout. But awaiting them, according to some experts, would be a nightmarish world of cold and darkness called **nuclear winter**.

Multiple atomic blasts would carry millions of tons of dust and soot into the atmosphere. Rising columns of smoke from thousands of fires created by the explosions would create large, dark clouds. Dust from the explosions and smoke from the fires would block sunlight. Temperatures would plummet, with devastating effects on the biosphere.

The nuclear-winter theory was first published in 1982 by Paul Crutzen. Others, among them an American group including the renowned Carl Sagan, have arrived at similar conclusions.

No one knows for sure, but experts believe that about 40 million metric tons of soot would be required to produce a nuclear winter. Should 100 cities burn in a nuclear exchange, the world might be immersed in a dark, cold glacial age. The more smoke, the longer and more severe the winter would be.

Eventually, studies suggest, the clouds would coalesce and form a dense pall stretching from Florida to Alaska. The cloud could very easily spread south across the equator and plunge the Southern Hemisphere into a chilly twilight. Making matters worse, the smoke-laden air could absorb the sun's energy. This would cause the cloud to rise higher and thus lengthen its stay in the atmosphere.

Many scientists believe that the pall of smoke from an all-out nuclear war would plunge surface temperatures 20° to 30° C (36° to 54° F). Widespread extinction would result. Crop production throughout the world could halt. A drop in the average daily temperature of 2° or 3° C during the growing season can cut wheat production in Canada and the Soviet Union in half. Grains make up some 70% of the world's food energy; nuclear winter would, therefore, cause widespread global starvation. Climatic changes would devastate the developed countries, which are mostly in the temperate zone. According to a recent study by the International Council of Scientific Unions, however, the major environmental impacts would occur in the tropics and subtropics, where most of the Third World nations are located. Reduced temperature and rainfall in these regions would wipe out thousands of species adapted to warm, constant climates.

Nuclear war by itself could also disrupt world trade of fertilizers, pesticides, and the like, resulting in further declines in farm productivity worldwide. Nuclear explosions could also create a toxic nightmare as millions of tons of potentially toxic and carcinogenic substances spewed into the atmosphere from burning cities. Asbestos, dioxins, and PCBs—among others—would spread to surrounding areas, making them even more inhospitable to life. Finally, an all-out nuclear war would also reduce the protective ozone layer, resulting in further damage. (See Chapter Supplement 15-1.)

Some scientists have criticized proponents of the nuclear-winter theory for underestimating the effects. Others, such as Edward Teller, known as the father of the hydrogen bomb, argue that certain factors would "thaw" the nuclear winter, and they suggest that this new theory of doom is exaggerated. Teller and his associates, for instance, contend that fire storms would probably not generate columns of smoke that rise into the upper atmosphere; thus, the duration of the nuclear winter would be lessened. Most recently (1986), scientists at the National Center for Atmospheric Research in Boulder, Colorado have for the first time included the effects of the oceans in the computer models for a nuclear winter. These new calculations indicate that the fall in temperatures worldwide would last days, or at most weeks. In such a case, the effects of the cold would be considerably less drastic. Some scientists are speaking of a **nuclear fall**. Undoubtedly, the debate will continue for years to come.

Predicting the Effects of Nuclear War

No one can accurately predict the effects of nuclear war, because no one knows how many atomic weapons would be hurled between warring nations, where they would explode, what the weather conditions would be, or other imponderables. Clearly, a fraction of the total nuclear arsenals of the United States and the Soviet Union could devastate the major population centers of every country in the world.

Of this we can be certain: Any nuclear explosion over an urban area would have catastrophic effects. Widespread nuclear war would have environmental and health consequences beyond our comprehension. Health care, transportation, and public sanitation would be crippled. Widespread climatic changes might grip the world in a winter or fall-like cold that would destroy plants and wildlife and grind agriculture to a stop.

Paul Warnke, former US coordinator for the Strategic Arms Limitation Talks (SALT II), noted: "In this the fourth decade of the nuclear age, it is tempting to assume that a nuclear exchange won't take place and that, in any event, there is nothing the average human being can do about it. But the fact is that a nuclear war could happen and very well may happen unless we, as citizens of a threatened world, decide that we will do something."

Suggested Readings

Crutzen, P. J. (1985). The Global Environment after Nuclear War. *Environment* 27 (8): 6–11, 34–37. Excellent discussion of the nuclear-winter theory.

Ehrlich, P. R., Sagan, C., Kennedy, D., and Roberts, W. O. (1984). *The Cold and the Dark: The World after Nuclear War*. New York: Norton.

Goodwin, P. (1981). *Nuclear War: The Facts on Our Survival*. New York: Rutledge Press. Excellent coverage.

Leaf, A. and Ohkita, T. (1988). Health Effects of Nuclear War. *Environment* 30 (5): 36–38. Brief article outlining health concerns.

Lifton, R. J. and Falk, R. (1982). *Indefensible Weapons*. New York: Basic Books. Eloquent analysis of the nuclear age.

Riordan, M., ed. (1982). *The Day After Midnight: The Effects of Nuclear War*. Palo Alto, CA: Cheshire Books. Detailed account worth reading.

Russett, B. (1983). *The Prisoners of Insecurity: Nuclear Deterrence, the Arms Race, and Arms Control*. San Francisco: Freeman. Excellent discussion of the nuclear arms race.

Sagan, C. (1985). Nuclear Winter: A Report from the World Scientific Community. *Environment* 27 (8): 12–15, 38–39. Excellent overview of scientific reports on the nuclear-winter theory.

Schell, J. (1982). *The Fate of the Earth*. New York: Avon. A stirring account.

Turco, R.P. and Golitsyn, G. S. (1988). Global Effects of Nuclear War. *Environment* 30 (5): 8–16. Excellent overview of the possible effects of nuclear war.

POPULATION

5

Population: Measuring Growth and Its Impact

One generation passeth away, and another generation cometh, but the earth abideth forever.

ECCLESIASTES 1:4

On May 24, 1985, a hurricane struck the small nation of Bangladesh, which lies just east of India along the Bay of Bengal. Winds clocked at over 160 kilometers per hour (100 miles per hour) struck the shoreline. A 5-meter (15-foot) tidal wave crashed inland, devastating homes and farms in its way. When the weather had cleared, officials surveyed the damage: A quarter of a million people were homeless; their rickety homes, built on silt islands off the shore, had been demolished by the storm. An estimated 4000 to 15,000 people had perished. Countless livestock had died. In 1988, a similar disaster struck. Heavy rains flooded two-thirds of the land, leaving 25 million people homeless.

On the surface the disasters in Bangladesh look like especially severe natural disasters. On closer examination, however, it is clear that much of the blame can be pinned on an underlying human problem—too many people.

Bangladesh is a small country with big troubles. No larger than Wisconsin, it houses 118 million people. Rural and poor, Bangladesh is the most densely populated nation in the world and is also among the fastest growing.

The people of Bangladesh and their livestock have created an environmental disaster. Cattle have overgrazed the land. Hordes of desperate peasants have, over the years, stripped away much of the vegetation in search of food, fuel, and shelter. When the rains come, floods invariably follow, with devastating consequences. Denuded hillsides cannot hold the rain back; thus, even moderate storms produce floods. Making matters worse, soil from these lands washes into streams. Brown rivers flow to the sea and deposit their silts in deltas. In search of farmland, people flock to the deltas by the tens of thousands.

Almost any ecologist will tell you that deltas belong to rivers and the sea, not to people. When the monsoons come, as they do most years, high waters flood these muddy lands, driving people away, and destroying their farms and homes. Violent hurricanes wreak havoc too as they sweep in from the Bay of Bengal, claiming the lives of thousands who have either no way of knowing of the storm's impending arrival or nowhere to go.

In Bangladesh people are dying in a vicious circle triggered largely by **overpopulation**—too many people for the existing resources. This chapter chronicles the toll of overpopulation. It also discusses why the human population has exploded, ways to better understand the population crisis, and, finally, the future prospects for world population.

Dimensions of the Population Crisis

The population crisis can be summarized in six words: **too many people reproducing too quickly**. This creates a myriad of problems, which are examined in the following sections.

Too Many People

"Hell is a city much like London," Shelley wrote in 1819. If he were alive today, the English poet might have put it differently: "Hell is a city much like Cairo, Calcutta, Shanghai, Bangkok, London, Los Angeles, and Mexico City." By the end of the century at least 22 cities worldwide are expected to have populations in excess of ten million people. Sixty others will probably have exceeded the five-million mark. Each of these urban centers will have its own unique problems, many of which may be linked to overpopulation—too many people.

The Plight of the Cities In both the rich and poor nations of the world, overpopulation brings incredible despair. In Cairo, Egypt, for example, a giant graveyard is home to several hundred thousand people who cannot afford or cannot find housing. They live in makeshift cardboard shanties among the many tombstones. Those of Cairo's 12 million inhabitants who are lucky enough to find housing don't count their blessings. Apartments are tiny. The municipal water system breaks down frequently. One-third of the population lives in housing without sewage.

In Calcutta, India, conditions are worse. More than 70% of the city's ten million people live at or below the poverty level. One water tap supplies, on the average, 25 slum

Figure 5-1. Los Angeles rush-hour traffic: a nightmare on a good day.

dwellings; and about half of the homes in this city have no indoor toilet. The streets are littered with trash and feces, and at least 600,000 people roam the city, homeless.

Shanghai, China, a well-to-do city by comparison, houses nearly 12 million people. Housing is provided by the government, but most citizens have only 2 to 3 square meters to call home—not much more than standing room. Shanghai's Huangpu River is foul with the smell of raw sewage and industrial waste. Jobs are in short supply.

Cities in the developed world are much better off, but the signs of overpopulation are still prevalent. The slums of New York and Chicago house the poor and disadvantaged. The living conditions in these areas are deplorable. Crime runs rampant. Cities in America are badly polluted as well. In Los Angeles, air pollution from vehicles kills trees in the neighboring mountains, burns the eyes of visitors and residents, and blocks views. Rush-hour traffic clogs the highways (Figure 5-1). Teenage gangs rob, brawl, and vandalize. Jobless and homeless men and women rummage for food in dumpsters and garbage cans.

Crowding in urban centers has been implicated in a variety of social, mental, and physical diseases. Many

Figure 5-2. Overpopulation, desertification, and drought have diminished the resources available to these nomadic people, forcing them to live in shanty towns.

social psychologists assert that social instability, divorce, mental illness, drug and alcohol abuse, high prenatal death, and rising crime rates in densely populated cities are due, at least in part, to stress from overcrowding. They call this the **inner city syndrome**. Research on humans and animals supports their contention. The most notable study was performed by the psychologist John Calhoun. In his research rats were confined to a specially built room and allowed to breed freely. As population snowballed, Calhoun observed increased violence and aggression, abnormal sexual behavior, cannibalism of young, and disruption of normal nesting and maternal behavior. The physiologist Hans Selye performed similar experiments. He observed hormonal imbalances brought about by stress. These imbalances could lead to ulcers, hypertension, kidney disease, hardening of the arteries, and increased susceptibility to other diseases. Other researchers argue that factors other than stress, such as lack of education, poverty, poor nutrition, and inadequate housing, contribute to the inner city syndrome.

Scientists are still unsure how crowding in cities affects human populations. But this we do know: too many people, especially in high-density urban centers, can greatly exceed the capacity of the environment to assimilate wastes. Large urban populations also place considerable demands on the outlying countryside for resources such as fuel, water, and food.

In 1989, about 45% of the world's people lived in cities; by 2000, it is estimated, this figure will jump to slightly over 60%. Crowding, and the many problems that result from it, will probably worsen.

Rural Despair A surplus of people strains the limits of rural areas as much as it does those of cities. Bangladesh, Kenya, Ethiopia, Mexico, and other less developed nations reel under the oppressive burden of burgeoning rural populations (Figure 5-2). In Kenya, for instance, women have an average of eight children. Because of malnutrition and disease, hundreds of thousands of these children die each year.

The story is much the same in other parts of rural Africa and in Central America, where overpopulation leads to unsanitary living conditions, water shortages, food shortages, and disease. In Africa, trees are cut down for firewood and grasses are overgrazed, turning once-fertile land into desert.

In an ironic twist, rural overpopulation feeds the urban crisis. Dismayed peasants and their families, unable to survive on their farms, migrate in large numbers to the cities in search of jobs and security. In Nigeria between 1962 and 1985 the proportion of people living in urban areas increased from two in ten to over six in ten. In Mexico City an estimated 1.5 million rural farmers and their families arrive each year. Nearly every major city in the world serves as a magnet for disheartened rural residents. What awaits them is often worse than what they have left.

Reproducing Too Quickly

Hunger, starvation, disease, poverty, illiteracy, pollution, unemployment, and barren landscapes are, to many

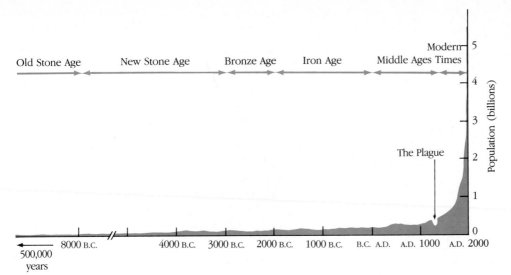

Figure 5-3. Exponential growth curve depicting world population. Note the rapid upturn in world population in the last 200 years.

observers, signs that the human population is much too big for the earth's resources. Efforts to eliminate these problems have been frustrated by an immediate problem: the continued rapid growth in population. Nowhere is this trend more evident than in Africa.

Ten years ago few African nations would admit to the need to control their rapidly growing population. Growing at a rate of 3%—doubling every 24 years—the African continent fell behind in food production and economic development. Food-exporting African countries suddenly became importers. Economic development fell in arrears. Disease continued to spread. Fortunately, many leaders now recognize their plight. The predicament they face is described by Paul Ehrlich in his classic book, *The Population Bomb*. Ehrlich writes:

In order to just keep the standard of living at the present inadequate level, the food available for the people must be doubled every 24 years. Every structure and road must be duplicated. The amount of power must be doubled. The capacity of the transport system must be doubled. The number of trained doctors, nurses, teachers, and administrators must be doubled. . . . This would be a fantastically difficult job in the United States— a rich country with a fine agricultural system, immense industries, and access to abundant resources. Think of what it means to a country with none of these.

The world's fastest-growing areas are, in decreasing order, Africa, Latin America, and Asia. In Latin America and Asia, at least, the rate of growth is declining; further declines are still necessary if countries are to stop the mounting despair. In many African nations, however, the rate of growth is increasing; the future of this great continent is in peril.

This survey of the overcrowded world shows a fundamental truth: that we generally live as we breed. Ecologists warn that unless we dramatically reduce population growth, many people will fall further and further behind.

Robert McNamara, president of the World Bank for 14 years, urged that "short of nuclear war itself, population growth is the gravest issue the world faces. . . . If we do not act, the problem will be solved by famine, riot, insurrection, and war." Solving the population, pollution, and resource "trilemma" requires a substantial and immediate downturn in population growth rate. Chapter 6 discusses growth control in more detail.

The Population Explosion

In the 60 seconds it takes you to find this book on your bookshelf and turn to this page, the world population increases by 172 people, a rate of 2.9 people every second!

At this rate of growth about 1.7 million people join the human population every week, approximately the number of people in Detroit. Overall, the world's population increases by 90 million people each year. In 1990 world population reached 5.3 billion. By the year 2000 it will very likely exceed six billion and may be growing by 100 million a year. Most of this growth takes place in the poorer nations, where three-quarters of the world's people now live.

The rapid growth of world population is a recent phenomenon. Throughout most of human history our population was quite small (Figure 5-3). During the time of Jesus, for example, there were probably fewer than 200 million people in the world. By 1850, though, world population had reached the one-billion mark, after which it began to grow much more rapidly (Table 5-1). In the following 80 years it doubled, reaching two billion by 1930. Forty-five years later it had doubled again, reaching four billion in 1975. By 2020 it is expected to reach eight billion.

What caused the rapid upsurge in population growth starting in the 1800s?

Figure 5-4. Birth rates (solid lines) and death rates (dashed lines) for two countries. In Sri Lanka a continued high birth rate and a drastic drop in the death rate resulted in rapid population growth. In Sweden the birth rate declined faster than the death rate, resulting in much slower population growth.

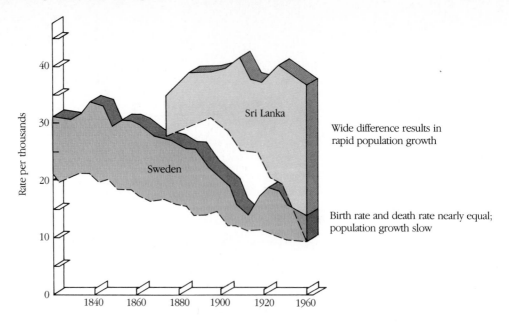

The Survival Boom

Throughout most of human history population was kept small by disease, famine, and war. These checks on population growth are some of the reduction factors that form part of environmental resistance (see Chapter 4). In the course of human history, though, several important changes have occurred, enhancing human survival and reproduction. These include the development of tools, the agricultural revolution, the Industrial Revolution, and the opening of the New World for settlement, most of which were discussed in Chapter 2. Each has helped to unleash human population growth.

Tools played a big part in the survival of our early ancestors. New tools and technologies had an enormous effect on human population during the Industrial Revolution in the 1700s and 1800s. Population began its rapid upturn at that time.

The modern industrial society, however, has made even greater strides in increasing survival. Modern medicine stands as one of the major advances of our time. New drugs, such as penicillin, dramatically lowered the death rate. Better health care and sanitation decreased infectious diseases and increased chances of survival. The pesticide DDT was used in the tropics to combat malaria-carrying mosquitoes. As a result of these developments death rates in many countries dropped precipitously (see the example of Sri Lanka in Figure 5-4) and life expectancy increased.

In 1900 the average white American female had a life expectancy of 50 years. By 1987, however, her life expectancy had climbed to 78.3 years. Life expectancy in white American males born in 1900 was about 47 years, but by 1987 it had increased to 71.5. Surprising at first, the increase in life expectancy is largely due to decreased infant mortality. Put another way, Americans live longer than their ancestors largely because more of them survive past the first year of life (Figure 5-5). Modern medicine has really helped more Americans survive those dangerous early years. Consider these statistics: between 1900 and 1978 the life expectancy for an infant increased 26 years, because more were saved in the first year; the life expectancy of a 60-year-old increased only 5.2 years.

As life expectancy increased and death rates dropped, birth rates continued at high levels in many countries (Figure 5-4). This fueled the rapid population growth that continues today.

A Double-Edged Sword: Expansion of the Earth's Carrying Capacity

Each of the advances cited above—tools, agriculture, medicine, and technological developments—increased the earth's **carrying capacity**, that is, the number of organ-

Table 5-1 World Population Growth and Doubling Time		
Population Size	**Year**	**Time Required to Double**
1 billion	1850	All of human history
2 billion	1930	80 years
4 billion	1975	45 years
8 billion (projected)	2017	42 years

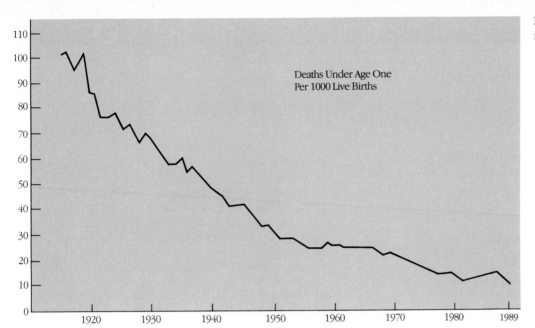

Figure 5-5. US infant mortality rate, 1915–1989.

Deaths Under Age One
Per 1000 Live Births

isms (in this case, humans) the earth can support.

All ecosystems have a specific carrying capacity for each population. Carrying capacity fluctuates from year to year depending on climate and other factors, as discussed in Chapter 3, but most organisms do little to change it. Humans are an exception to that rule: advances in tool making, agriculture, industry, and medicine allow us to extend the limits set by nature—to expand the carrying capacity.

To increase carrying capacity humans tend to develop to the maximum extent possible, or maximize, single variables such as energy and agricultural production. This is an approach that creates considerable problems. Modern agriculture and disease control, for example, increase the earth's carrying capacity. The result is a rise in human population, and more waste and pollution, which the earth cannot assimilate. Surely, we could feed even more people with further advances in agricultural technology, but the supplies of fresh water and clean air would be taxed even further. As Garrett Hardin notes, "Maximizing one [variable] is almost sure to alter the balance in an unfavorable way."

A survey of human history shows that increasing the earth's carrying capacity for our own kind has often been accomplished at the expense of the rest of the biosphere. The price has been depleted resources, increased pollution, and species extinction. The effects of such actions on ecosystems are the subject of much of this book.

Exponential Growth

Figure 5-3 showed human population growth over time. The curve is flat throughout most of human history; it is not until the last 200 years that population has veered upward. In that period population has followed a J-shaped curve called an **exponential curve**.

Exponential growth takes place when something increases by a fixed percentage every year. For example, a savings account with 5% compounded interest grows exponentially. The remarkable characteristic of all exponential growth is that early growth (in absolute terms) is quite slow. Once growth "rounds the bend," however, the thing being measured, whether money in a bank account or population, begins to increase more and more rapidly. For example, suppose that your parents opened a $1000 savings account in your name on the day you were born. This account earned 10% interest, and all the earned interest was applied to the balance, where it also earned interest. When you were 7, your account was worth $2000. By age 14, it had increased to $4000. By age 42 you would have $64,000 (Figure 5-6, page 113). If you left the money in a little longer, you'd find it growing faster and faster. At age 49 the account would be worth $128,000; at age 56 you'd have a quarter of a million dollars. Seven more years and you'd have half a million dollars, but if you waited until you were 70, your account would be worth over $1 million. Looking back over your account records, you'd find that during the first 49 years it grew from $1000 to $128,000, but during the last 21 years it increased by nearly $900,000. The rate of growth was constant over the entire period, and the account doubled every seven years, but it was not until it rounded the bend of the growth curve that things began to happen quickly.

Worldwide, population has been growing in the recent past at a rate of 1.8% per year. At this seemingly slow rate of growth world population doubles in just under 40 years. Making our situation worse, the numbers involved have become so large that each doubling means an

>◄ *Point/Counterpoint*

The Population Debate
The Case for More People

Julian L. Simon

The author is a professor of business administration at the University of Maryland at College Park. He has written several important books on population and economics, including The Economics of Population Growth, The Ultimate Resource, *and* The Economic Consequences of Immigration.

Many technological advances come from people who are neither well educated nor well paid—the dispatcher who develops a slightly better way of deploying taxis in his ten-cab fleet, the shipper who discovers that garbage cans make excellent, cheap containers for many items, the retailer who discovers a faster way to stock merchandise, and so on. Even in science one need not be a genius to make a valuable contribution.

In the past century there have been more discoveries and a faster rate of growth of productivity than in previous centuries, when fewer people were alive. Whereas we develop new materials almost every day, it was centuries between the discovery and use of, say, copper and iron. If there had been a larger population, the pace of increase in technological practice might have been faster.

Classical economic theory concluded that population growth must reduce the standard of living: the more people, the lower the per capita income, all else being equal. However, many statistical studies conclude that population growth does not have a negative effect on economic growth. The most plausible explanation is the positive effect additional people have on productivity by creating and applying new knowledge.

Because technological improvements come from people, it seems reasonable to assume that the amount of improvement depends in large measure on the number of people available.

Data for developed countries show clearly that the bigger the population, the greater the number of scientists and the larger the amount of scientific knowledge produced.

There is other evidence of the relationship between population increase and long-term economic growth: an industry, or the economy of an entire country, can grow because population is growing, because per capita income is growing, or both. Some industries in some countries grow faster than the same industries in other countries or than other industries in the same country. Comparisons show that in the faster-growing industries the rate of increase of technological practice is higher. This suggests that faster population growth, which causes faster-growing industries, leads to faster growth of productivity.

The phenomenon economists call "economy of scale"—greater efficiency of larger-scale production where the market is larger—is inextricably intertwined with the creation of knowledge and technological change, along with the ability to use larger and more efficient machinery and greater division of labor. A large population implies a bigger market. A bigger market is likely to bring bigger manufacturing plants, which may be more efficient than smaller ones and may produce less expensive foods.

A bigger population also makes profitable many major social investments that would not otherwise be profitable—railroads, irrigation systems, and ports. For instance, if an Australian farmer were to clear a piece of land far from neighboring farms, he might have no way to ship his produce to market. He might also have trouble finding workers and supplies. When more farms are established nearby, however, roads will be built that link him with markets in which to buy and sell.

We often hear that if additional people have a positive effect on per capita income and output, it is offset by negative impacts such as pollution, resource shortages, and other problems. These trends are myths. The only meaningful measure of scarcity is the economic cost of goods. In almost every case the cost of natural resources has declined throughout human history relative to our income.

Conventional wisdom has it that resources are finite. But there is no support for this view. There is little doubt in my mind that we will continue to find new ore deposits, invent better production methods, and discover new substitutes, bounded only by our imagination and the exercise of educated skills. The only constraint upon our capacity to enjoy unlimited raw materials at acceptable prices is knowledge. People generate that knowledge. The more people there are, the better off the world will be.

Is More Always Better?

Garrett Hardin

The author is a renowned environmentalist, writer, and lecturer. He has taught at several universities, and in 1980 he served as chairman of the board and chief executive officer of the Environmental Fund. He has written numerous books and articles on environmental ethics and is best known for his article "The Tragedy of the Commons."

To get at the heart of the "population problem," you should make careful measurements of the daily flow of water over Niagara Falls. You will discover that twice as much water flows over the falls during each daylight hour as during each nighttime hour. There in a nutshell you have the population problem.

Puzzled? You should be. The connection between Niagara Falls and population is not obvious. Before we can understand it we need to review a little biology.

For every nonhuman species there is an upper limit to the size of a population. Near the maximum, individuals are not so well off as they are at lower densities. Starvation appears. Crowded animals often fight among themselves and kill their offspring. Game managers and wildlife advocates agree that the maximum is not the optimum. If animals could talk, we suspect they would agree.

What about humans? Will we be happiest if our population is the absolute maximum the earth can support? Few people say so explicitly, but some argue that "more is better!"

Admittedly, we need quite a few people to maintain our complex civilization. A population the size of Monaco's, with about 25,000 people, could never have enough workers for an automobile assembly line. But Sweden, with some 8 million people, turns out two excellent automobiles. Eight million is a long way from nearly 5 billion, the approximate population of the world today.

Some say, "More people—more geniuses." But is the number of practicing geniuses directly proportional to population size? England today is 12 times as populous as it was in Shakespeare's day, but does it now boast 12 Shakespeares? For that matter does it have even one?

Consider Athens in classical times. A city of only 40,000 free inhabitants produced what many regard as the most brilliant roster of intellectuals ever: Solon, Socrates, Plato, Aristotle, Euripides, Sophocles—the list goes on and on. What city of 40,000 in our time produces even a tenth as much brilliance?

Of course, the free populace of Athens was served by ten times as many slaves and other nonfree classes. Forty thousand people free to apply themselves to intellectual and artistic matters is evidently quite enough to produce a first-rate civilization. Slaves gave 40,000 Athenians freedom to create. We are given creative freedom by labor-saving machines, certainly a more desirable form of slavery. But where are our geniuses?

Business economists are keenly aware of "economies of scale," which reduce costs as the number of units manufactured goes up. Communication and transportation, however, suffer from diseconomies of scale. The larger the city, the higher the monthly phone bill. More automobiles mean more signal lights per auto. More researchers mean more time spent in the library finding out what other investigators have done. Crimes per capita increase with city size. So do the costs of crime control. All these suggest that more may not be better.

Democracy requires effective communication between citizens and legislators. In 1790 each US senator represented 120,000 people; in 1980 the figure was 2.3 million. At which time was representation closer to the ideal of democracy? To communicate with his or her constituents each senator now has an average of 60 paid assistants (versus zero in 1790). As for the president, Franklin D. Roosevelt had a staff of 37 in 1933, when the population was 125 million, whereas Ronald Reagan had a staff of 1700 in 1981, when the population was 230 million. We have to ask whether democracy can survive unrestrained population growth.

Let's look at another aspect of the more-is-better argument. An animal population is limited by the resources available to it. With humans, a complication arises. Though the quantities of minerals on earth are fixed, improved technology periodically increases the quantity of resources available to us. In the beginning we had to have copper ores that tested out at 20%; now we are using ores with less than 1%. Available copper has increased but not the total amount of copper on earth.

Yet it is a mistake to speak of "running out" of mineral resources. When we use copper, we don't "use it up." Abrasion of our copper artifacts ultimately disperses the metal in tiny fragments over the face of the earth, but none of the copper is destroyed. It can all be gathered together again, *but* recon-

⧓ *Point/Counterpoint (continued)*

centrating it uses energy. All resource scarcity finally translates into a problem of energy. Is energy limited?

This is not an easy question to answer. The rate of energy input from the sun is strictly limited, but atomic energy is something else. In principle, atomic energy seems almost unlimited. The key question is this: can we get this energy at an acceptable environmental cost? Can we dispose of the lethal waste products safely? Can we prevent nuclear sabotage and blackmail? These questions are less scientific than trans-scientific—beyond science.

Let's return to Niagara Falls. Less water flows over the falls at night because more water is diverted to generate electricity when people aren't looking at the falls. It would be possible to use all the water to generate electricity, but then there would be no falls for us to look at. As a compromise, the volume of water "wasted" falling over the falls is reduced only at night. Therefore, the turbines and generators are not fully used 24 hours of the day, which means that local electricity costs are just a bit higher.

If the population continues to grow, the day may come when electricity is so scarce and expensive that the public will demand that Niagara Falls be shut down so that all the water can be used to generate electricity. Similar dangers face every aesthetic resource. Wild rivers can be dammed to produce more electricity, and estuaries can be filled in to make more building sites for homes and factories.

The maximum is never the optimum. With human populations, *quantity* (of people) and *quality* (of life) are tradeoffs. Which should we choose—the maximum or the optimum? This is a trans-scientific question.

Is our population now below or above the optimum? In trying to answer this I suggest that you make two lists. On one list write down all the things that you would expect to be *better* if the population doubled; on the other, all the things that would be *worse*. On which list would you put the availability of wilderness? Of theatres and museums? What about the noise level? Amount of democracy? Amount of pollution? Per capita cost of pollution control? Availability of parking spaces? Personal freedom?

When you are through, compare your list with your friends'. What value judgments account for the differences? Can these differences be reconciled? How?

Critical Thinking

1. Carefully outline each of the authors' arguments. Are their views consistent with reality? What critical thinking rules do they violate? Do the poor, developing countries benefit from more and more people? What about West Germany and Sweden, where population growth has stabilized? Are they suffering from the lack of growth?

2. How does Garrett Hardin answer Simon's assertions? Which man's views do you agree with? Why? Are your reasons based on feelings or facts?

3. What points does Simon make about resources? How does his economic training affect his views? How does Hardin respond? What do you think about Hardin's views?

---

extraordinary increase in the number of people. Resource demand and pollution also increase exponentially. So fast has the population begun to increase in sheer numbers that civilization is hard pressed to keep up with it. As we will see in Part 3 of this book, the limits of many resources are in sight. As Part 4 shows, the atmosphere, oceans, and rivers cannot assimilate all the wastes we produce.

Understanding Populations and Population Growth

The newspapers and magazines sometimes present a confusing array of statistics on population, making it difficult for the public to understand a global issue that ranks in importance with nuclear war. With a little effort you can learn enough **demography**, or population science, to weed out the fallacies from the facts of the population debate.

Measuring Population Growth

One of the most important population measurements is growth rate, which is frequently expressed as a percentage. What does it mean when demographers report that a population is growing at a rate of 3% per year? Is this something to be thankful for or concerned about? To answer this question, let's first see how **population growth rate** is determined.

Growth Rate World population growth rate is calculated by subtracting the number of deaths in a population from the number of births in any given year. Multiplying by 100 converts the growth rate into a percentage, as shown here.

Growth Rate (%)
$$= (\text{Crude Birth Rate} - \text{Crude Death Rate}) \times 100$$

Crude birth rate is the number of births per 1000 people in a population. **Crude death rate** is the number of deaths per 1000 people in a population.

The relationship between birth rates and death rates determines whether the world's population grows, shrinks, or remains the same. Some important growth rates are listed in Table 5-2.

What do the numbers in Table 5-2 really mean? A look at the growth rate of the developing nations, the earth's poorest, is instructive. The annual growth rate of these countries is 2.1%. A 2.1% growth rate means that a population is adding 2.1 persons each year for every 100 in the population. On the surface this seems rather small. These Third World nations, however, have, 3.9 billion people. A 2.1% growth rate, therefore, translates into 82 million *additional* people every year!

Birth Rates The balance between birth rates and death rates in various areas determines global population growth. Each of these rates, however, is affected by a variety of elements. The birth rate in a given population, for instance, depends on (1) the age at which men and women get married, (2) their educational level, (3) whether the woman works after marriage, (4) whether the couple uses reliable contraceptives, (5) the number of children the woman and her husband want, (6) the couple's religious beliefs, and (7) their cultural values.

Death Rates Death rates are equally important in determining population growth. A few decades ago death rates in the developing nations of the world were quite high. Few children lived past the first year. Therefore, even though birth rates were high, population growth remained low. Today, modern medicine, pesticides, and better sanitation, among other things, have reduced death rates. Without an accompanying fall in birth rates, the reduction in death rates has led to an unprecedented population growth in the poorer nations. Many developed nations have stabilized their growth by lowering birth rates.

Fertility and Zero Population Growth Demographers employ dozens of measurements to assess the dynamics of the populations they study. One of the most important is the total fertility rate, which is used to predict the growth of populations. The **total fertility rate** (TFR) is the number of children women are expected to have in their lifetime. The total fertility rate in the United States is 1.8; that means that ten women are expected to have 18 children. Canada's TFR is 1.7. In India, a much poorer nation, women have a TFR of 4.3, or 43 children born to every ten women.

One of the most encouraging population trends is that many countries have reached **replacement-level fertility**, the point at which couples produce exactly the number of children needed to replace them (Tables 5-3 and 5-4).

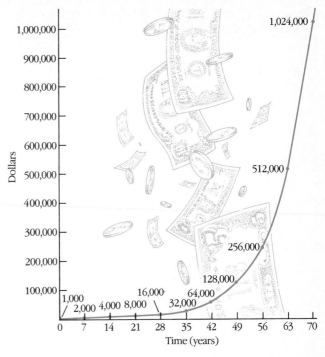

Figure 5-6. Exponential growth curve of a savings account with an initial deposit of $1000 and an interest rate of 10%.

Table 5-2	Growth Rate and Doubling Time in 1989	
Region	**Growth Rate (%)**	**Doubling Time (years)**
World	1.8	39
Developed countries	0.6	117
Developing countries	2.1	33
Africa	2.9	24
Asia	1.9	38
North America	0.7	99
Latin America	2.1	33
Europe	0.3	233
Soviet Union	1.0	70
Oceania	1.2	55

Source: Population Reference Bureau, *World Population Data Sheet, 1989.*

In the United States and other developed countries replacement-level fertility is a TFR of 2.1 children. That means that ten women must have 21 children to replace themselves and their husbands. Why the additional child? Because one of every 21 children dies before reaching reproductive age. Replacement-level fertility is higher in nations with higher death rates.

The press is forever confusing the world population picture, and for good reason: an astounding number of measurements have to be understood before one can assess a population's future. Consider the confusion that

Table 5-3 Total Fertility Rate in Mid-1989

Region		Total Fertility Rate	Population Size (Billions)
World		3.6	5.234
Developed countries		1.9	1.206
Developing countries		4.1	4.028
Africa		6.3	
Northern	5.6		
Western	6.6		
Eastern	6.9		
Middle	6.0		
Southern	4.7		
Asia		3.6	
Western	5.3		
Southern	4.8		
Southeastern	3.8		
Eastern	2.3		
(incl. China, Japan)			
North America		1.9	
US	1.7		
Canada	1.9		
Latin America		3.6	
Central	4.1		
Caribbean	3.0		
Tropical	3.6		
Temperate	2.9		
Europe		1.7	
Northern	1.8		
Western	1.6		
Eastern	2.1		
Southern	1.6		
Soviet Union		2.5	
Oceania		2.7	

Table 5-4 Countries at or Near Replacement-Level Fertility in 1989

Country	Total Fertility Rate
Sri Lanka	2.5
China	2.4
Reunion	2.4
Chile	2.4
Soviet Union	2.4
Ireland	2.3
Uruguay	2.3
Guadeloupe	2.3
Cyprus	2.3
Romania	2.3
Poland	2.2
South Korea	2.1
Puerto Rico	2.1
Martinique	2.1
Netherlands Antilles	2.1
Bulgaria	2.1
New Zealand	2.0
Czechoslovakia	2.0
Iceland	2.0
Portugal	2.0
Yugoslavia	2.0
Barbados	1.9
Australia	1.9
United States	1.9
Sweden	1.9
Bahamas	1.9
Cuba	1.8
France	1.8
United Kingdom	1.8
Hungary	1.8
Norway	1.8
Antigua and Barbuda	1.7
Japan	1.7
East Germany	1.7
Canada	1.7
Greece	1.6
Malta	1.6
Singapore	1.6
Netherlands	1.6
Finland	1.6
Spain	1.5
Belgium	1.5
Switzerland	1.5
Denmark	1.5
Austria	1.4
Luxembourg	1.4
West Germany	1.4
Italy	1.3
Hong Kong	1.3

Source: Population Reference Bureau, *World Population Data Sheet*, 1988.

arises over total fertility rate. Even though a population reaches replacement-level fertility, it does not mean that it has stopped growing. A population stops growing only when the death rate equals the birth rate (and the net migration is zero). Demographers call this **zero population growth**, or ZPG. The United States provides a good example. The TFR in the United States fell below replacement level in 1972, thanks in large part to modern contraception and widespread public awareness of the population dilemma. Even though the TFR has remained below replacement-level fertility since that time, the US population has continued to swell. And, demographers project, at least 70 years will pass before our population stops growing. Why?

The answer is really quite simple. The population continues to grow for two reasons: First, each year numerous people move to the United States from other countries. Some of the people come legally; others, however, arrive

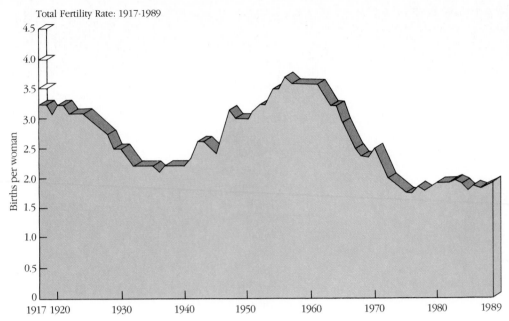

Total Fertility Rate: 1917-1989

Figure 5-7. Total fertility rate is a predictive measure; it tells us the number of children women are expected to have based on the current age-specific fertility rates (fertility rates in the current age groups). The total fertility rate in the United States has fluctuated widely with economic conditions. In the decade of the Great Depression women had a low total fertility rate. In the several decades following World War II, the total fertility rate shot up; in 1972 it fell below replacement level, where it has remained.

illegally. Legal and illegal immigration adds at least one million people (probably much more) each year to our population and is responsible for about 40% of our annual growth.

The rest of the growth rate comes from within. Figure 5-7 shows the total fertility rate for the United States going back 70 years. After World War II, the total fertility rate climbed dramatically and remained high for several decades, creating a **postwar baby boom**, a rapid rise in fertility which resulted in a fairly sustained increase in US population. The children of the baby-boom era are a tidal wave in the American population. Many of these children are now having children. Even though they are having, on average, only 1.8 children each, the number of women actually giving birth is still on the rise. In other words, because of the post-war baby boom, the reproductive age group (ages 15–44) continues to expand. More women are having children.

Demographers believe that the US population, fast approaching 250 million, will increase by 50 to 60 million before stabilizing. This delayed slowing is called the **lag effect**. Like an oil tanker that glides for a mile or two before coming to a stop, US population continues to grow long after replacement-level fertility was reached.

Stabilization is possible only if the number of women entering the reproductive age group levels off, if the total fertility rate remains below replacement level, and if immigration does not climb substantially. Small changes in any of these factors could have drastic effects on US population. For example, Susan Weber, executive director of a nonprofit organization called Zero Population Growth, says that if abortion becomes illegal the total fertility rate could climb to 2.2, slightly above replace-

ment-level fertility. In that case, US population would never stabilize.

A *Comparison of Growth Rates* The world is crudely divided into the haves and the have-nots—the **developed countries** and the **less developed, or developing countries** (Table 5-5). On the whole the developed countries, such as the United States, Canada, Great Britain, and the Soviet Union, grow fairly slowly, on the average about .5% per year (doubling time = 140 years). A number of developed countries, such as Sweden and Austria, have stopped growing altogether. Others are experiencing negative growth rates, meaning their populations are shrinking. The developing nations, in contrast, are growing rapidly—at an average of 2.1% per year (doubling time = 33 years). But these averages hide some dangerous trends in countries such as Kenya and Uganda, which are growing well over 3% per year.

Doubling *Time* Growth rates, total fertility, and replacement-level fertility are important measurements, but a better perspective on population growth can be gained by using the growth rate to calculate the time it takes a population to double. The following formula is used to determine **doubling time**:

$$\text{Doubling Time} = \frac{70}{\text{Growth Rate (\%)}}$$

Overall, the human population is growing at 1.8% per year. This gives a doubling time of just under 40 years. What is the doubling time of the developed nations (growth rate of .6%)? What is the doubling time of the less developed nations (growth rate of 2.1%)?

Table 5-5 Comparison of Developed and Developing Countries

Feature	Developed	Developing
Standard of living	High	Low
Per capita food intake	High (3100–3500 cal/day)	Low (1500–2700 cal/day)
Crude birth rate	Low (15/1000 population)	High (33/1000 population)
Crude death rate	Lower (10/1000 population)	Higher (12/1000 population)
Growth rate	Low (0.6%)	High (2.1%)
Doubling time	High (120 years)	Low (33 years)
Infant mortality	Low (20/1000 births)	High (90/1000 births)
Total fertility rate	Replacement level (2.0)	High (4.6)
Life expectancy at birth	High (72 years)	Lower (57 years)
Urban population	High (69%)	Low (26%)
Wealth (per capita GNP) (1985 US dollars)	High ($9,930)	Low ($660)
Industrialization	High	Low
Energy use per capita	High	Low
Illiteracy rate	Low (1%–4%)	High (25%–75%)

Source: Population Reference Bureau.

Migration

The previous section dealt primarily with global population, determined by the balance of death rates and birth rates. To calculate the growth of individual countries, states, or regions, demographers must take into account the number of people moving out of and into the region as well as the balance (or imbalance) between births and deaths.

Migration, technically speaking, is the movement of people across boundaries to set up a new residence. The term **immigration** refers to movements into a country; **emigration** refers to movements out. **Net migration** is the difference between immigration and emigration. Population growth in a country will stabilize if the growth rate and net migration are zero.

Immigration is one of the hottest topics today in the United States, for many reasons. One of the most important is that legal and illegal immigration into the United States accounts for 40% of the annual population growth. Efforts to stabilize US population will fail unless the coun-

try brings immigration and emigration more closely into balance.

Recent public opinion polls in the United States show that a majority favors reducing immigration of all kinds. Accordingly, Congress passed a law in 1986 levying penalties against those who knowingly employ illegal immigrants and requiring job seekers to provide proof of citizenship or legal immigration status. The law is controversial, and solutions to this problem are not easy or clear-cut.

People also move *within* the boundaries of a country, often in large numbers. Such migrations can dramatically alter local economies. Figure 5-8 shows the recent migration flow in the United States. The primary flow is from the northeastern and north-central regions to the South (mostly Florida) and West (Arizona, Wyoming, Utah, Alaska, Idaho, Colorado, New Mexico, and Texas). The migration has been so dramatic in recent years that for the first time in US history more than half the population lives in the West and South. This movement, called the

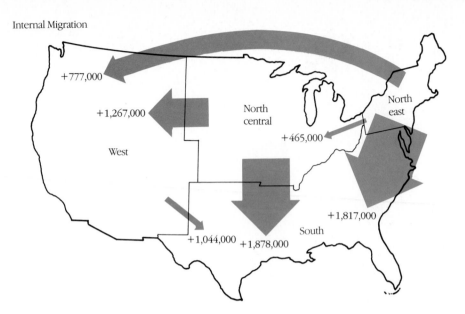

Internal Migration

Figure 5-8. Internal migration in the United States, 1975–1980, a period of extremely frequent migration. The migration has slowed somewhat but still creates many problems.

"sunning of America," is caused by (1) expanding (sunrise) industries in the West and Southwest, especially in electronics and energy production; (2) declining (sunset) industries of the northeastern and north-central states, especially auto and steel manufacturing; (3) a desire for a warmer climate; (4) a desire for a lower cost of living; (5) a preference for abundant recreation; (6) an aspiration for a less hectic, less crowded life-style; and (7) the growth of retirement communities.

The massive migration to the South and West has had many important economic benefits for these regions. Overall, income has increased and unemployment has fallen. New opportunities have opened up for builders and for restaurant owners, bankers, and other members of the service sector of the economy.

Ironically, the influx of people into the Sunbelt has destroyed many of the values the migrants had sought. With growth rates in the range of 3% to 5% per year, many western cities and states have been swamped by new residents. Some major western ski areas have become so overcrowded that skiers must make reservations days in advance. Air pollution in Sunbelt cities has worsened with increasing population; the clean air that many came for has disappeared. Traffic has become congested. The rapid rise in demand for housing has sent the cost of homes skyward. The breakneck speed of growth has made it difficult for local governments to provide water, schools, sewage treatment facilities, and the like. Spreading cities engulf smaller outlying communities and change the tempo of life. (See Case Study 5-1 for a discussion of problems facing San Diego.)

Internal migrants affect the places they come from as well. Since many of the out-migrants are young, educated, and skilled workers, they have created a significant drain on human resources in the northeastern and north-central United States.

Seeing Is Believing: Population Histograms

Demographers use growth rates, fertility rates, and doubling times to explain the future of the human population, but few techniques shed as much light as a nicely drawn graph. The old saying "A picture is worth a thousand words" is as true in demography as it is in anything else.

You have already seen a graph of exponential growth and pondered the consequences of such growth. What you learned was that rounding the bend of an exponential growth curve yields a remarkable growth in the population even though the growth rate remains small. Let's look at another graphic tool, the population histogram. This is a bar graph that tells a little about the history and a little about the future of a population (Figure 5-9).

The **population histogram** displays the age and gender composition of a population. The area of each horizontal bar on the histogram represents the size of a certain age and gender group. As illustrated in Figure 5-10, three general profiles exist: expansive, constrictive, and stationary. Morocco, a country which is expanding, has a large number of young people. If they produce more offspring than their parents did, the population will continue to expand at the base. If, on the other hand, family size decreases, the base of the histogram will begin to constrict. This is what is happening to the US population. Sweden presents an entirely different picture. For many years Swedish couples have been having the same number of children as their parents had; as a result, Sweden's population is stationary.

Population histograms can change over time and, therefore, are not always reliable for making long-term predictions of future growth. Expansive populations can become constrictive, constrictive populations can become expansive, and so on.

Histograms give a snapshot view that can be used for

Figure 5-9. US population by age and gender.

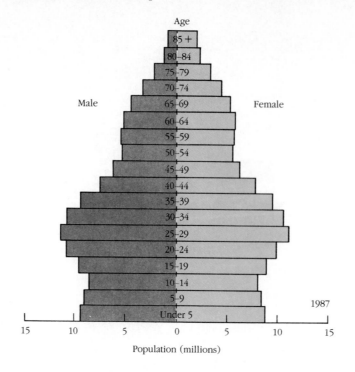

Figure 5-10. Three general types of population histogram. Expansive populations have a large percentage of young people and are expanding at the base. Constrictive populations have a tapering base, resulting from lower total fertility. Stationary or near stationary populations show no expansion or constriction. This results from couples having only the number of children that will replace them.

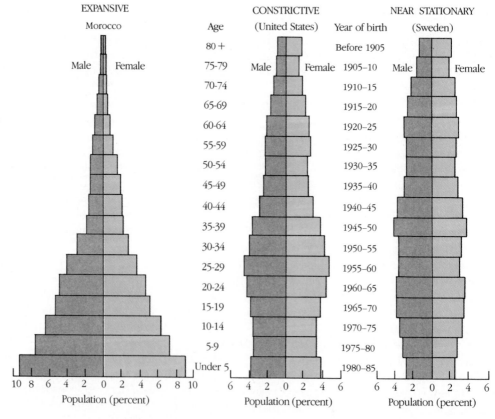

Case Study 5-1

Frontierism in San Diego: Seeking the Good Life in Southern California

This is a story of San Diego from 1970 to 1989. But you could change the name to that of a dozen or more cities and the story line would stay the same. What's happening in San Diego is happening in Tucson, Phoenix, Los Angeles, Miami, Boston, southern Maine, and Denver. It's called growth and it's creating headaches. Enormous headaches.

In twenty years, San Diego County in southern California has experienced a growth surge. The population has swelled from 1.4 million to 3.2 million. Today 180 people move into the county every day. Many believe that the growth will taper off some, but will continue for years to come. Developers and business interests view the growth with excitement, but many residents think differently. Why?

Increasing population has brought with it many problems. One of the most significant changes is a dramatic increase in drugs and guns. Drive-by shootings were virtually unheard of 20 years ago. In 1988, there were nearly 100 such killings. In 1970 there were 1500 reports of aggravated assault. Today there are nearly 11,000 a year. Auto thefts are on the rise too. In 1988 one of every 40 automobiles was stolen, a rate three times higher than in 1970. The number of jails and prisoners has doubled in the past 20 years and, because jails are over-crowded, many criminals are set free.

Consider another problem: traffic. Twenty years ago there were 360 kilometers (215 miles) of highways and traffic was a breeze. Today, 1.6 million registered vehicles cram 480 kilometers (290 miles) of highways, making commuting a nightmare.

All these people and all of their automobiles create significant air pollution problems as well. Although the number of days the city violates the federal smog levels is down considerably, the county still violates federal standards 45 days of each year. What is more, automobile travel is expected to increase substantially in the next decade, further degrading air quality. Officials believe that the present downward trend in automobile emissions will reverse itself by 1995. Officials

are considering tighter emission controls, mandatory car-pooling, and staggering work hours to help battle this difficult problem. Making matters worse, much of the pollution that plagues the city is imported from nearby Los Angeles, another area that's growing rapidly.

Wildlife has suffered enormously as the human populations swell. Twenty-six animals and plants in the county are on the endangered species list.

The rapid growth of population has resulted in an equally dramatic increase in trash. In the past twenty years, trash production has more than doubled. Living more extravagant life-styles, San Diego residents now produce, on average, 1.5 tons of garbage a year—up from under one ton per person in 1970.

Perhaps the most startling fact about San Diego is that it is in a desert, where it has virtually no water. Only 10% of San Diego's water comes from local sources; the rest is imported through an extensive set of pipes and canals. With new sources of water increasingly difficult to find, city officials are looking at ways of reclaiming sewage—purifying sewage—to make it usable for domestic consumption.

One final item of interest is the cost of housing. Twenty years ago the average new home cost $23,000. Today, residents camp out overnight for the chance to bid on a $300,000 town-home. The average new home is priced at $152,000—hardly affordable for most people. By 2010, a new house is expected to cost $1.3 million. What all this means is that real estate developers and builders will get rich, and that the rapid influx of people and the rising demand for limited resources will continue to fuel the spiraling increase in housing prices.

San Diego is clearly a city under siege. It's suffering from the stresses of excessive population size and rapid population growth. For many residents, the situation seems out of control. Many of the amenities that brought them to San Diego are rapidly falling by the wayside as the city and county swell with people seeking the good life in Southern California.

---

short-term planning of schools, hospitals, retirement homes, and so on. Mexican officials, studying their his-tograms, might make special efforts to expand educational facilities and maternity wards. The United States might look for ways to increase retirement homes and services for the elderly.

Shifts in the population histogram can have dire consequences if ignored. In the United States the postwar baby boom, caused by an increase in fertility (seen as a bulge in Figure 5-10, a population histogram) greatly

increased the number of school-age children in the 1950s and 1960s. New schools couldn't be built fast enough to keep up with the pace. Teachers were in short supply. Housing also became a serious problem as members of this age group married and searched for homes. Still larger problems loom in the early 21st century, when those people reach retirement age. By 2030 individuals 65 and older will make up about 20% of the US population, compared with 11% today. Health costs are almost certain to go higher, since people over 65 today account

Figure 5-11. (a) Future of world population. Constant path indicates population growth continuing at the current rate. The middle line indicates the future of world population if fertility drops off at a slow rate. The bottom line shows what might happen with a rapid decrease in fertility. As discussed in the text, the future may lie somewhere between the lower two lines. (b) The future of US population depends on future total fertility rate and annual net migration.

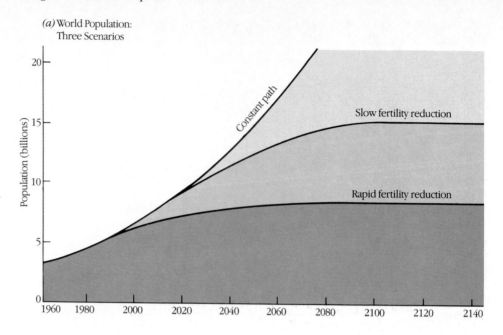

(a) World Population: Three Scenarios

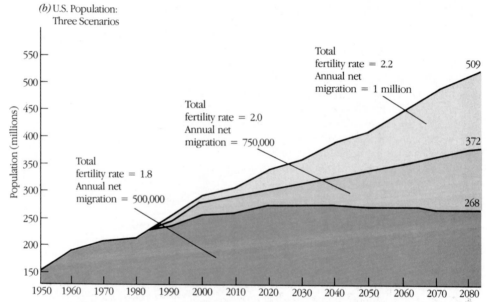

(b) U.S. Population: Three Scenarios

for nearly one-third of the nation's health bill. Gerontologists, new hospital facilities for the aged, and retirement homes will be needed to accommodate the elderly. Making matters worse, a smaller proportion of the population—including those in college today—will be supporting this large elderly group through the Social Security program.

The aging of the population also means an aging of the work force, which has positive aspects. It means that competition for entry-level positions will lessen, and it means that the labor force will, on the whole, be more experienced and will require less training. This could increase overall labor productivity—the amount of output per unit of labor.

The Future of World Population: Some Projections

Perhaps the questions most often asked of demographers are "What is the future of the world population? How big will it get? Will it really double in the next 40 years?"

We don't know and can't know the answers, for it is impossible to predict the future. The factors that affect future population growth, such as birth rates, fertility, and death rates, can change dramatically.

Regardless of the uncertainties involved in making estimates of future growth, many scientists have forecast growth. The world population growth rate is currently about 1.7%. If this rate continued into the future, the

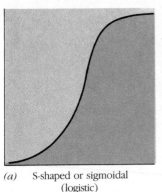 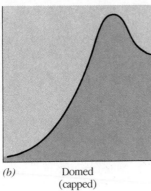 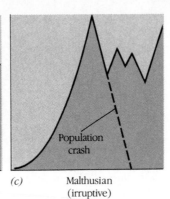

(a) S-shaped or sigmoidal (logistic)

(b) Domed (capped)

(c) Malthusian (irruptive)

Figure 5-12. Patterns of population growth. (a) Stabilization of population growth with a smooth transition to a stable population size. (b) Gradual dropoff caused by population exceeding the carrying capacity. (c) Population crash caused by irreparable change to the ecosystem. No doubt all three patterns will be seen in individual countries.

population of the world would reach one trillion (1,000,000,000,000) around 2300.

Long before then, demographers believe, the population growth curve will level off as a result of dropping birth rates, rising death rates, or both. Figure 5-11 plots this and several other possibilities for the United States and the world. A slow reduction in fertility would result in a gradual decrease in population growth and a stable population by the year 2100. The population size would be about 15 billion, three times what it is today. Still another projection, based on a rapid decline in fertility, shows stabilization around 2060 at 8 billion, 3 billion more than today. Exactly where the curve levels off will depend on the persistence and success of efforts in family planning and economic development in the Third World, as discussed in the next chapter.

Some demographers predict that shortages and pollution brought about by exponential growth will eventually cause the human population to begin to decrease in size, producing a dome-shaped population curve instead of a **sigmoidal curve** (S-shaped, characteristic of a stabilizing population), as shown in Figure 5-12. The decrease in population size in this scenario, ecologists believe, would occur when populations exceeded the earth's carrying capacity. The introductory section of this chapter pointed out many ways in which human populations have already pushed their numbers beyond the carrying capacity. Further pollution of the air and water; further shortages of food, water, wood, grazing land, and other natural resources; or continued extinction of plants and animals could cause the population size to decrease over time—or even quite suddenly.

Still another, ominous pattern may occur. Instead of leveling off or declining smoothly the human population might suffer sudden die-offs followed by spurts of growth and further plunges. This roller coaster ride is an unstable, **irruptive** pattern caused by severe ecosystem imbalance (Figure 5-12). Finally, the human population may plummet quite suddenly in a population crash caused by critical changes in global climate or, possibly, resource supplies. This sudden catastrophic die-off may be so severe that the human population becomes extinct or is reduced to only a fraction of its present size.

No one knows what will happen to world population. The only opinion on which the experts approach unanimity is that world population cannot grow indefinitely. Some countries will inevitably make a smooth transition to a stable state; others may experience periodic crashes caused by epidemics in crowded urban populations; others may fall to low population levels because of continued starvation and disease.

H. G. Wells once wrote that "human history becomes more and more a race between education and catastrophe." The population–resource–pollution bind we're in clearly illustrates this fact. What we decide today will have far-reaching effects that determine the kind of lives our children and theirs will have.

The doors we open and close each day decide the lives we live.

Flora Whittemore

Summary

Population growth is at the root of virtually all environmental problems, including pollution and resource depletion, and indirect social disruptions such as housing shortages, malnutrition, and inadequate health care. Rapid growth in population creates difficulties in meeting the basic needs of people. Crowding may cause mental illness, drug abuse, and various forms of antisocial behavior.

The rapid growth of world population is a recent phenomenon. Throughout most of human history our population was small, controlled by war, disease, and famine. But advances in agriculture and industry allowed us to indefinitely expand the **carrying capacity**, the number of people the earth could support.

Exponential growth, or fixed-percentage growth, follows a **J curve**. Initially, growth is slow, but once the bend of the curve is rounded, the absolute growth becomes remarkable. The environmental crisis is largely the result of our beginning to grow exponentially in recent history and then rounding the bend of the population curve.

World population growth is determined by subtracting the **crude death rate** from the **crude birth rate**. These are the number of deaths and births per 1000 people, respectively. **Doubling time**, the time it takes a population to double in size, is also used as a measure of population growth. The **developing countries**, those with low per capita income, inadequate education, and poor nutrition, contain three-fourths of the world's population and double about every 33 years. **Developed countries**, the rich, well-fed, and well-educated nations, double on the average every 140 years. Population growth in individual nations is determined by calculating the number of births minus deaths plus the **net migration**—the difference between **emigration** and **immigration**.

The most important measure of reproductive performance is the **total fertility rate** (TFR), the average number of children a woman will have. In developed countries a TFR of 2.1 is **replacement-level fertility**. Replacement-level fertility is one of the first steps toward achieving **zero population growth**.

Population histograms are an important tool for studying populations, because they show the proportion of males and females in a population in various age groups. The shape of the histogram suggests whether populations will grow, shrink, or remain stable.

The future of world population is difficult to determine, because many factors such as total fertility rate can change. Many demographers believe that world population will stabilize somewhere between 8 and 15 billion and that the exponential growth curve will be converted into a **sigmoidal curve**. However, other possibilities are also likely. The exponential growth curve may be transformed into a **dome-shaped curve** as population levels fall due to overextension of the carrying capacity. Or the pattern may become **irruptive**, a series of peaks and chasms. Finally, severe damage to the environment caused by exceeding the carrying capacity could result in a devastating crash, virtually eliminating humans from the face of earth. The pattern that population will assume is largely dependent on our efforts to control it, especially in the developing nations.

Discussion Questions

1. What is the population of the world? What is the population of the United States?

2. Is the rapid growth in world population of concern to you? Why or why not? How does population growth in the United States affect your life?

3. How many years did it take the world to reach a population of one billion people? How quickly did we reach the second, third, and fourth billions?

4. What factors kept world population in check for so many years? Discuss the advances that have unleashed population growth in the last 200 years.

5. Define the term *exponential growth rate*.

6. What is a population histogram? Describe the three general profiles. What information do population histograms give us?

7. How is the world population growth rate calculated?

8. Define replacement-level fertility and zero population growth.

9. Discuss the pros and cons of a lenient policy toward legal and illegal immigration in the United States. Do you favor strong immigration quotas?

10. Describe internal migration patterns in the United States. What factors account for them, and what are the impacts?

11. Debate the statement "Population is the cause of all the world's problems."

Suggested Readings

Brown, L. R. and Jacobson, J. L. (1986). *Our Demographically Divided World*. Worldwatch Paper 74. Washington, DC: Worldwatch Institute. Detailed analysis of world population. Valuable reading.

Conservation Foundation (1987). State of the Environment. A View Toward the Nineties. Washington, DC: Conservation Foundation. Well-documented report on the environment. Covers US population growth in Chapter 1.

Ehrlich, P. R. (1971). *The Population Bomb*. New York: Ballantine Books. A book that startled the world, telling of the hidden dangers of overpopulation. Dated but worth reading.

Grant, L. (1983). The Cornucopian Fallacies: The Myth of Perpetual Growth. *The Futurist* 17 (4): 18–22. Important article rebutting optimists who hold that there are no limits to growth in population and resource acquisition.

Gupte, P. (1985). *The Crowded Earth*. New York: Norton. Extraordinarily readable account of overpopulation and the effects of population control. A must!

Hardin, G. (1981). An Ecolate View of the Human Predicament. *Alternatives* 7: 241–262. Superb paper on overpopulation and carrying capacity. Exceptional insights.

Haub, C. and Grant, L. (1983). *Whatever Happened to the Population Bomb?* Washington, DC: The Environmental Fund. Interesting answer to an important question.

Haupt, A. and Kane, T. T. (1985). *Population Handbook*. Washington, DC: Population Reference Bureau. A superb primer.

Mayer, J. (1984). Food, Nutrition, and Population. In *Sustaining Tomor-*

Case Study 5-2

An Eye on the Experts: Sharpening Your Critical Thinking Skills

The future holds terrific promise and frightening uncertainty. Most of us are filled with questions about the future. How many people will be alive in the year 2000? How will we live? What will happen to wildlife, oil supplies, nuclear power, and the global climate?

The most honest answer is that no one knows for certain. This is especially true for population. Conditions change so rapidly and often so unpredictably, that no one can say for certain what the future holds.

Critical thinking demands that we be wary of projections. In many cases they turn out to be dead wrong. Nevertheless, environmental science deals at considerable length with the future, trying to determine how today's actions will affect our descendants and the many other species that share this planet with us. A lot of what you will read about—even in this book— is based on projections. We need projections to help society chart a path that is sustainable.

Perhaps one of the leading second-guessers is the US Census Bureau, a federal agency that keeps track of important demographic data. Unfortunately, the experts at the Census Bureau, who spend their days studying the present and predicting the future of our population, are often quite wrong. For example, the bureau projected that during the 1970s 40 million to 49.3 million children would be born in the United States. The actual number of births was far below this at 33.3 million, a sign that US population growth is slowing. The Census Bureau over-estimated deaths during that same period.

The experts missed again on projections of average life expectancy. Life expectancy at birth, how long a person will live on average, was expected to increase in the 1970s by about one half of a year. Instead, it rose 3.4 years because of success in reducing deaths from heart disease and strokes.

The Census Bureau also projected a slight decrease in the number of people per household. Its estimates showed that by 1980 the average household would contain slightly more than 3 people. The actual value, however, was 2.76 people per household, considerably lower than estimated.

Another area of speculation was internal migration in the United States. The US Bureau of Economic Analysis, an agency similar to the Census Bureau, made projections on regional growth in a seven-volume report. It projected that the Northeast would grow by 7.6 million people between 1970 and 1980; in fact, it grew by only 1.1 million because of a much larger than expected out-migration. The West, on the other hand, was expected to grow by four million during the decade of the '70s. It grew by 8.3 million. The South also fooled the experts, growing by 12.5 million instead of a projected 7.8 million.

The government isn't alone in its inability to glimpse accurately into the future. Utility companies that supply us with power are similarly fallible. A simple mistake in projecting electrical demands in the Pacific Northwest resulted in the country's largest bond default—the failure of a bonding company to make good on bonds it issued to build a series of nuclear power plants, discussed in Chapter 12.

The need for projections is bound to continue, making it even more important that we project accurately. City planners need to know whether or not populations will grow or decline to make intelligent plans. New housing, power plants, water-supply systems, sewage treatment plants, and other facilities needed by a growing region require lead time: time to plan, finance, and build. A new power plant may have a lead time of ten years. City planners are not alone in their need for accurate projections. Businesses also like advance knowledge of population trends to better meet market demands.

But as it is, our system plods along basing important decisions on often faulty predictions. Critical thinking suggests that we deal with projections with a healthy dose of skepticism. Examine the underlying suppositions and biases that go into projections and you can often find out why they go wrong. A rigorous examination of these factors can help us fine-tune our powers of projection.

row: A Strategy for World Conservation and Development, eds. F. R. Thibodeau and H. H. Field. Hanover, NH: University Press of New England. Good summary of overpopulation and its effects.

Myers, N. (ed.). (1984). *Gaia: An Atlas of Planet Management.* New York: Anchor Books. Good reference on population and its impacts.

Population Reference Bureau (1982). U.S. Population: Where We Are; Where We're Going. *Population Bulletin* 37 (2): 1–51. A fact-filled treatise on US population—past, present, and future.

Population Reference Bureau (1989). *1989 World Population Data Sheet* and *United States Population Data Sheet.* Washington, DC: Population Reference Bureau. Published annually with current statistics on US and world population.

Simon, J. L. (1983). Life on Earth Is Getting Better, Not Worse. *The Futurist* 17 (4): 7–14. Recommended reading, along with Grant's rebuttal (above).

Webb, M. and Jacobsen, J. (1982). *U.S. Carrying Capacity: An Introduction.* Washington, DC: Carrying Capacity. Good overview of ways human populations have exceeded carrying capacity.

World Resources Institute and the International Institute for Environment and Development. (1988). *World Resources 1988–89. An Assessment of the Resource Base that Supports the Global Economy.* New York: Basic Books. Excellent reference book.

World Commission on Environment and Development. (1987). *Our Common Future.* Oxford: Oxford University Press. Good reference.

6

Population Control: Key to a Sustainable Society

To rebuild our civilization we must first rebuild ourselves according to the pattern laid down by life.

ALEX CARREL

Most experts are convinced that population growth cannot continue indefinitely. This conclusion gained wide acceptance after an extensive computer study of the human future by researchers from the Massachusetts Institute of Technology, sponsored by the Club of Rome, and published in the book *The Limits to Growth*. The authors found from their analysis of population growth, resources, food, pollution, and industrial output that the human population would exceed the carrying capacity of the planet within a century if exponential growth continued (Figure 6-1). According to this forecast, food supplies, industrial output, and population will grow until rapidly diminishing resource supplies put a halt to industrial growth. Population will fall later because of food shortages.

The researchers next doubled their estimated available supply of nonrenewable resources, assuming that we might be able to expand our resource base through discoveries and new technologies that improve mining efficiency (Chapter 13). They found that the human population would still overshoot the carrying capacity and crash, only a couple of decades later (Figure 6-2). The rising death rate would be due in part to increased pollution from expanded industrial output.

In still another scenario the authors assumed that world resources were unlimited. In that study population growth was halted by rising levels of pollution. (See discussions of the greenhouse effect, depletion of atmospheric ozone, and acid rain in Chapter 15.) The

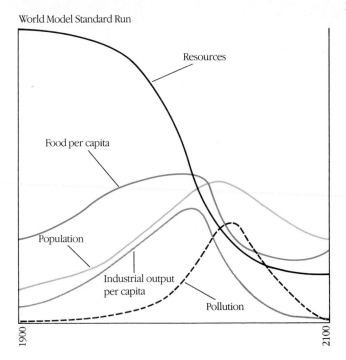

World Model Standard Run

Figure 6-1. Computer analysis of world trends. Because of natural delays in the system, both population and pollution continue to increase for some time after the peak of industrialization.

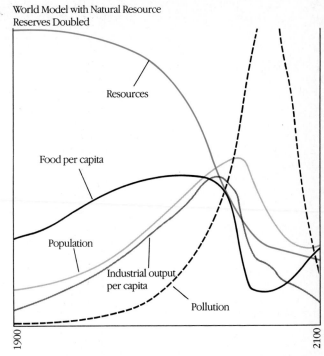

World Model with Natural Resource Reserves Doubled

Figure 6-2. The same computer analysis as Figure 6-1, except that a doubling of natural resources in the world is assumed.

conclusions of the MIT study were unequivocal. Any way you look at it, if human civilization continues on the same path, we will exceed the carrying capacity of the biosphere with perilous results. Can anything be done to change course?

Several computer models in the study showed that a stable world system could be reached by (1) immediately achieving zero population growth, (2) recycling, (3) pollution control, (4) soil erosion control, (5) soil replenishment, and (6) an emphasis on food production and services rather than industrial production (Figure 6-3).

The Limits to Growth study created a storm of controversy worldwide. Blinded by the frontier notions of unlimited resources, many critics simply couldn't believe its conclusions. Since the publication of *Limits to Growth*, however, 17 independent computer studies have all come to a similar conclusion: infinite growth in a finite system is impossible.

Despite the warnings of scientists, many political leaders remain blind to the consequences of exponential growth in population and industrial output. According to Professor D. H. Meadows and her colleagues, who collaborated on *The Limits to Growth*: "Every day of continued exponential growth brings the world system closer to the ultimate limits of . . . growth. A decision to do nothing is a decision to increase the risk of collapse." The longer we allow exponential growth to continue, the

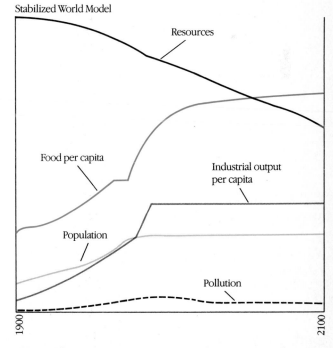

Stabilized World Model

Figure 6-3. The same computer analysis as Figure 6-1, with global zero population growth, pollution control, soil management, and sustainable agriculture.

Third World Population Growth: Why Should We Worry?

Daniel D. Chiras

The Third World countries are growing at a remarkable rate. In the next 30 years, the four billion Third World residents will double, if present-day growth continues. This creates domestic problems of colossal proportion, as noted in this chapter. To many Americans Third World growth is an issue of little importance; however, a careful study of the impacts shows that the rapid growth of population in Third World nations could have profound effects on our own welfare and our own economy. Consider the following examples.

The rapid growth of the human population south of the US border is resulting in a tremendous annual increase in the size of the work force in Mexico and other Latin American nations. Many of the new workers cannot find jobs at home, and as a result, they look to the United States as a way out. Many migrate illegally, crossing the borders at night to find work. But critics say that these individuals compete for low-level jobs and strain US hospitals, schools, and social welfare.

Illegal immigrants also strain our environment. At least 500,000 illegal workers and their families annually enter the United States. Some say that the number may be as high as 1.5 million. The United States is thus adding another Los Angeles (excluding the suburbs) to its population every year and a new California every 10 years. Unfortunately, say some environmentalists, this places additional burdens on already stressed land, air, and water.

Population growth abroad creates other less tangible impacts as well. For instance, 54 million people lived in the Middle East thirty years ago. Today, the population has climbed to over 135 million. The rapid growth of population fans the present religious, ethnic, and political turmoil—and the bloodshed, say some observers. From a purely selfish viewpoint, this turmoil threatens access to one of the world's largest oil reserves, from which the United States, Japan, and Europe import much of their oil.

Growing population abroad increases US food production. While that may be good for the economy, critics point out that it also places additional stresses on badly abused soils (see Chapter 7).

Expanding population abroad also wipes out species, a concern of many environmentalists in this country, as elsewhere. The loss of species is an aesthetic loss with many practical implications, discussed at length in Chapter 8. Many modern drugs, for instance, come from plants in the forests now falling to make way for more people. Population growth also threatens global ecological stability. The loss of rain forests, for example, contributes significantly to global warming (see Chapter 15).

Rapid population growth contributes to poverty, making it difficult for nations to pay back loans to developed countries. Billions of dollars in international debt now remain unpaid because countries cannot cope with their population growth. When countries fail to pay back loans to US banks, American consumers suffer. Banking services invariably become more costly.

Population growth throughout the world creates hardship and pain. To better the lives of our fellows on this planet and for purely selfish reasons, we must put a stop to the rampant growth. If we don't we will surely all suffer.

smaller the chance of achieving a sustainable future, and the greater the likelihood that we will damage the biosphere.

How Do We Control Population Growth?

Many experts agree that to avoid overshooting the carrying capacity and to prevent the likely population crash we should take measures now to control population growth.

Setting Our Goals

But what should our goals be? Hard decisions have to be made on two related fronts: population growth and population size. The immediate question is what to do with population growth. Should we let population growth continue unabated, should we try to reduce it, or should we stop growth altogether? And what about population size? Should we shoot for ten billion? Should we level off, then try to decrease population size? The answers to these questions differ depending on who is asked, partly because many people disagree on the fundamental issue: the number of people the world can support. Some observers contend that the world can comfortably accommodate only 500,000 people; others say that a population of 100 billion would be possible. Many ecologists argue that we have already exceeded the earth's carrying capacity (ample evidence is cited in Chapter 5); thus, growth should be halted as soon as possible.

Consider the ecologists' case: Today, about one-fifth of the world's people, or about one billion, live in a state of extreme poverty. They are inadequately fed and sheltered. They wander the streets of Calcutta, Bangkok, Cairo, and elsewhere begging for food and stealing what they can. At night they sleep in alleyways, under bridges, or in cardboard shelters. Another two billion people live on the edge, with barely adequate food and shelter and few amenities. Four families live in a two-room apartment and share a water tap with 25 other families. They have no sewage disposal systems.

All told, nearly three-fifths of the world's population is in bad shape. Strenuous efforts to improve the economic condition of the world's poor, in hopes of increasing personal wealth, have failed to keep up with growth. More and more people fall into the trap of poverty each year. Thus, many observers believe that putting a stop to growth now is the best way to reduce further suffering, environmental pollution, and resource depletion. Stopping growth, reducing population size, and pursuing sensible economic plans, as discussed below, can help break the vicious cycle of poverty, but the road ahead will be long and difficult.

Even if we could miraculously reach replacement-level fertility today, the enormous momentum present today ensures further growth to about eight billion, most of it occurring in poor nations. Many experts believe that the human population should be reduced through attrition—keeping birth rates below death rates. Reducing the size of human population throughout the world would help us live within the limits of nature and maybe ensure a better life for billions of poor people.

Population Control Strategies

How do we go about controlling population growth? For a number of years many population experts advocated a simple way: economic development. The logic was undeniably attractive: make jobs by increasing industry, farming, crafts, and services. More jobs meant more personal wealth. People could then afford decent housing, food, and education. Poverty and disease would be eliminated. As an added benefit, people would choose to have fewer children.

Demographic Transition The idea that economic improvement in the developing nations would solve the population crisis was an appealing one for many years, and there was plenty of evidence from developed countries that it could work. The evidence came in the form of a **demographic transition**, described below. Experts reasoned that if the demographic transition worked for Europe and North America, it could work for the rest of the world as well.

The demographic transition takes place in four definable stages (Table 6-1). Countries begin in Stage 1 with high birth rates and high death rates. The population is

Table 6-1	The Stages of the Demographic Transition
Stage 1—	Preindustrial High birth rate and high death rate Little or no increase in population Example: Finland 1785–1790 　　　　　　　Crude birth rate = 38/1000 　　　　　　　Crude death rate = 32/1000 　　　　　　　Natural increase = 0.6%
Stage 2—	Transitional High birth rate and falling death rate High population growth rate Example: Finland 1825–1830 　　　　　　　Crude birth rate = 38/1000 　　　　　　　Crude death rate = 24/1000 　　　　　　　Natural increase = 1.4%
Stage 3—	Industrial Falling birth rate and low death rate Slower growth Example: Finland 1910–1915 　　　　　　　Crude birth rate = 29/1000 　　　　　　　Crude death rate = 17/1000 　　　　　　　Natural increase = 1.2%
Stage 4—	Postindustrial Low birth rate and low death rate Slow population growth Example: Finland 1970–1976 　　　　　　　Crude birth rate = 13/1000 　　　　　　　Crude death rate = 10/1000 　　　　　　　Natural increase = 0.3%

Source: Haupt, A. and Kane, T. (1978). *Population Handbook.* Population Reference Bureau.

stable at this time. Because of improvements in health care and sanitation, however, death rates begin to fall. Birth rates remain high. This is called Stage 2; it is the most dangerous stage, because the large difference between birth and death rates leads to rapid population growth. As the country develops economically, however, birth rates begin to decrease. Population growth slows. This is Stage 3. Finally, over time, birth rates and death rates come into balance. Population growth is stopped in Stage 4.

The decrease in birth rates associated with economic development can be attributed to several factors. Perhaps the most important is the shift in people's attitudes toward children. Preindustrial farmers view children as an asset. Children help with farm work and can support their parents in old age. With industrialization and the migration of families to the city, however, children become an economic liability. Given competition for living space, each child means that more money has to be devoted to food and housing. If the children do not work, they create an additional financial drain on the family. As a result, smaller families generally prevail.

Figure 6-4. Family planning in India. A doctor lectures to a group of Indian women on various methods of birth control at a family planning clinic in New Delhi.

The idea that economic development can lead to zero population growth has been controversial. If factories are built and jobs are created in the developing countries, some argue, economic development and rising personal wealth will eventually reduce population growth. Supporters of economic development as a means to control population growth point to dozens of countries that have experienced demographic transitions. If it worked for the United States and Finland, why won't it work for Ghana, China, India, or Chad?

There are at least four reasons why not. First, the economic resources of many of the developing countries are too limited to build the type of industry needed for demographic transition. Second, the classic transition did not take place overnight: it took Finland over 200 years to approach a balance between birth rate and death rate. The developing countries with rapid doubling times do not have this kind of time. Third, population growth in many countries outstrips economic growth. Recent studies show, for instance, that a 1% growth in the labor force

requires a 3% economic growth. Think of the economic growth needed to sustain populations growing at 3% and 4% per year. For many countries it is hard enough to keep up with population growth. Getting ahead is a pipe dream. The fourth reason is that the fossil fuel energy sources that were essential to demographic transition in the developed countries are diminishing and becoming ever more costly. Without the rich mines of England or the great oil deposits of Arabia and North America, poor nations will probably never witness economic growth like that seen in the Western world.

Many of the world's countries have entered Stage 2 of the demographic transition, with high birth rates and low death rates. For the reasons given above, a demographic transition brought on by industrialization is unlikely. Thus, the demographic transition must come about in other ways, most likely through family planning and small-scale economic development.

Family Planning **Family planning** allows couples to determine the number and spacing of offspring. For countries stuck in Stage 2 of the demographic transition it can speed up the decline in the birth rate.

For family planning to work, information on birth control must be readily available. Birth control must also be accessible and inexpensive. People must be motivated to practice birth control, too. Strictly defined, **birth control** includes any method to reduce births, including contraception and **induced abortion**, the intentional interruption of pregnancy through surgical means or through drug treatments. A **contraceptive** is any chemical, device, or method that prevents fertilization.

Family planning programs lie on a continuum from voluntary to compulsory. **Voluntary programs** are those that make birth control available to the public at low cost. There is no pressure from the government. People choose the type of birth control and family size. Some voluntary programs are sponsored by governments (Figure 6-4), and some are run by private organizations, such as Planned Parenthood in the United States.

Family planning programs may be promoted by governments; such programs are euphemistically called **extended voluntary programs**. In these cases governmental agencies may hand out information on birth control and sterilization or sponsor posters, newspaper ads, television and radio announcements, and billboards (Figure 6-5). In Egypt, for instance, a song promoting birth control that was played with a government-sponsored commercial became so popular that it was a national hit. Payments or incentives are frequently offered by governments to couples practicing birth control or undergoing sterilization. Transistor radios, for example, are handed out in rural Egypt to couples who adopt some form of birth control. One of the key components of govern-

Figure 6-5. A familiar poster seen everywhere in India, with a smiling family and the slogan, "two or three children . . . is enough." India's family planning program is closely meshed with the government's health services. All doctors, nurses, and other health workers are trained and expected to provide family planning advice and assistance as part of their regular duties.

mental programs is an effort to change people's thinking about family size. Varinda Vittachi, a writer from Sri Lanka, reminds us, "The world's population problem will be solved in the mind and not in the uterus." Posters in Vietnam, for example, extol the virtues of a one-child family. Informational campaigns may also attempt to change stereotypical sex roles, persuading men that masculinity and self-importance are not related to the number of their children and convincing women that the childbearing role is not the only one that makes them valuable.

Forced family planning programs involve strict governmental limitations on family size and include punishment for those who exceed quotas. These rare programs may impose sterilization after a family reaches the allotted size, limit food rations for "excess" children, or tax couples who exceed the allowed number of children. China has what many observers consider a forced family planning program. It has adopted a one-child policy to halt its rapid growth (China has one-fifth of the world's population, 1.1 billion people). The one-child policy is intended to reduce China's population size to 700 or 800 million. Female workers meet regularly in small groups to discuss birth control. Government workers reportedly urge men and women with one child to undergo sterilization. Pregnant women who already have one child have reportedly been "forced" to have abortions, although authenticated cases are hard to come by, and occurrences may be rare. The Chinese government also supplements the monthly income of couples who pledge to have only one child. If a couple that has pledged to have one child has a second one, however, the government often requires repayment of the monthly bonus and cuts the monthly salary by 15%.

Many programs do not fit neatly into one of these categories. The government of Singapore, for instance, uses radio, billboards, and school curricula to suggest that two children per family are enough. This extended voluntary program is complemented by several coercive measures: The first two children of any family are allowed to attend local schools, but additional children must often be bused elsewhere. Furthermore, hospital costs increase for "surplus" children, and government employees who are not sterilized are given a low priority for public housing.

Small-Scale Economic Development "Family planning cannot exist in a vacuum. You can't just distribute contraceptives and tell people to go ahead and start lowering the birth rate," says Aziz el-Bindary, head of Egypt's Supreme Council for Population and Family Planning. "To have an effective family-planning program," he adds, "you also have to have an effective economy—where jobs are available, where health facilities are adequate."

Throughout the world, governments are finding that packaging together jobs for women, small-scale economic development, improved education, better health care, and contraception can reduce birth rates. The critical thinking skills you learned in Chapter Supplement 1-1 suggest that solutions often require many approaches. Family planning is a case in point.

Family planning will work if people want fewer children. With this in mind the United Nations Fund for Population Activities recently made substantial investments in clothing factories in which Egyptian women can work. Other programs are under way. The logic behind these activities is that working women often delay marriage and childbearing and thus have fewer children.

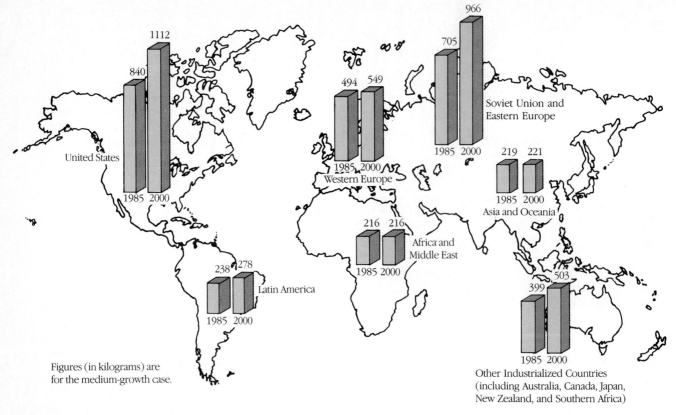

Figure 6-6. Current and projected per capita grain consumption in the world (including grain used for livestock).

Developed Countries—What Can They Do?

Many people think that population control pertains only to the developing countries, because many of the richer, developed countries are in Stages 3 and 4 of the demographic transition. Observers point out, however, that the high per capita consumption of the developed countries has put enormous strains on the earth's carrying capacity. Widespread pollution and resource depletion are two signs that many of the developed countries have exceeded the carrying capacity. Therefore, many ecologists argue, population growth must be reduced in the developed nations if we are to build a sustainable society. Today, the populations of West Germany, Hungary, and Denmark have begun to shrink because total fertility rates have fallen below replacement-level fertility. Their populations are declining. Nearly 20 European countries are close to zero population growth.

Population growth continues in a number of major industrial nations. Canada, the United States, and the Soviet Union comprise about half of the population in the developed world. Growth in these nations is of great concern to those interested in protecting the global environment because of the resource-intensive life-styles of their people.

According to some estimates, a single American or Canadian uses 25 to 38 times more resources than a citizen of India. Impact, therefore, is a result of population size *and* appetite—or consumption level. As shown in Figure 6-6, US residents consume nearly four times as much grain as residents of Third World nations in Africa, the Middle East, Asia, and Oceania. By using less, many people argue, developed countries can make possible a more equitable sharing of the earth's resources. This redistribution may help reduce world tensions.

Sharing resources is controversial. Garrett Hardin, author of *Filters Against Folly*, contends that global sharing of resources is not the answer to global stability, as some suggest. The rate of growth and needs of the nearly four billion residents of the Third World far exceed our capacity to help, he argues. Developed countries can assist the developing nations by sharing their knowledge of birth control, agriculture, health care, and appropriate technology. Financial assistance to help achieve a moderate rate of industrialization, using **appropriate technologies**—that is, industry that is labor intensive and uses local resources to meet local needs—can also help. Table 6-2 lists some suggestions for developed countries.

William and Paul Paddock propose a triage system to determine who gets financial and technical aid. They say

Table 6-2 Population Control Strategies for Developed Countries

Strategy	Rationale	Benefits
Stabilize population growth by restricting immigration, and by spending more money and time on sex education and population awareness in public schools.	High use of resources taxes the environment. Immigrants create serious strain on the economy and create social tension in conditions of high unemployment. Education helps citizens realize the importance of population control.	Limiting resource use leaves more for future generations and developing countries.
Provide financial assistance to developing countries for agriculture and appropriate industry. Aid should come from government and private sources.	Economic growth in developing countries will raise the standard of living and aid in population control.	The rich–poor gap would narrow. A decrease in sociopolitical tension and resource shortages would result.
Provide assistance to population control programs.	Better funded population programs can afford the increased technical assistance and community outreach programs necessary to provide information to the public.	Could result in faster decrease in population growth.
Make trade with less developed countries equitable and freer.	Freer trade will increase per capita income and raise standards of living with little effect on home economy.	A higher standard of living and increased job opportunities could result.
Concentrate research on social, cultural, and psychological aspects of reproduction.	Techniques available today are effective and reliable. What is needed is more motivation for population control, especially among poor countries.	Money will be better spent; research of this nature may help facilitate family planning in less developed countries.

that countries could be categorized in three groups: (1) those that have an adequate resource base and could survive hard times without aid, (2) those impoverished nations that would probably not survive drought and food shortage even with aid, and (3) those that can be helped. Group 1 needs no assistance. Group 2 must be given up as hopeless, for no amount of aid will help them. Group 3, if aided, could pull through a drought or other difficult period. Therefore, they suggest, we should maximize our aid for population control, and food, agricultural, and technological support and development by concentrating our efforts on Group 3. To some, this seems like an unethical approach to world assistance. Others see it as the only answer to apportioning the limited aid.

Developing Countries—What Can They Do?

The less developed countries recognize the need for population control. Today 93% of the world's population lives in countries with population control policies. The International Planned Parenthood Federation, established in the 1950s, has disseminated information on birth control and provided assistance to many countries. Since the 1960s the World Bank has given financial aid to developing countries with population control policies.

Unfortunately, having a population control policy and funding that policy are entirely different matters. Few governments today spend more than 1% of their national budget on family planning services. In addition to this monetary problem, illiteracy, rural isolation, and local taboos have hampered success. For these and other reasons, population control in developing nations has been limited. Results have been especially poor in India, Mexico, Brazil, Pakistan, and Indonesia.

Table 6-3 lists population control strategies for less developed countries. Accompanied by appropriate development, these suggestions could improve the lives of many people.

Making Strategies Work

Almost any venture encounters obstacles. Population control is no exception. In fact, it is bound to face more obstacles than most projects, because it strikes at the heart of what is to many people a personal issue: the right to

Table 6-3   Population Control Strategies for Developing Countries

Strategy	Rationale	Benefits
Develop effective national plan to ensure better dissemination of information and availability of contraception and other methods of population control. Do not rely on one type of control.	Each country better understands its people and thus can design better programs to spread population control information and devices.	More effective dissemination of information and, probably, a higher rate of success.
Finance education in rural regions, emphasizing population control and benefits of reduced population growth.	Education can help make population control a reality.	Slower population growth, more effective use of contraceptives, and more incentive.
Seek to change cultural taboos against birth control and cultural incentives for large families.	Changes in culture and psychology may be needed to make population control programs effective.	Such changes will help programs succeed.
Develop appropriate industry and agriculture, especially in rural areas to reduce or eliminate the movement of people from the country to the city.	Appropriate agriculture and industry will create jobs and better economic conditions for families. A higher standard of living could translate into better health care and greater survival of young, thus destroying need for large families.	This will result in higher standard of living, better health care, and impetus for control of family size.
Seek programs of development that attain a maximum spread of wealth among the people.	Development must not just help a select few, because benefits may not trickle down to needy.	Plans of this nature yield good distribution of income and help the needy rather than select few.
Integrate population policy with economic, resource, food, and land-use policy to achieve a stable state.	Finite resources require wise allocation and use; success in the long run depends on attempts to achieve a sustainable future.	Longevity and permanence are attainable if policies are integrated and take into account the requirements of a sustainable society.
Seek funding from the United Nations and developed countries.	Developed countries have a stake in stabilizing world population growth.	Developed countries could provide significant financial support.

have children. At least three major obstacles stand in the way of population control programs: psychological inertia, lack of education, and religious beliefs.

Psychological Barriers

Large families are an asset in the developing countries, since children help with the chores and later care for their parents. Given the high mortality rates in these countries, having many children ensures that some will sur-

vive. Even though death rates have fallen swiftly throughout the world and more children are now surviving, the traditional value of children has not been abandoned, and birth rates remain high.

Traditional views of family size often change slowly after a decline in death rates. Nowhere is this more evident than in India. In some economically developed regions, for instance, sons are still viewed in high esteem as a source of security in their parents' old age. The present government promotes small families of two or

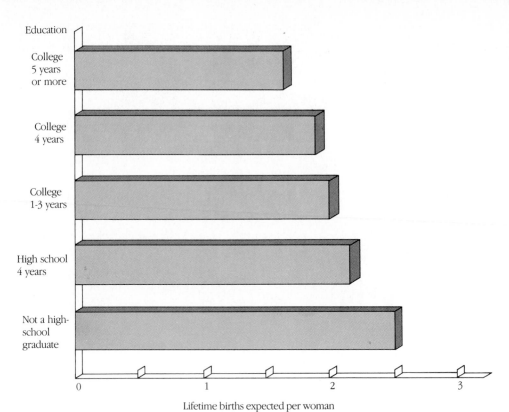

Figure 6-7. Total fertility rates of US women, by level of education.

Lifetime births expected per woman

three children. Many parents see the ideal family as two sons and a daughter, to hedge their bets should one son die. In trying to reach this goal, however, couples in 1990 averaged 4.3 children. "One son is no sons," Indians argue, still unsure of the lowered death rate. "To be sterilized is to tempt fate," another proclaims, arguing in favor of the insurance that two-son families provide. Until people begin to realize that one son is enough and that that son will probably survive, India's population growth will continue at its current rapid rate.

Having children is a self-fulfilling activity for males and females in most countries. Sociologists report that men and women are admired in many societies for the number of children they have, a view that is present in developed nations as well. The humorist and educator Bill Cosby pokes fun at this view, saying, "If a chimpanzee can have a baby, the human female [and male, I would add] should realize that the feat is something less than an entry for the *Guinness Book of World Records*."

Unfortunately, social acceptance and numerous other psychological factors result in the birth of many children who will never have adequate food, clothing, shelter, and education. In the face of such possibilities how can people continue to have large families? Citizens of developed countries, even those who love children and put stock in family life, tend to view children as an economic drain. Recent studies show that it costs low-income families in the United States about $60,000 to raise a child, including education through a publicly subsidized college. Middle-

income families spend about $90,000 per child. In the poor countries a child deprives the family of virtually nothing. In fact, since children represent a form of material wealth (free labor and support in old age) and satisfaction, they probably represent a net gain.

Education Barriers

US Department of Commerce figures show that the higher the educational level, the lower the total fertility rate (Figure 6-7). The reason for this is that educated women often pursue careers, postponing marriage and childbearing. Since the childbearing years are from 15 to 44, a woman who graduates from college at age 21, marries, but delays children until she is 30 has decreased her childbearing years by half.

Inadequate education in developing countries has several important effects on population growth. Men and women who lack educational opportunities or who choose not to go to school generally marry young and do not pursue careers that interfere with childbearing. Thus, the period of childbearing is much longer than that for couples who pursue higher levels of education. A lack of education also makes it more difficult for people to learn about alternatives to childbearing and proper use of contraceptives. It is no wonder, then, that the birth rate is still high in rural India, where 80% to 90% of the women cannot read or write.

Figure 6-8. Indira Gandhi, the late prime minister of India, speaks to a crowd in Calcutta about her governmental policies.

Religious Barriers

Religion may also be a powerful force in reproduction. In the West the Roman Catholic church has been cited as a prime cause of overpopulation. Strictly forbidding all "unnatural" methods of birth control, the Catholic church takes a dim view of the pill, the condom, the diaphragm, and abortion. In 1968 Pope Paul VI's encyclical *Humanae Vitae* condemned the use of contraceptives except the rhythm method, one of the least effective measures. The church's position today remains much the same. Theoretically, the church provides the guidelines for the sexual practices of approximately 600 million people. Recent surveys show, however, that the use of contraceptives by Catholic women, especially in Western nations, is nearly as high as that among non-Catholics. In Latin America, many priests speak out against the Vatican's official policy.

Birth control is a generally undiscussed subject among other religions. And since many of the Eastern and Mideastern religions compete with one another for followers, birth control is not advocated. In fact, the total fertility rates in Iran and Iraq are 6.3 and 7.2, respectively.

Overcoming the Obstacles

For family planning to work, its proponents must understand the culture, education, and religion of a people. Programs should be tailored to meet the needs of each group. People must have access to socially acceptable birth control, and in poor countries this must be inexpensive or free and must be easy to understand and simple to use, to reduce the risk of failure. Trained individuals must be available to counsel those seeking contraceptives, abortion, or sterilization. Most importantly, attitudes toward family size must be changed if programs are to

work. Perhaps the most effective way of changing attitudes is to provide work for women; to increase personal wealth; to reduce illiteracy; and to provide better health care, sanitation, and contraception.

Ethics of Population Control

Perhaps no other issue in environmental science is so laden with ethical problems as population control. To deny the right to reproduce is to deny one of the most basic and important of all human activities. To some, population control is a violation of deep religious beliefs. To others, it is an intrusion into a private matter; and for minorities, population control has overtones of genocide. Two important ethical questions are discussed in this section as an introduction to this complex ethical issue.

Is Reproduction a Personal Right?

Some people argue that the right to reproduce at will should be curtailed when the rights of the individual interfere with the welfare of society, the collective rights of all the people. For example, in 1975 India's government, under the late Prime Minister Indira Gandhi, began a program of forced sterilization (Figure 6-8). Changes in India's Constitution placed the rights of the whole of society above the rights of the individual. This short-lived and extreme program illustrates that an important end—the welfare of an overpopulated nation's people—can prompt governments to adopt tyrannical means to achieve that end.

The opposite point of view was well stated in 1965 by Pope Paul in a speech to the United Nations: "You must strive to multiply bread so that it suffices for the tables of mankind, and not, rather, favor an artificial control of birth, which would be irrational, in order to diminish the number of guests at the banquet of life."

Individuals who support the right to reproduce freely often argue that denying such rights takes away personal freedom. Paul Ehrlich argues that we "must take the side of the hungry *billions* of living human beings today and tomorrow, not the side of potential human beings. . . . If those potential human beings are born, they will at best lead miserable lives and die young." He argues that we cannot let humanity be destroyed by a doctrine of individual freedoms conceived in isolation from the biological facts of life.

Garrett Hardin argues that the integrity of the biosphere should be the guiding principle in the debate over population control. Recognizing that the welfare of the biosphere determines the welfare of all living things, including humankind, and that we have obligations to future generations to protect the biosphere, human population control becomes a biological imperative.

Is It Ethical Not to Control Population?

If we do not control population and instead let it run its course, will uncontrolled growth improve or worsen the world for future generations? If our actions rob from the future, can they be considered ethical?

To predict how our actions will affect the future, we can use the computer to forecast major trends in population, resources, and pollution. Such exercises, as we've seen in earlier examples, may be subject to error but nonetheless help us arrive at ethical decisions based on scientific information. Current research suggests that continued population growth is an invitation to disaster. The computer projections remind us that humankind comes with the same warranty that the dinosaurs had. Werner Fornos, director of the Population Institute, notes, "National leaders are recognizing that population growth without adequate resources and services results not in national strength but in national disaster."

Not controlling population, many experts agree, is unethical because it sidesteps our obligation to future generations. Critics of this viewpoint reply that technology and substitutes will be found to feed the hungry new mouths and that more people will lead to a better tomorrow. Julian Simon, an economist from the University of Maryland, is a leading advocate of this outlook. He argues that more people mean more knowledge and that knowledge is the key to a better future. This view ignores the finite nature of many resources, the ways renewable resources such as forests can be irreparably damaged, and the poverty that afflicts many of the world's people. (For a debate on this issue see the Point/Counterpoint in Chapter 5.)

The Status of Population Control

"The world of the mid-eighties is a world of stark demographic contrasts," writes Lester Brown, President of the Worldwatch Institute. Never have the differences among countries been greater. "Some populations change little in size from year to year or decline slightly," Brown notes, "while others are experiencing the fastest growth ever recorded." This section discusses the demographic contrasts, looking at areas of progress and setbacks in an effort to guide our future efforts.

Encouraging Trends

Developing Nations　Most of the world's developing nations see the need for programs to control population growth. Ten years ago this was not the case. Especially reluctant were Third World nations, which rallied behind the cry "Development is the best contraceptive." A decade of falling per capita food production and income and rising debt, however, has changed many leaders' minds. Today, even the World Bank, a private lending institution that has long ignored the problem, has called attention to the need for population control throughout much of the developing world. The first step in solving the population problem is awareness. The Third World countries are now aware of the problem and are trying to do something about it. In fact, for every dollar of foreign aid received to combat population growth the developing nations spend four dollars.

Many nations have successful population control programs and have made phenomenal progress. The largest reductions in growth rate have been witnessed in China, Taiwan, Tunisia, Barbados, Hong Kong, Singapore, Costa Rica, and Egypt.

China's decline in birth rate in the 1970s is the most rapid of any country on record. It may be family planning's greatest success story. China's 2.5% annual growth rate in the 1960s (doubling time 28 years) is now (1989) at 1.3% (doubling time 54 years).

Developed Nations　In the early to mid-1970s many developed countries updated their laws and policies governing family planning. Eighteen countries in Europe now have near stationary populations. Three European countries are shrinking. The United States and Japan, two of the most populous developed countries, show signs of slowing population growth. According to some demographers, the decline in growth in the developed nations is largely responsible for the large drop in the world's growth rate from over 2% in 1974 to 1.7% in 1989.

The United States exemplifies what has happened throughout much of the developed world. US abortion laws were liberalized in the early 1970s, access to birth control was increased, a growing number of women

On Immigration, The U.S. Must "Know When to Say When"

Brooke A. Martič

Ms. Martič graduated from U.C. Berkeley's International Political Economy program in 1988. She is now a Senior Analyst for the Center for Immigration Studies in Washington, D.C.

The earth's population expands by 230,000 people daily, mounting formidable political pressures on governments and pressing lagging Third World economies to the wall. War and civil strife in these countries is making matters worse. Simultaneously, the U.S. is seen as a haven with ample opportunities for everyone. Many people are abandoning their homelands for a chance at the American way of life.

In 1989 over 3.2 million people applied for 20,000 U.S. immigrant visas in a special visa "lottery" created by Congress. Another 643,000 legal immigrants and refugees moved to the U.S. under other immigrant categories. Hundreds of thousands more risk their lives to cross our borders illegally.

Some lawmakers believe we can address high demand by increasing admissions numbers. Congress is now considering reforms to increase the annual allotment of employer-sponsored immigrant and independent immigrant visas (for those without family or job offers in the U.S.). Some are also trying to make the so-called Second Family Preference Immigrant category—covering spouses and children of U.S. permanent residents—a numerically unrestricted category. (Under current law, only immediate relatives of U.S. citizens are able to enter the U.S. without a long wait.)

The Center for Immigration Studies, a Washington-based, nonprofit agency, has studied some of the possible effects of these changes. Using government projections, we estimate that, if enacted, these provisions could expand net U.S. immigration by as much as 100% over the first half of the 1990s. The U.S. would admit over one million immigrants *per year*. (See graph on page 137.)

Numbers could be much higher if Congress continues to pass special interest immigration legislation. In the summer of 1989, both houses voted to grant automatic refugee status to anyone from selected Soviet, East European, and Southeast Asian groups. The State Department estimates that over 21 million people could qualify.

Resettlement programs require money the U.S. doesn't have. Each legal refugee, for example, costs about $7,000 to resettle. No such figure can be tallied for illegal immigrants, but once they are granted permanent resident status, they are eligible for a wide range of federal, state, and local welfare programs. The three million illegal aliens who received amnesty in 1986 will cost $4 billion over the next four years.

The majority of U.S. immigrants congregate where large domestic minorities are pressing for better schools and housing. In Los Angeles, where an estimated 20% of all immigrants

Thousands

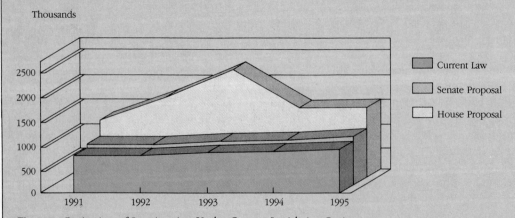

Five-year Projection of Immigration Under Current Legislative Options.

Source: Center For Immigration Studies "Backgrounder" October, 1989 (IRCA Legalized Aliens Excluded.)

settle, a 1987 *L.A. Times* survey estimated that as many as 200,000—most of them foreign-born—are living in converted garages, many without plumbing, heating or windows. A 1989 U.S. Department of Education study noted that "the presence of large numbers of recent immigrants creates a significant burden for some of the nation's school districts." Another 1989 study by the U.S. Labor Department says that recent immigrants are the group most likely to suffer from job displacement and wage stagnation caused by large immigrant influxes into the job market. These trends are likely to continue.

The U.S. was once big enough to ignore such problems. This is no longer the case.

The only way out is to restrain demand. Congress should approve a permanent but flexible ceiling on *all* U.S. immigration. Refugees should be included under the cap but should have the highest priority since they are by definition the most in need of resettlement. "Family reunification" immigration should also be limited to spouses and minor children of U.S. permanent residents. Current preferential arrangements for

brothers and sisters, married sons and daughters, etc., pushes up current and future numbers through "chain migration."

Finally, Congress and the Bush Administration must curb special interest immigration legislation. Such proposals usually reward those with the strongest lobbies here—often at the expense of the most needy. If such legislation is unavoidable, new admissions must be fitted under the next year's ceiling.

It is our opinion that any overall ceiling in immigration should be set to produce a net immigration of 450,000 per year over the next four years. In five years, Congress and the Administration could adjust the figure in light of prevailing fertility rates, emigration, and illegal immigration.

These moves will send a message to every intending immigrant and refugee—the U.S. will not harbor everyone. Although this may seem like a draconian stance to some, it will enable us to resettle some of the most needy while ensuring *everyone* a higher quality of life, for now and for future generations.

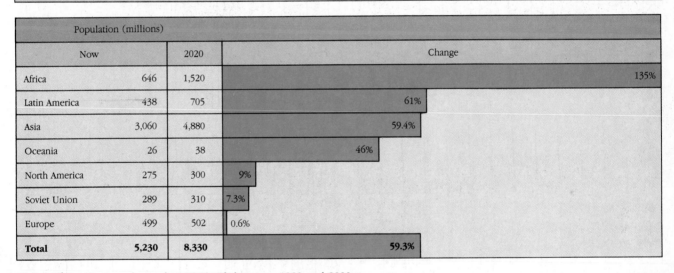

Figure 6-9. Projected population growth between 1989 and 2000.

entered the work force, educational opportunity for women increased, the feminist movement articulated a way of life in which motherhood was but one option, and many couples decided to remain childless. In addition, inflation and recession in the 1970s and early 1980s and soaring hospital costs no doubt helped reduce birth rates. The total fertility rate in the United States dropped to slightly below 2.1 in 1972 and since then has remained below replacement level (Figure 6-7). Should abortions be restricted, total fertility is expected to increase.

Discouraging Trends

Unfortunately, the world population dilemma is far from solved; there remain some rather discouraging signs. For instance, 35% of the world's population was under the

age of 15 in 1986. These children will provide a great deal of momentum for further growth as they marry and reproduce. During the 1970s death rates turned upward in many poorer countries, primarily because of hunger and malnutrition. Furthermore, food demand has now outstripped production, and world food reserves have dropped from 90 days in 1970 to 30 days in 1974 and have remained at this low level ever since. (For more on this topic see Chapter 7.)

Rapid growth is expected to continue in Africa, Latin America, and Asia, as shown in Figure 6-9. Expanded efforts are needed in these regions to avert widespread starvation, pollution, poverty, and resource depletion. Figures 6-10a and b show the relative growth of the developing countries and the developed world. With the rapid expansion of population in the Third World will come a

Figure 6-10. (a) Graph showing that most of the growth in the world population will occur in the developing countries. (b) Another way of showing the same phenomenon, this graph indicates the relative proportion of new residents in the world population in developed countries and developing countries. Both graphs point out a dangerous trend.

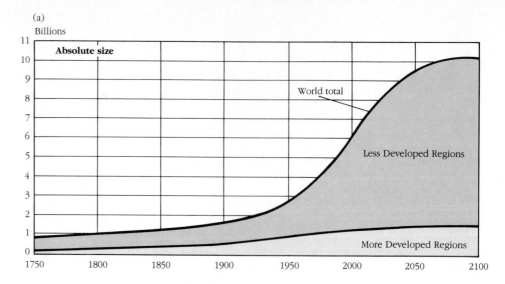

(a)

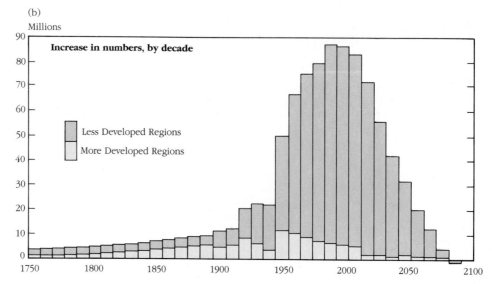

(b)

whole host of problems, many of which could affect the world in significant ways. (For a discussion of the connections between Third World growth and developed nations see Viewpoint 6-1.)

Another discouraging trend is that contraceptive availability is low. Studies suggest that many women of reproductive age throughout the world do not want additional children but that about half of them are not using effective methods of birth control. Furthermore, whereas more than 80% of all married women in developing countries have heard of contraception, in some places such as Pakistan only one in ten women has ever used it. Another discouraging sign is the steady decline since 1960 in nonmilitary, developmental foreign aid to poor countries.

One of the most visible failures in population control is India's, even though a national birth control program has been in effect since 1951. India's program was the first of its kind in the world. The rate of growth has slowed from 2.6% in 1969 to 2.1% in 1989. India has a population of over 800 million (1989), however, and will soon exceed China. In the next ten years India's population is expected to swell by over 210 million, almost equal to the current population of the United States.

Population growth is a social problem. It begins with people as the cause, and ends with people as the victims. Human civilization need not decay in an explosion of people, but, many experts agree, we can improve only if we learn to live within the limits of nature. This is the lesson of ecology and demography.

In a society with a relatively short attention span, in which wars and space voyages capture our attention, we may have difficulty sustaining our interest in controlling population growth. Long-term action, lasting through the rest of this century and well into the next, can ensure a balance between human beings and the planet that will allow us to explore our full potential.

I cannot believe that the principal objective of humanity is to establish experimentally how many human beings the planet can just barely sustain. But I can imagine a remarkable world in which a limited population can live in abundance, free to explore the full extent of man's imagination and spirit.

PHILIP HANDLER

Summary

Many experts believe that the growth of population and use of resources cannot continue indefinitely without overtaxing the earth's carrying capacity. This belief is supported by elaborate computer modeling studies, which suggest that the only route to a stable, sustainable society is global zero population growth, recycling, pollution control, soil erosion control, soil replenishment, and emphasis on food production and services rather than industrial production.

Since about one-fifth of the world's population lives in a state of extreme poverty without adequate food and shelter, some population experts believe that we have already exceeded the earth's carrying capacity. Family planning and limited economic development are ways to bring the human population back in line.

Family planning allows couples to determine the number and spacing of offspring. Programs may be voluntary, extended voluntary, or forced. Many experts believe that family planning should be part of an overall program that promotes economic development, jobs for women, health, and education.

Developed nations can contribute to a solution by reducing consumption and population size. They can assist the less fortunate with population control, agriculture, health care, and appropriate technology through financial aid and, especially, information sharing.

Many developing nations have population control programs, although funds are often inadequate. Increasing expenditures on such programs could have many long-range benefits. To be effective, programs in such countries must take into account the effects of religious beliefs, psychological factors, and educational levels.

No other issue in environmental science carries with it so many ethical ramifications. To some, population control implies denial of one of the most basic and important of all human freedoms, the right to bear young. To others, it violates deep-seated religious beliefs. Still others believe that the right to reproduce should be curtailed when the rights of the individual interfere with the welfare of society; these people ask whether it is ethical *not* to control population.

Many encouraging signs of our increasing control over population have been visible in recent years. Growth rates have dropped, and world population is expected eventually to stabilize somewhere between 8 and 15 billion. Progress in population control has been remarkable in many countries, especially China, Taiwan, Tunisia, Barbados, Hong Kong, Singapore, Costa Rica, and Egypt. Unfortunately, the population problem is far from being solved. In 1986, 35% of the world's population was under 15 years of age; this group provides a great deal of momentum for further growth as it enters reproductive age. Per capita food production has barely kept pace with food consumption. In many countries only a small portion of the people practice birth control. One of the most frustrating situations is in India, which after more than three decades of attempted population control has failed to bring its population growth rate below 2%.

Discussion Questions

1. Debate this statement: "The world cannot support the people it currently has at a decent standard of living, so we should help the developing nations become industrialized. Population will fall as a result, so population control programs are not necessary."

2. Define family planning. Make a list of the three major types of family planning programs. Give some specific examples of each.

3. The United Nations appoints you as head of population control programs. Your first assignment is to devise a population control plan for a developing country with rapid population growth, high illiteracy, widespread poverty, and a predominantly rural population. Outline your program in detail, justifying each major feature. What problems might you expect to encounter?

4. Describe ways in which developed countries might aid developing countries in solving the population crisis.

5. Discuss the "value" of children in less developed countries. How do these views differ from those of the developed countries? Are they similar to or different from your views?

6. Discuss reasons why the total fertility rate tends to be lower among more educated women.

7. Discuss general ways to ensure a high rate of success in population control programs.

8. Do we have the right to have as many children as we want? Should that right be curtailed? If so, under what conditions?

9. Discuss some of the encouraging and discouraging news regarding world population growth. What progress has been made? Where do we need to concentrate our efforts in the near future?

Suggested Readings

Brown, L. R., Chandler, W. U., Flavin, C., Jacobson, J., Pollock, C., Postel, S., Starke, L., and Wolf, E. C. (1987). *State of the World 1987*. New York: Norton. See Chapters 2 and 9 for an overview of population growth and population control.

Cole, H. S. D., Freeman, C., Jahoda, M., and Pavitt, K. L. R. (1973). *Models of Doom: A Critique of "The Limits to Growth."* New York: Universe Books. A rebuttal of the computer study.

Crooks, R. and Baur, K. (1990). *Our Sexuality* (4th ed.). Redwood City, CA: Benjamin/Cummings. Excellent coverage of birth control.

Gupte, P. (1984). *The Crowded Earth: People and the Politics of Population*. New York: Norton. Superb! You must read this wonderful book on the effects of population control projects.

Hales, D. R. and Williams, B. K. (1989). *An Invitation to Health: Taking Charge of Your Life* (4th ed.). Redwood City, CA: Benjamin/Cummings. Excellent coverage of contraception.

Hardin, G. (1972). *Exploring New Ethics for Survival: The Voyage of the Spaceship Beagle*. New York: Viking. A beautifully written discussion of population ethics.

Hardin, G. (1982). Some Biological Insights into Abortion. *Bioscience* 32 (9): 720–727. Important reading.

Hardin, G. (1985). *Filters Against Folly*. New York: Viking. Important reading.

Jacobson, J. (1983). *Promoting Population Stabilization: Incentives for Small Families*. Worldwatch Paper 54. Washington, DC: Worldwatch Institute. Excellent paper.

Jacobson, J. L. (1988). *Environmental Refuges: A Yardstick of Habitability*. Worldwatch Paper 86. Washington, DC: Worldwatch. Excellent.

Meadows, D. H., Meadows, D. L., Randers, J., and Behrens, W. W. (1974). *The Limits to Growth* (2nd ed.). New York: Universe Books. Excellent study of population, resources, and pollution.

Mumford, S. D. (1982). Abortion: A National Security Issue. *The Humanist* 42 (5): 12–42. Some interesting and controversial insights.

Population Reference Bureau (1989). World Population Data Sheet. Washington, DC: Population Reference Bureau. Excellent source of information.

Repetto, R. (1987). Population, Resources, Environment: An Uncertain Future. *Population Bulletin* 42 (2): 1–44. Excellent survey of the impacts of future growth.

RESOURCES

7

Feeding the World's People: Food and Agriculture

It is not in the stars to hold our destiny but in ourselves.

SHAKESPEARE

Thomas Jefferson wrote that "civilization itself rests upon the soil." The first towns, early empires, and powerful nations can all trace their origins to the deliberate use of the soil for agriculture (Chapter 2). But as R. Neil Sampson wrote, in most places on earth "we stand only six inches from desolation, for that is the thickness of the topsoil layer upon which the entire life of the planet depends."

This chapter discusses hunger, malnutrition, and many of the problems facing agriculture. It also suggests solutions that could help us develop a sustainable world agricultural system.

The Dimensions of Hunger

By various estimates hunger afflicts between 17% and 40% of the world's people, mostly in Asia, Africa, and Latin America. These people may suffer from **undernourishment**, a lack of calories, or **malnourishment**, a lack of the proper nutrients and vitamins. In the United States 10% to 15% of the population suffers from hunger, according to some experts.

Diseases of Malnutrition

Worldwide, an estimated 40 million people die each year of starvation and diseases linked to malnutrition. This is the equivalent of 300 jumbo jets, each holding around

400 passengers, crashing every day of the year with no survivors. Many of the victims are children.

Medical scientists recognize two major types of malnutrition: **kwashiorkor** (resulting primarily from a lack of protein) and **marasmus** (resulting from an insufficient intake of protein and calories). Kwashiorkor and marasmus are two extremes of protein–calorie deficiency, and most individuals have symptoms of both diseases.

Kwashiorkor In the rural villages or city streets of Latin America, Asia, or Africa, children lie in their mothers' arms. Their legs and arms are wasted away, and their abdomens are swollen with fluids (Figure 7-1). They look up with sleepy eyes, moving only occasionally. These children are suffering from kwashiorkor, a protein deficiency common in children one to three years of age. This disease generally begins soon after children are weaned, when they are deprived of their mothers' protein-rich breast milk and are often fed a low-protein, starchy diet.

Marasmus Other children are thin and wasted (Figure 7-2). Their ribs stick out through wrinkled skin. They suck on their hands and clothes to appease a gnawing hunger. Unlike victims of kwashiorkor, these children are alert and active. Suffering from marasmus, they lack protein and calories, a situation which often occurs in children who are separated from their mothers during breast-feeding as a result of maternal death, a failure of milk production (lactation), or the use of milk substitutes promoted by multinational corporations. Through slick advertising campaigns, many women in the Third World were once persuaded to bottle-feed children, using powdered milk substitutes. After starting their children on the supplements, however, many women found that they could not afford them. By then, their breast milk had dried up. To compensate, some women diluted what milk they could afford with water from contaminated streams; their children developed diarrhea, which further reduced food intake and worsened their malnourishment.

In the Third World countries, for every clinically diagnosed case of marasmus and kwashiorkor there are hundreds of children with mild to moderate forms of malnutrition, a condition much more difficult to detect. Like the more severely malnourished children, these children are prone to infectious diseases.

Effects of Severe Malnutrition Growing evidence shows us that malnutrition early in life often leads to a permanent physical retardation. Furthermore, the more severe the deficiency, the more severe the stunting. Several studies indicate that childhood malnutrition may also impair brain development, perhaps permanently. This is because 80% of the brain's growth occurs before the age of two. Victims, therefore, may be mentally retarded and lack physical coordination.

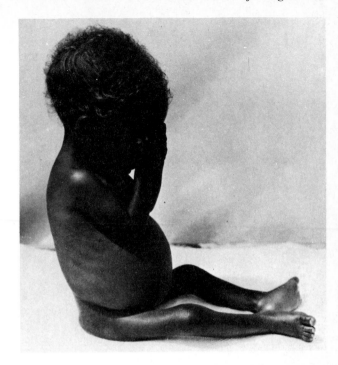

Figure 7-1. Kwashiorkor leads to swelling of the arms, legs, and abdomen. Children are stunted, apathetic, and anemic.

Malnourished children who survive to adulthood remain mentally impaired. They become the working citizens of the poor countries. Often plagued by malnutrition their whole lives, they remain prone to infectious diseases and provide little hope for improving their nation's agriculture, literacy rate, or economic level.

Declining Food Supplies

From 1950 until 1970 improvements in agricultural production and expansion of the land area under cultivation made significant inroads into world hunger. During this period world per capita grain consumption increased by approximately 30%, resulting in a substantial improvement in the diet of many of the world's people. From 1971 to 1984, however, world food production only kept pace with population growth. Between 1984 and 1988, food production per capita fell a surprising 15%. Should global warming, with its accompanying drought, and population growth continue, widespread hunger will worsen dramatically.

In the last two decades more and more countries have lost the ability to feed their people and have become dependent on wealthy agricultural nations such as Canada, Australia, and the United States. Food imports by the developing countries, which amounted to only a few million tons a year in 1950, are now over 100 million tons. But the safety net provided by major food-producing

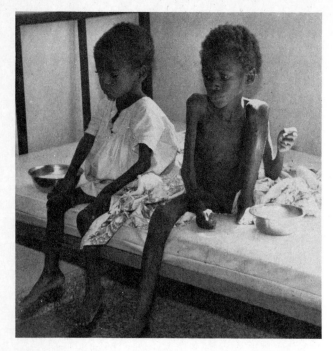

Figure 7-2. Victims of marasmus await medical attention at a hospital in Angola. Survivors of malnutrition may be left with stunted bodies and minds.

nations, such as the United States and Canada, may be failing. In 1988, US grain production fell by 35%. US farmers annually produce about 300 million tons of grain. Two-thirds of this grain is consumed domestically, another third is sold overseas or stored for lean years. Because of the hot, dry summer of 1988, grain production in the United States plummeted to only 190 million tons. That wasn't enough to supply domestic needs, let alone foreign demand. Thanks to previous surpluses, domestic and foreign demands were met, but now the vast surpluses are gone. Should subsequent years prove as unfruitful, foreign exports could slow to a trickle, leaving food-importing countries in danger.

Long-Term Challenges

Many food and agricultural experts believe that unless decisive, far-reaching steps are taken, widespread starvation is inevitable; millions will perish in the poor countries. The famine in Ethiopia, Chad, and the Sudan, in which hundreds of thousands of people have died in recent years, may foreshadow what is to come. Experts agree that the rich nations, to whom the Third World nations are highly indebted, will not be immune to the upheaval that results.

Three interrelated challenges face the world today: (1) feeding malnourished and undernourished people (the immediate problem), (2) meeting future needs for food

(the long-term problem), and (3) preventing deterioration of soil and water (the continuing problem). Building a sustainable worldwide agricultural system, coupled with vigorous population control programs and political and economic changes, could go a long way toward solving all of these problems. Before examining ways to build such a system, however, we will look at the major problems facing world agriculture.

Problems Facing World Agriculture

World agriculture can be summed up this way: In the poor nations many people cannot afford food. The cropland in many of these countries is often badly managed. Torrential rains and fierce windstorms carry the soil away, reducing agricultural output. Poor farming practices also slowly rob the soils of their nutrients. Because of the decay of farmland and the rising population, food supplies are falling. These countries must turn to the outside for help.

In the rich nations, in contrast, farmers generally turn out an abundance of food in most years, enough to feed their own people and anyone else who can afford to buy it. However, in the United States many farmers are going out of business, partly because of overproduction thanks to modern technology, fertilizers, and pest controls. Overproduction leads to lower prices, which ironically stimulate farmers to produce more in hopes that they can make their profit by high-volume production. Some government programs may also hurt farmers in the long run. For instance, government purchases of surplus grain encourage overproduction.

Modern agriculture has become a system that works best for large, wealthy corporations or those who hold vast acreage. "Get big or get out" was the message of the 1970s. To stay in the business, farmers bought expensive farm machinery—tractors that cost $60,000, for instance—and land at inflated prices. High interest rates on loans in the 1970s came back to haunt American farmers in the 1980s. Crop prices plummeted, as did land values, yet the cost of the things to run farms—fuel, fertilizer, and pesticides—continued to rise. Farming, in many cases, became a losing proposition. In fact, the US Department of Agriculture determined that in 1986 one farm went out of business every eight minutes. In 1988, 17,310 farms closed down operations, or about 50 farms every day.

The economics of farming also spelled trouble for the land and the long-term future of agriculture. Soil erosion, depletion of soil nutrients, and desert formation were the price unwittingly paid for mass producing cheap food. The following sections discuss many of the major problems plaguing agriculture throughout the world.

Soil Erosion

Soil erosion is the most critical problem facing agriculture today, both in the poor developing countries and in the much wealthier industrial nations. **Erosion** is the process by which rock and soil particles are detached from their original site by wind or water, transported away, and eventually deposited in another location. **Natural erosion** generally occurs at a slow rate, and new soil is usually generated fast enough to replace what is lost. **Accelerated erosion**, resulting from human activities such as overgrazing, decreases soil fertility, causing a decline in agricultural production. In the long term accelerated erosion can destroy land permanently (Figure 7-3). In the United States one-third of our topsoil has been lost to erosion since agriculture began. Soil erosion affects distant sites. For instance, pesticides may adhere to soil particles and be transported to nearby waterways. Sediment deposited in waterways increases flooding, destroys breeding grounds of fish and other wildlife, and increases the need for dredging harbors and rivers.

The economic damage from soil erosion in the United States is estimated to be more than $6 billion a year. Erosion annually destroys about 500,000 hectares (1.25 million acres) of US cropland. According to recent estimates, erosion exceeds replacement on nearly half of the country's farmland. All told, about three billion tons of topsoil are lost from American farms each year. Experts predict that if these trends continue, crop production will fall by 10% to 30% in the next 30 years. In the long run the US agricultural system could falter. Restoring topsoil on abandoned land could take 300 to 1000 years. (See Chapter Supplement 7-1 for more on soil erosion.)

Unfortunately, little information is available on soil erosion rates throughout the world. Experts agree, however, that the problem is significant in many regions, especially in the Third World nations. In China, the Yellow River annually transports 1.6 billion tons of soil from badly eroded farmland to the sea. The Ganges in India carries two times that amount. One conservative measure of topsoil loss puts the global figure at nearly 21 billion tons. Unfortunately, those countries with the fastest growing populations usually have the least money for soil conservation and the worst problems. As explained in Chapter 5, the rapid growth of population often precludes long-term care and proper management of natural resources, in the rich as well as the poor nations of the world.

Desertification: Turning Cropland to Desert

The United Nations Environmental Program recently predicted that by the end of the century one-third of the world's cropland will have turned into desert, a process

Figure 7-3. Soil erosion on farmland.

called **desertification**. Few countries are immune, because virtually all countries mistreat their soils. Desertification afflicts the United States, Africa, Australia, Brazil, Iran, Afghanistan, China, and India. Worldwide, an area the size of Belgium (about six million hectares or 15 million acres) becomes barren desert each year (Figure 7-4).

Desertification results from many factors: drought, the inherent fragility of the ecosystem in arid and semiarid (and subhumid) regions, overgrazing, overcropping (planting too many crops per year), deforestation, and plowing marginal lands. Overuse stems from overpopulation and the failure of existing socioeconomic systems to properly manage vulnerable lands.

Desertification is not new to humankind. In the ancient Middle East, for instance, the destruction of forests, overgrazing, and poor agricultural practices reduced local rainfall. Coupled with a long-term regional warming trend, the drop in rainfall turned once-productive pastureland and farmland in much of the Fertile Crescent, where agriculture had its roots, into desert (Chapter 2). In more recent times the United States found itself immersed in a major environmental crisis in the famous

Figure 7-4. Worldwide spread of deserts. Over 37 million square kilometers (14 million square miles) are now threatened. Six million hectares become desert each year.

dust bowl era of the 1930s. Farmers' fields turned dry, and huge dust storms swept the topsoil away. The dust bowl crisis was made more serious by extensive planting to supply Europe with food in the early years of World War II. When a natural drought struck, the crops failed, and heavy winds carried the soil away. Only through extensive conservation measures in the postwar years were farmers able to rebuild their soils.

Desertification, while widespread today, is especially bad in Africa. Beginning in 1968 a long-term drought in the Sahel region, coupled with overpopulation, overgrazing, and poor land management, began the rapid spread of desert southward in Ethiopia, Mauritania, Mali, Niger, Chad, and Sudan (Figure 7-5). The Sahara is also spreading northward, squeezing the people of North Africa against the Mediterranean. An estimated 100,000 hectares (250,000 acres) of rangeland and cropland are lost in this region each year.

Desertification and erosion, among other problems facing world agriculture, are already taking their toll. In Africa, a continent straining under the pressures of 650 million people, about 100 million do not have enough food to maintain normal health and physical activity. In Ethiopia, nearly one of every three people is undernourished. Nigeria is in a similar bind; one of every ten people is undernourished. In Chad, Mozambique, Somalia, and Uganda, 40% of the people are undernourished. Many are bound to die. Already, the death rate is

showing a substantial increase in many of these countries as famine worsens. In Madagascar, an island nation off the east coast of Africa, infant mortality, which serves as a ready indicator of nutritional deficiencies, rose from 75 per thousand in 1975 to 110 per thousand in 1989.

Deteriorating food supplies are being experienced elsewhere. In Latin America per capita grain production is on the decline. The number of undernourished preschool children in Peru now stands at nearly 70%. Infant mortality in Brazil is on the rise. Authors of the *State of the World*, published by the Worldwatch Institute, believe that "Latin America's decline in food production per person will almost certainly continue into the nineties."

Depletion of Soil Nutrients

In Chapter 3 you learned that farming can severely disrupt nutrient cycles. Excess fertilizer, for example, washes into lakes and streams, upsetting the balance. Harvesting crops also has a deleterious impact on the environment. Ten percent of the dry weight of plants is mineral matter picked up from the soil. Current farming practices remove most of the biomass of a crop. Thus, farmers essentially mine the soil; that is, they remove the minerals faster than they can be replaced. This problem is especially noticeable on lands planted two or three times a year and land fertilized only with artificial fertilizers.

Case Study 7-1

Stopping the Spread of Desert in China

China is a nation in trouble. With a growing population already well over one billion people, China's land is quickly falling into ruin. Centuries of overgrazing, poor agricultural practices, and deforestation have resulted in severe erosion and rapidly spreading deserts that gobble up the once-productive countryside.

The spread of deserts affects the lives of millions of peasants in China. In the Loess highland of northern China, for example, the land is cut with gullies, some hundreds of meters deep. Erosion from the raw, parched earth is an astounding 30–40 tons per hectare (12–16 tons of topsoil per acre) per year— far above replacement level. In all, some 1.6 billion tons are carried into the Yellow River annually, making it one of the muddiest rivers in the world.

According to one estimate, an area larger than Italy has become desert or semidesert in China in the last 30 years. Although most of the desertification is occurring in northern China, few areas are immune to the rampant destruction of the land.

In 1978, the Chinese government launched a reforestation project to stem the tide. By planting trees, shrubs, and grasses, they hope to form a giant green wall across the northern reaches of the nation. The wall will extend 6,700 kilometers (4,000 miles) and will be 400 to 1,700 kilometers (250 to 1000) miles wide. Moreover, it will return the land to productive use.

In 1985 the first phase of the program was completed. More than 5.9 million hectares (23,000 square miles) of barren land had been replanted with trees and shrubs, many of which have survived and put a stop to the growing desert. The second phase of the project will last another ten years. Government officials hope by the end of that period to have reforested nearly 6.4 million hectares (25,000 square miles).

The Yulin District is one of China's success stories. Before 1949, more than 400 villages and six towns had been invaded or completely covered by the encroaching sand. Today four major tree belts have been planted in the area, decreasing the southward push of the desert by 80%. Towering sand dunes now peep through poplar trees, and rice paddies sparkle in the sunshine. Grain production has been replaced by a diversified agricultural system, including animal husbandry, forestry, and crop production.

The trees provide shade and help reduce the shifting sand dunes. Shrubs and grasses now thrive on land once stripped of its rich vegetative cloak. Trees and shrubs grow in gullies and grasses carpet slopes, helping hold the soil in place and reversing the local climate change.

Local residents have built a diversified desert economy in what was once simply a desert. Juice from the desert cherry tree, which thrives in the desert climate and is extremely rich in vitamin C and amino acids, is now used to produce soft drinks, preserves, and beer. Twigs of the desert willow are used to make wicker baskets and trunks that earn local residents two million dollars in US currency every year.

Despite the encouraging signs in China, a report by the Shanghai-based World Economic Tribune says that, while nearly 10 million hectares (4,000 square miles) are planted every year, twice that amount is still being lost. In the northern province of Heilongjiang, home of China's largest concentration of virgin forest, loggers have reduced the tree cover from 50% to 35% in just 30 years. Government pricing policies promote overcutting.

The reforestation project is also plagued by a shortage of money and technical expertise. Because of short-sighted land use practices, some experts believe that China's Yangtze River, the nation's longest, could turn into a second Yellow River. Each year, its tributaries are turning muddier.

Reforestation is needed throughout the world to help reverse centuries of land abuse that have led to the spread of deserts and the gradual deterioration of the earth's soil. Forest replanting can also reduce global warming that now threatens the world climate.

Adapted from: Ming, L. (1988). Fighting China's Sea of Sand. *International Wildlife* 18 (6): 38–45.

Soil erosion also depletes the soil of nutrients. The annual cost of replacing nutrients lost from erosion in the United States is about $30 billion (in 1989 dollars).

High Energy Costs and Diminishing Supplies

Few relationships are as clear as that between the cost of food and the cost of fossil fuels, such as oil and natural gas. Oil byproducts power farm equipment. Oil and natural gas run food processing plants and factories that produce farm equipment. Natural gas is converted into nitrate fertilizers. Byproducts of oil are converted into pesticides that are sprayed onto fields from airplanes and tractors. Trucks, trains, and ships, powered by oil, move food across countries and across oceans. Even the stores that sell the food to the public require oil and natural gas for heating and lighting. So high is the energy

Figure 7-5. Desertification in the Sahel region of Africa.

demand, that modern agricultural societies invest 9 kilocalories of energy for each kilocalorie of food produced. Agriculture's dependency on fossil fuels became evident in the United States and elsewhere when petroleum prices increased from $3 a barrel to over $30 a barrel in the 1970s. Food prices climbed rapidly in tandem. Rising food prices hurt consumers in the developed nations. They also made things worse for developing countries by raising domestic production costs and raising import costs.

Oil is quickly being used up. By some estimates fewer than 60 to 70 years remain until world oil supplies are depleted. In the near term, expect rising prices. By 2000 or thereabouts, the demand for oil is expected to exceed supplies, a change that could fuel a spiral of inflation throughout the world. World population and food demand are expected to be at least 50% above current levels by 2005. Unless changes are made, more of the world's people will be unable to afford food and will suffer further hunger, starvation, and disease.

Water Mismanagement

Much of the world's food production is dependent on irrigation. In the United States, for instance, one-eighth of all cropland is irrigated; this land produces approximately one-third of the nation's food. Since 1950 the amount of irrigated cropland has doubled. Irrigated agriculture faces several problems: (1) depletion of groundwater, (2) competition for water supplies, (3) salinization, and (4) waterlogging of soil.

Agricultural Groundwater Depletion   In several regions in the United States groundwater is being depleted rapidly from **aquifers**, zones of porous material, such as sandstone, that contain water. Aquifers are naturally replenished by water from rain and snow that percolates through the soil and rock from the surface in **aquifer recharge zones**.

Aquifers are an important source of irrigation water. But many are being depleted (Figure 7-6). Of major concern is the depletion of groundwater in the Ogallala aquifer, a major prehistoric water deposit that is only very slowly being recharged by natural processes. (For more details on groundwater, see Chapter 10.) The Ogallala aquifer lies under the highly productive irrigated land of Nebraska, Kansas, Colorado, Oklahoma, Texas, and New Mexico. To date over 150,000 wells have been drilled into this aquifer, mostly for agricultural irrigation. Some parts of the aquifer, which took approximately 25,000 years to fill, have been depleted in only a few decades; water levels are falling 1 meter per year in heavily used areas, compared with a 1-millimeter replenishment rate. As a result many farm wells have already been abandoned. By the end of this century, groundwater experts predict, many parts of the aquifer will be drained. Irrigated agriculture in these regions will inevitably fail.

Competition for Water   Domestic and industrial use of groundwater can also deplete aquifers and reduce water available for agriculture. Already, municipal water demands in Arizona and California have made deep cuts into agricultural water use and put an end to much irrigated farming. In many western states agriculture competes with energy companies for water.

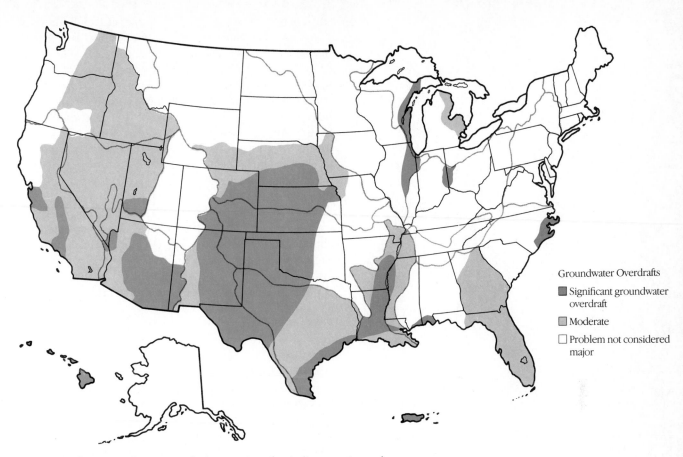

Figure 7-6. Groundwater overdraft. Areas in color indicate regions where water is being removed from aquifers by farmers, industries, and cities faster than it can be replenished.

Another source of competition is internal: the wasteful use of water by one sector of agriculture, which theoretically deprives others. In California, for example, approximately 450,000 hectares (over 1 million acres) of grassland are irrigated for cattle production. Grassland irrigation in this relatively dry climate consumes one-seventh of the state's water. But revenues from the cattle industry are only 1/5000 of the state's total annual earnings. Some critics argue that cattle production is a wasteful drain of water resources that deprives fish and wildlife as well as other more fruitful agricultural uses.

In the long run the loss of irrigated agricultural land—whatever the cause—will lower US food production, raising prices here and abroad.

Waterlogging and Salinization Irrigation has aided agriculture in many semiarid regions, but it has also created some problems. Irrigating poorly drained fields, for example, often raises the **water table**, the upper level of the groundwater. Two consequences of this practice are waterlogging and salinization (Figure 7-7). **Waterlogging** occurs if the water table rises too near the surface, filling the air spaces in the soil and suffocating the roots

of plants. It also makes soil difficult to cultivate. Second, water evaporates from the soil, leaving behind salts and minerals. The accumulation of these substances, called **salinization**, impairs plant growth and may make soil impenetrable.

Worldwide, one tenth of the irrigated cropland suffers from waterlogging. As a result productivity has fallen approximately 20% on this 21 million hectares (52.5 million acres) of cropland, an area slightly smaller than Idaho. In the United States salinization occurs to some degree on an estimated 300,000 hectares (800,000 acres). It is a growing problem in many other nations, including Pakistan, Iran, Iraq, Mexico, and Argentina. In Argentina alone, about 2 million hectares (5 million acres) of irrigated land has experienced a decline in productivity because of salinization. Worldwide, an estimated 40 million hectares (100 million acres) of irrigated farmland is suffering from salinization.

Conversion to Nonagricultural Uses

Farmland throughout the world is being taken out of production by urbanization, energy production, trans-

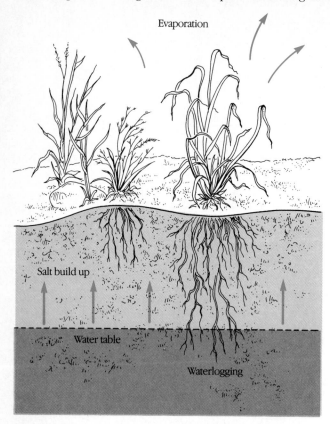

Evaporation

Salt build up

Water table

Waterlogging

Figure 7-7. Salinization and waterlogging. Salts and other minerals accumulate in the upper layers of poorly drained soil (salinization) when irrigation waters raise the water table and water begins to evaporate through the surface. The rising water table also saturates the soil and kills plant roots (waterlogging).

portation, and other nonfarm uses (Figure 7-8). Every hour about 90 hectares (220 acres) of actual or potential farmland in the United States is converted to nonfarm uses. This is equivalent to 2160 hectares (5400 acres) per day! In addition, about 600 hectares (1500 acres) of pastureland and rangeland is lost each day. Annually, the United States loses farmland equivalent to a strip about 1 kilometer (.62 miles) wide extending from New York City to San Francisco.

Farmland conversion is a worldwide phenomenon. West Germany loses about 1% of its agricultural land by conversion every four years, and France and the United Kingdom lose about 1% every five years. Little is known about the rate of agricultural land conversion in the developing world, but it is believed to be great.

Conversion of Cropland to Fuel Farms: A Future Problem

Some experts predict that when petroleum supplies begin to fall, around the year 2000, certain crops such as

sugar cane and corn will be used to make liquid fuels to power automobiles and other machinery. Microorganisms convert the sugar in corn, sugar cane, and other crops into burnable ethanol. This process is called **fermentation** and is basically the same process used to make wine, whiskey, and other alcoholic beverages. Alcohol can be mixed with gasoline in a ratio of one part of alcohol to nine parts of gasoline, creating **gasohol**. It can also be burned in pure form. In Brazil, for instance, cars and trucks now burn 100% ethanol.

Alcohol is a renewable energy resource, but tapping this largely ignored energy supply could have serious impacts on world food production. For example, Brazil eventually hopes to achieve complete self-sufficiency in automotive fuel with domestic ethanol production. Such a feat would require Brazilians to convert one-half of their farmland to fuel production, greatly reducing food exports. Should ethanol replace oil worldwide, major food producers such as Canada, the United States, and Australia will divert grain used for food to fuel production. One-fifth of the exportable corn produced in the United States would be needed to produce 2 billion gallons of ethanol, a fraction of total US liquid fuel needs. As fuel farms rise in prominence, exportable surpluses could fall. Food prices will inevitably soar, which will affect all peoples, rich and poor. Today, most countries import some grain, and nearly a dozen countries import more than half of the grain they consume each year. Unless more countries become self-sufficient, the world's hungry will be forced to compete with the automobile fuel industry and, ultimately, the automobile users of the world for precious grain (Figure 7-9). Few people doubt who will win.

Politics and World Hunger

Many of the world's food problems are political. Some agricultural experts maintain that plenty of food is produced each year but that about half of this output in the less developed countries never reaches the table. One reason is that governments, with good intentions or fraudulent ones, often become involved in the production and distribution system, reducing food availability.

In the Soviet Union, for example, government officials control the large farms. On a regional basis they decide how much fertilizer should be applied, rather than individually tailoring fertilizer demands to local needs. Such decisions ultimately reduce productivity. The failure of centralized governmental controls is underscored by the fact that private production, which takes place on about 1% of the country's farmland, yields about one-fourth of the yearly crop.

Political decisions in one country can also have tremendous impacts in other countries. In 1954, for example, the US Congress enacted legislation that drove

Figure 7-8. Urban sprawl, as shown here in Des Moines, Iowa, swallows up farmland at an alarming rate.

thousands of Third World farmers off their land. This law—the Farm Surplus Disposal Act, often called the Food for Peace program—authorized huge surplus grain shipments from the United States to Third World nations. This free or nearly free surplus grain entered countries with devastating consequences. Perhaps the most important effect was that Third World farmers, unable to compete with the cheap grains, were forced to abandon their farms. Ironically, Food for Peace stimulated US production, but as mentioned earlier in this chapter, government support had far-reaching effects on domestic agriculture, most notably increased production and decreased prices.

Government policies also have many subtle long-range impacts on agricultural production. A US program called payment in kind (PIK), enacted during President Ronald Reagan's first term, encouraged farmers to curtail production. The ultimate goal was to reduce grain gluts that drive prices down. The program worked this way: Farmers who had cultivated a piece of land for more than two years were allowed to take it out of production; to offset possible losses the government gave them a grain payment from federal stockpiles. Grain payments amounted to 80% to 95% of their expected production. The farmer was then free to sell the grain on the market. The offshoot of this program was that land speculators got into the farm business, bought marginal land, planted it for two years, and then took it out of production so that they could collect the government grain payments. This practice was euphemistically known as "sod busting." However, much of this land was rangeland that was not well suited to farming. The already meager topsoil quickly eroded away. Through payment in kind, critics argue, the government encouraged soil erosion on good rangeland or land that could have been used for grazing or crop

production in the future in conjunction with good soil conservation.

Politics and economics have had a great deal to do with the declining ability of many nations to feed their people. In Africa, for example, many countries must produce cash crops (coffee and tea, for instance) for export to the more developed countries to generate money to pay off their huge foreign debts. The net effect is that cash crops (such as strawberries) for export are produced in place of staples (such as corn) for domestic consumption. Governmental policy strongly influences this unsustainable agricultural policy. In Mexico, for instance, most credit for irrigation systems and roads is given to farmers who produce cash crops, such as tomatoes, for export to the United States. Cattle grazing is also given a high priority in Mexico and is supported by the government. But cattle pastures usurp farmland once used to produce crops for domestic consumption, and most cattle are destined for the United States.

Countless examples of government interference in agriculture exist. One final example illustrates the complex problem facing Ethiopia. There, farmers traditionally left land fallow for seven-year periods so that nutrients from the highly weathered, poor soil could be replenished by natural vegetation. This practice is now condemned by the Ethiopian government, which is interested in increasing farm production. If land is not cultivated within three years, it is confiscated. But not leaving it fallow will result in rapid deterioration and a long-term decrease in productivity.

Loss of Genetic Diversity

Before the advent of modern agriculture, grains and veg-

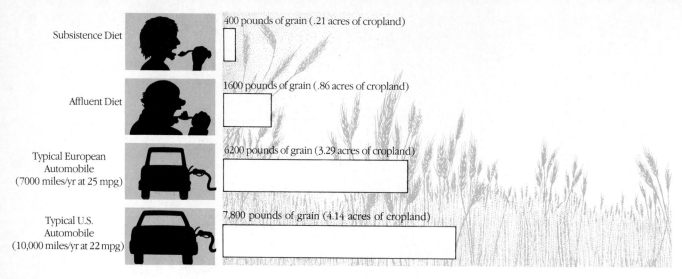

Subsistence Diet — 400 pounds of grain (.21 acres of cropland)

Affluent Diet — 1600 pounds of grain (.86 acres of cropland)

Typical European Automobile (7000 miles/yr at 25 mpg) — 6200 pounds of grain (3.29 acres of cropland)

Typical U.S. Automobile (10,000 miles/yr at 22 mpg) — 7,800 pounds of grain (4.14 acres of cropland)

Figure 7-9.   Grain appetites: people versus cars. Competition for "food" is about to begin, with the wealthy car owners in a better position to pay.

etables existed in thousands of varieties. Now, only a few of these varieties are commonly used (Table 7-1). (The same trend is occurring on ranches throughout the world as ranchers adopt breeds developed for maximum yield.) Many experts believe that this loss of variety, or **genetic diversity**, could have a significant impact on world agriculture.

Before looking at those effects, consider some of the main reasons why genetic diversity is dwindling. First and foremost, the new varieties often have a higher yield on mechanized farms. Between 1903 and 1976, for instance, new varieties of wheat allowed American farmers to double their yield; new varieties of corn allowed them to quadruple output per acre. These varieties were also better suited for machine harvesting and responded well to irrigation. Finally, seed companies benefited economically by concentrating their efforts on a few varieties.

The development of high-yield varieties in the 1960s is part of a worldwide agricultural movement often called the **Green Revolution**. It began in 1944, when the Rockefeller Foundation and the Mexican government established a plant-breeding station in northwestern Mexico. The program was headed by Norman Borlaug, a University of Minnesota plant geneticist, who developed a high-yield wheat plant (Figure 7-10). Before the program began, Mexico imported half of the wheat it consumed each year, but by 1956 it was self-sufficient in wheat production. By 1964 it was exporting half a million tons. Borlaug was later awarded a Nobel Prize for his work.

The success in Mexico led to the establishment of another plant-breeding center, in the Philippines. High-yielding rice strains were developed there and introduced into India in the mid-1960s. Again, the results were spectacular. India more than doubled its wheat and rice

production in less than a decade and has become self-sufficient in wheat production.

Important as it was, the Green Revolution contributed greatly to the decrease in agricultural diversity. One of the most important concerns is the loss of genetic resistance to disease. Local varieties of plants are acclimated to their environment; natural selection has ensured this. New varieties, on the other hand, often have little resistance to insects and disease. Moreover, planting of **monocultures**, expansive fields of one genetic strain, facilitates the spread of disease and insects. As discussed in Chapter 4, simplifying ecosystems removes environmental resistance that normally keeps potential pest populations in check. The potato famine in Ireland in the 1840s is one famous example of the effects of reducing diversity. Only a few varieties of potatoes were planted in Ireland. When a fungus (*Phytophthora infestans*) began to spread among the plants, there was little to stop it and no backup supply of resistant seed potatoes. Within a few years two million Irish perished from hunger and disease, and another two million emigrated. In addition to their susceptibility to disease, high-yield hybrids are also generally less resistant to drought and flood. In 1970 the southern corn leaf blight wiped out nearly one-fourth of the US corn crop. The American peanut crop, consisting of two varieties, was almost entirely destroyed in 1980 by drought and disease.

Reduced genetic diversity is a trend that began thousands of years ago during the agricultural revolution as humanity turned from hunting and gathering to an agricultural way of life (Chapter 2). That trend is worsened by a relatively new phenomenon: the destruction of tropical lands where many of the ancestors of our modern crops grow. Their fate is crucial to agriculture. Why? The future will require plants that can survive climatic

Table 7-1   Limited Diversity in American Agriculture

Crop	Varieties Available	Major Varieties in Use	Percentage of Total Production
Corn	197	6	71
Wheat	269	10	55
Soybeans	62	6	56
Rice	14	4	65
Potatoes	82	4	72
Peanuts	15	9	95
Peas	50	2	96

Source: Reichart, W. (1982). Agriculture's Diminishing Diversity. *Environment* 24(9): 6–11, 39–44.

changes, drought, disease, and insects. These ancient ancestors could provide the genetic material needed to improve our crops. Thus, reducing genetic diversity through deforestation and farming could have far-reaching effects on the future of agriculture. (For more on this problem see Chapters 7 and 8.)

The agricultural expert R. Neil Sampson once noted that "the progress of civilization has been marked by a trail of wind-blown or water-washed soils." Today, the destruction continues in the form of depleted soils, falling groundwater supplies, waterlogging, salinization, cropland conversion, political mismanagement, and declining agricultural diversity.

Much of the world's agricultural land is poorly managed. If the trend continues, we could conceivably destroy much of the renewable soil on which civilization rests. The population crash discussed in Chapter 5 could result from the widespread mistreatment of our agricultural land. In the following section we look at ways to convert our failing agricultural system into a sustainable system.

Building a Sustainable Agricultural System

Hunger and starvation are as much a problem of too many people as a problem of food shortages. One of the most important solutions to famine, then, is population control (see Chapter 6). Beyond that, we can (1) increase agricultural land; (2) grow more on it—that is, increase productivity; (3) develop alternative foods; (4) reduce food losses to pests; (5) increase the agricultural self-sufficiency of developing nations; and (6) find political solutions to ensure better distribution of food and to eliminate the production of cash crops instead of food crops for domestic consumption.

Increasing the Amount of Agricultural Land

Increasing the amount of cropland and rangeland available to us can be achieved primarily in two ways: (1) by tapping farmland reserves, that is, land not being used, and (2) by preventing loss of cropland from desertification, soil erosion, and farmland conversion.

Exploiting Farmland Reserves   For many years the United States and other agricultural nations have solved the problem of rising food demand by opening up new lands to the plow. The abundance of untapped farmland in the frontier days fostered reckless attitudes that resulted in poor land management. But the days of abundant reserves are quickly coming to an end.

US farmers currently cultivate 167 million hectares (413 million acres). The cropland reserve—land that can be farmed if necessary—is a paltry 51 million hectares (127 million acres). Projected increases in foreign and domestic demand will use up this land by the year 2000. In most of the other major agricultural nations, farmland reserves are also small. If land destruction continues at the current rate, therefore, the 21st century will be marked by falling production, food shortages, and soaring prices. With fuel farms taking over farmland, the outlook for agriculture becomes dimmer. For Third World nations one inescapable conclusion can be drawn: they will have to become increasingly self-sufficient. The developed nations won't have the food reserves to support themselves and very many developing nations. Agricultural self-sufficiency may require the developing nations to slow down and eventually stop population growth. Third World nations will also have to develop their own land, improve agricultural efficiency, reduce soil erosion and nutrient depletion, improve transportation, and conserve irrigation water.

According to the United Nations Food and Agriculture Organization, Africa and South America have large sur-

Figure 7-10. Comparison of an old variety of wheat (left and right) with a new short-stemmed high-yield variety (center). The first of the short-stemmed varieties was developed under the direction of Norman Borlaug.

pluses of land that could be farmed. In Africa, for example, only 21% of the potentially cultivatable land is in use. In South America only 15% is being farmed. Some experts believe that South American and African nations should develop this land, but only if the currently farmed land is protected against erosion and nutrient depletion. However, few experts believe that tropical rain forests should be developed as farmland. Soils there are poor in nutrients, easily erode in intense rains, and may become hardened when exposed to sun. (See Chapter Supplement 3-1 for more details.)

Tapping other unfarmed land might be an option, but it would severely affect many wildlife populations. Protecting wild species is important for many reasons, as discussed in Chapter 8. One of the key economic reasons for preserving wildlife is that wildlife itself can become a major source of income. Tourists spend considerable amounts of money to view African wildlife. Some wildlife may be "harvested" on a sustainable basis, further supporting local economies. Brazil, for example, has set aside a large tract of rain forest to protect native species and also to provide sustainable income for local people. Fruits, nuts, berries, and rubber can be harvested from the forests without destroying the trees or the wildlife that depend on them (see Chapter 9). Environmentalists hope that the first reserve will lead to many others throughout the tropics and could help halt the devastating destruction of this important biological resource.

In other areas of the world agriculture strains the limits of land availability. In Southeast Asia, for instance, 92% of the potential agricultural land is being farmed. In southwestern Asia more land is currently being used than is considered suitable for rain-fed agriculture. Per capita food production in Asia has begun to fall as population continues to increase. Population control and improved farming practices are needed to preserve this land and ensure a sustainable system.

Reducing the Spread of Deserts Developing new farmland is one answer to the imminent shortages of arable land. More important are measures to reduce the destruction of land currently farmed or grazed by livestock. Throughout the world people are working to stop the spread of deserts. In China agricultural officials have begun to plant a 6,900-kilometer (4,300-mile) "green wall" of vegetation to stop the spread of desert in the northern region (see Case Study 7-1). In Australia huge semicircular banks of earth are created in the windswept plains to catch seeds and encourage regrowth in areas denuded by livestock. In Iran migrating sand dunes are sprayed with a residue left over from oil production. The liquid dries on the surface and forms a gray mulch that retains soil moisture and facilitates the growth of drought-resistant plants. Within six years oil-mulched dunes are green with vegetation. Additional efforts worldwide include improving rangeland management, revegetating barren land, controlling wind erosion (Figure 7-11), preventing the use of marginal lands, and irrigating.

Soil Conservation Just as important as stopping the march of deserts are efforts to cut back on soil erosion. The rich nations as well as the poor will benefit. To date, however, farmers in most developed and developing countries have done little to reduce erosion. In the devel-

Figure 7-11. Among the vegetation to the left, protected from overgrazing, are 7-year-old trees, which in time will protect the land from the severe erosion pictured to the right.

oping world farmers struggle to meet their most basic needs and have neither the time nor the means to care properly for the land. Furthermore, few can see the benefits of soil conservation, because the gains tend to materialize slowly and usually take the form of a decrease in losses rather than an increase in actual food output.

Economics impairs soil-erosion control in the developed nations (discussed in Chapter 20). Caught between high production costs and low prices for grains, farmers may ignore the long-term effects of soil erosion while synthetic fertilizers artificially help them maintain yield in the short term. Governments can promote conservation through a variety of measures, many of which are discussed in Chapter Supplement 7-1. In 1985 the US Congress passed a farm bill that contains some important controls on soil erosion. The law denies any federal farm loans, subsidies, or even crop insurance to farmers who plow highly erodible land. The law also creates a land conservation program in which the federal government will pay farmers to remove highly erodible land from crop production for ten years and plant trees, grasses, or cover crops to stabilize the land and rebuild the soil. Farmers had "retired" an estimated 12 million hectares (30 million acres) by 1989, cutting erosion by 800 million tons per year.

Other Protective Measures Protecting land from desertification and soil erosion are two important steps in preserving farmland and, therefore, increasing our supply of arable land. But efforts to prevent the spread of cities, the proliferation of highways, and other nonfarm uses of arable land are also needed. Careful city planning and new zoning laws could help reduce farmland conversion by ensuring that homes, roads, airports, and businesses are not built on agricultural land. (For more on urban planning, see Chapter 21.)

Increasing the Yield of Cropland

Preserving Genetic Diversity In a forest in southwestern Mexico in 1979, four scientists discovered a few tiny patches of a wild, weedy-looking grass that could have enormous impact on corn production throughout the world. *Zea diploperennis*, a primitive near relative of modern corn whose known population numbers only a few thousand stalks, sets itself apart from all modern corns because it is a perennial; that is, unlike an annual it grows from the same root structure year after year rather than from seeds. To many people the rare corn may not seem very impressive, but to geneticists, the engineers of new corn breeds, it's the discovery of the century.

Corn, like any other crop, is only as good as the genetic "boosts" it receives, in other words, the infusions of fresh genes, which provide its resistance to disease, drought, insects, and so on. Valuable new genes come from wild plants or early varieties still grown by peasants in remote corners of the world. The new genetic vigor they confer to corn results in large harvests, and because corn accounts for one-fourth of the world's cereal grains, a large harvest is always crucial.

Genetic boosts have an enormous impact on corn production. In the last 60 years, for instance, corn harvests have more than quadrupled—from 20 bushels per hectare to 100 to 250 bushels—because of successful genetic improvements. Hence, breeders are always on the lookout for the very rare varieties that could help produce hardier, more resistant plants.

The primitive corn found in Mexico is also special because it is highly resistant to several diseases. The genes that provide this resistance could be introduced into conventional corn to produce a resistant perennial hybrid that grows in fields much like grass. This new plant would reduce erosion and save farmers annual costs of plowing, sowing, and cultivating. The potential savings are enormous.

New Plant and Animal Varieties Preserving the ancient relatives of modern crops is only one way to increase the yield of existing farmland. Another is to develop new plants and animals. New high-yield varieties of rice and wheat developed during the Green Revolution, for example, can produce three to five times as much grain as their predecessors when grown under proper conditions. New varieties of plants produced by breeding closely related plants to combine the best features of the parents are called **hybrids**.

As the new hybrids were introduced into many poor nations, the hopes of the Green Revolution dimmed, for farmers soon found that the hybrids required large amounts of water and fertilizer, unavailable in many areas. Without these, yields were not much greater than those of local varieties; in some cases they were even lower. The cost of the new varieties prevented many small farmers from buying them; as a result, they often went out of business as larger farms converted to the high-yield varieties. Also, new plants were often more susceptible to insects and disease.

The Green Revolution, written off by its critics as a failure, was the first step in a long, tedious process of plant breeding aimed at improving yield. Today, plant breeders throughout the world are developing crops with a higher nutritional value and greater resistance to drought, insects, disease, and wind. Plants with a higher photosynthetic efficiency are also in the offing. Efforts are even under way to incorporate the nitrogen-fixing capability of legumes (discussed in Chapter 3) into cereal plants such as wheat, a change that would decrease the need for fertilizers and reduce nitrogen depletion.

One exciting improvement announced in 1988 is a new variety of corn. Corn is a staple for 200 million people worldwide, many of whom are chronically malnourished. Because corn is such an important source of calories and protein throughout the world, researchers sought to find ways to improve it. The result of nearly two decades of work is a product called quality-protein maize (QPM). Studies show that only about 40% of the protein in common corn is ultimately used by humans. In contrast, roughly 90% of QPM's protein can be digested and used. In countries where corn is a staple, such as Africa and Mexico, QPM could help curb malnutrition. It could also help these countries become more self-sufficient in food production if costs could be lowered to make it affordable.

Some researchers are exploring the use of perennial crops for agriculture. **Perennials**, briefly described above, are those plants that produce flowers, fruit, or seeds year after year from the same root structure. Today, most agricultural crops are **annuals**, plants that must grow anew from seeds each year. Preliminary research suggests that productivity from perennials may be equal to or slightly lower than conventional annuals such as wheat, but the benefits from soil conservation, soil-nutrient retention, and energy savings may overwhelmingly favor them.

Just as new varieties of plants help increase yield, so do fast-growing varieties of fowl and livestock. Efforts are being made to improve plants and livestock by **genetic engineering**. This is a complex process involving several steps. For instance, scientists first identify genes that give plants resistance and other important properties that might increase yield. Next they isolate the gene and chemically analyze it so they can make copies of it. Finally, the gene is inserted into seeds, which then sprout and produce cells each containing the advantageous gene. The gene can be transferred to a plant's offspring.

Genetic engineering may be used in other ways that improve agriculture. A new strain of bacteria developed by scientists at the University of California inhibits the formation of frost on plants, thus potentially offering farmers a way of reducing crop damage and increasing yield. Other scientists are working on genes that give plants resistance to herbicides used to control weeds. A group of scientists at the Monsanto Company has developed a strain of bacteria that grows on the roots of corn and other plants. When eaten by insects, the bacteria release a toxic protein that kills the pest. Animal geneticists are now working on ways to improve livestock, combining genes from one species with those of another to improve efficiency of digestion, weight gain, and resistance to disease.

Genetic engineering is a much quicker way to produce new varieties than conventional plant breeding. However, it has generated a lot of public criticism. Much fear has been raised over its potential to create hazardous life forms that could escape into the environment, disrupting ecosystems and possibly killing people. Many scientists believe that these and a host of other concerns should not be dismissed lightly. Despite new standards for laboratory testing of genetically engineered organisms, overall controls on genetic engineering in the United States are, by many standards, inadequate. For example, five different agencies currently regulate the environmental release of genetically engineered microbes. Unfortunately, there is little coordination of efforts. Furthermore, these agencies operate under outdated laws, designed principally to control toxic chemicals and not genetically engineered organisms that can reproduce, mutate, and multiply when released. Critics argue that there is too little review of proposals for release. They also say that the federal agencies are inadequately prepared to handle the rising number of applications for release expected in the near future. For more on genetic engineering, see Chapter Supplement 7-2.

(a)

(b)

Figure 7-12. Increasing the efficiency of irrigation. (a) Elevated pipes reduce water loss. (b) Trickle systems deliver water to roots, cutting evaporation losses.

Soil Enrichment Programs Soil conservation helps preserve farmland but can also preserve soil fertility. Other methods for maintaining soil fertility include the use of artificial or organic fertilizers and crop rotation. These methods are covered more thoroughly in Chapter Supplement 7-1, on soil management.

Improving Irrigation Efficiency Improving irrigation efficiency could help reverse the inevitable decline in irrigated farmland caused by groundwater overdrafting. Fortunately, many techniques are available to reduce water consumption on farms and in cities. (Urban water conservation is discussed in Chapter 10.) On farms, lining irrigation ditches with cement saves a lot of water (and money), as does the use of pipes (Figure 7-12a). Farmers can also use drip irrigation systems to deliver water directly to the roots of the plants (Figure 7-12b). Computer systems can help conserve water by monitoring soil moisture content. These and other techniques are replacing currently wasteful practices and will probably grow in use as water supplies drop and prices for water rise.

New Foods and Food Supplements

In many countries scientists have been working to develop food supplements, using such things as algae and fish that are often rich in protein, carbohydrate, and other important nutrients. Because of consumer resistance, high costs, and other factors, supplements only slightly increase the food supply. A much more promising food source is native species.

Native Species The biosphere is an untapped biological reservoir of plants and animals, many of which may have a higher yield and better nutritional value than those we currently use. The winged bean of the tropics, for example, could be a valuable source of food, because the entire plant is edible: its pods are similar to green beans, its leaves taste like spinach, its roots are much like potatoes, and its flowers taste like mushrooms. Food scientists are looking for other plants with similar potential.

Native animals may also provide an important, sustainable food source in years to come. In Africa, for instance, native grazers are far superior to cattle introduced from Europe and America. They carry genetic resistance to disease and rarely overgraze grasslands, unlike cattle. Native grazers also generally convert a higher percentage of the plant biomass into meat and may be cheaper to raise.

Fish from the Sea By some estimates, fish provide about 5% of the total animal protein consumed by the world's population. Although three-quarters of the fish catch is consumed in the developed nations, fish protein is important to the diets of many poorer countries, in many cases supplying 40% of the total animal protein consumed.

(a)

(b)

Figure 7-13.   (a) Commercial catfish farm near Monticello, Arkansas. These feeders release food pellets when a catfish nuzzles an extended rod. (b) Fish raised in irrigation ditches and ponds in China and other countries supply needed protein.

Until 1971 increasing the world fish catch was viewed as an important way to increase the world's food supply. Between 1950 and 1970 the catch tripled, reaching approximately 70 million metric tons per year. Since the early 1970s, though, the world catch has stabilized between 66 and 74 million metric tons a year, despite intensive efforts to increase yield. On a per capita basis the fish catch declined in the 1970s by about 15% and continued to fall in the 1980s. Biologists warn that any increases could lead to the collapse of ocean fisheries. Maintaining current yields requires global cooperation to prevent overfishing and water pollution.

One method to increase fish supplies is to reduce spoilage by improving refrigeration in the entire production–consumption cycle. Some scientists believe that this step by itself could increase the supply of fish in the developing nations by 40%.

Commercial Fish Farms: Mariculture and Aquaculture

Commercial fish farms might also help increase fish protein supplies. **Fish farms** are forms of aquatic agriculture, called **aquaculture** in fresh water and **mariculture** in salt or brackish water.

There are two basic strategies in fish farming. In the first, fish are grown in ponds; population density is high and is maintained by intensive feeding, which is costly and, therefore, not suited to developing countries (Figure 7-13a). Fish (and shellfish) can also be maintained in enclosures or ponds where they feed on algae, zooplankton, and other fish naturally found in the aquatic ecosystem (Figure 7-13b). This system requires little food and energy and is suitable for poorer countries. Worldwide, fish farms produce about 4 million metric tons of food a year. Intensified efforts could double or triple this amount.

Eating Lower on the Food Chain

Many environmental science textbooks propose that wealthier citizens of the world can contribute to solving world hunger by eating lower on the food chain. The reasoning behind this idea is as follows: By consuming more grains, vegetables, and fruits—and less meat—citizens of the developed nations would free grain for the developing nations. A 10% decrease in beef consumption in the United States, for instance, would release enough grain to feed 60 million people. Chapter 3 showed that many more people could be sustained on 20,000 calories of corn if it were consumed directly than if it were first fed to cattle and the meat was then fed to people. The advantages of eating lower on the food chain are clear. Beef cattle produce about 19 kilograms (43 pounds) of protein per acre per year, whereas soybeans produce nearly 200 kilograms per acre.

The problem with this apparently logical answer to world hunger is that few people respond to moral appeals. Even more important, if Americans and other wealthy world citizens were to cut back on beef consumption, thus freeing tons of grain, the developing nations still might not be able to afford it. Sacrifices here would most likely not translate into gains abroad. Another problem is that land suitable for grazing is often not arable. The lesson to be learned from our study of ecology is that to feed their people, the Third World nations should concentrate on grain production rather than meat.

Reducing Pest Damage and Spoilage

Rats, insects, and birds attack crops in the field, in transit, and in storage. Conservatively, about 30% of all agricultural output is destroyed by pests, spoilage, and diseases. In the developing nations this figure may be much higher, especially in humid climates where crops are grown year-round and conditions are optimal for the spread of crop diseases and insects. Better controls on crop pests could increase the global food supply. For a detailed discussion of pest control see Chapter 17.

Inefficient transportation can delay food shipments. Meanwhile, rats, insects, birds, or spoilage may claim part of the supply. Spoilage can be prevented by refrigeration and other improvements. Grain, for instance, can be stored in dry silos or sheds to prevent the growth of mold and mildew and hold down rat populations. Technical and financial assistance from the developed countries could go a long way toward improving food storage and transportation, potentially increasing world food supply by 10% or more.

Increasing Self-Sufficiency

Increasing self-sufficiency among poorer nations is an essential part of decreasing hunger. Recommending increased self-sufficiency in today's "global economy" may seem out-of-step. For many countries, however, agricultural self-sufficiency may be one of their only hopes for survival. Many people in the Third World are exceedingly poor and few can afford imported grain or other foodstuffs. In the near term, shortages created by global warming could cut exports, or eliminate them. Rising fuel prices are bound to make food production costs in exporting nations increase dramatically in the coming years. Food crops could be diverted to liquid fuel production as well, forcing prices higher, and making food unaffordable. In some respects, then, individual nations might be better off feeding their own people, rather than depending on imports.

Self-sufficiency could be achieved by technical and financial aid aimed at building a sustainable agricultural system that (1) uses nonrenewable energy efficiently, (2) relies primarily on human labor, (3) conserves soil and soil nutrients, (4) uses water efficiently, (5) minimizes water pollution, (6) relies on a high level of species diversity, (7) reduces soil disruption from plowing and cultivating, and (8) uses perennial rather than annual crops.

Political and Economic Solutions

Solving world hunger will require dramatic changes in governmental policy throughout the world. Laws and policies that promote unsustainable farming practices must be abandoned or changed. A few of these were discussed earlier. To help ensure long-term production, water laws that encourage waste should also be changed. In the West, for example, farmers or ranchers who cut back on water they are allocated lose the rights to that water, thus eliminating incentive to be frugal. Simple changes in water laws could free up enormous amounts of water for future use. On another front, some farmers may find it advantageous to abandon crops that use large amounts of irrigation water. California's Central Valley, for instance, receives only about 10 centimeters (4 inches) of rainfall a year but supports cotton and rice fields. Rice requires the equivalent of 2.5 meters (100 inches) of rainfall. Cotton also requires large amounts of rain. Shifting production to more suitable climates could free up enormous amounts of water.

An Integrated Approach

This chapter began by listing three problems in world agriculture: the near-term need to feed malnourished and undernourished people, the long-term need to provide food for the coming generations, and the continuing need to preserve the soil on which our civilization depends. It should be clear that there are many ways to solve these problems. But which one is right? Critical thinking suggests that the answer is that not one but many of these ideas must be adopted and integrated into a policy to build a sustainable agricultural system. No single solution will suffice. That means we must find political solutions to hunger as well as practical changes in the way we till the land and use water. Taken together the solutions discussed in this section can help us build a sustainable system. Anything short of this is, according to some observers, doomed to fail in the long run.

To a man with an empty stomach, food is God.

Gandhi

Sharpening Your Critical Thinking Skills: Analyzing the Costs and Benefits of Agricultural Water

In this chapter we saw that agricultural irrigation water is being depleted from the Ogallala aquifer faster than it is being replenished. Annually, the net water loss is about 14 million acre-feet (1 acre-foot covers 1 acre to a depth of 1 foot). One of the proposed solutions to this problem is to divert water from the Missouri River at St. Joseph, Missouri, 850 feet (255 meters) above sea level, and pump it in a canal and pipeline to agricultural areas in western Kansas (elevation 2480 feet), eastern Colorado (3000 to 4000 feet), and Texas (3000 to 3500 feet).

Question

Do a cost-benefit analysis of this agricultural problem.

Step 1: Costs and Benefits

To balance costs and benefits, first make a list of all possible costs and benefits involved in this project. What are the major economic and environmental costs? What are the major benefits? How would you assign a numerical value to costs and benefits?

Step 2: Calculating Annual Pumping Costs

As you might have guessed, one of the chief economic costs would be incurred by pumping water uphill from St. Joseph, Missouri, to farms in Colorado, Kansas, and Texas. Let's assume that the water is going to be pumped to an average elevation of 3000 feet to make calculations easier. The electricity to pump 1 acre-foot of water up 188 feet costs about $25 (in 1979 dollars). How much will it cost to pump 14 million acre-feet from St. Joseph to 3000 feet?

Unfortunately, the cost of electricity is probably only about half the total annual cost. Repayment of loans, interest, maintenance, and personnel make up the other half. With this in mind, what is the total annual cost for pumping this water?

Step 3: Gross Returns on Corn Sales

Now that you've approximated the annual pumping costs, let's look at the approximate economic benefits. First, 14 million acre-feet of water will irrigate 8.38 million acres of land, yielding 754 million bushels of corn. If corn sells for $3 per bushel (1979 dollars), what would the gross return from corn sales be? Remember, the $3 selling price does not take into account any production costs. To be accurate, you'd have to subtract the cost of seed, fertilizer, herbicides, pesticides, labor, farm

equipment, land rental or mortgage, and other costs. You might call some farmers in your area or contact an agricultural specialist at your school and find out how much profit a farmer makes per bushel of corn so you can calculate the net return.

From an economic standpoint, how feasible does this project look? What alternatives might you suggest to meet the demand for irrigation water?

Answers

Step 1: Costs and Benefits

Costs
pumping cost
canal and pipeline construction costs
canal and pipeline maintenance
environmental costs
 reduced water flow in Missouri River
 effects on fish and wildlife
 effects on pollution concentrations
 effects on water supply for cities and farms
Benefits
continuance of irrigated agriculture
economic benefits
 profit
 jobs for farm laborers
 spinoff economic benefits—community services such as stores and farm equipment sales

Step 2: Calculating Annual Pumping Costs

2150 feet = total elevation water must be pumped
Cost to pump 1 acre-foot 2150 feet:

$$2150 \text{ feet} \times \frac{\$25}{188 \text{ feet}} = \$286/\text{acre-foot}$$

Annual cost to pump 14 million acre-feet:

$$\$286/\text{acre-foot} \times 14 \text{ million acre-feet} = \$4 \text{ billion}$$

Total cost for pumping, assuming pumping costs are only one-half of annual costs:

$$\$8 \text{ billion}$$

Step 3: Gross Returns on Corn Sales

$$754 \text{ million bushels} \times \$3/\text{bushel} = \$2.3 \text{ billion}$$

Summary

Throughout the world 40 million people die annually of starvation and diseases worsened by malnutrition. The major nutritional diseases worldwide are **kwashiorkor** (protein deficiency) and **marasmus** (protein–calorie deficiency). Growing evidence suggests that malnutrition early in life leads to a permanent retardation of mental and physical growth.

Satisfying future food demand and feeding those people alive today in a sustainable manner are three of the major challenges facing agriculture. But many problems stand in the way of the goals: soil erosion, nutrient depletion, desertification, dwindling supplies of fossil fuels, groundwater depletion, waterlogging, salinization, and farmland conversion. Politics compound many of these problems.

Food experts are concerned also about the loss of species diversity, arguing that it makes our agriculture system more vulnerable to insects, disease, drought, and other natural calamities. Today, a few high-yield species have replaced a variety of species that once provided food.

Numerous strategies must be employed to meet current and future food demands. Population control is first in importance. Beyond that, we can increase the amount of agricultural land in production by exploiting currently unused farmland, by decreasing erosion on existing land, by controlling the spread of deserts, and by reducing farmland conversion.

In conjunction with these efforts, food production (yield) can be increased by developing new plant and animal varieties, by soil conservation and enrichment, and by irrigation. New plant and animal varieties can be developed through artificial selection or by genetic engineering, but in order for these efforts to be successful, new varieties must not lose their resistance to disease, insects, and other natural factors that tend to reduce yield. Some agricultural researchers are exploring the use of perennial crops for agriculture. Such a system would reduce soil tillage, energy demands, and soil erosion. Careful soil enrichment programs in which natural organic wastes are used to replenish lost nutrients can be successful in retaining productivity. Finally, since much of our agricultural production comes from irrigated land and since many supplies of irrigation water are now threatened, more efficient use of water could extend production on land that otherwise might have to be abandoned.

For the most part, manufactured food supplements have proved unacceptable to people. Native species, especially grazers, can provide protein and have many advantages over introduced species. Fish provide about 5% of the total animal protein consumed by the world's people. Increasing the fish catch, however, is unlikely and could probably be carried out only on a short-term basis before populations were depleted. A better strategy would be to reduce spoilage. Commercial fish farms might make it possible for rich and poor nations to double or triple their fish production.

Each year a large amount of food is destroyed by insects, bacteria, fungi, and rodents during production, transportation, and storage. Simple, cost-effective measures could be devised to reduce this loss.

Limits to the world's petroleum supply suggest the need for more agricultural self-sufficiency among the poor as well as the rich. Such independence can be developed by employing many of the strategies outlined in this chapter. Most important, rich and poor nations need to devise a sustainable agricultural system that uses a minimum amount of nonrenewable energy, relies on human labor, conserves soil and soil nutrients, uses water efficiently, minimizes water pollution, relies on a high level of species diversity, reduces soil tillage, and uses perennial rather than annual crops.

Discussion Questions

1. What percentage of the world's population is malnourished? What are the short- and long-range effects of malnutrition?

2. What has happened to world per capita food supply in the last two decades?

3. What is desertification, what factors create it, and how can it be prevented?

4. Discuss the statement "Soil erosion control is too expensive. We can't afford to pay for it, because our crops don't bring in enough money."

5. How does the long-term outlook for oil affect world agriculture?

6. Describe waterlogging and salinization of soils. How can they be prevented?

7. Describe the decline in agricultural diversity. How could this trend affect world agriculture? Give some examples.

8. List and discuss the major strategies for solving world food shortages. Which are the most important? How would you implement them in this and foreign countries?

9. Debate the statement "Simply by practicing better soil conservation and replenishing soil nutrients we can reduce the need for developing new farmland."

10. Describe the Green Revolution, its successes and its failures. What improvements might be made?

11. Describe trends in the world fish catch. What happened to per capita yield in the 1970s? Did the 1980s show any improvements? Is it likely that the annual fish catch will increase in the coming years? Why or why not?

12. Should the developing countries become more self-sufficient in agricultural production? What are the advantages of self-sufficiency for less developed countries? What are the disadvantages?

13. You have been appointed head of a UN task force. Your project is to develop an agricultural system in a poor African nation, which imports more than 50% of its grain and still suffers from widespread hunger. Outline your plan, giving general principles you would follow and specific recommendations for achieving self-sufficiency.

Suggested Readings

Berry, W. (1977). *The Unsettling of America. Culture and Agriculture.* San Francisco: Sierra Club Books. Extraordinary discussion of American society, agriculture and ecology.

Brown, L. R., Chandler, W. U., Flavin, C., Jacobson, J., Pollock, C., Postel, S., Starke, L., and Wolf, E. C. (1987). *State of the World 1987*. New York: Norton. Chapter 7 summarizes important gains and further needs in agriculture.

Brown, L. R., Durning, A., Flavin, C., Heise, L., Jacobson, J., Postel, S., Renner, M., Shea, C. P., Starke, L. (1989). *State of the World 1989*. New York: Norton. Chapters 1–3 provide an excellent overview of problems and solutions.

Erhlich, A. (1988). Development and Agriculture. In *The Cassandra Conference. Resources and the Human Predicament*. P. R. Ehrlich and J. P. Holdren, eds. College Stations, TX: Texas A & M Press. Important essay in an excellent collection of articles on the environmental crisis.

Jackson, W., Berry, W., and Colman, B., eds. (1984). *Meeting the Expectations of the Land. Essays in Sustainable Agriculture and Stewardship*. San Francisco: North Point. Extraordinary collection of writings on sustainable agriculture.

Ming, L. (1988). Fighting China's Sea of Sand. *International Wildlife* 18 (6): 38–45. Superb discussion of the problems of and solutions to desertification.

Paddock, J., Paddock, N., and Bly, C. (1988). *Soil and Survival. Land Stewardship and the Future of American Agriculture*. San Francisco: Sierra Club Books. Worthwhile reading.

Pimental, D. (1988). Industrialized Agriculture and Nature Resources. In *The Cassandra Conference. Resources and the Human Predicament*. P. R. Ehrlich and J. P. Holdren, eds. College Stations, TX: Texas A & M Press. Important reading.

Reichert, W. (1982). Agriculture's Diminishing Diversity. *Environment* 24 (9): 6–11, 39–44. A superb discussion of the loss of diversity among crops and domestic livestock.

Reisner, M. (1989). The Emerald Desert. *Greenpeace* 14 (4): 6–10. Extraordinary exposé of water policy in the West.

Sampson, R. N. (1981). *Farmland or Wasteland: A Time to Choose*. Emmaus, PA: Rodale Press. Well-written book on world agriculture.

Schneider, S. H. (1988). Climate and Food: Signs of Hope, Despair and Opportunity. In *The Cassandra Conference. Resources and the Human Predicament*. P. R. Ehrlich and J. P. Holdren, eds. College Stations, TX: Texas A & M Press. Worthwhile reading.

Solkoff, J. (1985). *The Politics of Food*. San Francisco: Sierra Club Books. Excellent account of US agricultural policy.

Tanji, K., Lauchli, A., and Meyer, J. (1986). Selenium in the San Joaquin Valley. *Environment* 28 (6): 6–11, 34–39. Detailed account of the problems caused by irrigation.

Soil and Soil Management

> *To build may have to be the slow and laborious task of years. To destroy can be the thoughtless act of a single day.*
>
> WINSTON CHURCHILL

Throughout the world nations treat soil as if it were an inexhaustible resource. In the United States, for example, 500,000 hectares (1.25 million acres) of farmland are lost each year because of soil erosion. On some farms erosion is so great that two bushels of soil are lost for every bushel of corn produced (Figure S7-1).

To build a sustainable agriculture, erosion must be halted. This supplement discusses soil and soil formation and presents some cost-effective ways to reduce erosion and create a sustainable agricultural system.

Figure S7-1. Areas of critical erosion in the United States. Peaks represent the highest rates of erosion.

What Is Soil?

Soil is a complex mixture of inorganic and organic materials with variable amounts of air and moisture. Clay, silt, sand, gravel, and rocks are the inorganic components of soil. Detritus, organic wastes, and a multitude of living organisms are the organic components. Soils are described according to six general features: texture, structure, acidity, gas content, water content, and biotic composition.

Soil Formation

Soil formation is a complex and often slow process, even under the best of conditions. The time it takes soil to develop depends partly on the type of **parent material**, the underlying substrate from which soil is formed. To form 2.5 centimeters (1 inch) of topsoil from hard rock may take 200 to 1200 years, depending on the climate. Softer parent materials such as shale, volcanic ash, sandstone, sand dunes, and gravel beds are converted to soil at a faster rate (in 20 years or so) if conditions are favorable.

A number of physical processes contribute to soil formation. Daily heating and cooling may cause the parent rock material to split and fragment, especially in climates such as deserts where daily temperatures can vary widely. Water entering cracks in rocks expands in freezing, causing the rock to fragment further. Rock fragments formed by heating and cooling and by freezing are slowly pulverized into smaller particles by streams or landslides, by hooves of animals, or by wind and rain.

Soil formation is facilitated by a wide range of organisms. Chapter 4 described how lichens "gnaw" away at the rock surface by secreting carbonic acid. Lichens also capture dust, seeds, excrement, and dead plant matter, which help form soil. The roots of trees and large plants reach into small cracks and fracture the rock. Roots also serve as nutrient pumps, bringing up inorganic nutrients from deeper soil layers. These chemicals are first used to make leaves and branches, which can fall and decay, thus becoming part of the soil.

Grazing animals pulverize rock and gravel under their hooves and drop excrement on the ground, adding to the soil's organic

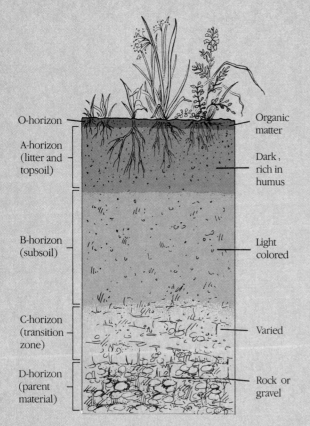

O-horizon

A-horizon (litter and topsoil)

B-horizon (subsoil)

C-horizon (transition zone)

D-horizon (parent material)

Organic matter

Dark, rich in humus

Light colored

Varied

Rock or gravel

Figure S7-2. Soil profile, showing the five horizons. Not all are found in all locations.

matter. The white rhinoceros, for example, produces about 30 tons of manure each year, which is deposited in its territory. A variety of insects and other creatures, such as earthworms, also participate in soil formation.

The Soil Profile

Soil is often arranged in layers of different color and composition. These layers are called **horizons**.

Soil scientists recognize five major horizons (Figure S7-2). The uppermost **O horizon**, or litter layer, a thin layer of organic waste from animals and detritus, is a zone of organic decomposition characterized by a dark, rich color. Plowing mixes it in with the next layer.

The **A horizon**, or **topsoil**, varies in thickness from 2.5 centimeters (1 inch) in some regions to 60 centimeters (2 feet) in the rich farmland of Iowa. This horizon is generally rich in inorganic and organic nutrients and is economically important because it supports crops. The A horizon is darker and looser than the deeper layers. It holds moisture because of its organic material, the **humus**, and is quite porous. The A horizon is also known as the "zone of leaching," since nutrients are leached out of it as water percolates through it from the surface.

The **B horizon**, or **subsoil**, is also known as the "zone of accumulation," because it receives and collects minerals and nutrients from above. This layer is lightly colored and much denser than the topsoil because of a lack of organic matter. The **C horizon** is a transition zone between the parent material

below and the layers of soil above. The **D horizon** is the parent material from which soils are derived. Not all horizons are present in all soils; in some, the layering may be missing altogether.

The soil profile is determined by the climate (especially rainfall and temperature), type of vegetation, parent material, age of the soil, and the organisms. Soil profiles tell soil scientists whether land is best for agriculture, wildlife habitat, forestry, pasture, rangeland, or recreation. They also tell us how suitable soil might be for various other uses such as home building and highway construction.

Soil Management

Soil management has two major goals: erosion control and nutrient preservation.

Erosion Control

Erosion can be controlled by a variety of techniques. This section discusses six major strategies: minimum tillage, contour farming, strip cropping, terracing, gully reclamation, and shelterbelts. These measures may raise the cost of farming in the short term, but, as Chapter 20 points out, they make good economic sense in the long run.

Minimum Tillage As the name implies, **minimum tillage** is a way of farming that reduces the physical disruption of the soil caused by plowing or cultivating crops for weed control. Typically, farmers plow their fields before planting a new crop. With special implements, however, they can plant right over the previous year's crop residue.

According to the US Department of Agriculture, in 1986 minimum tillage was practiced on over 27 million hectares (67 million acres) of land in the United States (Figure S7-3). In 1987, the USDA reported a decline to 19 million hectares (46 million acres). Because fields are protected much of the year by crops or crop residues, soil erosion can be decreased substantially—in some cases by as much as 90% (Figure S7-4). Minimum tillage also reduces energy consumption by as much as 80% and conserves soil moisture by reducing evaporation. Crop residues can increase habitat for predatory insects that help hold pest populations in check. This may reduce pesticide use and contamination of the environment (Chapter 17).

Despite its benefits minimum tillage has several drawbacks. For example, herbicides are often used in place of mechanical cultivation to control weeds. In addition, crop residues may harbor harmful insects that damage crops. Minimum tillage also requires new and costly farm equipment (Figure S7-4).

Contour Farming On hilly terrain, crops can be planted along level lines that follow the contour of the land. This is called **contour farming**. As illustrated in Figure S7-5, rows are planted across the direction of water flow on hilly or sloped land. This reduces the flow of water across the land, resulting in a tremendous reduction in erosion and a marked increase in water retention. In a Texas experiment water runoff from a contoured field was only 4 centimeters (1.6 inches) per year, compared with 11.6 centimeters (4.6 inches) per year for uncontoured fields. Soil erosion can be reduced by 60% to 80%.

Strip Cropping   As illustrated in Figure S7-6, **strip cropping** involves planting strips of alternate crops in fields that are subject to soil erosion by wind or water. Strip cropping can be combined with contour farming to further decrease erosion. As an example, farmers may alternate row crops such as corn with cover crops such as alfalfa. Water flows more easily through row crops and begins to gain momentum, but when it meets the cover crop, its flow is virtually stopped.

Terracing   For thousands of years many peoples have grown crops on terraces in mountainous terrain. Terraces have also been used in the United States for over 40 years on land with less pronounced slope (Figure S7-7).

Terracing today involves the construction of small earthen embankments on sloped cropland. These are placed across the slope to check water flow and minimize erosion. Terraces are expensive to construct and often interfere with the operation of large farm equipment; thus, farmers may prefer other, cheaper forms of soil conservation.

Gully Reclamation   Gullies are a danger sign of rapid soil erosion. Some gullies can work their way up hills at a rate of 4.5 meters (15 feet) a year.

To prevent gullies from forming, farmers must reduce water flow over their land. Contour farming, strip crops, and terraces all help. Special diversion ditches can be dug in channels where water naturally flows off fields. Already formed gullies can be stopped by seeding barren soils with rapidly growing plants, including trees. Small earthen dams can be built across them to reduce water flow, retain moisture for plant growth, and capture sediment, which will eventually support vegetation as erosion is reduced. Too often land with severe gullies is abandoned or haphazardly reclaimed, only to suffer worse erosion in time.

Shelterbelts   In 1935 the US government mounted a campaign to prevent the recurrence of the disastrous Dust Bowl days. One part of this program involved planting long rows of trees as windbreaks, or **shelterbelts**, in a north–south orientation along the margins of farms in the Great Plains (Figure S7-8). Today, thousands of kilometers of shelterbelts have been planted from Texas to North Dakota.

Shelterbelts block the wind and, therefore, decrease soil erosion and damage to crops caused by windblown dirt particles. In the winter they reduce the amount of blowing snow, thus allowing snow to build up in fields. This insulates the soil and also increases soil moisture and groundwater supplies. They can also improve irrigation efficiency by reducing the amount of water carried away from sprinklers. In addition, shelterbelts provide habitat for animals, pest-eating insects, and pollinators. Shelterbelts protect citrus groves from wind that blows fruit from trees. They have the added benefit of saving energy by reducing heat loss from homes and farm buildings.

A University of Nebraska study showed that terracing and contour farming each reduce erosion by 50%. Combined, they can reduce soil loss by 75%. Minimum tillage is even more effective. By itself it can reduce erosion rates by 90%, but when it is combined with terracing and contour farming, soil erosion rates can be reduced by 98%.

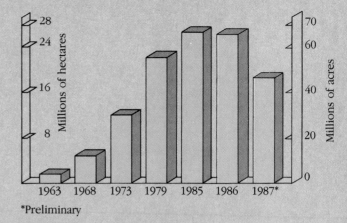

*Preliminary

Figure S7-3.   Growth in minimum (conservation) tillage in the United States. Source: USDA.

In a sustainable agricultural system erosion control assumes a position of great importance. Inexpensive techniques that are available now can slow down the rate of erosion and help us ensure food for future generations. Regardless, many farmers are reluctant to invest in such practices. An economic explanation of this phenomenon is given in Chapter 20.

Preventing the Depletion of Soil Nutrients

Preventing soil erosion saves topsoil and also preserves soil nutrients needed to maintain food output. Several ways to prevent the depletion of soil nutrients were mentioned in Chapter 7 and are discussed in more detail below.

Use of Organic Fertilizers   Organic fertilizers such as cow, chicken, and hog manure and human sewage are excellent soil supplements which replenish organic matter and important soil nutrients such as nitrogen and phosphorus. Other organic materials may be used to build soil; especially helpful are leguminous plants such as alfalfa that are grown during the off-season and plowed under before planting crops. Referred to as "green manure," they add organic matter to the soil, retain moisture, and also reduce soil erosion during fallow periods.

Soil enrichment with organic fertilizers (1) improves soil structure, (2) increases water retention, (3) increases fertility and crop yield, (4) provides a good environment for the bacterial growth necessary for nitrogen fixation, (5) helps prevent shifts in the acidity of soil, and (6) tends to prevent the leaching of minerals from soil by rain and snowmelt. In addition, the use of human wastes on farmland could significantly reduce water pollution by waste treatment plants.

Organic wastes have been successfully applied in many areas, but this form of fertilizing, like almost anything else we do, can create some problems. One of the leading ones is transporting waste to farms, which requires pipelines or trucks. Initial investment may be high, but the long-term benefits of applying

Figure S7-4.   Minimum tillage planter, designed to dig furrows in the presence of crop residue.

Figure S7-6.   Strips of corn protect small grain plants from the effects of wind.

Figure S7-5.   This land is farmed along the contour lines to reduce soil erosion and surface runoff, thus saving soil and moisture alike.

Figure S7-7.   This Iowa corn is grown on sloping land with the aid of terraces, small earth embankments that reduce water flow across the surface. The corn is planted in the stubble of last year's crop.

Figure S7-8.   Shelterbelts used to protect farmland in Michigan from the erosive effects of wind.

organic fertilizers could outweigh these costs. Another problem is that waste from municipal sewage treatment plants may be contaminated with pathogenic organisms such as bacteria, viruses, and parasites. Theoretically, some of these could be taken up by crops and reenter the human food chain. Toxic heavy metals, such as mercury, cadmium, and lead, are also present in municipal wastes and could enter crops. Better controls at sewage treatment plants or at the factories that produce these materials could alleviate the problem.

Use of Synthetic Fertilizers   In the United States synthetic fertilizer is the greatest source of nitrogen used by farmers. It is made by chemically combining nitrogen and hydrogen to form ammonia and ammonium salts. Some nitrates and urea are also made. Artificial fertilizers can be applied directly to the soil as a liquid, or they can be mixed with phosphorus and potassium in a dry, granular fertilizer.

Synthetic fertilizers containing nitrogen, phosphorus, and potassium—and made from natural gas—partially restore soil fertility but do not replenish organic matter or micronutrients necessary for proper plant growth and human nutrition. As a result soil is slowly degraded over years of agriculture. Excess fertilizer may be washed from the land by rains and end up in streams, causing a number of problems. (These are addressed in Chapter 16, on water pollution.)

To prevent the slow depletion of nutrients and to help develop a sustainable agricultural system, synthetic fertilizers should be supplemented by organic fertilizers. Synthetic fertilizers can probably be eliminated entirely in the developing nations or on small farms near abundant sources of organic fertilizer. Combined with programs of soil conservation and organic enrichment, synthetic fertilizers will inevitably play an important role in feeding the world's people. As fossil fuels become scarcer, however, synthetic fertilizers may become more costly.

Crop Rotation   In modern agriculture synthetic fertilizers and pesticides have allowed farmers to grow the same crop year after year on the same plot. This way farmers can concentrate their efforts on one crop that they know well. However profitable this practice might be, it is ecologically unsound, for it gradually depletes the soil of nutrients, can increase soil erosion, and often worsens problems with pests and pathogens. Crop rotation, once a common practice among farmers in the United States, could return us to a more ecologically sound and sustainable agricultural system.

Using the age-old practice of rotating crops, a farmer may plant a soil-depleting crop (corn) for one to two years and then follow it with a cover crop such as grasses or various legumes such as alfalfa. The cover crop reduces soil erosion, and legumes replenish soil nitrogen. Often, cover crops are not harvested but are simply plowed under to replenish organic matter and return nutrients to the soil.

Putting It All Together: Organic Farming   Today, thousands of farmers are running profitable farms either without chemical fertilizers and pesticides or with only small amounts of them. Called **organic farming**, this system avoids or largely excludes synthetic chemical compounds such as fertilizers, pesticides, growth regulators, and livestock food additives, relying instead on crop rotation, green manure, animal wastes, off-farm organic wastes, mechanical cultivation, and biological pest control to the maximum extent possible. Believing that a sustainable and productive agricultural system is more important than high yields alone, organic farmers concentrate on keeping their soil in top condition.

The success of organic farms baffles those who have come to think of modern agriculture as the only way to farm. The ecologist Barry Commoner compared organic and conventional farms of similar size and found that the organic farms produced nearly as well. Further research has supported this conclusion: organic farms produce about 11% fewer vegetables than a comparable "modern" farm, but the net income per hectare is about the same because of the lower costs. Organic farms require 60% less fossil fuel energy to produce corn crops with only 3% lower yields.

Not all organic farms perform as well as their modern counterparts. But they offer other benefits that are worth serious consideration. In two adjacent farms near Spokane, Washington, researchers compared the impacts of organic and conventional farming on soil. Both farms were begun in 1909. One farm, however, still relies on "old" practices, such as crop rotation. The second farm was converted to "new" practices in 1948. It uses inorganic fertilizers and pesticides and is planted year after year on the same land, returning little organic matter back to the soil.

To determine how these different forms of agriculture affect the land and its long-term productivity, researchers compared the topsoil thickness of the two operations. Their study showed that organic farming dramatically reduced soil erosion. Let's look a little more closely at the operations. The organic farm in the study operates on a three-year cycle, producing winter wheat the first year, then a crop of spring pea the following year. In the third year, Austrian winter pea is planted, but this crop is plowed under to provide nutrients and organic matter for the soil. The conventional farm, on the other hand, alternates between crops of winter wheat and spring pea. Although crop yields are similar for both farms, the organic farm produces a cash crop on a given field only two out of every three years. To some that may seem like a serious problem. But that's because agriculture is viewed narrowly by many people as a matter of production.

Earlier studies showed that organically farmed soil had higher levels of organic matter, a larger mass of organisms, and more soil enzymes. In many respects, then, organic farming ensures a healthier topsoil. The most recent study, published in 1988, showed that an organic farm's topsoil was, on the average, 15 centimeters (6 inches) thicker than its neighbors'. In addition, organically farmed soil had more moisture and a softer crust. From this perspective, organic farming wins hands down.

When analyzing agriculture, many so-called experts talk solely in terms of yield per hectare—the output of the land. It's a little like an employer who looks only at the amount of work an individual can produce per day, without any concern for the worker's health or the quality of the workmanship. Critical thinking suggests a broader look at agriculture, for example, the long-term prospects for the land, the effects of farming on the surrounding environment, and the health of the farmers and their families. When researchers computed the loss of topsoil on the two Washington farms, they found that the conventionally farmed soils, with slopes similar to those on the organic farm, could lose their topsoil within 50 years, exposing the dense, less fertile clay subsoil and, ultimately, destroying the land's productivity. The organically farmed land, on the other hand, could produce forever. Organic farming ensures a long-term supply of food. One of the chief goals of agriculture is to feed people. But if it is done at the expense of the land, it is surely an invitation for disaster.

There are some limits to organic farming: (1) organic fertilizers may not replace all of the phosphorus and potassium drained from the soil by plants, (2) nitrogen replacement by legumes may not always be enough to give high yields of crops, such as corn, that have a high nitrogen requirement, (3) total production may be lower than on a conventional farm because of land devoted to soil conservation, and (4) there is a three- to five-year transition period from conventional to organic farming during which yields may be low and losses to insects and weeds may be high.

Despite these problems, organic farming could become the wave of the future here and abroad. Increasingly, consumers are becoming wary of fruits and vegetables that are grown with pesticides. The market demand for organically grown produce is increasing each year.

To be sustainable, modern agriculture must undergo a radical shift. But the wholesale adoption of organic farming techniques will not come quickly. Farming has become a big business which, like so many others, is dominated by corporate giants and an industrial mentality. Modern agriculture subscribes to the philosophy that bigger is better. Although profits per bushel are small, farmers make up for them in volume. That leads to extensive monoculture, a practice with considerable problems. One of the most serious consequences is the loss of productive topsoil.

In developing countries, opportunities to shift to sustainable farming practices are numerous. For most countries this is their only hope of preventing disaster. Given the high cost of fossil fuels and the abundance of manual labor, organic farming is ideally suited to many Third World nations. The rapid growth of population, unfortunately, has shifted attention to the immediate problem of hunger, ignoring the continuing problem: how to farm without destroying the resource that makes all life possible.

Achieving Cost-Effective Conservation

Controlling soil erosion and soil fertility are as much economic problems as they are resource conservation problems. Because of rising production costs, short-term profits often supersede long-term matters of fertility and erosion control. What is most urgent to the farmer is the profit needed to keep up payments on land, fertilizer, machinery, pesticides, and seed, not the long-term stability of world agriculture.

Many farmers do not view soil erosion as a major problem, because fertilizers and new high-yield seeds have boosted crop yields and masked the effects of excessive erosion. In taking the shortest path between the top and bottom line on their balance sheets, farmers ignore practices that preserve the soil at its maximum productivity.

How can we achieve cost-effective soil conservation? Below are some suggestions:

1. Target areas where erosion is the most severe (see Figure S7-1). Currently, half of all cropland erosion in the United States comes from about 10% of the total farmland. Concentrated efforts on this land could be highly cost-effective in reducing soil erosion.

2. Expand programs by governmental agencies, such as the US Soil Conservation Service, to familiarize farmers with the most cost-effective ways of controlling erosion.

3. Enact agricultural zoning regulations locally to prohibit highly erodible land from being farmed.

4. Offer tax breaks to farmers to help defray the cost of soil conservation.

5. Require farmers applying for government loans, crop insurance, or aid to have an approved soil conservation program already in operation.

6. Strengthen the role of the states in soil conservation. Increase cooperation among state, local, and federal agencies involved in soil conservation.

Suggested Readings

Brock, B. G. (1982). Weed Control versus Soil Erosion Control. *Journal of Soil and Water Conservation* March/April: 73–76. A thorough study of weed control by tillage and herbicides.

Brown, L. R., Durning, A., Flavin, C., Heise, L., Jacobson, J., Postel, S., Renner, M., Shea, C. P., Starke, L. (1989). *State of the World 1989*. New York: Norton. Chapters 1–3 provide an excellent overview of problems and solutions.

Ehrlich, A. (1988). Development and Agriculture. In *The Cassandra Conference. Resources and the Human Predicament*. P. R. Ehrlich and J. P. Holdren, eds. College Stations, TX: Texas A & M Press. Important essay in an excellent collection of articles on the environmental crisis.

Jackson, W., Berry, W., and Colman, B., eds. (1984). *Meeting the Expectations of the Land. Essays in Sustainable Agriculture and Stewardship*. San Francisco: North Point. Extraordinary collection of writings on sustainable agriculture.

Jeffords, J. M. (1982). Soil Conservation Policy for the Future. *Journal of Soil and Water Conservation* January/February: 10–13. A view of the needs of soil conservation policy in the United States.

Owen, O. S. and Chiras, D. D. (1990). *Natural Resource Conservation*. New York: Macmillan. Excellent sections on agriculture, soils, and land management.

Postel, S. (1985). *Conserving Water: The Untapped Potential*. Worldwatch Paper 67. Washington, DC: Worldwatch Institute. Detailed study of agricultural water conservation.

Sampson, R. N. (1981). *Farmland or Wasteland: A Time to Choose*. Emmaus, PA: Rodale Press. Well-written book on agriculture and soil conservation.

US Department of Agriculture. (1980). *America's Soil and Water: Condition and Trends*. Washington, DC: Author. Valuable reference.

Vietmeyer, N. (1986). Lesser-Known Plants of Potential Use in Agriculture and Forestry. *Science* 232 (June 13): 1379–1384. Superb article on alternative food sources.

Wolf, E. C. (1986). *Beyond the Green Revolution: New Approaches for Third World Agriculture*. Worldwatch Paper 73. Washington, DC: Worldwatch Institute. Detailed assessment of agricultural prospects in developing countries.

Biomes

The biosphere is divided into geographically distinct land parts called biomes, each having its own specialized climate, plants, and animals. These diverse regions, made up of intricately balanced ecosystems, offer us a kaleidoscope of information on evolution and adaptation, which increases our understanding of the way the world works within the constraints of nature.

And yet, whenever we think we understand the workings of a biome (its climate, its topography, its predictability), we experience a Mt. St. Helens—a reminder from nature that life on earth is constantly in flux; changing, adjusting, adapting. Like nature, humans have altered biomes and ecosystems—but unlike nature, many of our changes have become irrevocable.

Through an understanding of the biomes we can learn about structure and function of the environment, which in turn will give us an awareness of how our world works—and, just as importantly, how it doesn't.

1 Atypical of the desert biome, the Great Sand Dunes National Monument in southern Colorado is lifeless and virtually barren.

2 The rim of the Grand Canyon in Arizona shows yet another face of the desert biome.

3 Cactus abound in the desert, blossoming briefly after the spring rains.

4 The tropical rain forest is perhaps the most diverse of all biomes. It is estimated that over 90% of the living organisms in these rain forests remain unclassified. Venezuela.

5 Rain forest in the Olympic Peninsula, Washington. Receiving over 200 inches of rain each year, this area maintains a lush growth of mosses, ferns, epiphytes, and trees.

6 The towering Alaska range overlooks the coniferous forests of the taiga, a biologically rich biome that extends across North America and much of Europe and Asia. Denali National Park, Alaska.

7 The taiga consists of large stands of one or two species of coniferous tree. Meadows occur naturally in mountain valleys, as shown here. Jasper Park, British Columbia.

8 An autumn view of the temperate deciduous forest biome.

9 Golden aspens herald autumn in the Maroon Bells-Snowmass Wilderness area, Colorado. High altitude yields a mix of coniferous and deciduous trees.

4

2

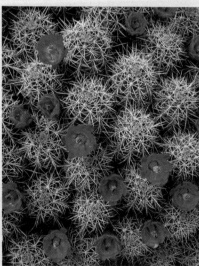

3

5

6

7

8

9

10 Alpine tundra is characterized by a short growing season, prohibiting the growth of trees or shrubs. Wildflowers, like the columbine shown here, grow densely in these areas.

11 Summery wildflowers and lichens add color to arctic tundra in the Soviet Union. Despite minimal precipitation, tundra areas remain damp from the presence of permafrost—permanently frozen soil that resists water absorption.

12 Flint Hills Preserve in Alena, Kansas, is one of the last remaining examples of the tall-grass prairie.

13 The African savannah is richly populated by grazing species that feed off trees, grass, and herbs. Kenya, Masai Mara Game Reserve.

10

11

12

13

The Promises and Perils of Genetic Engineering

No technological advance offers as much hope for a better future as genetic engineering, and none promises to revolutionize as much of our society—how we treat diseases and how we grow plants and rear livestock. Consider some of the promises:

The Promises

Scientists have successfully implanted the gene for human growth hormone into livestock embryos in hopes of producing marketable cattle much faster than by conventional means.

To help combat infections, researchers have isolated a bacterial gene that produces a naturally toxic chemical and have successfully transplanted that gene into another bacterium that grows on plant roots, giving plants a new way of warding off pests.

To help combat salinization, geneticists have introduced into oats certain genes that allow them to thrive in salty soils. Researchers are also experimenting with genetically engineered bacteria that could help plants absorb nutrients more efficiently, thus increasing crop yields.

On another front, genetic researchers have developed a mutant bacterium that retards frost development on plants. This development promises to save farmers millions each year.

In an attempt to help protect soils from nutrient depletion, genetic engineers are experimenting with ways to transplant the genes that would allow nonleguminous plants to fix atmospheric nitrogen. If successful, this could help farmers reduce fertilizer needs and water pollution created by excess fertilizer use.

Through genetic engineering, plant scientists are also developing a bacterium that produces a natural antifungal agent, protecting wheat from the devastating infestation of the take-all fungus, a disease which causes millions of dollars a year in crop damage.

The successes of genetic engineering have fostered extraordinary enthusiasm in the business community. Dozens of new companies have formed in recent years, and billions of dollars have been invested in the fledgling industry.

Nevertheless, safety questions remain. Will the genetically engineered bacteria escape into the environment, upsetting ecological balance? Experts agree that, once unleashed, a new form of bacterium or virus would be impossible to retrieve. Controlling it could prove costly and damaging.

Some individuals have criticized genetic engineering as a means of tinkering with the evolutionary process. Deliberate genetic manipulations, such as the transfer of chromosomes from one species to another, are different from anything that ordinarily occurs during evolution. Is it right, critics ask, to interfere with the genetic makeup of living organisms and, thus, the evolution of life on earth?

The Risk Debate

Recent research suggests that the dangers of genetic engineering have been blown out of proportion and that genetically engineered bacteria are not generally a threat to ecosystem stability. Most scientists agree.

At least two studies now indicate that genetically engineered bacteria applied to seeds, so that the bacteria take up residence on the roots, migrate very little from the site of application in the short term. Critics, however, are concerned with long-term consequences—ecological backlashes that may occur months or even years after application.

Many supporters of genetic engineering oppose attempts to halt its progress. To them, the promises of this revolutionary technology far outweigh any potential damage. The benefit of feeding the world's hungry with a genetically engineered strain of supercattle, they argue, far exceeds any possible threats. Even many leading environmental scientists who joined Rifkin and others at the outset of the battle in the 1970s have switched sides. But ecologists want to see the industry properly monitored and want careful testing before release. Genetically altered species, they say, are analogous to alien species introduced into new environments. The history of such introductions has been fraught with difficulties (see Chapter 8).

Environmental organizations also recommend caution. The federal government, they say, has poured enormous amounts of money into research on genetic engineering, but has spent very little to assess its potential risks. In *An Environmental Agenda for the Future*, a list of recommendations for environmental improvements in the United States, ten leading environmental groups called for a no-release policy until the government develops a full understanding of potential effects. They also called for tighter standards on designing and operating genetic engineering facilities.

Genetic engineering stands at a threshold. Still in its infancy, it offers an unparalleled opportunity for improving agriculture, animal husbandry, and medicine. At the same time, this revolutionary science carries with it an unknown potential for environmental harm that worries many of its critics.

8

Wildlife and Plants: Preserving Biological Diversity

The worst sin toward our fellow creatures is not to hate them, but to be indifferent to them; that's the essence of inhumanity.

GEORGE BERNARD SHAW

Stalking through the underbrush, two African peasants halt when they hear a rustling in the bushes ahead of them. Pushing back a branch, they spot the bull black rhinoceros, browsing on low branches. One of the men lifts his rifle and fires. The rhino stumbles a little, regains its footing, then races off. Another shot rings out; the rhino stumbles again, runs half a mile, and collapses in a cloud of dust.

Jubilant, the men run to the fallen beast. Their hearts beating hard, they saw off the rhino's horn and sneak into the trees, heading back to the village, where a middleman will buy the horn. The men will get $100 each, a considerable sum, for $200 will support a family for a year. Several more hunts this year and they will be wealthy men.

From Africa, rhino horns may go to Yemen or China. In Yemen they will be carved into expensive dagger handles. In China rhino horns are ground and sold as an aphrodisiac and as a fever-depressing drug costing $11,000 per kilogram (2.2 pounds).

African farmers, whose villages are doubling in size every 20 years, are plowing up grassland that was once grazing land for rhinos and other African wildlife. As the farmland spreads outward, the African rhino population is fragmented into many small groups, often with as few as 30 members. Poachers have further reduced some of these tiny groups to as few as eight animals. Interbreeding among rhinos has become a major problem.

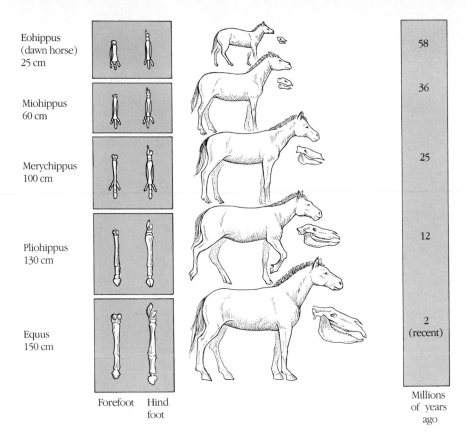

Eohippus
(dawn horse)
25 cm

Miohippus
60 cm

Merychippus
100 cm

Pliohippus
130 cm

Equus
150 cm

Forefoot   Hind
foot

58

36

25

12

2
(recent)

Millions
of years
ago

Figure 8-1. Stages in the evolutionary history of the horse.

"The rhino's demise," writes David Western, a Kenyan wildlife biologist, "stems from a jarring mix of prosperity and poverty. Asia is getting richer by the year and can afford astronomical prices for horn, while Africa is getting poorer and more desperate." Poachers, farmers, and, indirectly, wealthy Yemenis and Chinese have reduced the once-large rhino population to almost nothing. Numbering in the hundreds of thousands 50 years ago, the population of black rhinos was estimated to be less than 8000 in 1986. Fewer than 3800 were alive in 1989. The northern white rhino is now virtually extinct. If poaching and human population growth continue, there is little hope for the African rhino.

This chapter examines the extinction of plants and animals, brought on by a mixture of underlying causes. It also looks at the countless benefits of wildlife to human societies and suggests the need for urgent action throughout the world.

The Vanishing Species

As many as 500 million kinds of plants, animals, and microorganisms have made this planet home since the beginning of time. Today there are five to ten million species, two-thirds of which live in the tropics. Thus, 490 million species have become **extinct**, lost forever. Extinc-

tion is an evolutionary fact of life. But natural extinction differs considerably from the impending doom now facing the rhino and tens of thousands of plants and other animals. There are two reasons why natural extinction differs from the accelerated extinction now taking place.

First, during the course of millennia old species evolved into new ones (Figure 8-1). Consequently, many of the 490 million extinct species are represented today by their descendants. Modern extinctions, on the other hand, eliminate species entirely. If the rhino should vanish, it would leave no evolutionary legacy. It would be gone forever.

Second, the rate of extinction varies considerably. Even though some species did vanish because of severe climatic changes or increasing environmental resistance created by excess predation or disease, the rate of natural extinction was slow compared with today's **accelerated extinction** (Figure 8-2). Currently, one vertebrate (backboned) species becomes extinct every nine months, compared with a natural rate of one species every 1000 years. When plants, insects, and microorganisms are added, the extinction rate climbs to one species a day. Many experts fear that we have entered a new era of extinction unparalleled in the history of the earth.

If the world's population continues to grow and nations continue to destroy wildlife habitat at current rates, many experts predict, an average of 40,000 to 50,000 species will be destroyed per year over the next 20 years. By the

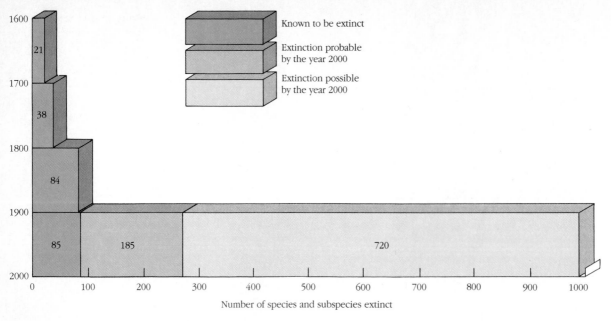

Figure 8-2. Extinction rate of vertebrate species past and projected. Notice the rapid increase in the 20th century.

end of the century we may be losing one species per hour!

Pandas, blue whales, tigers, and chimpanzees are the endangered species that make the headlines, largely because they are the most appealing or visible victims. But these species are, in fact, only a small part of the picture. Today, about 25,000 species of plants are threatened with extinction—one of every ten plant species on earth. Far more insect species teeter on the brink of extinction. Interest in the less appealing species is often difficult to stir.

Unless we curb population growth and properly manage our resources, the biological world at the end of the century will differ radically from the world we now live in. As many as one million species may have vanished between 1980 and 2000. The consequences of mass extinctions, and of the habitat destruction that is largely responsible for them, will be felt throughout the world.

What Causes Extinction?

Plant and animal extinction results from many underlying forces, such as economics, politics, and psychology. Figure 8-3 shows the specific activities that cause extinction and the relative importance of each. The top five, in decreasing order of importance, are: (1) habitat alteration, (2) hunting for commercial products, (3) introduction of alien and domestic species, (4) hunting for sport, and (5) pest and predator control. In Africa, extinction of the rhino stems from several of these factors, most

importantly habitat destruction and hunting for profit. In fact, multiple causation is the rule in animal and plant extinction.

Alteration of Habitat

Humans have always altered habitat, but today our numbers and our high demands place inordinate strain on habitat essential to many species (see Chapters 2 and 4). Alteration of habitat is the most significant single factor in extinctions. Habitat is destroyed by human civilization spreading into fields, forests, oceans, and waterways. Roads, strip mines, housing projects, dams, airports, farms, and cities usurp wildlife habitat. Even recreational areas have many adverse effects on wildlife. Dams built for boating and water skiing, for instance, release cold water from the bottom that threatens native fish species adapted to warmer waters. Dams impair salmon migration and have eliminated the Atlantic salmon and greatly reduced the Pacific salmon. On the Columbia River 10 to 15 million salmon once spawned on thousands of miles of streambed now blocked by dams. Today, the Columbia River system supports only 2.5 million salmon a year. Ski resorts and their attendant condominiums and restaurants wipe out winter feeding grounds for deer, elk, and other species.

Nowhere is the loss of habitat more noticeable than in the tropical rain forests. Tropical forests house about half of the earth's species. Once covering an area the size of the United States, the tropical rain forests have been cut by at least one-third. Countless species have perished as

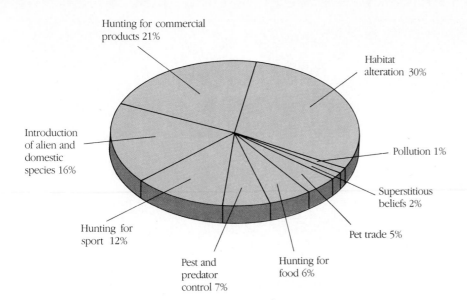

Hunting for commercial products 21%

Habitat alteration 30%

Introduction of alien and domestic species 16%

Pollution 1%

Superstitious beliefs 2%

Hunting for sport 12%

Pet trade 5%

Pest and predator control 7%

Hunting for food 6%

Figure 8-3. An approximate breakdown of the human activities that lead to extinction. More than one activity is often involved, however.

a result. (For more on tropical rain forests and their destruction see Case Study 9-1.)

Coral reefs, wetlands, and estuaries (the mouths of rivers) are other critical habitats now rapidly declining because of human development. Wetlands and estuaries are the home of many species (see Chapter Supplement 10-1) but are also highly prized by humankind. Their destruction threatens the future of waterfowl throughout the world.

Commercial, Sport, and Subsistence Hunting

Animals are hunted today for profit, for food, and for sport. Collectively, these activities play a major role in species extinction.

Commercial hunting represents the biggest threat. Whale hunting is one of the most widely publicized examples. Throughout the history of whaling, men have hunted species to the brink of extinction and then moved on to other species. The result has been a severe reduction in the number of whales of many species (Table 8-1). Thanks to recent efforts by the International Whaling Commission, commercial whaling has been greatly reduced. In its stead is a new industry: whale watching, with annual revenues that exceed those of the commercial whaling industry itself.

Despite bans on whaling, several nations continue to hunt whales under the guise of scientific research. Japan, South Korea, Iceland, and Norway all participate in such killing, an effort, say many critics, to keep their whaling industry alive. Japan is a principal perpetrator of this fraud. It even tried to reclassify one of its whaling ventures as a subsistence hunt, despite the fact that 90% of the whale meat from the coastal minke whale was sold on the open market. Iceland businesses have also profited

handsomely from whale meat from "research" slaughter, reaping over $8 million in revenue in a single year. New proposals from Japan, Norway, and Iceland seek to kill more whales in an attempt to determine the relationship between global fisheries and these remarkable creatures. By any other name, such a proposal is a license for continued slaughter.

Commercial hunting is systematically reducing the populations of many other endangered species. In parts of Africa, for example, endangered male gorillas are killed by tribes seeking parts of their bodies believed to possess magical properties. The hands and head are sold to tourists. Similar poaching has caused drastic declines in elephants, rhinos, jaguars, tigers, and cheetahs. In 1973, 130,000 elephants lived in the African nation of Kenya. Today, fewer than 20,000 are alive—a reduction of 85%.

Table 8-1 Whale Populations—Then and Now

Species	Number Before Commercial Whaling	Current Estimate
Blue	166,000	7,500–15,000
Bowhead	54,680	3,600–4,100
Fin	450,000	105,000–122,000
Gray	15,000–20,000	13,450–19,200
Humpback	119,000	8,900–10,500
Minke	250,000	130,000–150,000
Right	50,000	3,000
Sei (includes Bryde's)	108,000	36,800–54,700
Sperm	1,377,000	982,300

Figure 8-4. Sport hunting can endanger wildlife populations that are not carefully managed.

Even in protected areas, the herds are declining because of poachers. Ivory prices have soared in recent years and poachers have responded with a mass slaughter. An estimated 70,000 elephants will be shot for their ivory tusks. Another 10,000 young elephants will perish because their mothers have been killed. The US currently consumes 30% of the African ivory. One wildlife group, the African Wildlife Federation, hopes to convince Congress to ban US ivory imports and is working through the media to convince Americans to stop purchasing ivory. Recognizing the plight of the African elephant, the Kenyan government banned further export of ivory in 1989. Their announcement was underscored by burning a huge pile of ivory confiscated from poachers. Fortunately, some of the African nations have found that wildlife can greatly enhance their appeal to tourists and are attempting to crack down on poaching.

Hunting for sport is also a factor in wildlife extinction (Figure 8-4). However, the hunting of properly managed, nonendangered species frequently benefits populations by keeping them within the carrying capacity of their habitat. It is another kind of hunting that threatens rare and sometimes endangered wildlife: trophy hunting of endangered species.

People also hunt endangered species for food. For example, Eskimos annually kill a number of endangered bowhead whales. Environmentalists, with the support of scientific evidence, claim that the whale population (about 2300) cannot support further hunting. The Eskimos, however, argue that their own survival depends on the hunt. In a compromise decision, the Eskimos agreed to cut back on their annual kill and to reduce the loss resulting from wounding. Nonetheless, some environmentalists fear that the bowhead whale population is doomed. The new quotas, they say, could eventually wipe out this magnificent creature.

Introducing Foreign Species

Foreign, or **alien**, **species** introduced accidentally or intentionally into new territories often cause the extinction of native species. The water hyacinth, discussed in Chapter 4, is such an example. The English sparrow fluttering about in the bushes outside your window is another. Deliberately introduced into this country in the 1850s, the sparrow quickly spread throughout the continent. It now competes for nesting sites and food once used by bluebirds, wrens, and swallows.

Islands are especially vulnerable to new species. In Hawaii, for example, 90% of all bird species have been wiped out by human inhabitants and organisms (for example, rats) that humans have introduced. In New Zealand, half the native birds are endangered or extinct.

Florida is a showcase of alien species gone wild. Australian pines, introduced as ornamentals, have spread rapidly along coastal beaches. Their shallow roots are so dense that they destroy sandy beaches where many sea turtles lay their eggs. The Brazilian pepper, another introduced species, is growing wild in cleared land and especially mangrove swamps, the habitat of dozens of species. Worst of all is a species called the punk tree. This thirsty plant grows in swamps, creating a dense tangle of vegetation impassable to many animal species.

In Montana's Glacier National Park, the bald eagle, which visits McDonald Creek in the hundreds each fall, has all but vanished in recent years. The eagles have disappeared because their food source, a land-locked salmon that migrates into the stream to spawn, has nearly

Case Study 8-1

Saving Canada's Troubled Fisheries

Canada is a nation of incredible beauty. Vast open plains and rich forests blanket her lands. Wildlife still abound in her unspoiled wilderness. But the rich natural resource base that supports Canada's 25 million people is slowly coming unraveled. In eastern Canada, for example, 10,000 lakes are so acidic that experts believe they no longer contain fish. In Lake Ontario fishermen now reel in salmon ulcerated by tumors.

In south-central British Columbia, the number of nonresident fishermen dropped by half between 1960–1980. In the Yukon anglers are catching 28% fewer arctic grayling, pike, and lake trout than they were in 1975. In southern Manitoba, anglers can land only one-tenth as many walleye as they did 20 years ago.

Canada's freshwater fisheries are in chaos. Fish populations are declining, and spawning and rearing areas are being lost each year to urban development and industrialization. In some areas fish are contaminated with pollutants and are unfit for human consumption.

The destruction of Canada's once-famous fisheries results from many different causes. Overfishing by commercial interests, native Canadians, and recreationists is a leading cause. Air pollution from Canadian and US smokestacks and cities contributes, as do water pollutants from Canadian cities and paper mills. Fertilizers and pesticides from farmlands are also a major factor. Unfortunately, say many Canadians, the management of recreational fishing is left to the provincial governments, which are generally underbudgeted and stretched too thin. That impairs control.

Although anglers still harvest 45,000 tons of fish per year and spend $4.7 billion (in Canadian dollars) in the process, fishing has deteriorated significantly in the past five years. But many Canadian anglers have decided to take matters into their own hands. In record numbers, they are finding ways to rescue their favorite streams and to influence political leaders to help protect this vanishing resource. They are also putting pressure on the federal government to tackle some of the more overwhelming problems that require national solutions.

Anglers turned out in record numbers to save the Credit River, a trout stream 70 kilometers (40 miles) from Toronto,

home of three million people. There, anglers labored hard to remove dead trees and other obstructions that have changed the stream flow, making it warmer and unsuitable for trout. And they have begun to repair riverbank erosion which increases the sediment load in streams and destroys spawning sites.

The Credit River is only one of many rivers that have been rescued by these ambitious Canadians. Fifteen years ago the Bow River, a large stream that flows through Calgary, was likened to an open sewer, but local anglers applied pressure to political leaders to ensure minimum-flow dam releases, better waste treatment, and tighter fishing regulations. As a result, the river is now a blue-ribbon trout stream. Brown trout of 50 centimeters (20 inches) or more are now common in its waters.

Jim Gourlay, editor of a Canadian fishing magazine, says, "Instead of just griping as in the past, people are out there cleaning and improving streams, planting eggs and fry, pushing for mandatory hook and release and other conservation measures." In Ontario, anglers lobbied to institute fishing licenses. As a result of their successful campaign, Ontario fishing licenses now produce millions of dollars used to protect fish and manage their fisheries better.

Another exciting development is the Community Fisheries Involvement Program (CFIP) in Ontario, a cooperative program between private interest groups and the government. The government provides technical advice and some money; the group provides labor. Thousands of individuals work on such projects as cleaning up streams, building incubation boxes, and anything else to help improve trout and salmon streams.

Ontario's CFIP is a model program. British Columbia has started a similar program to protect streams. In a single year, 250 fishery projects involving 8,000 volunteers from all walks of life are carried out in British Columbia. This idea is spreading fast to other parts of British Columbia and is a good example of what people can do to make a difference in their environment.

---

vanished. Ironically, the salmon is an introduced species. No one knows for certain why its numbers are declining, but many think that it is because of another species, a freshwater shrimp introduced to Flathead Lake where the salmon spend the rest of the year. The shrimp, biologists think, eat the same food that the young salmon require. Since the early 1980s, the shrimp population has skyrocketed, destroying salmon food sources and, ultimately,

the salmon. Now that they're gone, the eagles that delighted visitors have vanished as well.

Pest and Predator Control

Control of pests also influences wild populations of plants and animals. DDT and other pesticides have taken a huge toll on American wildlife (Chapter 17). The peregrine

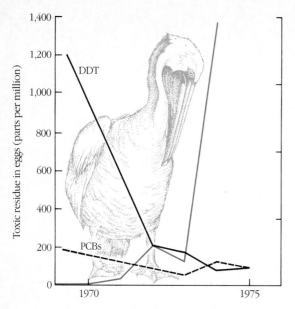

Figure 8-5. Banning of DDT and most uses of PCBs in the United States resulted in a dramatic decrease in DDT and PCB levels in brown pelican eggs and an equally dramatic increase in the number of eggs hatched.

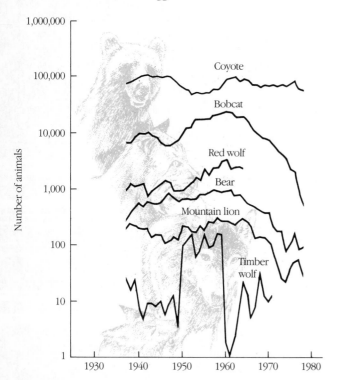

Figure 8-6. Number of predators killed by federal control programs. (States now handle most predator control, making data of total annual kill unavailable.)

falcon disappeared in the eastern United States by the 1960s as a result of DDT-induced reproductive failure. DDT caused eggshell thinning and gradually wiped out the entire population of falcons east of the Mississippi

River. Eagles and brown pelicans met a similar fate (Figure 8-5). Even the California condor suffered from eggshell thinning. DDT poisoning was, however, just one of many factors that spelled doom for this species. Among the others were lead poisoning (discussed later in this chapter), loss of habitat from home building and farming, and fire control, which eliminated takeoff and landing areas needed by these giant scavengers, whose wings span up to 3 meters (9 feet). Especially harmful to US migratory birds are the persistent pesticides, such as DDT and related compounds, that have been banned in the United States but are still used in Latin American countries. Birds in Latin America are also hard hit.

Predator control, once the cornerstone of wildlife management, has endangered or wiped out populations of wolves, bears, and other animals (Figure 8-6). Killing off predators creates an ecological backlash as the prey population balance is thrown out of kilter.

Collecting for Zoos, Individuals, and Research

Animals and plants are gathered in the millions throughout the world for zoos, private collectors, pet shops, and researchers. In 1988, more than 125 million fish, 1.2 million reptiles, and 1.5 million reptile skins were imported into the United States. Each year millions of tropical birds are brought to the United States, Canada, and Great Britain. However, for each bird that makes it into someone's home, 10 to 50 may die along the way. Survival of animals after they reach individual homes is also surprisingly low. Zoo animals fare much better, as a rule, especially where responsible officials have sought to modify enclosures, creating more "natural" settings. But not all zoos deserve commendation. In some, birds, apes, monkeys, reptiles, and others are housed in small cages without contact with other members of their species or adequate stimuli. They will live out their lives then die, requiring the zoo to replace them from the wild.

Among this influx of legally imported animals come countless members of endangered, threatened, and rare species. Smuggled in, many of these animals go to private collectors. Thankfully, zoos in the United States and many other nations have banded together to stop the illegal flow of animals.

Plants are also high-ticket items. Cacti and orchids, for instance, are in high demand today and support a growing industry. Over seven million cacti are imported each year into the United States from more than 50 countries. At home, collectors pillage Texas and Arizona deserts in search of salable cacti to adorn the lawns of their eager customers. In one area of Texas near Big Bend National Park, 25,000 to 50,000 cacti were uprooted in a single month. The international trade in exotic flowers, especially those originating in Turkey and the Mediterranean,

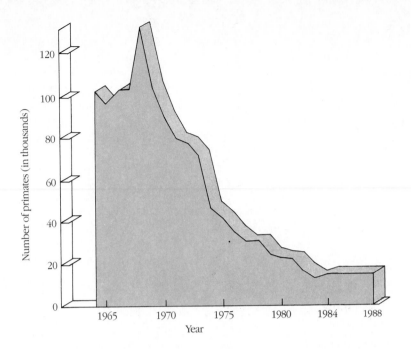

Figure 8-7. Decline in primates imported by the United States.

now threatens native populations in these areas. Although most flower bulbs sold in the United States are from domestic stock, tens of millions of bulbs are imported from Turkey to the Netherlands, where they are propagated or re-exported. The number of imports to the US doubled between 1982 and 1987.

To reduce the ravaging impact of commercial cactus rustling, Arizona has made it illegal to remove 222 different plant species. With penalties up to $1000 and jail sentences up to one year, Arizona has taken a small step to protect its native plants. Still, with only seven "cactus cops" to patrol the state, little can be done.

Researchers throughout the world use a variety of animals for their studies, many of which come from the wild. Demand is especially great for primates, monkeys, and the great apes such as chimpanzees. Taken from their homeland in Africa, as many as five chimpanzees die for every one that enters a laboratory. Fortunately, annual primate imports have dropped off rapidly in the United States and elsewhere (Figure 8-7). Nonetheless, some rare and endangered primates are still captured and sent to the United States. Approximately 200 great apes are imported into the United States each year, further diminishing the chances of their survival in the wild.

In 1975 the United States banned the importation of all primates for pets, but it allowed continued importation for zoos and research. Because research animals often do not breed in captivity and because they have a high mortality, continual replenishment from wild populations is likely to continue.

Today, 60 primate species are on the endangered list. Researchers have exploited many of these species with little concern for their declining population or ultimate survival in the wild. Most people agree that research must continue under humane conditions but that it should not be carried out at the cost of extinction. One solution is breeding captive animals to supply zoos and researchers. Another solution is the use of laboratory tests—for example, cell cultures—that replace live-animal tests. A surprising number of substitutes are already available. Actions such as these could stop the flow of these animals from their natural habitat and prevent the extinction of many primates.

Pollution

Pollution alters the habitat of plants and animals and thus contributes to extinction. Water pollution is especially harmful to organisms living in estuaries and coastal zones, where many economically important marine fish breed and spend their early years. Toxic wastes entering the food chain can have devastating effects, especially on the young, which are almost always more sensitive to pollution than adults. The sources and effects of pollution are discussed in Chapters 14–18. A few examples will suffice here.

In the semiarid farmland of California's San Joaquin Valley, well-intended irrigation projects have become a major problem for wildlife. Farmers installed tile drains in their irrigated fields to prevent salinization and waterlogging (Chapter 7); water drained from farmers' fields was diverted to a series of specially built evaporation ponds that also serve as nesting and feeding sites for waterfowl, whose numbers, for several reasons, have dropped by one-third in fewer than 30 years. The Kesterson National Wildlife Refuge contains 485 hectares (1200 acres) of ponds designed to do double duty—evaporation of agricultural water and waterfowl habitat

Figure 8-8. The Kesterson Wildlife Refuge was built to serve a dual purpose: to provide waterfowl nesting and feeding sites, and to provide farmers a place where irrigation runoff can be evaporated. Unfortunately, toxic metals leached from the soil poisoned the waterfowl in large numbers, forcing the shutdown of the refuge.

(Figure 8-8). In 1985 biologists found an unusually high incidence of abnormalities in chicks of waterfowl and wading birds at Kesterson. Chicks without eyes, beaks, wings, and legs were found, and many dead embryos were discovered. In natural conditions biologists expect a 1% deformity rate; at Kesterson the rate in some species reached 42% (Figure 8-9). Adult birds were also affected. Crayfish, snakes, raccoons, and muskrats that once flourished in the rich biological community also vanished. As the journalist Keith Schneider put it, "The Kesterson Refuge had become a place that killed the animals it was supposed to protect."

Water analyses showed the cause of the biological nightmare. Researchers found that levels of the toxic metal selenium in the water were a thousand times higher than was considered safe. The selenium (and other metals) had been leached from the soil by irrigation water and was concentrated in the pond by evaporation. Water quality was so bad that the Water Resources Control Board classified the refuge as a hazardous waste pit.

The unintentional poisoning of the ducks and geese at Kesterson violated terms of the Migratory Bird Treaty signed by the United States, Canada, and Mexico. For that reason the US Interior Department threatened to order the Federal Bureau of Reclamation to stop supplying water to 53 farms whose used irrigation water flows into Kesterson. Should the Interior Department do so, affected farmers would cease production of crops worth an estimated $500 million. The impacts on farm-related businesses and the general community would be staggering.

Because of the economic and social impact of such a measure, the state and federal governments have pursued other options to prevent waterfowl from being harmed further. First, they plugged up the inlets, stopping the inflow of polluted irrigation water. Next, ponds that received drainage were allowed to dry, then were filled with soil and reseeded, providing upland habitat for other species. As an interim measure, farmers have reduced irrigation on thousands of hectares, changed crops, and are recycling some of the water. An experimental treatment plant was operated for a short while, but had so many problems it was shut down.

Kesterson is only one of many examples of the threat that pollution poses to wildlife. Federal refuges in Utah, Wyoming, Texas, Nevada, Arizona, and other parts of California also show signs of toxic metal poisoning resulting from farm drainage. Several of these refuges have already confirmed the presence of deformities in chicks.

A recent survey by the US Fish and Wildlife Service found 84 hazardous waste sites at 39 wildlife refuges and related facilities. Ten of these sites have contaminated the wildlife areas. The remaining 74 could present problems in the future.

Outside of refuges, birds and animals often fare even more poorly. Already, acid precipitation has devastated fish populations in Norway, Sweden, Canada, and the United States. Oil spills can locally touch off an ecological disaster with far-reaching effects (see Chapter 16). The tragic spill of over 42 million liters (11 million gallons) of crude oil in Alaska's once-rich Prince William Sound has killed tens of thousands of birds and hundreds of sea otters. Its devastation could easily last for ten years or more.

Lead from shotgun shells is a fairly widespread toxic pollutant, which kills waterfowl by the tens of thousands each year. Hunters use more than 2400 tons of lead shot annually. Waterfowl pick up the lead from the bottom of rivers and lakes. Others may be wounded but survive with lead embedded in body tissues. Predatory birds, such as the bald eagle, that consume waterfowl containing lead shot are also dying.

Lead shot in animals that are wounded by hunters and then die without being retrieved is thought to be responsible, in part, for the loss of the California condor, which once ranged from British Columbia to the southern tip of the Baja peninsula in Mexico and across the entire southern United States. Two condors in the wild died in recent years from lead poisoning; lead levels in remaining birds, now all in the San Diego Zoo, are alarmingly high.

Lead shot is slated for a complete ban in the United States by 1991. It will be replaced by steel shot. While this is an important step, it won't eliminate the lead trapped in sediments on stream and lake bottoms. In 1988, low water levels in many lakes and rivers in the United States and Canada resulted in many additional cases of lead poisoning in waterfowl. Lower water levels allowed waterfowl to feed on sediment that had been out of reach in earlier years. Lead poisoning may have been a major factor in the dramatic loss of waterfowl in that year, decreasing from 50 million to 33 million.

In the Spring of 1988, disease struck the seals of Europe's North and Baltic Seas. Adult seals floated aimlessly in the water, too weak to eat or play, then died in record numbers. Pregnant females aborted. By the middle of the summer, seals were dying along hundreds of miles of coastline. By September, the disease had spread to the Atlantic coast of Ireland.

Some people called this scourge the "black death of the sea" because it recalled the epidemics of bubonic plague, or black death, that devastated Europe in the 1300s. Through this tragic turn of events, a population of harbor seals once numbering 18,000 has been cut to only 6,000. Biologists are certain that this, the largest die-off of seals in recorded history, is caused by the canine distemper virus, but that the virus is not working alone. Pollution in the seas, they say, may have greatly weakened the immune systems of the seals, making them more vulnerable to the virus.

Seals living in the waters off the coast of West Germany and the Netherlands are heavily contaminated with a toxic chemical called PCB (polychlorinated biphenyl), a substance once used as an insulator in electrical devices. PCBs and possibly other chemical contaminants are blamed for the seals' reproductive problems and for the suppression of their immune systems.

Figure 8-9. Defective killdeer embryo from the Kesterson Wildlife Refuge. Selenium and other toxic metals have resulted in a number of embryonic defects. This embryo has small eyes, a slightly twisted beak, and twisted and clenched legs and feet.

The seal plague is one of the latest manifestations of a chronic pollution problem in the North and Baltic Seas and a symptom of the global crisis now afflicting our wildlife.

Ecological Factors That Contribute to Extinction

Not all species are created equal. Because of peculiarities in behavior or reproduction, some are more vulnerable to extinction than others. Consider the passenger pigeon. At one time it inhabited the eastern half of the United States in flocks so large they darkened the sky. Probably the most abundant bird species to ever live, the passenger pigeon is now extinct because of widespread commercial hunting. Between 1860 and 1890 countless pigeons were killed and shipped to the cities for food. In 1878 the last nesting site in Michigan was invaded by hunters. When the guns fell silent, over one billion birds had been killed. By this time, the passenger pigeon was nearly extinct; only about 2000 remained. Broken into flocks too small to hunt economically, the birds were finally left alone. However, the number of birds dwindled year after year, until in 1914 the last bird held captive in the Cincinnati Zoo died.

This tragic, oft-told story has a lesson: some species have a **critical population size** below which survival may be impossible. The passenger pigeon population dropped below that level. The bird needed large colonies for successful social interaction and propagation of the species. Two thousand were simply not enough. This same problem faces the blue whale today.

Scientists know very little about the critical population size for many species. Overhunting, habitat destruction, and other activities discussed in this chapter can do irreparable damage before we even realize it.

Figure 8-10.   The California condor is on the verge of extinction. With only 27 birds remaining, all but one in the San Diego Zoo, the bird's future lies in the hands of scientists and zoo personnel. Hunting, habitat destruction, lead poisoning, and a low reproductive rate have resulted in the decimation of the condor population in the United States.

Organisms can be categorized as specialists or generalists, as we saw in Chapter 3. Specialists tend to become extinct more readily than generalists, who can exploit more food sources and can live in diverse habitats.

Animal size also contributes to extinction. Larger animals, like the rhino, are easier (and often more desirable) prey for hunters; they are also more likely to compete with humans for desirable resources, such as grazing land. Larger animals also generally produce fewer offspring, making it more difficult for reduced populations to recover. The large California condor, for example, lays a single egg every other year. Young condors remain dependent on their parents for about a year but are not sexually mature until age six or seven. Combined, these factors give the condor little resiliency to bounce back from pressures from human populations or natural disasters (Figure 8-10).

Another factor in extinction is the size of an organism's range. The smaller the range, the more prone the organism is to extinction. Finally, organisms often exhibit varying rates of tolerance to human presence. Bluejays and coyotes, for example, coexist with humans quite nicely, but grizzly bears move out when humans move into an area.

Keystone Species

According to a new ecological theory, in some ecosystems the extinction of one critical species, called a **keystone species**, may lead to the collapse of an entire ecosystem. To some, the idea of a keystone species is too simplistic, but to others it has become an important theme in conservation.

Consider some examples. The gopher tortoise is believed to be a keystone species of the southeastern United States. These magnificent animals dig long burrows in the sand, which they occupy for decades. Many other species also live in the burrow. For instance, the grey fox and diamond-back rattlesnake visit the burrow regularly. The Florida mouse and the gopher frog live exclusively in the gopher tortoise burrow. Mice excavate tiny side tunnels for their own living quarters; opossums and indigo snakes frequent the burrows as well. The gopher tortoise is so important, in fact, that in areas where it has been eliminated 37 species of invertebrates have disappeared.

The gopher tortoise's role as a keystone species in Florida prompted wildlife officials to list the animal as a "species of special concern." New regulations now ban the sale and hunting of tortoises and require a permit for people wishing to keep one as a pet.

Keystone species are also seen in complex ecosystems like tropical rain forests. Ecologist John Terborgh and a team of researchers from Princeton University have been studying plant communities in Peru for over ten years. They recently found that three-fourths of the birds and mammals of the Amazonian rain forest rely on fruit as their major food source. But fruit is available only nine months of every year. During the remaining period, monkeys, peccaries, parrots, and toucans live on figs. If figs are removed from the ecosystem, it could very well collapse.

The marine ecosystem in the US Pacific has a controversial keystone species, the sea otter. Sea otters are voracious eaters, feeding on sea urchins, abalone, crabs, and mollusks that dwell in kelp beds. The sea otter helps control sea urchin populations. Without predatory control, sea urchins devour kelp forests. In places where sea otters were eliminated, kelp beds were also destroyed, creating barren shorelines. As the sea otters repopulated the regions, they dramatically reduced urchin populations, allowing kelp and seaweed to grow again. The seaweed provides a habitat for fish and other fish-eating creatures like harbor seals and bald eagles. Thus, when the sea otter returns to an area, so do other species.

The existence of keystone species makes many scientists and environmentalists nervous about habitat destruction. Many little-known animals may be keystone species important to human societies. For instance, bats play a keystone role because they pollinate flowers and disperse seeds of jungle plants. Many tropical species, in fact, are entirely dependent on bats, including many important human food crops like avocados, bananas, cloves, cashews, dates, figs, and mangos. All these species rely on bats for survival in the wild.

Most scientists believe that we do not have enough information to categorize a species as a keystone in the ecosystem. Ironically, many of the species that attract conservationists and the lion's share of conservation funds are not keystone. Keystone species, in fact, are often unobtrusive, rare, or little-known species. Their survival, however, may ensure the survival of the more glamorous and well-known species.

Why Save Endangered Species?

Aesthetics

As Norman Myers has written, "We can marvel at the colors of a butterfly, the grace of a giraffe, the power of an elephant, the delicate structure of a diatom. . . . Every time a species goes extinct, we are irreversibly impoverished." Wildlife and their habitat are in many ways a rich aesthetic resource. The sight of a female trumpeter swan gently nudging her offspring into the water for their first swim, the eerie cry of the common loon at night, the lumbering grizzly bear on a distant grassy meadow, the sputtering of a pond full of ducks, the playful antics of sea otters, the graceful dive of the humpback whale—these enrich our lives in ways no economist could figure. For the weary urbanite home from the office, the sound of migrating geese stirs deep emotions, a satisfying sense that all is well. Destroying the biological world impoverishes all of us.

Ethics

Preserving endangered plants and animals is an ethical issue as well. What right, critics ask, do we have to tear apart the richly diverse biological world we live in? Don't other organisms have a right to live, too? Preserving life has become our duty because we have acquired the means to destroy the world. With that ominous power comes responsibility.

Economics

Economically, it makes good sense to protect the rich biological diversity we inherited from our forebears. "From morning coffee to evening nightcap," writes Myers, "we benefit in our daily life-styles from the fellow species that share our One Earth home. Without knowing it, we utilize hundreds of products each day that owe their origin to wild animals and plants. Indeed our welfare is intimately tied up with the welfare of wildlife. Well may conservationists proclaim that by saving the lives of wild species, we may be saving our own."

From the biosphere we reap fish from the sea; medicines and other products from plants; important plant and animal genes needed to improve domestic crops and livestock; a wealth of wildlife for hunters, anglers, and nature lovers; and research animals that provide valuable insights into human physiology and behavior. The economic benefits of these wild resources are enormous. By some estimates, for example, half of all prescription and nonprescription drugs are made with chemicals that came from wild plants. The commercial value of these drugs is around $20 billion per year in the United States and about $40 billion worldwide. The US Department of Agriculture estimates that each year genes bred into commercial crops yield over $1 billion worth of food. Similar gains can be documented for other major agricultural nations. About half of the increased productivity in corn over the last 50 years has resulted from "genetic transfusions" from wild relatives of corn or from corn's early ancestors grown now only in isolated regions.

"Wild species rank among the most valuable raw materials with which society can meet the unknown challenges of the future," writes Myers. Many developments from wild species now loom on the horizon and may offer us further financial gains and healthier lives. For example, the adhesive that barnacles use to adhere to ships may provide us with a new glue to cement fillings into teeth. A chemical derived from the skeletons of shrimps, crabs, and lobsters may help prevent fungal infections. An antiviral drug is now being developed from a Caribbean sponge. It has already proved effective against herpes encephalitis, a previously lethal brain infection that strikes thousands of people each year.

Plants and animals lost before they can be explored for possible benefits will diminish our opportunities to fight disease and increase productivity. A dozen regions located in the tropics and subtropics are the sources of virtually all commercially valuable plants and animals. They provide a reservoir of genetic material essential for the battle to fight disease, drought, and insects. Their loss would be a global tragedy with far-reaching effects on the food supply.

Ecosystem Stability

Finally, preserving species and their habitat helps ensure global ecosystem stability and, ultimately, our own future. The endangered biosphere provides us with many invaluable services free of charge. It controls pests. It recycles oxygen, carbon, and dozens of important nutrients. It maintains local climate. It helps control groundwater levels and reduces flooding. Without these hidden benefits humans would be an endangered species.

◀ *Viewpoint*

Playing God with Nature: Do We Have Any Other Choice?

Norman Myers

The author, a wildlife expert and renowned writer, has spent 25 years advising governments on park management. He has written The Sinking Ark *and* A Wealth of Wild Species.

I still hurt to recall the first time I went out on an elephant-cropping foray. It's hard to forget the screams of terror, the fountains of blood, and the sudden silence, broken only by the clinical talk of the technicians and scientists.

In South Africa's Kruger Park, where elephant cropping is a fact of life, officials work under the seemingly arrogant notion that only humans can keep a park wild in this human-dominated world. I spent months agonizing that there must be a better way to deal with nature.

Eventually, I clawed my way to the conclusion that Kruger's approach is the right one—this arrogance, this cold-blooded dominance of nature, is our only choice—but that we must do it with great caution. Despite my realization that such management is necessary, it sticks in my throat.

When I first went to Africa almost 30 years ago, human communities tended to be islands of settlement among a sea of wildlands. Wildlife could go about untrammeled, and I could bask in the unadulterated spectacle. Today, it is the wildlands that are islands. Africa is bursting at the seams with people; population pressures threaten parks from Ethiopia to Zimbabwe, from Kenya to Senegal, making intensive management a bitter but necessary reality.

The huge Kruger Park, two and a half times the size of Yellowstone National Park, is no exception. A 950-kilometer

(600-mile) fence surrounds the park, turning it into an island in an otherwise crowded land. That's where the trouble begins. Biologists theorize that islands, isolated from rejuvenating gene pools by water or fences or disturbed land, tend to have fewer species than a similar-sized portion of contiguous habitat. They are probably more susceptible to environmental change.

In fenced-off islands such as Kruger Park, wardens and scientists have concluded that they cannot allow the rich diversity of species to dwindle. To ensure diversity they have chosen to manage the park to the hilt, which is where this Viewpoint began.

The thought of the flesh of wild elephants ending up in cans in a supermarket may be incomprehensible or repellent. But that's what is happening, and for good reason—to prevent overpopulation and habitat destruction within the park. The alternative to control can be dangerous to parks, as I saw in Kenya. During the early 1970s the country was hit by drought. The elephants in Tsavo Park, already suffering from overcrowding, started to die in the thousands as food supplies shrank. People were starving, too. Yet park officials refused to allow anyone to touch the meat; they were aghast at the idea that the park's wildlife might be used to meet human needs.

Several years ago I returned to Tsavo and found that the local people were trying to acquire sections of the park for cultivation and grazing. One group of elders told me that their overall aim was to have Tsavo abolished altogether. "That park is an insult to us," one of them said. "We have a score to settle."

For their own survival, Africa's parks must take a lesson from Tsavo. Desperate, hungry people do not take pleasure in the pristine wilderness; they need food in their stomachs. Park managers must realize this. In protecting wildlife from the encroachment of humanity they can serve dual needs. Culling herds to eliminate overpopulation preserves habitat and ensures a sustainable ecosystem. In the process they can provide meat, mountains of it for neighboring peoples—reducing the animosity and making them more aware of the benefits of living side by side with wildlife.

But let us tread the path toward a human-dominated future very carefully. Once we have accepted such total management of a vast area, how long will it be before we go to extremes? Will we choose to eliminate species that aren't of any direct value to us? Will we begin to seed the savanna with "improved" grasses to increase yield, destroying native species uniquely adapted to the environment?

We established our parks as arks against a rising tide of humanity. Now we discover that it isn't enough to play Noah; willy-nilly, we are playing God. Let's do it carefully.

Human civilization thrives as the biological world thrives. In the short term, wildlife advocates argue, a species lost here and there may be of little consequence for overall ecosystem stability, but in the long term the cumulative effect of such losses threatens our own survival.

Opposing Views

Critics of the animal protection movement argue that too much energy and money is spent on saving endangered species. Researchers and wildlife departments have spent millions of dollars to save the California condor, which still teeters on the brink of extinction (Figure 8-10). A few critics argue that money would be spent more wisely on prevention. Other critics want to know why there is so much fuss over endangered species that block "human progress." Surely, a species lost here or there can have no great significance to us. The world will probably never miss the passenger pigeon or the peregrine falcon. And surely no irreparable ecosystem damage would result from the extinction of these and other species. Biologists argue, however, that if we take the attitude that each species by itself is dispensable, bit by bit we will destroy the rich biological world we live in. Somewhere the line has to be drawn: each endangered species is worth saving, because it stops the momentum toward widespread destruction. In growth-oriented societies this momentum may be difficult to slow down, much less to stop. Therefore, each hurdle put in its way becomes an important force in saving the living creatures that make up our web of life. In the words of the naturalist William Beebe, "When the last individual of a race of living things breathes no more, another heaven and earth must pass before such a one can be again."

How Can We Save Endangered Species?

Preserving species is not a simple matter; much work can be done on three overlapping levels: technical, legal, and personal.

Technical Solutions

Integrated species management is a diversified approach that attacks extinction on many fronts. Some general suggestions are:

1. Reduce habitat destruction by controlling development. Wetland and estuary destruction should be greatly curbed worldwide.

2. Establish preserves wherever needed to protect nesting grounds and other critical habitats.

3. Reduce commercial and trophy hunting when evidence shows that the hunted species is rare, threatened, or endangered and when synthetic products can replace those acquired from these animals and plants.

4. Improve wildlife management by upgrading habitat and protecting "nongame" species.

5. Strictly control the introduction of alien species, especially on islands.

6. Design careful predator and pest control management programs so as not to indiscriminately eliminate nontarget species. Be more selective in using poisons, and eliminate them wherever possible by selecting environmental control agents (see Chapter 17).

7. Reduce pollution of air, water, and land.

8. Increase public awareness of the value of wildlife and of factors causing extinction.

9. Increase public participation in habitat improvement, wildlife rescue efforts, and wildlife management (for example, by establishing hot lines and rewards for reporting poachers).

10. Increase private and governmental funding of captive breeding programs that raise endangered species for release and for habitat protection here and abroad.

11. Establish domestic breeding programs to generate research animals, rather than relying on imports. Eliminate unnecessary research on wild animals.

12. Toughen penalties and increase policing of animal and plant trade and poaching.

13. Promote international cooperation to curb the trade of endangered species.

14. Increase expenditures for all protective measures, possibly through new taxes or voluntary income tax programs under which citizens donate money to wildlife and protection.

15. Intensify research efforts to learn more about ecosystem stability and keystone species and to identify critical plant and animal habitats.

16. Help establish programs to protect keystone species.

17. Help fund extractive or productive reserves in Third World nations, allowing sustainable economic activity and species protection.

The following examples illustrate some technical solutions.

Zoos Lend a Hand Scientists at the Baltimore Zoo are engaged in a unique program to save the endangered lion-tailed macaque, a monkey that has disappeared from

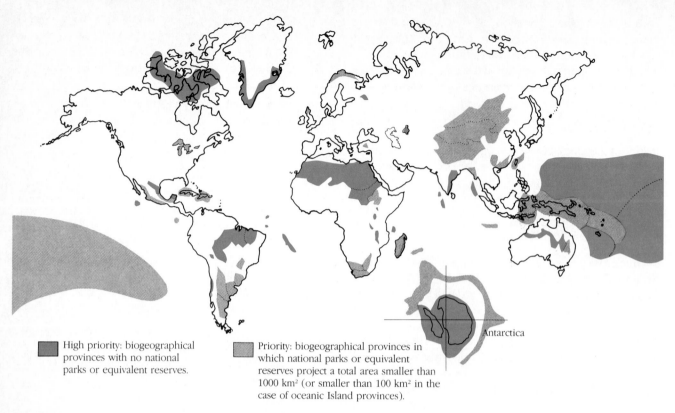

High priority: biogeographical provinces with no national parks or equivalent reserves.

Priority: biogeographical provinces in which national parks or equivalent reserves project a total area smaller than 1000 km² (or smaller than 100 km² in the case of oceanic Island provinces).

Figure 8-11.   Regions in need of protection to preserve important wildlife and plants.

the wild. Researchers first inject hormones that stimulate ovulation into female macaques. The hormones stimulate the release of a surplus of ova, which are flushed from the reproductive tracts of the monkeys and fertilized in petri dishes. The fertilized ova are next flown to the University of Wisconsin in Madison, where another team of researchers implants them into female rhesus monkeys, a common laboratory species. The use of surrogate mothers allows researchers to increase the number of captive monkeys and reestablish wild populations.

Zoos throughout the world have, for many years, been breeding endangered species in captivity to keep them from disappearing altogether and to establish populations that can be released into suitable habitat. One recent project was launched by zoos in the mainland United States in an effort to rescue three rare species of birds that are threatened by tree snakes on the island of Guam. Eventually up to 25 zoos may participate in special breeding programs to rebuild wild populations of the Guam rail, the Micronesian kingfisher, and Mariana crow.

The future of America's remaining California condors lies in the hands of wildlife biologists and personnel of the San Diego Zoo. The remaining wild condors were captured in 1986 and 1987 and transferred there, where workers are attempting to rebuild their numbers in the hope that they can be released into the wild. On April 29, 1988 the very first California condor conceived in captivity was hatched. Biologists hope to produce hundreds over the next ten years.

Helping Third World Nations Protect Priority Areas

The world map in Figure 8-11 shows high-priority areas, that is, regions with high species diversity that are in dire need of protection. Economically, they represent our best investment in wildlife protection. Most of these regions, however, are located in Third World countries without adequate resources to protect them. Thus, a fundamental problem arises: the bulk of the earth's species and their genetic resources are in the poor developing countries, but the economically wealthy, "biologically poor" developed nations are the ones that will benefit from the genetic resources. For this reason, many experts argue that the rich should share the cost of preserving the tropics. A .1% tax on internationally traded oil would net $1 billion a year and would go a long way toward establishing and maintaining large reserves in these high-priority areas.

Another way of protecting vital habitat in Third World nations is the **extractive reserve**, land set aside for native people to use on a sustainable basis. Huge tracts of rain forest in tropical countries, for example, could be set aside for harvesting berries, rubber, nuts, and fruits. While providing a sustainable income, the reserves also help protect native species. (For more on extractive reserves in tropical rain forests, see Case Study 9-1.) But how does a poor nation pay for a reserve? Conservation groups and others who support extractive reserves are asking lending agencies and lending nations to swap some of their debt for conservation programs. Dubbed "debt-for-nature swaps," they work like this: A nation that is forgiven a portion of its debt pledges to set aside land or to spend a certain amount of money on conservation programs. The World Bank, for instance, may forgive several million dollars of debt for a nation that pledges to spend half a million dollars on conservation.

Beyond Habitat Protection: Germ Plasm Repositories Preserving endangered plants and animals requires habitat protection. Wildlife, plants, and habitat are in many ways so ecologically bound together that one cannot be saved without the other. Realizing that much habitat is bound to fall to bulldozers and chain saws, biologists have been frantically searching through forests and fields to gather seeds for cold storage in **genetic repositories** (Figure 8-12). Here they can be held for future study and possible use. The US Department of Agriculture currently supports the National Germ Plasm System, which has over 450,000 plants in "stock" and plans to add 7,500 per year for many years to come. Third World members of the Food and Agricultural Organization of the United Nations voted in 1985 to establish a worldwide system of storing seeds, cuttings, and roots that can be used in agriculture. This system was created to thwart "genetic imperialism" by the developed nations, which have been collecting plants and seeds from Third World nations, altering their genetic composition, and then patenting them.

However important genetic repositories are, this strategy has some major drawbacks. First, despite storage at low temperature and humidity, many seeds rot and must be replaced. Others undergo genetic mutation when stored for long periods and are no longer useful. Finally, storage systems will not work for potatoes, fruit trees, and a variety of other plants. The problems faced in germ plasm storage underscore the need to protect land containing wild relatives of our most important crops: corn, wheat, rice, sorghum, and potatoes. Such efforts will help ensure geneticists a continual supply of new genetic material.

Legal Solutions

Integrated species management requires laws that protect rare, endangered, and threatened organisms and their habitats. Today, however, poverty caused partly by overpopulation and resource shortages hinders progress in habitat protection.

Figure 8-12. Inside a genetic repository laboratory. The lab is sponsored by the USDA to stock seeds from hundreds of thousands of plants from all over the world for future study.

In 1973, in response to the plight of wildlife and plants in the United States and abroad, the US Congress passed the **Endangered Species Act**. This act (1) requires the US Fish and Wildlife Service to list endangered and threatened species in the country, (2) creates federal protection of the habitat of listed species, (3) provides money to purchase this habitat, and (4) enables the United States to help other nations protect their endangered and threatened species by banning the importation of these species and by giving technical assistance.

Protection begins with the listing of an endangered or threatened species. Since 1973, 334 animals and 205 plants have been designated threatened or endangered in the United States. All federally funded or approved projects that might have an impact on endangered species must be reviewed by appropriate agencies, which can deny needed permits or ask for modifications that remove the danger.

Since the Endangered Species Act went into effect, thousands of projects have been through this process, and in most cases differences have been worked out amicably. The most renowned exception was the case of the snail darter and the Tennessee Valley Authority's Tellico Dam on the Little Tennessee River (Figure 8-13). Problems began in 1975 when an order came from a federal court to stop construction of the multimillion-dollar dam,

Figure 8-13. Measuring only 8 centimeters (3 inches), the snail darter created a big stir between environmentalists and industry. The impending destruction of the snail darter by the TVA's Tellico Dam brought the multimillion-dollar project to a standstill. After years of debate, Congress ordered the dam to be completed.

already 90% completed, which would flood the fish's only breeding habitat.

The order was upheld in the US Supreme Court. Congress soon established a committee to review requests for exemptions to the act. In 1979 the special committee refused to give the TVA an exemption, saying that the project was of questionable merit. The TVA applied more pressure on Congress, however, and later that year Congress authorized the completion of the dam. The snail darter was transplanted to several neighboring streams, where biologists believe it can live. Additional populations were discovered in several nearby streams.

The Endangered Species Act is one of the toughest and most successful environmental laws in the United States. "The real success story of the act," says Bob Davison, a National Wildlife Federation biologist, "is that there are species around today that would not have survived if the law had not forced agencies to consider the impacts of what they're doing while allowing development to proceed. . . . To a large extent, the law has succeeded in continually juggling those two competing interests." A battle over protecting old-growth forests and the spotted owl, which was listed as a threatened species in 1989, is raging in Oregon and Washington today. (For more on old-growth forests see Chapter 9.)

To protect rhinos, gorillas, and other endangered species, governments throughout the world have joined in an unprecedented legal effort to stop the illegal trade in rare and endangered species. But in many cases inadequate funding makes enforcement a joke. Inspectors can be paid off by illegal traffickers of endangered species. Governmental agents can patrol only a small fraction of the poachers' range and, at least in the United States, the courts have routinely been lenient toward poachers. In 1985, however, a Montana man was sentenced to 15 years in prison for killing and selling protected eagles and grizzlies. Conservationists are hoping that this sentence, the toughest to be handed down in US history, may mark a turning point.

Legal solutions can barely keep up with runaway population growth and burgeoning agricultural development. Nowhere are these trends more acute than in Africa. Without population control and strict laws to protect wildlife habitat, many of Africa's large herds will be wiped out. International wildlife organizations have stepped in to help raise personal awareness and settle the conflicts between humans and wildlife. Population control programs sponsored by a number of organizations can go a long way in stemming the tide of human population that threatens wildlife throughout the world.

Personal Solutions

Millions of us walk into fast-food restaurants every day and order a hamburger and fries. In Japan, people flock to fast-food restaurants for sushi and stir-fried vegetables, which they eat with disposable wooden chopsticks. Wealthy Californians head for Catalina Island for the weekend in their expensive hardwood yachts. In Central America, chain saws buzz in tropical rain forests; huge trees topple, and monkeys scurry for new homes in the outlying forests. Birds squawk and fly away, crowding into neighboring forests soon to be cut. The trees are hauled off on trucks to nearby mills, cut apart, and finally whittled down to make disposable chopsticks shipped by the millions to fast-food restaurants in Japan. Some of the wood will be fashioned into fine furniture, parquet floors, paneling, and high-quality coffins to bury the dead of the wealthy nations. On the barren forest ground ranchers plant grasses for their cattle, raised in large part to feed the hungry hamburger crowd of North America.

The hamburger and chopstick connection illustrates the part citizens unknowingly play in the extinction of this planet's rich biological diversity. Through excessive consumerism, apathy, and unchecked population growth, we become a part of the problem. But it need not be that way. You can find out which products come from tropical forests and find alternatives. Share the information with your family and friends.

You can take a more active role, too. Members of a Denver group called Volunteers for Outdoor Colorado are working with local wildlife officials to protect habitat and joining other groups to repair badly eroded hiking trails (Figure 8-14). From Monterey, California, to Alaska volunteers are spending their free time improving salmon streams badly damaged by sediment and debris from heavily forested areas. Similar groups exist in Washington, New Mexico, and Florida.

Join a group in your area. Contact your state fish and wildlife agency and find out what volunteer groups are up to. If there are no active groups in your area, why not start your own? Begin with a simple project. Your wildlife officials will surely know of a few inexpensive habitat improvement projects that you can tackle.

Since all resource extraction and processing affects wildlife, conservation can have an important effect. Shutting off lights when leaving a room, obeying the speed limit, and keeping the thermostat low in the winter will indirectly benefit wildlife. For further suggestions, check the EQ test at the beginning of the book. Your actions, combined with the actions of others like you, will cut down on pollution and land disturbance.

You can help educate others about protection of endangered species. You can join groups and spread the word through educational campaigns, lobbying, television ads, posters, books, pamphlets, and the like. Support organizations and politicians who fight against pollution, habitat destruction, commercial and trophy hunting, indiscriminate pest and predator control, and collection of animals and plants for research and home use.

Joining wildlife groups is one of the best ways to learn from dedicated experts with well-developed plans for wildlife protection. (See Environmental Action Guide for a list of organizations and their addresses.) Some organizations such as the Nature Conservancy and the Trust for Public Lands purchase habitat for rare and endangered species. Others, such as the National Wildlife Federation, Sierra Club, Audubon Society, and Wilderness Society, concentrate much of their effort in the legislative arena to promote sound environmental policy.

Wildlife Report

Wildlife advocates have a great deal to be happy about. In 1988, for example, scientists in Maryland spotted a breeding pair of bald eagles in the suburbs of Baltimore, something that has not been seen there for more than 50 years. The endangered black-footed ferret, trapped from the wild by conservation officials in an effort to protect the species, has fared well in captivity. It is reproducing rapidly, making scientists optimistic about reintroducing the animal into the wild within the next few years. The ferret population jumped from 18 in 1986, when the last known wild ferrets were put into captivity, to 60 in mid-1988. Thirty-six ferrets were born in 1988 alone.

In 1987, officials of the US Fish and Wildlife Service announced that they removed the American alligator from the endangered species list. After 20 years of protection, alligator populations in the southern United States have made a dramatic comeback. In Alabama and South Carolina, for instance, alligator populations have increased tenfold since the mid-1970s.

Figure 8-14. Volunteers head into an overused wilderness area in Colorado to repair badly eroded trails.

On another front, the first extractive reserve in the Amazon, in the state of Acre, Brazil, was recently endorsed by Acre's governor. It will be set aside for harvesting rubber, nuts, fruit, oils, and other products from trees in an attempt to reap sustainable economic benefits from tropical rain forests and to preserve natural diversity. Long advocated by US and international environmental groups, extractive reserves represent a major step forward in protecting the tropical rain forests from destruction.

In July, 1988, three peregrine falcon chicks were placed in special boxes atop a 23-story building in downtown Denver. The chicks were fed and monitored for several weeks until they were able to fly. It was hoped that the falcons, which migrate south in the winter, would return to Denver after two years to nest in the skyscrapers of the city. There they could prey on pigeons and eventually breed and maintain a steady population. Several more were released in 1989. Whether the birds will return is not yet known. Colorado is a leading state in peregrine recovery efforts. Over 500 peregrines have been released in Colorado and other states in the past ten years.

On another positive note, recent studies show that the range and population size of nearly two dozen songbirds in the United States have been expanding in recent years. Some believe that the large increase in the number of Americans who put out bird feeders during the winter months helps many species. Others believe that the slightly warmer climate may be partly responsible for the expansion. Still others believe that habitat changes that favor certain species may be the cause. Roger Tory Peterson, America's most renowned expert on birds, thinks that the northern cardinal, for example, is expanding its range because of the construction of bridges. The cardinal's range, Peterson says, was once bounded on the east by the Hudson river. The George Washington bridge, built across the river in 1931, may have enabled cardinals to cross the river more easily. Other species may also be expanding as a result of bridges. Regrowth of forests in certain areas of the country may also be helping songbirds. For instance, a hundred years ago Connecticut was 75% farmland. Today it is about 75% wooded. As farmland returns to forest, long absent species may be returning to the trees.

Finally, in news that thrilled many environmentalists, late in 1988 President Reagan signed into law a bill authorizing an increase in federal spending on endangered and threatened species. The bill, amending the original Endangered Species Act of 1973, increases the annual funding for plant and animal protection from about $30 million to $66 million by 1992. The bill also increases the fine for violating the act from $20,000 to $50,000 and authorizes the use of the first $300,000 collected in fines to be given out as rewards to informants who notify the US Fish and Wildlife Service of poaching or other criminal activities that affect endangered species. The amendment also makes it illegal to remove endangered or threatened plants from any property in the US without written consent of the landowner.

On balance, however, the prospects for wildlife are not bright. For most species, the situation is growing worse. Lead shot is believed to kill as many as two million waterfowl per year. Rhinoceros and elephant populations in Africa are fast on the decline. In 1987 newspapers announced the death of the last dusky seaside sparrow. These birds once flourished in Florida marshlands, but their habitat and food supplies were destroyed by housing development, business, pesticides, and fires.

Beach development in Florida has had an unfortunate impact on sea turtles. The lighting that accompanies new homes, condominiums, and businesses disorients newly hatched turtles. Hatchlings are endowed by nature with an instinct to head for water, and they find water by the glimmering of moonlight on the sea. In areas where beach development occurs, the turtles often become disoriented and, instead of finding their way to the sea, struggle toward land. Away from the ocean too long, the young turtles die. Greenpeace and other environmental groups are lobbying Florida for lighting ordinances and encouraging individuals to file complaints about light streaming onto turtle nesting areas.

The expanding human population and our growing demand for resources threatens to destroy hundreds of thousands of species. The US Fish and Wildlife Service added 12 species to the list of endangered species in 1987. Conservationists are concerned about the inability of the US Fish and Wildlife Service to protect threatened and endangered species. Approximately 1,000 species should be added to the endangered species list, say conservationists, but for one reason or another most of these will have to wait years to gain protection. Each year, hundreds of thousands of hectares of wetlands are destroyed here and abroad, and eleven million hectares of tropical rain forest are cut and burned to make way for human development.

What is civilized in us is not opera or literature, but a compassion for all living things and a willingness to do more than simply care.

DANIEL D. CHIRAS

Summary

Species extinction is a natural phenomenon which occurs during evolution as one species transforms into an entirely new one and also as a result of catastrophic climatic changes. Extinction has greatly accelerated in recent times because of human activities, including alteration of habitat; hunting for profit and sport; introduction of alien species; destruction of pests and predators; capturing species for zoos, private collections, and research; and pollution. Habitat destruction and hunting for profit are the leading causes. Today, one vertebrate species is believed to become extinct every nine months, compared with a natural extinction rate of one species per 1000 years. Species that become extinct are lost forever.

Numerous ecological factors also play a part in extinction. The **critical population size**, which is the number of organisms needed to ensure survival, varies from one species to the next. If a population is reduced below this level, survival may be impossible. Additional factors include the degree of specialization, position of an organism in the food chain, size of an organism's range, reproductive rate, and tolerance to human presence.

According to a new ecological theory, some species may be critical to the well-being of a great many others. Ecologists call them **keystone species** because their loss may lead to the collapse of an entire ecosystem.

Ecologists argue that there are many reasons for protecting endangered species. Aesthetic considerations, animal rights, ethical responsibility, economic benefits, and ecosystem stability are all compelling ones.

We can reduce the loss of plants, animals, and microbes through **integrated species management**. Some general suggestions are to reduce habitat destruction, establish preserves and extractive reserves, crack down on poaching and plant rustling, control the introduction of alien species, reduce pollution, fund captive-breeding programs, educate the public on the value of wild plants and animals, promote international cooperation, intensify research to learn more about ecosystem stability, and establish germ plasm centers to store seeds from wild plants that might be useful in improving the genetic stock of commercial crops.

One of the most effective tools for reducing the loss of endangered species in the United States, and to a certain extent abroad, has been the Endangered Species Act (1973). It prohibits importation of endangered species and sets out other guidelines to protect them. All federally funded projects that might have an impact on endangered species must be reviewed by the Fish and Wildlife Service or the National Marine Fisheries Service. The discovery of an endangered species only rarely results in the prohibiting of a project; in most cases only slight modifications need be made so the project can continue.

In addition to the many technical and legal solutions, numerous personal measures can be added: conserving resources, reducing waste and pollution, improving habitat, joining wildlife groups, and becoming politically active.

Discussion Questions

1. Debate the statement "Extinction is a natural process. Animals and plants become extinct whether or not humans are present. Therefore, we have little to be concerned about."

2. List and describe the factors that contribute directly to animal and plant extinction. Which ones are the most important?

3. Trophy hunters generally try to shoot the dominant males in a population. Natural predators, on the other hand, remove the sick, weak, and aged members of the population. How do trophy hunting and natural predators differ in their effects on the prey population? Use your knowledge of ecology and evolution.

4. Why are islands particularly susceptible to introduced species?

5. Discuss the "ecological" factors that contribute to species extinction.

6. Describe the concept of keystone species. What are its implications for the modern conservation movement?

7. Debate the argument "We must save endangered species of plants and animals."

8. Do we have a responsibility to preserve all living forms? Why or why not?

9. You are placed in a high government position and must convince your fellow executives of the importance of preserving other species. How would you do this? Outline a general plan for preserving species diversity.

Suggested Readings

Achiron, M. (1988). Making Wildlife Pay Its Way. *International Wildlife* 18 (5): 46–51. Elaboration on Norman Myers' Viewpoint in this chapter.

De Roy, T. (1987). When Aliens Take Over. *International Wildlife* 17 (1): 34–37. Eye-opening account of dangers of eliminating alien species once they have become established.

Domalain, J. (1977). Confessions of an Animal Trafficker. *Natural History* 87 (5): 54–57. Startling account of illegal practices in the animal trade.

Hansen, K. (1984). South Florida's Water Dilemma: A Trickle of Hope for the Everglades. *Environment* 26 (5): 14–20, 40–42. Thorough analysis of habitat alteration in southern Florida.

Jackson, P. (1986). Running Out of Room! *International Wildlife* 16 (5): 4–11. A moving tale of the effects of population pressure on Asian elephants.

Jackson, P. (1987). The Rhino's Fatal Flaw. *International Wildlife* 17 (1): 4–11. Alarming article on the fate of the rhinoceros.

Johnson, P. (1985). Smoothing the Way for Salmon. *National Wildlife* 23 (4): 31–35. Story of the work private individuals can do to help restore wildlife populations.

Laycock, G. (1966). *The Alien Animals.* Garden City, NJ: Natural History Press. A popular account of the troubles created by species introduction.

Morell, V. (1985). Masai. A Proud People in Kenya is on a Collision Course with Wildlife. *International Wildlife* 15 (3): 4–11. Superb account of conflicts between human civilization and wildlife.

Myers, N. (1985). The End of the Lines. *Natural History* (February): 2, 6, 10, 11. Some new thoughts on the new age of extinction and the effects it will have on evolution.

Ola, P. and d'Aulaire, E. (1986). Lessons from a Ravaged Jungle. *International Wildlife* 16 (5): 34–41. Graphic story about the destruction of tropical rain forests and our attempts to preserve species diversity.

Owens, M. and Owens, D. (1985). *Cry of the Kalahari.* New York: Houghton Mifflin. Touching account of two wildlife biologists in Africa. Important information on behavior, ecology, and conservation.

Ralph, C. J. (1982). Birds of the Forest. *Natural History* 91 (12): 41–44. Excellent look at endangered birds on the Hawaiian islands.

Raven, P. (1985). Disappearing Species: A Global Tragedy. *The Futurist* 19 (5): 8–14. Good overview of species extinction.

Robins, J. (1988). Grizzly and Man: When Species Collide. *National Wildlife* 26 (2): 20–27. Excellent case study.

Schneider, B. (1977). *Where the Grizzly Walks.* Missoula, MT: Mountain Press. Excellent discussion of how humans contribute to extinction, especially through habitat destruction.

Schwartz, D. M. (1988). Hog Havoc. *National Wildlife* 26 (4): 14–17.

Sunquist, F. (1988). Zeroing in on Keystone Species. *International Wildlife* 18 (5): 18–23.

Tanji, K., Lauchli, A., and Meyer, J. (1986). Selenium in the San Joaquin Valley. *Environment* 28 (6): 6–11, 34–39. Detailed account of the selenium problem at Kesterson caused by irrigation.

Wilcove, D. (1990). Empty Skies. *The Nature Conservancy Magazine* 40 (1): 4–13. Excellent overview of factors causing the decline of songbirds in the United States. Illustrates several key points discussed in this chapter.

Wilson, E. O. (1989). Threats to Biodiversity. *Scientific American* 261 (3): 108–116. Graphically illustrated discussion of habitat destruction here and abroad.

9

Rangeland, Forest, and Wilderness: Preserving Renewable Resources

Our duty to the whole, including the unborn generations, bids us restrain an unprincipled present-day minority from wasting the heritage of these unborn generations.

THEODORE ROOSEVELT

In the early 1970s scientists pored over satellite photographs of the drought-stricken African Sahel, a band of semiarid land that borders the southern Sahara. Little did they know that another important lesson was about to be revealed. One of them noticed a bizarre piece of land amid the spreading desert. On the surface it appeared much like an oasis in the parched desert landscape. Curious to find out the reason, Norman MacLeod, an American agronomist, flew to the site. There, surrounded by newly formed desert, was a privately owned ranch of 100,000 hectares (250,000 acres). Its grasses grew rich and thick even though vegetation in the surrounding fields had long since died, exposing the sandy soil.

The secret to the success of this ranch lay in several strands of barbed wire. Stretching around the perimeter of the ranch, this thin barrier held out the cattle of the nomadic tribes who had let them overgraze the surrounding communal property for decades. The ranch was divided into five sections, where a rigidly controlled number of cattle were grazed once every five years—further ensuring the ranch's survival.

The photographs of the drought-stricken Sahel show us the devastation wrought by mismanagement of renewable resources. "A continent ages quickly once we come," Ernest Hemingway observed. But this was not the only lesson. Perhaps even more significant was the discovery that this semiarid grassland could remain productive

Figure 9-1. Remnants of an ancient city in the Bamian Valley invaded by desert created by poor land management and drought.

despite the drought if it were well managed, a lesson crucial to humankind.

This chapter examines what happens to the world's **commons**, land that is shared by many people without control. It looks also at private lands, pointing out that in many cases they, too, suffer from poor management. Rangelands, forests, and wilderness areas are the key focal points of this chapter.

A Tragedy of the Commons

As far back as the days of ancient Greece, Aristotle recognized that property shared freely by many people often got the least care. Early civilizations toppled forests and carelessly overgrazed their cattle on their rangelands just as the tribesmen of the Sahel do today. History, however, shows that early civilizations paid dearly for their disregard. The skeletons of buildings from ancient cities stand out in deserts that were once rich forests and grasslands of the fertile crescent (Figure 9-1). Much of Iran and Iraq, now barren desert, once supported cattle, farms, and rich forests. Greece and Rome fell, historians believe, partly because of the misuse of their lands.

Economists have debated the fate of common resources such as air, water, and land for decades. It was not until 1968, however, that Garrett Hardin, a prominent US environmentalist, exposed the cycle of destruction in a classic paper called "The Tragedy of the Commons."

In England, Hardin noted, cattlegrowers grazed their livestock freely on fields called the commons. The commons fell into ruin, however, because of a lack of regulation. What doomed the commons was that the users became caught in a blind cycle of self-fulfillment. Individuals increased their personal wealth by increasing their herd size. Each additional cow meant more income at only a small cost, because the farmer did not have to buy new land or feed; the commons provided them. In his book *Filters against Folly*, Hardin notes that the cattlegrowers were rewarded for doing wrong. They realized that increasing their herd might lead to overgrazing and deterioration of the pasture, but they knew that the negative effects of overgrazing would be shared by all members of the community. Thus, each herdsman arrived at the same conclusion: he had more to gain than to lose by expanding his herd. This short-sighted thinking resulted in a spiraling decay of the commons. Over and over, growers increased their herd size, sharing the environmental costs with other users of the commons. Hardin summarized the situation as follows: "Each man is locked into a system that compels him to increase his herd without limit in a world that is limited." As each pursues what is best for himself, the whole is pushed toward disaster. "Freedom in a commons brings ruin to all," wrote Hardin.

The logic that compels people to abuse communal holdings has been with humankind as long as common property. Today, however, the process has begun to catch up with us. The deserts of the Middle East are a good example, as are millions of acres of desert in the US Southwest. The communal property of the Sahel is the most recent reminder. There overgrazing spawned environmental disaster. But economics played an important role in the problems of this area as well. In the 1960s, for example, loans to drill wells were made. Previously nomadic tribes became more sedentary. People no longer migrated south during the dry season. As a result, the land around the wells and human settlements deteriorated and the carrying capacity was soon surpassed.

Hardin's analysis of the tragedy of the commons, while important, is somewhat misleading, for it gives the impression that common resources are alone in being mismanaged. In many places private lands are no better cared for because short-term profit dictates management strategies (Chapter 20).

Today virtually all lands are gripped by the tragedy of overexploitation. A blindness to the concerns of the future is part of the frontier mentality discussed in Chapter 19. Short-term exploitation may have been permissible at one time, when the human population was small in relation to the earth's resources. Today such actions are intolerable. Too many people share this planet, and the cumulative effect of many small insults has become staggering.

Rangelands and Range Management

Rangelands are a vital component of global food, leather, and wool production. If properly managed, rangelands and the livestock they support can provide useful products indefinitely. If mistreated, rangelands, like other renewable resources, can be ruined. This section looks at problems facing rangelands and how they can be managed to ensure continued use. First, let's look at the problems created by mismanagement.

Rangeland Deterioration

Over 90 million hectares (225 million acres) of rangeland, much of it public land, have slowly turned to desert in the United States in the last 200 years. This is equivalent to an area one and a half times the size of Texas.

Figure 9-2. The Rio Puerco Basin of New Mexico, once a rich grassland, has turned to desert because of overgrazing.

The Navajo Indians, for example, live on a 6-million-hectare (15-million-acre) reservation in Arizona, New Mexico, and Utah. Theirs is a sun-parched land, dusty and dry. Eking out an existence, the Navajos live in destitution. Making matters worse, the population has begun to skyrocket; unemployment has worsened. To feed and clothe their people, the Indians have gradually increased the size of their sheep herds. Today, the herd size exceeds the carrying capacity by at least four times. Baked by the hot summer sun and swept by fierce winter winds, the overgrazed reservation is becoming an arid dust bowl.

Southeast of the Navajo reservation lies another tract of parched desert land, the Rio Puerco Basin. Lying northwest of Albuquerque, New Mexico, it was the breadbasket of the state in the 1870s, when the basin's rich grasslands supported huge herds of cattle. A century later, however, the land had crumbled under the strain of overgrazing (Figure 9-2). Erosion has formed gullies that widen by 15 meters (50 feet) a year. Wind and rain erode the soil five to ten times faster than it can be replenished naturally.

According to the Bureau of Land Management (BLM), which manages most of the federally owned rangeland, 62% of the public rangeland is in fair to poor shape. Fair rangeland currently supports one-half to one-fourth as much vegetation as the land is capable of supporting. Poor rangeland supports one-fourth or less. (Figure 9-3 shows rangeland degradation by region.) The Soil Conservation Service estimates that three-fourths of all rangeland and pastureland in the country needs better management. In general, private rangeland is in worse condition than federal rangeland. The major culprit in this deterioration is overgrazing. The Navajo lands and the Rio Puerco Basin, discussed above, show the effects: severe erosion, a drop in groundwater, desertification, the loss of wildlife, and the invasion of weeds.

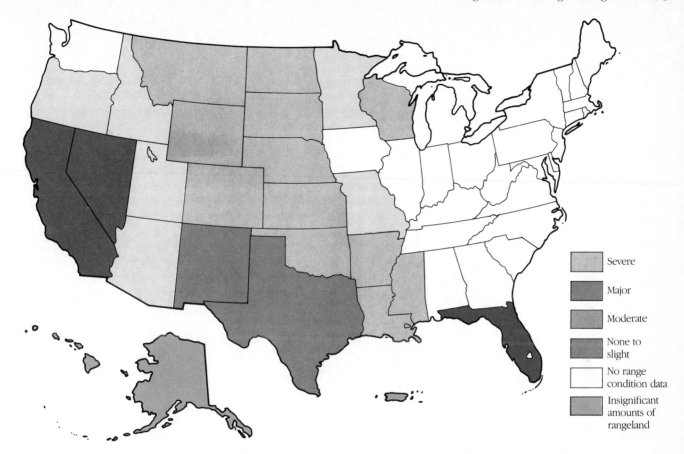

Severe

Major

Moderate

None to slight

No range condition data

Insignificant amounts of rangeland

Figure 9-3. Damage to nonfederal rangeland.

Range Management

The prospering island within the Sahel, discussed in the introduction to this chapter, provides living proof that rangeland can remain productive despite aridity. In fact, proper range management can benefit the land in many ways. Livestock, for instance, can help spread seeds and fertilize the soil.

Range management involves two basic techniques often employed simultaneously: grazing management and range improvement. Good **grazing management**, the first line of defense, requires careful control over the number of animals on a piece of land and the duration of grazing. **Range improvement** means controlling brush, revegetating barren areas, fertilizing impoverished soils, constructing fences and water holes, and similar measures to promote uniform grazing (Figure 9-4).

The key to proper range management is keeping livestock populations within the carrying capacity of the ecosystem while allowing them to remain profitable. This is often difficult, because carrying capacity varies from one year to the next depending on the weather. In dry periods the carrying capacity may be half of that in a normal year. Truly effective range management requires a willingness to cooperate with nature, benefiting from the good times

and cutting back during the bad.

The **Public Rangelands Improvement Act**, passed by the US Congress in 1978, promotes better range management on public land. It also calls for improvements of publicly owned rangelands currently managed by the BLM and the US Forest Service. Both agencies have guidelines for proper range management and attempt to manage their lands accordingly. However, some observers believe that the BLM's policy, formulated by a rancher advisory board, is not as sound as the Forest Service's, which is formulated by professional range management specialists and others.

This act also requires the BLM and Forest Service to reduce grazing where damage is evident. This strategy is not popular among ranchers, who either cannot see the benefits of improving range conditions or dispute the claims that they are overgrazing the land. One of the chief weaknesses of the Public Rangelands Improvement Act is that it does not pertain to Indian lands in the West, where grazing reductions are badly needed. As a result of private and federal actions, US rangelands seem to be gradually improving, but much more work is needed. Tremendous improvements are also needed in nations now gripped by drought.

Figure 9-4. Properly placed water holes help distribute cattle evenly on the range and avoid overgrazing of select areas.

Forests and Forest Management

Covering about one-third of the earth's land surface, forests provide many direct and indirect benefits. The most notable direct benefits are an estimated 5000 commercial products, such as lumber, paper, turpentine, and others, worth tens of billions of dollars each year. Forests also provide refuge from hectic urban life and opportunities for many forms of recreation. In many poorer nations, forests are a source of wood for cooking and heating.

Indirectly, the forests benefit us by protecting watersheds from soil erosion, thus keeping rivers and reservoirs relatively free of silt. Forests reduce the severity of floods and facilitate aquifer recharge. Forest lands also perform many important ecological functions: they assist in the cycling of water, oxygen, nitrogen, carbon, and other nutrients and provide habitat for many species.

The United States has about 300 million hectares (740 million acres) of forest land (Figure 9-5). About two-thirds is commercial timberland, and most of this is privately held. Each year US forest products sell for over $30 billion, and forestry employs 1.5 million people, making an important contribution to the economy.

Despite the great benefits of forests, only about 13% of the world's forest land is under any kind of management. In addition, only about 2% of the world's forests are protected in forest reserves. With world population growing, the demand for wood for fuel and goods rising, and the very real possibility of global warming, better management is badly needed to protect and expand this renewable resource.

Worldwide Deforestation

Forestry scientists estimate that somewhere between 30% and 50% of the world's forests have already been destroyed, mostly because of clearing for agriculture, firewood production, and commercial cutting. In Nepal population pressures force farmers higher and higher up the mountainsides in search of land to plant their crops. Coupled with a rising demand for fuelwood, both for residents and for climbing expeditions that regularly visit the Himalayas, this has meant the felling of large tracts of forest, creating a serious firewood shortage.

Nepal is not alone. Africa has lost about 30% of its forests, while Brazil, the Philippines, and Europe have lost 40%, 50%, and over 70%, respectively. In Third World countries ten trees are cut down for every tree that is planted. In Africa the ratio is 29 to 1. Besides destroying wildlife habitat, deforestation decreases sustainable fuel supplies needed for cooking and home heating. According to estimates of the Food and Agricultural Organization of the United Nations, 100 million people in 26 countries now face acute shortages of firewood. In rural Kenya some women spend up to 24 hours a week in search of wood. (See Case Study 9-1 for a discussion of deforestation in tropical rain forests.)

The destruction of forests is exemplified by the American experience. When the first colonists arrived, forests

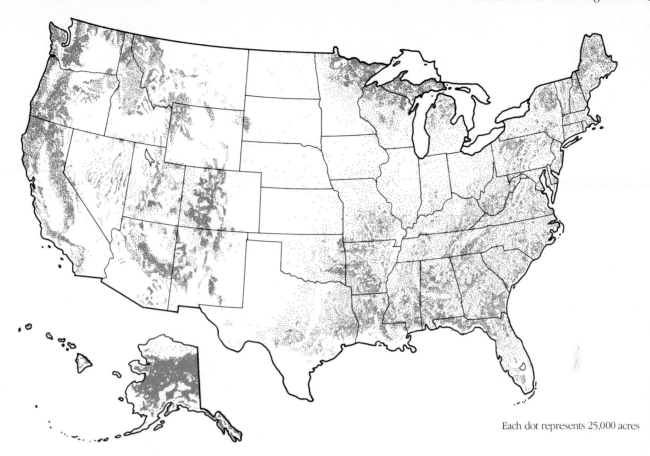

Each dot represents 25,000 acres

Figure 9-5. Distribution of forest land in the United States.

covered about half the land surface of the United States. Soon, though, they began to clear land for farms and towns and to build ships, homes, and highways.

White pines were especially hard hit by commercial harvesting, which started in the early 1800s in New England. Proceeding westward as these trees became depleted, commercial harvesters reached Minnesota and Wisconsin in 1870. By the early 1900s the white pine had been reduced to the point that it was no longer commercially profitable to harvest. Lumber companies moved into the South to cut the slash, loblolly, and longleaf pines, but within a few decades most of the profitable stands had been felled.

Fortunately, the southern pines bounced back from heavy harvesting. The trees grow rapidly in the hot, open areas created by previous tree harvests and in abandoned cotton and tobacco fields. Today, under better management, pines support a profitable timber industry in the South.

Forest Conservation in the United States

Forests have been exploited for most of human history with little concern for long-term productivity. Early com-

mercial interests in the United States took an especially narrow economic view of the public forests, seeking monetary gain with little concern for the future.

In 1891 President Benjamin Harrison established the first forest reserve, known as the Yellowstone Timberland Reserve, to help protect American forests. President Theodore Roosevelt added more land to the forest reserve system. By the end of his term over 59 million hectares (148 million acres) of forest had been saved from the commercial interests. Such interests had been running roughshod over much of the public land. Many of Roosevelt's actions were at the urging of the noted conservationist and forester Gifford Pinchot. Pinchot and Roosevelt recognized that forests could be harvested without permanent damage. Carefully managed, such lands could continue to produce valuable timber for future generations.

In 1905 Roosevelt established the Forest Service as part of the Department of Agriculture. At that time the forest reserves became known as the national forests. Pinchot became the first head of the Forest Service. He favored judicious forest development over strict preservation. His ideas laid the foundation for US forest management policy for decades.

Today the US Forest Service controls about 76 million hectares (187 million acres) of public land in 155 national forests and 19 national grasslands. About half of the Forest Service land is open for commercial harvesting. The Forest Service lands have many uses other than commercial timber cutting, such as livestock grazing, mining, hiking, skiing, hunting, camping, and other forms of recreation. The Forest Service also conducts numerous operations—including watershed protection, recreation enhancement, forest management, range management, pest and fire control, and wildlife habitat improvement—all geared to improve the national forests.

Forest Service lands are managed by guidelines set out in the **Multiple Use–Sustained Yield Act** (1960). This law requires that national forests and grasslands be managed to achieve the *greatest good for the greatest number of people in the long term* (the **multiple use** concept). With so many interests competing for the national forests' economic and scenic riches, numerous conflicts have arisen over this mandate. They generally involve environmentalists, on the one hand, and mining interests, timber companies, and resort developers, on the other. Central to the conflict is the question of how particular parcels of land should be used. Should a region be left for backpackers, hunters, and wildlife, or should it be leased to a ski resort developer? The controversy over wilderness designation discussed later in this chapter illustrates one important example of the conflict.

Sustained yield means that timber cutting should not exceed timber growth and should not destroy the forests' long-term productivity. Thus, the Forest Service's lands should be managed so that future generations can reap the same benefits as present generations. This requires special erosion-control measures and, in some cases, reseeding to help ensure regrowth. Special controls on clear-cutting, which are described below, can also increase the likelihood that forests will regrow after being cut. Critics of the Forest Service argue that this agency is violating its own policy by allowing overharvesting and timber cutting on steep slopes.

Forest Harvest and Management

Trees are commercially harvested by three basic methods: clear-cutting, selective cutting, and shelter-wood cutting.

Clear-cutting is a standard practice used primarily for softwoods (conifers), which grow in large stands with relatively few tree species. It is also used in tropical rain forests, which have tremendous species diversity. In clear-cutting operations loggers remove all the trees in 16- to 80-hectare (40- to 200-acre) plots, although on US Forest Service land clear-cuts are now limited by law to 16 hectares (40 acres), with some notable exceptions. For instance, in California, Oregon, and Washington Douglas fir stands have a clear-cut maximum size of 25 hectares

(60 acres). In the South clear-cuts can be even larger, up to 30 hectares (80 acres). In Alaska the limit is even higher, up to 40 hectares (100 acres).

On clear-cuts, loggers remove the commercial timber from a plot and often burn the remaining material. Burning this residual matter returns nutrients to the soil, facilitates regrowth, and reduces the threat of fires that could damage the regenerating forest. As the new stand grows, trees are thinned to eliminate overcrowding.

Clear-cutting has numerous benefits. Perhaps most important is that it is often faster and cheaper than other methods of harvesting trees. Clear-cutting can increase surface runoff, the flow of water over the ground's surface. This enhances stream flow and can increase the supply of water to cities, farms, and industry. Clear-cutting increases habitat for some species, such as deer and elk, which benefits hunters.

Critical thinking suggests, however, that the issue is not so simple. Not all clear-cuts are equal when it comes to elk habitat. A small clearing, for example, tends to be more beneficial than a larger one, because elk generally avoid open spaces of more than 8 hectares (20 acres). Elk prefer to remain at the edge of meadows, where they can race into the protection of nearby forests should a predator arrive. Thus, large square or rectangular blocks are less advantageous than smaller irregular cuts for elk. Another factor that determines whether a clear-cut increases or decreases elk habitat is the location of the cut. Winter range is a limiting factor in elk populations. Thus, clear-cuts in winter range, which make more food available to elk, are more beneficial than cuts in the more abundant summer range. In Rocky Mountain states, however, clear-cuts are generally made in elk summer range, high in the mountains.

Despite their many benefits, improperly sited clear-cuts can produce unsightly scars that may take years to heal (Figure 9-6). If not replanted or reseeded naturally, clear-cuts may suffer severe erosion. Erosion is especially troublesome if clear-cutting occurs on steep terrain. Eroded sediment fills streams and lakes, destroying fish habitat. Sediment also reduces the water-holding capacity of lakes and streams, which increases flooding, already more likely because of the elevated surface runoff. Erosion in clear-cut areas may deplete the soil of nutrients, thus impairing revegetation. In some instances, clear-cutting actually decreases surface runoff, which can deplete river water. Large open patches in mountainous terrain, for example, may accelerate a process called sublimation, the conversion of snow to water vapor. When this occurs, snow melts actually decrease. Routine burning in clear cuts can also damage soils by destroying bacteria needed for nutrient cycling. Burning can also volatilize soil nitrogen, robbing nutrients from the soil itself.

New studies in California suggest that widespread burning of forests generates three greenhouse gases, carbon

(a)

(b)

Figure 9-6. (a) Clear-cuts in the South Tongass National Forest, Alaska, and (b) in Kootenai National Forest, Montana.

dioxide, methane, and nitrous oxide, as well as other gases. Long after the flames have gone, burned soils continue to emit high levels of nitrogen oxides, particularly nitric oxide and nitrous oxide. Nitric oxide is converted to nitrogen dioxide. In the atmosphere nitrogen dioxide is converted to nitric acid, thus adding to acid deposition (Chapter Supplement 15-2).

Finally, clear-cutting destroys habitat and can contribute to the decline of many species, such as the ivory-billed and red-cockaded woodpeckers and numerous tropical species. In the Pacific Northwest, heavy cutting of **old-growth forests**, ancient forests more than 250 years old, with many sections from 500 to 800 years old, now threatens the spotted owl and several other species dependent on this rapidly vanishing habitat. Excessive cutting of the old-growth forests in the past century has devastated valuable salmon runs in Washington, Oregon, and California. Canadian old growth is also threatened. One of the hardest hit areas is Vancouver Island in British Columbia. This 12,408-square-mile island is rapidly being whittled away by clear-cuts. The southern and northern thirds of the island are covered with logging roads, and even the central region of the island, which houses the huge Strathcona Provincial Park, is being logged. Park officials, say critics, are making land trades with mining and logging interests that withdraw land from the park. In British Columbia, a national park creates a high degree of pro-

tection, whereas provincial park status means very little.

Vancouver Island has remarkable old-growth forests. Western hemlock is one of the main species, but massive red cedars also exist, some with a circumference of 20 meters (60 feet). Sitka spruce can climb to 100 meters (300 feet) or more. A century of feverish logging, though, has taken most of the best trees from the island.

In tropical forests clear-cutting is a prescription for ruin: soils become baked in the sun and too hard to support growth; others wash away in torrential rains. (For more on the effects of clear-cutting tropical forests see Chapter Supplement 3-1 and Case Study 9-1 in this chapter.)

New regulations by the US Forest Service are helping to reduce the impact of clear-cutting in national forests, but on private lands clear-cutting is largely unregulated. There are 34,000 privately owned tree farms in the United States, covering approximately 30 million hectares (75 million acres). Large commercial tree farms operate much like agribusiness. Seedlings are planted, fertilized from airplanes, doused with herbicides to control less desirable species, and sprayed with insecticides and fungicides to reduce losses. When the trees reach the desirable size, they are cut down, and the cycle begins again. Here and in the tropics, large clear-cuts are the rule.

Arguing for smaller clear-cuts on private as well as federal land, E. M. Sterling, an expert on forest management, writes, "The forested mountains of the Pacific Northwest ought to be as spectacular as any in the world. Tragically, they are not. . . . For unlike those of Europe, the great mountain forests of the Northwest are being scarred by ever-increasing clear-cut logging on both private and public lands." (See Figure 9-6.)

By comparison, Sterling notes, Austria harvests as much wood from its forests as does the Pacific Northwest. Yet Austrian forests show little evidence of clear-cutting, mainly because of strict forestry laws. What makes these laws unique is that they apply to public as well as private lands. Austrian law, for instance, forbids clear-cutting on all steep, erodible land. It also limits the size of clear-cuts. A private landowner may cut .6 hectares (1.5 acres) without permission, but must obtain a permit for larger clear-cuts. Seldom do clear-cuts exceed 2 hectares. Most clear-cuts are narrow strips that blend in with the terrain and ensure natural reseeding.

The lesson from Austria, environmentalists point out, is not that clear-cutting should be banned but that it can be improved in the United States to reduce erosion and the visual impact, for instance, by making cuts smaller and by blending them with the terrain. More efficient use of "waste" wood for paper or other products is needed. Restocking trees is also essential. A 1986 survey by the National Wildlife Federation showed that US foresters operating on public lands in Montana had violated Forest Service regulations that require restocking within five years of harvesting. Of 600 logged sites surveyed by the federation, about 200 had not been replanted even at a minimum level.

Selective cutting, as its name implies, is the removal of certain trees from a forest. Unfortunately, it is rarely used in the United States and Canada. The object of selective cutting is to preserve species diversity in forests, helping protect forests from disease and insects, while reaping a sustainable harvest. Properly done, a selective cut requires a periodic thinning of forests. Foresters harvest old, young, and middle-aged trees, retaining an equal percentage of each age class. Improperly done, a selective cut would remove only the best trees, an action viewed skeptically by some forest managers because it removes the genetically superior trees from the forest. Their seed is needed to keep the forest healthy. By selectively removing the strongest and healthiest stock, a forest, they say, may slowly degenerate, producing lower quality wood.

Selective cutting has several disadvantages, cost and time involved being the most critical. But this technique leaves no scar, causes little or no erosion, and does little damage to wildlife habitat. Despite these benefits, forestry experts point out, it cannot be viewed as a replacement for clear-cutting.

Shelter-wood cutting is an intermediate form of tree harvesting between clear-cutting and selective cutting. In this technique poor-quality trees are first removed. The healthiest trees are left intact. They reseed the forest and provide shade for their seedlings. Once the seedlings become established, loggers remove a portion of the commercially valuable trees. Enough are left in place to continue shading the seedlings. Finally, when the saplings are well established, the remaining mature trees are cut down.

Shelter-wood cutting has many of the advantages of selective silviculture. It leaves no unvegetated land, minimizes erosion, and greatly increases the likelihood that the forest will regenerate. However, it is more costly than either clear-cutting or selective cutting.

Shelter-wood and selective cutting can be economically competitive with clear-cutting when logging roads are present, even in forests consisting of one or two species, say some critics. Since shelter-wood and selective cutting prevent the scarring of forest land, they may provide additional economic and aesthetic advantages to regions that rely heavily on tourism, such as Washington, Oregon, Colorado, Wyoming, Alaska, and Canada. They are far less likely to create erosion and other problems associated with clear-cuts.

Prospects for the Future: Building a Sustainable System

Between 1970 and 2000 the demand for wood in the United States is expected to double because of population growth and increased demand. By the year 2020 the demand for wood and wood products is expected to exceed the supply; clearly, something must be done. Before considering solutions it is useful to look at the causes of deforestation.

What Causes Deforestation?

Deforestation in Canada, the United States, the tropics, and elsewhere stems from a great many factors. Analysts have traditionally viewed deforestation in Third World nations as a natural social response to poverty, unsustainable population growth, and landlessness. A new study by economist Robert Repetto of the World Resources Institute, however, puts much of the blame on governmental policies. Repetto notes that governments largely determine how a nation's forests will be used. Even governments that are committed to conservation, however, often have contradictory policies.

Unfortunately, many governments—the United States and Canada included—believe that forest protection can only occur at the expense of economic development. Repetto argues that the misuse of forests actually costs countries billions of dollars per year. Among the hardest hit are the poor countries of the developing world, many of which are saddled with immense foreign debt.

One problem is that governments have typically sold timber below market value to logging companies. Critics say that in the United States the Forest Service loses money on most, if not all, of its National Forests because it routinely auctions off timber rights to land that is unsuitable for logging. In such cases the highest bids are often lower than the government's cost for surveys, boundary markings, paperwork, and auctions. By not auctioning marginal or below marginal stands, Repetto believes, the US government could save $100 million per year. Below-cost timber sales are also a form of public subsidy to the lumber industry. Selling timber below cost discourages conservation by companies and individuals.

Many governmental policies here and abroad encourage excessive cutting of forests, profiteering, and an unsustainable exploitation of forest resources. One common problem in many Third World countries is that they restrict the export of raw wood by international companies in order to create jobs at home and encourage economic development of domestic wood processing industries. These nations hope that bans or heavy taxes that limit raw-wood exports will result in an increase in the export of finished wood products (for example, furniture), netting higher revenues than from trees that are processed elsewhere. Unfortunately, says Repetto, many of the small mills that open up are highly inefficient and use 50% more logs than the industry standard to achieve a given output of milled products. As a result, this policy results in an enormous increase in deforestation.

Another problem is short-term contracts. In many Third World nations, 35 years or more are required for a stand of trees to recover from logging, but contracts are written for only 20 years. This arrangement discourages companies from protecting forests because they have no long-term interest in their concession. Longer contracts, say critics, would encourage companies to harvest forests sustainably.

Economic policies can also influence deforestation. Heavy borrowing from international banks and industrialized nations has created enormous debt in the Third World (Chapter 8). To pay back the loans, countries often encourage unsustainable forest practices that result in widespread cutting.

Government tax policies also encourage deforestation. For instance, in Brazil, the government has offered income tax credits to investors in cattle ranches for up to 75% of the project's cost, and for up to 50% of a company's tax liabilities (the taxes owed in a given year). Cattle ranches have been a leading cause of deforestation in the Amazon. By 1980 they accounted for more than 72% of Brazil's deforestation. Ironically, many of the projects subsidized by tax credits could not turn a profit without these generous subsidies. Repetto estimates that this policy costs the Brazilian government about $3 billion a year.

Canadian forests suffer because of several legal loopholes. The Canadian government, in fact, has long encouraged deforestation with little regard for the environment. Making matters worse, Canadian citizens have little recourse when it comes to fighting harmful timber operations. As a result of these and other factors, the forests of British Columbia are falling at a rate far greater than the estimated annual sustainable yield. The pace of deforestation has nearly tripled since 1960. Ninety-four percent of the forest in British Columbia lies on public lands, and there are practically no institutional channels by which citizens can influence forest management. Citizens do not even have the right to sue to stop harmful forest cutting. As a result, many citizens have taken the law into their own hands and have turned out to block timber companies by forming human roadblocks or camping in the path of road construction. Local communities are worried about the loss of recreational opportunities and many people who turn out to block the bulldozers are anglers, artists, whale-watching guides, small business owners, and so on. They don't want an end to logging, just a voice in deciding how and where it should occur.

When conservationists argue for controls on deforestation, the timber industry responds with the threat of lost jobs. But in British Columbia, as in the American Pacific Northwest, wood products jobs have been steadily declining (because of automation) for years, while the annual cut has risen sharply. The timber companies wield an incredible amount of power and use the job issue as a smokescreen. Their power and influence are additional factors responsible for the widespread deforestation throughout the world.

Deforestation is a part of the frontier mentality, briefly described in Chapter 2. To solve our problems requires new policies, individual actions, and a new attitude toward natural resources (Chapter 19). In general, two strategies can be employed: increasing supply and decreasing demand.

Increasing Wood Supply by Protecting Forests
Protecting forests ranks among the best ways of ensuring a continued supply of wood and wood products. Clearcutting, especially in the tropics, could be more carefully regulated on both public and private land. In particular, the clear-cuts on steep terrain could be reduced to decrease erosion that impairs regrowth. Reseeding by logging companies should also be monitored more carefully, especially on public lands. In 1986 the US Congress passed an important law that prevents the US Agency for International Development (AID) from funding projects in the Third World, including dams and roadways, that could lead to the destruction of tropical forests. The law also directs the AID to help countries find alternatives to forest colonization and requires it to support preserves

Saving the World's Tropical Rain Forests

Rain forests cover much of the tropics in a thick, lush carpet. At one time, the tropical rain forests of the world covered an area twice the size of Europe. Today, the tropical forests have been reduced by half and logging continues at a feverish pace in many areas. Some experts think that tropical rain forests could be virtually obliterated by the end of the century if we do not effect strict measures to protect them.

The rate of deforestation varies considerably from one country to the next. In Thailand, Costa Rica, the Ivory Coast, and Nigeria the rate of deforestation is 3% to 6% per year. In other countries, such as Zaire and Cameroon, the rate of deforestation is much slower, but still of considerable concern in the long term.

All in all, many experts believe that about 11 million hectares (27 million acres) of tropical forests are cut down each year. That's an area about the size of Ohio. In 1988, satellite photographs of Brazil, just one of 80 countries endowed with tropical rain forest, showed that 8 million hectares had been recently cut and set afire. This discovery made many question whether the 11-million-hectare estimate was too low.

The loss of tropical rain forests is one of the most serious problems facing the world today, say Sandra Brown and Ariel Lugo, two US forestry scientists. It is a major cause of extinction since the tropical forests contain about two-thirds of the world's species, only about one-sixth of which have been named. The importance of preserving these and other species is described in Chapter 8.

Al Gentry, a researcher who studies tropical forests, says that the "loss of so many species is not only a tragic squandering of the earth's evolutionary heritage but also represents depletion of a significant part of the planet's genetic reservoir, a resource of immense economic potential." Genes from the tropical rain forest could help boost agricultural production, as discussed in Chapter 7. Wild species are also a source of new drugs to battle diseases.

Gentry points out another problem not often considered in the debate over tropical deforestation, that is, the effect of other methods of tree harvesting, such as selective cutting. Selective cutting, in fact, often alters the forest so much that it destroys as many species as clear-cutting.

Tropical forests play an important role in global cycling of oxygen and carbon dioxide. Global deforestation today accounts for about 25% of the world's annual increase in carbon dioxide. It is, therefore, a major factor in gradual global warming, as discussed in Chapters 1 and 15.

Clear-cutting tropical rain forests can reduce rainfall by 50% in the immediate vicinity, turning once-lush areas into barren deserts. When rains do come, they wash away the soil, filling streams and rivers with sediment. Tropical deforestation is already creating such problems in Pakistan, India, and the Amazon Basin of South America.

Tropical soils are poor agricultural lands. As noted in Chapter Supplement 3-1, they are generally nutrient poor and some may bake to a hard, bricklike consistency when exposed to sunlight. "Lush, high biomass tropical forests," says Al Gentry, "represent the end point of a very gradual accumulation of tightly held nutrients, continually recycled through the ecosystem over millennia." Cutting and burning the forests to make room for ranches and farms releases sufficient nutrients for a few years, but the nutrients are quickly taken up by crops or washed away, making the land useless.

Despite these and a number of other problems, tropical rain forests are receiving very little protection. The steps taken in recent years, however, are important in protecting forests from decimation. In 1988, the first extractive reserve in the Amazon Basin, in the state of Acre, Brazil, was established. Set aside for harvesting rubber, nuts, fruit, oils, and other products, the 40,000-hectare (100,000-acre) reserve will allow people to reap sustainable economic benefits from tropical rain forests and reduce overall deforestation. Rubber tappers, who

Case Study 9-1

make a living from the forests, worked with environmental groups to promote the concept of extractive reserves in Brazil, a much-needed alternative to deforestation, cattle-ranching, and agricultural colonization.

Rubber tappers and environmentalists have also been successful on other fronts. Consider an example. The Inter-American Development Bank (IDB) had loaned Brazil about half of the $60 million it needed to construct a highway into the heart of the Amazonian forest to open up land for the overcrowded state of Rondonia. Heavy pressure on the bank at the outset had caused officials to set aside $10 million of the highway loan money to guard against the adverse environmental impacts of road construction and to help create reserves for Indians who lived in the forests. Because Brazil had overlooked the agreement to protect the land, the rubber tappers, the National Wildlife Federation, and the Environmental Defense Fund pressured the IDB to cut project funding. Surprisingly, the bank agreed and suspended the highway loan, the first time a bank had halted a loan on the basis of environmental concerns.

Brazil petitioned the bank to restart the highway loan. In return, the government agreed to create 32 Indian areas in Brazil. The government also offered to protect 500,000 hectares of rain forest that could be affected by the highway and agreed to create four extractive reserves, totaling about 518,000 hectares.

Extractive reserves are much more sustainable than the typical uses of tropical rain forests. In the state of Rondonia, ranchers and farmers have cut down or disturbed 35% of the forests, creating millions of hectares of wasteland. On most soils, farmers can graze one head of cattle per hectare the first

five years. In the next two years, because of declining soil fertility, they can only graze one head of cattle per four hectares. After that the land is often destroyed. This waste contrasts with large tracts of rain forest that can be set aside for the sustainable harvest of latex, nuts, and other products. Research indicates that people can earn nearly twice as much money per hectare harvesting rubber as they can from cattle ranching.

On another front, in the small northern Central American country of Belize, a consortium of international environmental groups is working to set aside 100,000 hectares (240,000 acres) of rain forest. The Programme for Belize is asking individuals to purchase land for them at $50 an acre.

On still another front, scientists may have found a way to help start revegetation. In an experiment in December 1988, researchers showered deforested regions of the Brazilian rain forest with gelatin-coated seeds of tropical plants to promote revegetation of slopes and reduce landslides. Dropped from planes, each "seed bomb" contains approximately ten seeds, which fall directly to the ground. The gelatin absorbs moisture, causing the seeds to germinate quickly. If this experiment proves successful, it may be applied to other areas in years to come. Despite its promise, this technique will not duplicate the incredible diversity of forests.

Though positive steps are being taken, rain forests are being destroyed much faster than they can be protected. Millions of hectares of land are in ruin in the tropics. Individuals can help by joining groups that are fighting to protect the forests, by avoiding wood products derived from rain forests (disposable chopsticks, parquet floors, and many hardwoods), by recruiting friends, and by writing governmental representatives and urging action.

and other measures to save forests and promote biological diversity.

Forests suffer from a number of natural hazards, including diseases, insects, fire, droughts, storms, and floods. As shown in Figure 9-7, diseases, insects, and fires account for most of the damage. Sound management of forest land requires that foresters use an integrated plan that maintains trees in a healthy state. This means that trees should be thinned and that soil should be protected to ensure maximum fertility and water retention. Diversity should also be encouraged to reduce insect and disease outbreaks. Insect- and disease-resistant trees could be developed. Air pollution, which damages some trees and makes them susceptible to other environmental factors, should also be controlled. Finally, all imported trees and lumber should be carefully inspected to avoid accidentally introducing pests (Chapter 8).

Fire accounts for 17% of forest destruction, leveling .8 to 2.8 million hectares (2–7 million acres) of forest each

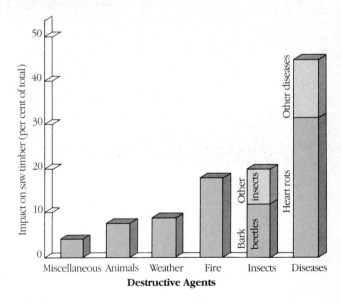

Figure 9-7. Causes of damage to US forest expressed as a percentage of the total annual damage.

(a)   *(b)*   *(c)*

Figure 9-8.   Benefits of forest fires. (a) Dense undergrowth in an Oregon pine stand results from the control of forest fires. (b) Controlled burning removes the undergrowth. (c) Periodic burning prevents disastrous fires, returns nutrients to the soil, and increases forage for cattle and wildlife.

year. According to the US Forest Service, 85% of all forest fires are accidentally or deliberately started by humans. The remaining 15% are ignited by lightning. The lightning fires, generally much larger than fires started by humans, account for about half of the forest damage each year.

To protect watersheds, timber, and recreational opportunities, the US Forest Service and state governments attempt to reduce forest fires by posting fire danger warnings and sponsoring television and radio announcements. Each year the Forest Service spends from $450 million to $600 million for fire fighting and surveillance.

Protecting forests from fire began with Gifford Pinchot in the early 1900s. It has no doubt saved billions of dollars worth of property and timber and countless animals. But ecologists and foresters now realize that strict fire control can actually be detrimental to forests. Fires are a natural event, with many benefits to the forest. For example, minor, periodic fires burn dead branches that have accumulated on the ground, returning the nutrients to the forest soil (Figure 9-8). Most animals can escape these minor **ground fires**, and living trees are generally left unharmed. Periodic ground fires also forestall intense, destructive fires.

In protected forests the story is quite different. If a fire breaks out in a forest that has been protected for long periods, the ample fuel supply may permit it to burn uncontrollably, spreading from treetop to treetop as a **crown fire**. Huge areas are destroyed in firestorms so hot that the soil itself is charred. Wildlife often perish in large

numbers. Generally, all living trees are severely burned and die as a result. One of the most devastating fires occurred in Yellowstone National Park in 1988. It began outside the park in Forest Service land that had been protected for many years, making it unstoppable once it crossed into the park.

Periodic fires protect our forests from devastation. In addition, ecologists now know that many forest species require occasional fires for optimum growth. For example, the cones of the jack pine open up and release their seeds during fires. Douglas fir, sequoia, and lodgepole pines also require periodic fires for seed release. As we have seen, fires return nutrients to the soil and remove brush that shades seedlings. They also help reduce disease and control insect populations.

Recognizing the benefits of periodic ground fires, forest managers now let many naturally occurring forest fires burn, provided they are not a threat to human settlements. The Forest Service also starts hundreds of fires each year to remove underbrush and litter, not only to reduce the chances of potentially harmful crown fires but also to improve wildlife habitat, soil fertility, and timber production and to increase livestock forage. These **prescribed fires** are set at times when the danger of their getting out of hand is low. (For a discussion of the forest fires in Yellowstone, see Case Study 9-2.)

The world's forests also require protection from pollution, which is taking a great toll. Acid deposition, discussed in Chapter Supplement 15-2, and ozone depletion,

Case Study 9-2

Controversy over Fire in Yellowstone National Park

In the summer and fall of 1988, Americans watched in horror as Yellowstone National Park, one of their oldest and most treasured parks, erupted in fire. Yellowstone has a history of fires burning in huge blocks of forest, often 100,000 acres at a time, but this one was different and in its wake is a storm of controversy.

Some critics of the Park Service, for example, blame the devastating fires on a controversial fire management policy begun in 1972. Known as the let-it-burn policy, it allows fires caused by lightning to burn as long as they do not endanger people, human property, special sites within the park, or endangered wildlife.

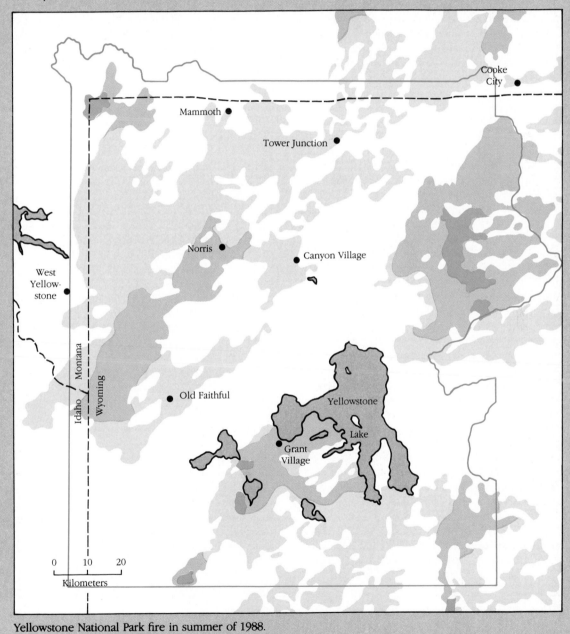

Yellowstone National Park fire in summer of 1988.
☐ On July 21 the fires were still relatively small
☐ On August 21 the fires expanded across large areas of the park
☐ By October 2, 720,000 acres of the park had burned
☐ Water areas

Case Study 9-2 (continued)

To blame the fires on park policy, however, is to ignore the facts. In 1988, record-breaking heat and drought turned the park into a tinderbox. Almost no rain fell on Yellowstone Park that summer, making it the driest summer in the park's history. Humidity on the forest floor was a startling 2% or 3%, much lower than kiln-dried wood, which contains 12% moisture.

Many of the fires started outside the park and were fought immediately, but fire fighters could do nothing to stop them. High winds and arid conditions conspired against them. The men and women who directed the fire-fighting efforts and who had never been beaten by a forest fire were helpless to stop the flames. Burning embers were carried 7 kilometers (4 miles), jumping valleys and fire lines with ease. At one point, nearly 10,000 men and women fought the blazes, but often they could only move out of the way of the rushing flames.

The news reported that the fires burned 450,000 hectares (1.1 million acres) inside the park—about half of Yellowstone. Satellite photographs, however, show that only about one-fifth of the park, or 180,000 hectares (440,000 acres) actually burned; about one-half of this area was badly burned.

The fires in Yellowstone had remarkably little effect on wildlife. By official estimates only 250 of the park's 40,000 elk perished. Only two grizzly bears and a dozen buffalo in the park's herd of 3,000 perished. Soil damage was also minimal.

The Yellowstone fires have provided an invaluable opportunity to study forest fires and their long-term effects. Yellowstone Park generally burns in a mosaic pattern: some areas burn while others are passed over by flames that jump from one location to another. The 1988 fire was no exception. A visit to the park shows charred stands in a checkerboard mosaic with live stands left untouched by flames. This checkerboard pattern increases the diversity of the Yellowstone ecosystem and also is thought to play a role in controlling fire in Yellowstone.

Forest researchers note that most patches of forest tend to go 200–400 years between large fires. Intense fires burn through an area and kill most of the trees. As a result, sun-loving lodgepole pines spring up. In a few decades the lodgepole pines form a solid canopy, creating shade. As the forest matures, shade-tolerant spruce trees and firs begin to sprout.

According to a widely accepted theory on forest fires, it is only after the forest develops a solid understory of spruce and fir that dangerous and uncontrollable fires can sweep through. Thus, a forest becomes naturally more susceptible to fire in about 200 years. Until that time it is somewhat protected because there is little dead wood on the ground. As a result, a fire on the ground cannot get up into the canopy. But in older forests with a significant amount of ground debris, fires climb into the crowns of the trees, spreading rapidly.

Young forests have long been thought to serve as natural firebreaks. Fires slow to a crawl or die in them. In fact, lightning striking a young stand often fails to start a fire. But in 1988, young forests burned much better than anyone had ever expected because of the dryness.

Scientists view the enormous fires in Yellowstone more as a renewal than a disaster. Large fires have swept through Yellowstone for thousands of years. Thus, most park plants have some sort of natural insurance against fire destruction. For instance, lodgepole pine, which comprises 77% of the forests in Yellowstone, produces two types of seed-bearing cones. The most numerous type develops on the trees for two years and then opens, dropping its seeds to the ground. The other is coated with a resin that seals the cone shut. These cones remain closed on the tree for decades, until the heat of a fire burns the resin, drying the cone, opening it, and letting seeds fall. After fires, tens of thousands of lodgepole pine seeds litter the forest floor, reseeding the forest.

Fire often kills conifer trees, opening up room for new seedlings. Other plants like willows, aspens, and many grasses and wildflowers can survive flames. These plants, in fact, depend on periodic fire to jolt them out of dormant periods. When their tops burn, new trees and plants sprout from the roots below. Fire also releases nutrients trapped in dead wood.

Some fish may benefit from the loss of tree cover around small streams and higher slopes. The loss of tree cover increases sun penetration, making streams warmer and lengthening the growing season both for insects and fish.

Over time, say park officials, most large animals will benefit from the summer's park fires. Nutrients in the ash could fertilize the soil and result in more nutritious food. In addition, forest fires temporarily open up areas, permitting grasses to grow and providing grazing habitat for elk.

The first winter claimed a large number of the elk and bison in the park. The lack of food caused by the fire was responsible for many of the deaths. But only 15% of the winter rangeland burned. Thus, ecologists believe that the drought, which lowered plant growth in Yellowstone, has had a greater effect on grazers than the fires themselves did. Record low summer rainfalls reduced grasses by as much as 60% in some areas.

The fire in Yellowstone may be a sobering warning of what is to come, given the global warming trend many scientists fear is already underway (see Chapter 15). In 1989 as well, forest and grass fires burned hundreds of thousands of acres in the West because of record-breaking heat and drought, deepening the worries of many.

covered in Chapter 15, are the principal agents of destruction. Forests are now threatened in the United States, Canada, and Europe.

Increasing Supply by Reducing Waste   Waste materials such as limbs, bark, and branches can be used for fuel, paper, and a variety of paper products. Some companies grind these materials into chips that are burned for energy or used for the production of paper, chipboard (particle board), and waferboard. The Masonite Corporation installed a $6-million system to generate energy from wastewood. It paid for itself in energy savings in a mere 18 months. Less desirable trees can be converted into paper and other products that require lower quality wood. As a result, forest yield can increase substantially.

Decreasing Demand: The Personal Connection

Recycling paper and more judicious use of wood products can help reduce future demands. Each year 1250 acres of Canadian forest are cut down to supply wood pulp for newsprint for the Sunday edition of the *New York Times*. Increased recycling of newsprint could reduce the amount of forest cut down each year.

The average American consumes over 272 kilograms (600 pounds) of wood per year in the form of lumber and paper. This is 4.5 times what the average European consumes and 40 times what citizens of the less developed countries use. Individuals could reduce overall consumption by recycling at home and at work, by using the backs of scrap paper for homework and notes, by carrying their own shopping bags to the store and by choosing not to use a bag for small items. The constant bombardment of advertising material can be stopped by writing companies and asking them to take you off their lists. To cut lumber use, smaller homes can be built, using 20% to 30% less wood. Earth-sheltered housing, discussed in Chapter 12, can also reduce our demand for wood while drastically cutting fuel consumption. Individuals, companies, colleges, and governments can help by purchasing paper products made from recycled paper (see Chapter Supplement 18-1).

Wilderness

Nineteen eighty-four was a banner year for wilderness protection in the United States. President Reagan signed into law a bill that added 3.5 million hectares (8.6 million acres) to the **National Wilderness Preservation System**. Since that time very little wilderness has been designated. Many areas proposed for wilderness designation in the United States are currently held up in Congress by a raging controversy over water rights. A key question being debated is whether wilderness is endowed with its own water rights. Many developers believe that wilderness areas should not be guaranteed any water rights. Environmentalists argue that this is ludicrous, that wilderness designation automatically assumes water rights. That means that rivers should have enough water to sustain healthy wildlife populations and riparian habitat. Wilderness without water rights is like a house without plumb-

ing. To some people—loggers and miners, especially—designating these lands as wilderness presented a roadblock to economic progress. To others, such as hikers and backpackers, it was a sign of hope.

Wilderness, as defined by US law, is "an area where the earth and its community of life are untrammeled by man, where man is himself a visitor who does not remain." Wilderness provides a temporary escape from modern society (Figure 9-9). Joseph Sax, author of *Mountains without Handrails*, writes that nature "seems to have a peculiar power to stimulate us to reflectiveness by its awesomeness and grandeur." It helps us understand ourselves and the world we live in, awakening us to the forgotten interdependence of living things. "Our initial response to nature," Sax writes, "is often awe and wonderment: trees that have survived for millennia; a profusion of flowers in the seeming sterility of the desert; predator and prey living in equilibrium. . . . [It] is also a successful model of many things that human communities seek: continuity, stability and sustenance, adaptation, sustained productivity, diversity, and evolutionary change."

But not everyone would agree. To many people, wilderness is the playground of the upper middle class, an elite group that is fighting to protect these lands and hinder others' opportunities to reap some economic benefit, especially through jobs created by mining and timber harvesting. Such a view may be typical of the way most people throughout most of human history have seen undeveloped lands; historically, wilderness has been viewed as something to subdue, something to exploit for short-term gain. In early colonial and postcolonial times, American lands represented untapped wealth—an unequaled opportunity to sustain a young, growing nation. The concept of wilderness preservation, had it arisen at that time, would have seemed absurd. Today some critics agree: wilderness designation locks up valuable resources (minerals and timber) needed by humans. Such conflicting views inevitably make compromise necessary.

Preservation: The Wilderness Act

The earliest efforts at wilderness preservation in the United States began in the 1860s. John Muir, founder of the Sierra Club and a longtime wilderness advocate, is credited with much of the early interest in saving wilderness for future generations. Further advances came in the 1930s, when the US Forest Service began to set aside large tracts of forest land, called **primitive areas**, for protection. Between 1930 and 1964 the Forest Service established over 3.7 million hectares (9.1 million acres) of primitive areas in the national forests.

In 1964 Congress passed the **Wilderness Act**, establishing the National Wilderness Preservation System. The Forest Service's primitive areas were renamed **wilderness**

Figure 9-9. Wilderness restores us. It is a vital resource in our world.

areas, tracts of land with restricted use. The Wilderness Act forbids timber cutting, motorized vehicles, motorboats, aircraft landings, and other motorized equipment (for example, chain saws), except to control fire, insects, and diseases or where their use was already established.

The Wilderness Act sought to create an "enduring wilderness," but many unwildernesslike activities were allowed to continue: livestock grazing and mining for metals and energy fuels, if claims were filed before the end of 1983. Wilderness areas throughout the United States are riddled with private inholdings, property owned by individuals and companies who control the mineral and water rights on the property. In Boulder, Colorado, for instance, a lawyer who owns a large section within the otherwise publicly owned Indian Peaks Wilderness Area cut through the public property with bulldozers to repair a dam on his property, causing incredible damage. Another man owns mineral rights in the Maroon Bells Wilderness Area near Aspen, Colorado and is

actively pressuring the Forest Service to allow him access so he can open a large marble quarry along the main path into the area.

The Wilderness Act directed the Forest Service, the Fish and Wildlife Service, and the National Park Service to recommend land within their jurisdictions for wilderness designation. By January 1989, 36.7 million hectares (90 million acres) of land were protected as wilderness. The Wilderness Act failed to consider BLM land for wilderness designation. The BLM controls vast acreages in Alaska and the West (a total of 180 million hectares, or 450 million acres). This serious omission was corrected by the **Federal Land Policy and Management Act** (1976). It calls on the BLM to submit recommendations on the wilderness suitability of its land. By January 1989 only 189,000 hectares (467,000 acres)—about 0.1%—of BLM land had been added to the system. An additional 9.8 million hectares (24.6 million acres) had been submitted to Congress by the BLM for wilderness designation.

Controversy over Wilderness Designation

Nothing heats the waters like wilderness designation. Environmentalists continually press for more land to be set aside; the mining and timber industries generally oppose wilderness designation because it locks up valuable resources. In the battles that ensue, environmentalists are accused of being elitists who want to exclude others from profiting from the earth's riches. Wilderness protection, critics say, costs Americans jobs. Ecologists and environmentalists note, however, that wilderness experiences and the ecological benefits provided by wilderness (erosion control, habitat protection, recycling of nutrients, and so forth) cannot be translated into dollars and cents, whereas the profit from lumber and mineral ores is easily put into economic terms. This makes it difficult for the nonenvironmentalists to understand their case.

The timber industry is one of the strongest opponents of wilderness designation. Since only slightly over 8% of all government land, or about 4% of all land in the United States, has been given wilderness protection, environmentalists argue that the timber industry's claims that wilderness is locking up our earthly riches are unfounded. Locally, however, wilderness tracts can tie up huge parcels of land, threatening the economic well-being of communities that have long made a living by timber harvesting.

Mining interests also argue that minerals are being locked up by the National Wilderness Preservation System. If anything, many environmentalists argue, the mining interests have been catered to in excess; by allowing some mining in wilderness areas, Congress has acted against the best interests of preservation, since mining conflicts with wilderness as much as any human activity could.

Wilderness advocates are pushing for more wilderness. But should we continue to set aside wilderness, especially if it contains oil, natural gas, or minerals that could be used today? Many environmentalists believe that wildlands are more valuable than these resources, because there are no substitutes for them once they have been destroyed. And, they note, there are many ways to expand our energy and mineral resources besides additional mining. (Some of these options are discussed in Chapters 12 and 13.) Environmentalists also argue that continued population growth necessitates an expansion of wilderness. But mining and timber interests validly point out that people also need minerals, wood, wood products, and energy. Locking up the wilderness limits economic development today and in the future. Controversy is raging today over oil and gas exploration in the Arctic National Wildlife Refuge in Alaska. Case Study 11-1 presents both sides of the issue.

The Wilderness Curse

Lured by the thought of quiet and solitude, backpackers pour into some US wilderness areas only to be dismayed by the crowds and special camping restrictions enforced by the Forest Service to protect lakes and streams from pollution. To many this overcrowding is a sign of the need for more wilderness, especially if the US population is to grow by 50–60 million people before it stabilizes, as many demographers now predict (Chapter 5). To others it is an example of one of the major problems caused by designating an area as wilderness: people are attracted, ultimately destroying the wilderness experience for many.

Wilderness crowding and the resulting environmental degradation can be reduced or eliminated by (1) educating campers on ways to lessen their impact, (2) restricting access to overused areas, (3) issuing permits to control the number of users, (4) designating where backpackers can camp, (5) increasing the number of wilderness rangers (even volunteers) to patrol areas and also pick up garbage and monitor use, (6) disseminating information about infrequently used areas to divert campers from overused areas, and (7) improving trails to encourage use of currently underutilized areas.

Wilderness, rangeland, and forest are all part of our lives. They cater to many different needs: eating, shelter, relaxation, and escape. A world without them is almost unimaginable. A group of scientists peering through the glass of their space shuttle 100 years from now will see the evidence of our actions today. Whether they see patches of ancient desert within rich, productive land or just the opposite depends on actions and decisions we make today.

The art of progress is to preserve order amid change.

A. N. WHITEHEAD

Summary

Our land, water, and air are in many ways akin to the commons where the English once grazed their livestock. Without regulation, the commons fell into a cycle of decay. Each cattle grower increased his herd to increase personal gain, bringing ruin to the unmanaged commons. The misuse of biospheric commons is the root of considerable environmental degradation. Resource mismanagement, however, also occurs on private lands.

Rangelands are a vital component of global food production. A history of overgrazing has resulted in the deterioration of both private and public rangelands, resulting in permanent loss of vegetation, erosion, desertification, wildlife extinction, invasion of weeds, and a drop in water tables. **Range management** helps us avoid these problems through two major strategies: **grazing management**, control of the number of animals on a piece of land and the duration of grazing, and **range improvement**, such as fertilizing, reseeding, and the use of techniques to encourage uniform grazing.

Forests benefit society directly by providing numerous commercially valuable products, and also by providing opportunities for recreation. Indirectly, forests benefit us by protecting watersheds from soil erosion; by reducing surface runoff; by recycling water, oxygen, carbon, and other important nutrients; and by providing habitat for a diversity of species.

Worldwide, millions of hectares of forest have been cut down. Overpopulation, shortsightedness, bad policy, and poverty are all to blame. In the United States forest protection began in the late 1800s. In 1905 President Theodore Roosevelt established the Forest Service. Its first head, Gifford Pinchot, promoted careful use of forests over strict preservation. His notions of multiple and sustained use have persisted. Today the Forest Service manages lands for commercial timber cutting, grazing of livestock, mining, and recreation.

The demand for wood and wood products in the United States is expected to exceed the supply by the year 2020. To ease the crunch two strategies can be used: increasing the supply and reducing the demand. Increasing the supply means taking better care of forests and reducing forestry wastes. Decreasing the demand means recycling and cutting down on unnecessary use of wood.

Wilderness is defined by US law as "an area where the earth and its community of life are untrammeled by man, where man is himself a visitor who does not remain." For some people wilderness provides an escape from hectic urban life, a chance to watch animals, exercise, or relax. To others it is an untapped resource with valuable minerals and timber. Because of such divergent views much controversy surrounds efforts to set wilderness aside.

In 1964 the US Congress passed the Wilderness Act, which established **wilderness areas** within land owned by the National Park Service, Fish and Wildlife Service, and Forest Service. The law prohibits timber cutting and motorized vehicles and equipment except in certain instances. Mining can take place in designated wilderness areas as long as claims were filed before the end of 1983.

Even though many commercial interests continue to fight wilderness preservation, conservationists note that rising population and the fact that more people are turning to outdoor recreation necessitates more wilderness to avoid overcrowding and damage. Additional solutions include educating campers on ways to minimize impact, restricting access to overused areas, controlling the number of users, increasing the number of wilderness rangers patrolling areas and picking up garbage, disseminating information about infrequently used areas, and improving trails to encourage the use of underutilized regions.

Discussion Questions

1. Discuss the tragedy of the commons. Give some specific examples of "commons" and how they are mistreated.

2. What are the major problems facing rangelands in the United States? What suggestions would you make to improve the condition of American rangelands and better manage them?

3. Define the following terms as they relate to forest management: *sustained yield, multiple use*, and *clear-cutting*.

4. Why are many foresters intentionally burning forests or letting some forest fires burn?

5. List and discuss ways to satisfy the growing need for wood and wood products in the coming years. Which of your ideas are the most ecologically sound? How would you carry out your ideas if you were an elected official?

6. In what ways can you personally reduce paper and wood waste and increase recycling?

7. What are the major arguments for and against wilderness preservation?

Suggested Readings

Brown, B. (1982). *Mountain in the Clouds*. New York: Simon and Schuster. Gripping account of deforestation practices in the Pacific Northwest and how they decimated salmon fisheries.

Brown, L. R., et al. (1988). *State of the World*. New York: Norton. See Chapter 5 on reforesting the earth.

Brown, L. R., et al. (1989). *State of the World*. New York: Norton. See Chapter 10 on how reforestation can be part of a global strategy to build a sustainable future.

Clawson, M. (1983). Reassessing Public Lands Policy. *Environment* 25 (8): 6–17. Thoughtful look at managing public lands.

Fearnside, P. M. (1989). A Prescription for Slowing Deforestation in Amazonia. *Environment* 31 (4): 16–20, 39–41. Detailed description of sustainable forest practices that can help save tropical rain forests.

Hales, L. (1983). Who Is the Best Steward of America's Public Lands? *National Wildlife* 21 (3): 5–11. Excellent reading.

Hardin, G. (1968). The Tragedy of the Commons. *Science* 162: 1243–1248. A classic paper.

Hardin, G. (1980). Second Thoughts on "The Tragedy of the Commons." In *Economics, Ecology, Ethics*, ed. H. E. Daly. San Francisco: Freeman.

Hughes, J. D. and Thirgood, J. V. (1982). Deforestation in Ancient Greece and Rome: A Cause of Collapse. *The Ecologist* 12 (5): 196–207. Detailed paper.

Maser, C. (1989). Life Cycles of the Ancient Forest. *Forest Watch* 9 (8): 11–23. Important for understanding forest ecology.

Mathews, O. P., Haak, A., and Toffenetti, K. (1985). Mining and Wilderness: Incompatible Uses or Justifiable Compromise? *Environment* 27 (3): 12–17, 30–36. Interesting look at a sticky issue.

Monastersky, R. (1988). After the Flames: Awaiting a Regeneration of Yellowstone. *Science News* 134 (19): 330–332. Interesting reading.

Monastersky, R. (1988). Lessons from the Flames. *Science News* 134 (20): 314–317. Worthwhile reading to further your knowledge on the benefits of the Yellowstone fire.

Nash, R. (1985). Proceed at Your Own Risk: Restoring Self-Reliance to the Wilderness. *National Parks* January/February: 18–19. Important thoughts on the value and protection of wilderness.

O'Dell, R. (1986). Alaska: A Frontier Divided. *Environment* 28 (7): 10–15, 34–37. Detailed look at preservation versus development.

Owen, O. S. and Chiras, D. D. (1990). *Natural Resource Conservation: An Ecological Approach* (Fifth edition). New York: Macmillan. See chapters 11 and 12 for more detailed information on rangeland and forest management.

Prescott-Allen, C. and Prescott-Allen, R. (1986). *The First Resource: Wild Species in the North American Economy.* New Haven: Yale University Press. A systematic analysis of the economic importance of wildlife in North America.

Sax, J. (1980). *Mountains without Handrails: Reflections on the National Parks.* Ann Arbor: University of Michigan Press. Discusses important aspects of park and wilderness preservation.

Sheridan, D. (1981). Western Rangelands: Overgrazed and Undermanaged. *Environment* 23 (4): 14–20, 37–39. Superb account.

Sterling, E. M. (1981). Forestry in Austria: Small Cuts and Grand Vistas. *Sierra* 66 (6): 40–42. Good comparison of US and Austrian forestry practices.

Watkins, T. H. (1986). The Conundrum of the Forest. *Wilderness* 49 (172): 12–24, 34–49. Comprehensive look at US management plans.

Watkins, T. H. (1988). Blueprints for Ruin. *Wilderness* 52 (182): 56–60. Extraordinary article on the fate of old-growth forests in North America.

Water Resources:
Preserving Our Liquid Assets

A river is more than an amenity—it is a treasure.

OLIVER WENDELL HOLMES

Water. We drink it. We wash with it. We play in, on, and underneath it. It finds a thousand uses in our factories and is a staple of food production. The abundance or lack of water often determines where we live and how well off we will be.

Despite its importance to human society water remains one of the most poorly managed resources on earth. Squandered and polluted by industry, agriculture, sewage treatment plants, and many other facilities, water is often treated as if it were of little importance to human survival. This chapter discusses the problems of water supply and flooding and presents solutions to the problems we now face from years of mismanaging our liquid assets. Chapter 16 addresses water pollution. Before looking at the problems of water supply, however, let's examine the water cycle.

The Hydrological Cycle

The global recycling of water is the **hydrological cycle**, or **water cycle**. It runs day and night, free of charge, busily collecting, purifying, and distributing water that serves a multitude of purposes along its path. At its heart are two processes driven by energy from the sun: evaporation and precipitation.

Water evaporates from waterways and lakes, from land, and from plants (Figure 10-1). In plants, its evaporation from leaves helps draw nutrients up the stem, much in the way that sucking on a straw draws water from a glass. The evaporative loss of water from leaves is called **transpiration**. The total loss of water from the soil and leaves is called **evapotranspiration**.

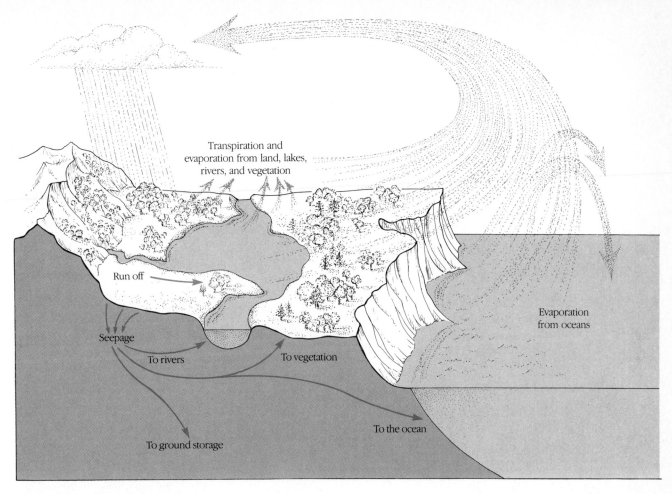

Transpiration and
evaporation from land, lakes,
rivers, and vegetation

Evaporation
from oceans

Run off

Seepage

To rivers

To vegetation

To ground storage

To the ocean

Figure 10-1. The hydrological cycle.

Water suspended in the clouds, except in polluted regions, is nearly pure. Collected as rainfall in rural areas, it can be used in steam irons in which distilled water is required. The reason for the purity of atmospheric moisture is that when water molecules evaporate, they leave behind dissolved impurities.

In the atmosphere, water is suspended as fine droplets (water vapor). The amount of moisture the air can hold depends on air temperature. The warmer the air, the more moisture it can contain. Atmospheric moisture content can be expressed as **absolute humidity**—the number of grams of water in a kilogram of dry air—or as relative humidity, the more common measurement. **Relative humidity** measures how much moisture is present in air compared with how much it could hold if fully saturated at a particular air temperature. At a relative humidity of 50%, for example, air has 50% of the water vapor it can hold at that temperature. If the relative humidity is 100%, the air is saturated.

When moisture exceeds saturation, clouds, mist, and fog form. Clouds, for example, form when moist air is raised by mountain ranges, when cold-air masses come in contact with moisture-laden air, or as warm air rises

to cooler levels (Figure 10-2). For rain to form, air must contain small particles, known as **condensation nuclei**, on which the water vapor collects. Condensation nuclei may be salts from the sea, dusts, or particulates from factories, power plants, and vehicles. Over a million fine water droplets must come together to make a single drop of rain. If the air temperature is below freezing, the water droplets may form small ice crystals that coalesce into snowflakes.

Clouds move about on the winds, generated by solar energy, and deposit their moisture throughout the globe as rain, drizzle, snow, hail, or sleet. Precipitation returns water to lakes, rivers, oceans, and land from which it came, thus completing the hydrological cycle. Water that falls on the land may evaporate again or may flow into lakes, rivers, streams, or groundwater, eventually returning to the ocean.

At any single moment 94% of the earth's water is found in the oceans, 4% is in inaccessible aquifers, and 1.5% is locked up in polar ice and glaciers. This leaves about .5% of the earth's water available for human use, but most of this water is hard to reach and much too costly to be of any practical value.

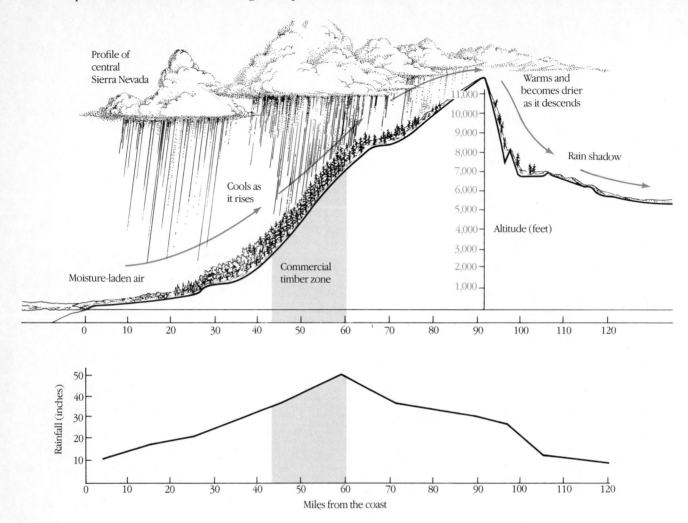

Figure 10-2.   Mountain ranges thrust air upward, causing it to cool. This increases the relative humidity of the air, forming clouds and precipitation. Most of the precipitation falls on the windward side of the mountain. As the air descends on the leeward side, it warms, and the relative humidity decreases. Leeward sides of mountain ranges are often arid.

Water Supplies and Usage

The Global Picture

Water problems face virtually every nation in the world. Despite the abundance of sparkling blue water around the Caribbean, for example, islands are plagued with freshwater supply problems. Rainwater for domestic use must be captured on rooftops and used sparingly. In many of the developing countries people spend a good part of their waking hours fetching water, frequently walking 15 to 25 kilometers (10 to 15 miles) a day to get water, often from polluted streams and rivers.

Three out of every five people in the developing nations do not have access to clean, disease-free drinking water. According to the World Health Organization, 80% of all disease in these countries results from the contaminated water that people drink and bathe in. Many coun-

tries are engaged in bitter rivalries over this precious resource. Argentina and Brazil, for instance, dispute each other's claims to La Plata River; India and Pakistan fight over the rights to water from the Indus. All told, over 150 river basins are shared by two or more countries.

The UN General Assembly proclaimed the 1980s as the International Drinking Water Supply and Sanitation Decade. The goal was to provide the world's population with clean drinking water and adequate sanitation by 1990. Unfortunately, when the decade ended the UN was a long way from meeting its goals. Work will continue, however, to get safe, clean drinking water to hundreds of millions of people.

Water shortages are commonplace today in many parts of the globe, including parts of Peru, central Chile, many parts of Mexico, Panama, parts of Africa, New Zealand, parts of Australia, Korea, Japan, Taiwan, western India,

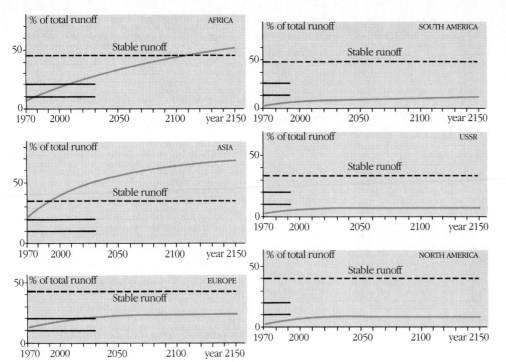

Figure 10-3. Graphic representation of water demand by continent. The bars on the left indicate 10% to 20% of the total runoff, which most countries can capture without major problems. When a country's demand (indicated by blue line) exceeds these levels, it may suffer severe shortages in dry years. These graphs hide local and even regional shortages.

Pakistan, most of Iran, southern Italy, Spain, and all of the Arab states except Syria. The future does not appear bright.

To many it may seem as if the world is running out of water. Far from it. Today's freshwater supply is the same as it was when civilization began. In fact, enough drinkable water falls on the land each year to flood it to a depth of 86 centimeters (33 inches), providing enough water to meet our needs several times over.

Water shortage arises from two principal problems: First, water is not evenly distributed across the face of the earth. Tropical rain forests are drenched with rain, whereas much of the US desert Southwest receives under 25 centimeters (10 inches) a year. The second reason is that human civilization has pushed well beyond the earth's carrying capacity, exceeding the readily available supply of this renewable resource, even in areas with abundant rainfall.

The world population withdraws about 9% of the potentially available freshwater runoff. Eighty-five percent of this water is used for crop and livestock production. Industrial use accounts for 7%, and domestic use makes up the rest. As a general rule it is economically feasible for most nations to withdraw 10% to 20% of the annual stream runoff, although some wealthy countries now draw off almost 30%. On individual rivers, water withdrawals may run much higher, causing some rivers to run dry.

Irrigated land is expected to double between 1975 and 2000. Domestic use is expected to increase fivefold. Industry's water demand is expected to increase twen-

tyfold. Rising demand almost certainly guarantees severe water shortages in the coming years. By the year 2000 water demand is expected to exceed water supply in at least 30 countries. Parts of Brazil, Mexico, many of the large Caribbean islands, East Java, Tasmania, the plains of central Thailand, half the United States, and most of Europe, India, and the Soviet Union will face this unwelcome prospect.

Figure 10-3 shows the projected water withdrawals throughout the world. The solid horizontal bars on the left mark 10% and 20% of the runoff, amounts that are economically feasible for most countries to withdraw. As the blue lines on these graphs show, Africa and Asia will require large withdrawals to meet their projected needs, far in excess of the stable runoff. **Stable runoff** is the amount of the total runoff that can be counted on from year to year. In most countries it is about 30% to 40% of the total average runoff. If the country's water demand exceeds the stable runoff, extreme water shortages can occur in dry or even moderately dry years.

Graphs and statistics can be deceiving. According to Figure 10-3 most continents should be in fairly good shape. None exceeds the stable runoff, except perhaps Asia. However, the continental averages are dangerously misleading, an important consideration for critical thinking. The graph for North America, for instance, suggests that enough water is available. Many parts of the United States, however, already tax their water supplies and suffer in dry years. The deception in these graphs stems from the fact that continental averages obscure local supply problems. North America also appears well off

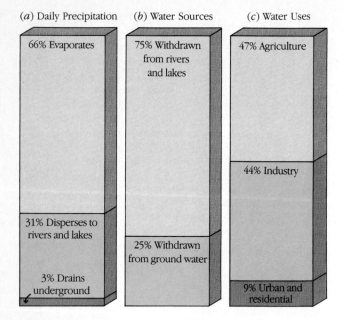

Figure 10-4. Disposition of precipitation, water sources, and users of water in the United States.

because the graph includes massive untapped water resources in Canada and Alaska that will probably never make their way to many water-short areas such as southern California, Arizona, Colorado, and New Mexico. It is important to consider the assumptions behind all projections to become a more critical thinker.

Water Use in the United States

On the average, over 15 trillion liters (4 trillion gallons) of precipitation fall on the United States every day. Two-thirds of all this precipitation (10.5 trillion liters) evaporates. Thirty-one percent (4.9 trillion liters) finds its way to streams, lakes, and rivers, and only 3% recharges groundwater (Figure 10-4a).

Each day, billions of liters of water (an average of 1540 billion liters, or 400 billion gallons) are withdrawn from surface water and groundwater in the United States. Three-fourths of this water comes from lakes, rivers, and streams, collectively called **surface waters**; the remainder comes from groundwater (Figure 10-4b).

Agriculture is the biggest water user in the United States. Nearly one of every two liters of water taken from surface and groundwater supplies is used to irrigate crops. Industry comes a close second (Figure 10-4c). Municipal water users consume relatively little.

The average American uses about 340 liters (90 gallons) of water for domestic use each day. But just how important water is in our lives is underscored by these facts: Growing food for a family of four requires 12,000 liters (3,200 gallons) of water a day. The steel in a washing machine requires 17,000 liters (4,500 gallons) to produce. Adding

these and other uses together, the per capita water demand in the United States comes to about 5,700 liters (1,500 gallons) per day.

Mismanaging Our Water Resources: Causes and Consequences

Virtually every community in the world has a problem with water (Table 10-1). In some there is too much water. In others water supply falls short of demand. Aquifers run dry because water is pumped out faster than it can be replenished. Streambeds dry up in summer months because of excessive demands for lawn watering. In many cases water is badly polluted. Resource managers believe, however, that much of our trouble stems from mismanagement. This section examines mismanagement and outlines some of the consequences.

The Numbers Game: Beyond Drought

Whether you live in New York, Alberta, Canada, Tulsa, Key West, or Bird City, Kansas, you've probably felt the impacts of water shortage at least once in your lifetime. Water shortages stem from natural factors, mismanagement, and overpopulation.

The chief natural cause of water shortage is drought. A **drought** technically exists when rainfall is 70% below average for a period of 21 days or longer. A severe drought results in a decrease in stream flow; a drop in the water table; a loss of agricultural crops; a loss of wildlife, especially aquatic organisms; a drop in the levels of lakes, streams, and reservoirs; a reduction in range production and stress on livestock; an increasing number of forest fires; and considerable human discomfort. In many parts of the world drought is likely to increase because of global warming and activities that reduce local rainfall, such as deforestation and overgrazing (Chapter

Table 10-1   Common Water Resource Problems
Inadequate surface water supply
Overdraft of groundwater
Pollution of surface water and groundwater
Quality of drinking water
Flooding
Erosion and sedimentation
Dredging and disposal of dredged materials
Drainage of wetlands and wet soils
Degradation of bays, estuaries, and coastal waters

9). Overpopulation throughout the world will also result in water shortfalls.

The years 1977 and 1978 stand out in the minds of many Californians as a time of great hardship. Normal snows and rains failed to materialize, leaving mountain reservoirs vulnerable to the heavy demand placed on them by the millions of people crowded into the area near San Francisco. Water was rationed out by the cupful, trucked in where possible. Lawns dried up. To many people the problem was a natural calamity brought on by drought. Others saw it for what it was: the hardship of a drought exacerbated by a population living well beyond the carrying capacity of the environment.

Drought has become painfully obvious in the 1980s as well. In 1987 severe drought once again struck the southeastern United States—the fourth year in a row of below-normal rainfall. Farmers in northern Alabama, Georgia, Tennessee, and the Carolinas were badly hit. Farm production fell drastically. The Tennessee Valley Authority reported that water supplies were at the lowest level ever recorded.

The West, particularly Oregon, California, Washington, and parts of Idaho and Montana, was also gripped by severe drought. The 1987 and 1988 drought, for example, cut water supplies for cities and farms. The drought hurt fisheries and the hydropower industry. Foresters worried as fires began to spring up throughout the West. In fact, 1987 had more forest fires than any year in the previous 30 years; 1988 was about the same.

Droughts are natural events, but global warming, overpopulation, and desertification may be combining to increase droughts or the severity of them. Changes are needed quickly.

Overdraft: Depleting Our Liquid Assets

Much of the water taken from streams and aquifers never returns to its source. For example, over 80% of the agricultural water applied to crops evaporates. Because of this, rivers flow at a fraction of their natural rates during high-use seasons. Aquifers often dry up completely, driving farmers out of business.

Rather than managing aquifers and other resources more responsibly or controlling population growth, the traditional approach has been to seek new water supplies. Expansion of water supply systems often triggers bitter conflicts among water users. New dams threaten recreational rivers (see Chapter Supplement 10-1). Land developers and energy companies often find themselves grappling with farmers over water rights, as each group attempts to meet its own needs. Today, many western states have locked horns over water.

Overexploitation of existing water supplies can severely limit growth. Consider the future of Wichita, Kansas. This growing city gets most of its water from deep

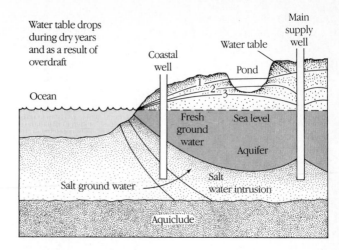

Figure 10-5. Saltwater intrusion into groundwater. Reduction of groundwater flow causes freshwater aquifers to retreat, allowing saltwater to penetrate deeper inland underground. Numbers indicate different positions of the water table and differential effects on surface water (pond). Coastal wells are first affected by falling groundwater. Inland wells pick up saltwater only after severe depletion.

aquifers, the Equus beds, which are recharged at a rate exactly equal to current water withdrawal. With water demand expected to double in the next 20 years, Wichita will need to develop new sources, conserve or ration water, curtail its growth, or a combination of these.

Overexploitation of groundwater in coastal regions may lead to **saltwater intrusion** into freshwater aquifers (Figure 10-5). Taking too much water from rivers and streams may allow salt water to intrude into their estuaries, upsetting the ecological balance in these important zones. This is a serious problem in Everglades National Park in Florida.

Ponds, bogs, and streams are sites that mark the intersection of aquifers and land surface. Many think of a pond as "exposed groundwater." Because of this link, groundwater overdrafting drains swamps and ponds, at times drying them up completely. Fish, wildlife, and recreation are often devastated.

To meet the growing demand, water departments around the United States wreak havoc on the natural environment. Denver, for example, has placed a number of dams throughout Colorado to capture snowmelt for summertime lawn watering, which doubles daily water consumption. Dams and reservoirs destroy the habitat of black bears, bighorn sheep, and many other animals. One dam, in Waterton Canyon near Denver, resulted in the death of one-half of the bighorn sheep population. The Denver Water Department is now actively planning a new reservoir on the same river. Many times larger, it would flood 33 miles of rapids and trout-fishing country. It would also reduce downstream flow of the South Platte River in

Figure 10-6. This large sinkhole developed quickly, swallowing part of a community swimming pool, parts of two businesses, a house, and several automobiles. It happened in Winter Park, Florida, on May 8, 1981.

Nebraska, a stopping-over spot for the endangered whooping crane and tens of thousands of sandhill cranes. At this writing the EPA had announced its intention to veto the project.

In Chapter 7 we saw that groundwater overdrafting threatened the long-term prospects for irrigated agriculture in the West. In Nebraska, Kansas, Colorado, Oklahoma, and Texas, for example, 81,000 hectares (200,000 acres) of farmland on which corn had been grown were taken out of production between 1977 and 1980 because of water shortages. (Estimates of cropland taken out of production since that time are not available.) By 2000, Texas may lose half of its irrigated farmland, 1.2 million hectares (3 million acres).

Groundwater fills pores in the soil and thus supports the soil above the aquifer; when the water is withdrawn, the soil compacts and sinks, a process called **subsidence**. The most dramatic examples of subsidence in recent years have occurred in Florida and other southern states, where groundwater depletion has created huge **sinkholes** that may measure 100 meters (330 feet) across and 50 meters (165 feet) deep (see Figure 10-6). Subsidence has occurred over large areas in the San Joaquin Valley of California, damaging pipelines, railroads, highways, homes, factories, and canals (Figure 10-7). Southeast of Phoenix over 300 square kilometers (120 square miles) of land has subsided more than 2 meters because of groundwater overdrafting. Huge cracks have formed, some 3 meters wide, 3 meters deep, and 300 meters long.

Are We Flooding Our Own Homes?

After shortages, the next major US water problem is flooding. Despite years of flood control work, costing in excess of $4 billion, property damage is continuing to rise. Today, floods cause damage valued at over $3 billion a year.

Causes of Flooding   A deceptively simple correlation can be drawn between floods and their apparent cause: heavy rainfall and snowfall. Closer examination reveals many causes that go unnoticed by most people. As illustrated in Figure 10-8, precipitation that does not evaporate must either run off or percolate into the soil. Whether it streams across the land's surface, possibly spilling over the banks of rivers, or sinks quietly into the soil to join groundwater and later discharge into streams is largely a matter of the surface features and temperature of the land, as shown in Figure 10-8. For example, major spring floods often result when snow melts before the ground has thawed. Vegetation also greatly influences surface runoff. Forest or grass cover retards water flow and promotes percolation. Heavily vegetated watersheds act as sponges. Light vegetation (for instance, in deserts) increases surface runoff and, hence, flooding. The fate of rainfall in many cases rests not so much in the hands of nature as the hands of farmers, urban planners, developers, and the like. Many activities strip vegetation from the land and increase runoff: farming, ranching, mining, off-road-vehicle use, hiking, construction, and clear-cutting on steep

Figure 10-7. Subsidence caused by groundwater overdrafting damaged this building and driveway in Contra Costa, California.

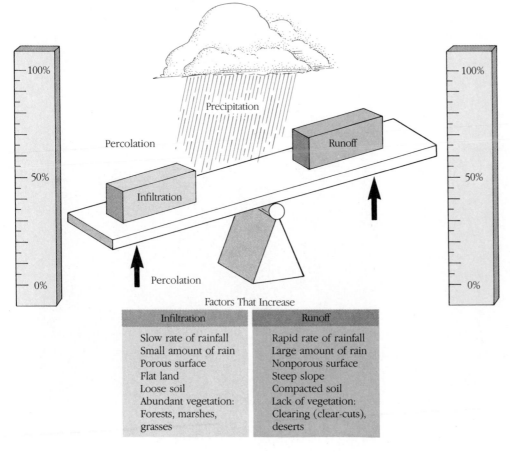

Figure 10-8. Percolation–runoff ratio. When 75% of the water flows over the surface, 25% percolates into the soil (ignoring evaporation).

Factors That Increase

Infiltration	Runoff
Slow rate of rainfall	Rapid rate of rainfall
Small amount of rain	Large amount of rain
Porous surface	Nonporous surface
Flat land	Steep slope
Loose soil	Compacted soil
Abundant vegetation: Forests, marshes, grasses	Lack of vegetation: Clearing (clear-cuts), deserts

Case Study 10-1

Ecological Solutions to Flooding and Water Supply Problems in Woodlands, Texas and Boston, Massachusetts

What do Woodlands, Texas and Boston, Massachusetts have in common, besides the distinguishing accents of their people? The answer is common sense. Both regions have made important decisions in land-use planning that will save millions of dollars by reducing flood damage.

Boston boasts a fine city park system that stretches from the center of the city into the outlying suburbs. Some consider it a landmark of park planning. But few realize that the park system with its meandering stream was built, in large part, to help control flooding. When the rains come, excess water flows into this 12-hectare (30-acre) basin where it is released to the sea, reducing property damage in the whole region.

In recent years, the city purchased 3,400 hectares (8,500 acres) of wetlands in outlying areas to reduce flooding. Rather than letting developers drain the wetlands and build on them, as is often the case, city officials purchased the land and set it aside for parks. Not only will it help reduce flooding the natural way, but it will also provide valuable habitat for fish and wildlife. The project cost one-tenth as much as a dam to control flooding.

Many housing developments in the United States are built with little concern for wildlife or for the developments' effects on flooding. Ian McHarg, a landscape architect who wrote a landmark book called *Design with Nature*, has begun to change the conventional way of building. McHarg suggests careful site analysis to look at wildlife needs, regions of natural flooding, unstable soils, and a half dozen other factors. From these, builders can design a site with much less impact. Unfortunately, many developers today don't make much of an effort to design in harmony with nature.

In Texas, developer George Mitchell decided to build a new town called Woodlands on an 8000-hectare (20,000-acre) forested tract north of Houston. He envisioned a city in harmony with the forces of nature and set out to build it. Mitchell and his staff of planners first analyzed the region they wanted to build on and found that they could leave the natural drainage system as open space. It would carry away water more effectively and more cheaply than a costly storm sewer system. That step alone saved $14 million in construction costs. Roads were built on high ground and buildings were not allowed on aquifer recharge zones, helping to protect an aquifer that supplies water for neighboring Houston. In 1979, rainwaters drenched the site. The streams swelled by 55%. In neighboring towns built with little regard for nature, water flows increased 180%. The towns suffered considerable flood damage while Woodlands managed fine.

Woodlands is an attractive community. Most of its trees still stand. The floodplains which were set aside for natural drainage and aquifer recharge harbor numerous birds and mammals, including bobcats and white-tailed deer.

Designing with nature requires wisdom and foresight. It's an approach to development that permits nature to direct the design of human settlement. It helps people live in harmony with nature, an important step in building a sustainable society.

An energy efficient house in Woodlands, Texas

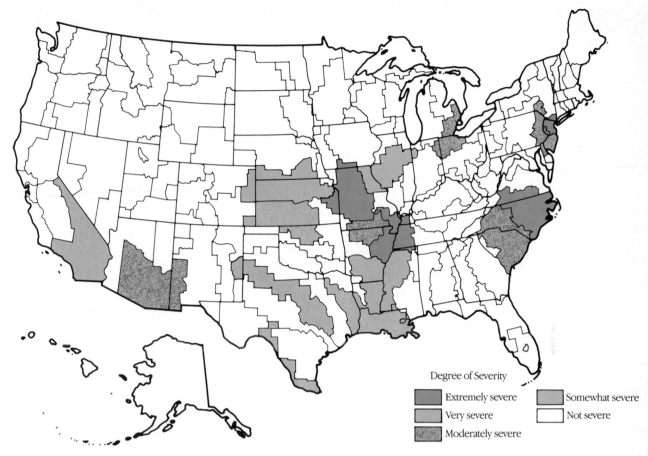

Figure 10-9. Flood-prone areas in the United States.

terrain. Water flowing rapidly over the surface of barren land into streams causes flooding and often transports a substantial amount of soil in the process. In a vicious circle eroded sediment fills bodies of water and reduces their holding capacity. This makes flooding more likely, even after moderate rainfall.

Highways, airports, shopping centers, tennis courts, office buildings, homes, and numerous other structures greatly reduce land permeability. The devastating flood in Kansas City, Missouri, in 1980 was the direct result of a heavy rainstorm falling on thousands of acres of impermeable surface.

Figure 10-9 shows the regions of the United States that are susceptible to flooding. Many of these areas are located along the Mississippi River and its tributaries. The **floodplains** of these rivers, regions along the flanks of rivers naturally subject to flooding, are popular sites for cities, towns, and farms. Taken together, habitation along floodplains and activities that increase surface runoff ensure humankind a future of flooded basements.

Controlling Flooding When rivers flow over their banks, workers rush in with sandbags to hold back the waters. When the floodwaters recede, city engineers plan ways to stop future devastation. Traditionally, potential floods have been fought by building dams and levees (embankments to hold the water back) along river banks. Sometimes of dubious merit, these approaches only treat the symptoms of a serious and costly problem, not the underlying causes. Watershed protection is another, perhaps more effective, approach. Watershed management is to flood control what preventive medicine is to health care. Some effective management tools include reducing deforestation and overgrazing in flood-prone regions and replanting trees, shrubs, and grasses on denuded hillsides. Consider, too, the effect of redesigning the urban environment in ways that slow down the rush of water during a rainstorm by diverting it to holding ponds or underground storage tanks or even pumping it to aquifer recharge zones. A lake in a city park could trap much of the surface flow in a rainstorm. Rooftops of downtown buildings can also be designed to hold water. Captured rainwater could then be released slowly to rivers or used to water lawns or recharge aquifers. Storm sewers could also divert water to special tanks that supply water to factories. Individual homeowners can divert gutter water to underground holding tanks to be used later for watering lawns and gardens and washing cars. (For an example of ways to avoid flooding, see Case Study 10-1.)

Undoing the Damage to Florida's Kissimmee River

Between 1964 and 1970 the Army Corps of Engineers hacked away at the winding Kissimmee River in Florida, hauling up mud and dumping it along the banks in a heap 5 meters (15 feet) high and 30 meters (100 feet) wide. When they had finished this huge flood-control project, a river that had once lazily meandered nearly 150 kilometers (100 miles) through the Florida landscape had been reduced to a 48-mile-long, 200-foot-wide, and 30-foot-deep canal to drain water quickly from the northern reaches of the watershed. The once-magnificent river is now referred to simply as "the Ditch." On the heels of the huge dredgers that planed the river into a straight-as-she-goes canal came the contractors, who threw up concrete dams and steel navigation locks approximately every 10 miles. They also built huge earthen dams extending for 2 miles on either side of the locks to hold back water drained from the upper section of the watershed. From here, water could be released at a controlled rate to avoid flooding below. The dams now form five long holding ponds along the length of the canal.

Once a rich habitat for bald eagles, deer, fish, waterfowl, and alligators, the Kissimmee River has become a sterile tribute to our tireless efforts to channelize rivers to control flooding. Most scientists condemn the channelization as a major environmental catastrophe. Gone are 12,000 hectares (30,000 acres) of the original 16,000 hectares of marsh, once a major breeding ground and stopping-off place for many dozens of species of water birds. Secondary canals built by landowners along the main canal drained another 80,000 hectares. Today,

the vast flocks of ducks that once rained down from the skies are gone. Gone, too, are the wading birds. By Florida Game and Fresh Water Fish Commission estimates, 90% of the waterfowl and 75% of the bald eagles have vanished from the region because of channelization. Also gone are the largemouth bass that attracted anglers from all over the nation.

Only two years after this enormous project had been completed, Florida biologists began noticing changes in Lake Okeechobee, into which the Kissimmee's clean waters once flowed. Dead fish and dying vegetation were the most blatant signs that something was awry in the lake, which provided drinking water for Miami and adjoining coastal cities. It didn't take biologists long to determine that the loss of marshlands, which purify waters and hold back sediment, was the reason for Lake Okeechobee's sudden change for the worse. The loss of the natural cleansing of the wetlands and the onslaught of pesticides, fertilizer, animal wastes, and sediment from cattle ranches and farms that sprang up along the river's banks created a water-quality problem for the lake.

The waters of the Kissimmee River flow south and once fanned out across southern Florida to nourish the multimillion-hectare wetlands known as the Everglades. On the southernmost tip of Florida is Everglades National Park. To make room for farms, much of the Everglades was drained, and the water from Lake Okeechobee and the Kissimmee River basin was shunted via canals to the coast. Reduced flows have disrupted the ecology of the Everglades, seriously threatening common species as well as a number of endangered species

Aerial view of the Kissimmee River canal and old river bed

Case Study 10-2

(Chapter 8). Reduced water flows also produced an ironic backlash. Because of a lowered water table, farmland once prized because of its rich soil began to subside at a rate that some experts think could hinder farming in the region. Lower water flows have also resulted in saltwater intrusion into surface water and groundwater.

Ironically, studies after the canal was completed indicate that it provides little or none of the expected flood control above Lake Kissimmee at the far northern end of the river. Making matters worse, the canal is now seen as a major threat to downstream areas. After heavy rains in central Florida, for instance, a slug of water travels rapidly southward along the canal. It can wipe out nesting waterfowl and drown unsuspecting wildlife. Under natural conditions the wetlands hold back water like a sponge, releasing it slowly with less damage.

Less than two years after the Army Corps of Engineers had trucked in the last load of cement, a special governor's conference committee released a report calling on the state to reflood the marshes that it had dried up in the channelization. Their report was only one of many to reach the same conclusion: channelizing the river had been a big mistake. And with a price tag of $30 million, it had been a costly one. Even the Army Corps of Engineers got into the action by commissioning a study to reevaluate the project and prepare recommendations for returning the river to its original state.

In 1983 Governor Robert Graham and supporters took steps to reverse the damage. In 1984 Phase 1 of the Kissimmee River Demonstration Project began. In an ironic twist, federal funds cannot be used to reclaim rivers—only to channelize them. Funding for restoring the Kissimmee comes from property taxes on residents in southern Florida.

The South Florida Water Management District, in charge of the project, built three dams to divert water back into the old river channel and flood the marshes along 20 kilometers (12 miles) of the river at a cost of $1.5 million. To satisfy the corps,

however, 38-foot-wide and 7-foot-deep notches were left in the diversion dams to allow boats to pass.

Studies show that Phase 1 has worked well. Wetland vegetation, waterfowl, and fish have returned to restored areas. M. Kent Loftin of the South Florida Water Management District estimates that full recovery may take eight to ten years, but that other changes are needed to fully restore the wetlands.

During the channelization, several reservoirs were built on the upper watershed of the Kissimmee River to help control floods. Water management upstream results in periods of high flow and long periods (four to five months) of zero flow. The "recovering" wetlands suffer serious setbacks because of these unusual flow regimens.

To help solve this problem the water district is buying back land and letting the system return to its natural state, a measure which will ensure water flows needed to maintain the wetlands. By the time the state is through buying land it could end up spending $40 to $50 million.

Phase 2 of the project began in late 1989. Efforts are under way to restore more wetland.

The water district is taking an ecological approach to restoration. Their goal is not to optimize one or a few valuable species, such as bass, but to restore the ecosystem so badly damaged by channelization. They are also working upstream to help prevent runoff from farms and dairies.

Restoring the Kissimmee River is part of a major ecological experiment aimed at saving Florida's fast-vanishing wetlands. By 2000 the complex wetlands of Florida will function more like they did 100 years ago. But at this very moment, despite the lessons learned in Florida and elsewhere, engineers and construction companies are hard at work draining enormous wetlands the world over. One of the largest lies along the Nile River. In the home of countless birds and wildlife, huge dredgers are now busily sucking up the mud and straightening the channel.

In some instances the wisdom of our predecessors helps in solving problems of today. But this is not always the case. Take streambed channelization as an example. **Streambed channelization** is a way, or so our ancestors thought, of reducing floods by streamlining rivers and streams. To do this, bulldozers rip the vegetation along a stream's banks and then deepen and straighten the channel, creating a glorified ditch that may be lined with concrete or rock. Over the years our experience with this dubious technique has shown that it generally eases flooding in the immediate vicinity. That minor benefit is outweighed by habitat destruction, increased erosion of stream banks, a loss of river recreation, and increased

flooding downstream. (For a detailed look at the consequences of channelization, see Case Study 10-2.)

Streambed channelization in the United States was begun seriously in 1954, when the **Watershed Protection and Flood Prevention Act** authorized the Soil Conservation Service to drain wetlands along rivers to make more farmland and to reduce flooding. To date, 13,000 kilometers (8000 miles) of US streams have been channelized. Another 13,000 kilometers have been targeted for similar "improvements." Critics argue that many projects are of dubious merit and should not be undertaken, especially when they destroy wetlands (see Chapter Supplement 10-1).

Figure 10-10. A kayaker paddles through turbulent white water. Dams can eliminate such recreational rapids.

Protecting Our Liquid Assets

Many parts of the world are heading for severe water shortages in the next 10 to 20 years unless we act quickly. But what can we do?

Population Control

Fortunately, solutions to the worsening worldwide water shortages are many. One of the most important is the control of population growth, particularly in Africa, Asia, and Latin America. Measures to control regional population growth in water-short areas in more developed countries, such as the desert Southwest of the United States, can also reduce water supply problems.

Supplementing population control are numerous technical, personal, and legal measures.

Technical Solutions: Costs and Benefits

Dams and Reservoirs Dams and reservoirs retain snowmelt and rainwater, increasing our supply of water, controlling floods, generating electricity, and increasing certain forms of recreation. They also allow the expansion of population.

Dams are no panacea, however. They are costly and may inundate towns, villages, good arable land, and wildlife habitat. Pakistan's recently completed Tarbela Dam alone displaced 85,000 people. China's mammoth Three Gorges Dam and reservoir, being built on the Yangtze River, will displace two million people and destroy nearly 41,000 hectares (100,000 acres) of farmland in the world's most heavily populated and agriculturally productive

river valley. Dams and reservoirs also destroy opportunities for certain forms of river recreation, including kayaking, rafting, and trout fishing (Figure 10-10). Exceptional floods can exceed the capacity of reservoirs and even destroy well-built dams.

Contrary to popular belief, dams and reservoirs do not always increase the amount of water available to some communities, because of high rates of evaporation. On Egypt's Lake Nasser, formed by the Aswan Dam, 10% of the total water supply evaporates each year.

Dams reduce stream flow into the ocean, resulting in changes in the salt concentration of estuaries, bays, and tidal waters. Reduced stream flow also reduces the nutrient inflow to coastal waters, with devastating effects on the producer organisms in the aquatic food web. Because of a reduction in algal growth at the mouth of the Nile, Egypt's annual sardine catch plummeted from 16,000 metric tons per year to 450.

Dams interrupt the natural flow of nutrient-rich sediment to floodplains, river deltas, and coastal waters. Good farmland must be abandoned or fertilized, often at a high cost. Sediments collected in reservoirs slowly fill them, making them useless. In the United States well over 2000 small reservoirs are totally clogged with sediment. Many larger reservoirs throughout the world will meet a similar fate. The $1.3-billion Tarbela Dam on the Indus River in Pakistan took nine years to build, but because of upstream soil erosion the reservoir could fill with sediment in 20 years. Lake Powell on the Colorado River will be filled in 100 to 300 years.

Dams also interfere with the migration of fish to spawning grounds. Salmon runs on many streams in the Pacific Northwest and the East Coast have been decimated by dams built without "ladders" that allow fish to pass the dam. Even when ladders are built, there's no guarantee

Figure 10-11. Desalination plant in Key West, Florida.

that fish populations will thrive. Salmon, for example, migrate downstream to the ocean after hatching. Reservoirs eliminate current, which normally helps them migrate. Many young salmon get lost in huge reservoirs, never reaching the sea where they mature and reach full size. In addition, dams generally release the coldest water from the bottom of the reservoir; this can make stream water exceedingly cold year round and decrease the spawning of native species. The Glen Canyon Dam on the Colorado River, for instance, releases chilly water from the reservoir's bottom and has converted what was once a warm-water fishery to a fine rainbow trout fishery. The cold water now threatens several native species, such as the humpback chub. To avoid changing stream temperature, some existing dams have been retrofitted with devices that combine warm water from the reservoir's surface with cooler deep waters. Newer dams are often fitted with multiple gates to ensure proper downstream water temperature.

Since the economic and environmental costs of dams and reservoirs are so high, caution is advised when undertaking a project of this nature.

Water Diversion Projects Denver, Los Angeles, Phoenix, New York, and many other cities and towns rely on water diversion projects to supply their needs. Often urban water departments build small dams on distant streams and then pipe the diverted water to neighboring reservoirs, from which it flows to water treatment plants.

Because up to 80% of the annual runoff of some streams may be taken, wildlife habitat and good recreational sites may be destroyed by diversion projects. In addition, these projects usually transfer water from rural areas of low population density, often agricultural regions, to cities. This practice creates bitter conflicts between urban and rural residents. Since the urban water users have a larger say in state policies, they usually win the legislative battles, leaving irrigated agriculture high and dry.

Diversion of clean, high mountain water in tributaries of the Colorado River ultimately diminishes the flow and quality of the Colorado, a source of water for about 20 million people. The lower sections of the river carry a heavy burden of sediment and dissolved salts. By removing clean water from the spring snowmelt, upstream diversions indirectly increase the concentration of salts downriver. By the time the Colorado reaches Mexico, its salt concentration is over 800 parts per million (violating a US–Mexico treaty), compared with 40 parts per million at its headwaters in the Rockies.

Water with a salt concentration over 700 parts per million cannot be used for agriculture unless it is desalinated or diluted. Drinking water must have a salt concentration of 500 parts or less. A bill passed by Congress in 1973 authorized the construction of three desalination plants along the Colorado River to remove over 400,000 tons of salt from the river each year. Costing an estimated $350 million, these projects help purify water to be used in California and Mexico.

Water diversion projects will continue to be built to meet the needs of industry, agriculture, and cities. The costs and benefits of proposed projects must be thoroughly analyzed.

Desalination The expense and ecological impact of dams and diversion projects often lead community planners to other approaches to increase their water supply. One of these is **saltwater desalination**, the removal of salts from seawater and brackish (slightly salty) water (Figure 10-11). The two main methods are evaporation and reverse osmosis. In **evaporation**, or distillation, the salt water is heated and evaporated, leaving behind the salts and minerals. The steam produced is then cooled, and pure water condenses out. In **reverse osmosis** water is forced through thin membranes whose pores allow the passage of water molecules but not the salts and minerals.

By either of these methods, seawater can be converted to drinking water or irrigation water. Since 94% of the water on earth is in the oceans, desalination might seem like the best answer to water shortages. Unfortunately, water produced by desalination is four to ten times more expensive than water from conventional sources.

Since 1977 the world's desalination capacity has increased dramatically. Nevertheless, desalination produces an insignificant proportion of the fresh water consumed by humankind. In the United States over 100 desalination plants now produce an excess of 1250 million liters (330 million gallons) per day. This is .001% of the total US freshwater requirement. Most plants are located in California, Texas, Florida, and the Northeast. Desalination plants are also in operation in Saudi Arabia, Israel, Malta, and a few other countries.

Even though costs have come down in recent years, energy requirements and construction costs of desalination plants may always remain prohibitively high. For agriculture, desalination seems even more unlikely, since water-short areas are usually located far from the sea.

Desalination plants permit a further extension of the earth's carrying capacity with potentially serious ecological impacts. For example, population growth in the Florida Keys, in part due to a new desalination plant, threatens coral reefs and native wildlife, such as the crocodile. Construction of houses and condominiums causes erosion that pollutes coastal waters. New residents produce an increasing amount of sewage and other pollutants that can affect water quality. Large quantities of salt and minerals from desalination plants could worsen water quality.

Groundwater Management Another technological fix for water shortages is to tap unused groundwater and to regulate withdrawal so it will not exceed recharge. In some areas groundwater can be replenished by disposing of wastewater in aquifer recharge zones. This water is cleansed as it percolates through the soil, although some of the organic waste cannot be broken down underground. Irrigation water from farm fields, storm water runoff, cooling water from industry, and treated effluents from sewage treatment plants could also be used.

Although aquifer recharge can help increase groundwater supplies, it can also present some problems. The most significant is the cost of disposal, since many cities are far from aquifer recharge zones.

Another cost-effective solution is careful design and planning of new housing and other development, preventing it from being built over aquifer recharge zones. For more on this subject, see Case Study 10-1.

Conservation and Recycling US water resource policy was defined in the past by the flow of federal dollars to dams and canals. That era, however, is ending for three principal reasons: First, federal support has nearly run dry. Second, the general public has shown a strong interest in preserving its rivers for recreation and wildlife. Third, many of the best sites for dams have been used, and those that remain will be more expensive to develop and are often more environmentally objectionable. Many experts, therefore, think that water policy will shift from water development to the efficient use of existing facilities and supplies. Conservation and recycling may be the wave of the future. Such a shift, they say, should take place as rapidly as possible because it is the most cost-effective solution to water supply problems and has the fewest environmental impacts. Since agriculture and industry are the biggest water users, efforts to recycle and conserve water in these sectors could do much to alleviate water shortages.

Agricultural water conservation is easily achievable. Open, dirt-lined ditches that deliver water to crops are only about 50% to 60% efficient, and they could be replaced with lined ditches, or, better yet, pipes. Sprinklers, which often waste up to half of their water, can be replaced with drip-irrigation systems (Chapter 7), which lose only 5%. However, drip irrigation can be used for only a limited number of crops, such as fruit trees, grapevines, and some vegetables. For the vast fields of wheat and corn, more conventional methods must be used. Still, water conservation efforts such as these can help reduce water demand.

Water conservation can help offset current shortages and can meet future demands—at a far lower cost than dams. In the agriculturally rich San Joaquin Valley of California, for example, irrigation water costs farmers about $5 per acre-foot (an acre of water one foot deep). Water is inexpensive to farmers for two reasons: First, federal and state taxes are used to pay for water projects. In other words, the public subsidizes the private use of water. Second, water is so cheap because the economic costs of the damage (for example, the loss of fish) created by water projects are not included in the project cost.

According to the Environmental Defense Fund, recycling water and scheduling water application according to crop and soil needs costs about $10 per acre-foot of water applied. Switching from irrigated crops, such as cotton, to dryland crops costs farmers about $40 per acre-foot in lost crop yield. Water conservation by more efficient drip and sprinkler systems costs about $175 per acre-foot. At first glance, farmers have little incentive to conserve water.

Critical thinking, however, suggests that such a look may be too narrow. For instance, by cutting back on water demand, farmers can reduce farm runoff and salinization (discussed in Chapters 7 and 8). Conservation also

The Third Stage of Environmentalism

Frederic D. Krupp

The author is the executive director of the Environmental Defense Fund.

I believe the late 1980s will see US conservationists embrace a major shift in tactics as we enter a newly constructive third stage in environmentalism's evolution.

The first stage of the conservation movement, represented by President Theodore Roosevelt and the early Sierra Club and National Audubon Society, was a reaction to decades of truly rapacious exploitation of natural resources, especially in the West. The early focus was on stemming the direct loss of wildlife and forest lands.

A key change occurred in the 1960s as people began to realize that they, too, were becoming victims of environmental abuses, that careless contamination of water, land, and air had sown seeds of destruction in the food chains of both wildlife and humans.

The environmental movement's response in this second phase was to work to halt abusive pollution, just as the early conservationists had tried to end the overexploitation of resources. The Environmental Defense Fund (EDF) was born in the forefront of the second phase, almost 20 years ago, in a victorious effort to stop the use of DDT, which threatened the osprey, bald eagle, and other species with extinction and had established an alarming presence in mother's milk.

The EDF's original vision—to present the evidence of environmental science in a court of law—proved to be an effective strategy to halt and even reverse environmental damage. Lawsuits, lobbying, peaceful protests and other direct efforts became the common expressions of environmental concern in this period.

Most environmental organizations still emphasize this form of reaction, but a new age of environmentalism may be dawning. The "New Environmentalists" say we cannot be effective solely by opposing environmental abuses. We must go the extra mile by finding alternatives to answer the legitimate needs that underlie ill-advised projects like destructive dams. Otherwise we are treating only the symptoms of problems that will surface again and again. When we answer the underlying needs, we perform a lasting cure.

If conservationists worry about the impact of a dam, for example, they had better address the water-supply or power-supply problem the dam was proposed to solve. They must concern themselves with the *science and economics* of environmental protection. Jobs, the rights of stockholders, and the needs of agriculture, industry, and consumers for adequate water and power—all of these issues must become part of the new environmental agenda.

For us to move beyond reactive opposition, to become a *positive* movement *for* better alternatives, means facing a difficult challenge. Our organizations will need to keep and recruit the best minds from every discipline, thinkers who can envision and persuasively lead the nation toward environmentally sound economic growth.

The EDF's experience in California utility regulation is one of several case studies in this New Environmentalism. In the late 1970s one of the country's largest utilities, Pacific Gas & Electric Company, had plans to build coal and nuclear power plants worth $20 billion. Forces on both sides were strong, and deeply entrenched. Then an EDF team—a lawyer, an economist, and a computer analyst—developed an unprecedented package of alternative energy sources and conservation investments and ultimately convinced PG&E to adopt the plan. Why? Because it not only met the same electrical needs but also meant lower prices for consumers, higher returns to PG&E stockholders, and a healthier financial future for the company itself.

The EDF's plan not only blocked construction of the polluting plants, it literally made the plants unnecessary and made obsolete an entire "bigger is better" mindset among electric power planners.

In the 1970s society badly needed the new institutions like environmental action groups and public-interest law firms that were created to meet the problems of that era. Today the need is for these institutions to become well equipped to envision solutions and to assemble new coalitions—even coalitions of former enemies—to bring about answers to environmental problems. But the third stage of environmentalism is in no sense a move toward compromise, a search for the in-between position. We will still need skillful advocacy—even in court—against narrow institutional vision or vested interest in the status quo. We must, however, become true advocates of a new course of action, not mere opponents of the old.

The EDF is the first environmental group to begin to fill this exciting new environmental niche, but even without its push in this new direction, the shift would have been inevitable. What the American public wants—some might say paradoxically—is both to expand our economic well-being and to preserve our natural resources and public health. It is up to us as environmentalists to prove this is no paradox and find the innovative ways to do both.

reduces the need for new water projects, which cost considerably more now that federal monies are becoming harder to come by—up to $500 per acre-foot. In the long run, water conservation makes good economic and ecological sense.

There are other reasons as well. Perhaps the most important is that water conservation can make farmers money. By investing in water conservation, farmers can cut water demand. Excess water could be sold to willing buyers. California cities and industries currently purchase water for about $200 per acre-foot per year. If a farmer could invest $50 per acre-foot to save water and sell the excess for $200 he or she would reap a $150 per acre-foot profit. This plan angers some people because the farmers would be making a profit on water that is subsidized by public funds. It seems only fair, they say, that the public be reimbursed as well. Farmers shouldn't be allowed to reap the full profit of a publicly subsidized resource. No matter what the outcome of the debate, water conservation makes good sense in the long run as well as the short run. A 10% decrease in agricultural water use, in fact, could double the amount of water available for cities and industries.

To avoid unsound water projects in the future, many think that federal and state governments should require those who benefit from a project to repay its full costs. "Demand for water projects would reflect the discipline of the marketplace, rather than the undisciplined pursuit of subsidies," say the authors of *An Environmental Agenda for the Future*, a collection of recommendations set forth by ten leading environmental groups. Making users pay the full cost, they argue, would make users focus on least-cost alternatives, particularly conservation. The cost of a water project should also include the costs of correcting environmental problems, making conservation and recycling even more attractive.

In January 1989, two major water supply agencies announced an historic agreement that may pave the way for the future. The Metropolitan Water District of Southern California agreed to pay for water conservation measures in the Imperial Irrigation District to the north. By lining irrigation canals and taking other measures, they will free up water for use in Los Angeles, San Diego, and neighboring cities at a cost far below that incurred by new dams and diversion projects.

In industry, water conservation can be achieved by redesigning processes and facilities. The use of water by steam-cooled electric power plants could be cut one-fourth by using dry cooling towers (Figure 10-12), although these require more energy and are more expensive to operate than wet cooling towers.

Municipal water systems throughout the United States lose, on average, 12% of their water. Boston records a loss of 35%. New pipes can reduce water loss enormously, saving streams and the wildlife that depend on them.

Problems with water supply can be eased or in some cases solved entirely by using water more efficiently. The potential for water conservation in industry, in homes, and in agriculture is immense. Water efficiency improvements offer the least cost and most environmentally acceptable solution to water supply problems. Water that is saved through efficiency measures can be made available for other uses immediately at a far lower unit cost.

Cities and industries can also recycle water. Wastewater from various industrial processes can be purified and reused over and over. Municipal sewage treatment plants are a common type of water purification system, though effluents are hardly potable. Further treatment, though, could make this water pure enough to drink. In Tokyo, for example, Mitsubishi's 60-story office building has a fully automated recycling system that purifies all of the building's wastewater to drinking-water purity.

As water shortages worsen and public and governmental awareness increases, recycling plants will become a necessity. Such plants decrease water demand and water pollution. Although the cost of building and maintaining large-scale recycling plants for large cities can be considerable, they may still be cheaper than new water diversion projects and dams and have added ecological benefits.

Doing Your Share: Personal Solutions

You and I can do many things to cut down on water use. Table 10-2 compares typical wasteful habits and shows how simple adjustments in our behavior translate into great water savings.

Education Solutions

Half the battle to ensure an adequate water supply can be won by education. Our children should learn the value of water and the growing demands on it; they should learn to conserve water. Such lessons may be as important as the math or history they learn. Our schools could go far toward making our youngsters resource-conscious citizens.

Legal Solutions

New laws and building codes can also help reduce household water consumption. In all areas, water-short or not, new homes could be required to use 25% to 50% less water than existing homes. This can be achieved by installing waterless or low-water toilets as well as flow restrictors for showers and faucets. Under a newly amended plumbing code, the state of Massachusetts now requires that all toilets installed after March 1, 1989 use no more than 6.2 liters (1.6 gallons) of water per flush, about one-half as much water as current models and one-third as much water as toilets in older homes. Toilets use 30% to 40% of the indoor water consumed by a family, so water-conserving models could generate significant savings.

Smaller lawns could be encouraged, and planted with low-water grasses. Drip and root-zone irrigation systems could be required. Systems that retain rainfall and gray water (used water from showers and faucets) could be installed. Gray water is suitable for lawn and garden irrigation. Finally, municipal water agencies could charge more for water used when evaporation is greatest, and less for water used when evaporation is low. This price incentive could help cut down on wasted water.

Perhaps the most serious deficiency in our regulation of water resources is poor coordination between water resource management and economic development. In many cases the two are discussed exclusively of each other. Builders put up tracts of new homes and then ask where the water will come from. City planners scramble to find water, throwing up dams and diversion projects and ruining rivers that contribute to the allure of the area.

Water development plans must be environmentally as well as economically sound. We cannot afford to assume that adequate water supplies will always be available no matter where we go and no matter what we do. In addition, we cannot blindly continue our extension of the earth's carrying capacity. Though we can make the deserts bloom, we must ask, "Should we?" Though we can pump water uphill and through tunnels under mountain ranges, we must ask about the consequences of these actions and whether there are other, ecologically sound, options. In some cases a proper analysis will cause us to abandon our plans. In others it may give us time to pursue options that make more ecological sense.

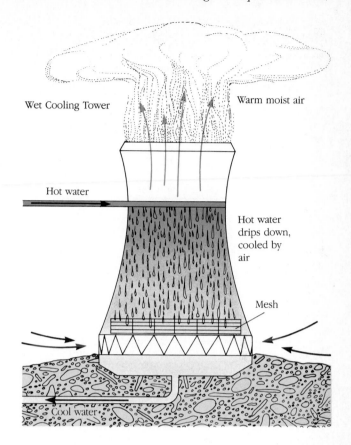

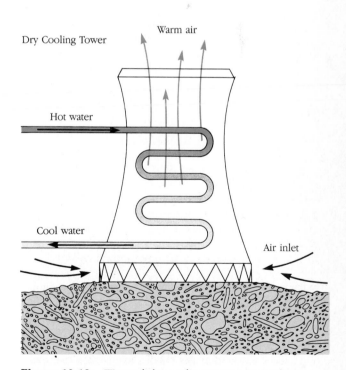

Figure 10-12. Wet and dry cooling towers. Water from electric power plants is cooled in them before being reused. Dry towers cost more to operate but conserve water.

Table 10-2   Water Savings Through Conservation

Activity	Typical Use (Liters/Use)	Efficient Use (Liters/Use)	Suggestions
Shower	77	2	Install flow restricters
		15	Shorter showers
Toilet	19	15	Bottles or bricks to displace water in tank
		13	Install low-flush toilet
		4	Install washdown toilet
Faucets	(liters/min.) 12	10	Low-flow faucets
		6	Flow restricters
Brushing teeth	20–40	2	Brushing with faucets off, then rinsing briefly
Shaving	24–48	2	Shaving with faucets off, followed by brief rinse
Washing dishes	60–80	10–20	Rinsing dishes in drainer or turning water on and off each time; using short cycle for dishwashers
Washing clothes	140	80	Front-loading washing machine
		100	Adjusting water level for each load
Outdoor watering	(liters/min.) 40	—	Nighttime watering can cut water demands in half

Let him who would enjoy a good future waste none of his present.

ROGER BABSON

Summary

The **hydrological cycle** is a natural system for collecting, purifying, and distributing water, which is driven by energy from the sun. At any single moment only .5% of the earth's water is available for human use.

Three-fourths of the water used in agriculture, industry, and our homes comes from **surface waters**, and the rest from groundwater. Two-thirds of the water withdrawn each day is returned to its sources, but the remaining portion is consumed, that is, it either evaporates or becomes incorporated in living tissue.

Major water supply problems include shortages, overexploitation of ground and surface water supplies, and flooding. Water shortages are brought about by a complex set of factors, including drought, overuse, waste, and overpopulation. Overexploitation of ground and surface waters occurs in many parts of the world today, creating shortages, conflict among water users, and severe economic impacts, especially for farmers. Groundwater depletion may also cause **saltwater intrusion** in coastal zones or in regions where freshwater aquifers lie near saltwater aqui-

fers. In some regions groundwater overdrafting has caused severe **subsidence**, or sinking of the land, which results in the collapse of highways, homes, factories, pipelines, and canals.

Flooding is also a major problem. Because of natural factors and poor land management, billions of dollars worth of damage is caused by floods each year. Any human activities that increase the surface runoff and decrease percolation can potentially increase flooding. Much of the damage from flooding is our own fault, for we tend to inhabit **floodplains**, river valleys that are subject to periodic flooding. To prevent flooding, **streambed channelization** is often carried out, involving the removal of vegetation and the deepening and straightening of channels. Such actions destroy wildlife habitat, increase the rate of streambank erosion, diminish recreational opportunities, alter the aquatic environment, and often increase flooding in downstream sites.

Water problems face virtually every nation in the world. Especially hard hit are some of the less developed nations, where water is far from villages and is polluted. Polluted water supplies are responsible for 80% of all illness in these countries.

The long-term prospects for meeting water needs for industry, agriculture, and individuals are dim for many parts of the world. Africa and Asia may find adequate water supplies difficult to come by, as may many other regions within countries, even in the developed world. The United States has been racked by a series of droughts in the 1980s and may face more if global warming continues.

There are many solutions to meeting future water demands and curbing flooding. Measures to control population growth, especially in water-short regions, are most important. In addition, numerous technological solutions can be used. The environmental impacts of each of these must be carefully considered along with long-term benefits.

Dams and reservoirs retain snowmelt and rainwater, help control floods, generate electricity, and increase certain forms of recreation. They are costly, however, often inundating wildlife habitat, farmland, and even towns. Dams also reduce stream flow into the ocean, resulting in changes in the salt concentration of estuaries, bays, and tidal waters. Dams also interrupt the natural flow of nutrient-rich sediment to coastal waters, with devastating effects on the aquatic food web. Fish migration is often impaired by dams.

Diversion projects increase water supplies in water-short regions but are often costly, reduce stream flow, destroy aquatic ecosystems, affect downstream salt concentrations, and create bitter conflicts among users.

Desalination of salt water is feasible in some places, but it is four to ten times more expensive than conventional freshwater development projects. Desalination plants produce salts that must be disposed of and encourage home and resort construction and population growth in some regions, resulting in ecosystem destruction.

In some areas it may be prudent to replenish groundwater by disposing of wastewater, industrial cooling water, storm water runoff, and water from agricultural fields in aquifer recharge zones or in special water disposal wells. Such water is purified as it percolates through the soil, but projects of this nature are feasible only if water supplies are near recharge zones.

Recycling and conservation can make important contributions to increasing water supplies in the near future. Since agriculture uses a large percentage of our water, devices to reduce irrigation losses that are inexpensive and easy to install are needed. Such measures are good short-term and long-term investments, both economically and environmentally.

New laws and building codes can help reduce water consumption in and around our homes. New homes can be fitted with water-conserving toilets and shower heads, special systems to reuse gray water for lawns and gardens, or recycling systems that actually purify water from all domestic uses for reuse. Individuals can assist by reducing water consumption.

Discussion Questions

1. What is the hydrological cycle? Draw a diagram showing how the water moves through the cycle.

2. Define *transpiration, evaporation, relative humidity, absolute humidity, saturation,* and *condensation nuclei.*

3. What sector is the largest water user in the United States?

4. Define the following terms: *groundwater, water table, saltwater intrusion,* and *subsidence.*

5. Discuss the problems caused by overexploitation of ground and surface water.

6. You are appointed by the governor of your state to study floods and flood control projects. Make a list of reasons why flooding is now severe and ways to correct these problems.

7. Discuss the benefits and costs (environmental, economic, and so forth) of dams and diversion projects.

8. How can desalination of seawater help solve water shortages? What are the limitations of and the problems created by this method?

9. Describe ways in which you and your family can help conserve water. Calculate how much water your efforts will save each day. How much will they save in a year?

Suggested Readings

Abbey, E. (1984). *Beyond the Wall.* New York: Holt, Rinehart, and Winston. See Chapter 5, The Damnation of a Canyon, for a poignant view of the controversy over Glen Canyon Dam.

Brown, L. R., Chandler, W. U., Flavin, C., Pollock, C., Postel, S., Starke, L., and Wolf, E. C. (1986). *State of the World 1986.* New York: Norton. See Chapter 3, on water resources, for Sandra Postel's excellent analysis of freshwater supplies and demands.

Kakela, P., Chilson, G., and Patric, W. (1985). Low-Head Hydropower for Local Use. *Environment* 27 (1): 31–38. An interesting look at the value of small dams.

Maranto, G. (1984). Saving South Florida. *Discover* 5 (4): 36–40. Excellent overview of the problems created by channelization and of how Florida officials are trying to reverse the damage.

McPhee, J. (1989). *The Control of Nature.* New York: Farrar Straus Giroux. Excellent reading on efforts to control floods and the backlashes resulting from these efforts.

Micklin, P. P. (1985). The Vast Diversion of Soviet Rivers. *Environment* 27 (2): 12–20, 40–45. In-depth look at major water diversion projects in the Soviet Union.

Postel, S. (1985). Thirsty in a Water-Rich World. *International Wildlife* 15 (6): 32–36. Excellent article on water supplies.

Postel, S. (1986). Water for the Future: On Tap or Down the Drain? *The Futurist* 20 (2): 17–21. Superb look at water conservation.

Postel, S. (1990). Saving Water for Agriculture. In *State of the World 1990.* L. Starke, ed. New York: Norton. Outlines current water shortages and strategies to avert them in the 1990s.

Reisner, M. (1989). The Emerald Desert. *Greenpeace* 14 (4): 6–10. Extraordinary article on California's profligate water demands.

Reuss, M. (1988). Along the Atchafalaya: The Challenge of a Vital Resource. *Environment* 30 (4): 6–11, 36–44. Detailed study of the politics of conflicting demands and needs.

Sheaffer, J. R. (1984). Going Back to Nature's Way: Circular vs. Linear Water Systems. *Environment* 26 (8): 10–15, 42–45. A look at ways to help solve both water shortages and water pollution.

Udall, J. R. (1986). Losing Our Liquid Assets. *National Wildlife* 24 (1): 50–55. Exceptional account of groundwater depletion in the southwestern US and its impacts.

Willey, W. R. Z. (1987). Economic Common Sense Can Defuse the Water Crisis. *EDF Letter* (March): 7. Excellent overview of the economics of water conservation.

World Resources Institute and International Institute for Environment and Development (1988). *World Resources 1988–89* New York: Basic Books. See Chapter 8.

Wetlands, Estuaries, Coastlines, and Rivers

From certain vantage points Chesapeake Bay resembles the vast, almost limitless ocean. For years Americans have treated it with the disrespect accorded to seemingly limitless resources. Today, however, the bay is in trouble. The rich abundance of organisms is diminishing, threatened by pollution, overfishing, and other activities.

The bay and its surrounding wetlands, in this case mostly swamps, are home to a variety of fish and shellfish, including blue crabs, oysters, and striped bass. Properly managed, the bay could provide enough food to feed Japan. However, Chesapeake Bay is much more than a food source for the 13 million people who live near it. It is just as much a recreational gold mine for hunters, anglers, nature enthusiasts, and boaters.

Chesapeake Bay is only 310 kilometers (195 miles) long, but its shoreline measures nearly 12,000 kilometers (7000 miles). In many ways the bay is a symbol of how wetlands and coastlines have been mismanaged for 200 years.

Old-timers maintain that the bay's waters once contained "wall to wall" oysters. Oyster populations have declined precipitously in the last 30 years. Bountiful harvests of striped bass may be a thing of the past, too. In 1970 the average number of bass netted per haul in seines was 30; in 1984 the count had fallen to only 4.2. Various strategies have been tried to reverse this downward trend. But the 1984 count prompted a moratorium on striped bass fishing in January, 1985. Fishermen were stunned by this action. Because of some recovery in bass populations between 1985 and 1990, the moratorium will be lifted. But tight restrictions will be imposed by the states.

Environmental Protection Agency (EPA) studies show that the bay's submerged vegetation, which is vital to fish populations, has dropped by 76% in the last 25 years. The EPA also notes that large **algal blooms** (bursts of algal growth resulting from certain kinds of pollution discussed in Chapter 16) now threaten the grasses. Mats of algae block sunlight from reaching the bottom.

Further damage to the bay is caused by aerobic (oxygen-requiring) bacteria, which deplete the oxygen supply when they decompose dead algae and organic pollutants in the water. Some oxygen depletion, or anoxia, is natural for the bay.

According to the Conservation Foundation, commercial development and population growth are the two biggest threats to the bay. The population by 2020 is projected to be 16 million. Most of the pollution comes from development—from oil spills, sewage, toxic chemicals, heavy metals, and runoff from the surrounding land. The bay's extensive drainage system, an area slightly smaller than Missouri, ensures that anything added to the land will probably end up in its waters. One of the worst pollutants is nitrogen (in the form of nitrates) from commercial duck farms, municipal sewage, and farm fields.

The tragedy of neglect witnessed in Chesapeake Bay reminds us that water management involves much more than protecting humans from floods and supplying water. This supplement examines the threat to wetlands, estuaries, coastlines, and rivers and looks at ways to better manage them.

Wetlands

Wetlands, perpetually or periodically flooded lands, fall into two types. **Inland wetlands** are found along freshwater streams, lakes, rivers, and ponds. Included in this group are bogs, marshes, swamps, and river overflow lands that are wet at least part of the year. **Coastal wetlands** are wet or flooded regions along coastlines, including mangrove swamps, salt marshes, bays, and lagoons (Figure S10-1).

The Hidden Value of Wetlands

Wetlands are an extremely valuable and productive fish and wildlife habitat. Many other animal and plant species also make the wetlands their home. Deer, muskrats, mink, beavers, and otters are a few of the species that live in or around wetlands. In addition, shellfish, amphibians, reptiles, birds, and fish also call these endangered places their homes.

Aside from their importance to wildlife, wetlands also play an important role in regulating stream flow. A study in Wisconsin showed that wetlands act like sponges, holding back rainwater and reducing natural flooding. They also filter out sediment eroded from the land and therefore help reduce sedimentation in streams. Wetlands also act as traps for nitrogen and phosphorus, two common pollutants washing from heavily fertilized land. Wetland plants absorb these nutrients, preventing water pollution downstream. The sponge effect reduces flooding and has the added benefit of recharging groundwater supplies. Coastal wetlands help protect human settlements by absorbing many storm surges, high waves that accompany high winds.

Wetlands are used to grow certain cash crops, including rice, cranberries, peat moss, and blueberries. Because their usefulness is not always apparent, however, they are often filled in or dredged to make way for housing, recreation, industry, and garbage dumps.

Declining Wetlands

By some estimates wetlands once covered an area the size of Texas (68 million hectares, or 170 million acres). Today, less than one-third of these rich wetlands remain, and what is left is fast heading toward oblivion. Estimates of wetlands destroyed each year range from 48,000 to 200,000 hectares (120,000 to 500,000 acres).

US wetlands are shown in Figure S10-1. No matter where you look, wetlands are in trouble. Coastal wetlands are victims of huge dredgers that scoop up muck in streams and bays to make them more navigable. Dredging drains adjoining swamps. Cities often fill in swamps to make room for homes, recreational facilities, roadways, and factories. Farmers fill in swamps to expand their arable land.

Protecting Wetlands

Concern for the loss of wetlands has stirred many state governments into action. For example, Florida passed legislation in 1972 to regulate all wetland development. Strict controls in Nassau County, New York, have also slowed destruction. The federal government has assumed an increasing role in wetland protection through executive orders that prohibit all its agencies from supporting construction in wetlands when a practical alternative is available. A farm bill passed in 1985 denies farmers federal benefits, such as loans, crop insurance, and subsidies, if they drain or fill wetlands to convert them to agricultural land. The federal **Coastal Zone Management Act** (1972) calls on states to develop plans to protect coastal wetlands. Virtually all states have either federally approved plans or state plans to regulate

their wetlands. However, only 16 states have measures to protect inland, freshwater wetlands. In 1988, New Jersey passed a law requiring buffer zones of 30 meters (100 feet) around important wetlands. Although a plan does not ensure protection, it is a step in the right direction.

The federal government also purchases wetlands that become part of the **National Wildlife Refuge System**. The system today contains 4 million hectares (10 million acres) of wetlands. The Fish and Wildlife Service also buys wetlands for protection. States own additional wetland acreage. All told, a little over one-fourth of our existing wetlands are protected by these various measures. In 1986 Congress passed a measure that will raise more money for the purchase of wetlands by increasing duck stamp fees and charging admission at certain wildlife refuges. Many environmentalists, however, believe that Congress should find additional sources of funding to purchase wetlands, sharply accelerating the current rate of acquisition.

Another important action aimed at protecting wetlands is the 1985 Farm Bill (Chapter 7). It included important "swampbuster" provisions—rules that deny federal benefits (low-interest loans and crop insurance) to farmers who drain and farm their wetlands. In 1987 the swampbuster provisions were tightened thanks to environmental groups. Under previous rulings large-scale wetland destruction was exempt if drainage had begun before the bill was passed. The new ruling eliminates the loophole and could save thousands of hectares of prairie potholes, small ponds that dot the landscape in Michigan, Minnesota, Wisconsin, and the Dakotas and are now under heavy pressure from farmers.

Unfortunately, many wetland protection laws and regulations are like toothless watchdogs. In many instances little is done to enforce these important laws. Each of us can write our representatives and find out what laws exist and how they are being enforced. Local action groups can help stimulate stronger enforcement when necessary. Personal action is needed more than ever in the face of deep cuts in the federal budget. On an international scale nations can cooperate to preserve wetlands.

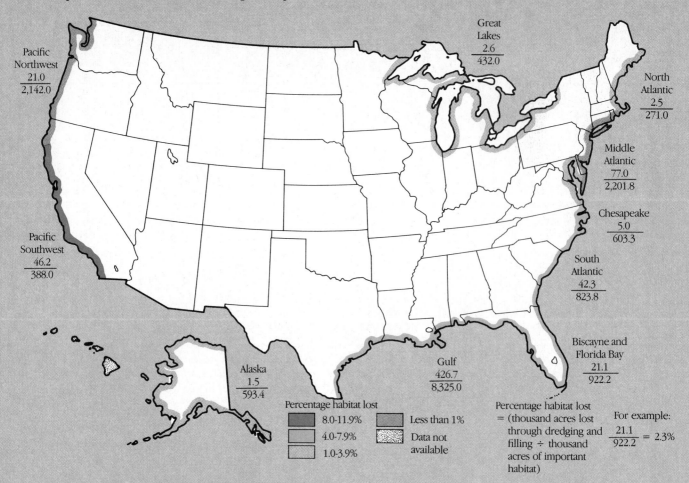

Figure S10-2. Destruction of the estuarine zone (estuaries and coastal wetlands) from dredging and filling.

In 1971 representatives of many nations met in Ramsa, Iran, to discuss the plight of wetlands and agreed to protect those lands within their jurisdiction. In 1986 the United States ratified the agreement. Four US wildlife refuges were added to a list of wetlands of international importance.

Estuaries

Estuaries are the mouths of rivers, where saltwater and fresh water mix. Estuaries, like wetlands, are critical habitat for fish and shellfish. Together, coastal wetlands and estuaries make up the **estuarine zone**. Two-thirds of all fish and shellfish depend on this zone during some part of their life cycle.

The estuarine zone gets its richness from the land—eroded land. Eroded sediment, rich in nutrients, is carried in streams to the ocean, where it supports an abundance of aquatic organisms, especially algae, the base of a large and productive aquatic food web.

Estuarine zones, like wetlands, also absorb water pollutants that flow from upstream sources (sewage treatment plants, industries, cropland, and so forth). Their importance as a water purifier is underscored by estimates that 1 acre of coastal wetland is the equivalent of an $85,000 sewage treatment plant.

Damaging This Important Zone

The estuarine zone is vulnerable to a variety of assaults: pollutants from sewage treatment plants or industries; sediment from erosion that buries rooted estuarine vegetation; oil spills; and dams that can cut off the life-giving flow of nutrients from the land. Cities may withdraw so much fresh water upstream that rivers run dry. In Texas, for example, drought and heavy water demands in past years have critically reduced water flow into estuaries. The Mexican delta of the Colorado River is a remnant of its former self. Freshwater inflows are critical to maintaining the proper salt concentration in brackish water of lagoons and coastal wetlands where mollusks and other organisms dwell. Salinity may be one of the most important factors determining shellfish productivity.

New research suggests that pollutants from the ocean can also concentrate in estuaries. Researchers from the Department of Energy's Oakridge National Laboratory, for example, have found that plutonium (a radioactive material deposited in the ocean from above-ground nuclear blasts) attaches to particles suspended in seawater and can be washed to estuaries. Particles can also carry heavy metals, organic pollutants such as PCB, and insecticides such as DDT.

Many organisms inhabit the estuarine zone, but the most

Endangered Species

In 1978, the Supreme Court blocked the completion of the nearly finished Tellico Dam in Tennessee in order to save a small, ugly fish called the snail darter.

While many applauded the decision, some asked how we can allow a fish to stand in the way of progress. The answer is twofold: (1) Everytime we destroy a species, we destroy an intricate web of interdependency. The near-extinction of the Guam rail, for instance, has had a dramatic effect on both its prey and its predators. The altered ecosystem struggles to adapt or ceases to exist. (2) We can be certain that plants and animals contain some of the answers to the myriad of problems we have not yet been able to solve. Is the cure for cancer hidden in one of the thousands of yet unclassified plants that are being destroyed at the rate of 100,000 acres per day in the tropical rain forest? Can the rapidly disappearing gorilla teach us what we need to know about our own evolution? We can be sure that, as the diversity of the gene pool shrinks, so do our chances for answering some of human-kind's more serious problems.

With the *Endangered Species Act* of 1973, the government assumed legal responsibility for vanishing plants and animals, but it has done little to fulfill this obligation. For instance, during the Reagan administration, over 3800 species were nominated, but only 36 received protection.

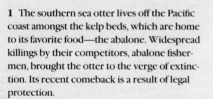

1 The southern sea otter lives off the Pacific coast amongst the kelp beds, which are home to its favorite food—the abalone. Widespread killings by their competitors, abalone fishermen, brought the otter to the verge of extinction. Its recent comeback is a result of legal protection.

1

2 The West Indian Manatee, now numbering only 1000, makes its home in the coastal waters of Florida. Its numbers have been reduced by habitat alteration, overhunting by humans, and injuries caused by collisions with power boats.

3 The largest living primate, the mountain gorilla has been hunted to the brink of extinction. Killed for its meat, collected for zoos, and slaughtered for body parts, the population has dwindled to about 1000.

4 The Panamanian Golden Frog lives in a three-square-mile region in Panama. Attractive to tourists in search of unusual pets, the endangered frog is now protected by law.

5 Once enjoying the largest geographic distribution on this continent, the timber wolf has been reduced to a handful of small packs in Alaska, Canada, and the Soviet Union. Probably one of the most feared and misunderstood of all animals, timber wolves were victims of widespread slaughter by humans who believed they were protecting their livestock. It is now known that wolves kill only the old or infirm in the herds.

6 The nation's symbol, the bald eagle, is endangered or threatened in most of its habitats in the United States. Habitat destruction coupled with pesticide poisonings have brought the birds' numbers down to an alarming 1300 pairs in the lower 48 states.

7 A shy animal cursed with a beautiful pelt, the snow leopard was first photographed in the wilds of Nepal in 1970. Studies indicate that as few as 16 remain today.

8 A sow grizzly will charge anything that threatens her cubs. At 1000 pounds, with a 30-inch neck and a 55-inch waist, it seems unlikely that this animal could run faster than 65 kilometers (40 miles) per hour. But she can—easily catching any human or beast that might unwittingly wander between her and her cubs. Only 800 grizzlies remain in the lower 48 states today.

9 Zoos, once contributing to the decimation of wildlife populations, have come to the assistance of endangered species. Here two zoologists introduce a motherless newborn African elephant to a lactating, and hopefully cooperative, female.

4

2

3

5

6

7

8

9

10 This California condor was hatched in the San Diego Zoo in 1983. Condor eggs are taken from nests in the Los Angeles mountains, and hatched and raised in captivity. Because they are victims of habitat alteration, the baby condors' chances of surviving to sexual maturity are less than 1 in 10.

11 A zooworker feeds a hatchling using a condor puppet. The use of the puppet teaches the baby to recognize its own kind and encourages it to feed normally. Careful avoidance of human contact will help to keep the birds wild and enable them to breed naturally once they are set free.

12 All wild condors were captured in 1986 and 1987 and placed in captivity in southern California. Successful programs at the San Diego Wild Animal Park and the Los Angeles Zoo, have helped to bring the number of California condors up to 38. So far, in 1990, 7 eggs have hatched, bringing the total number of eggs hatched in captivity up to 19. In the fall of 1991, up to four California condors are planned to be released into the wild.

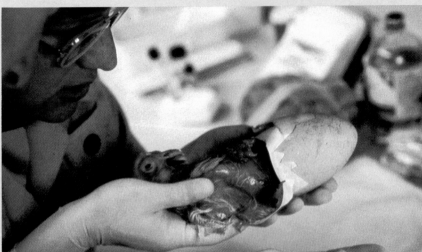

10

12

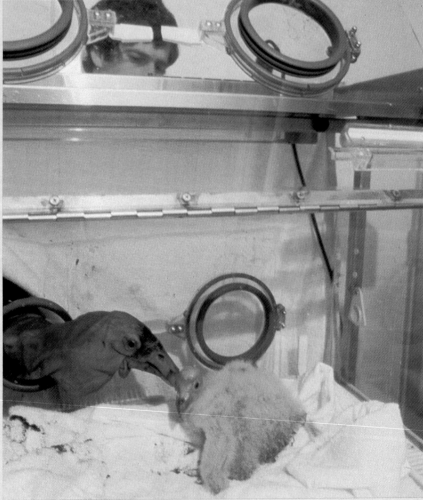

sensitive to pollution are clams, oysters, and mussels. Pollution is generally thought to be one of the major factors responsible for the decline in mollusk harvests in the last 20 years. Mollusks can concentrate toxic heavy metals, chlorinated hydrocarbons, and many pathogenic organisms, including those that cause typhus and hepatitis. These pathogens may not affect the mollusks' survival, but they make them unsafe for human consumption.

The estuarine zone is a battlefield of sorts, beleaguered by pollution, water loss, sedimentation, dredging, and filling, among other invaders. Compounding the damage is the widespread problem of overharvesting of fish and shellfish. It is generally agreed that the decline in US oyster production after 1950 was largely the result of overharvesting. Clams in the Northeast have likewise been severely overharvested. Chesapeake Bay is certainly a victim of heedless overfishing.

Over 40% of the US estuarine zone has been destroyed. The most severe damage has occurred in California, along the Atlantic coast from North Carolina to Florida, and along the entire Gulf of Mexico (Figure S10-2). Despite state and federal laws to protect this zone, destruction continues.

Protecting the Estuarine Zone

Protecting estuaries and coastal wetlands is largely a matter of common sense and good resource management, part of an overall management strategy whose beneficial effects ripple through the biosphere just as the adverse effects do now.

Improved water pollution control (Chapter 16) is a key element of the plan. Erosion control is equally important. Water conservation to preserve vital water flow into estuaries is also needed. Restraint is the final element. Restraint means restricting dredging and filling and ending the overharvesting of fish and shellfish.

Protecting the estuarine zone presents a unique challenge in the United States, because 90% of the coastal land in the 48 conterminous states is privately owned. In 1972 Congress responded to the plight of the estuaries and coastal wetlands by passing the **Coastal Zone Management Act**. This law set up a fund to provide the 35 coastal and Great Lakes states with assistance in developing their own laws and programs. It also provided them with funds to purchase estuarine zones and estuarylike areas in the Great Lakes states. These regions are to be set aside for scientific study; as of 1989, 18 national estuarine sanctuaries had been established. Eventually, 20 to 30 will be established. The Coastal Zone Management Act also allows for the establishment of national marine sanctuaries. Eight have been set aside off the coasts of California and Washington. These sanctuaries are established to protect vital habitat for a variety of marine mammals and fish.

Half of the US population lives in counties that are at least partly within an hour's drive of a coast, and a majority of the major cities are coastal. Discoveries of offshore oil, gas, and minerals pose new problems that require immediate answers. An abundant supply of cooling water makes the coastal zones prime candidates for new power plants and oil refineries. Because of current and potential problems the Coastal Zone Management Act is an important step in preserving US coasts.

Still, according to critics, the act leaves too much discretion in the hands of the states. Some states, in fact, either have not

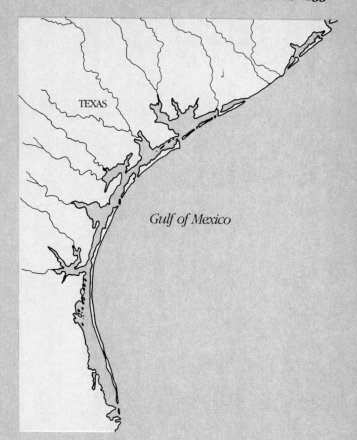

Figure S10-3. The Texas Gulf Coast is a barrier island coast. Barrier islands are moving bodies of sand that make poor sites for homes and resorts.

adopted programs or enforce their programs poorly, leaving their coastal waters open to misuse.

Fifteen additional federal laws have been passed to promote coastal zone management, but they are only as good as their enforcement. Lackadaisical enforcement is almost as bad as no enforcement at all. Without further, more serious efforts to improve coastal zone management, we will almost certainly lose more of this important habitat.

Coastlines and Barrier Islands

The eastern coast of the United States and Mexico is skirted by a chain of **barrier islands**. These narrow, sandy islands are separated from the mainland by lagoons and bays (Figure S10-3). An estimated 250 barrier islands lie along the Atlantic and Gulf coasts. Many of them are popular sites for recreation. Some have been purchased by the federal government and are used for recreation (as national seashores) and wildlife habitat, but most barrier islands are under private ownership.

In the last 40 years many of these islands have been developed for vacationers. Summer homes, roads, stores, and other structures have invaded the grass-covered dunes. According to estimates of the National Park Service, in 1950 only 36,000 hectares (90,000 acres) of barrier islands had been developed, but by 1980 a total of 113,000 hectares (280,000 acres) had been devel-

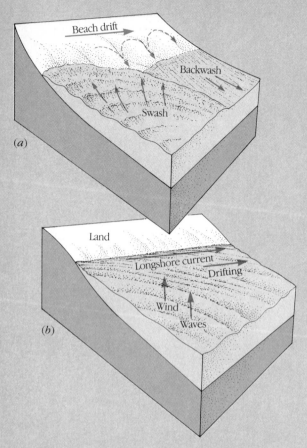

Figure S10-4. (a) Beach drift is caused by waves approaching obliquely. (b) Longshore currents are formed by offshore oblique winds and waves.

oped. If the current rate of development continues, by 1995 all of the remaining barrier islands will be developed.

Barrier islands and their beaches are part of a river of sand that migrates down the East Coast. The islands grow and shrink from season to season and year to year in response to two main forces: First, waves, which tend to arrive at an angle to the beach, erode the beaches, as illustrated in Figure S10-4. These waves create **beach drift**, a gradual movement of sand along the beach. In addition, the wind creates a **longshore current** parallel to the beach, which also moves sand along the beach (Figure S10-4b). Combined, beach drift and **longshore drift** are called **littoral drift**. Littoral drift causes the barrier islands to move parallel to the main shoreline. They shorten on one end and become longer on the other. Homes built on the upcurrent side of the island may collapse into the sea. Second, winter storms tend to wash over the low barrier islands and move the sand closer to the land, destroying houses, roadways, and other structures.

Home construction, attempts to stabilize the beaches and sand dunes of barrier islands, and road construction are a recipe for continued disaster. Federal actions and various relief programs have, in the past, encouraged development on barrier islands. When erosion and storm damage occurred, the government stepped in with money for disaster relief. Federal flood insurance paid for the damage, and federally subsidized construction

projects helped rebuild roads and stabilize the islands. A few years later other storms would devastate the islands, starting the rebuilding cycle all over again at considerable expense. Recognizing this fact, Congress passed the **Coastal Barrier Resources Act** in 1982 to prohibit the expenditure of federal money for highway construction and other development on barrier islands. It helped decelerate the destruction of barrier islands, saving the federal government substantial amounts of money.

Coastal beaches are the victims of human activities. Understanding the dangers requires a look at the natural processes that affect these beaches. Coastal beaches, like barrier islands, are eroded by longshore currents. Thus, US beaches are like great rivers of sand kept in constant motion by the major coastal currents. Sand lost in one area is subsequently replaced by sediment which is carried to the sea by rivers. Consequently, dams that trap sediment diminish the natural replacement of sand on coastal beaches. Pockets of coastal beach erosion plague California. According to one estimate, dams hold back nearly 40% of the sediment that once reached the mouth of the Santa Clara River north of Los Angeles. This robs California beaches of 15 million metric tons of sand each year. From New Jersey to Texas, the story is the same.

Some communities erect barriers, called jetties, to prevent erosion by longshore currents. These structures only slow down the process. Jetties are sometimes built to maintain navigable passageways in coastal harbors. In 1911, for instance, two 300-meter jetties were built on the New Jersey coast north of Cape May to prevent sand from filling in the harbor. While performing admirably in their appointed duty, the jetties have had a disastrous effect on downcurrent beaches. The beaches at Cape May grew thinner and thinner. By the 1920s the town was actively fighting back by building small jetties to keep the remaining sand from being washed away and to trap the sand flowing in the longshore currents. To the townspeople's dismay their efforts were fruitless. Beaches retreated by 6 meters (20 feet) a year. Lighthouses fell into the sea. The ocean threatened to swallow the airport. After years of anguish the town has turned to an expensive pumping system that will draw sand from above the two large jetties and move it down to the beaches.

To protect the shoreline the delicate balance between sediment flow and erosion must be maintained. Dams that retain sediment must be avoided, or rocky beaches will become common in the future. Beaches must be left alone to grow and shrink with natural cycles. Preventing their movement, which costs taxpayers millions of dollars a year, only seems to accelerate erosion and beach disappearance. Realizing that it makes more sense to cooperate with nature, many government officials on the Atlantic coast have begun to develop plans that would return developed shorelines to their natural state following destructive hurricanes. This step would begin a retreat from the shorelines, where nature probably intended humankind to be only a visitor.

Wild and Scenic Rivers

More and more people are flocking to free-flowing rivers to embark on a variety of sports: fly fishing, kayaking, rafting, inner tubing, and canoeing. This great recreational resource, however, is increasingly imperiled by dams and diversion projects (Chapter 10). Foreseeing the need to protect rivers for recreation and habitat for fish and other species, Congress passed the **Wild and Scenic Rivers Act** in 1968. It prevents the construction of dams, water diversion projects, and other forms of undesirable development along the banks of some of our remaining free-flowing rivers.

Undammed rivers offer much more than excitement for the river runner. Like wilderness, they provide unrivaled scenery and countless opportunities for relaxation and reflection. For the adventurous kayaker and rafter, they offer a taste of danger and the opportunity to test one's skills, strength, and endurance. They are home for fish and other forms of wildlife. Their canyons offer unequaled opportunity for geological study. Taken together, these opportunities make a river much more than a source of water.

Because rivers exist in varying states of development, Congress established a three-tiered classification scheme for the **Wild and Scenic River System**: **wild rivers** are relatively inaccessible and "untamed," **scenic rivers** are largely undeveloped and of great scenic value, and **recreational rivers** offer important recreational opportunities despite some development. By January 1, 1989 119 river segments had been included in the system, totalling over 15,430 kilometers (9,260 miles).

Few things are as likely to spark controversy as a wild or scenic river designation, because so many interests vie for a river's benefits: municipal water consumers, paper manufacturers, farmers, hunters, anglers, and whitewater boaters. Competing interests make compromise difficult or impossible. A dammed river provides water for a new paper mill, water skiing, and boating, but irretrievably floods the kayakers' rapids and the fly fisherman's favorite pools.

The fight to keep free-flowing US rivers free from development continues today. Much of the pressure to dam American rivers, however, has been taken off because of economic and legal forces. After years of paying for questionable dams, Congress has found that many projects return only a few cents for every dollar invested. As a result, federally subsidized water projects, often handed out as political favors, have fallen into political disfavor.

In the battle over water, compromises are inevitable, but commercial interests might better promote water conservation as a first step in meeting water needs (Chapter 10). Simple measures costing pennies can save millions of dollars. When possible, smaller dams that flood only a small portion of the river should be built. To prevent sedimentation, erosion controls in the watershed could be strengthened.

Many conservationists argue that valuable recreational rivers must be preserved just as endangered species are. When a scenic river gorge is dammed, it is gone forever. Conservationists can work with water developers to help select sites least likely to cause environmental damage.

A river is a vital resource, but dammed and diverted to water-hungry, often-wasteful consumers, that river becomes a tragic symbol of poor planning and undisciplined greed. Our goal should be to manage our rivers wisely and efficiently, minimizing waste and damage, ensuring future generations the use of treasures we now enjoy and too often take for granted.

Suggested Readings

Magder, R. (1989). Growth and the Environment Focus Paper. Washington, DC: Renew America. Survey of progress in controlling growth and development.

Energy: Winning a Dangerous Game

Our entire economic structure is built from and propelled by fossil fuels. We have invaded the long-silent burial grounds of the Carboniferous Age, appropriating the dead remains of yesteryear for the use of living today.

JEREMY RIFKIN

The year is 2000. Joanna Mills, mother of two high-school-age children, slips out of bed at 7 AM. Tiptoeing across the cold wooden floor she turns up the thermostat to a cool 18° C (64° F). Electric current begins to flow through the baseboard heaters.

Several states west, Jacob Cowens wakes with the rising sun, pads to the window, and throws open the heavily insulated curtains. The sun pours into the main room of his passive solar home. Despite the subfreezing Wisconsin weather his house stays pretty warm at night. Sunlight and heavy-duty insulation in the ceiling and walls have been his primary "heat sources" this winter. Stored in a thick cement slab under his tile floor and interior brick walls, the heat radiates into the room day and night, keeping it warm and comfortable (Figure 11-1). Last year the Cowenses' heating bill for the entire winter was $400, compared with the Millses' $3000 bill.

Joanna Mills. Jacob Cowens. One a victim of rising energy costs, the other barely affected. One struggling to pay bills that nearly equal the house payments, the other free to send his daughter to college and invest his money in new business ventures.

Between 1973 and 1985 Americans' natural gas bills rose over 600%. A $30 monthly home heating bill in 1973 became a $210-a-month monster in 1985. Energy analysts believe that winter heating bills could in the near future exceed monthly mortgage payments. Should this prediction come true, many families that have purchased cheaply made homes heated with natural gas or electricity will find themselves facing hard times.

The rise in energy costs is inevitable. Deregulation, rising demand, and declining supplies of nonrenewable fuels, especially oil, ensure a future of escalating costs. Few people will be immune to them. Making matters worse, many experts predict that sometime between 2000 and 2010 global demand for oil will outstrip supply. The cost of gasoline and other oil byproducts will soar as a result. Along with them the cost of food, clothing, and shelter will begin to climb, creating global inflation and economic stagnation unless something is done soon.

This chapter examines many aspects of energy. It explores our dependence on oil and other fossil fuels, the impacts of energy use, and how long conventional fossil fuels will last. Finally, taking a long-range look, it suggests ways to build the foundation of a wise energy strategy.

The Fossil Fuel Connection: Discovering Our Energy Dependence

Spilling up from deep wells or scooped up by the truckload in surface mines, energy in its many forms is the lifeblood of modern industrial society. At the same time, it is our Achilles' heel. Cut the supply off for even a brief moment, and pandemonium would ensue. Industry would come to a standstill. Millions would be out of work. Agriculture and mining would halt, with cataclysmic effects. Automobiles would vanish from city streets.

Important as it is, energy has been taken for granted for many years, and used wastefully. The prevailing mood was that there would always be more and it would always be cheap. The days of energy nonchalance came to an abrupt halt as a result of the 1973 oil embargo imposed by the Organization of Petroleum Exporting Countries, or OPEC, and a second embargo by Iran in 1979. OPEC nations used the 1970s to wield enormous economic power. Crude oil prices shot up from $3 a barrel to $34. Previously poor nations like Kenya rejoiced over their newfound wealth generated from oil. But in Europe and the United States the news was bleak. In step with this shocking price increase, the costs of consumer goods and farm products, produced and distributed with the aid of petroleum, shot upward. Inflation began a long climb that crippled the American and world economies; many countries have not yet recovered. In the United States long lines formed at gas stations. People became used to colder houses. All in all the 1970s were a rude awakening for a frontier society nurtured on the idea˚of unending resources. (For more on the frontier ethic see Chapter 19.)

But good often comes from disaster. The oil shortages of the 1970s were no exception. The OPEC tactics, while

Figure 11-1. Passive solar houses capture sunlight's energy in interior walls and floors and radiate it into rooms at night. Over the lifetime of the owner, passive solar energy can save tens of thousands of dollars in heating bills. The author's home (shown here) at 2500 meters (8200 feet) in the Colorado Rockies is heated for under $100 a year thanks to superinsulation, passive solar heat, and a 92%-efficient gas heating system that supplements the solar heat.

devastating to world economies, heightened awareness of the industrial world's dependency on foreign oil. Many cut back on energy waste, improved efficiency, and found new sources. Between 1973 and 1989, for example, American industry reduced its energy use by 30% and home energy consumption fell by 20%. Although a third of the decrease in industrial energy demand resulted from a shift from heavy industry to less energy-intensive service industries, the rest came from conscious efforts to cut back energy use. However, critics point out that energy conservation is a road a thousand miles long which we have only just begun to travel. (For more on the potential of energy conservation see Chapter 12.)

Energy conservation in the 1970s served a purpose few people recognize: it helped to revive the world economy. Political officials like to take credit for the economic recovery of the 1980s, especially in the United States, but critical thinking suggests that other reasons played a significant role. One of the chief reasons for economic recovery was conservation. How? Conservation created an excess of oil. Consumers had turned off the spigots, but producers had left the oil wells running. Temporary gluts stabilized prices of consumer goods, and, gradually, the economy began to pick up.

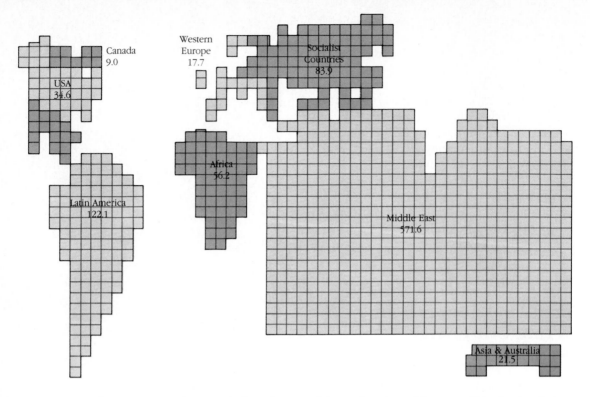

Figure 11-2. Proven oil reserves. An indication of where the United States, Canada, and Europe will be buying their oil. Numbers represent billions of barrels of oil in proven reserves. Each square equals one billion barrels of oil.

Stimulated by the still relatively high price of oil, many non-OPEC nations (Britain, for example) increased exploration and oil production. OPEC's share of the world oil market fell. In 1977 OPEC nations supplied two-thirds of the world's oil. By 1985 their share had dropped to one-third. To combat the loss in market share, OPEC announced in December 1985 that it would cut prices to undercut its competitors. For the consumer, in the near term, this was good news. By April 1986 crude oil prices had dropped by half, rising only slightly by 1990.

Lowered energy prices, although good for the economy, may have stifled energy conservation and have crippled the development of renewable resources, such as wind power and solar energy (see Chapter 12). Max Neiman, a political science professor at the University of California, Riverside, found that few local governments showed any sustained interest in the energy conservation programs they had drafted in the late 1970s. A potentially dangerous result of the lack of concern for energy conservation and alternative sources, says Christopher Flavin, a senior researcher at the Worldwatch Institute, is that the stage is now being set for an even stronger OPEC in the 1990s. With 56% of the world's oil, OPEC nations will be in position to control prices once again as the non-OPEC nations deplete their reserves. In other words, cheap energy today may mean a heavy dependence on OPEC oil in the future, with the potential for another

energy crisis in the 1990s.

These major global dramas played out in the international economy distract the public from the underlying problem of modern industrial society: oil, in many ways the lifeblood of such societies, is disappearing fast. The present oil "gluts" are only temporary—in many ways, the calm before the storm.

Energy Use—Then and Now

The history of energy use in the United States has been one of shifting dependency, as shown in Figure 11-2. This pattern may well continue in the future as today's energy sources run out and as new ones are developed.

One hundred years ago Americans had few choices for energy. Wood was the chief source. Today, our options are many: coal, oil, natural gas, hydropower, geothermal energy, solar power, nuclear power, and wind (Figure 11-3). American energy options began to expand in the late 1800s when factories shifted to coal as wood supplies were depleted. But many people found coal to be a dirty, bulky fuel, expensive to mine and transport. When oil and natural gas were made available in the early 1900s, therefore, coal use began to fall. These new fuels were much easier and cheaper to transport and had the additional advantage of burning more cleanly.

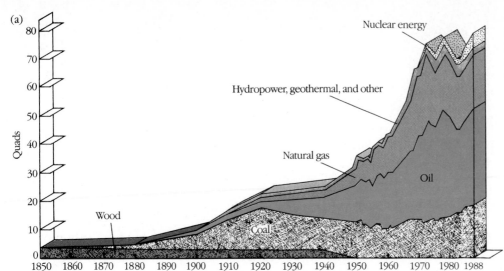

(a)

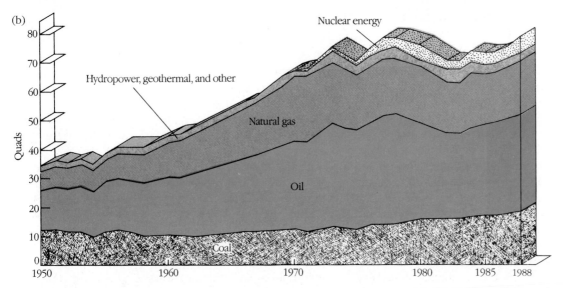

(b)

Figure 11-3. (a) Energy consumption in the United States by fuel type from 1850 to 1982. (b) Energy consumption from 1950 to the present.

Today, despite a far greater number of energy options, the United States depends primarily on three fuels: oil, natural gas, and coal. In 1989 oil accounted for 42% of total US energy consumption (Figure 11-4a). Natural gas provided 23%, and coal, 24%. Hydroelectric, solar, and geothermal power supplied 4%, and nuclear power provided 7%. Although it does not show up in the energy calculations, conservation provides us with a great deal of energy, too. (Figure 11-4b breaks down energy consumption by user.)

The United States is a major consumer in the international energy market. With only 6% of the world's population, the United States consumes about 30% of the world's energy (Figure 11-5). For many years the nation's energy appetite grew rapidly. Between 1965 and 1971, for example, energy consumption grew at 4.8% per year. After the oil embargoes of the '70s energy consumption took a dramatic downturn. Between 1979 and 1983, in fact, total annual energy consumption dropped by 10%. Higher prices, a poor economy, improved energy effi-

ciency, and conservation all contributed to the decline. With falling prices and a revitalized economy since then, however, energy consumption has climbed steadily upwards. In 1989, the US consumed nearly 81 quadrillion BTUs (quads) of energy, exceeding the all-time high of 79 quads in 1978.

The United States needs energy, and needs it badly. But each day thousands of tons of fuel are wasted. In fact, over half the energy consumed in the United States is wasted. The second law of thermodynamics states that energy is "degraded" when it changes form, as discussed in Chapter 3, so some waste is inevitable. But the amount of energy wasted in the United States far exceeds the inevitable loss. Given the nation's dependency on energy, its high cost, and the rapid decline in global resources, continued waste may become a detriment to Americans' future well-being. As Bruce Hannon put it, "A country that runs on energy cannot afford to waste it." No better advice was ever given to an industrial society. (Chapter 12 discusses ways in which energy waste can be cut.)

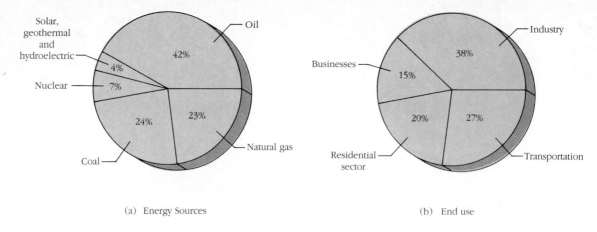

(a) Energy Sources   (b) End use

Figure 11-4. (a) Major energy sources in the United States. (b) Breakdown of energy use.

Impacts of Energy Production and Consumption

Energy does not come cheaply. In addition to the economic costs, which were discussed above, society pays a huge price for damage to the health of its people and its environment. Lung disease, acid rain, thermal pollution, and battered landscapes are but a few of the many costs we have paid, and will continue to pay, for energy use. Pollution, much of it caused by energy production and consumption, is the topic of Part 4 of this book. In the words of the economists, pollution and its impacts are **external costs**, because they are borne by society and not by the manufacturers directly responsible for them (Chapter 20). Making external costs known, reducing them whenever possible, and finding fair ways to pay for them are the main goals of the environmental movement, and the subject of much of this book.

For many people energy comes from the electric outlet or the gas station pump. Few people know much about the oil wells or coal mines from which our energy comes or the impacts they create. This section looks at some of the effects of the energy production–consumption cycle. This chain of events, called an **energy fuel cycle**, or **energy system**, is composed of many stages, as illustrated in Figure 11-6. The major phases are exploration, extraction, processing, distribution, and end use. The most notable environmental impacts occur at the extraction and end-use phases. Understanding fuel cycles helps us analyze the environmental and economic costs that can result from fuels and can help us find ways to eliminate or minimize the costs. All in all, understanding these cycles is an important tool for critical thinking, helping individuals look at the big picture.

Most of the electricity Americans use comes from four sources: coal, hydroelectric power, oil, and nuclear power. The impacts of hydroelectric power are discussed

in Chapter 10. For a discussion of the potential impacts of oil development on the Arctic National Wildlife Refuge in Alaska, see Case Study 11-1, and for information on the impacts of oil on oceans see Chapter 16. Nuclear power is discussed in Chapter 12. The following material discusses some of the impacts that occur during the production of electricity from the combustion of coal, which will help further your understanding of the energy fuel cycle.

Exploring for coal once caused enormous damage in the East because companies used bulldozers to carve away the sides of mountains in search of coal seams. Often, dirt and debris were pushed over the hillside, burying trees and vegetation. When coal was found, the bulldozers removed more material, creating additional waste that was discarded similarly. Today, these harmful exploration and mining practices have been largely stopped. Mining coal, however, still has considerable environmental impact. The rock and dirt over coal seams is hauled away, rather than pushed over the hillside, and is stored until the coal seam is exhausted. Then it is trucked back to the site and used to fill in the scar. Even though practices have changed, coal strip mines still create eyesores and can increase erosion. Roads into hill country can also erode, filling streams with sediment.

In the East, coal comes from underground mines. Coal mines are notorious for explosions and cave-ins, making underground coal mining the most hazardous of the major occupations in the United States. Since 1900 more than 100,000 Americans have been killed in underground coal mines, and one million have been permanently disabled. Thanks to stricter safety regulations deaths have dropped substantially in the last 50 years. Underground coal mines also cause **black lung**, or **pneumoconiosis**, a progressive, debilitating disease caused by breathing coal dust and other particles (Figure 11-7). Victims have difficulty getting enough oxygen because the tiny air sacs

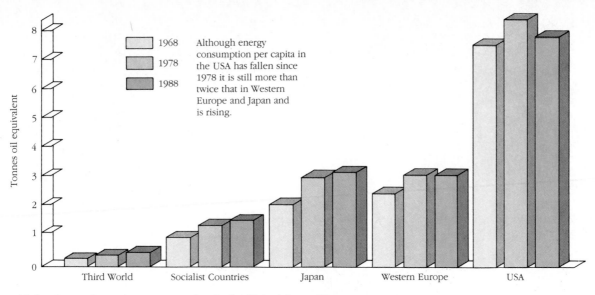

Figure 11-5. Per capita energy consumption in the United States, Western Europe, Japan, socialist countries, and the Third World.

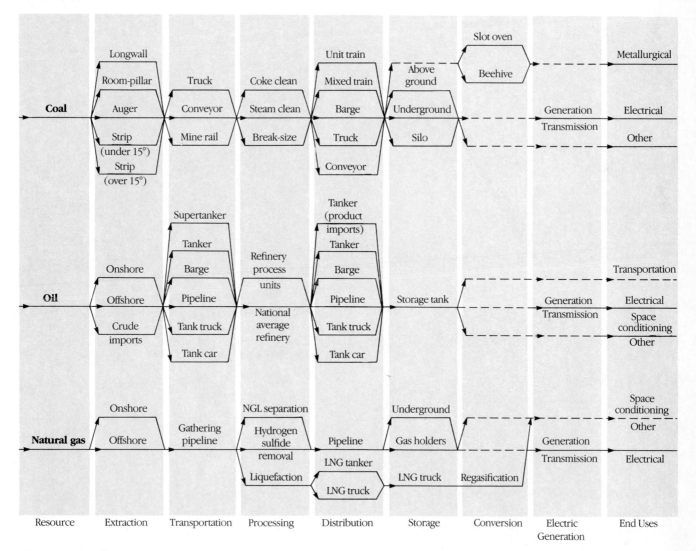

Figure 11-6. Energy systems, or energy fuel cycles.

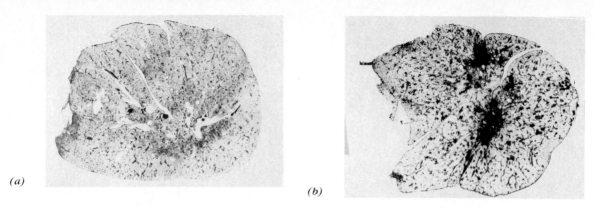

(a)　*(b)*

Figure 11-7. (a) A cross section of a normal lung. (b) A cross section of a lung from a retired coal miner with black lung.

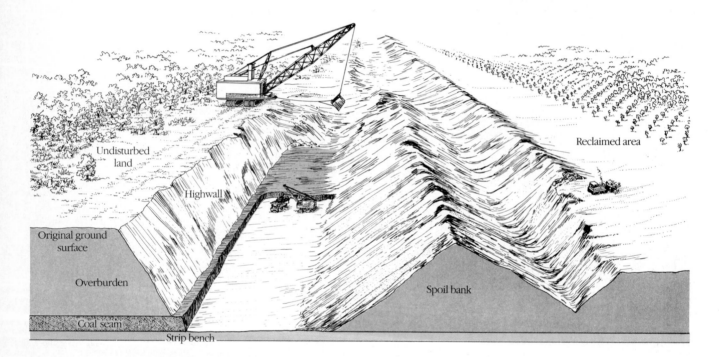

Figure 11-8. A strip mine (above). Dragline removes the overburden to expose the coal seam. Aerial view of a strip mine (below). If land is not carefully reclaimed, it could be permanently ruined.

Undisturbed land

Overburden

Overburden

Highwall

Bench

Coal bed

Pit

Spoil bank

Figure 11-9. Contour strip mining is common in the hilly terrain of eastern US coal fields. Until recently, overburden was dumped on the downslope, creating enormous erosion problems.

(alveoli) in the lungs break down. Exercise becomes difficult, and death is slow and painful. Despite safety improvements one-third of all US underground coal mines still have conditions conducive to black lung, a problem that costs taxpayers over $1 billion a year in federal worker disability benefits.

Collapsing mines also cause subsidence, a sinking of the surface. Cracks form on the surface, ruining good farmland. In some cases streams vanish in the fissures. Over 800,000 hectares (2 million acres) of land has already subsided in the United States from underground coal mining. For every hectare of coal mined in central Appalachia, over 5 hectares (12.5 acres) of surface becomes vulnerable to subsidence.

In the West coal is largely mined in **strip mines** (Figure 11-8). In surface mining operations the topsoil is first removed by scrapers and is set aside for later reapplication. Next, the **overburden**, the rock and dirt overlying the coal seam, is dynamited and removed by huge shovels, called **draglines**, to expose the coal seam. The coal is removed and hauled away, and then another parallel strip is cut. The overburden from the new cut is placed in the previous one, as illustrated in Figure 11-8. The overburden is then regraded to the approximate original contour, and the topsoil is replaced. Seeds are sown, increasing the likelihood that the area will revegetate.

Strip mines create eyesores, destroy wildlife habitat and grazing land, and may increase erosion. Destroying the natural vegetation and stockpiling overburden in soil piles in the hilly terrain of Kentucky, for instance, have

been shown to increase erosion from .4 metric tons per hectare to 2.4 to 145 metric tons per hectare (Figure 11-9). Proper reclamation can restore wildlife habitat and grazing land and eliminate the eyesores, but unless reclamation is carried out immediately after mining, erosion can become a major problem.

Surface mines can also disrupt and pollute groundwater supplies in the West, because many aquifers are located near or in coal seams. In Decker, Montana, extraction of coal from an aquifer seam resulted in a drop in the water table of 3 meters (10 feet) or more within a 3-kilometer (2-mile) radius of the mine. Residents who depended on the aquifer were forced to find new water supplies.

The dirt roads that transport the workers to the mines create enormous problems in hilly eastern regions if they have not been properly built. During heavy rains, for instance, sediment from the roadways washes into nearby streams, killing fish and other aquatic organisms. Sediment also fills streams, reducing their water-carrying capacity. When rain falls, water spills over the streams' banks, flooding farms and communities.

Many abandoned coal mines in the East leak sulfuric acid into streams. **Acid mine drainage**, as it is called, consists of sulfuric acid formed from water, air, and sulfides (iron pyrite) in the mine. A bacterium (*Thiobaccillus thioxidans*) facilitates the conversion, frequently making the waters highly acidic. Acid mine drainage kills plants and animals and inhibits bacterial decay of organic matter in water, thus allowing large quantities of organic matter

Figure 11-10. Streams affected by acid mine drainage in the United States.

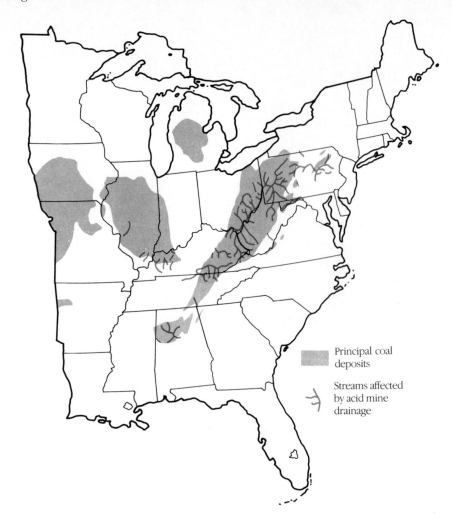

Principal coal deposits

Streams affected by acid mine drainage

to build up in streams. The sulfuric acid also leaches toxic elements such as aluminum, copper, zinc, and magnesium from the soil and carries them to streams.

Acid can render water unfit for drinking and lakes unsuitable for swimming. Municipal and industrial water must be chemically neutralized before use. Acid also corrodes iron and steel pumps, bridges, locks, barges, and ships, causing damage estimated in the millions of dollars each year.

US mines, most of them abandoned, produce about 2.7 million metric tons of acid a year. Acid mine drainage pollutes over 12,000 kilometers (7,250 miles) of US streams, 90% of which are in Appalachia (Figure 11-10). Further increases in coal production could increase acid mine drainage, although awareness of the problem has brought about many efforts to reduce it. In any case, cleaning up the abandoned mines could take decades and billions of dollars.

Underground mines produce enormous quantities of wastes, which are transported to the surface and dumped around the mouth of the mine. These wastes, called **mine tailings**, often wash into streams during heavy rains. The sediment, laden with heavy metals, acids, and other pol-

lutants, damages nearby streams. Coal-cleaning plants, which are designed to crush the coal and wash away impurities, also produce enormous quantities of waste. These wastes are often stored in holding ponds that sometimes leak and pollute nearby streams.

Once the coal is extracted, railroad workers load and ship it off to power plants, where it will be transported by conveyer into the furnaces. Diesel trains, which move most of the nation's coal, pollute the air with a black cloud of particulates and other potentially harmful pollutants, such as sulfur dioxide.

About two-thirds of US coal is burned to generate electricity. The remainder is burned in various industrial processes to create heat or to produce steel and other metals. Numerous environmental pollutants are produced during coal combustion. The most notable are waste heat (Chapter 16), particulates, sulfur oxides, nitrogen oxides, and oxides of carbon (carbon monoxide and carbon dioxide). These pollutants are responsible for most of the world's most threatening environmental problems, and are discussed in Chapter 15 and its supplements.

A 1000-megawatt coal-fired power plant, which produces electricity for about one million people, burns

Table 11-1 Major Environmental Impacts of Fossil Fuels

Fuel	Extraction	Transportation	End Uses
Coal	Destruction of wildlife habitat, soil erosion from roadways and mine sites, sedimentation, aquifer depletion and pollution, acid mine drainage, subsidence, black lung disease, accidental death	Air pollution and noise from diesel trains	Air pollution from power plants and factories—especially acid pollutants and carbon dioxide, thermal pollution of waterways
Oil	Offshore leaks and blowouts causing water pollution and damage to fish, shellfish, birds, and beaches; subsidence near wells	Oil spills from ships or pipelines	Air pollution similar to that from coal
Natural gas	Subsidence and explosions	Explosions, land disturbance from pipelines	Fewer air pollutants than coal and oil, but nitrogen oxides and carbon dioxide

about 2.7 million metric tons of coal each year. In the process it produces about 5 million metric tons of carbon dioxide, about 18,000 metric tons of nitrogen oxides, 11,000 to 110,000 metric tons of sulfur oxides, and 1500 to 30,000 metric tons of particulates.

Coal combustion produces millions of tons of potentially harmful chemical substances in the United States and abroad, causing billions of dollars in damage to fish, lakes, buildings, and human health. Pollution control devices aimed at reducing atmospheric contamination help reduce air pollution. But what they remove from a smokestack doesn't magically disappear. In fact, it becomes a solid waste that must be disposed of properly to avoid problems elsewhere. For example, a fine dust known as fly ash is formed during combustion. Fly ash is mineral matter that makes up 10% to 30% of the weight of uncleaned coal. It is carried up the smokestack with the escaping gases and, if it is captured by pollution control devices, becomes a hazardous solid waste. Sulfur dioxide gas is also emitted from the stack but can be removed by **smokestack scrubbers** (described in Chapter 15). Scrubbers produce a toxic sludge, containing fly ash and sulfur compounds, that must be disposed of. Some mineral matter that is too heavy to form fly ash remains at the bottom of the coal-burning furnace as bottom ash. It, too, is a hazardous waste that must be disposed of. Wastes from coal-burning furnaces and their pollution control devices are generally buried in the ground, but toxic substances from these sites may leak into groundwater supplies, polluting them.

A 1000-megawatt power plant may produce 180,000 to 680,000 metric tons of solid waste each year, including fly ash, bottom ash, and sludge from scrubbers. In 1987 US coal-fired power plants produced 76 million metric tons of solid waste per year. Approximately one-fourth of the waste is used to make road surfaces and other products.

A simple flick of a light or computer switch by millions of consumers indirectly kills fish, buries streams in sediment, increases flooding, destroys wildlife habitat, and fills the sky with hazardous pollutants. By understanding the hidden consequences of energy consumption, we can acknowledge that all of us are to blame for the long list of environmental problems discussed in this text. Knowing that the problems arise from our actions may stimulate us to take steps to reduce our contribution, for instance, by cutting back on waste. Knowing is not enough. Action is required. And individual action multiplied many times can have enormous impact.

Figure 11-6 showed that our three main fuels (oil, coal, and natural gas) have similar cycles. Table 11-1 lists some of the major impacts of energy production and consumption. More specifics are given in Part 4, on pollution.

Energy Trends: Supplies and Demands

Since 1860 energy use has grown worldwide at a rate of 5% a year, except for brief respites in the Great Depression years 1930–1935 and in the recent worldwide economic recession of the early 1980s. In 1989 global consumption had risen to about 350 quadrillion British

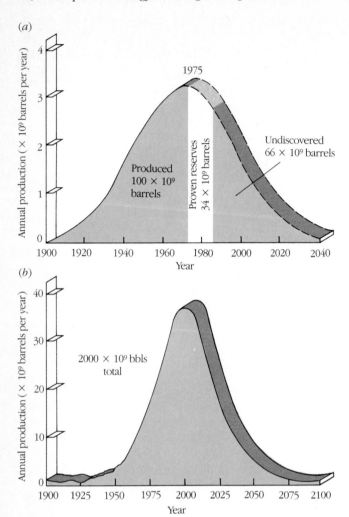

(a)

(b)

Figure 11-11. Petroleum production curves for the United States (a) and the world (b).

natural gas, and coal. To help sharpen your critical thinking skills, when examining the prospects for energy supplies keep in mind the many uncertainties outlined above.

Oil: The End Is Near

Most experts agree that when the last drop of crude oil has been burned, the world's oil fields will have yielded about 2000 billion barrels of oil. This figure is called the **ultimate production**. To date, 575 billion barrels have been produced and consumed globally. Of the remaining 1425 billion barrels, approximately 900 billion are **proven reserves**, that is, are known to exist. The remaining 525 billion barrels are undiscovered, that is, are thought to exist.

On the surface it appears as if the world is blessed with an abundance of crude oil. This illusion quickly shatters when we compute how quickly that oil may be used up by the energy-hungry world we live in. At the current rate of consumption the 900 billion barrels of proven reserves would last only about 40 years; the remaining 525 billion barrels, if indeed they exist, would last about 24 years.

Sixty-four years of oil—that's the good news. Now for the bad news: Any increase in the rate of oil consumption would drastically cut the lifespan of oil. With much of the Third World struggling to industrialize and the developed world still increasing in population and industrial output, an increase in oil consumption is very likely. The startling conclusion is that even a modest increase in energy use could cut the 64-year supply of oil in half.

Making matters even worse, long before the last drop is burned, signs of failure will be evident. You can understand this by studying Figure 11-11. These graphs show that petroleum production in the United States and the world is expected to follow a bell-shaped curve. Annual oil production rises to a peak and then begins to fall as reserves are depleted. As the proven reserves decline, oil companies are forced to work harder and harder to maintain production. But continued high demand will inevitably outstrip supplies, resulting in worldwide shortages and exorbitant oil prices. If alternative energy sources are not available, industry will suffer. Countries that have not developed other energy options will experience economic turmoil: inflation, economic stagnation, and widespread unemployment.

Global oil production is expected to peak sometime between 2000 and 2010. US oil production peaked in 1975. At that time half of the domestic supplies had been extracted and consumed. Despite greatly increased efforts the United States produced only 8.3 million barrels of oil per day in 1987, compared with 11 million in 1973. As a measure of how hard it is getting to find oil, in 1973 nearly 497,000 oil wells were producing more barrels of oil per day than 620,000 wells in 1987.

thermal units (BTUs), or 350 quads. The United States used one-third of that energy (not including biomass, such as wood). Assuming a growth rate in energy use of 4% to 5% a year, world energy consumption could be 550 to 600 quads by 2000, almost twice what it was in 1989. Will there be enough fossil fuel energy to meet these demands?

Answering this question is difficult for several reasons. Most importantly, no one knows how much fossil fuel lies within the earth's crust and is economically and environmentally feasible to extract. Making matters worse, no one knows how much fuel will be burned in the future. Will China's industrialization require massive inputs of fuel? Will the Soviet Union's attempts to improve its economy require the same? Will energy-efficient technologies cut demand and will people worldwide start living more energy-efficient life-styles, making searches for new energy unnecessary?

The next three sections examine the supply-and-demand picture for our three dominant fossil fuels: oil,

Case Study 11-1

Controversy over Oil Exploration in the Arctic National Wildlife Refuge

In the far northeastern corner of Alaska lies the Arctic National Wildlife Refuge, or ANWR. Set aside to protect wildlife, this refuge encompasses 7.7 million hectares (19 million acres). It has been described as the last great American wilderness. But if President Bush and the oil companies have their way, a major portion of this refuge—critical to wildlife—may soon be opened to oil exploration. If oil is found, the pressure will be on for widespread drilling in the coastal plain, a .6-million-hectare (1.5-million-acre) region—an area almost as large as Yellowstone Park.

Oil companies want to search for oil in the delicate arctic tundra of the coastal plain, a region 210 kilometers (115 miles) long and 50 kilometers (30 miles) wide. The *only* section of land along the entire 1800-kilometer (1100-mile) north coast of Alaska that is closed to oil exploration and drilling, the coastal plain is the annual calving ground of several large caribou herds. It also provides habitat for polar bears, grizzlies, wolves, moose, wolverines, and numerous small mammals. Many thousands of birds spend the summer there, raising their young and feeding off of insects. It is also the habitat of the musk ox, which was nearly hunted to extinction by 1969. Today the herd numbers over 400 thanks to conservation efforts.

In the wake of public protest over the tragic oil spill in Prince William Sound (Chapter 16), Congress and the oil companies have put off a decision to allow drilling in ANWR, but when the heat dies, most experts believe, the oil companies will be back, pushing to explore for oil within one of our most valuable wildlife refuges. Should oil be found, the pristine area would be turned into a nightmare of roads, oil platforms, waste ponds, buildings, and gravel pits. Four airfields and 50 to 60 oil platforms are proposed. A power plant and several oil processing plants would be built on the tundra and hundreds of kilometers of roads and pipelines would crisscross this delicate land.

While some proponents think that oil development and wildlife can coexist in harmony, critics are appalled at the prospects. They predict a 20% to 40% decline in one of the major caribou herds. They expect over half of the musk oxen to perish. Populations of grizzlies, polar bears, wolverines, and other animals are also likely to suffer substantial losses. Air pollution, water pollution, and hazardous wastes are all expected to have major impacts on wildlife and the long-term ecological health of the region. One exploratory well alone requires 35,000 cubic meters of gravel, which would be exca-

An oil drilling pad in Prudhoe Bay, Alaska

Case Study 11-1 (continued)

vated from streambeds to make pads and roads. Oil spills on the tundra and careless waste disposal, common in nearby Prudhoe Bay, would forever change this delicate landscape whose short growing seasons and harsh winters greatly impair the natural healing that normally takes place in the wake of human interference.

Proponents argue that we need the oil to cut reliance on foreign sources and that we need to start exploring in ANWR now because of the 10- to 15-year lead time required to fully "develop" the area. Geologic evidence, some say, indicates a high potential for oil discovery and oil is needed to help keep the Alaskan economy alive. They say that environmental regulations will ensure minimal impacts on wildlife and will help reduce environmental impacts.

Those in favor of preserving the refuge intact, however, argue that the environmental costs are too high. Experience in nearby Prudhoe Bay suggests that oil companies often ignore environmental regulations. Frequent violations have been cited. The Department of Interior, which promotes oil development, has recorded over 17,000 oil spills since 1973 in the Arctic. Where the oil saturates the soil, vegetation fails to recover. Opponents of exploration also argue that since the entire north coast of Alaska is already open to oil exploration and since ANWR is a wildlife refuge, exploration should not be allowed—now or ever. If there's oil there, let it lie. Let us find alternatives and let the wildlife live in peace. Furthermore, they point out that slight improvements in automobile gas mileage could easily "provide" as much oil as ANWR could generate over its lifetime—and at a much lower cost, both economically and environmentally. Moreover, critics say that the proponents have exaggerated the potential for finding oil that is economically feasible to recover. Based on the oil industry's own reports, they say, there's only a one-in-five chance of finding oil that could be recovered economically if oil prices were $33 per barrel. Today, oil sells for about $20 per barrel. Add these and the other negatives up and the cost of oil development in ANWR becomes a bargain with the devil.

Few of us will ever visit ANWR, but most of us take comfort in knowing there are places where wildlife are free of the torment of modern industrial society. We can keep it that way by insisting that the ANWR stay closed, by reducing our own energy consumption, and by convincing government officials to get serious about tapping conservation and the wide array of renewable energy resources now available to us.

Fortunately, trend need not be destiny. Restraint, through conservation, and our ingenuity can rescue the world from the impending disaster. To prevent widespread economic turmoil, however, we must act soon. If we learn nothing else from the bell-shaped curves in Figure 11-11 we should learn that to maintain our standard of living we must find something to replace oil, which powers our transportation system, heats our homes, and provides the raw materials for plastics and other synthetic materials, such as nylon. The replacements must come soon. Guidelines for wise energy choices are presented later in this chapter. Some energy options are discussed in Chapter 12.

Natural Gas: A Better Outlook

The outlook for natural gas is considerably brighter than the outlook for oil, although experts debate the size of our ultimate production. Estimates range from 5,000 to 12,000 trillion cubic feet. Taking 10,000 trillion cubic feet as the most reasonable estimate, let's examine the global outlook. About 2300 trillion cubic feet of natural gas have already been produced and consumed, leaving 7700. Of this, 4000 is proven reserve (we know it exists), and 3700 is yet to be discovered.

How long will the global supply of natural gas last? At the current rate of usage (70 trillion cubic feet per year), proven global reserves would last about 57 years. The as-yet-undiscovered reserves would last about 53 more years. As with oil, increased consumption could cut the lifespan of natural gas considerably, perhaps by as much as half. At least for the time being it appears that global supplies are adequate. In the United States, however, the picture is dimmer. The total known reserves of natural gas are about 187 trillion cubic feet. This is expected to last slightly less than 40 years. Since domestic gas production has already peaked and continues to drop, prices are likely to shoot upward. The rise in costs suggests a need for cheaper options and, especially, a need for more conservation.

Coal? The Brightest but the Dirtiest?

Coal is the world's most abundant nonrenewable fossil fuel. World proven reserves (estimated to be 786 billion tons) will last 200 years at the current rate of consumption. **Total resources** (all the coal in the ground) are estimated at 12,600 billion metric tons; half are believed to be recoverable. (Some energy analysts think that this estimate exaggerates the reserves by five or six times.) The recoverable reserves (an estimated 6300 billion tons) would last 1700 years. Growth rates of 4% to 5% would greatly reduce the lifespan of coal, but the large surplus suggests that coal could be with us for many years to

Hard Paths vs. Soft Paths—Opposing Views
If Energy Sources Are Thrown Away

A. David Rossin

The author was Assistant Secretary for Nuclear Energy, US Department of Energy (1986–87). He directed the Nuclear Safety Analysis Center at the Electric Power Research Institute (1981–86), and conducted research on nuclear power safety and materials for 16 years at Argonne National Laboratory.

By electing governors and legislators and by initiative referendums, the public can remove energy options. But decisions about investment in long-term capital-intensive projects, like power plants, are made by those with the responsibility to serve their communities and regions. They must choose among available options, then raise the money. They must have confidence that they can build and then operate the plant reliably for many years to pay back the investors.

New laws now encourage independent power producers and cogenerators. As a result, few utilities are even considering building new power plants at all. Some states have instituted a bidding system, and there seem to be plenty of non-utility bids to meet energy demand. The problem is that almost all of the bids involve burning natural gas to make electricity.

Some call natural gas the fuel of the future. But the future is now, and our decisions today may lock us into a very few options: conservation, improved efficiency, natural gas, and possibly insufficient amounts of electricity if conservation doesn't save enough.

In a free society, people, cities, and companies are permitted to make their own choices. In some countries that are energy-short, a centralized government makes the decisions about what uses have higher priority than others. This is the reality

in some eastern European and some developing nations, and may even come about in Sweden or Italy before the end of the century. But this would be a political disaster in the US.

During the last decade, the growth rate of solar and wind generators has been fast, but the base was small. The maximum contribution that these sources are capable of making to the nation's energy is 1% to 2%. This would be very valuable, because it would largely replace the burning of more natural gas, conserve gas for the future, and eliminate some carbon dioxide emissions which add to the greenhouse effect.

Conservation and improved efficiency have resulted in significant reductions in fuel use below the projections of the 1970s. The most dramatic cuts have been in oil use for gasoline and other transportation, and in electric power. Much of the remainder of the difference between the old energy use projections and actual use was a result of heavy industry closing in the US and moving overseas!

Bringing nuclear power back into the marketplace will not save the world from acid rain or from the greenhouse effect. No one claims that nuclear power (or any single source of energy) can do the job alone. We will need all of our options, and a monumental effort in conservation, to avoid shortages. Electricity production declines caused by government decisions to cut the burning of coal, oil, or gas to reduce carbon dioxide emissions cannot be made up by any strategy other than restricting growth and allocating electricity by priority.

Considerable capital investment is necessary for improving efficiency and recycling. Recycling requires energy: fuel for heat and collecting refuse, electricity to move it, separate it, and process it, to pump, clean up, and recycle solvents or wash water. There really is no free lunch. To do the things society wants to do will take energy.

Current trends indicate that an ever-increasing fraction of our energy demand will be supplied by electricity. It makes for a cleaner workplace, cuts pollutant emissions, and it can be controlled more precisely to make better and more economical products with less energy waste. Experience in the manufacturing sector reveals that as electricity replaces direct heating with fuel burning, unit production costs go down, prices become more competitive, sales go up, and the total amount of electricity used generally increases. But the efficiency of energy use has improved.

Nuclear power is not subsidized as some contend. The government has invested only $20 billion in nuclear power and radiation and safety research since 1950. But the utility investment in nuclear plants is well over $100 billion. The privately-owned utilities pay taxes, they pay the government to cover all the ultimate costs of waste disposal, they pay 40% of the

>|< *Point/Counterpoint (continued)*

budget of the Nuclear Regulatory Commission, and they pay into special funds to cover ultimate decommissioning of the plants, all out of the bills the ratepayers pay for electricity.

The US has reliable electricity because the utility system is dispersed. It has diversity of power plant type and fuel. It has reserve capacity against shutdowns, failures, storms and other natural catastrophes. It has dedicated and trained personnel whose job it is to get the system back into operation and restore service quickly.

There are those that talk of the "soft path," suggesting that individuals can have their own, small supplies of electricity and escape the monopoly power of the big utility company. Some do it. They are a few of the wealthy elite or rugged individualists who choose their life-style. But most people prefer to do other things, and count on their electricity to be reliable and affordable.

In fact, the terms "soft path" and "least-cost energy strategy" have become buzzwords for stopping construction of power plants by utilities. Some critics argue that by stopping utilities from building new power plants, the nation will have a more flexible energy future. Flexibility means that those with responsibility for society's energy supply have plenty of choices. The opposite is being forced by proponents. They are leaving fewer options, some of which may not even be politically acceptable, like energy allocation or rationing.

Most important is that the public, corporations as well as individual citizens, have the flexibility and the freedom to make their own choices: not just about energy supply, but about the kind of life they want to lead and the kind of society their children will inherit.

Critical Thinking

1. Summarize Rossin's key points, then analyze them using your critical thinking skills.

2. Rossin's main concern here is maintaining energy options, especially nuclear power. Do you agree or disagree that nuclear power should be given a second chance, given the recent accidents and the cost?

3. Does Rossin accurately portray the potential of conservation and renewable energy options? Which options does he leave out?

The Best Energy Buys

Amory B. Lovins and L. Hunter Lovins

Amory Lovins is a consulting physicist; his wife and colleague, Hunter, is a lawyer, sociologist, and political scientist. They have worked as a team on energy policy in over 20 countries, and are principals of the nonprofit Rocky Mountain Institute (RMI) in Old Snowmass, Colorado, which explores the links between energy, water, agriculture, security, and economic development. Most of RMI's income is from advising utilities on how to produce electricity more efficiently and more economically.

Raw kilowatt-hours, lumps of coal, and barrels of sticky black goo are more messy than useful. Energy is only a means of providing services: comfort, light, mobility, the ability to make steel or bake bread. We don't necessarily want or need more energy of any kind at any price; we just want the amount, type, and source of energy that will do each desired task at least cost.

In the United States about 58% of all delivered energy is needed as heat; 34%, as liquid fuels for vehicles; and 8%, as electricity (for motors, lights, electronics, smelters). Electricity is a premium form of energy. It is able to do difficult kinds of work but is extremely expensive—far too costly to provide economical heat or mobility. Yet over the next 20 years utilities want to build another trillion dollars' worth of plants—causing spiraling rates, bankrupt utilities, and unaffordable energy.

What's the alternative? First, we ought to use energy just as efficiently as is worthwhile. We're starting to do this. Making a dollar of gross national product (GNP) takes a quarter less energy now than it did ten years ago. Since 1979 the United States has gotten seven times as much new energy from savings (conservation and improving efficiency) as from all the new oil and gas wells, power plants, and coal mines opened in the same period. Yet it's worth saving even more. With our current technology we could double the energy efficiency of industrial motors or jet aircraft, triple that of steel mills, quadruple that of household appliances, quintuple that of cars, and improve that of buildings by tenfold to a hundredfold. Such increased "energy productivity" gives us the same services as now just as conveniently and reliably but at less cost to ourselves and the earth.

Electricity, being expensive, is especially worth saving. We are writing this under a new kind of light bulb that gives better light than the old kind, lasts thirteen times as long, uses a quarter as much electricity, and repays its $15 price in a year or two. Better motors and drivetrains in factories could save more electricity than all nuclear plants produce. The best refrigerator we can make uses 1/27th as much electricity as the one in use today. If Americans used all the best technologies now on the market, they could be using less than a quarter as much electricity as they use now. And they could be paying, for each kilowatt-hour saved, less than it would cost just to *run* a coal or nuclear plant to generate that kilowatt-hour. Thus, even if a new nuclear plant cost nothing to build, were perfectly safe, and didn't produce radioactive wastes or bomb materials, it would still save the country money to write it off, never run it, and buy efficiency instead.

Indeed, neither new fossil-fueled nor nuclear power plants would be necessary or economically feasible if we used electricity efficiently. Existing and small hydroelectric plants plus a bit of windpower (or, optionally, solar cells or industrial cogeneration—the simultaneous production and use of electricity and heat) would be enough. A government study showed that even if by the year 2000 the United States' GNP had increased by 66%, by buying the cheapest energy options Americans would use a quarter less energy and electricity than now and nearly 50% less nonrenewable fuel. The net saving: several trillion dollars and about a million jobs.

After efficiency, the next best energy buys are, as a Harvard Business School study found, the appropriate renewable sources. Such "soft technologies" include passive and active solar heating, passive cooling, high-temperature solar heat for industry, converting sustainably grown farm and forestry wastes to liquid fuel for efficient vehicles, present and small hydroelectric power, and windpower. Solar cells, too, will soon be generally cost-effective and join the list. Soft technologies tend to be smaller than huge power plants, so they can provide the cheapest energy where it's needed. Small isn't necessarily beautiful, but it usually saves money by matching the relatively small scale of most energy uses.

Careful studies in 15 countries show that the best soft technologies now available are cheaper than new power plants and could meet essentially all our long-term global energy needs.

Already, efficiency and renewables are sweeping the market, not because we say they should but because millions of people are choosing them as the best buys. The United States since 1979:

- has gotten more new energy from renewable sources than from any or all of the nonrenewables

- has ordered more new electric generating capacity from small hydroelectric plants and windpower than from coal or nuclear plants or both

- is now getting nearly twice as much delivered energy from wood as from nuclear power, which had a 30-year head start and direct subsidies of well over $100 billion.

Wood-burning, solar heat, and the like aren't always done well. People need much better information and quality control to choose and use the cheapest, most effective opportunities. But it is faster to build many small, simple technologies that anybody can use than a few huge, complex projects that take ten years and cost billions of dollars each. And that's what Americans are doing, to the tune of $15 billion worth in 1980 alone.

Every time you buy weatherstripping instead of electricity—because you can get comfort cheaper that way—you're part of the transition. And your part matters. The United States could eliminate oil imports in less than a decade, just by making buildings and cars more efficient. Conversely, each dollar spent on reactors can't be spent on faster, cheaper ways to save oil, and hence it delays energy independence. Power plants also provide fewer jobs per dollar than any other investment. Thus, every big plant built loses the economy, directly and indirectly, about 4000 net jobs, by starving all other sectors for capital.

In the United States 64% of the capital charge for new reactors is subsidized via taxes. The taxpayer also picks up much of the tab for nuclear fuel, decommissioning the worn-out plants, developing and regulating them, exporting them, coping with their hazards, and trying to fend off the nuclear bombs they spread. Despite the enormous government intervention, nuclear power is dying of an incurable attack of market forces throughout the world's market economies (and is in deep trouble even in the centrally planned economies, notably France and the Soviet Union). Wall Street won't pay for more reactors; over a hundred have been canceled; and most of the industry's best people have already left.

Fortunately, the same best energy buys that are vital to a healthy economy are also keys to national security. Centralized, complex, computer-controlled nuclear plants are sitting ducks for terrorists, accidents, or natural disasters. In contrast, a more efficient, diverse, dispersed, renewable energy system could be resilient. Major failures of energy supply simply couldn't happen. Hooking together decentralized electrical sources via the existing power grid, so that they could back one another just as giant power stations do now, would actually make electrical supplies more reliable, move supplies nearer users, and reduce dependence on fragile transmission lines.

People and communities are starting to solve their own energy problems. They've discovered that the problem isn't where to get 80 quadrillion BTUs a year, but how to seal the cracks around their windows. People are finding more to trust in local weatherization programs, community greenhouses, and municipal solar utilities. The energy transition is happening from the bottom up, not from the top down: Washington will be the last to know.

It's not for "the experts" to choose whether you need caulk or electricity. Pick your own best buys. The energy future is *your* choice.

Critical Thinking

1. Summarize the key points made in this viewpoint. Are they well supported? Can you detect any areas where bias may be tainting the authors' views?

2. The Lovins's argue that centralized power is ultimately more vulnerable than decentralized power. Why? Do you agree?

---

come. Coal will probably play a large role in the near term, for it can also be used to make liquid fuels to power our transportation systems and can be converted into a synthetic natural gas for home heating. (See Chaper 12 for a discussion of synthetic fuels from coal.)

The United States has about 30% of the world's proven coal reserves, or about 225 billion metric tons that can be recovered. This could last the country 100 to 200 years even at increased rates of consumption. It is believed that at least 360 billion additional metric tons of coal can be recovered in the United States, giving Americans a supply of coal that could last several hundred years.

Unfortunately, coal is not a clean-burning fuel. In fact, with the exception of oil shale, coal is the dirtiest fossil fuel known. When burned, as we have seen, it produces solid waste (bottom ash) that is buried in landfills. Particulates (fly ash) can escape up the smokestacks. Harmful gases are released into the air; some of them are converted into dangerous acids in the atmosphere. New and cleaner ways to burn coal are being developed, but progress has been slow. New technologies to burn coal and pollution control devices may help lessen the environmental damage, although some problems such as carbon dioxide pollution are inescapable. (Chapter 15 and Chapter Supplement 15-1 discuss the pollution problems associated with coal in more detail.)

Our Energy Future

The twin oil embargoes of the 1970s spurred many energy-dependent countries into action. Most cut back on consumption and sought to become energy independent by increasing their domestic fuel production. But the crisis also began a colossal debate on energy, particularly on which sources should form the foundation of our future. The debate has intensified because of the growing threat of global warming. Global warming is thought to be caused by carbon dioxide, which is released in large amounts from fossil fuel combustion in our homes, factories, and power plants.

Hard Paths and Soft Paths

Participants in this debate often line up in two camps. On one hand are the advocates of large, centralized sources such as the coal and nuclear power plants in operation today. Amory Lovins, an energy consultant and author, coined the term **hard path** to describe this option. The hard path relies primarily on nonrenewable energy resources, at least to meet near-term needs. Advocates of this path generally believe in the **technological fix**, arguing that new technologies or improvements in existing technologies are the answer to our energy problems.

Hard-path strategies generally involve extensive distribution networks to transport energy from sources to the end users and have many advantages, discussed in A. David Rossin's essay in this chapter. The alternate option is the **soft path**. Its advocates support conservation and renewable energy sources: active and passive solar heating, solar photovoltaics (which produce electricity from sunlight), wind, biomass, and hydropower. The soft path calls on both nontechnological solutions, such as individual actions to save energy, and technological solutions, such as solar panels and wind-powered electric generators. Advocates of this path generally promote decentralized energy, controlled by the consumers rather than the large corporations. They see their path as a way of ensuring a sustainable future in a world of finite fossil fuel supplies. Additional advantages are outlined in Amory and L. Hunter Lovins's essay in this chapter.

The debate between hard-path and soft-path advocates has had the important side effect of showing us the impressive number of energy options from which we can choose. At the same time, however, the sheer number of options and our lack of experience with them creates a paralysis of sorts. Poised between two proverbial roads, each offering benefits and risks, few of our governments have clear insights into which options are worth pursuing. The remainder of this chapter presents seven guidelines that can help nations and individuals choose the energy future that makes the most sense in the long run. These guidelines can help you think more critically about energy options.

Guidelines for Wise Decisions

Ensuring Positive Net Energy Production

The first rule of energy production is, quite simply, that the energy we get out of an energy resource must not be exceeded by the energy we invest in its exploration, extraction, transportation, and so on. In other words, energy input should not exceed energy output. This is sound energy policy as well as economic policy.

The energy we acquire from energy resources, such as coal, wind, and oil shale, minus the energy that is invested in producing the fuel is called the **net energy**. Net energy may be positive, meaning we get more energy out of the system than we put in, or it may be negative, meaning that we put in more energy than we get out. The higher the net energy yield, obviously, the better the energy source.

One way of improving net energy efficiency is to reduce transportation. Transporting energy in any form is expensive and also requires energy. The longer the distance, the more energy it takes to supply end users, and the greater the cost. Energy sources close to the end user, such as local power plants or even solar panels on individual homes and commercial buildings, can often be the

Table 11-2 Energy Quality of Different Forms of Energy	
Quality of Energy	**Form of Energy**
Very high	Electricity, nuclear fission, nuclear fusion*
High	Natural gas, synthetic natural gas (from coal gasification), gasoline, petroleum, liquified natural gas, coal, synthetic oil (from coal liquefaction), sunlight
Moderate	Geothermal, hydropower, biomass (wood, crop residues, manure, burnable municipal refuse), oil shale, tar sands

*Workable nuclear fusion reactors do not yet exist. Even if they were technically operable by the end of the first quarter of the twenty-first century, they would probably remain economically unfeasible.

most economical strategy for supplying energy where it is needed. Indeed, as you will see in the next chapter, Third World nations are now tapping solar energy and wind energy to supply electricity to rural villages isolated from centralized power plants.

The idea of net energy, however elementary, is frequently overlooked by advocates on both sides of the debate. Chapter 12 looks critically at net energy yield when considering our energy options.

Energy Matching

You wouldn't cut butter with a chain saw. Nor would you try to melt steel with sunlight. But for all intents and purposes that's just what people have been doing for years. We have mismatched energy sources with needs—by heating homes with electricity, for instance. Wise energy use requires a careful matching of energy sources with needs. As you well know, energy comes in a variety of forms. Each of these forms has a different **energy quality**, a measure of the amount of available work you can get from it (Table 11-2). Oil and natural gas, for instance, are highly concentrated and are said to be high-quality energy resources. When burned, they produce large amounts of heat. Sunlight is a lower-quality form. Streaming through the south-facing windows of a house, it produces a lower-level heat, more appropriate for home heating.

Using energy wisely means employing low-quality energy sources for tasks that call for it and high-quality energy for appropriate tasks. This approach reduces waste, conserves energy, and saves money. Energy matching should be applied to the planning of future needs, although many countries forget to do it. The rule is: determine what your energy needs will be, and then find sources that match those needs. If a community needs energy for home heating, don't build electric power plants, which are an inefficient source of home heat. Instead, develop energy conservation, passive solar energy, and other low-quality energy sources that match the end needs more precisely.

Converting Energy Efficiently

The third rule of wise energy use is to convert energy into work as efficiently as possible. **Conversion efficiency** is often called **first-law energy efficiency**. The first law of thermodynamics states that energy is neither created nor destroyed but merely transformed from one form to another (see Chapter 3). Thus, when natural gas is burned, it is converted into light and heat. To measure the conversion efficiency, simply divide the total amount of work you get out of a system by the total energy input. Multiplying this result by 100 converts it into a percentage. For example, suppose we burned 10 BTUs of natural gas to make a motor work and got 2 BTUs of actual work out of this fuel. The conversion efficiency would be 20%. As a practical example, conventional incandescent light bulbs have a conversion efficiency of about 5%. In other words only 5% of the electricity it takes to run the bulb is converted into light, and the rest is converted into heat. Fluorescent light bulbs, on the other hand, are about 22% efficient. First-law energy efficiencies can be determined for all of our machines. By increasing their efficiency, we can cut energy use and reduce environmental impacts.

Another way of looking at efficiency is through the second law of thermodynamics. The second law tells us that when energy is converted from one form to another, it is degraded; a certain amount of heat loss is inevitable (Chapter 3). First-law efficiency calculations do not take this law into account. To determine energy efficiency taking into account the unavoidable loss, we can make a **second-law efficiency calculation**. To do this, divide the minimum amount of work needed to perform a task by the actual amount used. This is a more precise measurement of energy efficiency. Reconsider the example given above. Here 10 BTUs are used to get 2 BTUs of work. Because of the second law, however, some energy is lost as heat; that's inescapable. Let's say that in this system we lose 5 BTUs. That means that from the remaining 5 BTUs we get only 2 BTUs of work. The second-law efficiency is only 40% by this reasoning when, in theory, the motor could have a second-law efficiency of 100%.

Reducing Pollution and Ensuring Safety

In many respects the opposing camps agree on the fourth rule of wise energy use: to produce energy safely and cleanly. Proponents of nuclear power, for instance, argue that this source of electricity is much cleaner than coal-fired power plants, which spew out a variety of harmful gases. They also note that uranium mining has caused fewer deaths than underground coal mining. Soft-path proponents support clean, low-risk energy resources but think that nuclear energy doesn't fit the bill. Accurately determining the risk becomes a major challenge (see Chapter 14).

Ensuring Abundance and Renewability

A careful look at the energy debate shows that hard- and soft-path advocates also often agree on the fifth rule of wise energy use: a desirable resource is one that is abundant and, ultimately, renewable. Developing renewable, long-term resources maximizes economic returns on money invested in research, development, and commercialization.

At the heart of the soft-path advocacy is a belief that renewable energy resources are the wave of the future. Hard-path advocates see fusion energy, which uses abundant hydrogen from the ocean, as the answer (Chapter 12).

Interestingly, the basis of energy production in the two pathways is the same. The sun and fusion reactors both produce energy from the fusion of atomic nuclei. The difference lies in the fact that hard-path advocates want to make their own miniature suns here on earth and try to contain the enormous amounts of heat and radiation they would give off. Solar advocates prefer to let the sun contain the fusion reactions and to capture the energy that beams down to earth.

Affordability

Energy proponents also agree on the need for an affordable energy supply. As you will see in the next chapter, conservation "provides" energy cheaper than any other source. Many experts agree that it must be considered the first line of attack in the struggle to provide a continuous inexpensive source of energy for all human societies. Comparing the costs of other energy sources, however, must be done carefully. Full economic costs, including all externalities, must be taken into account. In addition, government subsidies must be factored into the equation. Currently, coal, oil, and nuclear energy receive enormous subsidies from the federal government. They come in the form of research money, tax breaks, and inexpensive mineral rights on federal lands. In contrast, many renewable resources, such as solar and wind energy, have received little federal support over the years. As a result, they cost more and many appear less attractive. Critical thinking suggests that subsidies distort our economic decisions. Many think that our energy sources must be allowed to compete fairly and are asking the federal government to remove federal subsidies for many hard-path technologies now in use or to subsidize renewables equally. Creating a level playing field would help us discern the true costs of energy and could help us make better decisions for meeting future energy needs. (For more on subsidies see Amory and Hunter Lovins's essay in the Point/Counterpoint in this chapter.)

Abandoning the Old

A positive net energy balance (the bigger the better), energy matching, converting energy efficiently, reducing pollution, and abundance and renewability are the principles of an intelligent energy policy. The choice of our future energy resources rests on a fair comparison of our options. Our choices should be made with these criteria in mind.

To create an energy system that is economically and environmentally sound is an enormously complicated task, made more difficult by current controversies and the failure of many proponents to subject their "pet energy sources" to critical analysis. We must often abandon old allegiances and base our decisions on what is economically sensible, environmentally benign, and socially acceptable. Our goal is to design an energy system that makes sense given the world's limited resources and the need for a clean environment. Anything short of this is an invitation to disaster.

The trouble with our times is that the future is not what it used to be.

PAUL VALÉRY

Summary

The rising cost of energy is bound to become a major economic problem in the coming decades both in the United States and abroad. Energy is, in many ways, the lifeblood of modern industrial society. Without a steady supply modern industrial nations would face economic collapse.

Important as it is, energy was taken for granted for many years. The days of energy nonchalance came to an abrupt halt in the 1970s as a result of two oil embargoes. Although devastating to the economies of many nations, the embargoes heightened awareness of our energy dependency. Many nations took steps to conserve energy. Because of cutbacks in use and increased production by non-OPEC nations, the world seems to be awash in oil. The present oil "glut" is only temporary, however, and diverts public attention from an important problem: oil is fast on the decline.

The history of energy use has been one of shifting dependency, a trend likely to continue as conventional energy resources run out. Today, the United States depends primarily on three nonrenewable sources: oil, natural gas, and coal. The United States is the leading energy consumer in the world. With only 6% of the population, it uses about 30% of the world's energy. At least half of the energy Americans consume is wasted.

Energy does not come cheaply. In addition to the economic costs, society pays a price in the form of the destruction of health and the environment, effects that differ with each energy source. Because these costs are not part of production expenses, they are called **external costs**. Understanding the external costs requires us to understand **energy systems**, the long chains of events that span from energy exploration to extraction to consumption and, finally, to disposal.

One of the most perplexing problems facing society is whether there will be enough energy to go around in the future. A careful look at world oil supply and demand shows that the end is near for petroleum. Conservation and substitutes for this precious liquid fuel are badly needed. The prospects for natural gas are better, and the prospects for coal are much better. Unfortunately, coal is the dirtiest of the three.

Energy shortages stimulated a continuing debate on which energy resources should form the foundation of our future. Participants in this debate often line up in two camps. **Hard-path** advocates support large, centralized energy sources such as coal and nuclear power plants. **Soft-path** advocates support conservation and decentralized, renewable resources. This debate has exposed the impressive number of energy possibilities. The sheer number of options and our lack of experience with many of them, however, has created paralysis.

To develop a sensible energy future requires us to consider some basic guidelines: ensuring a positive net energy yield, matching energy quality with end use, improving the efficiency of conversion, using sources that do minimal environmental damage, developing inexhaustible sources, and finding sources that are affordable. These guidelines can help us achieve a sustainable energy system that minimally disrupts the environment.

Discussion Questions

1. What were the results of the oil embargoes in the 1970s? How did they affect your life?

2. Debate the statement "There is enough oil to go around for at least 70 more years."

3. Describe the term *energy system*. List all of the steps you can think of needed to produce gasoline from oil, starting with exploration, and discuss some of the impacts on the environment associated with each step.

4. Using your critical thinking skills, debate the statement "Coal is our energy savior. We must come to rely very heavily on it to achieve a stable future."

5. Describe the differences between hard energy paths and soft energy paths.

6. On what points do hard- and soft-path advocates agree?

7. You are studying future energy demands for your state. List and discuss the factors that affect how much energy you'll be using in the year 2000.

8. Using your critical thinking skills, debate the pros and cons of oil exploration in the Arctic National Wildlife Refuge.

Suggested Readings

Brown, L. R., Chandler, W. U., Durning, A., Flavin, C., Heise, L., Jacobson, J., Postel, S., Shea, C. P., Starke, L., and Wolfe, E. C. (1988). *State of the World 1988*. New York: Norton. See Chapters 2, 3, and 4 for information on energy alternatives.

Chandler, W. U. (1985). *Energy Productivity: Key to Environmental Protection and Economic Progress*. Worldwatch Paper 63. Washington, DC: Worldwatch Institute. Excellent look at energy costs and the role of conservation in stretching energy reserves.

Council on Environmental Quality. (1980). *The Global 2000 Report to the President*. New York: Penguin. Superb analysis of energy demands and supply.

Flavin, C. (1985). *World Oil: Coping with the Dangers of Success*. Worldwatch Paper 66. Washington, DC: Worldwatch Institute. A candid look at oil supplies.

Flavin, C. (1990). Slowing Global Warming. In *State of the World 1990*. L. Starke, ed. New York: Norton. Discusses strategies to use energy more efficiently.

Flavin, C. and Durning, A. B. (1988). *Building on Success: The Age of Energy Efficiency*. Worldwatch Paper 82. Washington, DC: Worldwatch Institute. Interesting analysis of the potential for energy conservation.

Gates, D. M. (1985). *Energy and Ecology*. Sunerland, MA: Sinauer. Excellent coverage of energy.

Gibbons, J. H., Blair, P. D., and Gwin, H. L. (1989). Strategies for Energy Use. *Scientific American* 261 (3): 136–143. Discusses possible role of several energy sources and conservation in meeting future energy demands without worsening global warming.

Hirsch, R. L. (1987). Impending United States Energy Crisis. *Science* 235 (March 20): 1467–1472. Important analysis of energy costs in the coming years.

Priest, J. (1984). *Energy: Principles, Problems, Alternatives*. Reading, MA: Addison-Wesley. Excellent introduction to energy.

World Resources Institute and the International Institute for Environment and Development. (1988). *World Resources 1988–89*. Excellent source of information for global energy supply and demand.

Future Energy: Making the Best Choices

I cannot say whether things will get better if we change; what I can say is they must change if they are to get better.

G. C. LICHTENBERG

You will probably see the end of conventional oil supplies in your lifetime. Your children may see the end of conventional natural gas. Over time, coal, possibly nuclear energy, and a variety of renewable energy resources could dominate the energy market.

The shift in energy dependency has already begun. France and the Soviet Union have opted to produce electricity with nuclear energy on a large scale. Brazil has chosen ethanol produced from sugar cane to power its fleet of trucks and cars. As world oil reserves decline and population and demand increase, more and more countries will make the shift away from oil to a variety of renewable and nonrenewable energy resources. To avoid major economic turmoil, experts believe, we must start developing replacement fuels immediately. But which of the many energy options should we turn to?

Answering this question requires an objective look at all possible sources—their supplies and, especially, their environmental impacts. This chapter broadly outlines the world's energy needs, establishing a "shopping list" to help us plan our energy future intelligently. It also looks at our many options, their benefits, and their environmental consequences. Finally, it suggests ways to achieve a sustainable system that will satisfy the energy demands of future generations.

Establishing a Shopping List

You should probably never go grocery shopping when you are famished; you will inevitably come back with a

Figure 12-1. Washington's Public Power Supply System's abandoned nuclear power plant.

trunkload of goodies that you don't really need and a bill that rivals the national debt. To avoid such calamities, you should prepare a shopping list so that you buy only what you need.

In many ways the United States and other major countries have been shopping for new energy without a list, investing blindly in projects that really were not needed. The exorbitant cost of this activity is now coming back to haunt us. Nowhere are the signs more evident than in the nuclear power industry. Spurred by projections of enormous electrical demand and the two oil crises in the 1970s, American utilities went on a rampage of sorts, planning and building nuclear power plants at a frenetic pace. But companies quickly found out that the demand projections were grossly inflated; we really didn't need the additional electricity. As a result, in the last decade American utilities have canceled plans to build over 100 nuclear reactors.

Canceling plans to build power plants is one thing, but abandoning plants still under construction is financially devastating. For example, the Washington Public Power Supply System (WPPS) (nicknamed WooPPS!) had hoped to build 7 to 15 nuclear power plants. Construction had started on five of them when officials realized that they had misprojected energy demand. They halted construction on four plants (Figure 12-1). Grossly exaggerating

its energy demands has cost the system plenty. WPPS defaulted on $2.25 billion worth of bonds, money lent to finance two of the plants that would have been paid back with interest from earnings from the generation of electricity. Money lent to build the other two was provided by bonds guaranteed by rate payers, who are now paying back the loans with interest. Washington is not alone. An Indiana company scrapped a plant that was 97% complete and had cost $2.5 billion, enough money to make you and 2499 friends millionaires.

Clearly, it is time to reassess our energy demands and establish some priorities—to draw up a shopping list for the near term and long term. Chapter 11 discussed the future of oil, natural gas, and coal, our key energy resources today. Several important trends were observed: (1) oil supplies are fast on the decline and in need of a substitute; (2) globally, natural gas supplies are adequate for the near term (the next 50 years), but domestic shortages and rising prices suggest the need for a replacement; and (3) coal is our most abundant fossil fuel, with supplies that could last hundreds of years.

Energy reserves are not our only concern when looking at a future energy supply system for the US, Canada, and other countries. The last chapter introduced a number of additional concerns—guidelines to help develop your critical thinking skills. One of the most important factors is the environmental damage caused by energy sources. Perhaps the most important concern is the contribution an energy source will make to global warming. In truth, all the energy in the world will do us no good if we disrupt our climate, creating widespread drought and starvation (Chapter 15).

Energy efficiency is also an important consideration. Conservation, discussed in more detail at the end of this chapter, can help us stretch existing fuel supplies, buying us time to develop alternatives and bring them on line. Conservation can also help us reduce carbon dioxide pollution, in large part responsible for global warming. Conservation is generally the cheapest and quickest "source" of energy available to humankind, and its potential has hardly begun to be exploited. Tapping this often overlooked source of energy must be our first line of attack in solving the world's energy demands.

Keeping in mind the factors mentioned above, our energy shopping list might look like this:

Short-term goals (within 20 years):
Item 1   Improve the energy efficiency of all machines, homes, appliances, buildings, factories, and so on.
Item 2   Find a replacement for oil (primarily used to make transportation fuel and plastics) that does not add to global carbon dioxide levels. Several replacements are available, including oil shale, oil from coal, hydrogen,

and biofuels (ethanol), but only the last two produce no increase in carbon dioxide.

Intermediate-term goal (within 50 years):

Item 3 Find a replacement for natural gas (which is primarily used for heating and industrial processes). Passive solar, synthetic gas from coal, and biofuels are possible replacements.

Long-term goal (within 100 to 200 years):

Item 4 Find a replacement for coal (which is primarily used to generate electricity). Solar voltaics, breeder reactors, fusion reactors, and hydropower are possible substitutes.

Nonrenewable Energy Resources

This section discusses future energy sources that are nonrenewable, that is, ones we can exploit only once and generally for a limited time, although that time may be considerable in the case of coal. Nonetheless, they may serve as important bridges to the sustainable energy system that must be in place a century from now. Oil is not covered here because of its limited potential.

Nuclear Fission

The nuclear reactors in use today in the United States and most of the rest of the developed world are fueled by naturally occurring uranium-235. This is a form of uranium (an "isotope") whose nuclei split, or **fission**, easily when they're struck by neutrons, giving off enormous amounts of energy (Figure 12-2). In fact, 1 kilogram (2.2 pounds) of this material, completely fissioned, could yield as much energy as 2000 metric tons of coal!

Fission reactors basically provide an environment in which uranium-235 nuclei can be bombarded with neutrons. Uranium-235 (U-235) is housed in **fuel rods** within the **reactor core** (Figure 12-3). The nuclei of U-235 naturally emit neutrons. These bombard other nuclei and cause them to split. Heat produced during fission is transferred to water that bathes the reactor core. As shown in Figure 12-3, the heated water, which circulates in a closed vessel, heats up water in another closed system. This water is converted to steam which drives a turbine, generating electricity. Most nuclear plants are cooled by water and are called **light water reactors** (LWRs). Other reactors use coolants such as liquid sodium but operate on the same principle.

The splitting of a U-235 nucleus produces two smaller nuclei, called **daughter nuclei**, or **fission fragments** (Figure 12-2). Over 400 different fission fragments can form during uranium fission; many of them are radioactive. (Radioactivity is described in Chapter Supplement 12-1.) When a U-235 nucleus fissions, it also releases neutrons,

which may strike other nuclei in the fuel rods and initiate a **chain reaction**. However, the chain reaction is controlled to regulate energy production and prevent the reaction from getting out of control. Runaway chain reactions could produce enough heat to melt the reactor core in a **core meltdown**. An atomic explosion in such a case would be unlikely because the fuel is not sufficiently concentrated.

The fission reaction is kept in check by water and the control rods. Bathing the core and carrying off energy, water absorbs some of the neutrons, reducing the rate of fission. The **control rods**, made of neutron-absorbing materials (boron), lie between the fuel rods. Raising or lowering the control rods regulates the rate of the fission reaction. When the control rods are completely lowered, the reactor is shut off.

The entire assemblage of fuel and control rods, bathed in coolant, is housed in a 20-centimeter- (8-inch-) thick steel container, called the **reactor vessel**. The reactor vessel is surrounded by a huge shield, and the entire unit is contained in a 1.2-meter- (4-foot-) thick cement shell, the **reactor containment building**.

Nuclear fuels, like fossil fuels, pass through a complex chain, from uranium extraction to waste disposal. At each stage, radioactive materials can escape, either by accident or during normal operation, creating environmental and health impacts. A full assessment of the entire energy system is necessary to evaluate the risks of nuclear power. Before doing so, let us look at the benefits of this source of energy.

Nuclear Power: Pros and Cons When nuclear power was first developed, proponents bragged that its energy would be so cheap that it wouldn't pay to meter houses. That dream has failed to materialize. In the United States nuclear power costs about 10 to 12 cents per kilowatt-hour, about twice the cost of coal. Building a nuclear power plant is two to six times more expensive than an equivalent coal-fired power plant. Nevertheless, nuclear power has become an important source of energy in the United States, providing approximately 95,000 megawatts of electricity, enough for 95 million people. It fits well into an electrical grid system that provides electricity to large numbers of people. Perhaps the most convincing argument for using nuclear power, as opposed to coal and oil, is that it produces very little air pollution. Studies show that the release of radioactive materials into the atmosphere from nuclear power plants is insignificant under normal operating conditions. In fact, one study published in *Science* found that coal-fired power plants released more radioactivity than nuclear-powered plants. Moreover, nuclear plants do not produce toxic gases such as sulfur dioxide and nitrogen dioxide, which are converted into acids in the air. Carried to the earth in rain, snow, and on particulates, these acids can cause consid-

Figure 12-2. (a) In a fission reaction a uranium-235 nucleus struck by a neutron is split into two smaller nuclei. Neutrons and enormous amounts of energy are also released. (b) A chain reaction is brought on by placing fissile uranium-235 in a nuclear reactor. Neutrons liberated during the fission of one nucleus stimulate fission in neighboring nuclei, which in turn release more neutrons. Thus, the chain reaction can be sustained.

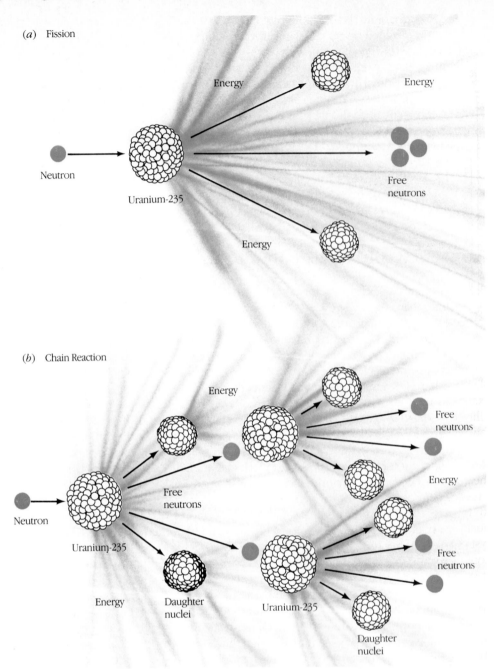

(a) Fission

(b) Chain Reaction

erable damage to the environment. (For more on acid precipitation see Chapter Supplement 15-1.) Another advantage of nuclear power is that it requires less strip mining than coal because the fuel is a much more concentrated form of energy. Reductions in mining mean less land disturbance and fewer impacts on groundwater, wildlife habitat, and so on. (See Chapter 11 for a description of some of the impacts of strip-mining coal.) The cost of transporting nuclear fuel is lower than that for an equivalent amount of coal. By using more nuclear power in the future we would also free up coal for making synthetic liquid and gaseous fuels, which can be used to power our transportation system and homes, respectively.

(Synthetic fuels are discussed later in this chapter.)

Despite its many advantages nuclear power has some substantial drawbacks, worthy of careful consideration. The most important are (1) waste disposal, (2) possible contamination of the environment with long-lasting radioactive materials from accidents at plants and during transportation, (3) thermal pollution from power plants, (4) health effects of low-level radiation, (5) limited supplies of uranium ore, (6) low social acceptability, (7) high construction costs, (8) a lack of support from insurance companies and the financial community, (9) questionable reactor safety, (10) vulnerability to sabotage, (11) the proliferation of nuclear weaponry from high-level reactor

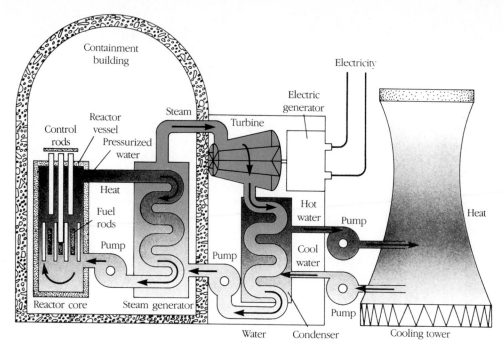

Figure 12-3. In a nuclear power plant, nuclear fission reactions in the reactor core heat up water to generate steam. The steam runs a turbine that generates electricity as in conventional power plants. Note that the water surrounding the reactor core circulates in a closed system, through which it heats up water in the steam generator. This double water heating system is used to prevent the possible escape of radioactivity in the steam.

wastes, and (12) questions about what to do with nuclear plants after their useful life of 20 to 25 years. This section examines reactor safety, waste disposal, social acceptability and cost, and the proliferation of nuclear weapons, four of the major areas of concern.

Reactor Safety Uncertainties regarding safety add to the list of problems with nuclear power. An accident at a nuclear power plant at Three Mile Island, Pennsylvania, in 1979 alerted Americans to the dangers inherent in nuclear power. A malfunctioning valve in the cooling system triggered a series of events that led to the worst commercial reactor accident in US history. Radioactive steam poured into the containment building. Pipes in the system burst, releasing more radioactive water, which spilled onto the floors of two buildings. Some radiation escaped into the atmosphere, and some was dumped into the Susquehanna River. The accident then took a turn for the worse. Hydrogen gas began to build up inside .the reactor vessel, threatening to expose the core and cause a complete meltdown. The gas bubble was slowly eliminated, but thousands of area residents had to be evacuated. Photographs of the core showed that a partial meltdown had occurred.

The accident at Three Mile Island had many long-term effects. It cost the utility (and its customers) many millions of dollars to replace the electricity the plant would have generated. Even more money (over $1 billion) was needed for the cleanup. By 1989 officials had not decided what to do with the radioactive water and sludge in the containment building. Bacteria proliferated in the contaminated water and might pose a significant health hazard.

The accident at Three Mile Island severely damaged the prospects of the nuclear power industry in the United States. However, nuclear advocates applauded the manner in which the accident had been handled and pointed to their success in preventing a major catastrophe as proof of the safety of nuclear reactors. Utility authorities said the accident might cause a few cancers. John Gofman and Arthur Tamplin, radiation health experts, contended that the exposure to low-level radiation that residents received for 100 hours or longer would cause at least 300, and possibly as many as 900, fatal cases of cancer or leukemia. (For more on the health effects of low-level radiation, see Chapter Supplement 12-1.)

A 1975 study on reactor safety, called the Rasmussen Report for its principal author, Norman Rasmussen of the Massachusetts Institute of Technology, showed that the probability of a major accident at a nuclear power plant was not more than one in 10,000 reactor-years. (A reactor-year is a nuclear reactor operating for one year; for example, 10 reactors operating for 20 years are 200 reactor-years.)

If 5000 reactors were operating worldwide, as some advocates propose for the year 2050, we could expect a major core meltdown every other year! *Each accident* might cause between 825 and 13,000 immediate deaths, depending on the location of the plant. In addition, 7500 to 180,000 cancer deaths would follow in the years after the accident. Radiation sickness would afflict 12,000 to 198,000 people, and 5,000 to 170,000 genetic defects would occur in infants. Property damage could range from $2.8 billion to $28 billion.

In January 1989, 429 nuclear reactors were operating throughout the world, and 117 were under construction

The Nuclear Disaster at Chernobyl

In April 1986 the world stood in horror as the news of a major nuclear accident at the USSR's Chernobyl nuclear power plant unfolded. Announcements of radiation spreading throughout Europe, falling on farms and cities, frightened Europeans. In the days that followed the first signs of the accident, pandemonium broke loose. Early news reports indicated that 2000 Soviets had died in the accident, but as time went by it was clear that the immediate death toll would be much lower. The property damage caused by the explosion would, however, turn out to be far greater than many people had anticipated. The Soviet and European lives lost to cancer over the years, while still in question, may make this the worst disaster in the history of industrial society.

What happened at Chernobyl? A report released by the Soviet government in August 1986 says that the plant operators were running some tests on the reactor. They reduced power output to test the turbines. To approximate an emergency, however, the men decided to deactivate several safety systems, a direct violation of plant regulations. During the test, the cooling water flowing through the reactor core fell rapidly. Without sufficient coolant the 200 tons of uranium housed in the reactor's fuel rods quickly heated up. The reactor temperature soared as high as 2800° C (5000° F), twice the temperature required to melt steel. An enormous steam explosion blew the roof off the building. Flames from 1700 tons of burning graphite, a neutron-absorbing agent in the core, shot 30 meters (100 feet) into the air. While Soviet fire fighters risked their lives to contain the disaster by spraying water down on the reactor core from the roofs of nearby buildings, the uranium fuel melted, spewing highly radioactive isotopes into the atmosphere. Swept upward by the heat, these materials circulated far and wide.

On the advice of Swedish and West German nuclear experts Soviet helicopters began to drop sand, lead, and boron on the molten mass of graphite and uranium. Officials announced that the core had melted down but had not melted through the thick cement slab below it. In the days that followed the explosion, workers tunneled under the molten core to install a cooling system. Cement was poured around the molten fuel to keep radioactivity from contaminating groundwater.

The plant was eventually entombed in concrete, where it will sit for several hundred years at least while the radiation levels dissipate. To prevent radioactive soil particles deposited near the plant from being blown away, the soil around the plant was sprayed with a special liquid plastic that hardens on contact.

The Chernobyl accident was made worse by the Soviet government's delays in reporting the accident and in evacuating people. Thirty-six hours after the government admitted that an accident had occurred, officials began to bus residents who lived near the power plant to distant communities. All told, 116,000 people living within a 30-kilometer radius of the plant were relocated. Many of them will never be able to return to their homes.

Besides losing their homes and their personal possessions, tens of thousands of Soviets may have been exposed to high levels of radiation. Estimates of human exposure near Chernobyl indicate that whole-body radiation for persons in the immediate area of the plant ranged from 20 to 100 rems (see Chapter Supplement 12-1). This is 4 to 20 times higher than the allowable exposure for US nuclear workers. Exposures below 100 rems cause nausea and vomiting and also increase the likelihood of cancer. Although no one knows for certain, it is likely that about one of every ten people, or about 15,000 people exposed to radiation from Chernobyl in the 30-kilometer radius, will succumb to cancer.

The accident at Chernobyl caused a number of immediate deaths from radiation poisoning. All told, 31 people died within the first four months. Workers who were saved by heroic procedures are likely candidates for cancer.

One hundred thirty kilometers (eighty miles) south of Chernobyl lies the city of Kiev, with a population of 2.4 million. Early reports indicated that Kiev had been unaffected by the accident, since prevailing winds had swept the radioactive cloud north and west. Unfortunately, the winds later shifted, sending radioactivity over the city. Residents were told to wash frequently and to keep their windows closed. Water trucks washed down city streets, and residents were warned not to eat lettuce. Soviet officials closed schools two weeks earlier than normal so that Kiev's 250,000 students could be evacuated. Confirming that Kiev had also been pelted with radiation, US technicians found that 14 American tourists who had visited the city two days after the accident had absorbed 1.5 rems, or about the equivalent of 50 chest X-rays. Residents may have received significantly higher doses. A 5-rem exposure would cause an additional 14 cases of cancer per 100,000 people, or about 335 additional cases of cancer in the total population.

In July of 1989, the *New York Times* reported that 100,000 additional Soviet citizens may have to be evacuated from nearby areas as many as 330 kilometers (200 miles) from Chernobyl. Scientists in the neighboring republic of Byelorussia have found dangerous levels of radiation in the soil in that region three years after the explosion. They are concerned about the long-term health of the region's people and are urging a five-year evacuation. Many fear high rates of cancer if the people are not removed.

Radiation also spread throughout Europe. One study shows that at least 13,000 people will die from cancer caused by the accident in the next 50 years. Radiation was greatly diluted by the time it hit the United States, and estimates show that only

Damaged Chernobyl
nuclear reactor
soon after the
1986 explosion.

10 to 20 Americans will die from cancer caused by the accident in the next 50 years. That's not a lot, unless you're one of the victims.

Besides exposing large numbers of people to potentially harmful radiation, the Chernobyl accident threatened crops, farmland, and livestock in the Ukraine, the Soviet Union's "breadbasket." By some estimates, up to 150 square kilometers (60 square miles) of land is so contaminated that it cannot be farmed for decades, unless the Soviets remove the contaminated topsoil. Agriculture outside of the Soviet Union also suffered from the ill effects of radiation. Soon after the acci-

dent, for example, Italian officials turned back 32 freight cars loaded with cattle, sheep, and horses from neighboring Austria and Poland because of abnormally high levels of radiation. Radioactive fallout from the accident also settled on Lapland, an expanse of land encompassing northern Sweden, Norway, Finland, and the northwestern part of the USSR. Lapland is occupied by a seminomadic people who raise reindeer for food. The reindeer feed on lichen, which was contaminated by radiation after the accident. So heavily contaminated was the meat that Laplanders were forced to round up and slaughter their herds.

Case Study 12-1 (continued)

To critics of nuclear power, Chernobyl was a painful reminder of the hazards of nuclear energy and the economic costs of pursuing this controversial option. It showed the potential for widespread health effects and the difficulties encountered in mass evacuations. It stands as a symbol of the Faustian bargain we have made with the atom to power modern society. Supporters of nuclear energy say that the accident was an anomaly. Adequate shielding would have minimized the widespread contamination. Regardless, public opinion of nuclear power has nosedived here and abroad. Increasingly,

many people are worried that nuclear power is a bargain with the devil. It may provide us energy, but the cost is too high. Voters in California showed their distrust of nuclear power in 1989 in a referendum calling for the closure of the Rancho Seco nuclear power plant. The first public vote of no confidence comes at a time when Americans are scrambling for alternatives to fossil fuels in order to slow down global warming. The vote of no confidence was especially startling since many Americans were beginning to reconsider nuclear power.

or soon to be built. If the Rasmussen report is correct, when all 546 reactors are functional, we can expect a major meltdown every 18 years.

The Rasmussen study was heavily criticized for failing to include such possibilities as sabotage and human error. The report was discredited later by the Nuclear Regulatory Commission, which argued that it had underestimated the real risk.

Perhaps the weakest point in the Rasmussen Report is its failure to consider the role of human error. Many experts believe that the accident at Three Mile Island was worsened by operator confusion. A small problem subsequently turned into a disaster. Many people argue that misjudgment and performance errors among personnel could negate technological improvements designed to make plants safer. Human errors and oversight can also occur during construction.

Critics are also concerned with unforeseen technical difficulties. A hydrogen bubble at the Three Mile Island plant, for example, took the experts by surprise. Fortunately, the utility was able to release the pressure before a catastrophe occurred. Critics wonder what would have happened if the utility had been unsuccessful.

Numerous backup systems in nuclear power plants are designed to prevent a core meltdown and the release of radiation. The accident at the Three Mile Island, near Harrisburg, Pennsylvania, showed that plants are not invulnerable. The April 1986 accident at Chernobyl in the Soviet Union reinforced deep concerns about the dangers of nuclear energy and awakened the world to the widespread environmental contamination that can result from a meltdown of a reactor without proper containment (see Case Study 12-1).

Clouding the issue of reactor safety is the possibility of terrorism. In 1975 two French reactors were bombed. Nuclear power plants could become targets of similar attacks. Damage to the cooling system could result in a

meltdown, with radiation leakage. Most plants are easily accessible and hence vulnerable to attack. Protection from ground and air assaults may be impossible. Even though security has been improved at many plants, the threat of well-planned terrorist actions cannot be ignored.

All told, the question of reactor safety remains open. Continued development of nuclear power in the United States and abroad will give us a chance to find answers, but the cost of this experiment, many critics say, could be astronomical.

Waste Disposal   One of the most notable sources of contamination from the nuclear energy system is the uranium mill, where ore is crushed and the uranium is extracted to make nuclear fuel (the **enrichment** process). In the United States **mill tailings** were indiscriminately dumped near mills and along rivers until the late 1970s (Figure 12-4). Some were even used for fill in construction of homes and buildings. In Grand Junction, Colorado, for example, tailings were spread over land before 4000 homes were built. Residents in these homes are exposed to radiation equivalent to ten chest X-rays per week. The leukemia rate in Grand Junction is twice that of the rest of Colorado. Workers are now busy removing the sandy radioactive waste, but will not be done until 1994. Removal has already cost over $23 million.

In the United States approximately 125 million metric tons of tailings have been haphazardly discarded on or near mill sites. Some of this waste has been buried and then covered with topsoil and vegetation. Nearly 11 million tons were dumped along the banks of the Colorado River and its tributaries. Stricter regulations should help prevent a recurrence of this problem.

Nuclear power plants produce an assortment of high- to low-level wastes. In the United States nearly 4 million cubic meters of low-level radioactive wastes from reactors and other facilities (hospitals, laboratories) have been buried in shallow excavations. Low-level wastes are hazardous for a relatively short time, usually no more than 300 years. In contrast, high-level wastes can be dangerous for *tens of thousands of years*. The common way of measuring the lifespan of a radioactive substance is by its **half-life**, the time it takes for half of the material to decay. The half-life of plutonium-239 (Pu-239), a highly radioactive waste product of nuclear reactors, is 24,000 years. It generally takes about eight half-lives for a material to be reduced to .1% of its original mass, at which point it is often considered safe. For Pu-239, this is about 200,000 years!

Because high-level waste disposal is such a tricky political issue, few US politicians have been willing to tackle it. Only recently has Congress passed a law that calls on the Department of Energy to find a suitable site to build a disposal facility for high-level waste. (See Chapter 18 for a discussion of this law.) Meanwhile, high-level wastes are building up at nuclear power plants and weapons facilities throughout the United States. In some locations wastes have begun to leak. At Richland, Washington, where fissile ores are enriched for atomic bombs, high-level wastes have been stored in large steel tanks since the plant opened in World War II. Since it began operating the plant has produced several hundred million liters of highly radioactive liquid waste. Gradually, the tanks deteriorated. Since 1958 at least 2 million liters of waste have leaked out of tanks into the soil and, possibly, the groundwater. Approximately 150 million liters of waste is still in storage tanks that could leak. Clearly, something needs to be done about these wastes and the 18,000 metric tons of waste currently held at nuclear power plants. (For more on radioactive waste and what is being done about it, see Chapter 18 and Case Study 18-2.)

Low- and medium-level wastes are a problem because of their sheer volume and potential for contaminating the environment. For many years these wastes have been improperly disposed of. In some cases they were deposited in steel drums that were stored underground. The drums can leak radioactive materials into groundwater and surface waters. Because of such problems half of the low-level waste sites in the United States have been closed. (Chapter 18 describes where low- and medium-level wastes go and what is being done about future waste disposal.)

Until the 1960s low- and medium-level radioactive wastes were often mixed with concrete, poured into barrels, and dumped at sea off the coast. Thousands of barrels of waste were dumped in Massachusetts Bay near Boston; some were dumped in the harbor itself. Approximately 47,000 barrels of radioactive wastes were dumped 48 kilometers (30 miles) off the coast of San Francisco, near the Farallon Islands, and numerous barrels were dumped in the ocean near New York City. Although this activity has been banned in US waters since 1970, it still occurs

Figure 12-4. Radioactive mill tailings from a uranium processing plant in Grants, New Mexico. Wastes like these have been dumped throughout the American West, sometimes along riverbanks where they can enter waterways.

in the Irish Sea. England annually dumps large quantities of radioactive wastes there without even placing them in barrels.

Social Acceptability and Cost Two of the most important factors controlling the future of nuclear energy are its social acceptability and its cost. The two are tightly linked.

The construction of a nuclear reactor costs $2 billion to $8 billion, compared with $500 million to $1 billion for an equivalent coal-fired power plant. Costs are high because of strict building standards, expensive labor, construction delays, and special materials needed to ensure that plants will operate safely. Because of the high cost and risk of damage, US banks are refusing to lend utilities the money needed to finance construction. Repair costs are many times higher than those for conventional coal-fired plants. For example, saltwater corrosion of the cooling system in a reactor owned by the Florida Power and Light Company cost over $100 million to repair. The utility paid $800,000 a day to make up for the lost electricity. A

Figure 12-5. Nuclear reactions in a breeder reactor. Neutrons produced during fission strike nonfissionable "fertile" materials such as uranium-238. U-238 is then converted into fissionable plutonium-239, which can be used in the reactor as fuel.

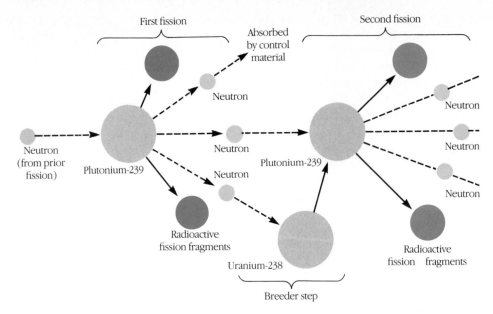

similar problem in a coal-fired power plant would have cost a fraction of this amount.

A lack of public support, exorbitant costs, and a rash of canceled plants in recent years have crippled the atomic power industry. Increasingly, critics argue that nuclear power may be costing us millions of dollars unnecessarily. France, which gets 60% to 70% of its electricity from nuclear plants, is thought to be paying 35% to 60% more for that electricity than it would have for electricity from coal.

What makes nuclear power so expensive and threatens to make it so costly in the future, besides high construction and maintenance costs, are poor performance, accidents, decommissioning costs, and now, loan defaults. Colorado's only nuclear reactor generated its first electricity in 1976. In its 13-year lifetime, the plant has been shut down half of the time because of various mechanical problems. The long "down time" cost that utility and its customers $20 to 30 million per year. In 1989, the company decided to shut the plant down. Accidents and routine maintenance cost utilities dearly and raise the costs of nuclear power, as we saw at Three Mile Island. Now that some nuclear power plants are reaching the end of their lifetime, a new and very costly problem faces utilities: decommissioning, dismantling plants at the end of their 20-to-30-year lifespan. Experts estimate that it may cost utilities $500 million to $1 billion to decommission a 1000-megawatt plant. This cost will inevitably add to electric bills. Finally, the rash of cancellations puts a huge burden on the US economy. Even though utilities halt construction of nuclear plants, they must still pay back the billions of dollars they have borrowed. Customers inevitably foot most of the bill. Bondholders, who sometimes finance municipal projects, can also be left in financial ruins, as was the case when the Washington Public

Power Supply System defaulted on $2.25 billion in bonds. Because of the default electric rates have increased dramatically. The region is now saddled with a debt that may climb to $8 billion or $9 billion with interest. Utilities have assumed the financial responsibility for three of the canceled reactors. The cost will be passed on to the customers. Bondholders have been stuck with the rest.

Proliferation of Nuclear Weapons At least 21 countries have the materials and the technical competence to build nuclear bombs. Many of these countries are politically unstable or are in volatile regions where war could easily erupt. Six countries—the United States, China, the Soviet Union, Britain, France, and India—have already test-fired nuclear weapons. The plutonium used in these bombs comes from special reactors designed to make it and from conventional nuclear reactors. Critics of nuclear power argue that the spread of nuclear power throughout the world will make fissionable materials more widely available.

Nuclear fuels could be stolen by terrorist groups. According to the energy analyst Denis Hayes, with careful planning an armed group could steal plutonium from any number of nuclear facilities. "No wizardry is required to build an atom bomb that would fit comfortably in the trunk of an automobile," he writes. Strategically planted, a terrorist bomb could prove disastrous.

Breeder Reactors The world's supply of the uranium-235 used in light water reactors will last about 100 years at the current use rate. Increases in the production of electricity by nuclear power plants, however, could greatly reduce the lifespan of uranium reserves. To counter the decline in fuel supply, the breeder reactor has been proposed. The **breeder reactor** is similar in many respects to the light water reactor described earlier. However, it performs an additional function: it makes the fissionable material plutonium-239 from uranium-238, an abundant isotope of the element uranium. In the breeder reactor fast-moving neutrons from the reactor core strike the nonfissionable U-238 placed around the core and convert it into Pu-239 (Figure 12-5). The neutrons come from small amounts of Pu-239 located in the fuel rods of the breeder reactor. Theoretically, for every 100 atoms of Pu-239 consumed in fission reactions, 130 atoms of Pu-239 are produced—hence, the name "breeder."

The attraction of breeder reactors is that they would be fueled by U-238 found in the wastes of uranium processing plants or in spent fuel from fission reactors. In the United States the estimated supply of fuel for breeder reactors would last 1000 years or more. In addition to providing a long-lasting supply of electrical energy, breeder reactors could reduce the need for mining, processing, and milling uranium ore. Fuel prices might remain stable because of the abundance of uranium-238. Breeder reactors also do not create chemical air pollution (if we ignore problems from mining and milling).

Breeder reactors have been under intensive development in the United States for over 30 years. The most popular design is the **liquid metal fast breeder reactor**, which uses liquid sodium (instead of water) as a coolant. Heat produced by nuclear fission in the reactor core is transferred to the liquid sodium coolant, which in turn transfers this heat to water. The water is then converted to steam and used to generate electricity.

This breeder reactor, on the surface, sounds like the answer to our electrical energy needs. It has numerous problems, however, some so great that they may make the technology impractical. The most significant problem is that it takes about 30 years for the reactor to break even, that is, to produce as much Pu-239 as it consumes, although scientists are now attempting to shorten this period. Clearly, the fate of breeder reactors hinges on a drastically shortened pay-back period. The second major problem is the cost: $4 billion to $8 billion. In addition, many of the problems of light water reactors are applicable to the breeder reactor.

Another problem would be the large quantities of plutonium located at breeder reactors. Plutonium is long-lived and extremely toxic if inhaled. The liquid sodium coolant can also be dangerous. It reacts violently with water and burns spontaneously when exposed to air. Leaks in the coolant system could trigger a catastrophic accident at a breeder reactor. Should the core melt down, experts warn, a small nuclear explosion equivalent to several hundred tons of TNT might occur. Rupturing the containment building, the explosion would send a cloud of radioactive gas into the surrounding area.

In 1983 the US Congress canceled further funding on an experimental breeder reactor, the Clinch River project

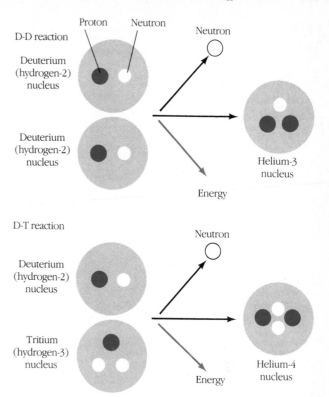

Figure 12-6. Two potentially useful fusion reactions.

in Tennessee. Having spent nearly $2 billion for planning, Congress decided that further investments were not wise. The prospects of the breeder reactor are currently being assessed by France now that its commercial breeder reactor, the Superphenix, is in operation. In 1987 the reactor was shut down for seven months due to a sodium leak. In 1988 the plant was shut down four times, but each time only for a short time. The French are now planning to build a second breeder reactor.

Nuclear Fusion

The sun, a solar furnace powered by a special type of nuclear reaction called fusion, has been the inspiration of humankind for millennia. **Nuclear fusion**, taking place in the sun and other stars, results when four hydrogen nuclei fuse to form a helium nucleus, a slightly larger nucleus (see Chapter 2). Fusion, however, requires extremely high temperatures to overcome the mutual electrostatic repulsion of the positively charged nuclei. When hydrogen nuclei fuse, they emit large quantities of energy in the form of high-energy radiation. Several fusion reactions are of interest today. Experimental reactors are powered either by the fusion of two deuterium nuclei (a hydrogen nucleus with one neutron) or deuterium and tritium nuclei (a hydrogen nucleus with two neutrons) (Figure 12-6). Some scientists believe that

Case Study 12-2

Cold Fusion: Science on Trial

In 1989, a US scientist and his British colleague made an announcement that shook the world. They claimed to have discovered a room-temperature fusion process, called **cold fusion**. Given the problems and high costs of fission energy and the decline in conventional fossil fuel supplies, many people were thrilled by the discovery. Cold fusion promised virtually unlimited energy.

Stanley Pons of the University of Utah in Salt Lake City and British co-worker Martin Fleischmann made their announcement after running experiments in which they passed an electrical current between two electrodes immersed in a liquid containing lithium and deuterium oxide. They reported that excess heat and neutrons given off in the aqueous system came from the room temperature fusion of deuterium atoms on one of the electrodes.

Troubles soon began, however. Pons and Fleischmann chose to announce their findings at a press conference, bypassing normal peer-review processes that weed out faulty experimental results before publication. As expected, their announcement created as much criticism as excitement.

The controversy that arose resulted in an exciting test of critical thinking skills. First, in order to determine whether the results were valid, many scientists tried to duplicate the experiments in their own labs, a requirement of all critical thinking. Most efforts failed. But some studies seemed to support the original findings. Excess neutrons detected in these experiments suggested that cold fusion may indeed be occurring. But even these results drew heavy fire.

Within weeks, what had been hailed as a scientific breakthrough had become back-page news. Pons and Fleischmann, in fact, retracted many of their claims.

MIT nuclear physicist Martin Deutsch called much of the research "garbage," claiming that many researchers had failed to run controlled experiments, needed for all good scientific research and described in Chapter Supplement 1-1. The most damaging testimony came from chemist Nathan Lewis of the California Institute of Technology, who explained all of the original assertions as the result of either faulty assumptions in calculations or poorly controlled experiments.

As the dust settled on the controversy it appeared that in the heat of a potentially important discovery several reputable scientists may have temporarily suspended their critical thinking skills. But several lingering questions remain. What produced the excess heat? And why did some carefully run studies indeed detect neutrons?

Most believe that the heat produced by the reaction may have come from a chemical reaction, not fusion. The neutrons detected by several researchers, some say, may have come from the fusion of some atoms on the electrodes, but in numbers too small to be of significance. The Pons–Fleischmann results, then, may have come from an unexpected chemical reaction that occurs along with miniscule amounts of cold fusion. The long-term benefit of this discovery is still unknown, but most scientists are skeptical about the once glowing prospects of cold fusion.

---

fusion can occur at room temperature under certain conditions. For a discussion of the controversy over this process, see Case Study 12-2.

Controlled fusion, if it proves successful, offers several advantages. The most important is the abundance of fuel. Deuterium, for example, is found in water, plentiful in the earth's oceans. Tritium does not exist naturally and must be made from lithium, but lithium supplies will not constrain energy production. The energy analyst John Holdren estimates that at current rates of energy consumption in the United States fusion would meet energy needs for up to 10 million years!

Unfortunately, fusion also has some drawbacks, which may forever make it unattainable commercially. The first of these is that fusion reactions take place at temperatures measured in the hundreds of millions of degrees celsius. The main obstacle, then, is finding a way to contain such an extremely hot reaction. No known alloy can withstand

these temperatures; in fact, metals would vaporize. To contain fusion reactions scientists have devised two possible reactors, which suspend tiny amounts of fuel in air within a metal reactor vessel. The most popular technique now used in experimental fusion reactors is called **magnetic confinement** (Figure 12-7). The superheated fuel forms a hot plasma—a hot gas of nuclei and electrons—that is suspended in an electromagnetic field set up by magnets inside a reactor vessel. Small-scale experimental reactors of this type have been developed and operated in the United States, Europe, the Soviet Union, and Japan. Despite 40 years of research, however, researchers have failed to reach the break-even point, that is, the point at which they get as much energy out of the system as they put into it. Achieving this point would be just the first step in a long, costly climb to commercialization. To be economical, fusion reactors must generate energy efficiently.

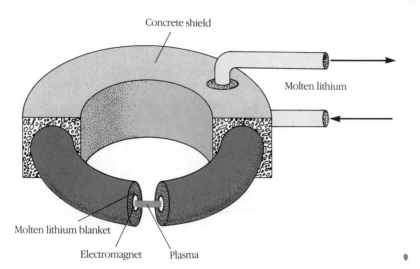

Concrete shield

Molten lithium

Molten lithium blanket

Electromagnet   Plasma

Figure 12-7. One type of fusion reactor. The fusion reactions occur suspended in an electromagnetic field.

In the proposed designs, heat released from the fusion reaction would be drawn off by a liquid lithium blanket. The heat would be used to boil water and create steam for electric generation. The lithium blanket would also capture neutrons, creating tritium, which would be extracted and then used as fuel.

Scientists estimate that a demonstration fusion reactor could be running by the year 2000, but commercial-scale plants would not be possible until 2020 or 2030 at the earliest. Scientists at Lawrence Livermore Laboratory (a government research lab in California) are asking Congress for over $1 billion to fund two experimental reactors, but, because of federal budget problems, they are not optimistic about funding. The cost of a commercial fusion reactor cannot be accurately assessed at this time, but it could cost three to five times more than a comparable breeder reactor, or about $12 billion to $20 billion.

The deuterium–tritium fusion reactor is the most feasible type of fusion reactor, but tritium is radioactive and is difficult to contain. Because of the high temperatures in fusion reactors tritium can penetrate metals and escape into the environment. There are other problems. Fusion reactors would produce enormous amounts of waste heat (see Chapter 16). Fusion reactions would also emit highly energetic neutrons, which would strike the vessel walls and weaken the metal, necessitating replacement every two to ten years. Metal fatigue could lead to the rupture of the vessel and the release of tritium and molten lithium, which burns spontaneously when it contacts air. A leak might destroy the reaction vessel and the containment facilities. Neutrons emitted from the fusion reaction would also convert metals in the reactor into radioactive materials. Periodic maintenance and repair of reactor vessels would be a health hazard to workers, and radioactive components removed from the reactor would have to be disposed of properly.

Coal

Coal is the most abundant fossil fuel, one not likely to be depleted soon. As we saw in Chapter 11, world coal supplies could last hundreds of years, and the United States has about one-third of world reserves. Coal's abundance is its chief advantage. In the United States coal is widely used to generate electricity for millions of homes and businesses. Some businesses also burn coal directly to generate heat and steam. The abundant supply ensures its continued use. Like nuclear energy, coal fits nicely into the electrical grid system, and it currently costs 5 to 7 cents per kilowatt-hour (about half the cost of nuclear energy). Energy conservation and cogeneration, discussed later, are the only energy sources presently cheaper than coal. Coal-fired power plants are an established and low-risk technology, and they are inexpensive to build (compared with nuclear power plants), giving coal an edge over nuclear power. Additionally, synthetic natural gas and oil can be made from coal, providing fuel for transportation and buildings. For countries with abundant supplies coal is clearly an economical fuel to burn.

On the other hand, coal is a dirty and environmentally costly fuel. At virtually every step in the coal energy system significant impacts may occur; air pollution, black lung, subsidence, and erosion are just a few. (Table 11-1 summarized many of those problems; see also Chapter 15 and Chapter Supplement 15-1.) Surely, the impacts of the coal production–consumption cycle could be lessened through tougher laws and enforcement of existing laws. In addition, technological developments could help us burn coal more efficiently and more cleanly (see Chapter 15).

Although the price of coal-generated electricity is bound to increase as tighter controls are placed on smokestacks, coal will undoubtedly serve as a major source of energy in the coming years. Our challenge is to control the many impacts or, if the cost of control becomes too high, to switch to cleaner, cheaper forms of energy production.

Natural Gas

Natural gas is a mixture of low-molecular-weight hydrocarbons, but the main one is methane. It is burned in homes, factories, and electric utilities and is often described as an ideal fuel because it contains few contaminants and burns cleanly. Like oil it is easy to transport, but only within countries where it is transported from one location to the next through pipelines. It has a high net energy yield, too.

Natural gas is extracted from wells as deep as 10 kilometers (6 miles). On land, drilling rigs generally have minimal impact unless they are located in wilderness areas, where roads and noise from heavy machinery and construction camps can disturb wildlife. However, natural gas extraction can cause subsidence, a sinking of the earth, within a considerable radius of wells. One notable example is in the Los Angeles–Long Beach Harbor area, where extensive oil and gas extraction beginning in 1928 has caused severe subsidence. Over well sites the ground has dropped 9 meters (30 feet). Natural gas is generally safe to transport in the gaseous form in pipelines. To transport it across oceans, however, it must be liquefied. In the liquid form natural gas is unstable and highly flammable. A ship containing liquefied natural gas could burn intensely.

Natural gas supplies, as described in the last chapter, are uncertain. The best estimates indicate that enough natural gas exists throughout the globe for 60 to 110 years of consumption at current rates. Like oil it, too, will be replaced in the coming years.

Synthetic Fuels

The world is running out of oil, essential for transportation, heating, and chemical production. Three nonrenewable substitutes are available: oil shale, tar sands, and coal. Each of these can be converted into liquid and gaseous fuels, known as **synthetic fuels**, or **synfuels**.

Oil Shale   Oil shale is a sedimentary rock that contains an organic material known as **kerogen**. Kerogen is driven from the rock by heating. In a liquid state this thick, oily substance is called **shale oil**. Like crude oil it can be refined and purified to make gasoline and other by-products.

Oil shale is also found in large quantities. Deposits lie under much of the continental United States, with the richest ones in Colorado, Utah, and Wyoming. Large deposits are also found in Canada, the Soviet Union, and China. The US Geological Survey estimates that shale deposits in the country contain more than 2 trillion barrels of oil, although not all of it would be recoverable.

The chief advantages of oil shale are its versatility and its large supply. Oil shale technology is not fully developed, however, and costs currently make it uneconomical to produce. The high cost of production stems from shale's poor net energy efficiency: about one-third of a barrel of oil (or an energy equivalent) is needed to mine, extract, and purify a barrel of shale oil. Net energy analysis shows that shale oil production creates only about one-eighth as much energy as conventional crude oil production for the same energy investment.

Oil shale brings with it some environmental costs, as well. Shale that is strip-mined, as coal is, disturbs large tracts of land, increasing erosion and reducing wildlife habitat. Mined shale is next crushed and heated, a process called **surface retorting** (Figure 12-8). Surface retorting also produces enormous amounts of solid waste, called **spent shale**. A small operation producing 50,000 barrels per day would generate about 19 million metric tons of spent shale each year. Since the shale expands by about 12% on heating, not all of it could be disposed of in mines. Dumped elsewhere, the spent shale may be leached by water, producing an assortment of toxic organic pollutants that could contaminate underground and surface waters.

Oil shale retorts require large quantities of water, too, but oil shale is typically found in arid country. Retorts would also produce significant amounts of air pollutants unless carefully controlled.

To bypass the solid waste problem, oil shale companies have experimented with a process called *in situ* **retorting**. Shale deposits are fractured by explosives. A fire is started underground and forced through the shale. The heat drives off the remaining oil, which is collected and then pumped above ground.

In situ retorts eliminate many impacts caused by surface mining but, for several reasons, have not worked well. One of the chief problems is keeping the fire burning. Operators have found that groundwater often seeps into the retorts and extinguishes the fires. Operators have also found it difficult to fracture the shale evenly, which is necessary for uniform combustion. In addition, *in situ* retorts produce more sulfur emissions than surface retorts, raising the cost of extracting shale oil.

A 1980 study by the US Office of Technology Assessment found that huge government investments would be needed to help the oil shale industry produce even small amounts of oil. In 1985 the federal government withdrew virtually all of its support from the industry because of the poor economics. Today, only one oil shale plant is operating in the United States. The plant cost $1 billion to build and is in trouble. It is barely covering operating costs and is producing oil at $42 per barrel, two to three times the cost of conventional oil.

Tar Sands Tar sands are sand deposits impregnated with a petroleumlike substance called **bitumen**. Found throughout the world, tar sands are another source of liquid fuels. Tar sands can be strip-mined and then treated in a variety of ways to extract the bitumen. Hot-water processing is the only method used commercially. *In situ* methods similar to those used in oil shale extraction are also being tested.

The largest deposits of tar sand are in Alberta, Canada, in Venezuela, and in the Soviet Union. In the United States six states have commercially attractive deposits.

Tar sands are plagued with many of the problems that face oil shale. They are expensive to mine. The sand sticks to machinery, gums up moving parts, and eats away at tires and conveyor belts. Tar sand expands by 30% after processing, and, as with oil shale, its production requires large amounts of water, which is badly polluted with oil in the process. Safe disposal is difficult and costly.

The most significant barrier to tar-sand development is poor economics. Like oil from oil shale, synthetic crude oil from tar sands has a poor energy efficiency: at least .6 of a barrel of energy is consumed per barrel produced. Worse, world reserves of tar sand oil are insignificant compared with world oil demands.

Coal Gasification and Liquefaction The abundance of coal in the United States and abroad has stirred interest in **coal gasification** and **liquefaction**. In both technologies coal is reacted with hydrogen to form synfuels. Coal gasification produces combustible gases. Coal liquefaction produces an oily substance that can be refined.

Unfortunately, coal gasification produces numerous air pollutants and requires large quantities of water. Many critics argue that gasification is a dirty alternative compared with natural gas and renewable energy options. Reducing pollution could make gasification too costly to be practical. The cost of synthetic natural gas is high enough without expensive pollution control equipment. For example, building a plant may cost $1.3 billion to $2 billion. An additional, and serious, problem is the low net energy production. Synthetic gas produced from surface mines is about 1.5 times more expensive than natural gas, and synthetic gas from coal taken from underground mines is 3.5 times more expensive.

Coal liquefaction is similar in principle to coal gasification. There are four major ways of making a synthetic oil from coal, but each involves the same general process: adding hydrogen to the coal. The oil produced by liquefaction must be purified to remove ash and coal particles.

Coal liquefaction could provide us with liquid fuels, but it would be costly. It produces air and water pollutants and requires large amounts of energy. Like coal gasification, it might be preempted by other energy sources that are cheaper and cleaner.

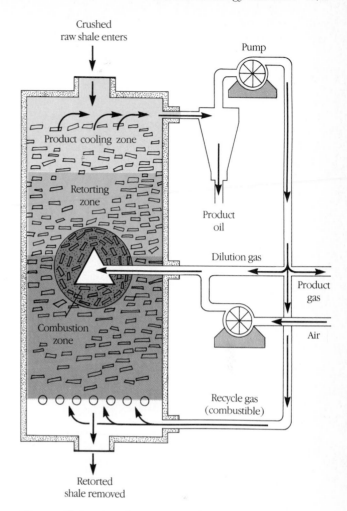

Figure 12-8. A surface retort used to extract the kerogen from oil shale. Raw shale is introduced at the top of the retort vessel. Retorted shale is removed at the bottom after being burned. Oil is driven off as a vapor in the hot gas at the top of the retort.

Renewable Energy Resources

Imagine a world powered by the sun, the winds, and other renewable forms of energy. It's a dream that may become a reality because of the rising threat of global warming. One hundred years from now, maybe sooner, our descendants may well live in such a world. Houses would be heated by the sun. Windmills and photovoltaics would provide electricity. Liquid fuels would come from crops. This vision of a soft-path energy future will be examined in this section.

Solar Energy

Oil, natural gas, oil shale, coal—all have limits. The sun, in contrast, is expected to last for several billion years. Even though only two-billionths of the sun's energy

Figure 12-9. Schematic representation of a passive solar house.

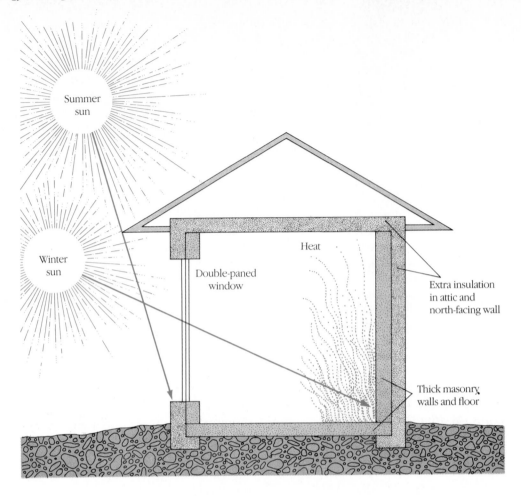

Summer sun

Winter sun

Heat

Double-paned window

Extra insulation in attic and north-facing wall

Thick masonry walls and floor

strikes the earth, it still adds up to an impressive total. The sunlight striking an area the size of Connecticut could provide enough energy to power the entire United States, including all homes, factories, and vehicles. Despite the enormous potential, solar energy provides only a fraction of US energy needs. Contrary to popular misconception, this poor showing is not because solar is limited to a few areas. In fact, significant sources of solar radiation are available across the nation.

Pros and Cons of Solar Energy The most notable advantage of solar energy is that the fuel is free. All we pay for are devices to capture and store it. Solar energy is a nondepletable energy resource available as long as the sun survives. It is a clean form of energy, although construction of solar units creates pollution and solid wastes, as does any manufacturing process. Over their lifetime solar systems produce much more energy than is needed to make them. Years of pollution-free operation offset the pollution created by production. Most solar systems can be integrated with building designs and therefore do not take up valuable land.

Solar energy offers the advantage of great flexibility. Current systems provide energy for remote weather-sen-

sing stations, single-family dwellings, and commercial operations. Solar energy can be collected to meet the low-temperature heat demands of homes or the intermediate- or high-temperature demands of factories. Solar electricity today provides energy to power radios, lights, watches, road signs, stream-flow monitors, and space satellites.

No major technical breakthroughs are required before we can use many existing solar systems, such as active solar water heating and passive solar space heating. Some improvements in design and costs could enhance the economic appeal of others, such as active solar space heating and cooling and solar voltaics.

Because rising prices of natural gas and oil will take a larger and larger chunk out of family and corporate budgets in the near future, those who invest in solar now could well enjoy an advantage over those who continue using costly fossil fuels. Over a lifetime solar energy can save a homeowner in cold climates $50,000 to $80,000 at current energy prices. As fossil fuel prices rise, solar homes may be the only affordable homes of the future.

The major limitation of solar energy is that the source is intermittent: it goes away at night and is blocked on cloudy days. Consequently, solar energy must be collected

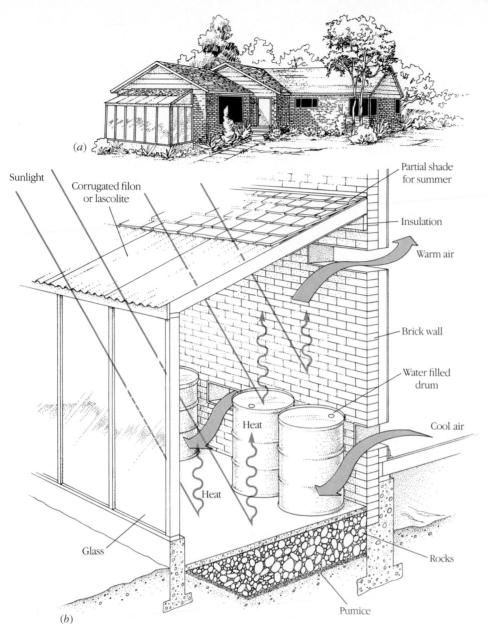

Figure 12-10. (a) Existing homes can add solar greenhouses and other passive solar ideas. (b) Sunlight penetrates the greenhouse glass and is stored in the floor or in water-filled drums. Warm air passes into the house.

(a)

Sunlight

Corrugated filon or lascolite

Partial shade for summer

Insulation

Warm air

Brick wall

Water filled drum

Heat

Cool air

Heat

Heat

Glass

Rocks

Pumice

(b)

and stored, but current storage technologies are limited. As a result, many solar users must have a backup system to provide heat during cloudy periods. Some forms of solar energy are currently uneconomical (for example, solar cells used to generate electricity).

There are three major types of direct solar energy systems: passive, active, and photovoltaics. Understanding each one can help us assess the potential of this largely untapped energy source.

Passive Solar Heating Passive solar heating is the simplest and most cost-effective solar system available. Often described as a system with only one moving part, the sun, it is designed to capture solar energy within a

building (Figure 12-9). Sunlight streams through south-facing windows and heats interior walls and floors of brick, tiles, or cement. The heat stored in these structures radiates into the rooms, heating the air day and night. On cloudy days solar homes are kept warm by heat that continues to radiate from heat-absorbent materials and by backup systems.

Passive solar design requires good insulation, internal heat storage (thermal mass), south-facing windows, and, usually, shutters or heavy curtains to block the outflow of heat at night. Overhangs block out the summer sun (Figure 12-10). Passive solar, unlike other forms of direct solar energy, may be difficult to install in existing homes without major modifications. Because of the problems of

Figure 12-11. An active solar heating system. The flat plate collectors shown here circulate a fluid that picks up heat captured by the black interior.

such retrofitting, passive solar is best suited for new homes.

Well-designed passive systems can provide 100% of a home's space heating. One passive solar home in Canada, built by the Mechanical Engineering Department of the University of Saskatchewan, had an annual fuel bill of $40, compared with $1400 for an average American home. The house was so airtight and well insulated that heat from sunlight, room lights, appliances, and occupants provided enough energy to keep it warm. The cost of this house was little more than that of a tract home. The author's superinsulated solar home at 2400 meters (8000 feet) in the Colorado Rockies is heated for less than $100 a year. As a rule of thumb, solar houses today cost about 10% more than conventional houses of similar size. Rising energy costs, however, could easily offset the high price.

Thousands of American homeowners have selected another solar option, the **earth-sheltered** house, built partly or entirely underground to take advantage of the insulative properties of soil. Properly designed earth-sheltered homes are well lighted, dry, and comfortable. They require less external maintenance. Partial sheltering—say, along the back of the house only—saves energy and renders the house virtually indistinguishable from a conventional solar home.

Active Solar Active heating and cooling systems rely on solar collectors, generally mounted on rooftops. Most collectors are insulated boxes with a double layer of glass on one side (Figure 12-11). These are called **flat plate collectors**. The inside of the box is painted black. Sunlight

is absorbed by the black material and converted into heat. The heat is carried away by water (or some other fluid) flowing through pipes in the collector or by air blown in by a fan. The heated water or air is then carried to some storage medium, usually water, in a superinsulated storage tank. After transferring its heat to the storage medium, the water or air is returned to the collectors.

In many parts of the country active solar water and space heating are competitive with electric heating. As natural gas prices rise, active solar is likely to become a cheaper option.

Photovoltaics Photovoltaics, or **solar cells**, provide a way of generating electricity from sunlight. Solar cells consist of thin wafers of silicon or other materials. These materials emit electrons when sunlight strikes them; the electrons then flow out of the wafer, forming an electrical current (Figure 12-12).

Electricity from solar voltaics currently costs much more (five to ten times more) than electricity from conventional sources, but costs have fallen quickly in the last two decades. Experts predict that improvements in production could soon make photovoltaics competitive with electricity from coal and nuclear power plants. In 1988, researchers at Sandia National Laboratories in Albuquerque, New Mexico, reported an efficiency rate of 31% in a new line of photovoltaics. The best commercially available cells have an efficiency rate of 18% to 20%. The improvement, along with more government support, could help photovoltaics become cost competitive sooner.

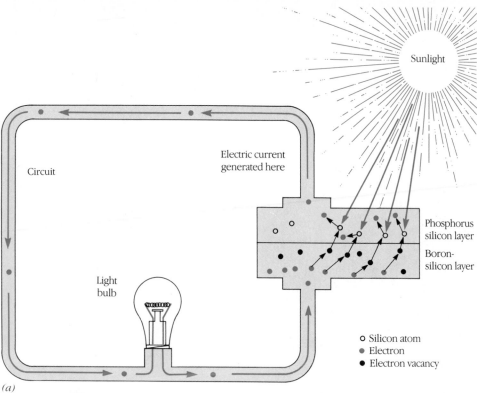

Figure 12-12. (a) Photovoltaic cells made of silicon (and other materials). When sunlight strikes the silicon atoms, it causes electrons to be ejected. Electrons can flow out of the photovoltaic cell through electrical wires, where they can do useful work. Electron vacancies are filled as electrons complete the circuit. (b) Array of solar voltaic cells. These cells are being used to power a railroad switching station in Alaska.

Circuit

Electric current generated here

Phosphorus silicon layer

Boron-silicon layer

Light bulb

○ Silicon atom
● Electron
● Electron vacancy

Sunlight

(a)

(b)

Photovoltaics may have their first significant market in the less developed nations. Many of these nations lack fossil fuel resources and can't afford to import them, or they can't afford to build centralized fossil-fuel-burning plants. In addition, the needs of these countries are generally small and are concentrated regionally.

Wind

About 2% of the sun's energy striking the earth is converted into wind. Winds form in two major ways. First, because sunlight falls unevenly on the earth and its atmosphere, some areas are heated more than others. The

Figure 12-13. Windmill generators near Livermore, California.

warm air rises, and cooler air flows in from adjacent areas. The earth's most important circulation pattern develops as warm air near the equator rises, drawing cooler polar air toward the tropics. The earth's rotation then causes air to circulate clockwise in the Northern Hemisphere and counterclockwise south of the equator. The second major wind-flow pattern results from the unequal heating of land and water. Air over the oceans is not heated as much as air over the land. Therefore, cool oceanic air often flows landward to replace warm, rising air.

Table 12-1 Present and Estimated Electrical Generation Costs*

Source	Cents per Kilowatt-Hour	
	1989	2000
Nuclear	8–12	na
Coal	6	na
Gas and oil	6–9	na
Hydroelectric	3–6	3–6
Wind	7–9	4–5
Geothermal	4.5–5.5	4–6
Photovoltaic	20–40	6–18
Solar thermal	8–12	5
Biomass	5	5

*Generation costs are costs to companies. (na = estimates not available)
Sources: Public Citizen Critical Mass Energy Product and Worldwatch Institute.

The potential of wind energy is enormous. Tapping the globe's windiest spots could provide 13 times the electricity now produced worldwide. The Worldwatch Institute estimates that wind energy could provide 20% to 30% of the electricity needed by many countries. Today, however, wind-generated electricity accounts for only a tiny portion of the world's enormous energy needs.

Wind can be tapped to generate electricity and heat, pump water, or do mechanical work (grinding grain, for example). Wind developers generally see two possibilities: large wind farms and backyard models that supply individual needs. Small-scale generators are easier to mass-produce than large systems. They have small blades that are less subject to the stress that adds to the cost of building larger units (Figure 12-13). Small generators produce more electricity in light winds and, therefore, operate more efficiently. Small wind generators can be located close to the end user. A breakdown jeopardizes only the individual, whereas breakdowns in large-scale operations can affect a whole community or city. On the negative side, small generators create a greater aesthetic impact than large units on carefully selected wind farms. Individual units require a fairly substantial investment ($5,000 to $20,000), and many homeowners may not have enough knowledge of local wind resources to make intelligent decisions about the feasibility of a proposed system.

Wind energy offers many of the advantages of direct solar energy: It is clean and renewable, uses only a small amount of land, and is safe to operate. Moreover, wind technologies do not preclude other land uses; wind farms, for example, can be grazed and planted. The technology

Table 12-2 Estimate of Available Biomass in the US

Type	Total Resources (Million Tons)	Total Energy Potential (Quads)	Recoverable Energy (Quads)
Crop residues	340	5.1	1.0–4.6[1]
Forestry residues	300	4.5	0.9–4.0
Urban refuse	135	1.2	0.3–1.1
Manure	45	0.7	0.1–0.6
Total	820	11.5	2.3–10.3

[1]Recoverable energy probably lies in the middle to lower part of the range given here.
Source: Kendall, H.W. and Nadis, S.J. (1980). *Energy Strategies: Toward a Solar Future.* Cambridge, MA: Ballinger, p. 170.

is well developed, and the fuel is free. Among solar-related electrical generating options, wind is the closest to being ready for widespread adoption. The costs of wind-generated electricity are rapidly dropping, but they are currently greater than those of coal, hydroelectric power, and nuclear energy (Table 12-1). According to some energy experts, mass-producing wind generators could bring the prices down further, making wind-generated electricity one of the cheapest sources available. Like solar voltaics, wind may become a source of energy in many developing nations.

There are, of course, disadvantages to wind systems. The wind does not blow all of the time, so backup systems and storage are needed. Storage technologies seem to be one of the major weaknesses of the wind energy system. Second, many states haven't extensively surveyed their winds, making it difficult for businesses and homeowners to decide whether wind energy would be practical. Third is the visual impact. Individual windmills and wind farms can be eyesores. Fourth, large wind generators may be noisy and may impair television reception, although fiberglass blades reduce interference by half. Some generators may also impair the microwave communications used by telephone companies.

Biomass

Biomass is the organic matter contained in plants. It is produced by photosynthesis and is, therefore, a form of solar energy. Biomass supplies about 19% of the world's energy. In the United States and other developed countries it supplies a smaller portion of the energy needs, only about 3%.

Useful biomass includes wood, wood residues left over from the timber industry, crop residues, manure, urban waste, industrial wastes, and municipal sewage. Some of these can be burned directly, and others are converted to methane and ethanol. The simplest way of getting energy from biomass is to burn it, but it may make more

sense to convert it to gaseous and liquid fuels and raw materials for the chemical industry to replace declining oil and natural gas supplies.

The US Office of Technology Assessment recently projected that biomass could supply 15% to 18% of the nation's energy needs by 2000 with aggressive research and development, government sponsorship, tax breaks, and other incentives. Crop and forestry residues, urban refuse, and manure could produce 3% to 14% of US energy demands (Table 12-2). Fuel farms (discussed in Chapter 7) could add additional energy. One of the most important contributions from biomass would be ethanol, a liquid fuel that can be burned in vehicles. Wood burned in factories and homes can provide large amounts of energy. Burnable municipal trash could likewise help produce heat, electricity, and steam, supplementing coal and other fossil fuels, but incinerators are a poor substitute for recycling programs when one looks at the energy each saves.

By using wastes and converting cropland to fuel farms, many countries, such as the United States and Canada, could make biomass a significant, renewable energy resource in the future. Biomass can help us reduce our dependence on nonrenewable energy resources, and it offers many other advantages. The most notable advantages are its high net energy efficiency, when it is collected and burned close to the source of production, and its wide range of applications. Biomass does not pollute the atmosphere with carbon dioxide, long implicated in the greenhouse effect (Chapter 15), as long as the plant matter burned equals the plant matter produced each year. Burning some forms of biomass, such as urban refuse, reduces the need for land disposal, as discussed in Chapter Supplement 18-1.

Biomass has some drawbacks, too. Improper management of fuel farms and forests that produce biomass could lead to soil erosion, sedimentation, destruction of reservoirs, and flooding. Removing crop and forestry resi-

Figure 12-14. This view of Mono Dam in California shows how the reservoir was filling with silt. In succeeding years the dam filled completely and gradually was reclaimed by the surrounding forest.

dues may reduce soil nutrient replenishment. Increasing reliance on fuel farms and forests could increase competition for their products, raising the prices of food, wood, and wood products. Biomass can create large amounts of air pollution, for example, smoke from wood stoves. Finally, transportation costs for biomass are higher than traditional fossil fuels, because biomass has a lower energy content.

Sugar cane, corn, and grain could all be grown to produce alcohol, helping to fill the need for an alternative to oil. Certain nonfood crops could also be grown to produce liquid fuel. For example, a desert shrub (*Euphorbia lathyris*) found in Mexico and the southwestern United States produces an oily substance that could be refined to make liquid fuel. In arid climates the shrub could yield 16 barrels of oil per hectare on a sustainable basis. The copaiba tree of the Amazon yields a substance that can be substituted for diesel fuel without processing. Sunflower oil can also be used in place of diesel. Farmers could convert 10% of their cropland to sunflowers to produce all the diesel fuel needed to run their machinery. Eventually, the entire transportation system could be powered by renewable fuels, which would help us reduce global warming.

Hydroelectric Power

Humankind has tapped the power of flowing rivers and streams for thousands of years. The hydrologic cycle is driven by sunlight, making hydropower yet another form of indirect solar energy. Falling water, propelled by gravity

and replenished by rainfall, offers many advantages. It is renewable, creates no air pollution or thermal pollution, and is relatively inexpensive. Furthermore, the technology is well developed.

On the opposite side of the coin are numerous problems. Sediment fills in reservoirs, giving them a typical lifespan of 50 to 100 years (although large projects may last 200 to 300 years). Thus, even though hydroelectric power is renewable, the dams and reservoirs needed to capture this energy have a limited lifetime (Figure 12-14). Once a good site is destroyed by sediment, it is gone forever. Dams and reservoirs create many additional problems, which were discussed in detail in Chapter 10, on water resources.

Brazil, Nepal, China, and many African and South American countries have a large untapped hydroelectric potential. In South America, for instance, hydroelectric generating potential is estimated at 600,000 megawatts. By comparison, the United States, the world's leader in hydroelectric production, has a present capacity of about 70,000 megawatts and an additional capacity of about 160,000 megawatts.

Estimates of hydroelectric potential can be deceiving, because they include all possible sites. Such estimates do not take into account whether dams would be economical or technically possible. Critical thinking requires a careful consideration of these factors. For example, half of the US hydroelectric potential is in Alaska, far from places that need power. The potential for additional large projects in the United States is small, because the most favorable sites have already been developed. In addition, the

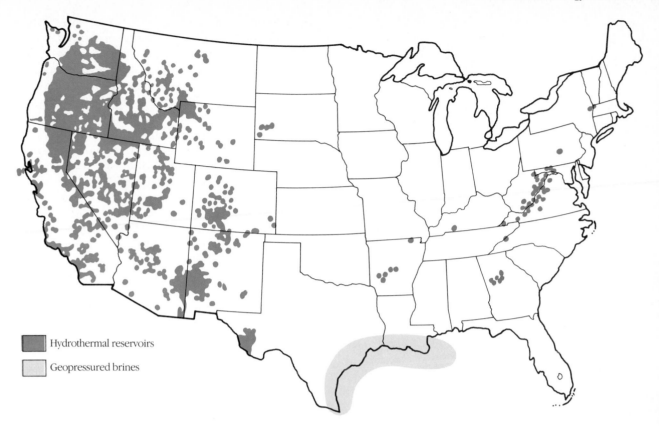

Figure 12-15. Geothermal resources in the United States.

high cost of constructing large dams and reservoirs has increased the cost of hydroelectric energy by 3 to 20 times since the early 1970s.

For the United States the most sensible strategy may be to increase the capacity of existing hydroelectric facilities and install turbines on the over 50,000 dams already built for flood control, recreation, and water supply. In appropriate locations small dams could provide energy needed by farms, small businesses, and small communities. But all projects must be weighed against impacts on wildlife habitat, stream quality, estuarine destruction, and other adverse environmental effects.

In the developing nations small-scale hydroelectric generation may fit in well with the demand. In China over 90,000 small hydroelectric generators account for about one-third of the country's electrical output.

Geothermal Energy

The earth harbors an enormous amount of heat, or **geothermal energy**, which comes from the decay of naturally occurring radioactive materials in the earth's crust and from magma, molten rock beneath the earth's surface. Geothermal energy is constantly regenerated, but because the rate of renewal is slow, overexploitation could deplete this resource regionally.

Geothermal resources fall into three major categories. The map in Figure 12-15 locates zones where the two most practical forms of geothermal energy can be tapped.

Hydrothermal convection zones are places where magma penetrates into the earth's crust and heats rock containing large amounts of groundwater. The heat drives the groundwater to the earth's surface through fissures, where it may emerge as steam (geysers), or as a liquid (hot springs).

Geopressurized zones are aquifers that are trapped by impermeable rock strata and heated by underlying magma. This superheated, pressurized water can be tapped by deep wells. Some geopressurized zones also contain methane gas.

Hot-rock zones, the most widespread but most expensive geothermal resource, are regions where bedrock is heated by magma. To reap the vast amounts of heat, wells are drilled, and the bedrock is fractured with explosives. Water is pumped into the fractured bedrock, heated, and then pumped out.

Geothermal energy is heavily concentrated in a so-called ring of fire encircling the Pacific Ocean and in the great mountain belts stretching from the Alps to China. Within these areas hydrothermal convection zones are the easiest and least expensive to tap. Hot water or steam from them can heat homes, factories, and greenhouses. In Iceland, for example, 65% of the homes are heated

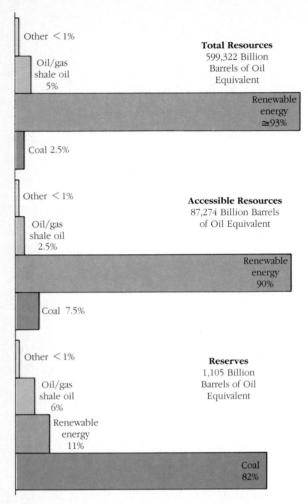

Figure 12-16. Estimates of the potential energy resources available to the United States. (a) *Total resources* takes into account all energy, irrespective of the economic or technological feasibility of tapping it. (b) *Accessible resources* is the energy that is available with current technology, irrespective of costs. Note that renewable energy makes up 90% of the accessible energy. As fossil fuel resources decline or fall into disfavor, renewables will be available and in much larger quantity than the finite nonrenewable energy resources. (c) *Reserves* includes energy that is technologically and economically feasible to tap. Coal and other fossil fuels currently top the list, but as they become depleted, the United States will invariably turn to renewable sources.

this way. Iceland's geothermally heated greenhouses produce nearly all of its vegetables; the Soviet Union and Hungary also heat many of their greenhouses in this way. Steam can be used to run turbines to produce electricity. The Philippines hopes to make geothermal energy its second-ranking source of electricity soon. Although still in the early stages of development, geothermal electric production is growing quickly in the United States, Italy, New Zealand, and Japan. By 2000, some experts believe,

the United States could produce 27,000 megawatts of electricity from geothermal energy, enough for 27 million people, about one-tenth of the population.

Hydrothermal convection systems have several drawbacks. The steam and hot water they produce are often laden with minerals, salts, toxic metals, and hydrogen sulfide gas. Many of these chemicals corrode pipes and metal. Steam systems may emit an ear-shattering hiss and release large amounts of heat into the air. Pollution control devices are necessary to cut down on air and water pollutants. Engineers have also proposed building closed systems that pump the steam or hot water out and then inject it back into the ground to be reheated. Finally, because heat cannot be transported long distances, industries might have to be built at the source of energy.

Hydrogen Fuel

Hydrogen could help replace oil and natural gas. Hydrogen gas is produced by heating or passing electricity through water in the presence of a **catalyst**, a chemical that facilitates the breakdown of water into oxygen and hydrogen without being changed. When hydrogen burns, it produces water and energy. Thus it is a renewable fuel with an essentially limitless supply. Hydrogen burns cleanly, too, producing nitrogen dioxide and water vapor. What makes it so appealing is that it does not add carbon dioxide to the air. Hydrogen is also easy to transport and has a wide range of uses, such as automobiles, gas ranges, and furnaces.

Unfortunately, it takes considerable energy to produce hydrogen fuel. This low net energy yield could make it an expensive form of energy. Hydroelectric power, wind, solar power, and other renewable energy resources, however, could be used to generate the electricity needed to make hydrogen. They are "free" energy sources and have unlimited supplies. Proponents argue that when power demand is low, renewables could generate electricity to make hydrogen economical. Still, the prospects for hydrogen are questionable today because of its low net energy yield. An efficient technology aimed at breaking down water by sunlight may not be available until the year 2000 or later.

The Renewable Energy Potential

Many people will be surprised to learn of the vast potential of renewable energy at human disposal. Robert L. San Martin, who heads the US government's renewable energy program, estimates that renewable energy sources provide about $18 billion worth of energy each year, or about one-twelfth of our total energy demand.

San Martin examined total energy resources available to the US. **Total resources** include all energy, whether or not it is practically or economically recoverable. In his study, San Martin found that renewable energy sources accounted for about 93% of the total resources (Figure 12-16). Coal, oil, natural gas, and oil shale accounted for less than 7.5%.

Making some reasonable allowances for what is accessible using current technology, San Martin found that renewable energy sources still came out on top; they made up 90% of all accessible resources, as shown in Figure 12-16. **Accessible resources** is a measure that looks only at technological feasibility but does not take into account economics. As such, it is much more realistic than total resources, but not a true reflection of current circumstances. Accessible resources is a measure of promise. It indicates energy that will be available as the economic picture changes and as finite resources become depleted.

As shown in Figure 12-16, accessible renewable energy that is available to us *each year* is equivalent to 70 to 80 billion barrels of oil. In contrast, nonrenewable energy resources under US control are equivalent to nine billion barrels of oil, but these resources are not renewable. When they are used, they are gone. Renewable energy, therefore, far outstrips the nonrenewable sources. In addition, renewable sources are available to the United States in large amounts year after year.

What is keeping people from tapping renewable energy resources? Right now, it's mostly economics and a long history of oil, gas, and coal use. The United States and a great many other countries were founded on renewable energy, mostly wood, but shifted to nonrenewables, such as oil and natural gas. Today, as shown in Figure 12-16, coal, oil, and gas make up the bulk of our current reserves—energy supplies that we can tap economically. These energy sources are generally cheaper than most renewable energy resources, but that's largely the result of heavy subsidies from the government (discussed in Chapter 11) and because economic externalities (environmental damage) are not figured into their total cost.

As prices rise and as countries try to stave off global warming, renewable energy resources could become more important sources of energy. They can provide energy for many thousands of years to come. All that's missing is the will to make the shift to a sustainable system.

Conservation

Conservation must be at the heart of all energy strategies. Nations that rely on energy cannot afford to waste it. Conserving energy offers numerous advantages: First, it can significantly reduce the cost of producing goods, giving industries an economic advantage in the marketplace, reducing inflation, and saving consumers millions of dollars. Second, it helps us stretch our fossil fuel supplies and thus gives us more time to find substitutes. Third,

energy conservation can reduce environmental pollution, land destruction, and waste disposal. Fourth, conservation can be brought "on line" rapidly, much faster than all other major sources of energy.

The economic savings from energy conservation can be enormous. For example, improving the efficiency of machines and appliances, that is, reducing their energy demand, costs about 1 to 2 cents per kilowatt-hour of energy saved. In contrast, coal-fired power plants produce electricity for 5 to 7 cents per kilowatt-hour. Nuclear power plants, on average, produce electricity for 10 to 12 cents per kilowatt-hour.

Slightly more expensive than reducing the energy demand (improving efficiency) of machines and appliances is a process called **cogeneration**, in which waste heat from one industrial process is captured and used for other purposes, usually electrical production. Cogeneration is an emerging source of energy. In 1989, 2300 cogeneration plants were operating, producing 27,000 megawatts of energy that otherwise would have been wasted. Plans to increase cogeneration are under way. Another 42,000 megawatts are under development. By 2000 cogeneration could provide 100,000 megawatts, and its estimated full potential is seven times greater than this estimate. Two hundred more projects, which will generate an additional 6000 megawatts of electricity, are soon to be completed. For years, most American industries produced their own steam using natural gas or oil. They purchased electricity from local utilities. The overall efficiency of this scheme was between 50% and 70%. Cogeneration, however, boosts the efficiency to 80% to 90%. The cost of electricity from cogeneration is 4 to 6 cents per kilowatt-hour.

Conservation does not mean "freezing in the dark," as some would have us believe. For factory owners it means installing equipment that captures waste heat to generate inexpensive electricity. Ultimately, it translates into enormous savings. It may mean redesigning manufacturing processes to cut waste energy, so that goods like steel or aluminum can be produced with considerably less energy and at a fraction of the cost. It can mean more profit or economic survival in the international market. For the homeowner, it may mean adding insulation to reduce heat loss, resulting in significant savings on utility bills, money that can be spent on summer vacations. It means installing storm windows that cut drafts and make rooms more comfortable. Both insulation and storm windows are small investments that are paid back in short periods. For the commuter, conservation may mean driving within the speed limit, keeping the car tuned, driving an energy-conserving car, or using mass transit whenever possible. Certainly, personal efforts require a little sacrifice, but in the end they can save us hundreds of dollars a year and can fill us with pride for doing something about the condition of the world's resources and its environment.

Table 12-3 Energy Conservation Suggestions

1. Water Heating.
   Turn down thermostat on water heater.
   Use less hot water (dishwashing, laundry, showers).
   Install flow reducers on faucets.
   Coordinate and concentrate time hot water is used.
   Do full loads of laundry, and use cooler water.
   Hang clothes outside to dry.
   Periodically drain 3 to 4 gallons from water heater.
   Repair leaky faucets.

2. Space Heating
   Lower thermostat setting.
   Insulate ceilings and walls.
   Install storm windows, curtains, or window quilts.
   Caulk cracks and use weatherstripping.
   Use fans to distribute heat.
   Dress more warmly.
   Heat only used areas.
   Humidify the air.
   Install an electronic ignition system in furnace.
   Replace or clean air filters in furnace.
   Have furnace adjusted periodically.

3. Cooling and Air Conditioning
   Increase thermostat setting.
   Use fans.
   Cook at night or outside.
   Dehumidify air.
   Close drapes during the day.
   Open windows at night.

4. Cooking
   Cover pots, and cook one-pot meals.
   Turn off the pilot lights on stove.
   Don't overcook, and don't open oven unnecessarily.
   Double up pots (use one as a lid for the other).
   Boil less water (only the amount you need).
   Use energy-efficient appliances (crock pots).

5. Lighting
   Cut the wattage of bulbs.
   Turn off lights when not in use.
   Use fluorescent bulbs wherever possible.
   Use natural lighting whenever possible.

6. Transportation
   Car-pool, walk, ride a bike, or take the bus to work.
   Use your car only when necessary.
   Group your trips with the car.
   Keep car tuned and tire pressure at recommended
     level.
   Buy energy-efficient cars.
   Recycle gas guzzlers.
   On long trips take the train or bus (not a jet).

The United States has made some strides in energy conservation, but its conservation potential has hardly begun to be tapped. The same can be said about virtually all industrial nations. Canada and the Soviet Union, which use energy more wastefully than the United States, are examples. The World Resources Institute reported in 1987 that the world could meet 90% of its energy needs between now and the year 2020 simply by making more efficient use of the energy we now use. Even though the world population is expected to double between 1980 and 2020, they say, only a 10% increase in energy production would be needed if existing energy-efficiency technologies were put in place.

Enormous opportunities to conserve exist in buildings, industry, and transportation. Several auto manufacturers, for example, now have test models that get 98 miles per gallon. The average new car rolling off the assembly line in the United States in 1990 got only 27.5 miles per gallon, one of the lowest averages in the developed world. By increasing the average mileage to 40 or 50 miles a gallon, we could double the lifespan of our oil supplies. Similar gains could be made in home heating and industrial processes.

But how can people be compelled to use energy more efficiently? At least four options are available. The first is a tax on fuels. By raising the taxes on fuels, governments could reduce energy consumption. In the United States, the average consumer pays about 30 cents per gallon in tax. In Denmark, the tax is nearly $3.00 per gallon. It is aimed at discouraging the use of gasoline. Increased taxes on gasoline in the United States and Canada could help cut consumption and could free up money for building efficient mass transit or for funding alternative energy sources for the near future.

Efficiency standards could also be applied to help cut back on energy waste. The National Appliance Conservation Act passed in 1987, for instance, sets tight energy standards for all new appliances. By 1992 all major household appliances, such as refrigerators, will consume 20% less electricity. The act will reduce peak US demand for electricity by nearly 22,000 megawatts by the end of the century, the equivalent of 20 large nuclear power plants. California passed a similar law calling for a 50% cut in electrical usage by new appliances. Efficiency standards could be applied to all new homes and factories as well. Automobile mileage standards, already in place, could be tightened. Pressure from Ford and General Motors to relax standards must be ignored.

Changes in pricing can also help reduce energy waste. For example, some utilities now charge customers more for electricity used during peak hours. Why? Meeting peak demand is very costly for utilities. It often requires construction of additional power plants, needed only a few hours a day to meet the peak demand. If utility companies can reduce peak demand through pricing, they can avoid having to build expensive new facilities.

Another way to save energy is through least-cost planning. Utilities are regulated by states. When they want to build a new power plant, they must seek state approval.

Over half of the public utility commissions now require utilities to prove that a new power plant is the most cost-effective way of providing electricity before construction begins. Utilities forced to do least-cost planning frequently find that new plants are far more expensive than other approaches. Improvements in generating efficiency, peak-pricing schemes, purchasing electricity from other companies, cogeneration, and promoting energy conservation measures in homes, factories, and businesses save customers money and often prove to be more profitable for utilities. Along this line, some companies are offering cash rebates to consumers who purchase energy-efficient appliances. These incentives are two to three times cheaper than the cost of producing energy at a new plant and the payback period is only two or three years, not twenty. Other utilities provide free energy audits or low-cost loans to invest in insulation or storm windows.

Energy inefficiency speeds up the depletion of valuable nonrenewable energy supplies. Many experts think that it is time to change our ways. Each of us can make small changes in life-style, cutting back a little waste here or there. Together individual actions add up to enormous change. A tree or two planted by a house, for example, provide shade, cutting down summer cooling bills. Conifers can even provide protection from the winter wind, helping to reduce annual energy bills. Studies show that three trees planted by light-colored houses throughout an entire residential neighborhood in Phoenix could cut cooling demand by 18%. In Sacramento and Los Angeles, the savings would be even greater—34% and 44%, respectively. Nationwide, tree planting near homes and businesses in the United States could cut energy demand by nearly 1 quad, or about one-eightieth of our total energy demand. Doubling the insulation of a new home adds 5% to the cost but pays for itself in five years in reduced energy costs. Collectively, our actions can add up to enormous savings, both economically and environmentally. (For some suggestions to save money, see Table 12-3.)

Despite the many advantages of conservation, it has not been as widely practiced as many would like. Why? Part of the reason is that Americans, Canadians, and others have been blessed in the past with abundant energy. Today, many suffer under the delusion that abundant energy will be ours indefinitely. Another reason conservation is not as widely practiced is that fossil fuels and nuclear energy have been subsidized by federal programs, which makes them cheaper to the customer. High-efficiency products also cost slightly more than less energy-efficient ones, and many people neither calculate the long-term savings nor think about the environmental consequences of waste. Many governments have not been committed to energy conservation programs either. In 1987, the US government spent only $162 million on energy conservation, down from $344 million in 1980.

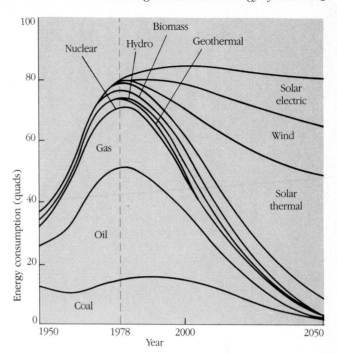

Figure 12-17. One projection of possible energy resources in the United States. These figures are based on full commitment to renewable resources.

The current outlay for conservation is equal to about five hours of military spending. This poor showing is hard to imagine when the investment-to-savings ratio for conservation can be as high as 1:1000—a thousand dollars saved for every dollar invested.

Building a Sustainable Energy System

Many observers see coal and nuclear power as the mainstays of the American energy diet in the coming decades. But others see a solar energy transition in the making. They believe that solar energy in its many forms could supply the bulk of US energy demands, along with conservation (Figure 12-17).

A renewable energy system will bring many changes. One major shift is that energy production will become a more personal matter. Conservation, photovoltaics, active solar heating, passive solar heating, and windmills will be built, replacing huge centralized nuclear- and coal-powered electric plants. Placing energy sources closer to the consumer could be a good thing; knowing where our energy comes from may help us respect its importance and use it more efficiently.

But how will renewable energy supplant the non-renewables that currently power US society? Where will the fuel come from to power the transportation system?

Table 12-4   Meeting Energy Needs of a Solar-Powered Society

Demand Sector	Sources	Application	Percentage of Total Energy Use
Residential and Commercial	Passive and active solar systems, district heating systems	Space heating, water heating, air conditioning	20–25
	Active solar heating with concentrating solar collectors		
	Solar thermal, thermochemical, or electrolytic generation	Cooking and drying	~5
	Biomass		
	Photovoltaic, wind, solar, thermal, total energy systems	Lighting, appliances, refrigeration	~10
			Subtotal ~35
Industrial	Active solar heating with flat plate collectors, and tracking solar concentrators	Industrial and agricultural process heat and steam	~7.5
	Tracking, concentrating solar collector systems	Industrial process heat and steam	~17.5
	Solar thermal, thermochemical, or electrolytic generation		
	Solar thermal, photovoltaic, cogeneration, wind systems	Cogeneration, electric, drive, electrolytic, and electrochemical processes	~10
	Biomass residues and wastes	Supply carbon sources to chemical industries	~5
			Subtotal ~40
Transportation	Photovoltaic, wind, solar thermal	Electric vehicles, electric rail	10–20
	Solar thermal, thermochemical, or electrolytic generation	Aircraft fuel, land and water vehicles	
	Biomass residues and wastes	Long-distance land and water vehicles	5–15
			Subtotal ~25
			100

Source: Kendall, H. W. and Nadis, S. J. (1980). *Energy Strategies: Toward a Solar Future.* Cambridge, MA: Ballinger, p. 262.

Increasing efficiency is the first and perhaps most important source. Cars that routinely get 50 to 80 miles per gallon could double or triple the existing oil supply. Mass transit, many times more efficient than the automobile, would further stretch energy supplies. To power automobiles, trucks, and buses, ethanol would be used. Hydrogen might be a supplement, but its economic prospects are not very bright. Electricity generated from inexpensive photovoltaics could power cars, trains, and buses.

Low- to intermediate-temperature thermal energy in the sustainable society would come from solar energy. Solar sources could be used for home space heating, water heating, and many industrial processes. Electricity would come from solar voltaics and windmills. Table 12-4 gives a further breakdown of energy demand by sector, showing us what kind of energy is needed and how it would be used.

Shifting to a Sustainable Transportation System

Americans are engaged in a dangerous love affair with their automobiles, an affair that many experts think cannot be sustained much longer. Pollution, declining oil supplies, and crowding on urban highways are the forces that could spell doom for the two-car family.

Today, nearly 30% of the energy Americans consume is used by the transportation sector and much of that powers our automobiles. The American passion for automobiles has spread throughout the world. The global automobile fleet, in fact, has expanded from 50 million in 1950 to over 400 million in 1990. In the United States, there is one car for every two people. In Western Europe the ratio is one to three. Automobile travel accounts for 90% of the motorized passenger transport in the United States and 78% in Europe. Each year, Americans travel nearly 2000 billion miles in their automobiles—the equivalent of more than ten round trips to the sun, 93 million miles away.

The love affair with the automobile will probably end in your lifetime, chiefly because of declining oil supplies. Liquid fuels from coal, oil shale, and fuel farms will probably not be able to save the auto from its slow extinction. In the next decade, as fuel demand outstrips supply, fuel prices are bound to climb, forcing people to more and more efficient automobiles. Many people, especially urban residents, will give up their second and third automobiles and turn to public transportation and other more efficient means of transportation. Eventually, nations may be forced to ration their gasoline and diesel fuel. Agriculture, cooking, and mass transit will probably receive the highest priority. Rationing, if it happens, will cause a further decline in automobile transportation.

In Stockholm, the Office of Future Studies has proposed that the city work to phase out the private automobile. They recommend considerable expansion of the existing mass transit system, because it is much more efficient, and also expansion of the fleet of rental vehicles to be hired for vacations and other special occasions.

Improving Efficiency Improving automobile efficiency can prolong the lifespan of the automobile. In 1982 the average new American automobile got about 9 kilo-

Table 12-5	Fuel Efficiency by Passenger Transportation Mode in Western Europe
Mode of Transportation	Kilojoules per Passenger Kilometer/Mile[1]
Van pool	400/640
Rail	400/640
Bus	450/720
Car pool	650/1,040
Automobile	1,800/2,880
Airline	3,800/6,080

[1]A kilojoule is 1000 joules, a unit of energy or work.
Source: Worldwatch Institute.

meters per liter of gasoline (22 miles per gallon). By 1989, mileage had climbed to 11.4 kilometers per liter (27.5 miles per gallon), still a long way from what experts think is achievable. Ironically, the best gas mileage was achieved by a German vehicle, the diesel Volkswagen Rabbit. It averaged 25 kilometers per liter (60 miles per gallon) on the highway. The British Leyland, a four-passenger prototype vehicle, leaves the Rabbit in the dust when it comes to gas mileage. It gets 34 kilometers per liter (83 miles per gallon). A Japanese vehicle currently gets the same mileage.

Rapid improvements in gasoline mileage could stretch world oil supplies. New foams and plastics, which make cars lighter and more efficient, could increase the safety of the new fleet of smaller, more energy-efficient vehicles. The alleged dangers of smaller cars could also be mitigated by tough drunk-driving laws, enforcement of speed limits, and better driver education.

From Road to Rails and Buses No matter how much the automobile fuel economy improves, the car can never compare to bus and train transportation (Table 12-5). In urban centers buses and trains can achieve a fuel efficiency of about 62 passenger kilometers per liter (150 passenger miles per gallon) of fuel. Intercity train and bus transport increases efficiency to 82 passenger kilometers per liter of fuel (200 passenger miles per gallon)—seven times better than the average new car today.

Faced with declining fossil fuel reserves, many cities will turn to mass transit. It is just a matter of time. The bicycle could supplement buses and high-speed trains. Investments that promote bicycle commuting represent one of the cheapest options available to cities and towns.

For decades, the bicycle has been a major means of transportation in many European and Asian countries. Following suit, some cities in the United States have laid out extensive bike paths for commuters (Figure 12-18). Davis, California, is a leader in promoting bicycle trans-

Figure 12-18. The bicycle is an efficient and economical alternative to the automobile for some commuters.

portation. Today, one-fourth of all commuter transport within the city is by bicycle. Some streets are closed entirely to automobile transport, and 65 kilometers (40 miles) of bike lanes and paths have been established. Bicycles won't replace cars, buses, and trains, but they can augment these forms of transportation. Because of vast differences in the climate, layout, and topography of cities, the bicycle won't find a place in all of our cities and towns.

Redesigning the transportation systems of cities and towns is one of the largest challenges facing modern industrial cities. The transition will probably come incrementally. As people abandon their automobiles, they're most likely to shift to buses and car pools. As the demand for buses climbs, cities are likely to install high-speed rail systems, perhaps down the median strip of existing highways. High-speed rail is incredibly expensive. To be profitable, it requires high participation, high density population in outlying areas, and a large central business district.

Economic Changes Accompanying a Shift to Mass Transit To be successful, transportation systems in a sustainable society must be efficient, yet flexible. Americans won't be happy about the shift, but in the long run we haven't got much choice. The shift to more efficient forms of transportation is likely to lead to significant shifts in our economy as well.

The automobile industry is the world's largest manufacturing industry and supports a number of other economically important industries. Manufacturers of rubber, glass, steel, radios, and numerous automobile parts will therefore also feel the impacts of the shrinking automobile market. So will the service sector: gas stations, automobile dealerships, and repair services.

Today, 20 cents of every dollar spent in the United States is directly or indirectly connected to the automobile industry and its suppliers. Eighteen cents of every tax dollar the federal government collects comes from automobile manufacturers and their suppliers.

Shifting toward a sustainable transportation system will not only dramatically change our life-style but will also create dramatic shifts in our economy. But don't hit the panic button. Steel and glass will be needed to build buses and trains. Automobile workers will find jobs in plants that produce buses and commuter trains. Mechanics will shift as well to service the new fleet of more efficient vehicles. Some workers, however, will inevitably be forced to find employment in new areas. Helping them adjust to the changes is an important task.

A wise energy future is economically, ecologically, and socially acceptable. Great inertia now lies in the way of the transition to a sustainable system: Industrial nations have been built on fossil fuels. The power plants that serve 245 million Americans and 24 million Canadians are in place. Switching to alternative fuels is no easy task. One of the chief problems with shifting to renewable energy resources and conservation is that it will require a massive amount of investment capital. Undoubtedly, this change will create new economic opportunities but will also threaten many powerful political and economic interests: the large energy companies, power companies, and the public representatives from energy-producing states.

The shift to renewable energy resources will take many years, even given the new urgency created by global warming. The length of the transition will depend on our political will and our willingness to build a system that is sustainable.

There are many ways of going forward, but only one way of standing still.

Franklin D. Roosevelt

Summary

The world is fast depleting oil and natural gas, two mainstays of its voracious energy appetite. It is also suffering the early symptoms of global warming. Both trends suggest the need for alternative energy sources. Two major groups exist to pick from: nonrenewables and renewables. But wise decisions require a careful look at all of our options and their impacts.

Some nonrenewable energy sources may have a place in the immediate future. Nuclear reactors are fueled by uranium-235, whose nuclei split when they are struck by neutrons. This process, **fission**, releases an enormous amount of energy. Nuclear fuels pass through a complex cycle from mining to waste disposal; at each stage of the cycle radioactive materials can escape, either by accident or through normal operations.

Nuclear power offers many advantages over coal and oil. Its light water reactors produce very little air pollution. Less land is disturbed by mining. The cost of transporting nuclear fuels is lower than for an equivalent number of BTUs of coal. Using nuclear fuels to generate electricity also frees coal for the production of synthetic oil and gas.

The major problems with nuclear power are disposal of radioactive wastes, contamination of the environment, thermal pollution, health impacts from radiation, limited supplies of uranium ore, low social acceptability, high construction costs, questionable reactor safety, lack of experience with the technology, vulnerability to sabotage, proliferation of nuclear weaponry from high-level wastes, and loss of individual freedom as security is tightened to protect citizens from sabotage and theft of nuclear materials.

Supporters of nuclear power have proposed the **breeder reactor** to get around the problem of limited fuel supply. Besides producing electricity, the breeder reactor makes fissionable plutonium-239 from the abundant uranium-238. Breeder reactors could use U-238 from waste piles, spent fuel rods, and uranium ore, providing electricity for several hundred years. **Liquid metal fast breeder reactors**, the technology of choice today, have many problems. The rate of conversion of U-238 to plutonium-239 is quite slow, requiring about 30 years for fuel production to equal consumption. A liquid metal fast breeder reactor could cost $4 billion or more. The presence of large quantities of fissionable materials packed in the core could lead to meltdowns or very-low-magnitude nuclear explosions in case of accidents. In addition, all the problems of light water reactors are applicable.

Another proposed energy system is **fusion power**. Fusion is the uniting of two or more small nuclei to form a larger one, a process accompanied by the release of energy. After three decades of research fusion is still a long way from being commercially available. But optimism prevails, for the forms of hydrogen needed to fuel the fusion reactor are abundant and could provide energy for millions of years.

Fusion reactions occur at extremely high temperatures. Safely containing such reactions is a major challenge. The cost of a commercial fusion reactor could reach $12 billion to $20 billion, many times more than that of conventional fission reactors, which currently face financial difficulties. The emission of highly energetic neutrons from the fusion reaction would weaken metals and necessitate their replacement every five years. Metal fatigue might lead to rupture of the vessel and the release of radioactive materials or highly reactive lithium.

Coal could be a major source of energy for years to come, but unless it can be burned more cleanly, the environmental cost of continued use may become astronomical. Natural gas supplies are greater than oil reserves, but they, too, will be depleted within the next century. Rising prices and the eventual decline in supplies suggest the need to find a replacement. Substitutes for oil and natural gas could come from oil shale, tar sand, and coal, which can be converted into liquid and gaseous fuels, known as **synfuels**.

Oil shale contains an organic material known as **kerogen**. Kerogen can be extracted by heating. Widespread development of this resource would require mining of extensive regions and large energy inputs, and it would result in numerous air pollutants and the production of large quantities of solid waste that would have to be disposed of safely to prevent groundwater and surface water contamination. Currently, the costs of shale oil compared with those of crude oil impede the development of this industry.

Tar sands are sand deposits impregnated with a petroleumlike substance called **bitumen**. Found throughout the world, tar sands can be mined and treated with heat or chemical solvents to remove the bitumen. Only a small proportion of the bitumen can be recovered, however, and global supplies are insignificant compared with energy demand. A low net energy yield also plagues this industry.

Coal can be converted to gaseous and liquid fuels. **Coal gasification** produces large quantities of air pollution and solid waste. Natural gas produced in such operations is expensive because of the low net energy yield. Surface mining destroys vegetation and habitat. **Coal liquefaction** is similar in principle, and it has many of the same problems.

Solar energy is abundant, but it provides only a fraction of our energy needs. **Passive solar** systems are the simplest and most cost-effective. Buildings are designed to capture sunlight energy and store it within **thermal mass**, walls and floors; the stored heat is gradually released into the structure. **Active solar** systems rely on collectors that absorb sunlight and convert it into heat, which is then transferred to water or air flowing through them. Pumps generally move water or air to a storage unit, where heat can be drawn off as needed. **Photovoltaics** are made of silicon or other materials that emit electrons when struck by sunlight, thus producing electricity.

Solar systems provide many advantages over conventional power sources. The fuel is free, nondepletable, and clean. When operating, systems produce no pollution and pay back the energy invested in their production. The major limitations are that the source is intermittent, making it necessary to store energy overnight or on cloudy days.

Winds can be tapped to generate energy. Wind energy offers many of the advantages and disadvantages of solar energy.

Biomass, a form of indirect solar energy, has some potential. Forest and crop wastes and fuel farms could be used to supply large amounts of energy. Useful biomass includes wood, wood residues, crop wastes, industrial wastes, manure, and urban waste. The simplest way of getting energy from biomass is to burn it, but many believe that a more sensible strategy would be to convert it to gaseous and liquid fuels and chemicals needed by the chemical industry.

Hydroelectric power, another indirect form of solar energy, is renewable, creates no air pollution, and is relatively inexpensive. Sediment fills in reservoirs, however, giving them an average lifespan of 50 to 100 years. The potential for hydroelectric power is limited in the developed countries, because the best sites have already been developed or are located far from population centers where the energy is needed. In the developing nations sites capable of producing large amounts of energy are available, but high construction costs may impair their development.

The earth harbors a great deal of energy from the decay of naturally occurring radioactive materials in the crust and from **magma**, molten rock. The most useful geothermal resource is the **hydrothermal convection zone**, where magma penetrates into the crust and heats rock formations containing large amounts of groundwater. The heat pressurizes the groundwater and drives it to the surface through fissures. Currently these zones are exploited for space heating and electricity. Such systems produce steam and hot water laden with toxic minerals, salts, metals, and hydrogen sulfide. Noise pollution is also a problem.

Hydrogen fuel is produced by heating or passing electricity through water in the presence of a catalyst. Water breaks down into hydrogen and oxygen. Hydrogen is a clean-burning fuel that could replace gaseous and liquid fuels. It is easy to transport but is explosive. Electricity needed to make hydrogen could be generated from solar energy, wind energy, or hydroelectric facilities. The prospects for hydrogen are questionable today because of its negative energy yield.

Renewable energy resources now technologically accessible far outstrip nonrenewable energy resources, including coal. Renewable resources are not more widely used because of economics and a history of nonrenewable energy dependency. As prices of oil and other fossil fuels rise and as countries fight to reduce global warming, renewable energy resources will become more widely used.

Conservation is one of the key untapped energy resources for tomorrow. By reducing energy waste in homes, factories, and transportation, we could inexpensively unleash an enormous supply of energy. The largest gain can be had when new homes and offices are built with passive and active solar systems and heavy insulation. Energy efficiency can be increased by fuel taxes, efficiency standards, pricing changes, least-cost planning, and individual actions.

The transportation sector currently consumes about 30% of US energy. A sustainable transportation system could help reduce that demand, but it would look sharply different from the current wasteful network based primarily on automobiles and jet aircraft. Mass transit by bus and train will probably replace much of the automobile traffic, both in cities and between them. Improved efficiency in vehicles can help lengthen the lifespan of the auto, but in the long term the auto will probably go the way of the dinosaur.

Energy experts have shown that we can substitute renewable energy resources such as solar energy for nonrenewables such as oil, natural gas, coal, and nuclear power without drastically changing society. A smooth transition can be made into a sustainable future, but it will require an immediate investment in renewable energy resources by governments and individuals.

Discussion Questions

1. Describe how a light water fission reactor works. What is the fuel, and how is the chain reaction controlled?

2. What are the advantages and disadvantages of nuclear power?

3. What is a breeder reactor? How is it similar to a conventional fission reactor? How is it different? Discuss the advantages and disadvantages of the breeder reactor.

4. What is nuclear fusion? Discuss the advantages and disadvantages of fusion energy.

5. What is oil shale? Discuss the benefits and risks of oil shale development.

6. Discuss the potential of tar sands in meeting our future energy demands.

7. Define coal gasification and liquefaction.

8. Discuss the advantages and disadvantages of solar energy.

9. Describe the difference between passive and active solar systems. What features are needed in a home to make passive solar energy work?

10. What are photovoltaic cells? In your opinion, should we develop photovoltaic cells in preference to nuclear energy? Why or why not?

11. Wind energy is close to being competitive with conventional electricity. Should we develop this energy resource in preference to nuclear power, coal, or shale? Why or why not?

12. What is biomass? How can useful energy be gained from it?

13. How is geothermal energy formed? How can it be tapped? Describe the benefits and risks of geothermal energy.

14. Using your critical thinking skills, debate the statement "Hydroelectric power is an immensely untapped resource in the United States and could provide an enormous amount of energy."

15. What are the major problems facing hydrogen power? How could these be solved?

16. Using your critical thinking skills, debate the statement "Conservation is our best and cheapest energy resource."

17. Discuss ways in which you could conserve more energy at home, at work, and in transit. Draw up a reasonable energy conservation plan for you and your family.

18. Make a list of criteria (cost, pollution, and so on) to judge energy resources, and list them in decreasing importance. Which energy technologies discussed in this chapter would be most suitable, according to your criteria?

Suggested Readings

Nonrenewable Energy Systems

Flavin, C. (1985). *World Oil: Coping with the Dangers of Success.* Worldwatch Paper 66. Washington, DC: Worldwatch Institute. Detailed study.

Flavin, C. (1987). *Reassessing Nuclear Power: the Fallout from Chernobyl.* Worldwatch Paper 75. Washington, DC: Worldwatch Institute. Important look at nuclear power and its future.

Gofman, J. W. and Tamplin, A. R. (1979). *Poisoned Power: The Case against Nuclear Power Plants before and after Three Mile Island.* Emmaus, PA: Rodale Press. Well-written analysis.

Hohenemser, C., Deicher, M., Ernst, A., Hofsass, H., Linder, G., and Recknagel. (1986). Chernobyl: An Early Report. *Environment* 28 (5): 6–13, 30–43. Comprehensive preliminary report on the accident.

Hohenemser, C. and Renn, O. (1988). Shifting Public Perceptions of Nuclear Risk: Chernobyl's Other Legacy. *Environment* 30 (3): 4–11, 40–45. Discusses the changing attitudes toward nuclear power following the Chernobyl accident.

Jungk, R. (1979). *The New Tyranny.* New York: Warner Books. An important, well-written book.

Kaku M. and Trainer, J., eds. (1982). *Nuclear Power: Both Sides.* New York: Norton. Excellent, balanced coverage.

Kunreuther, H., Desvousges, W. H., and Slovic, P. (1988). Nevada's Predicament: Public Perceptions of Risk from the Proposed Nuclear Waste Repository. *Environment* 30 (8): 16–20, 30–33. Outlines problems with efforts to place a high-level radioactive waste disposal site in Nevada.

Manning, R. (1985). The Future of Nuclear Power. *Environment* 27 (4): 12–17, 31–37. Excellent overview of the industry view of changes needed to revive nuclear power in the United States.

Renewable Energy Systems

Brown, L. R. (1981). *Building a Sustainable Society.* New York: Norton. General, optimistic survey of sustainable energy technologies.

Brown, L. R. et al. (1989). *State of the World.* New York: Norton. See Chapter 10 for a climate-sensitive energy strategy.

Bungay, H. R. (1983). Commercializing Biomass Conversion. *Environ. Sci. and Technol.* 17 (1): 24A–31A. Technical overview of biomass conversion.

Chandler, W. U. (1985). *Energy Productivity: Key to Environmental Protection and Economic Progress.* Worldwatch Paper 63. Washington, DC: Worldwatch Institute. Detailed study of energy conservation.

Deudney, D. and Flavin, C. (1983). *Renewable Energy: The Power to Choose.* New York: Norton. Fact-filled, highly readable book.

Elridge, F. R. (1980). *Wind Machines.* New York: Van Nostrand Reinhold. Excellent overview.

Flavin, C. (1981). A Renaissance for Wind Power. *Environment* 23 (8): 31–41. Interesting look at wind energy.

Flavin, C. (1983). Photovoltaics: International Competition for the Sun. *Environment* 25 (3): 7–11, 39–43. Good survey of economics and history of solar voltaics.

Flavin, C. (1984). *Electricity's Future: The Shift to Efficiency and Small-Scale Power.* Worldwatch Paper 61. Washington, DC: Worldwatch Institute.

Flavin, C. and Durning, A. B. (1988). *Building on Success: The Age of Energy Efficiency.* Worldwatch Paper 82. Washington, DC: Worldwatch Institute. Important reading.

Flavin, C. and Pollock, C. (1985). Harnessing Renewable Energy. In *The State of the World,* ed. Linda Starke. New York: Norton. Superb overview of the potential of renewable energy.

Group, L. (1978). *Solar Houses: 48 Energy-Saving Designs.* New York: Pantheon. Filled with interesting information on solar houses.

Hempel, L. C. (1982). The Original Blueprint for a Solar America. *Environment* 24 (2): 25–32. Excellent look at some highlights in solar history.

Kakela, P., Chilson, G., and Patric, W. (1985). Low-Head Hydropower for Local Use. *Environment* 27 (1): 31–38. Realistic look at the potential of small dams for generating electricity.

Kendall, H. W. and Nadis, S. J., eds. (1980). *Energy Strategies: Toward a Solar Future.* Cambridge, MA: Ballinger. Detailed survey of energy sources and their prospects for the future. Superb!

Pollock, C. (1986). *Decommissioning: Nuclear Power's Missing Link.* Worldwatch Paper 69. Washington, DC: Worldwatch Institute. Authoritative coverage of the costs involved.

Renner, M. (1988). *Rethinking the Role of the Automobile.* Worldwatch Paper 84. Washington, DC: Worldwatch Institute. Detailed coverage of the growth in automobile use and alternative transportation systems.

Shea, C. P. (1988). *Renewable Energy: Today's Contribution, Tomorrow's Promise.* Worldwatch Paper 81. Washington, DC: Worldwatch Institute. Excellent resource.

Radiation Pollution

Chapter 12 discussed nuclear energy, but this is only one potential source of radiation exposure. This supplement describes radiation—its sources and its effects.

As we saw in Chapter 2, the **atom** is composed of a **nucleus** and an **electron cloud**. The nucleus contains **protons** and **neutrons** and constitutes 99.9% of the mass of an atom. The much lighter **electrons** orbit in a cloud around the positively charged nucleus.

Atoms of a given element, such as carbon or uranium, all have the same number of protons in their nuclei. But they may contain slightly different numbers of neutrons. For example, uranium atoms all contain 92 protons. Some uranium atoms may have 146 neutrons, however, and others have 143. These alternate forms are called **isotopes**. To distinguish them scientists add up the protons and neutrons and tack the sum of these two onto the name of the element. The form of uranium containing 146 neutrons is called uranium-238 (92 protons + 146 neutrons = 238). The form containing 143 neutrons is called uranium-235.

Excess neutrons in some isotopes sometimes make them unstable. To reach a more stable state, they emit radiation. Unstable, radioactive nuclei are called **radionuclides**. They occur naturally or can be produced by various physical means. For the most part the naturally occurring radionuclides are isotopes of heavy elements, from lead (82 protons in the nucleus) to uranium (92 protons in the nucleus). There are three major types of naturally occurring radioactive emissions: alpha particles, beta particles, and gamma rays. X rays, which are also considered in this chapter, are artificially produced as described below.

Alpha particles consist of two protons and two neutrons, the same as a helium nucleus. They are positively charged. Alpha particles have the largest mass of all forms of radiation. In air they travel only a few centimeters. They can be stopped by a thick sheet of paper, so it is easy to shield people from them. In the body, alpha particles can travel only about 30 micrometers (about the width of three cells) in tissues. They cannot penetrate skin and are therefore often erroneously assumed to pose little harm to humans. But if alpha emitters enter body tissues, say, through inhalation, they can do serious, irreparable damage to nearby cells and their chromosomes.

Beta particles are negatively charged particles that are emitted from nuclei. They are equivalent to electrons found in the electron cloud, except that they contain more energy. Beta particles arise when neutrons in the nucleus are converted into protons, a process that helps stabilize radionuclides. A small amount of mass and energy is lost; this is the energetic beta particle that is ejected out of the nucleus.

The beta particle is much lighter than the alpha particle and can travel much farther. It can penetrate a 1-millimeter lead plate and can travel up to 8 meters (27 feet) in air but only 1 centimeter in tissue. Beta particles from some radionuclides have enough energy to penetrate one's clothing and skin but generally do not reach underlying tissues. They can, however, damage the skin and eyes (causing skin cancer and cataracts).

Gamma rays are a high-energy form of radiation with no mass and no charge, much like visible light but with much more energy. Gamma rays are emitted by nuclei to achieve a lower-energy, more stable state. They are often emitted after a nucleus has ejected an alpha or beta particle, because the loss of these particles does not always allow the nucleus to reach its most stable state. Some gamma rays can travel hundreds of meters in the air and can easily penetrate the body. Some can penetrate walls of cement and plaster or a few centimeters of lead.

Unlike the three previously discussed forms, the **X ray** does not originate from naturally occurring unstable nuclei. Rather, X rays are produced in X-ray machines when a high voltage is applied between a source of electrons and a tungsten collecting terminal in a vacuum tube (Figure S12-1). When the electrons are ejected, they strike the collecting terminal. Colliding with tungsten atoms, they are rapidly brought to rest. The energy they carried in is released in the form of X rays, which behave like gamma rays but have considerably less energy. They cannot penetrate lead.

All the forms of radiation described above are called **ionizing radiation**, because they possess enough energy to rip electrons away from atoms, leaving charged ions. Ions are the primary cause of damage in tissues.

How Is Radiation Measured?

Radioactive elements lose mass over time because of the emissions from their nuclei. Each radionuclide gives off radiation at its own rate, called the **radioactive decay rate** and measured

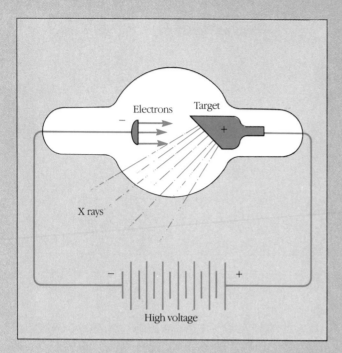

Figure S12-1. An X-ray machine.

in disintegrations per second. For example, 1 gram of radium decays at a rate of 37 billion disintegrations per second! The rate of radioactive decay determines the **half-life** of a radionuclide, that is, the time it takes for half of a given mass of a given radionuclide to decay into more stable isotopes.

Radiation exposure in humans is expressed in several different ways. One of the most widely used measures is the rad. **Rad** is the **radiation absorbed dose**, or, simply, the amount of energy that is released in tissue (or some other medium) when it is irradiated. One rad is equal to 100 ergs (a unit of energy) deposited in 1 gram of tissue.

As radiation travels through tissue, it loses its energy. The rate of energy loss is called the **linear energy transfer**, or LET. Put another way, LET is the amount of energy lost per unit of distance the radiation travels. Because of their mass, alpha particles travel only short distances through tissue and, therefore, lose their energy rapidly. They are said to have a high LET. The energy is transferred to the tissues. X rays, gamma rays, and beta particles travel farther through tissues and lose their energy more slowly; they have low LETs. Consequently, 10 rads of energy from beta particles would do less damage than 10 rads from an alpha particle, because the energy from an alpha particle is lost in a shorter distance.

The term **rem** takes into account the linear energy transfer and thus indirectly indicates the damage that a given amount of radiation will cause in tissue. For X rays, gamma rays, and beta particles 1 rem is essentially equivalent to 1 rad, but for alpha particles, because of their high LET, 1 rad is equivalent to 10 to 20 rems.

As a point of reference, a medical X ray may be equivalent to about .1 to 1 rem, depending on the type. The safety standard for workers in the United States is 5 rems per year. Background radiation is measured in thousandths of rems, or millirems (mrems).

Sources of Radiation

Radiation comes from two sources: natural and anthropogenic. Both contribute to our daily radiation exposure.

Natural Sources

Radiation is all around us. It is in rocks, in the air we breathe, and in the water we drink. Even the sun and distant stars bombard us with radiation. For many years experts believed that Americans received, on average, 160 to 200 millirems per year from natural and anthropogenic sources, with the highest exposures in high-altitude states such as Colorado, Montana, Wyoming, and others. In 1988, however, the National Council on Radiation Protection and Measurements announced that the average American may be exposed to nearly twice as much radiation each year as previously thought. The revised estimate takes into account the exposure resulting from **radon**, a radioactive gas found throughout the country. Radon is emitted from a naturally occurring radioactive element, radium, found in rock and soil. It seeps into homes and buildings and is breathed into the lungs, where it gives off damaging radiation. (For more on radon see Chapter Supplement 15-3.)

Table S12-1 shows that radon probably accounts for 55% of the annual radiation exposure. All told, natural sources are responsible for slightly more than 80% of our exposure to radiation. Slightly under 20% of our exposure comes from human sources, such as X rays and consumer products (televisions and luminous-dial watches).

Anthropogenic Sources

Anthropogenic radiation sources are many: (1) medical therapy (X-ray treatment for cancer) and diagnosis (X rays for bone fractures), (2) detonation of nuclear weapons in testing and the Second World War, (3) nuclear energy, (4) television sets, (5) luminous dials on watches, and (6) air travel.

Table S12-1   Estimated Radiation Exposure in the US

Sources of Radiation	Exposure and Percent of Annual Dose
Natural	
Radon	200 millirems (55%)
Cosmic	27 millirems (8%)
Rocks and soil	28 millirems (8%)
Internal exposure	40 millirems (11%)
Anthropogenic	
Medical X-rays	39 millirems (11%)
Nuclear medicine	14 millirems (4%)
Consumer products	10 millirems (3%)
Others	Less than 1%

Source: National Council on Radiation Protection and Measurements.

Effects of Radiation

How Does Radiation Affect Cells?

All forms of radiation ionize and excite biologically important molecules in tissues. Positively charged alpha particles, for example, draw electrons away from atoms in body tissues. Negatively charged beta particles in tissues may repel electrons of various atoms, causing them to be expelled from their atoms. They, too, produce positively charged ions. Gamma rays and X rays, on the other hand, are uncharged, but they possess lots of energy, which may be transferred to electrons as they pass through tissue. This energy excites the electrons and may cause their expulsion from the atoms, forming ions. Alternatively, the energy imparted to the electrons may make chemical bonds in molecules unstable and more easily broken. (Recall from Chapter 2 that all molecules are made up of atoms.)

Ionization of atoms in water and other molecules in tissues is responsible for much of the damage caused by radiation. Water molecules become positively charged when electrons are ripped from their atoms, as shown in Figure S12-2. Electrons that are freed from water molecules may combine with uncharged water molecules, forming negatively charged water molecules. Both positively and negatively charged water molecules rapidly break up into highly reactive fragments called **free radicals** (Figure S12-2).

Free radicals react almost instantaneously with biologically important molecules. When they react with oxygen, for example, hydrogen peroxide is formed. This powerful oxidizing agent damages or destroys proteins and other molecules, causing cell death. If it is extensive enough, cellular destruction can kill the organism. In some instances damage may be quickly repaired by the cells without any long-term effects. In other cases the damage may not be expressed until years after the exposure, in the form of mutations and cancer (Chapter 14).

Health Effects of Radiation

The effects of radiation on human health depend on many factors, such as the amount of radiation, the length of exposure, the type of radiation, the half-life of the radionuclide, the health and age of the individual, the part of the body exposed, and whether the exposure is internal or external.

Numerous studies of radiation have revealed some interesting generalizations: (1) Fetuses are more sensitive to radiation than children, who are, in turn, more sensitive than adults. (2) Cells undergoing rapid division appear to be more sensitive to radiation than those that are not. This is especially true in regard to cancer induction. Thus, lymphoid tissues (bone marrow, lymph nodes, and circulating lymphocytes) are the most sensitive of all the body's cells. Epithelial cells, those that line the inside and outside of body organs such as the intestines, also undergo frequent cellular division and are highly sensitive to radiation. In sharp contrast, nerve and muscle cells, which do not divide, have a very low sensitivity and rarely become cancerous. (3) Most, if not all, forms of cancer can be increased by ionizing radiation.

Health experts divide radiation exposure into two categories. Exposures over 5 to 10 rems per year are considered high-level. Below this, the exposure is low-level.

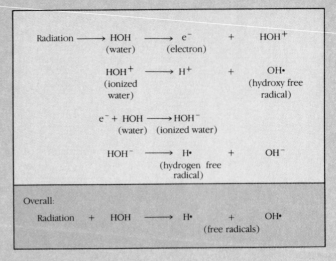

Figure S12-2. Ionization of atoms in water molecules.

Impacts of High-Level Radiation The most important information on high-level radiation comes from studies of the survivors of the two atomic bombs dropped on Japan at the end of World War II (Figure S12-3). Studies of these and other groups have led to several important findings. First, the lethal dose for one-half the people within 60 days is about 300 rads. Second, a dose of 650 rads kills all people within a few hours to a few days. Third, sublethal doses, or doses that do not result in immediate death, range from 50 to 250 rads. Victims suffer from **radiation sickness**. The first symptoms, which develop immediately, are nausea and vomiting; 2 to 14 days later, diarrhea, hair loss, sore throat, reduction in blood platelets (needed for clotting), hemorrhaging, and bone-marrow damage occur. Fourth, sublethal radiation has many serious delayed effects, including cancer, leukemia, cataracts, sterility, and decreased lifespan. Fifth, sublethal radiation also has profound effects on reproduction, increasing miscarriages, stillbirths, and early infant deaths.

High-level radiation exposure is rare today. We could anticipate such exposures only in workers at badly damaged nuclear power plants or munitions factories or in nearby residents. Nuclear war, even on a limited scale, would also expose large segments of the human population to dangerously high levels of radiation (Chapter Supplement 4-1). Accidents during the transporting of nuclear fuels and high-level wastes might also result in dangerously high exposures.

Impacts of Low-Level Radiation The effects of high-level radiation are well-established, but as is the case with low-level exposure to any toxic agent, health effects are harder to discern with low-level radiation. A growing body of evidence shows that low-level radiation increases the likelihood of developing cancer (Figure S12-4). For example, studies of individuals who years earlier were treated with radiation for acne, spinal disorders, and even syphilis show elevated levels of cancer and leukemia. In addition, children whose necks have been irradiated by medical X rays have an elevated rate of thyroid cancer. Lung cancer rates are elevated in uranium and fluorspar (calcium fluoride) miners, both of whom are exposed to radon gas. Factory work-

Figure S12-3. Hiroshima survivors.

ers who painted watch dials with radium early in this century developed bone cancer and a serious disease of the bone marrow called aplastic anemia. Several studies show that the rates of leukemia, tumors of the lymphatic system, brain tumors, and other cancers are 50% higher in infants whose mothers have been exposed to ordinary diagnostic X rays (2 to 3 rads) during pregnancy. One study showed that a 1-rem exposure to fetuses causes an 80% increase in mortality from childhood cancer. According to some estimates, low-level radiation from testing of nuclear weapons in the 1950s and 1960s caused 400,000 deaths in children because of cancer. Studies of workers at a plant in Hanford, Washington, the US government's source of radioactive materials for nuclear weapons, showed that the death rate from cancer was 7% higher than expected and that workers who died of cancer had received on the average only about 2 rems per year, well below the supposed safe level of 5 rems a year. As a final note, a new study of 27,000 Chinese radiologists and X-ray technicians showed that exposure to low levels of radiation increased the risk of developing cancer by 50%. Leukemia (a kind of blood cancer), breast cancer, thyroid cancer, and skin cancers were the prevalent types. This study confirms research done in the United States and Europe but is important because of the large number of subjects studied.

Low-level radiation is probably much more harmful than many scientists estimated a decade ago. The noted radiation biologist Dr. Irwin Bross, for example, estimates that low-level exposure is about ten times more harmful than previously calculated. He and others have called for major revisions of the maximum allowable doses for workers.

But is there a threshold level below which no damage occurs? No one can say for certain. Some health officials believe that no level is safe, because the effects of continued low-level exposure are cumulative.

According to two radiation experts, John Gofman and Arthur Tamplin, exposure to the current public health standard set for anthropogenic sources—.17 rad per year—will result in a 30-year exposure of 5 rads. This would cause 14 additional cases of cancer each year for every 100,000 people exposed, or about 14,000 additional cancer cases per year in adults over 30 and at least 2000 cases of cancer in individuals under 30 years of age. Studies by the National Academy of Sciences and the National Research Council suggest that a .17-rad exposure to anthropogenic radiation will probably increase the rate of cancer by 2% and the incidence of serious genetic diseases by about 1 birth in every 2000. Thus, some experts suggest lowering the health standard to 0.017 rad per year.

Some geneticists warn, however, that genetic disorders caused by radiation may be passed from generation to generation and may increase in incidence with subsequent generations. For example, if the incidence of genetic disorders and birth defects with a genetic basis were 1 in 2000 in the first generation, it would be 5 in 2000 in their offspring. A major consideration, then, is what impact radiation exposure today will have on subsequent generations. Are current policies and radiation standards posing a danger to future generations?

Low-level effects are small and hard to detect. The long latent period between exposure and disease makes it difficult for researchers to link the cause and the effects. Thus, studies such as those cited above have stirred a considerable amount of controversy. Although the results and conclusions of individual low-level radiation studies are debatable, on the whole they seem to consistently point to one conclusion: a substantial risk is created by subjecting people to low-level radiation. The question becomes what level of risk is acceptable. At what point do the benefits of X rays, nuclear power, and other uses outweigh the risks?

Bioconcentration and Biological Magnification Table S12-2 lists some radionuclides emitted from nuclear weapons and nuclear power plants. Some of these radionuclides are absorbed by humans and organisms and may become concentrated in particular tissues.

For example, iodine-131 is released from nuclear power plants, both during normal operations and in accidents. Fallout on the ground may be incorporated in grass eaten by dairy cows. It is then selectively taken up (bioconcentrated) by the human thyroid gland, where it irradiates cells and may produce tumors. Milk contaminated with I-131 is especially harmful to children.

Strontium-90 is released during atomic bomb blasts. It may also be released from reactors in small amounts under normal operating conditions but in large quantities in accidents. Strontium-90 is readily absorbed by plants and may also be passed to humans through cow's milk. It seeks out bone, where it is deposited like calcium. With a half-life of 28 years, it irradiates the bone and can cause leukemia and bone cancer.

Accumulation of radionuclides within tissues has important implications for human health, as seemingly low levels may become dangerously high in localized regions. Some radio-

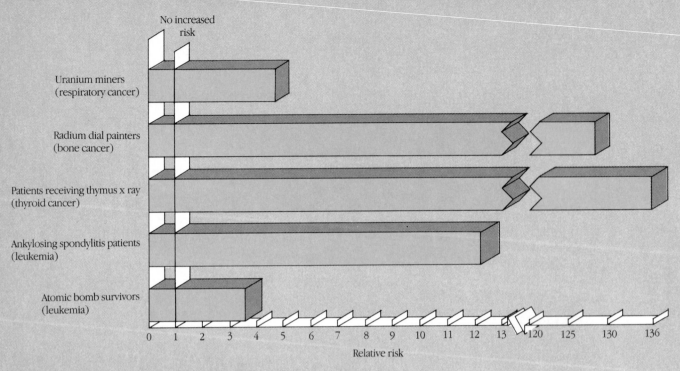

Figure S12-4. Relative risk of cancer in people exposed to various kinds of radiation. The relative risk tells us the probability of contracting cancer. In other words, a uranium miner is four times more likely to develop respiratory-tract cancer than someone who is not a miner. Atomic bomb survivors, in general, are 3.5 times more likely to develop leukemia than nonirradiated people.

nuclides may be biologically magnified in the higher trophic levels of food chains.

Minimizing the Risk

Radiation can be reduced in several ways. Since X rays are the most significant anthropogenic source of exposure, prudence dictates a cautious use of them. High-dose exposures especially warrant thorough discussion with the doctor.

Important ways to reduce X-ray exposure include (1) asking the physician if previous X rays or diagnostic procedures you've had would provide the same information; (2) reducing X-ray exposure in children; (3) informing physicians and dentists that you are pregnant (don't wait to be asked); (4) if you are pregnant, avoiding all X rays of the pelvis, abdomen, and lower back unless they're absolutely necessary; (5) avoiding mobile X-ray units, because they tend to give higher-than-necessary doses; (6) if you are a woman under the age of 50 and have no family history of breast cancer, avoid routine mammographs; (7) questioning the necessity of preemployment X rays; (8) if you must be X-rayed, requesting that a full-time radiologist do it; (9) asking if the X-ray machine and facilities have been inspected and set to minimize excess exposure; (10) requesting that a lead apron be placed over your chest and lap for dental X rays and that a thyroid shield be placed around your neck; (11) cooperating with the X-ray technologist (do not breathe or move during the X ray); and (12) making sure the operator exposes only those parts of the body that are necessary.

Radiation exposure from medical diagnosis and treatment is by far the easiest to control on an individual level. Controls on exposures from nuclear weapons testing, possible nuclear war, and possible catastrophic accidents at nuclear power plants may seem out of the average citizen's hands. Nevertheless, individual citizens can have a significant cumulative impact on nuclear

Table S12-2	Radionuclides from Nuclear Weapons and Reactors
Nuclear Weapons	**Nuclear Reactors**
Strontium-89	Tritium (Hydrogen-3)
Strontium-90	Cobalt-58
Zirconium-95	Cobalt-60
Rubidium-193	Krypton-95
Rubidium-106	Strontium-85
Iodine-131	Strontium-90
Cesium-137	Iodine-130
Cerium-141	Iodine-131
Cerium-144	Xenon-131
	Xenon-133
	Cesium-134
	Cesium-137
	Barium-140

policies by educating themselves and others, voting responsibly, becoming involved in political organizations, and writing letters to their representatives in Congress. For an opinion on the influence of such letter writing and for some suggestions on how to write letters on political issues, see the Viewpoint "The Right to Write" in Chapter 21.

Suggested Readings

Conservation Foundation (1987). State of the Environment. Washington, DC: Conservation Foundation. General coverage of radiation issues.

Gofman, J. W. (1981). *Radiation and Human Health.* San Francisco: Sierra Club Books. Comprehensive survey of the effects of low-level radiation on health.

Hobbs, C. H. and McClellan, R. O. (1980). Radiation and Radioactive Materials. In *Toxicology: The Basic Science of Poisons* (2nd ed.), ed. J. Doull, C. D. Klaassen, and M. O. Amdur. New York: Macmillan. Excellent review.

Johnson, C. J., Tidball, R. R., and Severson, R. C. (1976). Plutonium Hazard in Respirable Dust on the Surface of Soil. *Science* 193: 488–490. Interesting case study.

Laws, P. W., and Public Citizen Health Research Group. (1983). *The X-Ray Information Book.* New York: Farrar, Straus, and Giroux. General coverage of X rays, their effects, and ways to minimize exposure.

Mancuso, T. F., Stewart, A., and Kneade, G. (1977). Radiation Exposures of Hanford Workers Dying from Cancer and Other Causes. *Health Physics* 33: 369–385. Excellent study.

13

The Earth and Its Mineral Resources

Conservation is humanity caring for the future.

NANCY NEWHALL

To a person gazing out on a vast expanse of land—the prairies of Oklahoma or Texas, for instance—the earth seems stable and permanent. But churning deep within its interior is molten rock that can spew out in frightening displays, burying villages and farmland, leveling trees, and killing wildlife. Huge land masses fold and twist as they crash together. Others move like pieces of a jigsaw puzzle as if possessed by a fiery devil in the earth's interior.

Evidence of the earth's restlessness is all around us in earthquakes, volcanoes, and mountain ranges. Nowhere is this more apparent than in California. Consider, for instance, that in 1.5 million years Los Angeles will lie where San Francisco is. My publisher, now just south of San Francisco, will be well on its way to Canada. (Bundle up, gang.)

What's happening is that a narrow strip of land, bounded on the east by the San Andreas fault and on the west by the Pacific Ocean, is inching its way northward. It is part of a huge plate in the earth's crust encompassing nearly the entire Pacific Ocean. On its slow northward creep it is taking along its tiny strip of exposed land—and millions of unsuspecting passengers. The rumbling and earth-shattering quakes that now plague the West Coast give evidence of the Pacific plate grinding against the American plate, which includes North and South America (Figure 13-1).

The Earth and Its Riches

To understand this process and to understand the important resources we gather from the earth's restless crust,

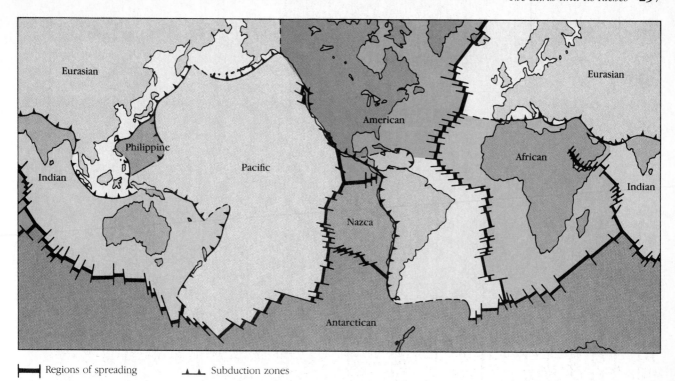

Regions of spreading Subduction zones

Figure 13-1. The earth's crust is broken into thin moving plates on the surface. Parallel lines indicate regions of spreading. Lines with solid triangles indicate subduction zones where one plate slides under another. Triangles indicate movement of the plates.

let us turn back to the time the earth began to form. Five billion years ago the earth began to cool; the surface of this molten mass gradually formed a thick rocky crust (Chapter 2). As the earth cooled, water vapor in the atmosphere condensed and rained down, forming oceans, lakes, and rivers. Today 29% of the earth's surface is land, and 71% is water. Beneath the crust is the mantle, and beneath that, the core.

A Rocky Beginning

After the earth had cooled, its crust consisted of solid rock. Over time, this rock was subject to many altering forces. Wind and rain, for example, created tiny particles that would give rise to soil. Rock was compressed and ripped apart by the earth's internal forces, creating new types of rock. These and other forces are active today, helping to reshape the planet.

Despite these dramatic changes, the earth's crust contains the same inorganic compounds, called **minerals**, that were present during its fiery beginning. These minerals are made up of elements (Chapter 2). The most abundant elements in the earth's crust are oxygen, aluminum, iron, and magnesium. Others, like gold and platinum, are extremely rare.

Rocks are solid aggregates of minerals. As a rule, several different minerals are found in each of the many types of rock. To simplify matters, however, geologists divide rocks into three major classes: (1) **Igneous rocks**, such as basalt and granite, are those formed when molten minerals cool. (2) **Sedimentary rocks**, such as shale and sandstone, are formed from particles eroded from other types of rock. (3) **Metamorphic rocks**, such as schist, are formed when igneous or sedimentary rocks are transformed by heat and pressure during mountain-building processes.

This chapter is concerned with the nonfuel minerals that are used to make numerous products (automobiles, computers, and ballpoint pens) required by modern industrial societies. Most nonfuel minerals come from igneous rocks. These minerals are often concentrated in igneous rocks by geological processes. A concentrated deposit of minerals that can be mined and refined economically is called an **ore**. Most ores are mined and then treated to produce their final product, for example, metals such as aluminum and zinc.

The Movements of Continents

Unknown to many, the earth's crust is continuously being recycled. As parts of it are gobbled up and turned into

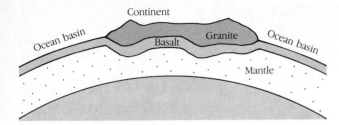

Figure 13-2. The earth's crust consists of two layers: the continental (granite) and subcontinental (basalt).

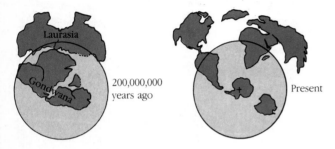

Figure 13-3. View of the earth 200 million years ago and today, showing the drift of continents resulting from movement of tectonic plates. The supercontinent shown on the left is called Pangaea.

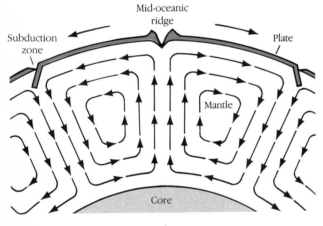

Figure 13-4. Movement of molten rock in the mantle (arrows) propels the tectonic plates.

molten rock, new rock is formed elsewhere. Geologists call this process the **rock cycle**. To understand it we must take a closer look at the earth's surface.

The earth's mineral crust consists of two layers: the outer **continental layer**, made mostly of granite, and the inner **subcontinental layer**, made mostly of a black rock called basalt. The continental layer is found only where land masses have formed. The subcontinental layer, lying beneath it, extends along ocean bottoms (Figure 13-2).

As we saw in Figure 13-1, the crust is broken into a number of plates, called **tectonic plates**. Some are oceanic, having little or no exposed continent riding on them. They consist mainly of a huge slab of basalt. Other tectonic plates consist of both land masses and ocean floor. Both the continental and the subcontinental layers are present in these plates.

Approximately 200 million years ago, the land masses we call continents were all part of a large supercontinent called **Pangaea** (Figure 13-3). Some evidence indicates that a similar supercontinent existed approximately 400 million years ago as well. Pangaea was probably centered where Africa lies today. But it broke up, geologists speculate, because of the buildup of heat in the underlying mantle. The heat accumulated beneath the continent for many years and eventually split Pangaea into many pieces, the continents. They migrated away from the center toward cool areas in the mantle.

Scientists believe that, over time, the continents may act as insulators, holding heat in. Eventually, then, the land masses may shift and join again, forming a new supercontinent and allowing the cycle to begin again.

But what causes the plates to move? Geologists think that the underlying mantle is a semisolid material that rises along one end of the plate and sinks along the other. This movement propels the plate slowly but surely. As shown in Figure 13-4, two adjoining plates pull apart at midoceanic ridges. Some of the rock pours out into the crack and solidifies. This formation of new crust under the ocean is called **sea-floor spreading**. Underlying molten rock sweeps the plate along. At the other end the plate crashes against an adjacent plate and is pushed under it. This process is called **subduction**. The continental and subcontinental layers are thrust into the mantle and melted, thus becoming part of the slow-flowing molten rock that sustains the cycle. The Nazca plate (see Figure 13-1) clearly shows the relationships. The eastern border of this plate is a subduction zone. Here it is pushed under the American plate. The western border is a region of sea-floor spreading.

The rock cycle shows that rocks of the earth's crust are reformed and implies that mineral resources must be renewable. However, new mineral deposits are formed primarily along the midoceanic ridges, inaccessible to humankind. We must make do with the finite ore deposits within our reach.

Mineral Resources and Society

More than 100 nonfuel minerals are traded in the world market. These materials, worth billions of dollars to the world economy, are vital to modern industry and agriculture. The major minerals used in the United States are

Table 13-1 Major Metals and Mineral Consumption in the US[1]

Metal/Mineral	1986[2]	1988
Aluminum	5,143,000	5,400,000
Antimony	34,433	41,400
Chromium	406,000	526,000
Cobalt	7,818	8,640
Copper	2,136,000	2,280,000
Gold	3,270,000 troy ounces[3]	3,300,000 troy ounces
Iron ore	52,560,000	67,500,000
Lead	1,148,000	1,170,000
Magnesium	115,000	135,000
Manganese	657,000	742,000
Mercury	44,582 flasks	not available
Molybdenum	19,303	16,818
Nickel	162,900	180,000
Platinum	3,537,000 troy ounces	2,700,000 troy ounces
Silver	182,000,000 troy ounces	132,000,000 troy ounces
Tin	51,535	
Vanadium	5,700	
Zinc	1,017,000	

[1]Measured in metric tons (2200 pounds) unless indicated otherwise.
[2]Estimates published in 1987 based on the first 9 months of consumption.
[3]1 troy ounce = 31.1 grams.
Source: US Bureau of Mines

shown in Table 13-1. Several dozen are so important that if any one of them was suddenly no longer available at a reasonable price, industry and agriculture would be brought to a standstill.

Who Consumes the World's Minerals?

The developed countries are the major consumers of minerals. With one-fourth of the world's population, they consume at least three-fourths of its mineral resources. These resources come from domestic mines and from many distant developing countries. The United States, with only about 6% of the world's population, consumes about 20% of its minerals. Japan, Canada, Europe, and the Soviet Union also consume large quantities.

The developing nations use mineral resources as well, but their per capita consumption is much lower than Americans, Canadians, Europeans, and others in developed countries. Latin America, Africa, and Asia have nearly three-fourths of the world's people but use only 7% of the aluminum, 9% of the copper, and 12% of the iron ore. Continued slow economic growth will probably not change these figures much by the year 2000, although there are notable exceptions, such as South Korea and Taiwan, which are quickly becoming industrialized.

Growing Interdependence and Global Tensions

As shown in Figure 13-5, the United States, Japan, and Europe depend heavily on mineral imports. This dependency, growing rapidly in recent years, results primarily from three factors: the depletion of high-grade ores in many industrial nations, rich deposits in developing countries, and lower labor costs in the developing countries. Because of the shift to foreign producers and because of domestic conservation and recycling, the American mining industry suffered drastic declines in the early 1980s. Between 1980 and 1986, for instance, mining revenues declined substantially. Employment in American mines dropped by 18% between 1981 and 1986—a decline equivalent to 600,000 jobs. After 1986, however, revenues and employment began to rise. From 1986 to 1988, for instance, revenues increased over 33%.

The vast majority of the United States' mineral resources come from fairly reliable sources, countries that are politically stable. What frightens some US analysts is that some crucial minerals, such as chromium and platinum, come from unreliable sources (Figure 13-6). To protect against embargoes or sudden cutoffs resulting from political upheaval in mineral-exporting nations, the United States stockpiles a three-year supply of all strategic minerals. If these supplies ran out before imports

Figure 13-5. Reliance of the United States, Japan, the European Economic Community, and the Eastern bloc on foreign and domestic mineral supplies.

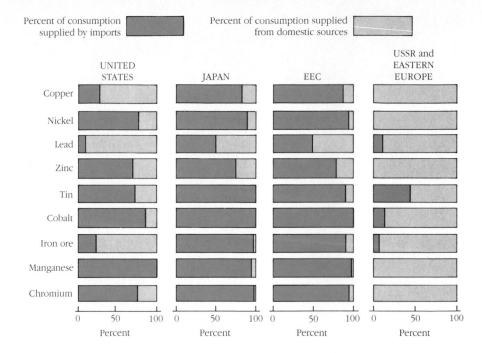

resumed, however, the economic foundation of the Western world would crumble.

Many poor, mineral-exporting nations complain about the low prices they receive for their exported raw minerals. They also feel cheated by the Western world, which buys minerals cheaply and converts them into products that reap them trillions of dollars a year. Making matters worse, many mineral-exporting nations have borrowed heavily from the West. Interest payments amount to billions of dollars, yet the rich, developed countries continue to exploit the vast mineral resources of the developing countries, profiting at the expense of the poor.

Following the lead of the OPEC nations, mineral-exporting countries may unite to form OPEC-style cartels to raise the price of mineral exports. Since over half of the copper and nearly all of the tin and aluminum reserves are in developing nations, cartels could form around these commodities. Developed countries would be forced to pay higher prices for imported minerals. Worldwide inflation could follow, crippling industry and sending millions of workers home.

Will There Be Enough?

This chapter concerns itself primarily with matters of supply, answering the critical question "Will there be enough?" Finding an answer to this question is no easy task. We will look at the question in both the short term and the long term.

In the short run, strategic supplies can help us weather sudden embargoes. That problem seems well taken care of. In the long run, however, the outlook is mixed. Some mineral supplies are adequate for many years to come, even at an increased rate of use. But other important minerals, for which no known substitutes exist, are fast on the decline. Gold, mercury, and silver are examples of such minerals. Something must be done, and done quickly, to bridge the gap.

Remember also the effect of exponential growth. Should mineral consumption increase rapidly, minerals thought to be in abundance would quickly become depleted. A resource with a billion-year lifespan would last only 580 years at a 3% growth rate.

Meeting Future Needs

Slowing the growth of the world population, perhaps stopping growth altogether, would help reduce the depletion of minerals. Beyond that, future demand can be met by at least four strategies: expanding reserves, finding substitutes, recycling, and conserving. This section looks critically at each of these options.

Can We Expand Our Reserves?

In the short term developed countries, such as Canada and the United States, have many opportunities to expand

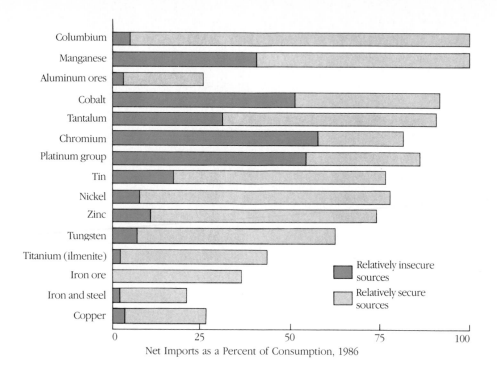

Columbium
Manganese
Aluminum ores
Cobalt
Tantalum
Chromium
Platinum group
Tin
Nickel
Zinc
Tungsten
Titanium (ilmenite)
Iron ore
Iron and steel
Copper

■ Relatively insecure sources
▨ Relatively secure sources

0   25   50   75   100

Net Imports as a Percent of Consumption, 1986

Figure 13-6. US imports of some major minerals as a percentage of total consumption. The complete bar shows the total imports as a percentage of US consumption; the colored part of the bar indicates the percentage of the minerals consumed which comes from insecure sources. The difference between the total and the colored bar represents the imports from secure sources as a percentage of total consumption.

reserves. In the long term, however, some insurmountable barriers lie in wait. This section looks at the factors that can help us expand reserves as well as factors that limit them.

Rising Prices, Rising Supplies The **law of supply and demand** has been the centerpiece of Western economic thinking for years. This law essentially says that when demand outstrips supply, prices rise. Rising prices stimulate production, which in turn increases supply. (For more on supply and demand see Chapter 20.)

Economists use this law to support their belief that the developed countries can continually expand the supply of minerals to meet ever-increasing demands. Their logic follows this path: First, rising mineral prices will make it economical for companies to mine deposits of minerals that were previously uneconomic to extract. Rising prices will also stimulate exploration. Companies will find new reserves, which can be mined as needed. This line of reasoning works well in the short term, but beyond that it falls apart. It is conceivable that as supplies dwindle, the price of a mineral would rise so high that people would simply stop buying it. In other words, when a mineral supply falls below a certain point, mining becomes too costly. For all intents and purposes the mineral is depleted. All this means that rising prices will expand our supplies, but only up to a point. Beyond that other measures must come into play.

Using Technology to Expand Reserves Technological advances will help us find ways to meet the rising

demand for minerals. Over the years technological improvements in the mining and processing of ore, for example, have increased our reserves of numerous minerals. Figure 13-7 shows the expansion of reserves in recent times.

Scientists and engineers are working on techniques to improve mining efficiency, allowing companies to exploit low-grade ore economically. One group of biologists, for example, recently found that certain algae bind gold ions; it is believed that they could help extract gold from mine wastes or even natural waters.

Technological advances are badly needed to increase mining efficiency. But technology is cost-driven. When mineral supplies fall below a certain point, costs could conceivably become too great for consumers to bear. Mining would be curtailed. Technology is not *the* answer. It can help, but it is only part of the solution.

Factors that Reduce Supplies Higher prices can stimulate production and new technologies can expand our reserves; a look at the bigger picture shows that there are many forces that can have the opposite effect. Rising energy and labor costs, for example, tend to cut our reserves. When these costs rise faster than the cost of minerals, reserves that were once economically attractive may become uneconomical to mine.

Exploration, mining, and production are also influenced by interest rates. High interest rates on capital needed for exploration and mining may slow these activities. Some economists predict that competition for capital will escalate in coming years, driving up interest rates

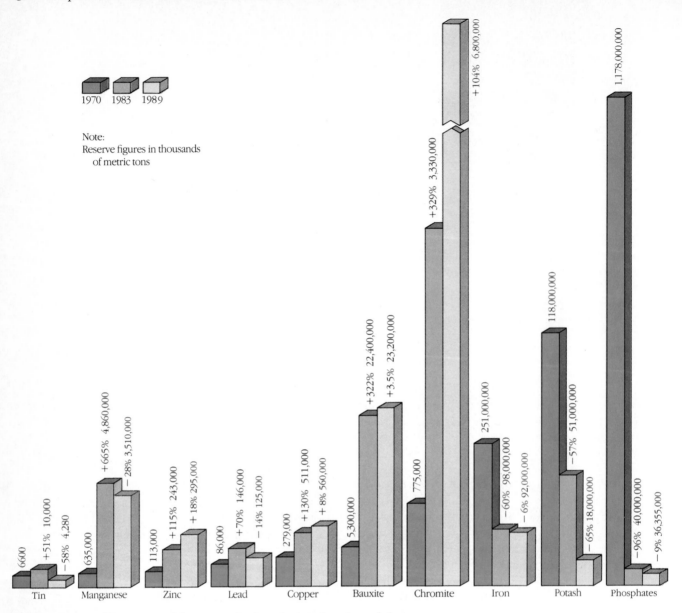

Figure 13-7. Changes in world reserves of selected minerals and metals between 1970 and 1989.

and slowing the expansion of our reserves. Before we turn to other ways of expanding reserves, let's look at energy and environmental impediments that will come into play in the near future.

Rising Energy Costs: A Key Factor Mineral resources will eventually be **economically depleted**, that is, mined until they can no longer be produced at a profit. Energy prices will play a big role in determining when mining a particular mineral becomes uneconomical and stops. As lower- and lower-grade ores are mined, the amount of energy required for mining and refining is fairly constant up to a certain point, beyond which it

increases dramatically (Figure 13-8). At this point these resources become too expensive to mine; their costs exceed what the market will bear. For all intents and purposes they are depleted.

According to the US Department of the Interior, the costs of minerals will remain constant until the year 2000. After 2000, though, they are expected to increase 5% a year, the same rate at which energy costs are predicted to increase. Rising energy prices combined with declining ore deposits could very well bring an end to some of our important mineral resources within 30 to 40 years. Runaway inflation could grip the world economy unless something is done.

Antarctica: Protecting the Last Frontier

New lands have always meant riches to explorers. Precious metals and gems were the rewards for those who ventured into uncharted territories far from their homelands. When French explorer Yves-Joseph de Kerguelen-Tremarec first saw the new continent in the southernmost reaches of the Indian Ocean in the 1700s, his mind raced with thoughts of hidden wealth, among them timber, rubies, and diamonds. Two hundred years later this vast expanse of land, called Antarctica, has offered up virtually none of the promised mineral riches.

Making up one-tenth of the earth's surface, Antarctica would be an attractive land holding—were not most of it covered with a sheet of ice a mile thick. Ironically, the coastlines and seas around this magnificent land support a rich and varied biological community. Despite the formidable cold, fierce winds, and long winter darkness, this remote wilderness and the frigid but nutrient-rich waters around it support large populations of penguins, seals, whales, sea birds, algae, krill, and fish—all uniquely adapted to the severe weather.

The promise of riches brought humanity to the frozen tip of the world where a summer day of −38° C (−35° F) is considered warm and where temperatures as low as −89° C (−127° F) have been recorded. But the wealth reaped from this land so far has been limited to information. Fighting frequent storms with winds often reaching 320 kilometers per hour (200 miles per hour), scientists are unravelling the mysteries of Antarctic life. They hope that knowledge they gain about ocean currents and Antarctic weather will improve their understanding of global weather. The officials who send them, while interested in advancing science, have a different agenda. They are hoping for mineral resources and oil that might be locked beneath the ice.

The first real impetus for Antarctic mineral exploration came after the perilous climb in oil prices in the 1970s. Reports from the US Geological Survey speculated that huge oil reserves might lie beneath Antarctica's continental shelf. But evidence of oil, many experts say, is sketchy. Similar studies of the Atlantic coast of the United States, for instance, have suggested the presence of oil beneath the continental shelf, but so far virtually no oil has been discovered there.

Sizable mineral deposits may also lie beneath the ice and snow of Antarctica, say geologists. Traces of minerals have been located in the Pensacola Mountains, which rise out of the ice of this frozen land, giving some even more hope of striking it rich. Like the early rumors of gold in the American West, these discoveries have accelerated research activity in Antarctica. The world's nations have begun to line up for their share of the envisioned wealth. Conflict is now arising over ownership of minerals and oil, should they be found.

Between 1908 and 1943 Great Britain, France, Norway, Australia, New Zealand, Chile, and Argentina claimed 85% of Antarctica. The United States and the Soviet Union, which had done their share of South Pole exploration and research, did not establish specific claims. Finding it preferable to retain an interest in all of the continent, the superpowers refused to acknowledge the territorial claims of the others, creating a sticky international situation.

In 1959 a cooperative agreement, the Antarctic Treaty System, was established to govern Antarctica. The treaty had two objectives: to maintain the continent for peaceful uses by prohibiting all military activities and weapons tests, and to promote freedom for scientific research. The treaty set aside arguments over land claims so that the sixteen nations who signed the agreement could carry out research. International cooperation for science became the basis for a whole new political order. Should minerals and oil be discovered, however, the seven original claimant countries could once again lay claim to their territories.

Hints of Antarctic minerals and oil now threaten to destabilize the climate, pitting one nation against another. The confrontation arises between the haves—the seven claimant nations—and the have-nots—those who signed the treaty and have participated in Antarctic research and management but have no legal claim to the land or resources. The have-nots also include a growing number of Third World nations, which have done no research in Antarctica but would like a share in any potential wealth. Added to the growing fray are several international organizations, such as Greenpeace International and the International Union for the Conservation of Nature and Natural Resources. They are interested in protecting Antarctica from exploration and extraction and from adverse impacts resulting from the growing tourist trade as well.

In 1988, the race for resources began when the signatories of the treaty agreed to adopt regulations that would permit oil and mineral exploration. When these regulations are approved, oil and mineral prospecting by seismic testing and other techniques with relatively minor environmental impacts could begin immediately, anywhere on the continent. Full-scale development, such as mining and drilling, however, is barred until a new agreement is made.

Many critics are afraid that in the rush to find riches, Antarctica's environment will be destroyed. The concern for environmental protection is well founded. Offshore oil rigs, for instance, would operate in some of the roughest seas in the world. Huge icebergs could rip apart platforms and drilling rigs. Ice flows could trap and crush ships. Mining would require huge amounts of energy just to melt the mile-thick icecap. Mining is made more problematic by physical constraints. Overland transport of ore, for example, would be difficult because of fierce winds and cold temperatures. Ice extending 1000 miles from land would make it difficult and costly to transport ores to awaiting ships.

Environmentalists are also concerned about the potential

Case Study 13-1 *(continued)*

impacts of resource development on the rich marine ecosystem. Oil spills, they assert, could be carried ashore and wipe out huge breeding colonies of penguins, birds, and seals. Oil slicks might also kill algae, which are responsible for replenishing 20% of the region's oxygen. Caught under the ice and in the chilly waters, oil could persist 100 times longer than in warmer waters. Finally, cleaning up an oil spill would be impaired by the short "warm" season. And, if a well blew out, winter might set in before workers could plug it up, thus allowing oil to collect under the ice for up to nine months, with potentially devastating effects.

Mineral extraction would require enormous amounts of energy, create pollution, damage the ice pack, and severely disrupt the exposed land, with its notoriously short growing season. A footprint scar takes a decade to heal. Pollution from mining operations would also disrupt scientific research on air pollution. Most damaging to wildlife, however, would be the use of ice-free coastal areas for processing and shipping ores to faraway countries.

Already, the activities of researchers and their daily maintenance have created significant air and water pollution. Solid waste has become another major problem. The National Science Foundation (NSF), which oversees US research in Antarctica, has outlined a multimillion-dollar plan to minimize pollution. The cleanup and prevention program for NSF's field camps and three year-round stations in Antarctica may cost more than $30 million over the next three or four years.

NSF officials note that there is a serious problem at McMurdo Station, which houses more than 1,000 people in the southern hemisphere during the summer. The cleanup plan calls for a sewage treatment system to help prevent the further deterioration of water quality.

Many environmental observers, however, see these efforts as inadequate. They would like to see the NSF speed up the process, and note that the National Science Foundation has been talking about many of these things since 1980. In addition, NSF has yet to address other issues such as the impact of uncontrolled vehicle emissions.

On February 16, 1989, an Argentinean ship ran aground in the harbor at McMurdo, spilling hundreds of thousands of gallons of diesel fuel. No one knows the impact this will have on aquatic life, because nothing like it has ever happened before. But the diesel fuel could kill many fish, penguins, and seals in the area.

One observer called Antarctica a political chess table. Research activity has taken precedence over environmental protection. Should oil and minerals be discovered, this relatively pristine wilderness could be sullied beyond imagination.

Some experts believe that dreams of minerals and oil are just that—dreams. Only enormous oil fields could ever be economical. Mineral extraction is even more economically preposterous, they say. But others are quick to point out that political pressures for development outweigh economic logic and the numerous ecological constraints to development. If world supplies of strategic minerals, for example, were threatened by political instabilities, world leaders might overlook the environmental and economic barriers to mineral extraction.

The Antarctic presents an unparalleled global challenge. Owned by no one, it is a harsh and fragile land where damage to the rich marine ecosystem could easily be irreparable. For years Antarctica has stood as a symbol of world peace and cooperation, but with hints of its mineral and oil riches reaching the outside world, that unique and harmonious relationship has begun to deteriorate. The challenge, many say, is to balance the costs and benefits of resource exploitation. When the costs to wildlife and the economic costs of extracting resources in the world's harshest environment are stacked against the benefits, the whole idea of development may seem preposterous.

Environmental Costs   A final problem that dims the prospects of long-term expansion of mineral reserves is environmental costs. Mining lower-grade ores will produce increasing environmental damage: larger surface mines will be needed to produce the same amount of ore, more material will be transported to smelters for processing, more waste will be produced at the mines and smelters, and more air and water pollution will result. Because of this, environmental protection costs will escalate, adding to the cost of production and ushering in the economic depletion of some minerals.

Minerals from Outer Space and the Sea   The mineral deposits on earth are finite and in many cases they have been heavily exploited, leaving little for future gen-

erations. As a result, some companies are considering three untapped frontiers—outer space, Antarctica, and the oceans—as potential sources for new minerals. (Case Study 13-1 discusses Antarctica.)

Outer space contains a wealth of mineral matter in planets and asteroids. Can't we send space ships up and haul back the riches? Critical thinking requires a look at the bigger picture, especially the potential costs of this appealing scheme. Daniel Deudney and Ben Bova debate this and other related issues in the Point/Counterpoint in this chapter.

The ocean is a vast resource of minerals, many of which are dissolved in the water itself. However, the concentrations of most dissolved minerals are generally too low to be of economic importance. More important are min-

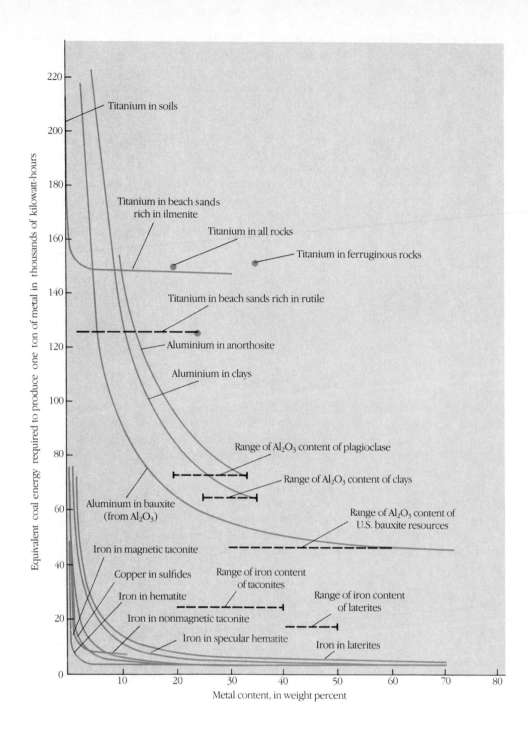

Figure 13-8. Energy investment for recovery of minerals related to the concentration of ore.

eral deposits on the seafloor. Most noteworthy of these minerals are small lumps called **manganese nodules**, which lie along deep ocean bottoms (Figure 13-9). Abundant in the Pacific Ocean, manganese nodules contain several vital minerals, notably manganese (24%) and iron (14%), with smaller amounts of copper (1%) and cobalt (0.25%).

Several mining companies have explored the possibility of dredging the seabed to mine these odd mineral deposits. The environmental impacts of such an action are largely unknown. Preliminary studies show that sea-

floor mining may be technically feasible and may be profitable. The biggest impediment is a legal one. No international agreements spell out who owns the deep seabed. Western countries have the wealth to exploit the resource. Developing nations, contending that the seabed belongs to all nations, believe they should receive a portion of the proceeds.

At a United Nations Conference on the Law of the Sea the developing countries proposed an international tax on seabed minerals, which could provide millions of dollars a year for agricultural and economic development.

Figure 13-9. Manganese nodules on the ocean's floor.

A comprehensive Law of the Sea Treaty, worked out with US negotiators during the Carter administration, would have included this plan, but the United States refused to sign the treaty under President Reagan. The UK and West Germany also refused to sign. In 1984 the US joined several nations in an agreement to work out conflicts over mining minerals on the ocean floor. This agreement will do nothing to help Third World nations benefit from the possible riches on the ocean floor.

Many people believe that seabed taxation to benefit the developing nations is fair. Others see it as an unfair way of cutting into profits. As a result of the growing conflict, progress toward mining the seabed has come to a halt.

Can We Find Substitutes?

The substitution of one resource for another that has become economically depleted has been a useful strategy for industrialized nations. Shortages of cotton, wool, and natural rubber, for example, have been eased by synthetic materials derived from oil. Synthetic fibers have found their way into American clothes. Synthetic rubber has replaced most natural rubber from trees. Substitution will unquestionably play an important role in the future, too. But is substitution a crutch we can lean on forever?

Critics argue that substitutions have created unreasonable faith among the public in the ability of scientists to come up with new resources to replace those that are being depleted. They also note that many substitutes have limits themselves. Plastics have replaced many metals, for example, but the oil from which plastics are made is a limited resource.

Substitutions will play an important role in the future, but they are not a cure-all for pending mineral shortages. Some resources may have no substitutes at all. For instance, it may be impossible to find substitutes for the manganese used in desulfurizing steel, the nickel and chromium used in stainless steel, the tin in solder, the helium used in low-temperature refrigeration, the tungsten used in high-speed tools, or the silver used in photographic papers and films.

Finding substitutes has become a race against time. Since we have rounded the bend of the exponential curve of demand, the time of economic depletion is fast approaching, perhaps more quickly than substitutes can be found. A wise strategy would be to identify those resources that are nearest to economic depletion and then promote widespread conservation and recycling. Research to find substitutes should begin immediately or, if research is already under way, it should be greatly accelerated.

Can Recycling Stretch Our Supplies?

Recycling of materials can alleviate future resource shortages, greatly reduce energy demand, cut pollution, and reduce water use. Instead of being discarded in dumps, valuable minerals can be returned to factories, melted down, and reshaped. In the United States and many other countries, however, recycling efforts have fallen short of their full potential except in a few instances such as the automobile. Approximately 90% of all American cars are recycled. The recycling of aluminum, steel, and many other metals could be doubled.

Recycling: Only a Partial Answer Recycling helps increase the time a mineral or metal remains in use, or its **residence time**. It also helps us save an enormous amount of energy. For example, manufacturing an aluminum can from recycled aluminum uses only 5% of the energy required to make it from aluminum ore (bauxite). Recycling, however, will not permit our current exponential growth in mineral use to continue indefinitely. Why is this so?

First, during the production–consumption cycle some minerals and metals enter into long-term uses, for example, aluminum used for wiring or bronze for statuary. Also, some materials are lost through processing inefficiencies, are lost accidentally, or are thrown away on purpose. For these reasons it is impossible to recycle 100% of a given material. A more practical goal would be 60% to 80%. Thus, recycling can slow down the depletion of a mineral resource but cannot stop it. In theory, recycling can double our mineral resource base, but continually rising demand and inevitable losses will eventually deplete that reserve.

Is Outer Space the Answer to Our Population and Resource Problems?

No Escape from the Population Bomb

Daniel Deudney

The author is a Hewlett fellow in Science, Technology, and Society at the Center for Energy and Environmental Studies at Princeton University. He writes on international security, space, and global commons issues. He is the author of several papers on space resources and coauthor of Renewable Energy: The Power to Choose *(Norton, 1983).*

Much of the recent writing about humanity's future in space has been dominated by outlandish proposals for large-scale space operations that aim to bypass the earth's resource limits, either by exporting people from the planet or by importing energy and materials from space. These include space colonies, solar-powered satellites, and asteroid mining operations. At first glance these massive undertakings have a logical appeal: the earth is limited, space is infinite.

Even though space contains vastly more "living" room, energy, and materials than the earth, this abundance cannot be brought to bear meaningfully on the earth's problems. Whatever the long-term prospect for human colonies in space, the earth's population and resource problems will have to be solved on earth in the near term.

Long a staple of science fiction writers, space colonies were among the many visionary ideas proposed by the pioneers of rocket technology in the early years of this century. In recent years space colonies have been advanced as a near-term solution to the earth's population and environmental problems by Professor Gerard O'Neill, a Princeton University physicist. In his writings he details plans to build colonies with first 10,000 inhabitants, and later a million. Manufactured out of materials from the earth, and then the moon and asteroids, the colonies would be made completely self-sufficient by harnessing the sun for all sources of energy. Such colonies would float freely in space, circumventing the need to tame the harsh environments of other planets. Space colonization is specifically promoted as a solution to overcrowding and environmental degradation on earth. By the exporting of ever greater numbers of people into these orbiting cities, the wildlife and wilderness qualities of earth could be protected, or perhaps even expanded.

Life in space is envisioned as pastoral, pollution-free, and pluralistic—like floating garden cities. Yet in reality space habitation would probably be bleak. Thick metal shielding would be necessary to block the lethal cosmic and solar radiation. Life would be like that in a submarine—cramped, isolated, and uneventful. For at least the next several decades people will go into space to perform various specialized missions— or perhaps briefly as tourists—but they will not live there in significant numbers.

No scientific laws forbid large space colonies, but the technology to build and maintain such structures remains conjecture. Structures of the size envisioned are thousands of times larger than anything yet built for space; no doubt there will be unforeseen and even insurmountable technical problems. The ecologist Paul Ehrlich points out that scientists have no idea how to create large, stable ecosystems of the sort that would be needed to make space colonies self-sufficient. The key to such knowledge is, of course, much more study of the ecosystems on earth, many of which are becoming less diverse and less stable. It could be many decades before scientists know enough to understand—let alone recreate—ecosystems as complex as those now being degraded. There are also unanswered questions of human biology: Can babies be born and grow up in a weightless environment? Would the various forms of cosmic and solar radiation make the mutation and cancer rate unacceptably high?

Space colonies are not even a partial answer to the population and environmental problems of earth. At a time when at least one billion people live in "absolute poverty" on earth, it makes little sense to think about building fabulously expensive habitats in space. Simply transporting the world's daily increase of about 240,000 people into space would consume the annual gross national product of the United States. Each launch of the space shuttle costs $300 million, and that doesn't even include the $25 billion in research and other initial costs. Maintaining the complex life-support systems in orbit costs thousands of dollars an hour.

Population stabilization is a difficult social challenge, but it is certainly less complex than the organizational and political skills needed for large-scale colonization of space.

Solar-powered satellites, or Sunsats, are often seen as complements of large-scale space colonization and as a source of

⟨✕⟩ *Point/Counterpoint (continued)*

energy for earthlings. The Sunsat's principal appeal is its ability to collect virtually unlimited amounts of solar energy day and night without polluting the earth's atmosphere. The construction of Sunsat, however, would be an undertaking of unprecedented size and cost. A 1980 NASA and US Department of Energy study estimated that 60 satellites, each as big as Manhattan Island, would be needed to produce the current US electrical usage. The cost estimates ranged from an optimistic $1.5 trillion to a more probable $3 trillion. If two Sunsats were built each year, one heavy-lift rocket—seven times the largest rocket ever built—would have to be launched each day for 30 years to build and service them.

More troubling are planetary-scale environmental risks. Beaming trillions of watts of microwaves through the atmosphere for extended periods is almost certain to alter the composition of gases in unpredictable ways. Launching millions of tons of material into orbit would also release large quantities of exhaust gases into the upper atmosphere, perhaps disrupting the ozone layer which screens out ultraviolet light. An operational Sunsat system big enough to make a difference in the terrestrial energy equation would be a shot-in-the-dark experiment with our atmosphere.

The other mirage of abundance in space that has recently received attention is asteroids. These irregularly shaped rocks, which orbit the sun, range in size from as small as a grain of sand to as big as the state of Texas. Although there is probably enough metal in the asteroid belt to meet world needs for many centuries, getting them to the earth's surface would be costly and energy-intensive and would risk an accidental collision and ecological disruption on a colossal scale. Long before it becomes feasible or economical to bring rare metals from space, scientists should be able to turn the abundant clay, silicon, alumina, hydrocarbons, and iron in the earth's crust into materials that can be used to meet global needs.

In summary, large-scale space colonization and industrialization is an unworkable attempt to escape from the problems of the earth. In the struggle to protect the earth from overpopulation, ecological degradation, and resource depletion, outer space has a great, largely unfulfilled role to play. It can be valuable not as a source of energy or materials or as a place to house the world's growing population, but rather as a tool both for learning more about our planet and for assisting problem solving here on earth.

Toward a New World

Ben Bova

Ben Bova is the author of more than 75 futuristic novels and nonfiction books about science and high technology. Among them are Welcome to Moonbase, *an examination of the economic, social, and scientific benefits to be reaped from a permanent settlement on the Moon. He is the former editor of* Omni *and* Analog *magazines, and president emeritus of the National Space Society.*

We live in a solar system that is incredibly rich in energy and raw materials. In interplanetary space more wealth than any emperor could dream of is available for every human being alive. Instead of thinking of our world as a finite pie that must be sliced thinner and thinner as population swells, we must go out into space and create a larger pie so that everyone can have bigger and bigger slices of wealth.

What can we gain from the New World that begins a couple of hundred miles above our heads? Of immediate concern, there is energy in space—enormous energy from unfiltered sunlight. The sun radiates an incomprehensible flood of energy into space—the equivalent of 10 billion megatons of H-bomb explosions every second. The earth intercepts only a tiny fraction of this energy, less than .2 billionths of the total energy given off by the sun.

For the purposes of our generation, a few dozen solar-powered satellites in geostationary orbit above the equator may be sufficient. Each would beam 5 billion watts of energy to us, making a substantial contribution to our energy supplies.

Natural resources are also there in superabundance. Thanks to the Apollo explorations of the moon, we know that lunar rocks and soil contain many valuable elements such as aluminum, magnesium, titanium, and silicon.

Looking further afield, there are asteroids that sail through the solar system, especially in a region between Mars and Jupiter called the Asteroid Belt. To a miner or an industrialist they are a bonanza. An MIT astronomer, Tom McCord, estimates that there are hundreds of millions of billions of tons of nickel–iron asteroids in the belt. The economic potential of these resources is incalculably large. And mixed in with these are many other valuable metals and minerals.

This "mother lode" of riches is a long way from earth—hundreds of millions of miles—but in space, distance is not so important as the amount of energy you must expend to get where you want to go. Like sailing ships, spacecraft do not burn fuel for most of their flight; they simply coast after achieving sufficient speed to reach their destination. Space is not a barrier, it is a highway. The biggest and most difficult step in space flight is getting off the earth's surface.

The opportunity offered by space is not to export people from earth, but to import valuable resources to earth.

We will move outward into space because it is biologically necessary for us to do so. Like children driven by forces beyond our understanding, we head into space citing all the good and practical and necessary reasons for going. But in actuality we go because we are driven.

The rocket pioneer Krafft Ehricke has likened our situation on earth today to the situation in a mother's womb after nine months of gestation. The baby has been living a sweet life, without exertion, nurtured and fed by the environment in which it has been enclosed. But the baby gets too big for that environment, and the baby's own waste products are polluting that environment to the point where it becomes unlivable. Time to come out into the real world.

Our biological heritage, our historical legacy, our real and pressing economic and social needs are all pointing toward the same conclusion: It's time to come out into the real world. Time to expand into the new world of space.

Critical Thinking

1. Summarize the views of both authors on using outer space to reduce population pressure, acquire energy, and find mineral resources. Using your critical thinking skills, analyze each argument.

2. Based on your analysis, which author do you most agree with? Why?

Can Conservation Stretch Our Supplies?

The author Nancy Newhall once observed that "conservation is humanity caring for the future." Conservation is often the cheapest and easiest strategy for stretching our mineral resources. In the long run our consumption of mineral resources will inevitably fall as prices rise. Prudence dictates strategies that delay the economic depletion of minerals. This book takes the position that conservation must begin with individuals like yourself and spread through society to the highest levels, where it will be reflected in national policies. Combined with recycling, conservation measures can greatly extend the lifetime of many valuable mineral resources. As shown in Figure 13-10, continued exponential growth—the track we have been on until recently—is the fastest route to depletion; recycling will slow down, but not stop, this depletion. Recycling and conservation measures combined will give us more time to develop new mining technologies and find substitutes.

Some Suggested Personal Actions Tolstoy wrote that "everyone thinks of changing the world, but no one thinks of changing himself." Personal actions go a long way toward increasing our mineral supplies, and they can save you money. Recycling and conservation are two such actions. Careful buying can also reduce waste. By purchasing food in recyclable containers or renewable wrappers (paper), for example, you and millions like you can

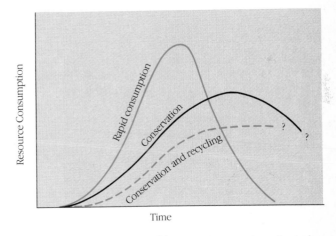

Figure 13-10. Three possible scenarios on a hypothetical time scale. Which one will we take?

curb the depletion of various minerals and oil, from which plastics are made.

When you buy soft drinks and other beverages, choose returnable bottles (although they are increasingly hard to find) in preference to recyclable aluminum cans, and aluminum cans above steel cans. Returnable bottles can be reused with a minor expenditure for transportation and washing; recyclable ones have to be transported, melted down, and then reformed, requiring more energy.

As for transportation, any measure that saves energy also saves valuable minerals. Buy smaller cars, and use mass transit so your family can get by with only one car. Build smaller homes. Buy solar homes.

When buying something, look for quality; a well-made product, by outlasting its inexpensive imitations, will be well worth the extra labor and material that was put into it. In the long run high-quality materials save resources. Write letters supporting durable products, and complain to manufacturers whose junk falls apart on you soon after you get it home. Remember that consumers are to economics what voters are to elections.

You can also help by supporting various legal solutions, for example, recycling programs on a local, state, or national level. Recycling bins in neighborhoods should be encouraged: volunteer groups can pick up the materials regularly and sell them for profit. (The city might even assist by providing trucks.) Waste recovery systems at local dumps might also be helpful in conserving recyclable materials. Palo Alto, California, for example, gives city residents a free dump pass if they bring recyclable materials to the recycling center just outside the dump gates.

Recent polls show that three-fourths of all Americans questioned favor increased recycling. However, only 11% of US waste is recycled. "Between saying and doing," the author F. W. Robertson wrote, "is a great distance." Blessed with cheap energy resources and abundant food sources, the United States has become the most wasteful country in the world. Used to the no-deposit, no-return appeal of modern products, the nation is unnecessarily throwing away its future. Clearly, Americans are robbing their descendants of their future. Conservation is not a roadblock to progress. For a society that has always viewed uncontrolled, ever-escalating production and consumption as evidence of prosperity, conservation is a sign of progress.

Tomorrow's growth depends on the use we make of today's materials and experiences.

ELMER WHEELER

Summary

Five billion years ago the molten earth began to cool, forming a crust of rock. This rock was subject to many altering forces. Soil formed from particles eroded by rainfall. New types of rock formed in the earth's crust.

The crust contains inorganic compounds called minerals. There are usually several types of mineral in any one type of rock. Concentrated deposits of minerals that are economical to mine and process are called **ores**.

The earth's crust participates in an enormous rock cycle, in which new rock is continuously formed while parts of the crust are melted. The crust is broken into huge **tectonic plates**, which move about over millions of years, a process called **continental drift**. The tectonic plates pull apart at midoceanic ridges. Molten rock flows up into the gap, creating new crust. In other parts of the plate the crust is thrust under adjoining plates and is melted.

Minerals are worth billions of dollars in the world market. Some are so important that industry and agriculture in developed nations would come to a standstill if they suddenly became unavailable. The developed countries are the major consumers of minerals. With one-fourth of the world's population, they consume three-fourths of its minerals. Most of these nations import large quantities from developing nations.

Developing nations have become increasingly unhappy with the low prices they receive for their raw minerals. Some analysts fear that they will form cartels to control prices. This would result in an upward trend in the cost of metals, with serious consequences for the economies of industrial nations.

One of the key issues regarding minerals is whether there will be enough to meet future needs. In the short term, strategic supplies will help us weather any sudden drops in supply. In the long term, however, some important minerals are being used up.

Some economists believe that rising prices will stimulate exploration and new technologies, which will increase our reserve of nonfuel minerals. They also argue that substitutes for economically depleted mineral resources will be made available through advances in science and technology. As mineral resources of lower and lower concentration are exploited, however, energy investments in mining and processing will become higher and higher. After 2000 or 2010 the real cost of minerals may begin to rise by 5% a year, making many minerals uneconomical to mine and process. Competition for capital may raise interest rates, which would make it economically unprofitable to mine and process lower-grade ores. Mining low-grade ores increases environmental damage as well.

Minerals from the sea may help expand our reserves, but economic, environmental, and legal questions have yet to be worked out. One of the most promising resources is **manganese nodules**, which are found on the seabed. They primarily contain manganese and iron.

Substitution of one resource for another that has become economically depleted has been a useful strategy in the past, but it will be of limited value in the future. Substitutes for scarce minerals may have limits themselves. Some minerals have no adequate substitutes. Finally, in some cases economic depletion may occur before substitutes can be developed.

Recycling can alleviate future resource shortages, cut energy demand, reduce environmental pollution, and create jobs, but it is not a panacea. Many materials are put into permanent use. For short-term goods, such as cans and bottles, recycling recovers 60% to 80% of total production. In short, recycling can slow down the depletion but cannot stop it.

Conservation is a highly favored approach. Combined with recycling, it can greatly extend the lifetime of many valuable minerals and give us more time to find substitutes and develop new mining strategies.

Discussion Questions

1. Describe the earth's crust using the following terms: *continental* and *subcontinental layers, midoceanic ridges, seafloor spreading, subduction zones, tectonic plates,* and *continental drift.*

2. Define nonrenewable and renewable mineral resources, and give examples of each. How are they different?

3. Why is it difficult to determine the level of future mineral demand? How would you go about calculating the demand for minerals in the year 2000?

4. Debate the statement "Economic forces will ensure us a continual supply of mineral resources. As prices rise, we'll find new resources, develop new technologies, and find substitutes for minerals currently used."

5. What are reserves? Explain how mineral reserves can expand.

6. The trend over the past few decades has been increasing mineral reserves. Why has this occurred? Will it continue? Why or why not?

7. Outline a plan to meet the future mineral needs of our society. Describe your plan. What are its most important components?

8. Draw a graph of the energy required to mine an ore as the grade decreases. What are the long-term implications?

9. Debate the statement "There is no need to worry about running out of minerals; we will find substitutes for them."

10. List the advantages of recycling and resource conservation.

11. Make a list of ways in which you can cut down your resource consumption by 20%.

Suggested Readings

Barry, P. (1985). Cities Rush to Recycle. *Sierra* 70 (6): 32–35. How American cities are responding to shrinking land sites for waste disposal.

Bogart, P. (1988). On Thin Ice: Can Antarctica Survive the Gold Rush? *Greenpeace* 13 (5): 7–11. Telling account of the troubles in Antarctica.

Chandler, W. U. (1984). Recycling Materials. In *State of the World,* ed. Linda Starke. Washington, DC: Worldwatch Institute. Superb review.

Council on Environmental Quality. (1980). *The Global 2000 Report to the President.* New York: Penguin. Chapter 12 provides an excellent survey of nonfuel mineral resources.

EDF. (1988). *Coming Full Circle: Successful Recycling Today.* Washington, DC: Environmental Defense Fund. Superb survey of recycling programs.

Pollock, C. (1987). *Mining Urban Wastes: The Potential for Recycling.* Worldwatch Paper 76. Washington, DC: Worldwatch Institute. Fact-filled reading.

US Bureau of Mines. (1989). *Mineral Commodity Summaries, 1989.* Washington, DC: US Government Printing Office. Excellent source of information on minerals published every year.

US Bureau of Mines. (1990). *Mineral Facts and Problems.* Washington, DC: US Government Printing Office. Superb reference published every five years.

IV

POLLUTION

Toxic Substances: Principles and Practicalities

Life is a perpetual instruction in cause and effect.

RALPH WALDO EMERSON

Shortly after midnight on December 3, 1984, a 35-meter-high (100-foot) cloud of methyl isocyanate gas escaped from the Union Carbide Corporation's chemical plant in Bhopal, India. Within minutes, chaos erupted. Frightened residents—coughing and screaming—ran helter-skelter through city streets to escape. Few people knew what was going on. Fewer still knew what to do.

A current report puts the death toll near 3000. Nearly 2000 more will die by 1995. All told, 200,000 people had been injured (Figure 14-1). Of these, 17,000 have been permanently disabled, largely with lung ailments. Several thousand dead and decaying animals littered the streets. In 1989, Union Carbide agreed to pay $470 million in damages to the victims and their families. The newspapers announced the end of the case, but reporters did not study the facts carefully. The payment was only for interim relief and represents only a fraction of the $3.3 billion being sought by 500,000 plaintiffs.

What makes the Bhopal accident more tragic is that any number of simple steps could have prevented it or lessened its severity. Had citizens been advised to breathe through wet towels, for example, many victims would be alive today. Had one of the backup systems in the plant operated, the accident could have been prevented altogether.

Focusing world attention on the risk of toxic chemicals, Bhopal remains a stark symbol of one of the possible consequences of the marriage between modern society and the chemical industry. But it is just one of many examples worth noting. Case Study 14-1 shows growing evidence of the impact of electricity on our health. This chapter looks at toxic substances—their effects and their control. It concludes with an assessment of the risks and benefits of life in a world dependent on technology and the products of the chemical industry.

Principles of Toxicology

Sixty thousand chemical substances are sold commercially in the United States. By various estimates, 8600 food additives, 3400 cosmetic ingredients, and at least 35,000 pesticides are in use. Chemical production and use have skyrocketed since World War II (Figure 14-2). Of the commercially important chemicals, however, only a small number—perhaps 2%—are known to be harmful. Still, this small percentage amounts to hundreds of potentially dangerous chemicals, hazardous mainly to workers but also to the general public, as the Bhopal tragedy makes clear. According to the Department of Transportation, more than 62,000 accidents during the transportation of toxic wastes occurred in the United States between 1980 and 1987. The human cost in comparison with Bhopal was small—2756 people killed and injured. Many more accidents occur in the workplace and in toxic waste facilities.

Perhaps the biggest problem with toxic substances is our lack of knowledge about their effects. The National Academy of Sciences notes that fewer than 10% of US agricultural chemicals and 5% of food additives have been fully tested to assess chronic health effects. Testing potentially harmful substances is a costly and time-consuming task, made more difficult by the 700 to 1000 new chemicals entering the marketplace each year.

Biological Effects of Toxins

Toxic substances, or **toxins**, are chemicals that adversely affect living organisms. **Toxicology** is the study of these effects. Our concern in this chapter is primarily with the effects on humans.

Average citizens are exposed to toxic substances at home, at work, or while moving about outdoors. In many cases they have little control over exposure. Polluted air from nearby power plants or highways exposes them to dozens of potentially harmful substances. In some cases, however, we intentionally expose ourselves to harmful substances, such as smoke from cigarettes.

Chemical substances exert a wide range of effects, depending on the amount we receive. This amount, in turn, is determined by the concentration (or dose) of the toxin and the duration of the exposure—known as the two D's (dose and duration). Some effects may be subtle, such as a slight cough or headache caused by air pollution. Others can be pronounced, such as the violent convulsions brought on by exposure to certain insecticides. **Toxicologists**, the scientists who study toxins, classify effects as acute or chronic.

Figure 14-1. Indian woman being treated for exposure to deadly MIC that escaped from Union Carbide's Bhopal, India facility, causing widespread sickness and death.

Annual Production of Synthetic Organic Chemicals, 1920–1987

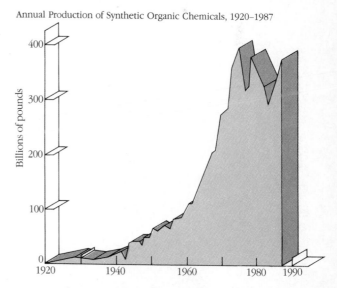

Figure 14-2. Growth in the production of synthetic organic chemicals in the United States.

Acute Effects Acute effects are those symptoms that appear right after exposure. In Bhopal, for example, many people complained of chest pains and severe eye irritation. Many children died soon after being exposed. Acute

The Dangers of Asbestos

Asbestos is the generic name for several naturally occurring silicate mineral fibers. Asbestos is useful because of its resistance to heat, friction, and acid; its flexibility; and its great tensile strength. Over 3 million metric tons of asbestos are used worldwide each year for thousands of different commercial applications.

Two-thirds of all asbestos produced is added to cement, giving it a better resistance to weather. Asbestos also insulates steel girders in buildings and serves as heat insulation in factories, schools, and other buildings. In addition, it can be found in brake pads, brake linings, hair driers, patching plaster, and a multitude of other products.

Asbestos is dangerous because its fibers are easily dislodged. Floating in the air, these fine particles may be inhaled into the lungs, where they are neither broken down nor expelled but remain for life. Three disorders may result: pulmonary (lung) fibrosis, lung cancer, and mesothelioma.

Pulmonary fibrosis, or **asbestosis**, is a buildup of scar tissue in the lungs that may occur in people who inhale asbestos on the job or in buildings with exposed asbestos insulation. The disease takes 10 to 20 years to develop after the first exposure.

Exposures to asbestos at low levels, even for short periods, can cause lung cancer. The death rate from lung cancer in asbestos insulation workers in the United States is four times the expected rate. The incidence of lung cancer in asbestos workers who smoke is 92 times greater than in asbestos workers who don't smoke, providing a striking example of synergism.

Asbestos is the only known cause of **mesothelioma**, a cancer that develops in the lining of the lungs (the pleura). Highly malignant, this cancer spreads rapidly and kills victims within a year from the time of diagnosis.

Scientists have long wondered how asbestos causes cancer. New research suggests that asbestos fibers attach to DNA outside cells and carry it into them, there creating mutations that lead to cancer. Measurable amounts of DNA are normally found in tissue fluids surrounding cells of the body. After binding to the DNA, the research shows, the asbestos fibers pierce the cell membrane. What happens to the DNA inside the cell is still unknown. Several possibilities exist. One possibility is that the DNA disrupts or turns off genes that control a cell's growth. With the control mechanism paralysed, the cell begins to duplicate wildly. Another possibility is that the DNA may carry cancer-causing genes into the cell. Once inside, the genes become activated, triggering the cell to divide.

An estimated 8 million to 11 million American workers have been exposed to asbestos since World War II. Studies show that over a third had lung cancer, mesothelioma, or gastrointestinal cancer. The expected death rate in the population for these diseases is roughly 8%.

The use of asbestos in the United States for insulation, fireproofing, and decorative purposes was banned in 1978. In 1979 the EPA began to assist states and local school districts in identifying and removing hazardous asbestos crumbling from pipes and ceilings. Since that time, Johns Manville, a major supplier of asbestos products, has been inundated with personal damage suits amounting to over $2 billion. In 1983 the corporation filed for bankruptcy and reorganization under federal law.

In 1986 the EPA proposed a ban on all remaining asbestos products, completely phasing them out by 1996. To protect workers in the meantime, the Occupational Safety and Health Administration toughened its rules.

Incidence of cancer in asbestos workers in the United States and Canada. (Ratio of the number of observed to the number of expected deaths times 100.)

effects often disappear shortly after the exposure ends and are generally caused by fairly high concentrations of chemicals during short-term (acute) exposures.

Chronic Effects Chronic effects are delayed, but long-lasting, responses to toxic agents. They may occur months to years after exposure and usually persist for years, as in the case of emphysema caused by cigarette smoke or pollution. Chronic effects are generally the result of low-level exposure over long periods (chronic exposures). It is important to note, though, that short-term exposures may also have delayed effects. In Bhopal, for example, methyl isocyanate may have caused a host of long-term effects, symptoms such as paralysis that appeared weeks after the exposure.

Chemicals can affect virtually every cell in the body. Their hidden effects, such as cancer, mutations, birth defects, and reproductive impairment, pose the most serious challenge to society.

Cancer Cancer annually kills 500,000 people in the United States. **Cancer** is an uncontrolled proliferation of cells that forms a mass, or **primary tumor**. Cells may break off from the tumor and travel in the blood and other body fluids. The spread of cancerous cells is called **metastasis**. In distant sites the cancerous cells may form **secondary tumors**.

Every cancer starts when a single cell goes haywire, a process that occurs most often in tissues undergoing rapid cellular division, for example, the bone marrow, lungs, lining of the intestines, ovaries, testes, and skin. Nondividing cells, such as nerve cells and muscle cells, rarely become cancerous.

Despite years of intensive research scientists remain uncertain about the causes of some types of cancer. Many cancers, they know, begin after **mutations**, or changes in the genetic material, DNA. Ninety percent of all chemicals known to cause cancer also cause mutations in bacterial test systems. The causative agents include viruses, a variety of chemical substances, and physical agents such as X rays and ultraviolet light.

New studies indicate that emotions may also play an important role in the development of cancer (and other diseases), possibly by acting through the immune system. Researchers at the Johns Hopkins University, for example, studied the incidence of cancer in medical students who took the Rorschach test, which measures personality, between 1948 and 1964. The research showed that students who suppressed emotions were 16 times more likely to develop cancer later in life than students who vented their emotions. More research is needed to determine if the cause-and-effect relationship between mental health and cancer is real. Critical thinking demands a closer look.

Although the chromosomes of cancer cells are typically abnormal in structure or in number, a number of **carcinogens** (agents that cause cancer), such as asbestos, certain plastics, and certain hormones, apparently do not directly alter the DNA or cause mutations. Other mechanisms must be involved.

In the 1970s many experts believed that about 90% of all cancers were caused by environmental pollutants and other agents such as X rays, ultraviolet light, and viruses. The public was shocked at this revelation. However, more careful research has shown that only 20% to 40% of all cancers are caused by workplace and environmental pollutants. The rest presumably arise from smoking, dietary factors, and natural causes. In the 1970s it seemed as if cancer was sweeping the United States. Many people began to believe that cancer rates were rising rapidly. As shown in Figure 14-3, however, most cancer rates have remained fairly constant for 50 years, except for lung cancer, which has risen dramatically in both men and women; testicular cancer, which has also risen; and stomach cancer, which has fallen. Overall, the incidence of cancer in the United States has increased slightly since 1962. Despite an outpouring of private and public funding on research, however, very little if any improvement has been made in treating the disease. (See John Bailar and Elaine Smith in the Viewpoint in this chapter for a discussion of this situation.)

By one commonly cited estimate made by two Oxford University scientists, 8000 Americans die each year of cancer caused by environmental factors, such as air pollution. Another 8000 cancer deaths are attributed to food additives and industrial products, such as pesticides used around the house; 16,000 deaths result each year from occupational exposure to harmful substances. The researchers note, by comparison, that tobacco causes at least 142,000 lung cancer deaths each year.

Mutations Agents that cause mutations are called **mutagens**. In general, three types of genetic alteration are seen: (1) changes in the DNA itself, (2) alterations of the chromosomal structure that are visible by microscope (deletion or rearrangement of parts of the chromosome), and (3) missing or extra chromosomes. For our purposes the term *mutation* encompasses all three.

Mutations can be caused by chemical substances, such as caffeine, or physical agents, such as ultraviolet light and other high-energy radiation (see Chapter Supplement 12-1). In humans, mutations can occur in normal body cells, or **somatic cells**, such as skin and bone. Such mutations occur quite frequently but are usually repaired by cellular enzymes. If a mutation is not repaired, it may lead to cancer.

The reproductive cells, or **germ cells**, in the male and female gonads are also susceptible to mutagens. Unrepaired germ-cell mutations may be passed on to offspring.

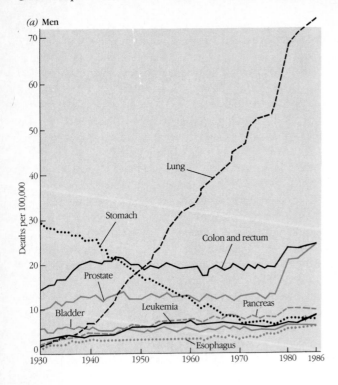

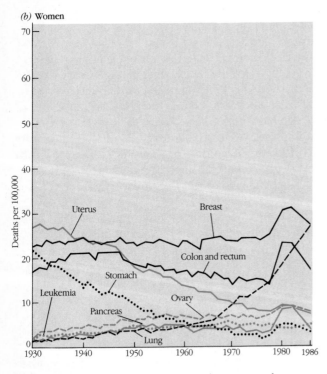

Figure 14-3.   Cancer rates in men and women in the United States. Most cancer rates have stabilized except for lung cancer, which is rising, stomach cancer, which has fallen, and testicular cancer, which has risen (but is not shown on the graph).

If a genetically damaged ovum, for example, is fertilized by a normal sperm, the mutation is passed on to every cell in the offspring. The defective gene may prove lethal, or it may manifest itself as a birth defect or a metabolic disease (a biochemical disorder). However, some germ-cell mutations may not evidence themselves in the first generation but may be expressed in the second and third generations. This delayed effect makes it difficult for scientists to pinpoint the causes of various diseases.

Genetic mutations are present in about 2 of every 100 newborns. The causes of mutations in humans are not well understood. Abnormal chromosome numbers, responsible for some diseases such as Down's syndrome, are related to maternal age (Figure 14-4). Broken and rearranged chromosomes are also related to maternal age. As women enter their 30s, their chances of having a baby with an abnormal number of chromosomes increase; after age 40, the chances skyrocket. Geneticists believe that the older a woman is, the greater the chance that she has been exposed to mutagens, hence the greater the likelihood that her child will have a mutation.

Other diseases, associated with actual structural defects in the DNA molecule itself, seem to increase in incidence as the father gets older but are not related to the mother's age (Figure 14-5). These defects may be caused by mutagens and may result in birth defects, cancer, and other diseases.

Birth Defects   Seven percent of children born in the United States have a birth defect: a physical (structural), biochemical, or functional abnormality. The most obvious defects are the physical abnormalities such as cleft palate, lack of limbs, or spina bifida (a disease characterized by an imperfect closure of the spinal cord, often resulting in paralysis). According to many scientists, the incidence of birth defects is greater than 7%, perhaps as high as 10% to 12%, because many minor defects escape detection at birth. For example, mental retardation and certain enzyme deficiencies are commonly missed by physicians.

Agents that cause birth defects are called **teratogens**; the study of birth defects is **teratology** (from *teratos*, Greek for "monster"). In humans, teratogenic agents may be drugs, physical agents such as radiation, or biological agents such as the rubella (German measles) virus (Table 14-1). No one knows for sure what percentage of birth defects is caused by chemicals in the environment.

Embryonic development can be divided into three parts (1) a period of early development right after fertilization, (2) a period when the organs are developing (*organogenesis*), and (3) a period during which the organs have formed and the fetus mainly increases in size. Teratogenic agents have a pronounced effect on development during organogenesis (Figure 14-6). The organs are most sensitive early in their development; as time passes, they become less and less sensitive.

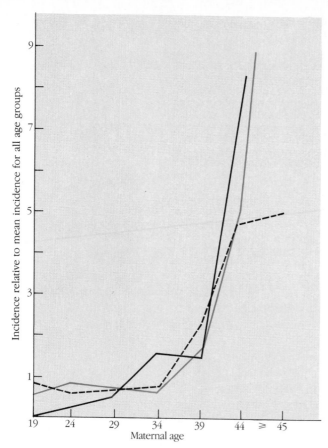

Figure 14-4. The incidence of several chromosomal abnormalities in newborns, involving the wrong number of chromosomes, is related to the mother's age. All abnormalities are trisomies of various chromosomes. These trisomies result when chromosomes fail to separate during the development of the ovum. When fertilization occurs, the offspring ends up with an extra chromosome for certain chromosome pairs.

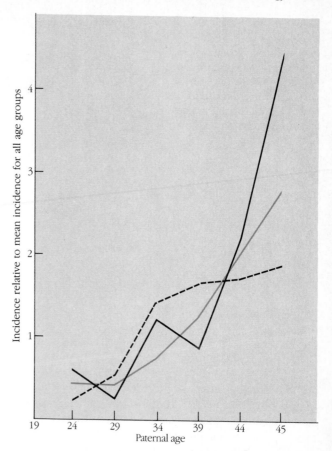

Figure 14-5. The incidence of several cartilage and bone diseases in newborns, caused by DNA damage, is related to the father's age.

The effect of a teratogenic agent is related both to the time of exposure and the type of chemical. Certain chemicals affect only certain organs; for example, methyl mercury damages the developing brains of embryos. Other chemicals, such as ethyl alcohol, can affect several systems; for instance, children born to alcoholic mothers exhibit numerous defects, including growth failure, facial disfigurement, heart defects, and skeletal defects.

Reproductive Toxicity Reproduction is a complex process, involving many steps. An ovum and sperm must be formed and successfully united. The zygote, the product of this union, must divide by mitosis and become implanted in the wall of the uterus, where it acquires nutrients from the mother's blood. Tissues develop from the ball of cells, and then organs develop from these tissues. Meanwhile, the mother is undergoing metabolic

Table 14-1 Some Known and Suspected Teratogens in Humans

Known Agents	Possible or Suspected Agents
Progesterone	Aspirin
Thalidomide	Certain antibiotics
Rubella (German	Insulin
measles)	Antitubercular drugs
Alcohol	Antihistamines
Irradiation	Barbiturates
	Iron
	Tobacco
	Antacids
	Excess vitamins A and D
	Certain antitumor drugs
	Certain insecticides
	Certain fungicides
	Certain herbicides
	Dioxin
	Cortisone
	Lead

Figure 14-6. Schematic representation of human development, showing when some organ systems develop. Sensitive periods are early in development. Exposure to teratogens during these times will almost certainly cause birth defects.

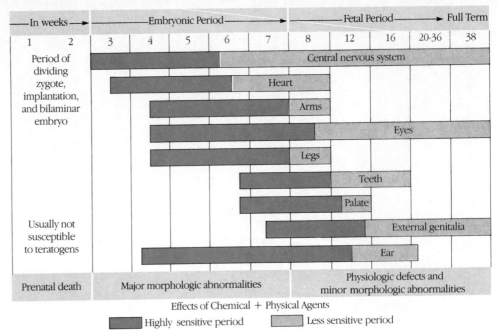

and hormonal changes. At the end of the developmental period birth takes place, requiring hormones that contract the uterus and expand the cervix (the opening between the vagina and uterus). Next, the breasts begin to produce milk, a hormonally regulated process called lactation. Chemical and physical agents may interrupt any of these complex processes, interfering with reproduction. The field of study that examines the effects of physical and chemical agents on reproduction is called **reproductive toxicology**.

The effects of drugs and environmental chemicals on reproduction have become a major health concern in recent years. Studies have shown that male factory workers temporarily become sterile when exposed on the job to DBCP (1,2-dibromo-3-chloropropane). Men who routinely handle various organic solvents often have abnormal sperm, unusually low sperm counts, and varying levels of infertility. A wide number of chemicals such as diethylstilbestrol (DES), borax, cadmium, methyl mercury, and many cancer drugs are toxic to the reproductive systems of males and females.

Some examples will help illustrate the effect of chemical toxins on reproduction. Two researchers from Laval University in Quebec examined the records of 386 children who had died of cancer before the age of 5. Their study showed that many of these children's fathers had been working at the time the children were conceived in occupations that exposed them to high levels of hydrocarbons. Some were painters exposed to paint thinners, and some were mechanics exposed to car exhaust. This study suggests that hydrocarbons had entered the bloodstream, traveled to the testes and there damaged the germ cells. The resulting genetic defect (mutation) was passed to the offspring.

Another example occurred when pregnant women were given the synthetic estrogen DES in the 1950s and 1960s. DES was administered to women who either had a history of miscarriages or had begun to bleed during pregnancy. Bleeding is an early symptom of miscarriage, and DES was given in hopes of preventing it. (It is now known that DES cannot prevent miscarriage.) Years later, uterine and cervical cancers began to appear in the daughters of DES-treated women. Research is uncovering reproductive damage in their sons, too.

How Do Toxins Work?

Toxic substances exert their effects at the cellular level in three major ways: First, they can affect **enzymes**, the cellular proteins that regulate many important chemical reactions. A disturbance of enzymatic activity can seriously alter the functioning of an organ or tissue. As examples, mercury and arsenic both bind to certain enzymes, blocking their activity. Second, some toxins can bind directly to cells or molecules within the cell, thereby upsetting the chemical balance within the body. Carbon monoxide, for example, binds to hemoglobin in the blood; this interferes with the transport of oxygen and can lead to death if levels are high enough (see Chapter 15). Third, some toxins can cause the release of other naturally occurring substances that have an adverse effect. Carbon tetrachloride, for example, stimulates certain nerve cells to release large quantities of epinephrine (adrenaline), believed to cause liver damage.

Waterbed Heaters and Power Lines: A Hazard to Our Health?

In 1979 two researchers from the University of Colorado reported a link between high-current electric power lines and the incidence of childhood leukemia. Their study suggested that extremely low-frequency (ELF) magnetic fields produced when electricity flows through wires may be the cause of the increased incidence of cancer in children living nearby. The researchers found that the death rate from cancer in children was twice what was expected in the general public.

ELF magnetic fields are virtually everywhere. What is more, they easily penetrate walls of buildings and the human body. ELF fields are also found around power stations, welding equipment, subways, and movie projectors.

In 1986, researchers from the University of North Carolina announced the results of a study that supports the Colorado research. The new study showed a five-fold increase in childhood cancer (particularly leukemia) in residents living near the highest ELF fields 7 to 15 meters (25 to 50 feet) from wires that carry electricity from power substations to neighborhood transformers. Adding to the concern, a researcher from Texas recently found that ELF fields increase the growth rate of cancer cells. In addition, cancer cells exposed to ELF fields are 60% to 70% more resistant to the body's naturally-occurring killer cells.

In 1986, another group of scientists reported that ELF fields given off by electric heaters in waterbeds and electric blankets increased the likelihood of miscarriage. Three-fourths of the miscarriages in women who used electric blankets occurred from September to the end of January. In the group that used electric heaters in their waterbeds 61% of the miscarriages occurred during those months. By comparison, women using neither a waterbed heater nor an electric blanket had a 44% miscarriage rate during this same period.

Although researchers are not sure how the general public could be protected from ELF fields created by power lines, they agree that manufacturers could easily design waterbed heaters that do not create electric fields. Electric blankets are another story.

ELF fields may also be a cause of congenital birth defects in humans. Scientists know that ELF fields affect fetal development in pigs, chickens, and rabbits.

The power industry is concerned about the effects of ELF fields. Leonard Sagan, manager of the Radiation Sciences Program at the Electric Power Research Institute (EPRI) says that the real cause of the increased cancer may be something else. For instance, individuals living near ELF fields may also be exposed to increased pollution from traffic. Sagan agrees that electric blankets and waterbeds deserve more attention, however. The EPRI is now spending over $2 million per year to study ELF fields.

Many scientists and industry representatives remain skeptical about the link between cancer and ELF fields. Most think that there is not enough information to necessitate protective action. From a public health perspective, most agree, there is some reason for concern.

Factors Affecting the Toxicity of Chemicals

Predicting the harmful effects of chemicals is no easy task. Age, sex, health, and a variety of other factors contribute to the final outcome. Consider this case: a family of six was living near a Canadian lead and zinc smelter that released large quantities of lead. Each member was exposed to high levels of lead. Because of age and health differences, however, the symptoms were quite varied. For example, the father and a 4-year-old boy suffered from colic and pancreatitis. The mother developed a neural disorder. Two other children experienced convulsions, and the last developed diabetes.

Three of the most important factors influencing the effects of a given chemical are the dose, the duration of exposure, and the biological reactivity of the chemical in question.

Dose and Duration In general, the higher the dose and the longer the exposure, the greater the effect. To demonstrate the effect of dose, toxicologists expose laboratory animals to varying doses and determine the response. The resulting graph is called a **dose–response curve** (Figure 14-7). The dose that kills half of the test animals is called the LD_{50}, or the lethal dose for 50% of the test animals. By comparing LD_{50} values, scientists can judge the relative toxicity of two chemicals. For example, a chemical with an LD_{50} of 200 milligrams per kilogram of body weight is half as toxic as one with an LD_{50} of 100 milligrams. In other words, the lower the LD_{50}, the more toxic a chemical.

Biological Activity The toxicity of a chemical is a function of its biological activity, that is, how it reacts with

Figure 14-7. Dose–response graph for two chemicals (a and b) with differing toxicities. The LD$_{50}$ is the amount of chemical that kills one-half of the experimental animals within a given time. The higher the LD$_{50}$ value, the less toxic the chemical is.

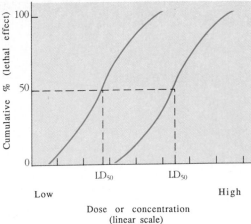

APPROXIMATE ACUTE LD50s
OF A VARIETY OF CHEMICAL AGENTS

AGENT	LD50 (mg/kg)
Ferrous sulfate	1,500
Morphine sulfate	900
Phenobarbital sodium	150
DDT	100
Picrotoxin	5
Strychnine sulfate	2
Nicotine	1
d-Tubocurarine	0.5
Hemicholinium-3	0.2
Tetrodotoxin	0.10
Botulinus toxin	0.00001

enzymes or other cellular components. The more reactive it is, the more effect it has. Inert substances—those that do not chemically react with cellular components—generally are not toxic, although there are notable exceptions such as asbestos.

Age Young, growing organisms are generally more susceptible to toxic chemicals than are mature adults. For example, two common air pollutants, ozone and sulfur dioxide, affect young laboratory animals two to three times more severely than they affect adults. Among humans, infants and children are more susceptible to lead and mercury poisoning than adults, because their nervous systems are still developing.

Health Status Poor nutrition, stress, bad eating habits, heart and lung disease, and smoking all contribute to poor health and make individuals more susceptible to certain toxins. Genetic factors may also determine one's response to certain toxic substances. Some individuals are genetically prone to heart disease, lung cancer, and other disorders brought on by environmental factors.

Synergy and Antagonism The presence of two or more toxic substances can alter the expected response. Different chemical substances can act together to produce a **synergistic response**, that is, a response stronger than the simple sum of the two responses. One of the most familiar examples of toxic **synergism** is the combination of barbiturate tranquilizers and alcohol; although neither taken alone in small amounts is dangerous, the combination can be deadly. Pollutants can also synergize. For instance, sulfur dioxide gas and particulates (minute airborne particles) inhaled together can reduce air flow through the lung's tiny passages; the combined response is much greater than the sum of the individual responses

(Case Study 14-1 discusses the synergistic effect of smoking and asbestos.)

Chemicals can also negate each other's effects, a phenomenon called **antagonism**. In these cases, a harmful effect is reduced by certain combinations of potentially toxic chemicals. In mice exposed to nitrous oxide gas, for example, mortality is substantially reduced when particulates are also present. Scientists are uncertain of the reasons for this phenomenon.

Bioconcentration and Biological Magnification

Two factors not mentioned in the previous discussion that profoundly influence toxicity are bioconcentration and biological magnification. **Bioconcentration** is the accumulation of certain chemicals within the body. For example, the human thyroid gland bioconcentrates iodide. The level of iodide in the thyroid is thousands of times higher than that in the blood. Scallops, marine bivalve mollusks that feed on material suspended in water, selectively take up certain elements from seawater, such as zinc, copper, cadmium, and chromium. The level of cadmium in scallops, for example, is 2.3 million times that of seawater.

When harmful chemicals become concentrated in organisms, trouble may begin. For example, certain persistent (nonbiodegradable) organic molecules, such as the pesticide DDT, concentrate in body fat. Bioconcentration opens the door for a phenomenon called **biological magnification**, the buildup of chemicals in organisms in a food chain. As shown in Figure 14-8, DDT in water is taken up by zooplankton, single-celled organisms in the water. Small fish ingest DDT when they feed on zooplankton. Higher-level organisms also accumulate this substance. Tissue concentrations become higher at higher levels of the food chain. Biological magnification occurs because DDT is a fat-soluble chemical that takes up a rather permanent residence in body fat. The more fish an osprey eats, the higher its DDT levels become. The concentration of DDT may be several million times greater in fish-eating birds than it is in the water (Figure 14-8). For humans, the magnification that occurs in our food chain may be as much as 75,000 to 150,000.

Biological magnification exposes organisms high on the food chain to potentially dangerous levels of many chemicals. Synthetic chemicals like DDT, some lead and mercury compounds, and even some radioactive substances are all biomagnified.

The Roots of Controversy

Human society has progressed considerably in the last 200 years. Its understanding of toxic chemicals, however, is still barely out of the Dark Ages. The reasons for this are many. One of the most important is that it is neither practical nor ethical to test toxic chemicals on human beings. As a result, toxicologists must rely on tests on rats, mice, rabbits, and other laboratory animals. The results of experiments on laboratory animals cannot always be extrapolated to humans. As a friend once reminded me, "Contrary to popular belief, the human is not a large rat." Lab animals frequently react differently to chemicals than humans do; they may be able to break them down better, or they may not be able to break them down as well. Physiological differences between humans and lab animals make it difficult to predict if a chemical harmful to an animal will be injurious to us.

Our ignorance of toxic effects also stems from the fact that humans are frequently exposed to many potentially harmful chemicals and may be exposed over long periods. For practical reasons most toxicity tests are performed on one substance at a time. Because of synergy and antagonism, extrapolating the results from single-chemical tests to the real world can be misleading.

Another problem is that most tests of toxicity, especially those for mutations and cancer, are performed at high exposure levels rarely if ever experienced by the average citizen or even by most workers. The fact that a large dose of a chemical induces cancer in a lab animal does not necessarily mean that the chemical will cause cancer in the typically low doses to which humans are exposed.

If scientists can't make accurate extrapolations from high doses to low doses, why do they perform their experiments that way? Researchers use such high doses to speed up their experiments. The time required to develop a noticeable cancer is quite long. Human cancers may develop 5 to 30 years after exposure. As a general rule the entire process from exposure to manifestation takes about one-eighth of the lifespan of an animal. Thus,

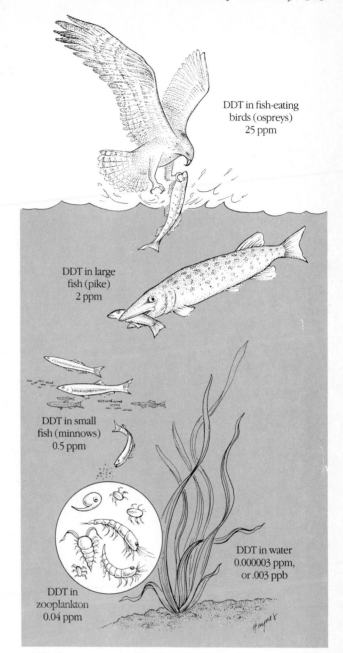

Figure 14-8. The biological magnification of DDT increases at higher levels in a food chain.

DDT in fish-eating birds (ospreys) 25 ppm

DDT in large fish (pike) 2 ppm

DDT in small fish (minnows) 0.5 ppm

DDT in water 0.000003 ppm, or .003 ppb

DDT in zooplankton 0.04 ppm

anything that speeds up the process, such as a high dose, helps cut costs. To test for low-level effects, scientists would need very large numbers of experimental animals to generate statistically valid results. High-dose studies, therefore, reduce the number of lab animals needed and can cut time and costs—a significant factor, since cancer studies can cost $500,000 to $1 million per chemical.

One of the greatest controversies in toxicology involves the **threshold level**, that is, the level below which no effects occur. The incidence of cancers in lab animals is

◄ *Viewpoint*

Are We Losing the War Against Cancer?

John C. Bailar III and Elaine M. Smith

John Bailar is a biostatistician at McGill's School of Public Health. Elaine Smith is a biostatistician at the University of Iowa Medical Center. This Viewpoint is adapted from an article published in the New England Journal of Medicine, *May 1986.*

Between 1950 and 1986, both private- and government-sponsored cancer research grew at a tremendous rate. Toward the end of that period substantial efforts were made to apprise physicians, patients, and the general public of the research that had gone on in the previous three decades. But what progress have we made in the fight against cancer? Have our efforts to find treatments for this disease been successful?

In 1962 cancer was the recorded cause of death for 278,562 Americans. In 1986, 24 years later, 469,330 Americans died of cancer, an increase of 68%. But this increase is somewhat deceptive. During that period the US population was growing. The relative proportion of people in older age categories also increased. When these two factors are considered, the real growth in the cancer rate turns out to be 10.1%.

But mortality data do not tell the whole story. We might ask not how many Americans die of cancer but how many contract the disease? From 1973–1974 to 1985–1986 the age-adjusted incidence rate increased by 12.3%. The conclusion: cancer is on the rise.

But have we made inroads in other areas, such as treatment? To look at the overall effectiveness of new cancer treatments, we might focus on long-term survival rates. Our data show that five-year survival rates for white patients with all forms of cancer ran from 50% in 1975 to 51.3% in 1981.

Our data show that cancer mortality rates have increased slowly but steadily over several decades. There is no evidence of a recent downward trend. There has been, in our view, little progress in treating most cancer as reflected in the survival rate data. In this sense, we are losing the war against cancer. Substantial increases in our understanding of the nature and properties of cancer have not led to a corresponding reduction in the incidence or mortality of this disease.

The National Cancer Institute recently announced a nationwide goal of reducing cancer mortality by 25% to 50% by the year 2000. It is unlikely, in our view, that we will achieve this goal.

These comments about a lack of progress are in no way an argument against the earliest possible diagnosis and the best possible treatment of cancer. The problem, as we see it, is the lack of any substantial recent improvements in treating the most common forms of cancer.

The main conclusion we draw is that 35 years of intense effort focused largely on improving treatment is a qualified failure. The results have not been what they were intended or expected to be. But we think that there could be much current value in an objective review of the reasons for this failure. Why were hopes so high? What went wrong? Can future efforts be built on more realistic expectations? And why is cancer the only major cause of death for which age-adjusted mortality rates are still increasing?

On the basis of past medical experience with infectious and other nonmalignant diseases, we suspect that the most promising route to cut cancer rates is prevention. Reducing smoking, indoor air pollution, and workplace exposure and other efforts could pay huge dividends in the long run. History suggests that savings in both lives and dollars would be great.

---

a function of dose: the higher the dose, the greater the incidence of cancer. The same relationship is known to be true in some human cancers; for instance, the incidence of lung cancer clearly increases with the number of cigarettes smoked. Some workers assume that if a chemical is harmful at high levels, it will also be harmful at the lowest levels. Others argue that there is a dose level—the threshold level—below which no harmful effect occurs (Figure 14-9); in other words, extremely low levels of certain chemicals are completely safe. It appears that a threshold may exist for some chemicals but not for others, such as asbestos.

Controlling Toxic Substances

The United States produces over 170 million metric tons (378 billion pounds) of synthetic chemicals and creates 250 to 300 million metric tons of potentially hazardous wastes each year. The need to control toxic substances has grown dramatically, and many laws have been passed to regulate them (Table 14-2).

Federal Control

In 1976 the US Congress passed the **Toxic Substances Control Act**. It is designed to screen new chemicals and ban or limit the use of those that present an unreasonable health risk.

The act has three major parts: (1) premanufacture notification, requiring all chemical companies to tell the EPA of new substances they want to introduce into the market; (2) requirements for testing new or existing chemicals that are believed to present a risk to the public and the environment; and (3) stipulations for the control of several existing hazardous chemicals.

Under the first part of the act, companies are required to notify the EPA 90 days before they import or manufacture a chemical substance not currently in commercial use. The EPA then has 90 days to decide whether the chemical can be introduced and whether any restrictions are necessary to minimize its risk.

Scientists at the EPA review existing toxicity data on the chemical and information provided by the manufacturer. However, most new substances have not been tested for toxicity, carcinogenicity, and other adverse effects, and manufacturers generally do little toxicity research because of its cost. Thus, the EPA must rely on toxicity data from chemicals with a similar structure. In many cases new chemicals belong to classes of compounds that have been adequately tested, so EPA officials can make decisions based on this information.

If the new chemical is believed to pose little risk, it is approved. If it might be hazardous, however, the agency asks the manufacturer to test its toxicity and report back.

Second, the Toxic Substances Control Act requires the EPA to take a look at chemicals that were in use before its passage. Those deemed risky are required to undergo toxicity testing.

Finally, through the act Congress also took specific actions against certain chemicals it believed were hazardous. The most radical controls were placed on polychlorinated biphenyls (PCBs), an insulating fluid used in electrical transformers. Because of PCBs' stability in the environment (persistence and resistance to biodegradation), the widespread contamination they had already caused, their ability to bioconcentrate, and their known toxicity to laboratory animals, Congress in May 1979 banned their manufacture and distribution except in a few limited cases.

Market Incentives to Control Toxic Chemicals

In 1988, one of the most controversial environmental laws in the United States went into effect. The state of California's Safe Drinking Water and Toxics Enforcement Act prohibits the contamination of drinking water by toxic chemicals. It also sets in motion a complex market strategy to control the exposure to toxic chemicals in foods and consumer products.

This experiment in toxic control owes its origin to a citizen-sponsored initiative, Proposition 65, which was passed overwhelmingly by voters in 1986. The state of

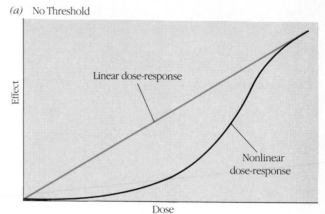

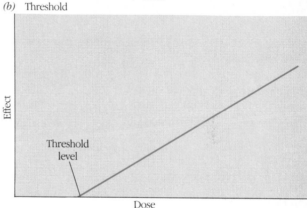

Figure 14-9. (a) A hypothetical dose–response curve indicating the absence of a threshold level, a level below which no effect occurs. (b) A hypothetical dose–response curve showing a threshold level.

California sets standards (acceptable levels) for potential toxins in various consumer products and foods. The law requires that manufacturers who violate these standards print warnings on their products, noting that the amount of potential toxin in the product exceeds the state's safe level. The proponents of this law believe that consumers will shun such products, creating a market force that will encourage manufacturers to reduce levels of toxins in their products.

California's regulators worked quickly to produce standards. In a single year, their regulators have produced more standards than the United States EPA has managed to create under the Federal Toxic Substances Control Act in over a decade. (For more on market incentives see Chapter 20.)

Determining the Risks

Ralph Waldo Emerson once wrote, "As soon as there is life there is danger." Every day of our lives we face many dangers, some obvious, some hidden. The study of the

Table 14-2 Federal Laws and Agencies Regulating Toxic Chemicals

Statute	Year Enacted	Responsible Agency	Sources Covered
Toxic Substances Control Act	1976	EPA	All new chemicals (other than food additives, drugs, pesticides, alcohol, tobacco); existing chemical hazards not covered by other laws
Clean Air Act	1970, amended 1977, 1990*	EPA	Hazardous air pollutants
Federal Water Pollution Control Act	1972, amended 1977, 1978, 1987	EPA	Toxic water pollutants
Safe Drinking Water Act	1974, amended 1977	EPA	Drinking water contaminants
Federal Insecticide, Fungicide, and Rodenticide Act	1948, amended 1972, 1973, 1988	EPA	Pesticides
Act of July 22, 1954 (codified as § 346(a) of the Food, Drug and Cosmetic Act)	1954, amended 1972	EPA	Tolerances for pesticide residues in food
Resource Conservation and Recovery Act	1976	EPA	Hazardous wastes
Marine Protection, Research and Sanctuaries Act	1972	EPA	Ocean dumping
Food, Drug and Cosmetic Act	1938	FDA	Basic coverage of food, drugs, and cosmetics
Food additives amendment	1958	FDA	Food additives
Color additive amendments	1960	FDA	Color additives
New drug amendments	1962`	FDA	Drugs
New animal drug amendments	1968	FDA	Animal drugs and feed additives
Medical device amendments	1976	FDA	Medical devices
Wholesome Meat Act	1967	USDA	Food, feed, and color additives; pesticide residues in meat, poultry
Wholesome Poultry Products Act	1968		
Occupational Safety and Health Act	1970	OSHA	Workplace toxic chemicals
Federal Hazardous Substances Act	1966	CPSC	Household products
Consumer Product Safety Act	1972	CPSC	Dangerous consumer products
Poison Prevention Packaging Act	1970	CPSC	Packaging of dangerous children's products
Lead Based Paint Poison Prevention Act	1973, amended 1976	CPSC	Use of lead paint in federally assisted housing
Hazardous Materials Transportation Act	1970	DOT (Materials Transportation Bureau)	Transportation of toxic substances generally
Federal Railroad Safety Act	1970	DOT (Federal Railroad Administration)	Railroad safety
Ports and Waterways Safety Act	1972	DOT (Coast Guard)	Shipment of toxic materials by water
Dangerous Cargo Act	1952		

CPSC = Consumer Product Safety Commission
DOT = US Department of Transportation
EPA = US Environmental Protection Agency

FDA = Food and Drug Administration
OSHA = Occupational Safety and Health Administration
USDA = US Department of Agriculture
*currently pending

Source: Council on Environmental Quality

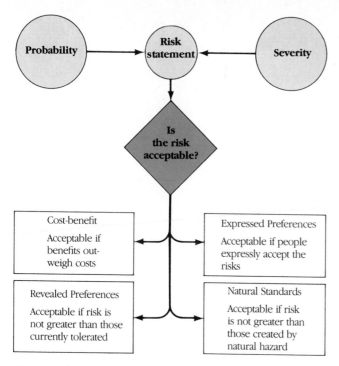

Figure 14-10. Determining the acceptability of a risk. Cost–benefit analysis is the most common method of determining risk acceptability, but three other methods can also be used, as shown here.

daily risks of modern technological societies has become an important policymaking tool. This section looks at risk and how it is assessed.

Risks and Hazards: Overlapping Boundaries

Two types of hazard are broadly defined by risk assessors: anthropogenic and natural. **Anthropogenic hazards** are those created by human beings. **Natural hazards** include events such as tornadoes, hurricanes, floods, droughts, volcanoes, and landslides. Natural hazards often have a human component. For instance, the damage from floods is, in large part, the result of our living along floodplains, channelizing streambeds, or changing the vegetative cover (Chapter 10). Similarly, earthquake damage can be greatly magnified by overpopulation and bad building practices. Hazards befalling human society exact an enormous price: human lives, human health, economic ruin, social disruption, mental illness, environmental destruction, and animal and plant extinction. Measuring the damage is never easy.

Citizens of the developed world have become increasingly aware of the hazards to which they are exposed. This increase in awareness of risk results from several factors. First, television and other advanced communication systems bring news of the hazards to our homes from all over the world. Second, our increased material wealth has given us more free time to consider the hazards around us. In poor societies, for example, people tend to ignore risks in the workplace because of their need to make a living wage. Third, we are more aware of hazards today because we are exposed to more hazards. As technology and population grow, more and more dangers are created.

Risk Assessment

Since the mid-1970s a new and rather imprecise science, called **risk assessment**, has been developed to help us understand and quantify risks posed by technology, our life-styles, and our personal habits (smoking, drinking, and diet).

Risk assessment involves two interlocked steps: hazard identification and estimation of risk. **Hazard identification** is both the recognition of dangers that exist today and the complicated art of predicting future dangers. **Estimation of risk** generally involves two processes (Figure 14-10). The first is determining the **probability** that an event will occur. This process answers the question "How likely is the event?" The second stage is determining the **severity** of an event, answering the question "How much damage is caused?" Determining probability and severity is complicated and fraught with uncertainty.

The next step, determining the overall level of risk, is often a difficult one. To understand how difficult it is to assess risk, consider nuclear power. The probability of a nuclear core meltdown is thought to be small. (Nuclear power advocates tell us that the probability is one chance in 10,000 years of reactor operation. See Chapter 12 for more information on nuclear power plant accidents.) Even though the probability of a meltdown may be small, the consequences of such an event would be severe. Thousands of lives might be lost, and billions of dollars in property damage would result. Thus, the probability and severity factors indicate distinctly different levels of risk. Assigning a risk value to nuclear power is nearly impossible because of the disparity between probability and severity. A combined scale is now being developed to help scientists solve this problem.

Risk Management: Decisions about Risk Acceptability

Risk assessment is ultimately designed to help society manage its hazardous environment. No matter what we do—whether it is screwing in a light bulb or flying cross-country in a jet—we put ourselves (and possibly our environment) at risk. Nothing is safe, or "entirely free from harm." The science of risk assessment recognizes that human life is haunted by hazards. Rather than talking in terms of safety, which is absolute, the risk assessor speaks in terms of risk, which is relative. Activities that we commonly consider safe are better seen as low-risk functions. "Unsafe" activities are better labeled as "high-risk functions."

Knowing the relative risk of a technology is one thing. Knowing the acceptability of that risk, what price society will pay for certain activities, is quite another story. **Risk acceptability** is one of the trickiest issues facing modern society. Why? Because we are fickle. What appears "safe" one day becomes suspect the next after a widely publicized accident. Irrational fears crop up and frighten us away from relatively low-risk activities.

The acceptability of risks is also determined by **perceived benefit**—how much benefit people think they will get from something. In general, the higher the perceived benefit, the greater the risk acceptability (Figure 14-11). As an example, the risks of a new steel mill might be overlooked by a community with high unemployment. Automobile travel provides the most telling example of the way in which perceived benefits affect our decisions. The risk of dying in an automobile accident in the United States is 1 in 5000 in any given year. (Incidentally, the risk of a fatal accident is greatly increased on Friday and Saturday evenings.) Over your lifetime, the risk of dying in a car accident is far higher if you don't wear seat belts, about 2 in 100, than if you do, 1 in 100. Meanwhile, substances believed to be far less hazardous than driving are banned from public use primarily because their benefit is not so highly valued or because a ban is involuntarily imposed on us.

Perceived harm, the damage people think will occur, also heavily influences our views of risk acceptability. In general, the more harmful a technology or its by-product is perceived to be, the less acceptable it is to society. Efforts to find a burial ground for high-level nuclear wastes clearly illustrate this point (see Chapter 18). Over two-thirds of Americans recently polled favor nuclear power, but few of them want a waste repository in their state—an inconsistent but entirely human reaction based on perceived harm. Many environmental disputes center around the issue of risk, with companies viewing potential risks differently from citizens and environmental groups. (For more on this topic see Case Study 17-1 and Chapter 20 on economics.)

Decisions, Decisions   Decisions on modern sources of risk—technologies, personal habits, and pollution—are becoming more and more commonplace as environmental scientists uncover harmful effects. Nowhere was this more evident than in Tacoma, Washington, south of Seattle. A copper-smelting plant owned by Asarco Incorporated annually contributed $20 million to $30 million to the local economy. Unfortunately, it daily spewed out 765 kilograms (1700 pounds) of arsenic into the air, 23% of the total arsenic emissions in the United States. Arsenic,

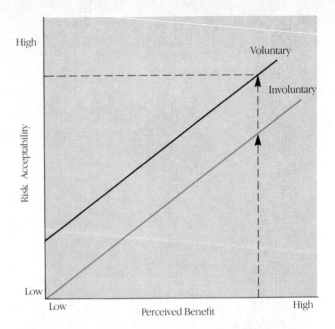

Figure 14-11.   Risk becomes acceptable, in general, as the perceived benefit rises. Voluntary risks, that is, ones people agree to, are generally more acceptable than risks imposed without consent.

which has been linked with both lung and skin cancers and neurological disorders, was found in alarmingly high levels in the blood of smelter workers and their children at a nearby school. In 1983 health officials estimated that residents nearest the smelter had a lifetime cancer risk of 9 in 100. Even if the best available control technology were put in place, health experts predicted that the cancer risk would be about 2 in 100. The company argued that reducing arsenic emissions would force it to close the factory, costing the community millions of dollars in tax revenue and income for its 1300 employees.

In this landmark case the EPA went to the people of Tacoma to explain the situation (Figure 14-12). William Ruckelshaus, then head of the agency, proposed that the residents be brought into risk–benefit analysis. This action angered many environmentalists, who accused the agency of asking residents to make a choice between clean air and jobs. The EPA held extensive hearings in Tacoma to outline the risks. The company hired a public relations firm to outline the benefits of the smelter. By the end of the hearings dozens of residents were wearing buttons that expressed their sentiment. The buttons said, quite simply, "Both." During the public clamor over the Tacoma smelter Asarco decided to close the plant because of the declining metals market, which made operations uneconomical.

Nonetheless, the Tacoma experiment points out the trade-offs that many countries must make in which environmental risk is weighed against economic risk and

Figure 14-12. Asarco parents at a town meeting in Tacoma, Washington. The sign reads, "Don't risk our children."

employment. Clearly, the choices are not easy. Uncertainty about the health effects of arsenic levels made the situation more confusing.

Several ways of making decisions about risk acceptability were shown in Figure 14-10. Each is riddled with problems. The most common decision-making tool is the **cost-benefit technique**, the method left to the people of Tacoma. While popular, this technique of weighing costs against benefits has serious weaknesses. For example, the benefits are generally easily measured: financial gain, business opportunities, jobs, and other tangible items. Many of the costs are less tangible. External costs, discussed in Chapter 20, are among the most difficult to quantify. Human health, environmental damage, and lost species come with no price tags attached. Cost–benefit analysis then suffers because many important costs are poorly documented, spread out, and unquantifiable, whereas the benefits are often clear and quantifiable. Experience shows that cost-benefit analyses of pollution control, for example, often overestimate compliance costs. Economists, say leading environmentalists, fail to consider technological innovation, which frequently decreases compliance costs, making the economics of pollution control more favorable.

Recent efforts by economists to assign a dollar value to environmental and health costs may help make decisions of cost and benefit more useful. In addition, the efforts of ecologists and environmental scientists to measure the impacts of technologies and their by-products on wildlife, the environment, recreation, health, and society in general may also help. Because of the lack of information on environmental and health costs, authors of *An*

Environmental Agenda for the Future (a list of recommendations made by ten leading environmental groups) suggest that society should adhere to the better-safe-than-sorry adage, even if predictable costs purchase benefits that cannot be calculated with certainty. In the long run, we may be better off.

Actual versus Perceived Risk   The main purpose of risk assessment is to help us create cost-effective laws and regulations to protect human health, the environment, and other living organisms. Ideally, good lawmaking requires that the **actual risk**, or the amount of risk a hazard really poses, be equal to the risk perceived by the public. When actual and perceived risk are equal, public policy can be formulated to yield cost-effective protection (Figure 14-13).

When the perceived risk is much larger than the actual risk, costly **overprotection** may occur. For example, laws and federal regulations regulating air pollution in the United States are believed by many, especially those in industry, to be too strict. These people argue that the damage caused by air pollution (actual risk) is less than the public thinks (perceived risk). Therefore, they assert that the cost of air-pollution control far outweighs the savings in damage to human health and the environment. Others argue that the perceived risk of air pollution is far smaller than the actual risk and that current policies do not adequately protect the public. In their eyes **underprotection** places a burden on society, affecting people's health, welfare, economics, and environmental quality.

Only through arduous efforts to identify and quantify risk accurately can we match perceived and actual risks.

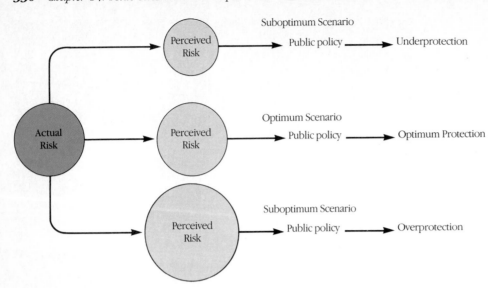

Figure 14-13. Matching the actual risk and the risk that a society perceives is essential to the formulation of good public policy. But perceived risk and actual risk do not always match, as shown.

The Final Standard: Ethics

Ultimately, our environmental decisions are based on our **ethics**, that is, the values we hold, or, simply, what we view as right and wrong. Values that affect our decisions come from our parents, relatives, friends, enemies, teachers, religious leaders, and politicians. These values shift over the years—sometimes subtly, sometimes dramatically—changing as we become older and as our priorities shift. Although our ethics are often never explicitly stated, they play an important role in our lives. They determine how we vote, what friends we associate with, how we treat one another, how we carry on business affairs, and, finally, how we act with regard to the environment.

Prioritizing Values

Values play an important part in decisions of risk acceptability. Benefits and costs will be incurred in virtually all of our decisions, and we must always weigh them against each other. This balancing of costs and benefits requires us to prioritize our values. If a coal mine were to be placed outside of your community, for instance, clear advantages might be realized: more jobs and a stronger economy. However, certain costs such as air and water pollution might be incurred. The decision to open the mine would be influenced by the priority of values.

Prioritizing values requires us to ask what we value the most. What is more important to us in environmental decision making? Economics? Health? Wildlife? A new reservoir will bring much-needed water to an area, thus allowing it to grow and prosper, but will destroy valuable wildlife habitat and recreation areas. How do we choose? Do we save recreation areas and wildlife habitat and find other solutions to the water shortage, or do we dam the river, destroying the wildlife and the recreation area?

Space–Time Values

Further insight into values comes from looking at **space–time values**. These are simply the concerns we have for other people and other living organisms in time (the present and the future) and space (you, your family, community, state, nation, and world).

As shown in the scatter diagram in Figure 14-14, individual interest can be identified by a single point that denotes one's space and time concerns. Most people's interests lie toward the lower end of the scales, tending toward self-interest and immediate concerns. Some people call this selfishness, but it can also be considered a natural biological tendency to be concerned with the self. Among animals, awareness of the needs of others is a feature only of social creatures like monkeys and lions; however, concern for the uppermost end of the space–time graph is a distinguishing feature of the human animal. The unique human ability to ponder the consequences of actions is indeed fortunate. It is fortunate because humans have reached a position of unprecedented power as molders of the world's environment. Our power to change the world to our liking has never been greater; nor has our power to destroy ever attained such heights.

Sound decision-making in a sustainable society requires that we know where our priorities lie on the scatter diagram. Three important space–time questions require answers: (1) Is our decision based primarily on self-interest? In other words, are we looking at the issue solely in terms of how we might benefit or be harmed? (2) Is our decision based primarily on concern for others who are alive now who might benefit or be harmed? And how far does our concern go? Are we concerned with the well-being of stockholders or citizens of the community, state, nation, or world? (In some cases, local

Figure 14-14. Various people's spatial and temporal interests are indicated by the points on this graph. Most individuals tend toward the lower end of the scales, being concerned primarily with self and the present.

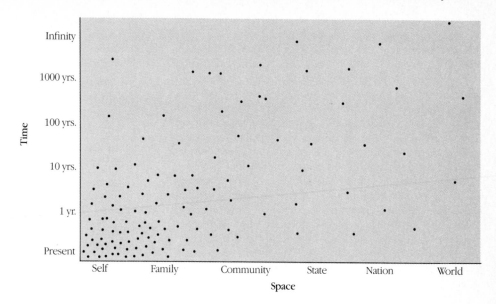

actions can have global impact.) (3) Finally, is our decision based on the good of future generations? Will our decisions today benefit or harm those who follow us?

Building the Future Incidents like the chemical accident at Bhopal, India, point out the Faustian bargain modern society has struck with manufacturing. The Toxic Substances Control Act and risk assessment are attempts to make that bargain with the devil as cost-free as possible.

In a general sense, society is in the business of protecting itself and ensuring an adequate future for generations to come. What we do with hazardous materials, in many ways, lays the foundation for the future. If we are reckless with them, future generations will pay the price. If we exploit them cautiously, our offspring will probably be better off. Ethics helps bridge the gap between the present and the future. As our concerns expand along the time–space continuum, we can begin to view our actions differently.

Slowly, modern society is learning what the American Indians have known for centuries. Unborn generations have a claim on the land, air, and water equal to our own. The need for a new, broader ethic that protects the future can no longer be ignored.

Chance fights ever on the side of the prudent.

EURIPIDES

Summary

Citizens of developed countries depend on a vast number of chemical substances, some of which may be harmful to human health and the environment. Toxic substances, or **toxins**, are those chemicals that cause any of a wide number of adverse effects in organisms. **Toxicology** is the study of these effects. Humans are exposed to toxins at home, at work, and while moving about in the environment. **Acute effects** are manifest immediately after exposure, are often short-lived, and are generally caused by high levels of exposure. **Chronic effects** are delayed, long-lasting, and generally the result of low-level exposure. Toxic effects include cancer, mutations, birth defects, and reproductive impairment.

Cancer, an uncontrolled proliferation of cells, is often caused by an alteration of the genetic material, although other mechanisms may be involved. The incidence of cancer has nearly leveled off since 1950 in most cases except for lung cancer, testicular cancer, and stomach cancer. Recent evidence shows that 20% to 40% of all cancers are caused by environmental workplace pollutants.

Mutations, or structural changes in the genetic material of cells, are caused by chemical and physical agents called **mutagens**. Mutations occur in humans at a fairly rapid rate but are usually repaired. If not repaired, they may lead to cancer. If they occur in **germ cells**, they may be passed on to offspring, leading to birth defects, stillbirth, spontaneous abortion, or cancer.

Birth defects—structural and functional ailments—are observed in 7% of all newborn American children. The study of birth defects is called **teratology**. Agents that cause them are called **teratogens**. In humans, there are few known teratogens, although many substances are suspected.

Reproductive toxicity is the study of the toxic effects of chemical and physical agents during the reproductive cycle. Numerous chemical agents impair human reproduction.

Toxins exert their effects at the cellular level by blocking enzyme activity, by binding to cells or molecules within cells, and by causing the release of naturally occurring substances in amounts harmful to organisms. Numerous factors affect toxic agents, making it difficult to predict the effects. Some of these include dose, age, health status, synergy, and antagonism. **Synergy** occurs when different substances act together to produce a response that is stronger than expected. **Antagonism** occurs when substances negate the effects of each other.

Toxicity is also profoundly influenced by **bioconcentration**, the ability of an organism to selectively accumulate certain chemicals within its body in certain tisuses. **Biological magnification**, the buildup of chemicals within food chains, also affects toxicity. Biological magnification occurs with chemicals that can be concentrated in certain tissues (such as fat) and are resistant to chemical breakdown.

The **Toxic Substances Control Act** was passed by the US Congress in 1976. It requires premanufacture notification to the EPA of all new chemicals to be produced or imported, calls for EPA-mandated testing of new and existing chemicals thought to be harmful, and establishes specific controls on several existing chemicals. The State of California recently passed a potentially powerful new law to control toxic chemicals in foods and other consumer products. The law requires the state to establish safety levels for potentially toxic chemicals and requires manufacturers to print toxin levels on food containers if they violate state standards, thus creating a market force to control toxins in consumer products.

Risk assessment is the science dedicated to understanding risk. Risk assessors first identify actual and potential hazards and then determine the **probability** (likelihood) and **severity** of the hazard. Once these are determined, a statement regarding risk can be made. It is then up to **risk managers**, usually our public officials, to determine how best to deal with the risk.

To regulate hazards better, risk managers must determine the **acceptability of risk**. Risk acceptability is determined by many factors, the most important being the **perceived benefit** (the benefit people think they will gain) and the **perceived harm** (the harm they expect to suffer).

Risk assessment is ultimately designed to manage risks in the most cost-effective manner. To do so, the **perceived risk** (the amount of risk people think is posed) must be equal or very close to the **actual risk**. The actual risk may be difficult to determine, especially in the case of new technologies with which society has had little experience.

Ethics is a code of what is right and wrong. All decision making entails ethical considerations, but ethics are often not clearly understood by individuals. Thus, it is helpful to prioritize our values to help us think about our decisions. Space and time are two important components of ethics. Most people tend to be concerned with the immediate future and self-interests. The interests of future generations, therefore, are often neglected.

Decision making requires a better understanding of the **space–time values**. Through education, more people can be made aware of the needs of future generations and the effects that current actions have on them. A value system that seeks to optimize the future by acting now is critically needed to build a sustainable society.

Discussion Questions

1. Define the terms *toxin, carcinogen, teratogen,* and *mutagen.*

2. Compare and contrast *acute toxicity* and *chronic toxicity* with respect to time to onset of symptoms, persistence of the effect, level of the toxic agent, and duration of exposure.

3. What is cancer? Discuss how it may form.

4. List some of the possible consequences of somatic and germ-cell mutations in humans.

5. What is teratology? Do teratogenic chemicals always create birth defects when given during pregnancy? Why or why not?

6. Make a list of factors that influence the toxicity of a chemical in a given individual.

7. Define the terms *synergism* and *antagonism.*

8. Define the terms *bioconcentration* and *biological magnification.* What factors can be used to predict whether a chemical will be biologically magnified?

9. Why is our knowledge of the effects of toxic chemicals on humans so limited?

10. Describe the major provisions of the Toxic Substances Control Act.

11. What are the two major types of risk? Give examples.

12. Describe the major steps in determining the level of risk posed by technology.

13. What factors determine whether a risk is acceptable to a population?

14. Many more people die in Montana and Wyoming from falls while hiking than are killed by grizzly bears. Why, then, are people so concerned about being killed by a bear when their chances of being killed in a fall are much greater?

15. What are space–time values? In general, where does your concern lie in space and time?

Suggested Readings

Bowonder, B., Kasperson, J. X., and Kasperson, R. E. (1985). Avoiding Future Bhopals. *Environment* 27 (7): 6–13, 31–37. Superb study of the Bhopal tragedy and ways to avert future accidents.

Goldbaum, E. (1987). Can Cell Cultures Predict Toxicity? *Industrial Chemist* January: 34–37. Interesting look at an alternative way to test toxicity.

Klaassen, C. D., Amdur, M. O., and Doull, J. (1986). *Casarette and Doull's Toxicology: The Basic Science of Poisons* (3rd ed.). New York: Macmillan. Superb reference.

Mausner, J. S. and Kramer, S. (1985). *Epidemiology: An Introductory Text.* Philadelphia: Saunders. Excellent reference.

Postel, S. (1988). Controlling Toxic Chemicals. In *State of the World.* Starke, L., ed. New York: Norton. Excellent overview of toxic chemicals and their control.

Waldbott, G. L. (1978). *Health Effects of Environmental Pollutants* (2nd ed.). St. Louis: C. V. Mosby. Good coverage of toxic effects of pollutants.

Willgoose, C. E. (1979). *Environmental Health: Commitment for Survival.* Philadelphia: Saunders. Good introductory text.

Risk

Center for Ethics and Social Policy (1977). *Ethics for a Crowded World.* Berkeley, CA: Graduate Theological Union. A concise, readable book that raises ethical consciousness regarding environmental issues.

Chess, C. and Hance, B. J. (1989). Opening Doors: Making Risk Communication Agency Reality. *Environment* 31 (5): 10–15, 38–39. Important discussion of the role of public agencies in risk management.

Chiras, D. (1982). Risk and Risk Assessment in Environmental Education. *Amer. Biol. Teacher* 44 (4): 460–465. A more technical presentation of risk and risk assessment.

Lowrance, W. W. (1976). *Of Acceptable Risk: Science and the Determination of Safety.* Los Altos, CA: Kaufmann. A fine book on risk that covers many important issues.

McKean, K. (1985). Decisions, Decisions. *Discover* 6 (6): 22–31. An interesting look at the psychology of risk assessment.

Opians, G. H. (1986). The Place of Science in Environmental Problem Solving. *Environment* 28 (9): 12–17, 38–41. Important look at the role of science in assessing risk.

Peterson, C. (1985). How Much Risk Is Too Much? *Sierra* 70 (3): 62–64. A lively but critical look at risk assessment.

Slovic, P. (1987). Perception of Risk. *Science* 236 (April 17): 280–285. Good discussion of risk perception.

Wilson, R. and Crouch, E. A. C. (1987). Risk Assessment and Comparisons: An Introduction. *Science* 236 (April 17): 267–270. Excellent introduction.

Global Lead Pollution

Lead is one of the most useful metals in modern industrial societies. Used by humankind for over 3000 years, lead is found in ceramic glazes, batteries, fishing sinkers, solder, and pipe. In gasoline, lead enhances combustion and helps reduce engine knocking.

Lead has long been known as a highly toxic poison. It affects many organs and enters the body in many ways. It has a special affinity for bone and brain tissue. High-level exposure in certain factory workers has caused neurological symptoms: fatigue, headache, muscular tremor, clumsiness, and loss of memory. If exposure is discontinued, patients may slowly recover, but residual damage—such as epilepsy, idiocy, and hydrocephalus (fluid accumulation in the brain)—often results. Continued high exposure may lead to convulsions, coma, and death. Some scientists believe that lead drinking vessels and lead pipes in water systems may have caused a decline in birth rates and increased psychosis in ancient Rome's ruling class, contributing to the fall of the Roman Empire.

Today, because of better controls on lead in the workplace and in commercial products, acute poisonings are rare. Nonetheless, many people throughout the world are regularly exposed to low levels of lead with serious consequences.

Sources of Lead

Lead contaminates our food, water, air, and soils. No one is free from this potentially toxic metal. Figure S14-1 shows the major sources of lead among the general public. For most of us, food tops the list. Some of the lead in food comes from lead-arsenate pesticides or from automobiles, power plants, and smelters. Deposited in the soil, it is taken up by food crops. About half the lead in the human diet comes from the solder in cans.

Until recently, most of the concern for lead exposure has centered on atmospheric lead. In a recent study the EPA estimated that 88% of the lead in the air we breathe comes from automobiles, except around lead smelters and steel factories, both of which release large quantities of this harmful metal into the atmosphere.

Lead is a discriminator. Its victims are primarily children and, among them, mostly children of poor black families (Figure S14-2). Lead is a poison of the poor. Even today children living in old, neglected buildings often ingest flakes of lead-based paint, which was applied before a ban was enacted in the 1940s. Children may also eat dirt contaminated with lead from passing vehicles, or may inhale lead in the atmosphere near highways. In 1986 a major study by the EPA revealed that lead levels in

drinking water in many cities exceeded federal standards, potentially threatening the health of millions of Americans. Lead is believed to come from solder in pipes and from lead pipes used in older homes.

Effects of High-Level Exposure

High-level lead exposure can cause a number of neurological disorders, including fatigue, headache, muscular tremor, lack of appetite, clumsiness, and loss of memory. These symptoms are a result of damage caused by inorganic lead in the brain and spinal cord (the central nervous system). If the damage is severe enough, death results. Organic lead (alkyl lead in gasoline, for example) causes a host of psychological disorders,

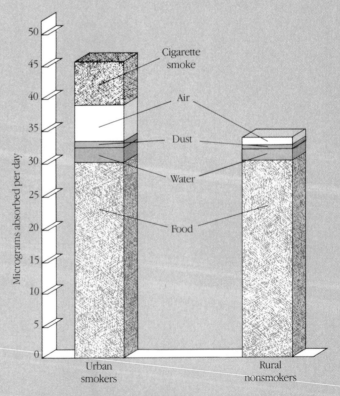

Figure S14-1. Sources of lead absorption in humans. The major source of lead is food: from lead arsenate pesticides, air pollution deposited on the soil, and food containers.

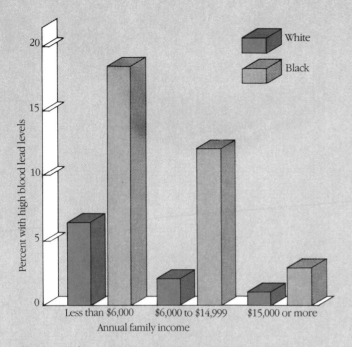

Figure S14-2. High lead levels in US children from 6 months to 5 years old, according to their parents' income and race (1976–1980).

including hallucinations, delusions, and excitement, and may lead to delirium and death.

Lead exposure can also affect the nerves that arise from the brain and spinal cord (the peripheral nervous system). The most common symptom in individuals exposed to high levels of lead is weakness of the extensor muscles, which cause the joints to open. On a cellular level, lead destroys the insulation (myelin sheath) of nerve cells. This may be responsible for the reduction in nerve-impulse speed commonly seen in patients who have been exposed to high levels of lead.

Lead also damages the kidneys, causing a disturbance in the mechanisms that help us conserve valuable nutrients (such as glucose and amino acids) that might otherwise be lost in the urine. Prolonged, high-level exposure causes a progressive buildup of connective tissue in the kidney and degeneration of the glomeruli, the filtering mechanism that separates wastes from the blood stream.

Lead has a profound effect on reproduction, in laboratory animals and humans alike. Numerous reports show that the rate of spontaneous abortion is much higher in couples either of whom has been exposed to high levels of lead in the workplace. Recent studies show decreased fertility and damaged sperm in male workers with high to medium levels of lead in their blood. According to one study, exposure of a pregnant woman to high levels of lead in household drinking water nearly doubles the risk of her having a retarded child.

Effects of Low-Level Lead Exposure

The toxic effects of large doses of lead have long been known, but only recently have we begun to understand what effects

low-level exposure may have in human populations.

About 8% to 10% of the lead ingested by adults is absorbed by the intestines, but children have a much higher absorption rate—perhaps as high as 40%. In addition, children are more sensitive to the effects of lead than adults. The developing brain seems to be the most sensitive organ. The toxicity of lead is increased in malnourished and iron-deficient children, who often come from poor urban families.

A number of studies have looked at the effects of lead levels on mental functions. Herbert Needleman and his colleagues performed a study of over 3000 children in the first and second grades in two towns near Boston. Children with high lead levels in their bodies (but still below toxic levels) had significantly lower IQ scores than those with low levels. Attention span and classroom behavior were also significantly impaired. Several other studies showed that lead levels in the blood of greater than 40 micrograms per 100 milliliters diminish intelligence and mental capacity in children under 6 years of age. A recent study in England showed that at an early age even marginally elevated levels of lead may have lasting adverse effects on intelligence and behavior.

In another important study, researchers found that exposure to small amounts of lead before birth, even at levels once considered safe, appears to seriously affect mental development. Low-level exposure slows important aspects of mental development during the first two years of life and possibly beyond that. If additional studies support these results, researchers believe that the federal standard for acceptable blood levels should be lowered for fetuses.

Lead contaminates livestock and wildlife as well as humans. Studies in Illinois, for example, show that lead levels in urban songbirds were significantly higher than those in their rural counterparts, although concentrations in urban birds did not approach toxic levels. A similar study of mice and voles showed that rodents living near major highways had significantly higher levels of lead than those living near less frequently used roads, but the elevated levels were apparently not toxic. Possible long-term effects on reproduction were believed to be minimal.

No one is free from lead exposure today, not even the residents of rural, nonindustrialized countries. Sergio Piomelli and his colleagues estimate that blood levels in humans before lead pollution became prevalent were about 100 times lower than the normal range found today in Americans. Clearly, the highest exposures occur in the citizens of technological societies, whose air is polluted by automobiles, power plants, and smelters and whose food is contaminated by lead solder and atmospheric fallout. However, even residents of Nepal have levels ten times higher than those estimated to be present before the widespread use of lead.

Controls on Lead

Alarmed by the mounting evidence regarding the effects of lead in children, the EPA in 1973 began a progressive restriction of the lead content of gasoline (Figure S14-3). Between 1974 and 1980 lead consumed in gasoline dropped by 62%, and ambient lead levels decreased by 54%. According to a study released in 1983, blood levels in over 27,000 Americans living in 64 areas have dropped from an average of 14.6 micrograms per 100

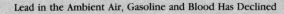

Lead in the Ambient Air, Gasoline and Blood Has Declined

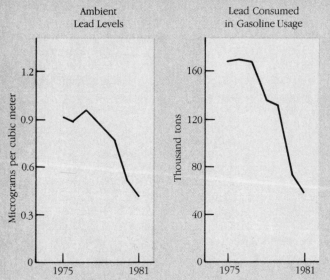

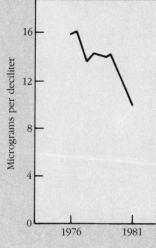

Figure S14-3. Reductions in lead in gasoline have resulted in marked decreases in lead in the air and blood of Americans.

Source: *National Air Quality and Emissions Trends Report, 1982,* USEPA

milliliters in February 1976 to 9.4 micrograms in February 1980.

Studies that found a strong statistical link between lead levels in blood and high blood pressure spurred the EPA to impose a 90% reduction in the lead in gasoline by the end of 1985. Because of new research showing the harmful effects of lead exposure to fetuses, the EPA recently announced a complete ban on leaded gasoline to be in effect by the mid-1990s.

While the United States has aggressively gone after lead in gasoline, most European nations have done little in this area, and it appears that regulations may be a long way off. There are several reasons for this. First, even though many nations favor the conversion to unleaded gasoline, a uniform policy is necessary, because Europeans make frequent border crossings. Cars that burn unleaded fuels cannot burn leaded fuels without destroying their catalytic converters (see Chapter 15). France and Italy are strongly opposed to unleaded fuels, largely for economic reasons. It would cost about 2 cents per gallon more to burn it and about $150 to $400 per car more to manufacture cars equipped with catalytic converters. Not until a European consensus can be reached will unleaded fuels be introduced there. Great Britain and West Germany, however, have at least begun to take positive steps to get the lead out of their gasoline and air.

Cities in Third World nations are even further behind. The lead content of their gasoline is, on the average, twice that of the developed countries. Malnutrition and high lead levels in the air will almost certainly have serious effects on their children.

As noted above, food is the major source of lead in the United States. The lead concentration in the average American diet is 100 times that of our prehistoric ancestors. In 1979 the Food and Drug Administration issued an advance notice of allowable lead levels in food, aimed at reducing the intake of lead from lead-soldered cans by one-half over a five-year period. These measures should go a long way toward reducing lead levels in the American diet. In 1986 Congress also banned the use of

lead solder in pipes. Since drinking water accounts for about 20% of the lead exposure in Americans, the EPA recently decided on new regulations to reduce lead in public drinking water supplies. The action may lower lead exposure in drinking water for about 138 million people.

The EPA will require public water suppliers where lead is a problem or where the water is slightly acidic to treat their water with alkaline additives. This will reduce lead leaching from pipes. Critics warn, however, that erosion controls will not be adopted quickly. A year or more of surveys may be necessary to establish where problems exist. The new regulations also permit water suppliers up to three years to develop treatment strategies.

The proposed ruling by the EPA would lower allowable levels of lead in the drinking water to about 5 parts per billion (ppb). The current level is 50 ppb. The reduction is not going to be ten-fold because the current limit of 50 ppb is water measured at the tap, while the proposed standards would measure water lead levels leaving treatment plants. Because most lead enters water after it leaves the treatment plant, some toxicologists believe that the new ruling is not going to lower public exposure very much. Only time will tell.

Suggested Readings

Klaassen, C. D., Amdur, M. O., and Doull, J. (1986). *Casarette and Doull's Toxicology: The Basic Science of Poisons.* New York: Macmillan. Good, technical resource.

Leyden, J. (1985). Nobody Wins with Lead. *National Wildlife* 23 (1): 46–48. Clearly written overview of the lead controversy.

Needleman, H. L. and Landrigan, P. J. (1981). The Health Effects of Low-Level Exposure to Lead. Annual Review. *Public Health* 2: 277–298. A critical look at the effects of lead in children.

Singhal, R. and Thomas, J. A. (1980). *Lead Toxicity.* Baltimore: Urban and Schwarzenberg. Excellent technical review.

Air Pollution: Protecting a Global Commons

Not life, but a good life, is to be chiefly valued.

SOCRATES

Springtime. That glorious time of year when birds return from their wintering grounds in colorful breeding plumage, full of song. Trees explode in foliage. Forests turn green overnight. Flowers poke through the earth's crusty skin, bending in warm sunshine. Spring rejoices in new promises—unless, of course, you make the water your home.

Along the pristine Tovdal River basin in southern Norway, far from urban centers, spring has arrived. With each day the sun climbs higher above the horizon. Squirrels, chattering high on tree branches, shake off a chill that has been with them over the long, gray winter months.

In the river the ice cracks with the rising heat, and huge chunks drop into the frigid waters. The snow becomes mushy. A small trickle percolates through the crystalline lattice, dripping to the ground. From here, tiny rivulets flow to the nearby river. As the days get longer, rivulets of melted snow swell. The Tovdal River rises in its banks. But something is awry. Fish dart fitfully along the bottom. Respiration becomes labored. Frantically, they flit from sunken log to rock. But there is no shelter from the invisible stranglehold. Within days the fish become listless, then die, turning belly up and floating to the sea.

What has happened to the fish of the Tovdal River? Toxic air pollutants deposited in the now-melting snow have suddenly flooded the river. The two most harmful are sulfuric acid and nitric acid. Besides being toxic, these acids leach harmful substances like aluminum from the soil and carry them to receiving streams. Aluminum irritates the gills and causes mucus to build up on them, eventually suffocating the fish.

Table 15-1 Natural Air Pollutants	
Source	**Pollutants**
Volcanoes	Sulfur oxides, particulates
Forest fires	Carbon monoxide, carbon dioxide, nitrogen oxides, particulates
Wind storms	Dust
Plants (live)	Hydrocarbons, pollen
Plants (decaying)	Methane, hydrogen sulfide
Soil	Viruses, dust
Sea	Salt particulates

The story of the Tovdal River is a tale of modern destruction brought on by air pollutants from an expanding industrial society. This chapter discusses air pollution, providing the principles needed to understand it. Four chapter supplements follow, giving details on some of the most important air pollution issues facing us today.

Air: The Endangered Global Commons

Air is a mixture of gases, including nitrogen (79%), oxygen (20%), carbon dioxide (.03%), and several inert gases: argon (almost 1%), helium, xenon, neon, and krypton. Water vapor exists in varying amounts.

Air is a finite resource capable of cleansing itself of many, but not all, pollutants. Satellite pictures show how huge air masses sweep across the earth's surface, picking up moisture and pollutants in one region and depositing them kilometers away.

Transparent, powerful, nurturing, air is a global resource. It is owned by no one. Many share in its use, but no one has sole responsibility for protecting it. This makes it easy to pollute and difficult to protect. As the example of the Tovdal River basin clearly shows, even those who do not pollute it suffer from the disregard of others.

The Trees Are Responsible

When he was President, Ronald Reagan argued that acidic pollutants, like those that caused the death of Tovdal River fish, come largely from natural sources, such as volcanoes. A former mayor of Denver contended that the city's summertime air pollution was, in his view, caused by pine trees. The message of both politicians is "Don't worry

about air pollution from human sources. The biggest source is natural events: volcanoes, dust storms, forest fires, and the like" (Table 15-1). Unfortunately, both men were telling only half the story.

In sheer quantity, natural pollutants often outweigh the products of human activities, the so-called **anthropogenic pollutants**. Nevertheless, anthropogenic pollutants create the most significant long-term threat to the biosphere. Why? Natural pollutants come from widely dispersed sources or infrequent events. Therefore, they generally do not raise the ambient pollutant concentration very much. In contrast, power plants, automobiles, and factories release large quantities in a restricted area, so their contribution to local pollution levels is often quite significant.

Air Pollutants and Their Sources

Take a deep breath. If you live in a city, the chances are you have just inhaled tiny amounts of dozens of different air pollutants, most in concentrations too small (we think) to be harmful. This chapter concerns itself primarily with six major pollutants: carbon monoxide, sulfur oxides, nitrogen oxides, particulates, hydrocarbons, and photochemical oxidants. Lead, an important air pollutant, was discussed in Chapter Supplement 14-1; radiation was examined in Supplement 12-1.

In 1980 the United States produced 160 million metric tons of air pollution. Thanks to conservation, better pollution control, and a ban on lead, by 1988 US production of the five major pollutants had fallen to 135 million metric tons.

Figure 15-1 shows that the major air pollutants come from three principal sources: transportation, stationary sources (factories and power plants), and industrial processes. Air pollutants are released from vaporization (or evaporation), attrition (or friction), and combustion. **Combustion** is by far the major producer.

Coal, oil, natural gas, and their refined products, such as gasoline, are organic fuels. They come from either plant or animal remains buried by sediments millions of years ago. For this reason they are called fossil fuels. Fossil fuels consist primarily of carbon and hydrogen atoms linked by covalent bonds. When this organic matter is ignited, an interesting thing happens. The initial source of heat, say, a match, breaks some of the covalent bonds. This releases energy in two forms: light and heat. Heat released in the process breaks other bonds, permitting the burning to occur until the fuel runs out. Oxygen reacts with carbon and hydrogen. Complete combustion, which rarely occurs, produces carbon dioxide (CO_2) and water (H_2O). Incomplete combustion produces carbon monoxide (CO) gas and unburned hydrocarbons (Figure 15-2).

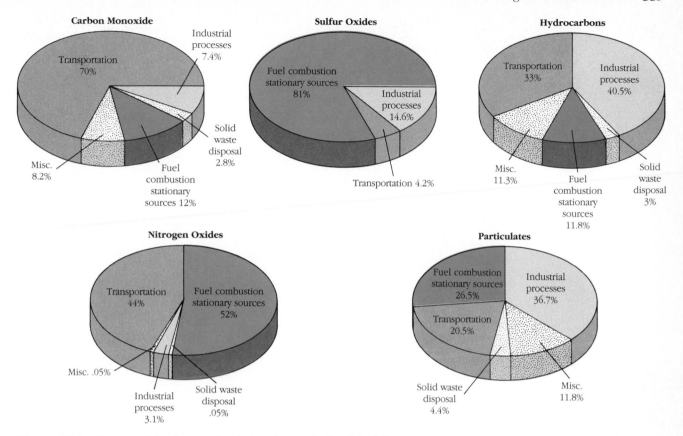

Figure 15-1. Sources of the five regulated air pollutants in the United States.

Most fuels contain some mineral contaminants. These unburnable contaminants may be carried off by hot combustion gases, escaping into the air as particulates. Other contaminants, such as sulfur, actually react with oxygen at high combustion temperatures, forming sulfur oxide gases, notably sulfur dioxide (SO_2) and sulfur trioxide (SO_3) (Figure 15-2). In the absence of pollution control devices these gases escape with the other smokestack gases.

Combustion must take place in air, for air provides a source of oxygen. But air also contains nitrogen. During combustion, nitrogen (N_2) reacts with oxygen to form nitric oxide (NO). NO is quickly converted to nitrogen dioxide (NO_2), a brownish-orange gas seen in many modern cities.

Primary and Secondary Pollutants

The atmosphere is, in many ways, a chemist's nightmare; it contains hundreds of air pollutants from natural and anthropogenic sources. These pollutants, called **primary pollutants**, often react with one another or with water vapor. A whole new set of pollutants, called **secondary pollutants**, is made in this way. Technically, secondary pollutants are chemical substances produced from the chemical reactions of natural or anthropogenic pollutants,

reactions powered by energy from the sun. These new pollutants may be more harmful than the chemicals that gave rise to them. For example, sulfur dioxide gas is released from a variety of sources such as coal-fired power plants and oil shale retorts. In the atmosphere, SO_2 reacts with the oxygen and water to produce sulfuric acid (H_2SO_4), a toxic pollutant with far-reaching effects (Chapter Supplement 15-2).

Toxic Air Pollutants

Health officials and environmental activists have long been concerned about the hundreds of potentially toxic pollutants released into the atmosphere each year in the United States. Although emitted in much smaller quantities than the five major pollutants discussed above, these *toxic air pollutants*, experts believe, may be responsible for numerous cancer deaths. A recent study by the EPA, for example, suggested that 45 of the toxic air pollutants may cause as many as 1700 cases of cancer each year.

By some estimates there are about 400 toxic air pollutants released into the atmosphere in the United States. Surprisingly, these chemicals are not currently regulated. No controls are required. Efforts to put a halt to their release have been under way for many years, but have been stymied by industry pressure, resulting, in large part,

Figure 15-2. Products of fossil fuel combustion.

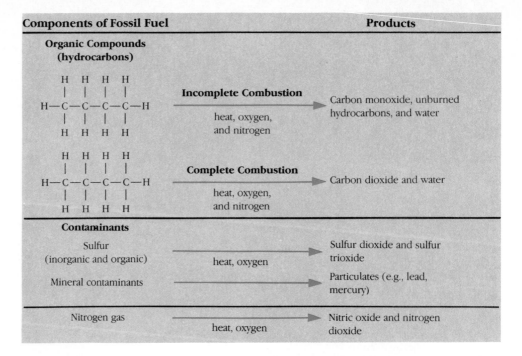

from the high cost of control. In 1989 President George Bush introduced legislation requiring factories to control toxic air pollutants. Critics, however, think that the requirement is too weak and subject to economic analysis that will sacrifice human health for profits.

The Effects of Climate and Topography on Air Pollution

Brown-Air and Gray-Air Cities

If you are like most Americans you live in a city. That city generally falls into one of two categories, based on the climate and the type of air pollution. Older, industrial cities like Nashville, New York, Philadelphia, St. Louis, and Pittsburgh belong to a group of **gray-air cities** (Figure 15-3a); newer, relatively nonindustrialized cities such as Denver, Los Angeles, and Albuquerque belong to the group of **brown-air cities** (Figure 15-3b).

Gray-air cities like New York are generally located in cold, moist climates. The major pollutants are sulfur oxides and particulates. These pollutants combine with atmospheric moisture to form the grayish haze called **smog**, a term coined in 1905 to describe the mixture of smoke and fog that plagued industrial England. The gray-air cities depend greatly on coal and oil and are usually heavily industrialized. The air in these cities is especially bad during cold, wet winters, when the demand for home heating oil and electricity is heavy and atmospheric moisture content is high.

Brown-air cities are typically located in warm, dry, and

sunny climates and are generally newer cities with few polluting industries. The major sources of pollution in these cities are the automobile and the electric power plant; the primary pollutants are carbon monoxide, hydrocarbons, and nitrogen oxides.

In brown-air cities atmospheric hydrocarbons and nitrogen oxides from automobiles and power plants react in the presence of sunlight. A number of secondary pollutants such as ozone, formaldehyde, and peroxyacylnitrate (PAN) are formed in this witch's brew. The reactions are called **photochemical reactions** because they involve both sunlight and chemical pollutants. The resulting brownish-orange shroud of air pollution is called **photochemical smog**. Ozone (O_3) is the major photochemical oxidant; a highly reactive chemical, it erodes rubber, irritates the respiratory system, and damages trees.

In brown-air cities early morning traffic provides the ingredients for photochemical smog, which reaches the highest levels in the early afternoon (Figure 15-4). Because the air laden with photochemical smog often drifts out of the city, the suburbs and surrounding rural areas usually have higher levels of photochemical smog than the city itself. Major pollution episodes in brown-air cities usually occur during the summer months, when the sun is most intense.

Today, the distinction between gray- and brown-air cities is rapidly disappearing. Most cities have brown air in the summer (when sunlight and automobile pollutants are prevalent) and gray air in the winter (when pollution from wood stoves and oil burners and the moist, wet air conspire to darken the skies).

Researchers have found some instances where natu-

(a)

(b)

Figure 15-3. Two types of air pollution: (a) Gray-air smog in Detroit. (b) Brown-air smog in Los Angeles.

rally occurring pollutants do affect air quality. In Atlanta, for example, the trees emit a number of highly reactive hydrocarbons. These react with nitrogen oxide, a gas emitted from automobiles and other combustion sources, producing ozone. Most air pollution analysts discount the contribution of trees to urban air pollution because they represent only a small fraction of the hydrocarbons present. Critical thinking suggests a more careful look. For example, new research shows that hydrocarbons from trees are 50 to 100 times more reactive than hydrocarbons from human sources. They react to form **ozone**, a potentially hazardous pollutant discussed later in this chapter.

Careful study can keep us from making foolish mistakes. The EPA argued that Atlanta could meet federal air quality standards for ozone by reducing the level of hydrocarbons from human sources by 30%. The new data, however, suggest that when the contribution from trees is added, the human-related sources would have to be cut by 70% to 100%.

Hydrocarbon reduction is the often-pursued strategy because it is technologically feasible. But in this case, a 70% to 100% reduction would be impossible. How can the city meet federal standards? By reducing nitrogen oxides from human sources. Although much more difficult, it is the only way to cut ambient ozone levels in the city and neighboring suburbs.

Factors Affecting Air Pollution Levels

Wind and Rain Pollution is something of an enigma to many people. One day is clear. The next finds the skies filled with ugly crud. Numerous factors contribute to this

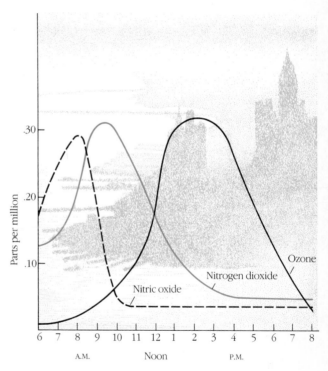

Figure 15-4. Nitrogen oxides and hydrocarbons (not shown here) react to form ozone and other photochemical oxidants. Because sunlight and time are required for the reactions to occur, maximum ozone concentration occurs in the early afternoon. Hydrocarbon levels would follow the same pattern as nitric oxide levels.

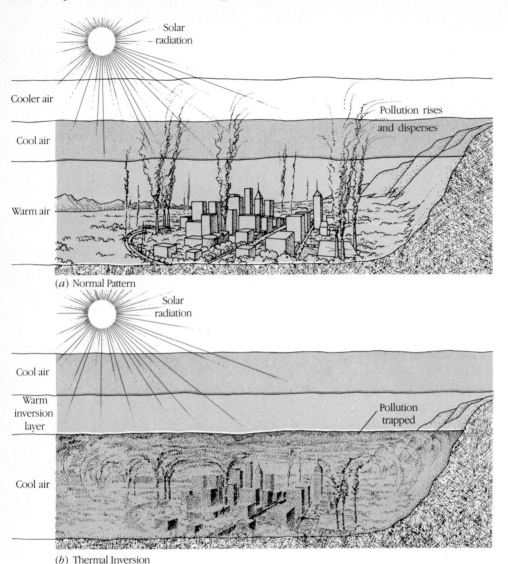

Cooler air

Cool air

Warm air

Solar radiation

Pollution rises and disperses

(a) Normal Pattern

Solar radiation

Cool air

Warm inversion layer

Cool air

Pollution trapped

(b) Thermal Inversion

Figure 15-5. (a) During normal conditions, air temperature decreases with altitude; thus, pollutants ascend and mix with atmospheric gases. (b) In a temperature inversion, however, warm air forms a "lid" over cooler air, thus trapping air pollution.

puzzle. For example, wind sweeps dirty air out of cities. Rain washes pollutants from the sky. Contrary to what many people think, the pollutants do not disappear. They are merely transferred from one medium to another, a process called **cross-media contamination**. Airborne pollutants can travel hundreds, perhaps thousands, of kilometers to other cities or unpolluted wilderness. The acidic pollutants responsible for the massive fish kills in the Tovdal River basin were blown into Scandinavia from industrial England and Europe (see Chapter Supplement 15-2).

Mountains and Hills Salt Lake City stretches out below a giant mountain range, covered with snow much of the year. Granite peaks and jagged cliffs make this one of the most scenic American cities—that is, when the mountains are visible through the air pollution. Residents of Salt Lake City would tell you that mountain ranges and

hills can be an asset to a city, but they can also be a curse. They often block the flow of winds and trap pollutants for days on end. Mountains also block the sun, which, as you shall soon find out, helps disperse pollutants.

Temperature Inversions Supposed you were riding a hot-air balloon into the sky on a normal day. You would find that warm ground air rises and expands. As it expands, it cools. Pollution rises with the warm ground air. Because of this, the atmosphere is gently stirred, and ground-level pollution is reduced (Figure 15-5a). Atmospheric mixing is brought about by sunlight. Striking the earth, sunlight heats the rock and soil. This heat is transferred to the air immediately above the ground. The warm air then rises, mixing with cooler air.

On your flight into the wild blue yonder you would find that the temperature dropped steadily as the balloon rose. Under certain atmospheric conditions, however, you would find a different temperature profile with tre-

Figure 15-6. This large subsidence inversion occurred in August 1969.

Effects of Air Pollution

In 1989, the EPA announced that 110 million Americans—nearly half of our population—were breathing unhealthy air. More than 60 cities failed to meet federal air quality standards, even though the standards had been eased and the deadline for meeting them was extended repeatedly in the past 17 years. Ozone is the biggest problem. According to the EPA, 76 million Americans are exposed to potentially dangerous levels of this pollutant. Particulates rank second, despite dramatic declines since 1970. All told, nearly 50 million Americans breathe air containing potentially harmful levels of particulates. According to a recent government report, by the year 2000 air pollution will be killing 57,000 Americans a year.

The American Lung Association estimates that air pollution costs Americans $40 billion a year in health costs, or about $160 per year for every man, woman, and child. Air pollution damages crops and buildings, adding at least another $10 billion to the price tag. Many other costs cannot be calculated: the loss of scenic view, the destruction of a favorite fishing spot, the erosion of an important statue. This section looks at the external costs of pollution, outlining a small fraction of the damage.

Health Effects

Acute Health Effects Considerable evidence has accumulated to show that air pollution affects us on a day-to-day basis (Table 15-2). In 1966, for example, one of the first studies of urban air pollution's effects showed that when levels of sulfur dioxide rose, a large proportion of the urban residents in New York City became ill. The incidence of colds, coughs, rhinitis (nose irritation), and other symptoms increased fivefold almost overnight. When the air cleared, the acute effects disappeared.

Many other acute effects come about from exposure to air pollution. Visitors to Los Angeles often complain of burning or itching eyes and irritated throats caused by photochemical smog. Commuters in heavy traffic are familiar with the headaches caused by carbon monoxide from automobile fumes. These acute effects, however, are generally ignored or simply seen as part of the price we pay for city living.

Chronic Health Effects Long-term exposure to air pollution may result in a number of diseases, including bronchitis, emphysema, and lung cancer.

One out of every five American men between the ages of 40 and 60 has **chronic bronchitis**, a persistent inflammation of the bronchial tubes, which carry air into the lungs. Symptoms include a persistent cough, mucus buildup, and difficulty breathing. Cigarette smoking is a major cause of this disease, but urban air pollution is also

mendous implications for air pollution levels (Figure 15-5b). For example, on some winter days you would find that the air temperature dropped as you rose, but only to a certain point. After that the temperature would begin to increase. This inverted temperature profile is called a **temperature inversion**. Temperature inversions create warm-air lids over cooler air (Figure 15-5b). Because the cool dense ground air cannot mix vertically, pollutants become trapped near the ground, often reaching dangerous levels.

Temperature inversions fall into two categories. A **subsidence inversion** occurs when a high-pressure air mass stalls and forces a layer of warm air down over a region. These inversions may extend over many thousands of square kilometers (Figure 15-6). A **radiation inversion**, on the other hand, is usually local and short-lived. It is a phenomenon many of us witness on cold winter days. Radiation inversions begin to form a few hours before the sun sets. As the day ends, the air near the ground cools faster than the air above it. Thus warm air lies over the cooler ground air. The cool ground air cannot rise. Pollutants accumulate. The inversion usually breaks up in the morning when the sun strikes the earth, beginning the vertical mixing. Radiation inversions are common in mountainous regions, especially in the winter when the sun is obstructed by the mountains and therefore unable to warm the ground enough to stimulate vertical mixing.

Table 15-2 Major Air Pollutants—Their Sources and Health Effects

Pollutant	Major Anthropogenic Sources	Health Effects
Carbon monoxide	Transportation industry	Acute exposure: headache, dizziness, decreased physical performance, death
		Chronic exposure: stress on cardiovascular system, decreased tolerance to exercise, heart attack
Sulfur oxides	Stationary combustion sources, industry	Acute exposure: inflammation of respiratory tract, aggravation of asthma
		Chronic exposure: emphysema, bronchitis
Nitrogen oxides	Transportation, stationary combustion sources	Acute exposure: lung irritation
		Chronic exposure: bronchitis
Particulates	Stationary combustion sources, industry	Irritation of respiratory system, cancer
Hydrocarbons	Transportation	Unknown
Photochemical oxidants	Transportation, stationary combustion sources (indirectly through hydrocarbons and nitrogen oxides)	Acute exposure: respiratory irritation, eye irritation
		Chronic exposure: emphysema

a contributing factor. Sulfur dioxide, nitrogen dioxide, and ozone are believed to be the major causative agents.

Emphysema, another chronic effect, kills more people than lung cancer and tuberculosis combined. Emphysema is the fastest growing cause of death in the United States. Over 1.5 million Americans suffer from this incurable disease. As they become older, the small air sacs, or **alveoli**, in the lungs break down. This reduces the surface area for the exchange of oxygen with the blood. Breathing becomes more and more labored. Victims suffer shortness of breath when exercising even lightly.

Emphysema is caused by cigarette smoking and may be caused by urban air pollution as well. One study, for instance, showed that the incidence of emphysema was higher in relatively polluted St. Louis than in relatively unpolluted Winnipeg (Figure 15-7). Studies in Great Britain showed that mail carriers who worked in polluted urban areas had a substantially higher death rate from emphysema than those who worked in unpolluted rural areas. Ozone, nitrogen dioxide, and sulfur oxides are the chemical agents believed responsible for this disease.

A number of studies have shown that lung cancer rates are higher among urban residents than among rural residents (even after the influence of cigarette smoking has been ruled out). Critical thinking recommends caution. Why? Health studies of human populations, or *epidemiological studies*, rely on statistical methods with some unavoidable shortcomings (see Chapter Supplement 1-1). In lung cancer studies, for example, researchers usually compare death certificates of urbanites with death certificates from rural residents. If the urban population

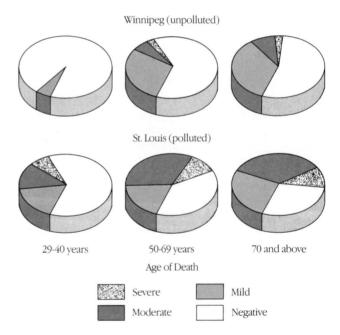

Winnipeg (unpolluted)

St. Louis (polluted)

29-40 years   50-69 years   70 and above

Age of Death

Severe   Mild

Moderate   Negative

Figure 15-7. Incidence of emphysema in St. Louis and Winnipeg. Note the increased incidence of emphysema in all three age groups in the more polluted urban environment of St. Louis.

shows a higher incidence of lung cancer, it is tempting to conclude that the cause was urban air pollution. But researchers must be careful to eliminate other causative agents, such as smoking and occupation. Most researchers eliminate smokers from the study, but not all take into

Table 15-3 Damage to Materials from Air Pollution

Material	Damage	Principal Pollutants
Metals	Corrosion or tarnishing of surfaces; loss of strength	Sulfur dioxide, hydrogen sulfide, particulates
Stone and concrete	Discoloration, erosion of surfaces, leaching	Sulfur dioxide, particulates
Paint	Discoloration, reduced gloss, pitting	Sulfur dioxide, hydrogen sulfide, particulates, ozone
Rubber	Weakening, cracking	Ozone, other photochemical oxidants
Leather	Weakening, deterioration of surface	Sulfur dioxide
Paper	Embrittlement	Sulfur dioxide
Textiles	Soiling, fading, deterioration of fabric	Sulfur dioxide, ozone, particulates, nitrogen dioxide
Ceramics	Altered surface appearance	Hydrogen fluoride, particulates

account urban occupations (factories), where men and women may have been exposed to high levels of cancer-causing pollutants. Thus, the slightly higher incidence of lung cancer in some urban settings may result from occupational exposure or some other factor. It could also result from pollution. Only time will tell.

High-Risk Populations Not all individuals are affected equally by air pollution. Particularly susceptible are the old and infirm, especially people with lung and heart disorders. Carbon monoxide is especially dangerous to people with heart disease, because it binds strongly to hemoglobin, the oxygen-carrying protein in red blood corpuscles. This binding reduces the oxygen-carrying capacity of the blood. For sufficient oxygen to be delivered to the body's cells, the heart must pump more blood during a given period. This puts a strain on the heart and may trigger heart attacks in individuals with weakened hearts.

Medical researchers recently announced that the health risk from air pollution is six times greater for children than for adults. Children are more susceptible than healthy adults because they are more active and therefore breathe more. As a result, they may be exposed to more pollution. In addition, children typically suffer from colds and nasal congestion and thus tend to breathe more through their mouths. Air bypasses the normal filtering mechanism of the nose, so more pollutants enter the lungs.

Effects on Other Organisms

Fluoride and arsenic poisonings have occurred in cattle grazing downwind from metal smelters. Acids produced from power plants, smelters, industrial boilers, and automobiles have been shown to be extremely harmful to wildlife, especially fish (see Chapter Supplement 15-2). Reports from Scandinavia, West Germany, and the United

States suggest that forest productivity may be reduced significantly by acid precipitation. In southern California millions of ponderosa pines have been damaged by air pollution (mostly ozone) from Los Angeles.

Ozone, sulfur dioxide, and sulfuric acid are the pollutants most hazardous to plants. Ozone, for instance, makes plants more brittle and likely to crack. Farms in southern California and on the East Coast report significant damage to important vegetable crops. City gardeners also report damage to flowers and ornamental plants. Sulfur dioxide damages plants directly, causing spotting of leaves. Recent studies show that air pollutants may also make some plant species more desirable to leaf-eating insects. According to botanists from Cornell University, air pollution and other stresses cause plants to produce a chemical called glutathione, which protects leaves from pollution but also attracts insects that normally have no interest in these plant species. All told, ozone and other pollutants may cause as much as $10 billion worth of crop damage (in reduced yield) in the United States each year.

Effects on Materials

Air pollutants may severely damage metals, building materials (stone and concrete), paint, textiles, plastics, rubber, leather, paper, clothing, and ceramics (Table 15-3). The two most corrosive, and therefore harmful, pollutants are sulfur dioxide and sulfuric acid.

The damage to human materials is both costly and tragic, for many of the structures attacked by air pollutants are irreplaceable works of art. The stone in the Parthenon in Athens, for instance, has deteriorated more in the last 50 years than in the previous 2000 years because of air pollution. The Statue of Liberty, which was recently restored, had been pitted by sulfuric and nitric acids. The Taj Mahal in India, like many other buildings in the world, is being defaced by air pollution from local power plants.

Sulfuric and nitric acids cause cosmetic damage to metals and reduce their strength. In the Netherlands bells that had been ringing true for three or four centuries have, in recent years, gone out of tune because of air pollution. Acidic pollutants have eaten away at them, lowering their pitch and rendering once familiar tunes indecipherable. Particulates blown in the wind erode the surfaces of stone, doing significant damage; hydrogen sulfide gases tarnish silver and blacken leaded house paints. Ozone cracks rubber windshield wipers, tires, and other rubber products, necessitating costly antioxidant additives.

The economic damage caused by air pollution is immense. Society pays for cleaning sooty buildings, repainting pitted houses and automobiles, and replacing damaged rubber products and clothing. The economic damage to statuary and other works of art cannot be calculated.

Global Warming/Global Change

Scientists have long known that air pollution can affect local weather. For example, smoke from factories can substantially increase rainfall in areas downwind. In recent years, however, scientists have debated the effects of air pollution on global climate. Before examining the issue, let us first look at the global energy balance, the basis for all climate.

Global Energy Balance

Each day the earth is warmed by the sun. Sunlight strikes the earth and heats the surface; this heat is then slowly radiated back into the atmosphere. Eventually this heat, or **infrared radiation**, escapes the earth's atmosphere and returns to space. Thus, an energy balance is set up: energy input is balanced by energy output.

This balance may be altered by air pollutants, notably CO_2. Naturally occurring CO_2 allows sunlight to pass through the atmosphere and heat the earth, but absorbs infrared radiation escaping from the earth's surface and radiates it back to earth. This process helps maintain the earth's temperature. Any increase in the concentration of CO_2 would slow down the escape of heat.

Upsetting the Balance: The Greenhouse Effect

Visitors from another galaxy circle our planet. Directions from an ancient transmission helped them make the long voyage. But upon arriving they find a barren, hot desert. Instruments on board indicate that life once existed on this planet, but there is nothing to show of it now. Only heat and parched land.

Scientists on board speculate that runaway carbon dioxide levels turned the earth into a planet much like Venus. A nearby planet, Venus may have met a similar fate at some early point in its history, which accounts for its daily temperature averaging well over 450° C (842° F).

Of course, all of this is speculation, but it points to a trend about which scientists have wondered for decades, the **greenhouse effect**: a rise in global temperature brought on by increasing atmospheric CO_2.

The thermometer in the little greenhouse that heats much of my house in winter reads 38° to 49° C (100 to 120° F) on a sunny winter day even when the outside temperature is well below freezing. Imagine what the global temperature would be if we were to encompass the earth in a sphere of glass. In many ways we may be doing just that, for CO_2 acts like the glass in a global greenhouse, slowing down the escape of infrared radiation from the earth's surface.

Scientists recently discovered that a variety of gases, found in increasing concentrations in the atmosphere, reradiate infrared radiation to earth more effectively than CO_2. These include methane, nitrous oxide, and chlorofluorocarbons (CFCs). Most of these gases are from human sources, although a few like methane come from natural sources. Adding just one molecule of freon gas is equivalent to 10,000 molecules of carbon dioxide. Freons are discussed in Chapter Supplement 15-1.

Between 1870 and 1989 global concentrations of CO_2 increased 21.5% (from 290 to 352 parts per million). This rise is attributed to increasing global consumption of fossil fuels. Many scientists believe that global CO_2 levels could double between 2030 and 2050 if fossil fuel consumption continues to increase at the current rate. What would be the effect of such a doubling? The best global climate computer models predict that a doubling of CO_2 will increase the average daily temperature by about 2° to 5° C (3.5° to 9° F). At first glance this increase seems insignificant. However, such a change could drastically alter global climate (Figure 15-8). Much of the United States and Canada would be drier than normal. If this happened, many midwestern agricultural states, now barely able to support rainfed agriculture, would suffer crippling declines in farming. The United States and Canada, now major food exporters, could become food-importing nations.

The greenhouse effect may sound like science fiction, but it is not. Scientists have known about it for over 100 years. Early atmospheric scientists noted that normal levels of carbon dioxide found in the atmosphere make the earth habitable. Without carbon dioxide, the earth would be so cool that life would probably not exist.

Like so many things in the world around us, a little carbon dioxide is good; too much may be devastating, disrupting the global climate. Some scientists think that the signs of disruption are already beginning. Consider

Case Study 15-1

Offsetting Global Warming: Planting a Seed

On October 11, 1988, a small electric utility announced plans to help finance the planting of 50 million trees in Guatemala. Applied Energy Services (AES) of Arlington, Virginia, is the first company ever known to take direct responsibility for offsetting the carbon dioxide produced by two fossil fuel power plants it plans to build. Oak Ridge scientist Gregg Marland commented, "To a certain extent, it's symbolic, but it's also courageous." Gus Speth of the World Resource Institute (WRI) called it "one of the most far-sighted and socially responsible decisions a private company has ever made."

AES currently operates three small coal-fired plants. When the company decided to build two additional plants, its chief executive officer, a former Energy Department official, decided that they ought to try to do something about emissions. They contacted WRI for suggestions. WRI found a suitable project in Guatemala proposed by CARE, Inc., an international relief and development agency. Over the next ten years CARE officials hope to plant enough trees to remove a total of 15 million metric tons of carbon from the atmosphere, approximately the same amount that AES's 180-megawatt coal-fired plant in Connecticut will emit over its 40-year life.

The price tag for this project is staggering. The utility will establish a $2 million endowment. CARE, the US Agency for International Development, and the government of Guatemala each will donate $2 million. Peace Corps volunteers will plant the trees, and their service and training are estimated to be valued at $7.5 million.

Some critics think that American companies would be more likely to invest in domestic tree-planting. US utilities could help offset carbon dioxide emissions from new coal-fired power plants by planting millions of hectares of trees. One approach might be to plant them on marginally productive, highly erodible farmland that is being taken out of use in the United States to reduce the nation's soil erosion problem. Daniel Dudek of the Environmental Defense Fund says that utilities could lease the land that is being set aside by farmers (with federal assistance). Here they could plant fast-growing trees like cottonwoods, providing habitat for wildlife. The cost would be about 70 cents per ton of carbon dioxide emissions removed. A utility wishing to offset emissions from a 1,000-megawatt coal-fired power plant would need to plant a forest with a 30-mile diameter. Individuals can help as well, by planting shade trees around their homes. Cities and states can help by planting trees along highways and in parks.

Tree planting will not save us from global climate change. Current emissions from fossil fuel combustion are far too great for this strategy to work by itself. Combined with measures to reduce deforestation and other approaches—like conservation, recycling, and renewable fuels—tree planting can play an important part in reversing the dangerous global climate change now under way.

these facts: Five of the hottest years in the past 100 years have occurred since 1980 (Figure 15-9). Drought and heat cut grain production in the United States by one-third. In a normal year, US farmers produce 300 million metric tons of grain. Americans normally consume 200 million metric tons of grain. The remaining 100 million metric tons is exported or put into storage. But in 1988, domestic production fell to 195 million metric tons, just below domestic consumption. Export obligations were met by emptying storage bins, and by the end of the year, US storage bins were all but empty for the first time in many years. Making matters worse, in 1989 the winter wheat harvest in Kansas failed for lack of spring moisture and had to be plowed under. Should the drought continue, the crippling declines in world food production described above could become a reality. America might experience another devastating dust bowl that could last for decades.

The drought of 1988 brought with it record forest fires and grassland fires, another harbinger of things to come

if global warming continues. The most widely publicized fires were those that swept through Yellowstone National Park (see Case Study 9-2). Wildfires raged out of control in 1989 as well. Hundreds of thousands of hectares burned in the summer of 1989 in California, Wyoming, Nebraska, Colorado, Idaho, and other states.

In the fall of 1987, an iceberg the size of Rhode Island broke off from Antarctica. Scientists think that this and other icebergs, chipping off in record number, may be another sign that global warming has begun. Satellite photographs also show that over the past 15 years the polar ice caps have shrunk 6%.

In June of 1988, James Hansen, a cautious and highly respected scientist at NASA, told the press that he was nearly certain that global warming had begun. But not all atmospheric scientists are convinced. The exceedingly hot years, they say, may be due to normal climatic variation. Droughts and heat waves are not uncommon. However, the probability of five hot years in a decade is fairly small.

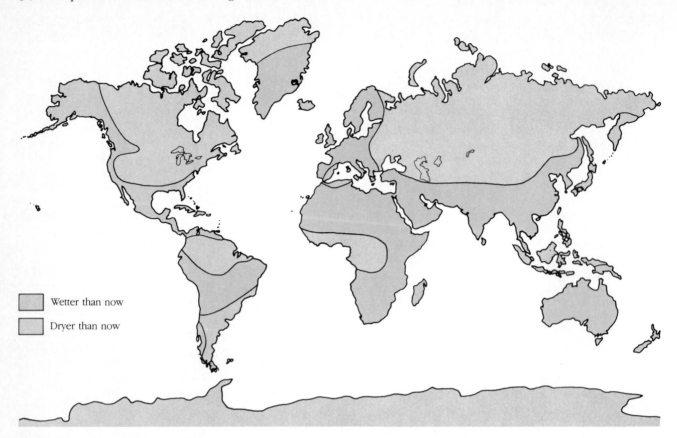

Wetter than now

Dryer than now

Figure 15-8.   Possible future changes in climate resulting from the greenhouse effect. This map is based on climatic conditions thought to have existed 4500 to 8000 years ago, when the average temperature was highest.

Should global warming continue, most of you will experience dramatic changes in your lifetimes. Your children will inherit a world far different from the one you know, a world racked with bizarre weather and potentially serious economic problems. What can you expect for yourself and your children should this trend continue?

Some experts think that in the next 40 to 50 years rainfed agriculture in the Midwest states will come to an end. But don't expect irrigation to take up the slack. River flows and groundwater are likely to fall substantially as the Midwest dries under the influence of global warming. The northern states may take up some of the slack as their climate becomes warmer, but the exchange probably won't be equal—we won't gain as much production as we've lost. Why? First, not all of the northern soils are as rich or as deep as the farmland in Kansas, Iowa, Indiana, Ohio, and Illinois. Shorter growing seasons on the northern grasslands for many hundreds of years have retarded the rate of soil formation. Second, even though northern regions may be warmer, they might not receive adequate rainfall to support farming (Figure 15-8).

You could also expect deserts to spread northward from New Mexico, Texas, Arizona, California, and Nevada

into neighboring states, destroying cropland and pastures. Snowfall in the Rockies may decline substantially, reducing the lucrative skiing and winter recreation industries. Rainfall may increase on both coasts, increasing flooding and devastation. But increased rainfall would not necessarily increase crop production. Heavy spring rains in western New York recently flooded fields and prevented farmers from planting their crops on time. By the Fourth of July, corn was only ankle high in many fields. Because of the crop failures in 1989, President Bush authorized a $900-million aid package to farmers.

Global warming will make cities unbearable in the summer. Global climate modelers predict that by 2030 the number of days above 32° C (90° F) will increase from 36 to 87. In Dallas, the number will increase from 17 to 78, greatly increasing utility costs for cooling, increasing water demand for irrigation, and generally making life unbearable for millions of people.

Ocean communities would also suffer enormously. By 2050, for example, sea level is expected to rise 50 to 100 centimeters (2 to 3 feet). By the end of the next century it could rise 180 to 210 centimeters (6 to 7 feet). The rise in sea level would result from two factors: (1) the melting of glaciers and of the land-based Antarctic ice pack and (2) an expansion of the seas resulting from warmer temperatures.

Most of the world's people live near the ocean. In the United States, for instance, over half the population lives within 83 kilometers (50 miles) of the ocean. Even a modest increase in sea level would flood coastal wetlands, low-lying fields, and cities. The rise in sea level would worsen the damage from storms. Waves produced during hurricanes and routine storms would sweep further inland, damaging more homes and cities than today. Many people would have to relocate or build new houses on stilts. Cities may be forced to build levees to hold back the seas or gradually move to higher ground as buildings are retired. The inland creep of the ocean is bound to usurp farmland and wildlife habitat and create more crowding as people compete for a limited land base. The changes would not come cheaply and they would not be easy.

Third World nations would suffer enormously as the oceans rise. In Asia, for instance, rice is produced in many low-lying regions that would be reclaimed by the sea. Storm surges could carry salt water onto some of the remaining fields, killing crops and poisoning the soil. Bangladesh, a country with 118 million people in 1990, would be especially hard hit. By some estimates, 17% of the land area will be under water by 2030, worsening crowding in a country the size of Wisconsin.

Humans are not the only species that would suffer during global warming. A great many plants and animals could face difficult times as well. If the change in temperature continues at the predicted rate, many species will be wiped out. Others will suffer incredible declines in their populations. Others may adapt or migrate to suitable habitat.

Professor Margaret Davis of the University of Minnesota used a computer to predict the effects of a global temperature increase on several tree species. If global carbon dioxide doubles by 2050, hardwood trees east of the Mississippi will have to shift 500 kilometers (300 miles) northward. Beech trees, for instance, would disappear from the southeastern United States, except in some mountainous regions. Suitable beech tree habitat would shift north to New England and southeastern Canada, which is now the extreme northern limit of its range. At the end of the last ice age, for example, beech trees "migrated" northward by dispersing seeds, but at a rate of only 20 kilometers per 100 years—far slower than the 500-kilometer migration needed to avoid destruction. Many other trees in the United States face a similar fate.

Animals, like plants, respond to warming trends by shifting to new habitats. But not all animals are capable of moving as far as necessary. Stanford University researcher Dennis Murphy studied the Great Basin Mountains, lying between the Cascades and Sierra Nevadas on

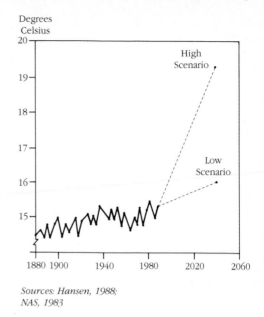

Sources: Hansen, 1988; NAS, 1983

Figure 15-9. Graph of average global temperature since 1880 with possible projections to the year 2040.

the west and the Rockies on the east. His studies indicated that 44% of the mammals, 23% of the butterflies, and a smaller percentage of birds would be lost by a 3° C increase in global temperature.

The changes now apparently in motion may result in a dangerous positive feedback (Figure 15-10). The oceans, for instance, are a major reservoir for carbon dioxide, storing 60% more than the atmosphere. Without the oceans, carbon dioxide levels in the air would be much higher than they already are. As the earth's temperature rises, however, the ocean's ability to dissolve and hold carbon dioxide falls. The oceans will then release much of the carbon dioxide they have been absorbing into the atmosphere, accelerating the rate of change.

Climate models also predict more violent and unusual weather throughout the world. Tornadoes in Colorado, practically unheard of before 1987, have wreaked considerable damage; flooding rains on the east coast and bizarre droughts in the water-rich Pacific Northwest may all be the result of a dramatic shift in global climate.

Global warming is also the result of deforestation: the rapid loss of trees, especially in the tropics. Trees absorb carbon dioxide, which they use to produce nutrients, woody tissue, and bark. Worldwide, forests are being cut much faster than they regrow. As a result, deforestation "contributes" about one-fourth of the annual global increase in carbon dioxide. To stave off or stop the rapid increase in global temperature will require a massive reforestation of the earth. Australia recently announced that it was going to plant 1 billion trees by the year 2000, in part to offset global warming. A few other countries are following suit. Individuals can help by recycling, building smaller homes, and supporting reforestation

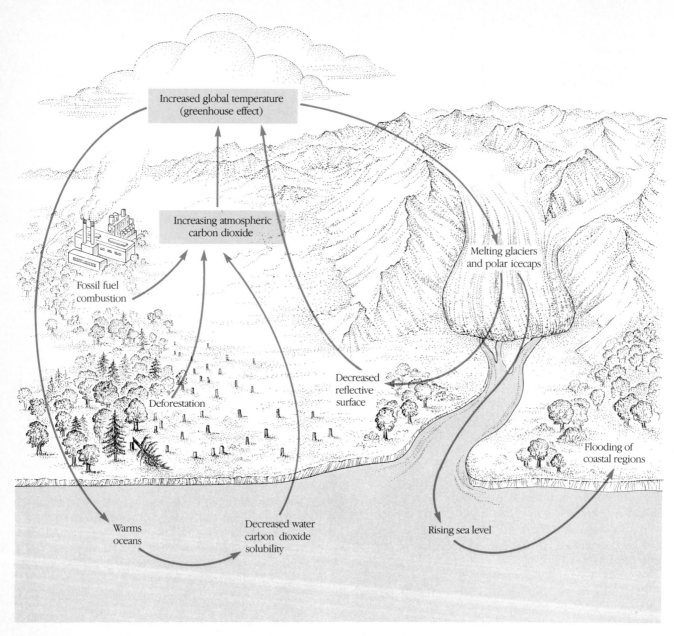

Increased global temperature
(greenhouse effect)

Increasing atmospheric
carbon dioxide

Fossil fuel
combustion

Melting glaciers
and polar icecaps

Deforestation

Decreased
reflective
surface

Flooding of
coastal regions

Warms
oceans

Decreased water
carbon dioxide
solubility

Rising sea level

Figure 15-10. The effects of particulates on global temperature depend on their color and their altitude, as well as on the amount of moisture in the atmosphere.

projects, even lending a hand to replant clear-cut areas, roadsides, abandoned fields, and backyards. To offset the carbon dioxide that your life-style creates would require you to plant 400 trees. A family of four would need to plant 2.5 hectares (6 acres) of fast-growing trees to offset its lifetime carbon dioxide production. Obviously, this is an impossibility for most people; other measures are needed. But what?

Energy conservation is a good candidate. By cutting energy demand 20% to 50%, we can make significant inroads into global warming. That means walking, bicycling, or riding a bus to school or work, building smaller,

energy-efficient homes, insulating existing homes, recycling, and using efficient appliances or doing some things by hand (for example, mixing by hand rather than using an electric mixer or drying clothes on a line rather than using a dryer). Using renewable resources and supporting family planning can also help.

In 1989, the EPA announced a plan to slow down the release of carbon dioxide and other greenhouse gases. The plan, they say, would cut the rate of increase to 0.6° C to 1.4° C per century. Energy conservation and renewable resources—long-time favorites of the environmental community—are prime candidates to accomplish this task. The EPA's approach will not eliminate global warming, only slow down the rate of change. It says to us that we must learn to live with global warming—to adapt to flooding and reduced land mass, to adapt to the loss of agricultural land and to let species die.

Many critics agree that this is a good start but argue that they would rather stop the increase altogether. We can do it by shifting to renewable energy resources, adopting much stricter energy conservation measures, using cogeneration, eliminating chlorofluorocarbons, increasing automobile efficiency and mass transit, sharply increasing home energy efficiency, halting deforestation, and improving energy efficiency in industry (see Chapter 12). The changes must come soon and must be substantial.

Air Pollution Control

Air pollution can be controlled at three interlocking levels: legal, technological, and personal. This section examines some of the major strategies we have used to control this pervasive problem and some still untapped solutions.

Cleaner Air through Better Laws

Society's laws are like the earth's creatures: evolving to fit the present better. This is true with clean-air legislation, now in its third decade of existence in the United States. Early clean-air laws were fairly weak and ineffective, but they laid the foundation for some of the most progressive environmental legislation in the world.

Today, the federal **Clean Air Act** (and its amendments) provides a wide range of protection through various measures. The first major advances came with a sweeping set of amendments to the act in 1970. They resulted in (1) emissions standards for automobiles, (2) emissions standards for new industries, and (3) air-quality standards applied to urban areas and aimed at protecting human health and the environment.

The 1970 amendments successfully reduced air pollution from automobiles and industry. They stimulated many states to pass their own air pollution laws, but they also created some major problems. For instance, in regions that exceeded ambient air-quality standards, the law prohibited new factories or the expansion of existing facilities. The business community objected, arguing that this stipulation prevented business expansion. In addition, some of the wording of the 1970 amendments was vague and required clarification. Of special interest were provisions dealing with the deterioration of air quality in areas that were already meeting federal standards.

Because of these and other problems the Clean Air Act was once again amended in 1977. To deal with the limits

on industrial growth in areas that were exceeding the ambient air-quality standards, or **nonattainment areas**, lawmakers presented a creative policy. It allowed factories to expand and new ones to be built in nonattainment areas only if three provisions were met: (1) new sources must achieve the lowest possible emission rates, (2) other sources of pollution under the same ownership or control in that state must comply with emissions-control provisions, and (3) newcomers must ask existing companies to reduce their pollution emissions, thus allowing the newcomers a portion of the air resource. The last provision, the **emissions offset policy**, proved quite successful. Combined with the stipulations described above it allows industrial expansion while ensuring a net decrease in pollution. In most cases the newcomers pay the cost of air-pollution control devices.

How to protect air quality in areas that were already meeting standards sparked considerable debate in Congress. Environmentalists felt that the ambient air-quality standards, in effect, gave industries a license to foul clean air, letting them pollute up to permitted levels. The 1977 amendments set forth rules for the **prevention of significant deterioration** of air quality in clean-air areas. These rules apply only to sulfur oxides and particulates, the pollutants viewed by Congress as the two most deserving of immediate action. Many air pollution experts disagree.

The 1977 amendments strengthened the enforcement power of the EPA. In previous years when the EPA wanted to stop a polluter, it had to initiate a criminal lawsuit; violators would often engage in a legal battle, since court costs were often lower than the cost of installing pollution control devices. Thanks to the 1977 amendments, the EPA can now initiate civil lawsuits, which do not require the heavy burden of proof needed for criminal convictions.

More important, the EPA was allowed to levy noncompliance penalties without going to court. These penalties are assessed on the ground that violators have an unfair business advantage over competitors that comply with the law. Penalties equal to the estimated cost of pollution control devices eliminate the cost incentive of polluting.

The Clean Air Act is working, there can be no question about it (Figure 15-11). Public support for clean air is as strong today as it was 10 to 20 years ago. Between 1970 and 1982 particulate emissions in the United States dropped by more than half. During the same period emissions of sulfur oxide gases, hydrocarbons, and carbon monoxide all fell (Figure 15-11). The only major air pollutant to increase was nitrogen oxide, which rose by 12%. Since 1982, annual emissions of these pollutants have leveled off. Further cuts are required to reduce or eliminate many problems.

Efforts to clean up the air in the United States continue today, despite deep economic cuts in environmental programs. The most recent advance came in 1986, when the EPA announced that it was drawing up regulations requir-

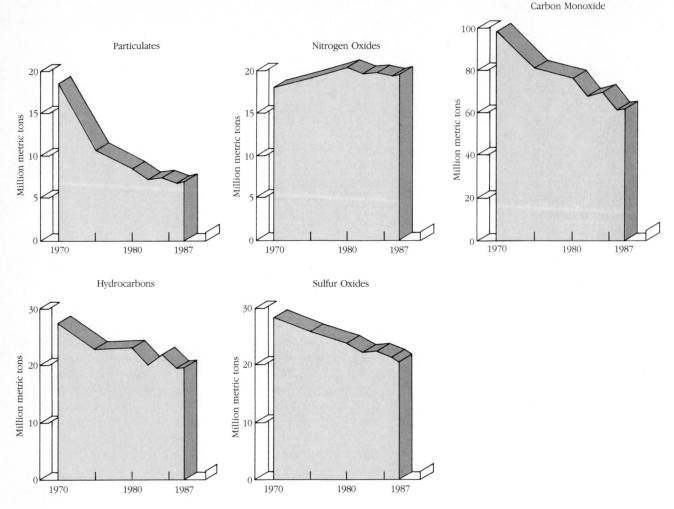

Figure 15-11. Because of tighter pollution controls and increased efficiency, emission of many pollutants has decreased. Nitrogen oxide levels have climbed because of a lack of US technology to eliminate the gas from combustion sources.

ing all wood-burning stoves to have pollution-reducing devices such as catalytic converters, which convert unburned hydrocarbons to water and carbon dioxide and also convert carbon monoxide to carbon dioxide.

As successful as it has been, the Clean Air Act still has its critics. Environmentalists, in particular, argue that the law needs to be broadened to include ways to cut back on acid precipitation, and to control many toxic air pollutants, which are currently unregulated. In 1989, President George Bush introduced a Clean Air Act Amendment that would help solve these problems, but the measures he proposed were far weaker than many people had hoped for (Chapter Supplement 15-2).

Cleaner Air through Technology

Legal solutions, in many cases, require technological solutions—new ways to cut down on pollutants. Two general approaches are usually pursued: (1) the removal of harmful substances from emissions gases and (2) the conversion of harmful pollutants in emissions gases into harmless substances. The first strategy is the most common for stationary combustion sources.

Stationary Sources In electric power plants, for example, **filters** separate particulate matter from the stack gases (Figure 15-12a). Smoke passes through a series of cloth bags; the bags filter out particulates. Filters often remove well over 99% of the particulates, but they do not remove gases.

Cyclones are also used to remove particulates, generally in smaller operations (Figure 15-12b). In the cyclone, particulate-laden air is passed through a metal cylinder. The particulates strike the walls and fall to the bottom of the cyclone, where they can be removed. Cyclones remove 50% to 90% of the large particulates

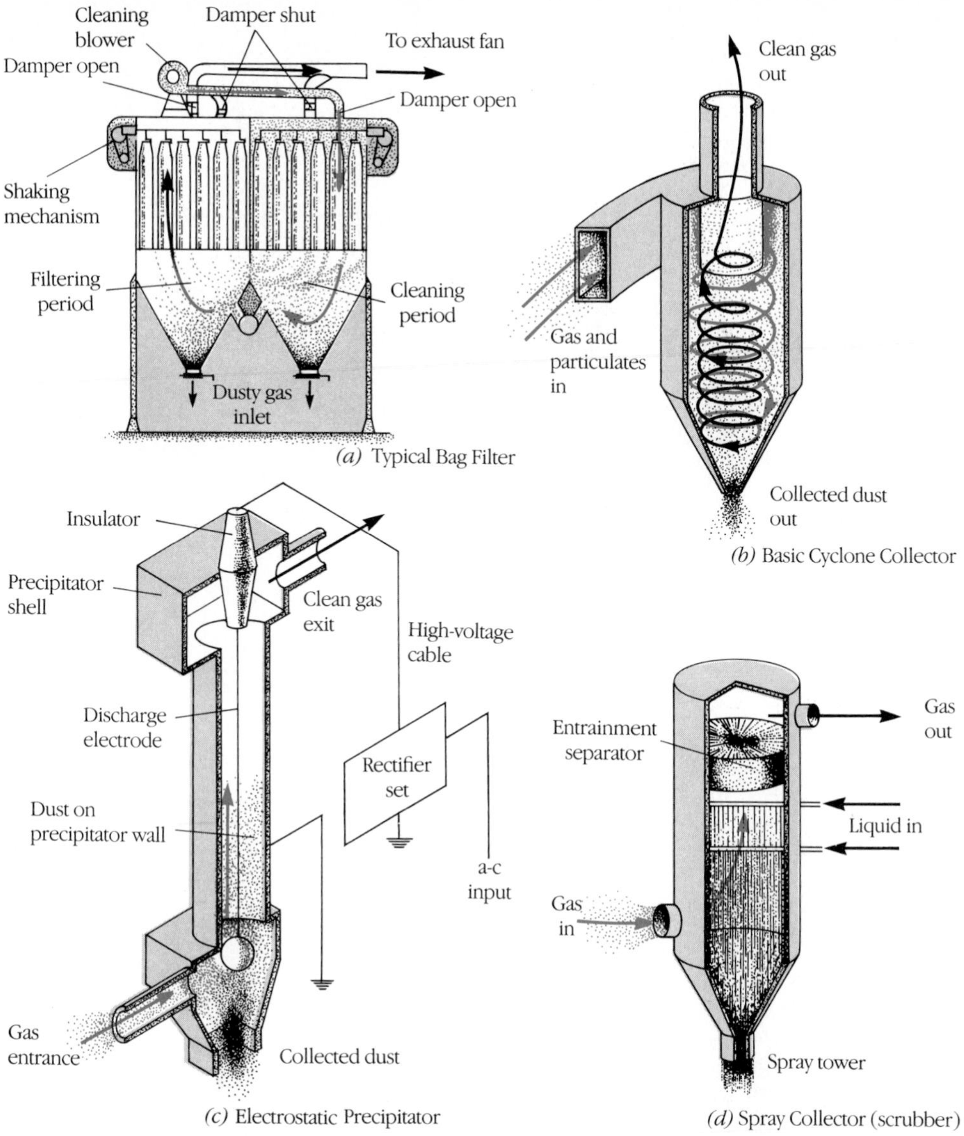

Figure 15-12. Four pollution control devices used for stationary combustion sources.

(a) Typical Bag Filter

(b) Basic Cyclone Collector

(c) Electrostatic Precipitator

(d) Spray Collector (scrubber)

but few of the small and medium-sized ones. Like filters, cyclones have no effect on gaseous pollutants.

Electrostatic precipitators, also used to remove particulates, are about 99% efficient; many of the major coal-burning facilities in the United States have installed them (Figure 15-12c). In electrostatic precipitators, particulates first pass through an electric field, which charges the particles. The charged particles then attach to the wall of the device, which is oppositely charged. The current is periodically turned off, allowing the particulates to fall to the bottom.

The **scrubber**, unlike the other methods, removes both particulates and gases such as sulfur dioxide (Figure 15-12d). In scrubbers, pollutant-laden air is passed through a fine mist of water and lime, which traps particulates (well over 99%) and sulfur oxide gases (approximately 80% to 95%).

Removing air pollution from stack gases helps clean up the air, but it creates a problem that many people fail to consider: hazardous wastes (see Chapter 18). Particulates from pollution control devices, for instance, contain harmful trace elements and other inorganic substances. Scrubbers produce a toxic sludge rich in sulfur compounds and mineral matter. Improper disposal can create serious pollution problems elsewhere.

Mobile Sources Reducing pollution from mobile sources can also be achieved by changes in engine design that reduce emissions. However, most new engine designs do not reduce emissions of carbon monoxide, nitrogen oxides, and hydrocarbons to acceptable levels. This makes it necessary to pass the exhaust gases through **catalytic converters**. Attached to the exhaust system, these devices convert carbon monoxide and hydrocarbons into water and carbon dioxide.

Figure 15-13. Magneto-hydrodynamics. Coal is mixed with an ion-producing "seed" substance, such as potassium, and burned. A hot ionized gas is given off and shot through a magnetic field. The movement of the ionized gas through the magnetic field creates the electrical current. Air or water is also heated and used to run an electrical generator.

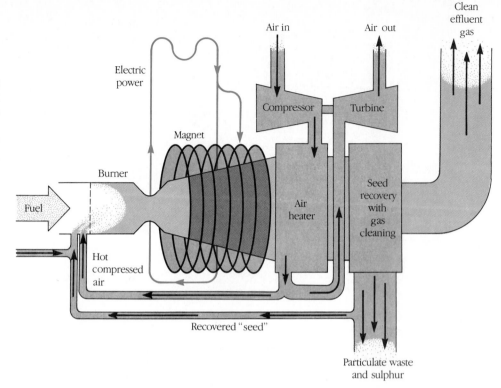

Cars with catalytic converters can meet emissions standards if they are kept well tuned. The use of leaded gasoline in these vehicles destroys the catalytic surface, often resulting in emissions that greatly exceed standards. Recent statistics show that one of every ten American vehicles equipped with a catalytic converter is run on leaded gasoline.

In the United States conventional catalytic converters do not remove nitrogen oxides, leaving this pollutant largely uncontrolled. American auto manufacturers contend that an affordable converter to do this job cannot be developed. But Volvo, the Swedish automobile manufacturer, introduced a catalytic converter in 1977 that lowered nitrogen oxide emissions to well below current US automobile standards. Researchers at Argonne National Laboratories are now experimenting with a process that will allow power plant operators to remove nitrogen oxides from smokestacks. A chemical added to scrubbers removes 70% of the nitrogen oxides and could cost much less than currently available technologies.

New Ways to Burn Coal Recent technological developments may help all nations use coal more efficiently and reduce air pollution in the process. One of these is **magnetohydrodynamics** (MHD) (Figure 15-13). Coal is first crushed and mixed with potassium carbonate or cesium, substances that are easily ionized (stripped of electrons). The mixture, burned at extremely high temperatures, produces a hot ionized gas—a **plasma**—con-taining electrons. The plasma is passed through a nozzle into a magnetic field, generating an electrical current. The heat of the gas creates steam, which powers a turbine.

MHD is about 60% efficient, compared with the 30% to 40% efficiency of a conventional coal-burning power plant. MHD systems remove 95% of the sulfur contaminants in coal, have lower nitrogen oxide emissions, and produce fewer particulates than conventional coal plants, but they release more fine particulates.

Coal may also be burned in **fluidized bed combustion** (FBC), a developing technology that is more efficient and cleaner than conventional coal-fired burners. In FBC finely powdered coal is mixed with sand and limestone and then fed into the boiler. Hot air, fed from underneath, suspends the mixture while it burns, thus increasing the efficiency of combustion. The limestone reacts with sulfur, forming calcium sulfate and reducing sulfur oxide emissions. Lower combustion temperatures in FBC reduce nitrogen oxide formation.

Cleaner Air through Conservation: A Framework for Personal Actions

Conservation can also help reduce air pollution. Recycling, increasing efficiency, and reducing demand are part of the conservation strategy. Through individual, corporate, and government conservation, energy demand can be substantially lowered. When energy demand falls, so does air pollution. Any segment of society, any activity big

What's Sacrificed When We Arm

Michael G. Renner

The author is a senior researcher at Worldwatch Institute and author of National Security: The Economic and Environmental Dimensions *(Worldwatch Paper 89).*

Our Common Future, the highly acclaimed report of the World Commission on Environment and Development, argues that Ethiopia could have reversed the steady advance of desertification threatening its food supply in the mid-seventies by spending no more than $50 million a year to plant trees and fight soil erosion. Instead, the government in Addis Ababa pumped $275 million per year into its military machine between 1975 and 1985 to fight secessionist movements in Eritrea and Tigre.

When famine struck in 1985, more than a million Ethiopians died. Emergency relief measures alone carried a price tag of $500 million.

This is only one example of how governments preoccupied with military dangers ignore the equally or perhaps even more serious threats to their national security posed by environmental degradation. Facing internal or external foes armed with increasingly larger and more sophisticated arsenals, most leaders feel they have little alternative but to keep up. The result is a never-ending arms race and a waste of precious resources. A seemingly limitless flow of weapons exports gives almost any nation or insurgency access to advanced conventional weaponry that is almost impossible to defend against—ironically, this makes all parties much less secure. For the nuclear-weapons states, the situation is even worse.

Compared to the imperative to guard against potential and real enemies at home and abroad, environmental protection is perceived as a luxury. National security is a meaningless concept, though, if it does not encompass the preservation of livable conditions within a country. Increasingly, states are finding their well-being undermined by environmental threats such as soil erosion on their croplands, pollution of their air and water, or cataclysmic floods unleashed by denuded watersheds.

Particularly in the last two cases, these problems can be exacerbated by activities beyond their borders and, therefore,

Trade-offs Between Military and Environmental Priorities

Military Priority	Cost	Social/Environmental Priority
Trident II submarine, F-16 jet fighter programs	$100 billion	One-third of estimated clean-up costs for US hazardous waste dumps, over 50 years
Stealth bomber program	$79 billion	80% of estimated costs to meet US clean water goals by 2000
German outlays for military procurement and R&D, FY 1985	$10.75 billion	Estimated clean-up costs for West German sector of the North Sea
Approx. 2 days of global mil. spending	$4.8 billion	Annual cost of proposed UN Action Plan to halt Third World desertification, over 20 years
MK-50 Advanced Light Weight Torpedo	$6 billion	Annual cost to cut sulfur dioxide emissions by 8–12 million tons/year in the US to combat acid rain, over five years

beyond their direct control. On a global scale, climate change and ozone depletion pose serious challenges to the safety and well-being of every nation.

Unfortunately, the Ethiopian famine is only one example of the fact that most leaders still see national security as primarily guaranteed by force of arms. The United States, for example, spent close to $300 billion on its military in 1988—even as US–Soviet relations were entering into a new detente—but only $85 billion (of which almost two-thirds were private funds) to deal with very concrete environmental pollution threats. This disparity in spending priorities is even greater in the Soviet Union, which, according to President Gorbachev, spends 77 billion rubles on the military but less than 10 billion rubles on environmental protection.

Because the pursuit of military power is such a costly endeavor, it drains resources needed to protect the environment and, thus, is beginning to lessen the security of many nations. This is made clear by a look at how fragments of military budgets applied elsewhere could do wonders to combat serious environmental problems (see table).

The Worldwatch Institute has estimated that a cumulative sum of about $774 billion would have to be expended worldwide during the final decade of this century to turn around adverse environmental trends in four priority areas: protecting topsoil on croplands from further erosion, reforesting the earth, raising energy efficiency, and developing renewable

sources of energy. This is at most 8% to 9% of current annual world military spending.

Many solutions to environmental problems may lie in developing new technologies. But worldwide spending on military research and development is sapping away funding in this critical area as well, growing from $13 billion a year in 1960 to an estimated $100 billion in 1986. That amount exceeds the combined governmental outlays on developing new energy technologies, improving human health, raising agricultural productivity, and controlling pollution.

National defense is a universally recognized and legitimized objective. Entrenched and powerful institutions guarantee a continuous flow of money in that direction. By contrast, the environment still does not have an adequate voice in most parliaments, cabinet rooms or ministries.

For instance, national leaders usually regard building public transportation systems that save energy and reduce pollution as a local task. Dealing with hazardous wastes is often viewed as a regional rather than a national responsibility, while developing renewable energy sources or new production technologies is left to the private sector, which may not have much incentive to get involved.

Thus, the tradeoffs presented in the accompanying table can be relatively invisible to national leaders, particularly those facing insurgencies or aggressive neighbors. As the example of Ethiopia shows, they are, nevertheless, real.

or small, can benefit from conservation.

The conservation strategy applies to every aspect of our lives. Each unnecessary product we buy, each bottle or can we toss out, and each gallon of gas we waste contributes to global pollution. This book promotes individual responsibility as an effective means of cutting down on resource demand and pollution. Individual actions, added together, can help us extend the lifetime of limited resources and can protect the global commons.

Cost of Air Pollution Control

Air pollution control costs money, but in many cases it can prove profitable. Today many companies are finding that pollution control can actually be financially rewarding; for instance, the Long Island Lighting Company recovers vanadium from particulates collected at its

power plants; in 1976, when the company began its program, it sold 362 tons of vanadium—about 9% of the total US vanadium production—for $1.2 million. The company continues to extract vanadium today but only makes $10,000 to $20,000 a year because it is using a cleaner fuel. The Chemical Division of the Sherwin-Williams Company installed pollution control systems at a Chicago plant, saving that company $60,000 a year.

The EPA estimates that an annual expenditure of $5 billion in the United States would eliminate or sharply decrease most air pollution damage, saving between $7 billion and $9 billion a year from this investment, improving the quality of life, bettering people's health, improving visibility, reducing damage to buildings and statues, and protecting wildlife. Chapters 19 and 20 explain why society is often reluctant to invest in pollution control even though it could save billions of dollars.

The power of man has grown in every sphere, except over himself.

WINSTON CHURCHILL

Summary

Air pollutants come from a variety of **natural** and **anthropogenic** sources. Anthropogenic pollutants—the products of human activities—represent the most significant threat to the environment and its inhabitants. The six major pollutants are carbon monoxide, sulfur oxides, nitrogen oxides, hydrocarbons, particulates, and photochemical oxidants. The major sources are transportation, power plants, and industry.

Cities fall into two categories. **Brown-air cities** are new and relatively nonindustrialized and are found in sunny, dry climates. Their major pollutants are carbon monoxide, ozone, and nitrogen oxides. In these cities hydrocarbons and nitrogen oxides react in the presence of sunlight to form secondary pollutants, major components of **photochemical smog. Gray-air cities** are older, industrialized cities, often in colder climates. Their major pollutants are particulates and sulfates; they combine with moisture to form **smog.**

Numerous factors such as wind, precipitation, topography, and temperature inversions affect regional air pollution. A **temperature inversion** occurs either when a high-pressure air mass stagnates over a region and forces a layer of warm air down (**subsidence inversion**) or when ground air cools faster than the air above it (**radiation inversion**). Both inversions result in the buildup of air pollution.

Air pollution affects human health in many ways. Short-term pollution episodes have numerous acute effects, including discomfort, burning eyes and throat, colds, coughs, heart attacks, and death (in extreme cases). Particularly susceptible are children and patients with heart and lung disease. Chronic effects are also possible. **Bronchitis**, a persistent inflammation of the bronchi, is characterized by a persistent cough, mucus buildup, and difficulty breathing. Cigarette smoking and air pollution are two major causes of this condition. **Emphysema**, the progressive breakdown of the air sacs in the lungs, is also caused by air pollution and smoking. Some studies have linked air pollution with cancer, but on the whole the evidence is contradictory.

Domestic animals are affected by many air pollutants, but the most noticeable impacts of air pollution are on materials, such as rubber, stone, and paint. Ozone, sulfur dioxide, and sulfuric acid are the most damaging. Crops and forests are also damaged by air pollutants, particularly ozone, sulfuric acid, and sulfates, although the exact magnitude of the effect is unknown.

The most potentially hazardous effects are those on climate. Carbon dioxide released from the combustion of fossil fuels acts as a reflector of infrared radiation. Thus, heat that normally escapes into space is reradiated to the earth's surface, warming the atmosphere. This so-called **greenhouse effect** may gradually change global climate, melting glaciers and Antarctic ice, raising the sea level, and disrupting agriculture.

Several other gases also contribute to greenhouse warming, including methane, nitrous oxide, and chlorofluorocarbons. Controlling these pollutants may be just as important as controlling carbon dioxide.

If trends continue, global temperature could be 2° C to 5° C warmer by 2050. Already signs of warming are present. Five of the hottest years in a century occurred in the 1980s. The drought of 1988 crippled American agriculture. But drought is not the only consequence of global climate change. Violent storms and unpredictable weather may also occur with greater frequency as the world heats up.

Air pollution is controlled at three interlocking levels: legal, technological, and personal. In the United States the most powerful tool is the Clean Air Act, its amendments, and its resultant regulations, which collectively provide for emissions standards for automobiles, national ambient air-quality standards for major pollutants, emissions standards for new stationary pollution sources, ways to prevent deterioration of pristine and less polluted air, and stronger EPA enforcement.

Reductions in pollution can also be brought about by energy conservation and by using alternative energy sources such as solar energy and wind energy. Harmful pollutants can be removed from smokestacks by cyclones, filters, electrostatic precipitators, and scrubbers, but each of these produces solid wastes that must be safely disposed of.

Air pollution control costs money, but many companies are finding that it makes good business sense to reduce pollution by recovering waste products. The EPA estimates that an annual expenditure of $5 billion would save $7 billion to $9 billion annually in reduced damage to crops, forests, buildings, and health.

Discussion Questions

1. Why do anthropogenic air pollutants generally have more impact than natural pollutants?

2. What are the six major air pollutants? What are the five major air pollution sources?

3. Toward what specific pollutants should most of our control efforts be directed? Why? Give specific examples of damage caused by these pollutants.

4. Describe the pollutants produced by burning fossil fuels. How is each product formed? Why is coal a "dirtier" fuel than gasoline?

5. Define the terms *primary* and *secondary pollutant.*

6. In what ways are brown-air and gray-air cities different? What are the major air pollutants in each city? Why does the distinction break down?

7. What is photochemical smog? How is it formed? Why are suburban levels of photochemical smog often higher than urban levels?

8. Describe some factors that affect air pollution levels in your community.

9. What is a temperature inversion?

10. What are the acute health effects of (1) carbon monoxide, (2) sulfur oxides and particulates, and (3) photochemical oxidants?

11. What are the chronic health effects of air pollution? Which pollutants are thought to be the cause of these effects?

12. Of all the impacts of air pollution, which are of the most concern to you? How have your education, home life, and religious beliefs affected your answer to this question?

13. Describe the greenhouse effect. What causes it? What are its effects? What can be done to lessen its impacts?

14. Make a complete list of your ideas for controlling air pollution in your city or town, and rank them according to their effectiveness and their feasibility. Which ones are practical?

15. Discuss the air pollution control legislation enacted by the US Congress, highlighting important features of the acts and amendments.

16. Discuss the control of air pollution from stationary pollution sources. In what general ways can pollution be reduced? Give specific examples of how you can help reduce air pollution.

Suggested Readings

Abrahamson, D. E. (ed.). (1989). *The Challenge of Global Warming*. Washington, DC: Island Press. Comprehensive analysis of global warming and its effects.

Cohn, J. P. (1989). Gauging the Biological Impacts of the Greenhouse Effect. *Bioscience* 39 (3): 142–146. Interesting look at some of the projected biological impacts of global warming.

Hardin, G. (1985). *Filters against Folly*. New York: Viking. See Chapter 14 for an interesting discussion of the greenhouse effect and the political difficulties in solving it.

Jacobson, J. L. (1989). Swept Away. *Worldwatch* 2 (1): 20–26. Exceptional account of the problems arising from flooding due to global warming.

Maranto, G. (1986). Are We Close to the Road's End? *Discover* January: 28–38. Excellent review of current information on the greenhouse effect.

Myers, N. (1989). The Heat is On: Global Warming Threatens the World. *Greenpeace* 14 (3): 8–13. Excellent overview of the threats of global warming.

Postel, S. (1986). *Altering the Earth's Chemistry*. Worldwatch Paper 71. Washington, DC: Worldwatch Institute. Detailed study of air pollution's effects.

Renner, M. (1989). *National Security: The Economic and Environmental Dimensions*. Worldwatch Paper 89. Washington, DC: Worldwatch Institute. Detailed discussion of the relationship between environmental protection and national security.

Sedjo, R. A. (1989). Forests: A Tool to Moderate Global Warming. *Environment* 31 (1): 14–20. Important survey of the ways forests can help us reduce global warming.

Schneider, S. H. (1988). Climate and Food: Signs of Hope, Despair, and Opportunity. In *The Cassandra Conference: Resources and the Human Predicament*. P. R. Ehrlich and J. P. Holdren, eds. College Stations, TX: Texas A & M University Press. Important reading from a leading authority on global climate.

Schneider, S. H. (1989). The Changing Climate. *Scientific American* 261 (3): 70–79. Excellent overview.

Stratospheric Ozone Depletion

Encircling the earth is a thin, protective layer of ozone gas (O_3), which screens out 99% of the sun's harmful ultraviolet light. The **ozone layer** occupies the outer two-thirds of the stratosphere, 20 to 50 kilometers (12 to 30 miles) above the earth's surface (Figure S15-1). The screening effect of the ozone layer protects all organisms from damage caused by ultraviolet light, which is known to be mutagenic and carcinogenic.

When ultraviolet light strikes ozone molecules, it causes them to split apart (Figure S15-2). The products, however, quickly reunite, reforming ozone and giving off heat. Thus, the ozone layer is a renewable layer that converts harmful ultraviolet light into heat.

Life on earth depends on this screening mechanism; without it, life would not exist. If the ozone layer suddenly disappeared, animals would be seriously burned and would develop cancer and lethal mutations, and plants would perish.

Activities That May Deplete the Ozone Layer

Many experts believe that some human activities are destroying the ozone layer. Three major activities have been singled out: (1) the use of spray cans and refrigerants that contain freon gas, (2) high-flying supersonic jets, and (3) the detonation of nuclear weapons.

Freons

In 1951 freon spray-can propellants hit the market in the United States. Freons, described in Table S15-1, are also known as **chlorofluorocarbons** (CFCs). Two CFCs were commonly used: (1) freon-11, a propellant now banned in several countries, including the United States, and (2) freon-12, still used in refrigerators, air conditioners, and freezers.

Until the early 1970s chemists listed freons as inert (unreactive) chemicals. Their release into the atmosphere was therefore of little concern. Conventional wisdom held that freon gases simply diffused into the upper layers of the atmosphere, where they were broken down by sunlight, without harm.

In the early 1970s, however, two US scientists reported that the breakdown products of freons could react with stratospheric ozone (Figure S15-3). They hypothesized that these products

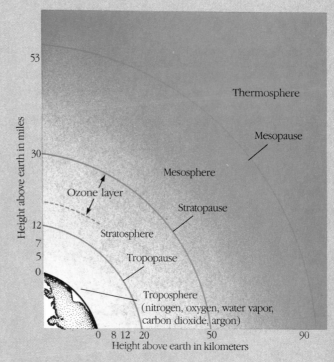

Figure S15-1. The earth's atmosphere is divided into layers. Ninety-five percent of the oxygen is found in the troposphere. The ozone layer occupies the outer two-thirds of the stratosphere.

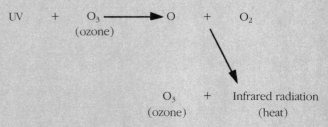

Figure S15-2. Screening effect of ozone in the stratosphere. The ozone molecule is split by ultraviolet (UV) light, but the reaction is reversible. When ozone reforms, infrared radiation or heat is given off.

Table S15-1 Commonly Used Freons

Generic Name	Use	Chemical Name	Chemical Formula
Freon-11	Spray-can propellant	Trichloromonofluoromethane	Cl | Cl—C—Cl | F
Freon-12	Coolant in refrigerators, freezers, and air conditioners	Dichlorodifluoromethane	Cl | F—C—Cl | F

could eventually deplete the protective ozone layer. Shortly after their announcement, three other research teams reported similar findings, showing that a single chlorine free radical (a highly reactive chlorine atom produced when freons are broken down) could react with as many as 100,000 molecules of ozone (Figure S15-4). By many estimates, this process would eventually deplete the ozone layer.

High-Altitude Jets and Nuclear Explosions

The ozone layer may be vulnerable to other human activities as well. For instance, scientists believe that aircraft such as the supersonic transport (SST) flying in the stratosphere may also destroy ozone through the release of nitric oxide produced by jet engines. The nitric oxide gas reacts with ozone to form nitrogen dioxide and oxygen (Figure S15-5).

In 1971 the US Congress killed plans to help finance the construction of 300 to 400 SSTs; however, the British–French Concorde is in use today on a limited scale. Because the Concorde flies lower and burns less fuel than the proposed US SST, it has less impact on the ozone layer. Ordinary commercial jets also produce nitric oxide. They help deplete the ozone layer, but to a smaller degree.

The detonation of nuclear weapons in the atmosphere also produces nitric oxide. The active period of atmospheric testing of nuclear weapons in the 1940s and 1950s caused a moderate and short-lived decrease in ozone, suggesting that nuclear war could cause a dangerous reduction (see Chapter Supplement 4-1).

Other Sources of Destruction

Nitrogen fertilizer that farmers apply to their fields may be converted into nitric oxide gas. It could gradually diffuse into the stratosphere, there reacting with ozone molecules. Although the use of fertilizers has risen dramatically in the last decade, no one knows what harm they may cause in the long run.

Other chemical pollutants such as methyl chloride and carbon tetrachloride may also diffuse into the ozone layer and eliminate ozone molecules. Natural pollutants such as nitrogen oxides from volcanoes and chloride ions from sea salt may also destroy the ozone layer, although these processes are presumably in balance with natural ozone replenishment.

Extent and Effect of Depletion

In 1974 two chemists, F. Sherwood Rowland and Mario Molina, startled the scientific world when they announced that previously considered safe CFCs could destroy the ozone layer. Their projections indicated that CFCs could eventually destroy 20% to 30% of the ozone layer, imperiling life on earth.

In the years following their reports scientists have grappled with the question of ozone depletion. Many people grew skeptical of the early projections because new estimates indicated that the effect would be much smaller than Rowland and Molina had suggested. By 1984 most experts on the issue believed that a decline of only 2% to 4% sometime in the next century would result from CFC release into the atmosphere.

As a result, stratospheric ozone depletion became a "dead" issue for a while. But in the mid-1980s EPA scientists delivered a shocking report, saying that a 60% decline in stratospheric ozone levels by 2050 was likely if the production of CFCs continued to grow by 4.5% a year. Even a 2.5% increase, they said, would deplete the ozone layer 26% by 2075.

These projections came in the wake of startling results from satellite studies to assess stratospheric ozone and CFC levels. One study, for example, showed that the CFC concentration in the stratosphere had doubled in ten years. But more frightening was a report on satellite measurements of the ozone layer over Antarctica. Much to the surprise of scientists, a mysterious hole in the ozone layer appeared each spring over Antarctica. About the size of the United States, this hole seemed to be growing worse each year. Scientists set out to determine the cause and found that CFCs and several natural climatic conditions were to blame. For example, a vortex of wind (a whirlpool in the atmosphere) circles the pole during the winter months and cools the stratosphere by blocking out warmer air currents. This contributes to the formation of polar stratospheric clouds. Scientists have found that ice crystals in these clouds facilitate the breakdown of ozone. Chemical reactions that destroy the ozone may occur on the surfaces of the ice crystals.

Researchers have found a similar hole in the ozone layer over the Arctic. Ozone levels there, however, are not as severely depleted because the vortex is not as strong as it is in the south. Therefore, invading winds can break through, keeping the Arctic air much warmer than its southern counterpart.

What about ozone levels elsewhere? Numerous studies showed a slight decline in the ozone layer throughout the world, but these results were thrown into question because ozone levels change from year to year. How could scientists be sure they were seeing a real decline?

In 1987, several government agencies, including NASA, convened an international panel of more than 100 scientists to study all of the satellite data and make a judgement once and for all. In March of 1988, the scientists confirmed that ozone had declined 1.7% to 3% over the northern hemisphere since 1969 (Figure S15-6). Over the heavily populated areas of North America and Europe ozone levels had fallen 3%. The greatest declines, ranging from 5% to 10%, were recorded over Antarctica and the southern tip of Argentina.

Ozone depletion will increase the amount of ultraviolet light striking the earth. In reasonable amounts, ultraviolet light tans light skin and stimulates vitamin D production in the skin. However, excess ultraviolet exposure causes serious burns and may induce skin cancer. It would also be lethal to bacteria and plants.

Medical researchers believe that a 1% depletion of the ozone layer would lead to a 2% increase in skin cancer. Given a 16% decrease in ozone by 2000, skin cancer would rise by 32%, resulting in 100,000 to 300,000 more diagnosed cases each year. Assuming a 4% mortality rate, about 4,000 to 12,000 additional deaths would occur each year in the United States alone.

Studies of skin cancer show that light-skinned people are much more sensitive to ultraviolet light than more heavily pigmented individuals. In addition, some chemicals commonly found in drugs, soaps, cosmetics, and detergents may sensitize the skin to ultraviolet light. Thus, exposure to sunlight may increase the incidence of skin cancers among light-skinned people and users of many commercial products.

Plants are also affected by excess ultraviolet light. Intense ultraviolet light is usually lethal to plants; smaller, nonfatal doses damage leaves, inhibit photosynthesis, cause mutations, or stunt growth.

(*a*) Photodissociation of Freon 12

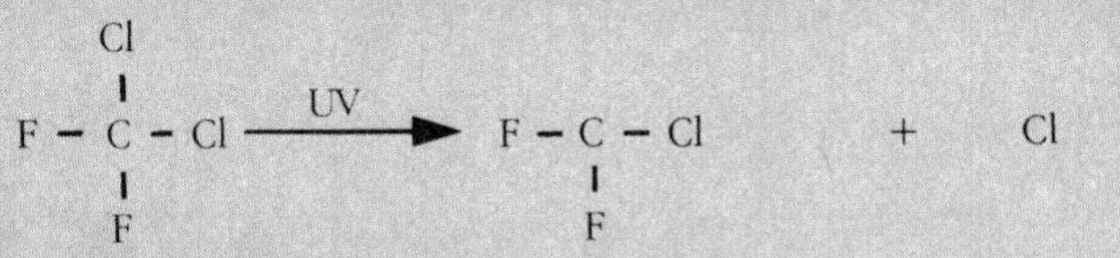

(*b*) Ozone Depletion

$$Cl + O_3 \longrightarrow ClO + O_2$$

Cl (Chlorine free radical); O_3 (ozone)

Figure S15-3. (a) The freons, or chlorofluorocarbons, are dissociated by ultraviolet light in the stratosphere. This produces a highly reactive chlorine free radical. (b) The free radical can react with ozone in the ozone layer, thus reducing the ozone concentration and eliminating the ultraviolet screen.

(*a*) $$ClO + O \longrightarrow Cl + O_2$$

ClO (Chlorine oxide); O (oxygen free radical); Cl (chlorine free radical); O_2 (molecular oxygen)

(b) $$ClO + O_3 \longrightarrow ClO_2 + O_2$$

Figure S15-4. (a) A single molecule of freon gas can eliminate many thousands of molecules of ozone, because the chlorine free radical is regenerated. (b) Chlorine oxide (formed when the chloride free radical reacts with ozone) can also react with ozone.

$$\text{Supersonic Transport Jet} \longrightarrow NO + O_3 \longrightarrow NO_2 + O_2$$

NO (nitric oxide); O_3 (ozone); NO_2 (nitrogen dioxide); O_2 (oxygen)

Figure S15-5. Supersonic and subsonic jets produce nitric oxide, which can react with ozone and reduce the ozone layer.

Preventing Ozone Depletion

The implications of a decline in the ozone layer are profound. The early projections of ozone depletion, in fact, moved several nations, including the United States, Sweden, Finland, Norway, and Canada, to cut back on CFC emissions. In 1978, for example, the United States banned freon used in spray cans. Freon-12, used as a refrigerant and coolant and also as a blowing agent in the production of plastic foam, was not affected by the ban. This ban helped cut global CFC emissions. Between 1974 and 1984, CFC emissions had dropped 21%.

Unilateral actions of the sort taken by the United States, Canada, and the Scandinavian countries were viewed by many as only a partial solution. Worldwide cooperation was needed. In 1985 the United Nations Environment Program negotiated an agreement with 50 nations calling for international cooperation in studying stratospheric ozone depletion. Very little came out of these talks.

In 1987 the UN sponsored new negotiations aimed at reducing CFC production worldwide. In September of that year, 24 nations signed a treaty, called the Montreal Protocol, which would cut CFC production in half by 1999. The agreement paved the way for a gradual decline in production in the industrial nations. Critics argued that it had too many loopholes. For one, it allowed Third World nations to increase production for ten years, then freeze production at that time. It also allowed the Soviet Union to complete construction of two large CFC plants.

David Doniager, a senior attorney with the Natural Resources Defense Council, argued that the agreement would slow down the depletion of stratospheric ozone but would not stop it. He called the treaty a major "half-step forward." EPA officials were disappointed as well. Their computer projections showed that an 85% reduction in CFC emissions was needed to stabilize CFC levels in the atmosphere.

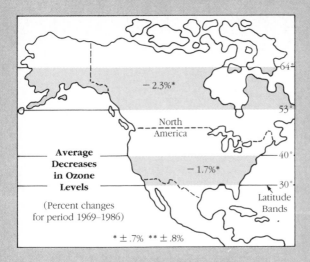

Figure S15-6. Map of North America showing ozone depletion.

The Montreal Protocol went into effect in January 1989. But before that, something unusual happened. In March of 1988, the international panel announced that ozone levels had fallen throughout the world. Two weeks later, DuPont, a major producer of CFCs, called for a total worldwide ban on CFC production. (Only two weeks earlier the company had said that it would not support a ban.) Then, in April 1989 the Food Service and Packaging Institute, an alliance of manufacturers that produce disposable foam products such as plastic cups and fast-food containers, announced that they would voluntarily stop using CFCs as blowing agents. They would replace them with less damaging agents.

The Montreal Protocol limits five CFC compounds: 11, 12, 113, 114, and 115. It also limits three compounds called *halons*. (In halons, an atom of bromine replaces some or all of the chlorine atoms.) Halons are consumed in much smaller quantities worldwide but they are far more effective in destroying ozone than CFCs. Their production levels will be frozen at 1986 values.

Manufacturers are pursuing several options: (1) production of less stable CFC compounds, (2) production of non-CFC substitutes, (3) conservation (using less to do the same job), and (4) recycling. Already, considerable progress has been made in these four areas.

Consider the first option and some of the roadblocks to bringing it about: By adding a hydrogen atom to the stable CFC molecule, researchers can make less stable CFCs that break up in the lower atmosphere. The chlorine atoms released during this process, they say, are less likely to reach the stratosphere.

Several less stable CFCs are already on the market. One of these, HCFC-22, is now used as a coolant in some home air conditioners. It is 20 times less destructive than the CFC-12 now used in refrigerators and automobile air conditioners.

HCFC-22 is not limited by the Montreal Protocol, but manufacturers are reluctant to switch because it would require substantial changes in air conditioning and refrigeration units. Why? HCFC-22 works at a higher pressure than CFC-12. Therefore, systems that are using this new chemical will require much heavier compressors and stronger tubes. That means more cost.

Because of this problem, refrigeration and air-conditioning industries are hoping that another compound, HFC-134A, which is not expected to be on the market for several years, will prove to be a good substitute for CFC-12. It does not contain chlorine or bromine, and may work in current cooling systems without substantial redesigns.

The second most widely used CFC is CFC-11. It is now used primarily as a blowing agent for foam and, outside of the United States and a few other countries, as a spray-can propellant. HCFC-123 has been touted as a possible replacement.

Perhaps one of the most difficult challenges is finding a replacement for CFC-113. This compound is an all-purpose cleaner for circuit boards produced for the computer industry. Because CFC-113 was not being considered for banning until a few years ago, the industry had not actively pursued replacements. At the signing of the Montreal Protocol, in fact, work to find a substitute had not even begun.

In January 1988, researchers announced the development of a compound called BIOCAT EC-7. It may be a partial replacement for CFC-113. This substance, isolated from orange peels, is very similar to kerosene and turpentine. EC-7 may replace a sizable share of the CFC-113 market, but it has its limitations. It is not versatile and is flammable. Industry representatives believe that no single compound will replace CFC-113 completely.

The ozone story is an encouraging one, illustrating how scientific knowledge can be used by business, government, and environmentalists for the common good of humankind. It also illustrates how disparate factions can work together and may serve as a model for other environmental disputes, such as global warming, where international cooperation is badly needed to avert potential disaster.

Suggested Reading

Shea, C. P. (1988). *Protecting Life on Earth: Steps to Save the Ozone Layer*. Worldwatch Paper 87. Washington, DC: Worldwatch Institute. Extraordinary coverage of the ozone controversy.

Acid Deposition: Ending the Assault

In the 1960s a forest ranger named Bill Marleau built the cabin of his boyhood dreams on Woods Lake in the western part of New York's Adirondack Mountains. Isolated in a dense forest of birch, hemlock, and maple, the lake offered Marleau excellent fishing. Ten years after Marleau finished his cabin, however, something bizarre happened: Woods Lake, once a murky green suspension of microscopic algae and zooplankton, teeming with trout, began to turn clear. As the lake went through a mysterious transformation, the trout stopped biting and soon disappeared altogether. Then the lily pads began to turn brown and die; soon afterward, the bullfrogs, otters, and loons disappeared, too.

What had happened to Woods Lake? What had destroyed the web of life at this small, isolated lake, far from any sources of pollution? Scientists from the New York Department of Environmental Conservation say that Woods Lake is "critically acidified." As a result, virtually all forms of life in and around it have perished or moved elsewhere. The lake became acidified from acids and acid precursors deposited from the skies in several forms.

Acid deposition—acid rain and snow and acidic precursors that are deposited in other ways—is becoming widespread, as are lakes like Marleau's. Woods Lake is only one of about 375 lakes and ponds in the western Adirondacks turned acidic and hazardous to virtually all forms of life by acid deposition. In eastern Canada 100 lakes have met a similar fate. In Scandinavia the death count is 10,000. Across the globe thousands of lakes

now lie in wait; according to researchers, their turn is coming unless something is done, and quickly.

Widely publicized as one of the most serious environmental threats facing us today, acid deposition is a phenomenon of great environmental and economic importance. Evidence is mounting to show that acid deposition turns lakes acidic, kills fish and other aquatic organisms, damages crops, destroys forests, alters soil fertility, and destroys statues and buildings. Moreover, scientists are finding that acid precipitation is more widespread than once thought and is taking a larger toll on our environment and pocketbooks than originally imagined. Recent reports indicate that it poses a universal threat, affecting the developed countries as well as many Third World nations. Earthscan, an international environmental group, reports that acid deposition is already damaging soils, crops, and buildings in much of the Third World. Rapidly growing urban centers with their poorly regulated industry and traffic congestion are largely the culprits. Ironically, tough pollution laws in developed countries have given multinational corporations incentives to set up operations in Third World nations, whose pollution laws are, if existent, certainly much weaker.

What Is Acid Deposition?

Acid deposition refers to the deposition of all forms of acids from the sky. To understand this phenomenon more clearly requires a look at acids: how they form in the atmosphere and how they are deposited.

Acidity is related to the amount of free hydrogen ions (H^+) in solution. The degree of acidity is measured on the pH (potential hydrogen) scale, which ranges from 0 to 14 (Figure S15-1). Substances that are acidic, such as vinegar and lemon juice, have low pH values, that is, less than 7. Basic (alkaline) substances, such as baking soda and lime, have high pH values on the scale—greater than 7. Neutral substances, such as pure water, have a pH of 7. A change of 1 pH unit represents a tenfold change in the level of acidity; thus, rain with a pH of 4 is 10 times more acidic than rain with a pH of 5, 100 times more acidic than rain with a pH of 6, and 1000 times more acidic than rain with a pH of 7.

In an unpolluted environment rainwater is slightly acidic, having a pH of approximately 5.7; the normal acidity of rainwater

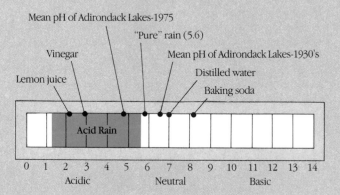

Figure S15-1. A pH scale to indicate acid–base level.

is created as atmospheric CO_2 is dissolved in water in clouds, mist, or fog and is converted into a mild acid, carbonic acid. **Acid precipitation** is rain and snow with a pH below 5.7.

Wet Deposition

Acid rain and snow are formed when two pollutant gases, the sulfur and nitrogen oxides, combine with water. Sulfur oxides form sulfuric acid; nitrogen oxide gases react with water to form nitric acid. Both are powerful acids. They may accumulate in clouds and fall from the sky in rain and snow. This process is called **wet deposition**. Even coastal fogs may contain droplets of acid that, when deposited on buildings or plants, can cause noticeable damage. A new study reports that moisture droplets in low-lying clouds (fog) tend to be much more polluted than rain or snow that falls from them. Fog and clouds bathe trees in highly acidic water. Making matters worse, recent studies suggest that the evaporation of recently deposited cloud water from forest canopies may result in acid concentrations on leaf surfaces that are higher than those found in the cloud droplets themselves.

Dry Deposition

Sulfur and nitrogen oxide gases also form sulfate and nitrate particulates. These may settle out of the atmosphere like fine dust particles. This process is called **dry deposition**. Settling onto surfaces, these particulates can combine with water to form acids. Sulfur and nitrogen oxide gases may also be adsorbed onto the surfaces of plants or solid surfaces, where they, too, combine with water to form acids. This is another type of dry deposition.

Where Do Acids Come From?

Acid precursors come from both natural and anthropogenic sources. The natural sources of sulfur oxides include volcanoes, forest fires, and bacterial decay. Anthropogenic sources of this pollutant are of major concern, however, because they are often concentrated in urban and industrialized regions, causing local levels to be quite high. About 70% of all anthropogenic sulfur dioxide comes from electric power plants, most of which burn coal.

Like the sulfur oxides, the nitrogen oxides arise from a wide variety of sources. The most important anthropogenic sources are electric power plants and motor vehicles (Figure S15-2).

American factories, cars, and power plants currently produce approximately 21 million metric tons of sulfur dioxide and about 19 million metric tons of nitrogen oxides a year. The EPA estimates that annual sulfur dioxide emissions may increase to around 26 million metric tons by 2000 and that annual nitrogen oxide emissions will increase to 25 million metric tons.

The Transport of Acid Precursors

It is not uncommon for rains in the northeastern United States to have pH values of 4 to 4.21; values similar to these have been reported consistently for 25 years. The acidic substances in the

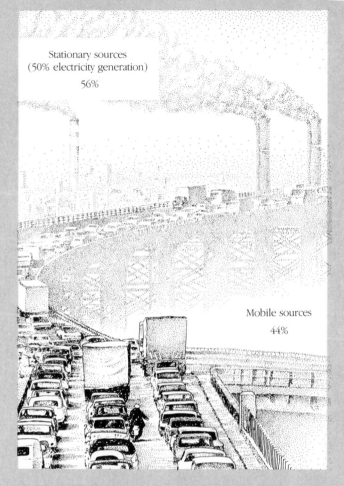

Total Human-caused Nitrogen Oxide Emissions in the U.S. in 1977

Stationary sources
(50% electricity generation)
56%

Mobile sources
44%

Figure S15-2. Anthropogenic sources of nitrogen oxides in 1987.

rain of this region often originate many hundreds of kilometers away.

Acid precursors and acids can remain airborne for two to five days and may travel hundreds, perhaps even thousands, of kilometers before being deposited. Scientists have found that the acid rain and snow in southern Norway and Sweden come from England and industrialized Europe. In the United States acid precipitation falling in the Northeast originates in the industrialized Midwest, primarily the Upper Mississippi and Ohio River valleys. Indiana and Ohio are the two major producers. Moving eastward, the mass of pollutants tends to converge on New York state and New England, where recent studies show that 50% of the lakes are in jeopardy because of low acid-neutralizing capacities.

Acid deposition is widespread. Montana, Florida, Colorado, New Jersey, California, Canada, the Amazon Basin, and the Netherlands, to name a few, have all documented acid precipitation in regions downwind of polluted areas.

The level of precipitation acidity at these locations can be quite high. In the White Mountains of New Hampshire, between 1964 and 1974, the average annual pH was about 4 to 4.21,

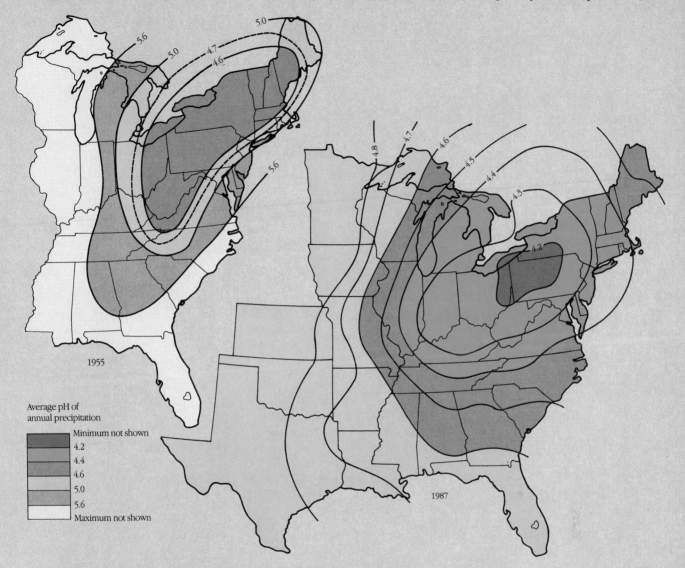

Figure S15-3. Acid precipitation in the eastern United States, 1955 and 1984. Note the worsening of acid rain and the wider area experiencing it.

Average pH of annual precipitation

Minimum not shown
4.2
4.4
4.6
5.0
5.6
Maximum not shown

nearly 100 times more acidic than normal precipitation. During that decade the level of acid increased 36%. In Europe rain and snow samples frequently have pH values between 3 and 5; in Scandinavia rain with a pH as low as 2.8 has been recorded. Rainfall samples collected in Pasadena, California, in 1976 and 1977 had an average pH of 3.9. One of the lowest pH measurements was made in Kane, Pennsylvania, where a rainfall sample with a pH of 2.7 was recorded—rain as acidic as vinegar. The grand prize for acidic rainfall, however, goes to Wheeling, West Virginia, where a rainfall sample had a pH of 2—stronger than lemon juice. More recent studies of acid fog downwind from the Los Angeles basin, however, show fog water with pH levels as low as 1.69.

In well-studied areas such as southern Norway and Sweden and the northeastern United States, two ominous trends have been observed. The first is that acid precipitation is falling over a wider area than it was 10 to 20 years ago; the second is that

the areas over which the strongest acids are falling are expanding (Figure S15-3).

Impacts of Acid Precipitation

Acidification of Lakes

Throughout the world, lakes and rivers and their fish are dying at an alarming rate. In the 1930s, for example, scientists surveyed lakes in the western part of the Adirondacks. Sampling the pH of 320 lakes, they found that most were fairly normal, with pHs ranging from 6 to 7.5. In 1975 a survey of 216 lakes in the same area showed that a large number had pH values below 5, a level at which most aquatic life perishes (Figure S15-4). Of the acidified lakes, 82% were devoid of fish life. A new study of 1500 lakes in New York's Adirondack Park found that 25% of the lakes

Frequency Distribution of pH in Adirondack Lakes

Figure S15-4. The pH level in Adirondack lakes.

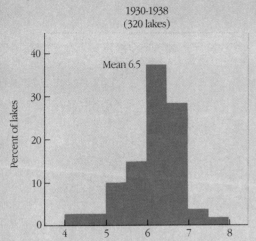

1930-1938
(320 lakes)

Mean 6.5

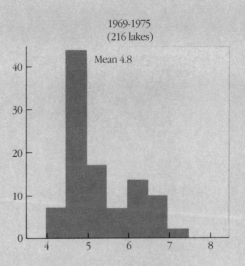

1969-1975
(216 lakes)

Mean 4.8

are so acidic that fish no longer live in them. Another 20% of lakes are acidic enough to be endangered.

In 1988 the National Wildlife Federation (NWF) published a list of US lakes that have high acid levels or are sensitive to acid deposition. The list includes 1700 lakes. The EPA estimates that another 14,000 lakes in the United States may already be acidified or have little capacity for neutralizing acid.

The NWF study shows that eastern lakes have been particularly hard hit by acid deposition. One of every five lakes in Massachusetts, New Hampshire, New York, and Rhode Island is acidic enough to be harmful to aquatic life. As acid deposition continues these lakes could become acidic cauldrons, lethal to virtually all forms of life. Recent studies have shown that acid deposition also occurs widely over the northern and central portions of Florida, with precipitation ten times more acidic than normal (Figure S15-5). One-third of the lakes in Florida are now acidic enough to be harmful to aquatic life.

In the mountains of southern Scandinavia the story is being repeated but on a much larger scale. Acidification of surface waters has occurred at a rapid rate for 40 years. In Sweden approximately 20,000 lakes are without or soon to be without fish. Salmon runs in Norway have been entirely eliminated because of the impact of acid precipitation on egg development. As a result, inland commercial fishing has ended in some areas.

In Canada the prospects for lakes and rivers are dimming. Nine of Nova Scotia's famous salmon-fishing rivers have already lost their fish populations because of acidity. Eleven more are teetering on the brink of destruction. In southern Ontario and Quebec acid precipitation has destroyed at least 100 lakes. By 2000, scientists predict, nearly half of Quebec's 48,000 lakes will have been destroyed. Because much of the acid reaching these lakes is believed to come from the United States, Canada has exerted considerable pressure on the US government to reduce air pollution. Its plea has fallen on deaf ears.

Parks and Wilderness Areas

Parks and wilderness areas in the United States are endangered by acid precipitation, because many lie downwind from major industrial centers and have thin soils and waters low in **buffers**,

chemical substances that allow aquatic systems to resist changes in pH. When H^+ levels increase, buffers combine with the free ions and eliminate them from the solution. When levels fall, they release them, thus maintaining a constant pH.

Preliminary data suggest that acid precipitation in the Great Smoky Mountains National Park has already put stress on existing trout populations in the poorly buffered lakes. The average annual pH for rainfall is 4.3, over ten times more acidic than normal rain. Also of particular concern is the Quetico–Superior Lake country of Canada and northern Minnesota. There are three major parks in this area: Quetico Provincial Park in Canada, and Voyageurs National Park and the Boundary Waters Canoe Area Wilderness in the United States. A recent EPA report stated that one-quarter to one-third of the lakes in the two US parks have so little buffering capacity that fish and other life forms are in danger. In the first comprehensive study of parks in the US West, scientists found that Yosemite, Sequoia, Mount Rainier, North Cascades, and Rocky Mountain national parks were already affected by acid precipitation. A number of others were not affected but could be if western precipitation continues to increase.

Effects on Aquatic Ecosystems

Many species of fish (brown trout and lake trout) die when the pH drops below 4.5 to 5 (Figure S15-6), although some species such as yellow perch and lake trout are slightly more resistant.

Acidity is only part of the reason fish die in acidified lakes. Scientists have found that when a lake's pH falls below the critical level, the concentration of toxic trace elements increases. Acidic rainwater or snowmelt dissolves elements like aluminum, mercury, and lead, which are naturally found in the soil and rocks. The acidic waters carry the metals to streams and lakes. Dr. Carl Schofield has shown that aluminum irritates the gills of brook trout, causing a buildup of mucus and, ultimately, death by asphyxiation (Figure S15-7).

Springtime Snowmelt: An Acid Bath Spring creates a special threat to fish and other aquatic organisms. Melting snow releases its acid in a sudden torrent, quickly elevating the acidity

Figure S15-5. Rainfall pH values in Florida.

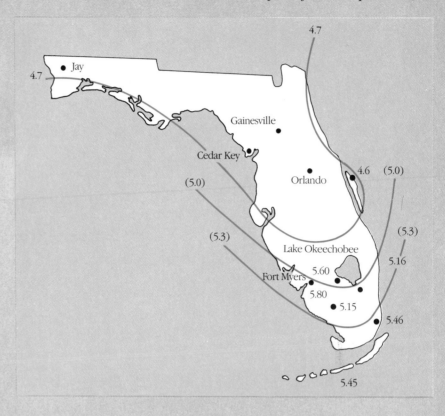

of lakes and streams. This surge of acids coincides with the sensitive reproductive period for many species of fish. What happens is that acids accumulate in the snow over the winter. When the snow begins to melt, the surface melts first. This water drains through the unmelted snowpack and leaches out the majority of the acids. The first 30% of the meltwater contains virtually all of the acid and typically has a pH of 3 to 3.5, which is toxic to eggs, fry, and adult fish as well.

Widening the Circle of Destruction

Fish are only one of the many organisms affected by acid precipitation. Professor Erik Nyholm of Sweden's University of Lund has found that songbirds living near acid-contaminated lakes lay eggs with softer shells than birds feeding around unaffected lakes. He also found elevated levels of aluminum in the bones of the birds that lived near acidic lakes, and he hypothesized that the aluminum had come from ingestion of aquatic insects living in acidified waters. The aluminum interferes with normal calcium deposition, resulting in defective (soft) eggshells and, ultimately, fewer offspring.

In 1987, the Izaak Walton League of America, a private environmental protection group, reported preliminary evidence that acid precipitation has contributed to a steep decline in the population of black ducks on the East Coast. The number of black ducks sighted has decreased by 60% in the last three decades. Although other factors have accounted for part of this decline, it appears that acids are killing the aquatic insects the ducks feed on. Declining insect populations rob female ducks and their offspring of protein at critical stages in the birds' life cycle. Baby black ducks in acidic areas also grow up to 60% more slowly than ducks raised in nonacidic wetlands because of the lack of protein in their environment.

Professor F. Harvey Pough of Cornell studies fertilized spotted salamander eggs in his laboratory and found that exposure to water with a pH of 5 prevented normal embryonic development and resulted in gross deformities that were usually fatal. The mortality of fertilized eggs was 60% at pH 6 but only 1% at pH 7.

Spotted salamanders breed in "temporary ponds" created by melted snow. These ponds are likely to be highly acidic in regions where acid precipitation is prevalent; as a result, the fate of the spotted salamander is bleak. The spotted salamander is as important as birds and small mammals in the food chain. "A drastic change in its population," Pough says, "would be likely to have repercussions throughout the entire ecosystem."

Some Human Consequences

In addition to damaging aquatic ecosystems, acids also damage crops, forests, buildings, and statues. Current estimates hold the economic costs of acid precipitation to be around $5 billion a year in the United States.

A report by the Ohio state government contends that if something is not done quickly to control acid precipitation, 2500 lakes a year will die in Ontario, Quebec, and New England throughout the remainder of the century. Many of the areas most susceptible to acid precipitation depend economically on recreation. In Ontario alone approximately 2000 fishing lodges contribute $150 million a year to the economy. A spokesperson for the Ontario Ministry of the Environment says that even at the current rate of acid precipitation, "there could be a $64-million loss and serious survival problems for 600 lodges over the next 20 years."

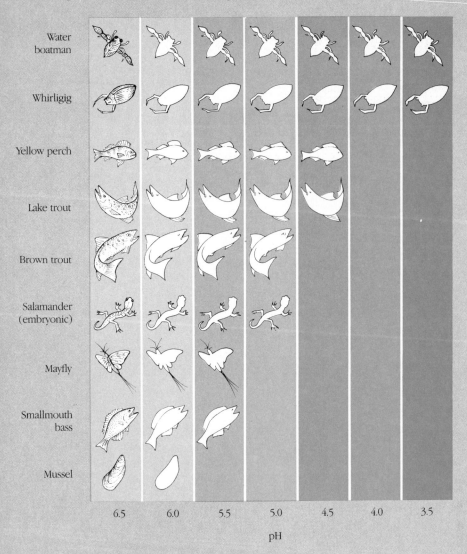

Labels (left side, top to bottom): Water boatman, Whirligig, Yellow perch, Lake trout, Brown trout, Salamander (embryonic), Mayfly, Smallmouth bass, Mussel

pH axis: 6.5 6.0 5.5 5.0 4.5 4.0 3.5

pH

Figure S15-6. The sensitivity of fish and other aquatic organisms to acid levels varies. The figure indicates the lowest pH (highest acidity) at which the organisms can survive. The yellow perch, for example, can withstand a pH of 4.5, while a mussel cannot live at pH levels lower than 6.0.

The same ominous future seems probable for parts of the United States. In New York state the destruction of lakes and the subsequent decrease in sport fishing is expected to cause a loss of about $1 million annually.

Forest Damage Numerous studies show that acid precipitation damages forests and may cause significant decreases in productivity. Acid precipitation causes foliar damage to birch and pines, impairs seed germination of spruce seeds, erodes protective waxes from oak leaves, and leaches nutrients from plant leaves. In Czechoslovakia researchers estimate that 300,000 acres of forest has been destroyed by pollution, mostly acid precipitation. In West Germany, 500,000 hectares (1.25 million acres) of forest is dying. Even the famous Black Forest is now severely damaged by acidic pollutants from industry and, especially, the automobile. In Vermont's Green Mountains half the red spruce, a high-elevation tree, have died from acid precipitation and acid fog. Lower-elevation sugar maples are also on the decline. With each passing day the stakes get higher. Nowhere is the impact felt more than in Canada, where the forestry industry contributes about 15% of the gross domestic product.

Swiss scientists believe that damage to trees may increase the likelihood of avalanches, because trees help retain snow on steep mountainsides. In the next few years 10% of the "barrier forests" may be lost, endangering the safety of mountain residents, skiers, and highway travelers.

In 1988, Robert Brock, a forest epidemiologist, reported preliminary findings from studies on Mt. Mitchell in North Carolina that may help explain why forests are dying. His results showed that low-lying clouds that often bathe spruce and fir trees on the mountain were considerably more acidic than vinegar. Two days after a two-day cloudy period, Brock found that needle tips looked singed. These needles contained 7 to 11 times more sulfate than healthy ones.

In 1988 a researcher from the University of Colorado proposed another hypothesis to help explain why the world's forests are dying. One of the chief assassins, he says, may be an acid-loving moss that grows on the forest floor. Professor Lee Klinger has studied 100 regions in 30 states where forests are dying. In each one he found a thick layer of moss carpeting the forest floor.

Mosses are natural sponges that hold so much water that the surface soils become saturated. The feeder roots and the trees

Figure S15-7. These fish were confined to a cage in a stream affected by acid rain. They died of asphyxiation.

die for the same reason that a houseplant dies when it is over-watered: water eliminates air from the soil. Plants literally suffocate. Mosses may also kill certain fungi that help trees absorb nutrients (mycorrhizal fungi). Mosses also acidify the water passing through them. Acidic water dissolves toxic trace metals like aluminum found in the soil, which can also kill the root system.

Crop Damage Concern for agriculture has also been raised by numerous researchers, but the results of many studies are inconclusive. Some researchers have reported that simulated acid precipitation decreases crop productivity, but other scientists have found increases, and still others, no effect. Acid precipitation with a pH less than 3.0 damages leaves on bean plants. Laboratory studies of tobacco plants show that simulated acid precipitation leaches calcium from leaves. Furthermore, timothy grass treated with simulated acid rain with pHs ranging from 2.2 to 3.7 died after extended exposure.

Acid precipitation is particularly harmful to buds; therefore, acids falling on plants in the spring might impair growth. In addition, acid precipitation appears to inhibit the dark reactions of photosynthesis, a series of chemical reactions in which plants produce carbohydrates and other important chemicals.

Acids may also damage plants by altering the soil. For example, acid rain may leach important elements from the soil, resulting in lower yield and reduced agricultural output. Acidification of soils may also impair soil bacteria and fungi that play an important role in nutrient cycling and nitrogen fixation, both essential to normal plant growth. Recent evidence shows that acids dissolve aluminum from the soil; aluminum damages cells in the water-transporting tubules of trees, especially closing off water transport. Trees die from thirst.

In some areas sulfur and nitrogen from acid rain may enhance soil fertility. However, direct damage to growing plants and damage to the soil could easily offset the fertilizing effect.

Acid-Sensitive Areas Acidification of lakes and streams occurs quite predictably in areas with a common geological denominator: thin soils with low acid-neutralizing capacities. Acid-sensitive areas in the United States, shown in Figure S15-8, include many mountainous regions. Ironically, the major producers, Indiana and Ohio, are the least vulnerable. Their thick topsoils contain abundant buffers.

Damage to Materials Acid precipitation also corrodes manmade structures and has taken its toll on some of special importance, such as the Statue of Liberty, the Canadian Parliament Building in Ottawa, Egypt's temples of Karnak, and the Caryatids of the Acropolis—not just architectural works but works of art, priceless treasures. John Roberts, Canada's former environmental minister, estimates that erosion of buildings in North America may be costing $2 billion to $3 billion a year. Acid rain may also damage house paint and etch the surface of automobiles and trucks. A recent US report claims that acid precipitation causes an estimated $5 billion in damage to buildings in 17 northeastern and midwestern states. The price tag includes the cost of repairing mortar, galvanized steel, and stone structures as well as the cost of repainting. It does not include damage to automobile paint, roofing materials, and concrete, potentially adding billions of dollars to the cost.

New Acid Rain Threat

The Environmental Defense Fund (EDF) recently uncovered another, potentially troublesome impact of acid deposition. They believe that nitrogen in the form of nitrates or nitric acid that is deposited in Chesapeake Bay stimulates the growth of algae and aquatic plants. As described in Chapter 16, excess plant growth can impair navigation. Moreover, surface plants can reduce light penetration, robbing deeper layers of sunlight needed to promote photosynthesis. Photosynthesis helps maintain oxygen levels in deeper waters. Without it, sealife may perish. When sea vegetation dies in the fall, it decays. The bacteria that break down this organic matter rob the water of oxygen, killing many aquatic organisms in the process.

According to the EDF's estimates, acid deposition contributes at least one-fourth of the nitrogen entering Chesapeake Bay each year from human activities. Acid deposition, in fact, ranks second only to fertilizer runoff as a source of nitrogen. EDF scientists believe that acid precipitation may be a significant source of nitrogen pollution along the entire eastern seaboard and could negate efforts to reduce surface runoff.

The EDF has recommended that the states in the Chesapeake Bay watershed find ways to cut nitrogen oxide emissions from power plants, factories, and motor vehicles. Improved energy

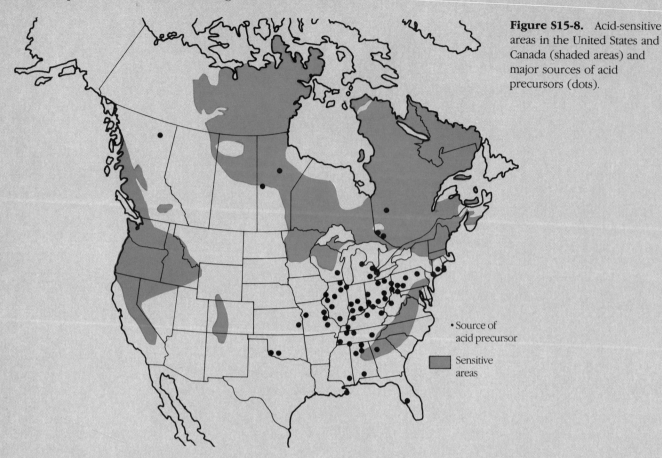

Figure S15-8. Acid-sensitive areas in the United States and Canada (shaded areas) and major sources of acid precursors (dots).

• Source of acid precursor

Sensitive areas

efficiency, they say, could go a long way toward cutting emissions.

Solving a Growing Problem

The first significant US governmental action against acid precipitation came in 1979 when Congress passed the **Acid Precipitation Act** to identify the sources and evaluate the effects of acid deposition. Congress promised to take steps to limit or eliminate sources of acid deposition. But by 1989 virtually nothing had been done at the federal level to put a stop to this menace.

In 1981 the prestigious and usually cautious National Academy of Sciences recommended a 50% cut in emissions of sulfur, as well as sharp cuts in nitrogen emissions. Continued emissions at present levels, the committee noted, represent a threat to human health and to the biosphere. Many other scientists and environmentalists agree with the academy. Strong support for control is gathering behind three strategies: (1) installation of scrubbers on all existing coal-fired power plants, (2) combustion of low-sulfur coal in all utilities, (3) combustion of coal that has been cleaned to remove sulfur. Individual conservation efforts, when taken together, could add up to significant cuts in sulfur emissions.

One of the few bits of encouraging news in the United States comes from New York state. In 1984 the legislature passed a bill calling for a 30% reduction in sulfur emissions by 1991. In

that same year, nine European nations and Canada signed an agreement to reduce sulfur emissions by 30% over a ten-year period. Minnesota also recently passed legislation to curb the growing problem.

In 1988, President Ronald Reagan, after eight years of dragging his feet on acid deposition controls, urged the EPA to ratify an international treaty that would help reduce acid deposition. The treaty will limit nitrogen oxide emissions worldwide to 1987 levels in ratifying nations. On the surface that may sound good. What it means is that further economic growth would have to occur without a net increase in nitrogen oxides. To do this requires that nations use the best available technology to reduce emissions from new vehicles and power plants and that they make cuts in existing plants.

Critics warn that the measure is too little too late. The current level of acid deposition is already costing the nations of the world billions of dollars a year and wreaking untold damage on the environment. Deeper cuts must be made. In 1989 President George Bush announced a bold new plan to cut annual emissions of sulfur dioxide by nearly half (10 million tons) and to cut nitrogen oxides by about 10% by the year 2000. But when his plan arrived in Congress in the spring of 1989, it had been watered down. The President only asked for a 9-million-ton-per-year cut in sulfur dioxide. He asked for a 2-million-ton cut in nitrogen oxides, but based on emissions in the year 2000. The net effect of this might be no cut whatsoever in annual emissions—and very likely an increase. At this writing environmental groups and concerned citizens are working to convince Congress to pass tougher legislation.

Many countries, impatient with the slow cuts in acid pollutants, have looked toward another strategy: attacking the symptoms directly. For example, in 1977 the Swedish government embarked on an expensive program to neutralize acidic lakes by applying lime. By 1988, 4000 lakes had been neutralized. Water quality in many of the lakes improved, and fish populations were saved. Sweden also undertook an ambitious stream-liming program in an effort to save inland salmon and other fish. In 1982 it built a liming plant on the banks of the Fyllea River and in the next two years added 2250 metric tons of lime to the waters, at a cost of $2.2 million. In 1988 and 1989, Sweden will spend over $18 million on liming lakes, rivers, streams, and forests. As of March 1988, 15,000 lakes require treatment. Critics argue that liming is an intermediate solution at best, a little like cardiopulmonary resuscitation administered to a heart attack victim. In Canada liming costs $120 per hectare ($50 per acre), so that treating a single lake can cost between $4,000 and $40,000. In five years, however, treated lakes turn acidic again.

Pursuing another approach, Cornell's Carl Schofield is developing a strain of acid-resistant brook trout. Despite its immediate logic, this approach is doomed to fail. Even if a strain of trout that could survive at pH 4.8 were developed, what would happen when the pH dropped to 4.5? And what about the trout's food supply?

Prospects for the Future

The United States is one of the few industrialized nations that has not taken steps to curb acid precipitation. Between 1979 and 1995, the US power industry projects, 350 new coal-burning power plants will go on line. The EPA predicts that sulfur dioxide emissions could increase from current levels of 21 million metric tons to about 26 million metric tons per year by 2000; nitrogen oxide emissions may increase from current levels of 19 million metric tons to 25 million metric tons by 2000.

Without decisive action, many experts believe, hundreds, perhaps thousands, of lakes will be destroyed in the next two decades. A 50% reduction in sulfur emissions would go a long way toward reducing acid precipitation in the East. In the West, restrictions on automobile use would help curb the nitrogen oxide emissions responsible for much of the acid precipitation. Energy conservation, solar energy, and wind energy could also become valuable allies in this battle.

Suggested Readings

Crocker, T. D. and Regens, J. L. (1985). Acid Deposition Control. A Benefit–Cost Analysis: Its Prospects and Limits. *Environ. Sci. and Technol.* 19 (2): 112–116. Good look at cost–benefit analysis.

Greenberg, D. S. (1985). Fast Cars and Sick Trees. *International Wildlife* 15 (4): 22–24. A short, well-written article on the connection between automobile pollution and the destruction of Germany's forests.

Havas, M., Hutchinson, T. C., and Likens, G. E. (1984). Red Herrings in Acid Rain Research. *Environ. Sci. and Technol.* 18 (6): 176A–186A. Excellent rebuttal of false claims made by opponents of acid-precipitation controls.

Johnson, A. H. (1986). Acid Deposition: Trends, Relationships, and Effects. *Environment* 28 (4): 6–11, 34–43. Comprehensive summary of the National Academy of Sciences report. Well worth reading.

Kahan, A. M. (1986). *Acid Rain: Reign of Controversy.* Golden, CO: Fulcrum. Balanced view of acid deposition.

Luoma, J. R. (1987). Black Duck Decline: An Acid Rain Link. *Audubon* 89 (3): 19–24. Extraordinary piece on the links between acid deposition and the dramatic decline in black ducks.

Mello, R. A. (1987). *Last Stand of the Red Spruce.* Washington, DC: Natural Resources Defense Council and Island Press. Highly readable account of forest damage and the underlying causes.

Postel, S. (1984). *Air Pollution, Acid Rain, and the Future of Forests.* Worldwatch Paper 58. Washington, DC: Worldwatch Institute. Detailed analysis of air pollution and forests.

Scheiman, D. A. (1986). Facing Facts. *The Amicus Journal* 7 (4): 4–9. Superb overview of the latest findings on acid deposition.

Wentworth, M. (1986). What's Wrong with Liming? *Outdoor America* 51 (1): 12–14. Wonderful survey of ameliorative attempts to offset the effects of acid precipitation.

Indoor Air Pollution

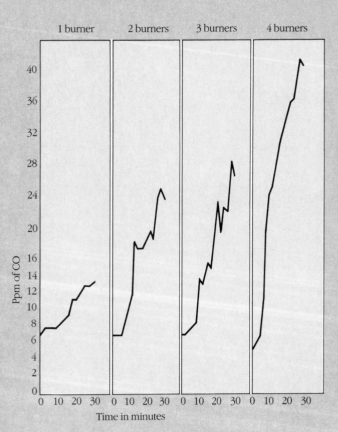

Figure S15-1. Carbon monoxide levels in a kitchen with one to four gas burners on.

Recent studies show that many Americans are exposed to high levels of toxic substances. Radioactive materials, formaldehyde, particulates, and dozens of other toxins in remarkably high levels enter our bodies through our lungs and skin. Where they come from may be a startling surprise.

Numerous scientific studies show that American homes and businesses are the source of potentially harmful substances called **indoor air pollutants**. A new rug, a new couch, a stove, or new paneling may all be emitting toxic chemicals into the air we breathe. In a recent report the EPA said that toxic substances from the home and office are much more likely to cause cancer than ambient air pollutants, for two reasons. First, indoor levels are often much higher than outdoor levels, in some cases up to 100 times higher. Even in pristine rural areas, indoor air can be more polluted than outside air next door to a chemical plant. Second, most people spend the bulk of their lives indoors. In 1986 the EPA announced that indoor air pollution may be causing several hundred US cancer deaths a year. In 1988, they revised that figure upward. One pollutant alone, radon (described later), may cause as many as 2000 to 20,000 cases of lung cancer each year. The three chemicals believed responsible for most of the deaths are benzene, from cigarette smoke, chloroform, a water contaminant given off during hot showers, and radon.

Indoor air pollutants come from a variety of sources. Some of the most important are wood and kerosene stoves, natural gas appliances, tobacco smoke, plywood, paneling, furniture, and rugs. Even making homes energy efficient, if not done correctly, can increase the danger.

Products of Combustion

Cigarettes, pipes, cigars, gas stoves and ovens, gas space heaters, water heaters, kerosene stoves, and wood stoves are the major combustive sources of indoor air pollution. Water heaters and furnaces are generally vented outside, so gaseous pollutants do not build up inside. But gas stoves and kerosene space heaters are not vented. Thus, as the fuel burns, carbon monoxide and nitrogen dioxide enter the room air.

As shown in Figure S15-1, carbon monoxide (CO) levels in the kitchen can increase from a few parts per million to over 40 parts per million when four burners are in operation for half an hour or so. CO levels increase appreciably in neighboring rooms, too. Nitrogen dioxide levels often follow the same pattern. (For a summary of the health effects see Table 15-2.)

Concentrations of indoor air pollutants fall after combustion sources are turned off, of course, but it may take several hours before normal levels are reached in conventional, poorly sealed homes (Figure S15-2). In well-sealed, energy-efficient homes that haven't made allowances for ventilation, it could take much longer. If several lengthy meals are cooked during a day, exposure to these pollutants can be quite high.

Researchers have found that in certain homes the nitrogen oxide and carbon monoxide levels can exceed the limits set to protect human health. What the long-term implications are, no one knows.

Sulfur dioxide is not generally a problem in homes unless kerosene stoves are used. Kerosene space heaters introduced in the late 1970s, for example, can release significant amounts of carbon monoxide, nitrogen dioxide, and sulfur dioxide if they are improperly functioning. For example, if the wick is damaged or covered with soot or if there is water in the fuel, pollution levels may become intolerable, creating headaches, coughing, and irritation of the throat. Even when operating normally they produce intolerable levels of pollution and, therefore, have been banned for indoor use in many states.

Particulates are generally not found in homes unless smokers are present. One study showed that the average particulate level was 40 micrograms per cubic meter in a home without smokers. Smokers raised particulate levels in some cases to 700 micrograms, over ten times greater than the level allowed by air-quality standards.

Carbon monoxide binds to blood hemoglobin and reduces the oxygen-carrying capacity of the blood, an effect especially harmful to individuals with heart and lung diseases. Sulfur oxides and nitrogen oxides are lung irritants and are responsible for emphysema and chronic bronchitis. In the long term they may also cause lung cancer. Cigarette smoke contains a number of carcinogens.

To reduce the buildup of these toxic chemicals, all combustion sources should be properly vented. With gas stoves, the overhead fans should be used when houses are closed up. Kerosene stoves should be avoided altogether. Cigarette smoking should be prohibited.

Some scientists have suggested that gas stoves should be eliminated from homes and replaced with electric stoves, even though electric cooking is an inefficient way of using this energy source. Passive solar heating can help reduce gas space heating. In tightly sealed, energy-efficient homes, air-exchange systems should be installed (Figure S15-3). These periodically replace room air with outside air; heat exchangers transfer heat to the incoming air, so little heat is lost to the outside.

One scientist working for the National Aeronautics and Space Administration found that certain plants help reduce indoor air pollution. Spider plants appear to be the best at gobbling up pollution. One plant per room is sufficient to prevent the buildup of nitrogen oxide.

Formaldehyde

Formaldehyde is a familiar preservative for biological specimens. It is also a common indoor air pollutant. About 2.7 million metric tons of formaldehyde is used in the United states each year. Because it is in so many consumer products, formaldehyde

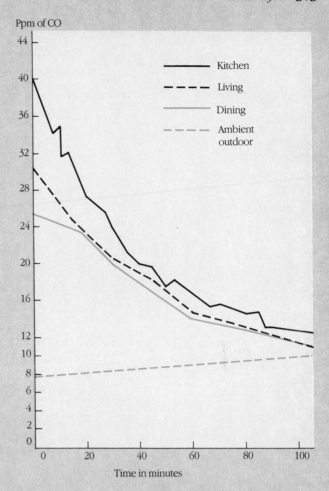

Figure S15-2. Drop in carbon monoxide level in the kitchen, dining room, and living room after the stove is turned off.

has become a chemical that few of us can escape. It is found in the adhesive in plywood, particle board, and wood paneling, and it leaks into houses and trailer homes, often in high concentrations. It is hard to avoid these days. Foam insulation (urea-formaldehyde foam that is injected into walls) contains it, as do permanent-press clothes, paper products, carpets, toothpaste, shampoos, waxed paper, grocery bags, and some medicines, to which it is added to kill bacteria, fungi, and viruses.

The levels of formaldehyde in homes insulated with foam are four times higher than those in homes with other types of insulation. Furniture containing formaldehyde in the wood and fabric has been shown to increase the levels in a previously unfurnished house by three times.

People with the greatest risk are those living in trailers or in homes with newly installed foam insulation, particle board, plywood, or paneling (Table S15-1). The better sealed a home or trailer is, the higher the levels of formaldehyde. In conventionally built homes with their abundance of cracks and poor insulation, air exchange between the inside and the outside occurs roughly once every hour. In airtight homes the turnover is much slower, perhaps once every five hours. The tighter the home, the greater the concentration of formaldehyde unless special

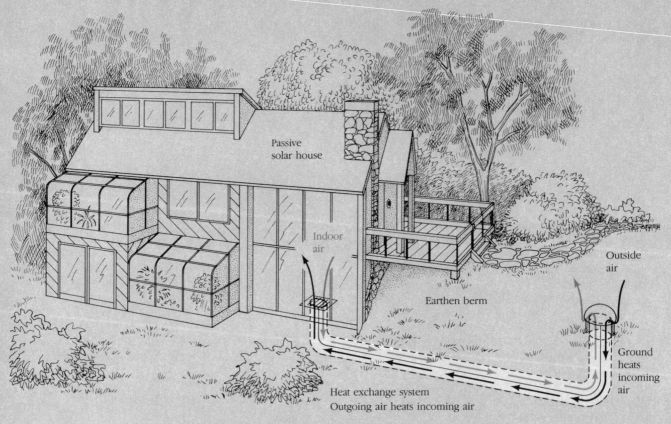

Figure S15-3. Air-exchange mechanism in a tightly sealed, energy-efficient house can prevent the buildup of toxic substances and greatly cut down on heat loss and energy use.

precautions are taken to avoid products containing this harmful substance.

Formaldehyde irritates the eyes, nose, and throat, but sensitivity among people varies. Some are sensitive to levels of 1.5 to 3 parts per million, but others who have been exposed to formaldehyde for long periods become sensitized and respond to levels as low as .05 parts per million.

Formaldehyde causes nasal cancer in rats and possibly mice at high levels. Given to monkeys at levels to which humans are typically exposed at work and at home, formaldehyde causes cellular changes believed to be the early stages of cancer within the linings of the respiratory tracts. It has also been shown to cause mutations in bacteria and many other organisms; many mutagens are also carcinogens. One epidemiological study showed a possible link between formaldehyde exposure and skin cancer in humans, but evidence is sketchy.

The US Consumer Product Safety Commission banned ureaformaldehyde insulation on February 22, 1981, but removed the ban in 1983. So far the EPA has taken little action; its only regulation is one that requires pesticides containing formaldehyde to disclose this fact on the label.

The EPA's refusal to regulate formaldehyde reflects a new attitude. In previous regulatory decisions the fact that a chemical had caused cancer in any laboratory animal was enough to warrant controls. Under the Reagan administration, though, the EPA took a more conservative approach, maintaining that with-

out conclusive proof from statistical studies on humans exposed to formaldehyde, regulations were unwarranted.

Many critics believe that this approach could weaken protection of human health. Norton Nelson, a highly regarded health scientist, contends: "Epidemiological studies must be regarded as a crude and insensitive tool. Only the most violent and intense carcinogens are likely to be detected by epidemiological techniques." Many critics, satisfied with imposing bans based on positive results in animal carcinogenicity studies, argue that the EPA should ban the use of formaldehyde in plywood, particle board, paneling, and textiles. The Department of Housing and Urban Development now requires mobile-home builders to disclose any use of formaldehyde-containing products.

Radioactive Pollutants

A more difficult indoor air pollutant to control is the naturally occurring radioactive gas radon. A daughter product of radium, radon is found in rocks and soils. It enters homes through cracks in the foundation or from the soil in homes without foundations. In some cases stone, brick, and cement contain small quantities of radium that emit radon into homes and other buildings.

Indoor concentrations of radon may be several times the natural, or background, level. Homes built on tailings from uranium mines have levels two to five times higher than normal background, as do the more energy-efficient, airtight homes without proper ventilation.

Inhaled, radon gas may emit radiation in the lungs that can cause mutations and lung cancer. Of greater concern, however, are the radioactive decay products of radon, particles like radioactive lead that can become lodged in the lung, providing long-term internal exposure to radiation. These decay products may also adhere to airborne particles that can be breathed into the lungs and also become lodged.

In 1987, the EPA surveyed radon levels in houses in ten states and found that one of every five homes exceeded the "action level," that is, the level at which mitigation efforts are recommended to lower radon exposure. The action level (4 picocuries per liter of air) poses the same lung cancer risk to individuals as smoking half a pack of cigarettes per day or receiving 200 to 300 chest X rays per year. In 1988, a new survey of 11,000 homes in Arizona, Indiana, Massachusetts, Minnesota, Missouri, North Dakota, Pennsylvania, and several midwestern states found that one of every three homes exceeded the action level. As a result, Lee Thomas, then Administrator of the EPA, recommended that virtually everyone in the United States test their residence for radon.

Bernard Cohen, a health physicist from the University of Pittsburgh, notes that radon in houses may cause more deaths than all other types of radiation exposure—natural and anthropogenic—combined. As with all forms of cancer, the link between cause and effect is difficult to prove. The risks associated with low-level radon exposure in American homes are generally extrapolated from studies of workers who have been exposed to very high levels. These include survivors of the atomic bomb blasts, recipients of high-dose X rays, and workers in uranium mines.

Researchers, however, still wonder whether there is a threshold to radiation effects, that is, a level below which no hazard exists. Erring on the conservative side, EPA officials have assumed that even tiny exposures present some risk. However, controversial new radon studies by Cohen challenge this view, suggesting that there is a threshold level below which radon is of no harm. Similar studies in Scandinavia, Florida, South Carolina, New Jersey, and New York have shown the same result, suggesting that the dangers of low levels of radon may be exaggerated. It's now up to the EPA to determine where that threshold is and to determine if the action level is appropriate.

Radon can be controlled in existing and new buildings. In existing homes and offices, for example, owners can install air-to-heat exchangers that vent indoor air while saving heat. Owners can also seal cracks in the foundation to prevent radon from entering. In new construction, builders can take simple steps to prevent radon from becoming a problem by laying gravel under the foundations of new homes. Porous pipes can be placed in the gravel to draw the radon away, keeping it from entering the house.

Most of the radon that seeps into American homes comes from the underlying soil and rock. New research, however, shows that this is not the only source. In some instances, researchers point out, soil radon emission is low but radon levels in buildings can still be high. In these cases, radon may be coming from groundwater. Turn on your taps and out comes water and radon gas that escapes into the room. To control such

Table S15-1 Formaldehyde in Mobile Homes and Houses Insulated with Urea-Formaldehyde Foam (UFF)

Type	Average Level of Formaldehyde (Parts per Million)
Homes without UFF	0.03
Homes with UFF	0.12
Mobile homes with UFF	0.38
Background[1]	0.01

[1]The background level is the normal atmospheric concentration of formaldehyde.

emissions is difficult. Homeowners may have to connect to public water supplies, if they are available, or drill new wells.

Chloroform and Trichloroethylene

In recent years two toxic and highly volatile chemicals, chloroform and trichloroethylene (TCE), have been identified in many municipal drinking-water supplies. These substances come primarily from groundwater polluted by industry and by hazardous waste dumps. Julian Andelman of the University of Pittsburgh estimates that drinking water tainted with these chemical contaminants causes an estimated 200 to 1000 cancer deaths each year. However, Andelman believes that toxic vapors given off from showers and baths could be more dangerous than poisoned drinking water. His work, for example, shows that a hot shower gives off half of the dissolved chloroform and 80% of the TCE, exposing the bather as well as other family members. Dishwashing and laundry water may also increase the indoor concentrations of these substances.

Controlling Indoor Air Pollutants

Indoor air pollution is a relatively new problem. No US laws directly address it. Legal experts argue that the **Clean Air Act** could be applied. The EPA, which enforces the act, has used it only once to regulate an indoor air pollutant, asbestos.

One way to address indoor air pollutants would be to develop indoor air standards, but the technical and legal problems in enforcing indoor standards would undoubtedly be enormous.

A provision of the Clean Air Act that could be applied to indoor air pollutants authorizes the EPA to draw up emissions standards for hazardous air pollutants. The EPA could develop formaldehyde emissions standards for plywood, particle board, paneling, and other household products.

A second approach might be to use the **Toxic Substances Control Act** (Chapter 14). It gives the EPA broad authority to control the production, distribution, and disposal of potentially hazardous chemicals. Bans on plywood, carpets, furniture, and other products containing formaldehyde could be applied. To date, the EPA Toxic Substance Office has been too burdened with other duties to take such actions. Outright bans would seem less desirable than emissions controls.

A final potential weapon is the **Consumer Product Safety Act**. It gives the Consumer Product Safety Commission the authority to regulate consumer products deemed hazardous to the public. Products that generate indoor air pollution could certainly qualify. The commission can, by law, develop safety standards for various products. Standards for stoves, for example, could indicate the permissible emissions of carbon monoxide, and standards for plywood, textiles, and furniture could set acceptable formaldehyde emissions. The commission can also require manufacturers to warn the public of potential dangers associated with the use of their products.

Indoor air pollution is an emerging problem that has only recently been brought to the attention of the American public. More work is needed to find out how indoor air affects our health. Creative solutions are needed to reduce exposure to the potentially harmful substances found in the air in homes and office buildings.

Suggested Readings

Environmental Protection Agency. (1988). *The Inside Story: A Guide to Indoor Air Quality.* Washington, DC: United States Environmental Protection Agency. General overview of indoor air quality problems.

Lipske, M. (1987). How Safe Is the Air Inside Your Home? *Natural Wildlife* 25 (5): 34–39. Excellent reference.

Sheldon, L. S., Handy, R. W., Hartwell, T. D., Whitmore, R. W., Zelon, H. S., and Pellizzari, E. D. (1988). *Indoor Air Quality in Public Buildings.* Vol. I. Washington, DC: United States Environmental Protection Agency. Technical report for those interested in more detailed information.

Sheldon, L. S., Zelon, H. S., Sickles, J., Eaton, C., and Hartwell, T. D. (1988). *Indoor Air Quality in Public Buildings.* Vol. II. Washington, DC: United States Environmental Protection Agency. Technical report for those interested in more detailed information.

Smith, K. R. (1988). Air Pollution: Assessing Total Exposure in the United States. *Environment* 30 (9): 10–15, 33–38. Interesting coverage of efforts to study indoor air pollution and personal exposure.

US House Committee on Science and Technology. (1985). *Radon and Indoor Air Pollution.* Hearing before the Subcommittee on Natural Resources, Agriculture Research, and Environment. Ninety-Ninth Congress. October 10, 1985.

Vietmeyer, N. (1985). Plants That Eat Pollution. *National Wildlife* 25 (5): 10–11. A look at plants thought to lower pollution.

Water Pollution: Protecting Another Global Commons

It's a crime to catch a fish in some lakes, and a miracle in others.

EVAN ESAR

"If there is magic in this planet, it is in water," wrote Loren Eiseley. Covering 70% of the earth's surface and making up two-thirds or more of the weight of living organisms, water is indispensable to life. Despite its crucial role in our lives water is one of the most badly abused resources. Chapters 7 and 10 described how overexploitation of groundwater and surface water creates regional shortages that disrupt agriculture and society. Pollution of estuaries was discussed in Supplement 10-1. This chapter covers water pollutants: where they come from; how they affect living organisms; and, finally, legal, technological, and personal measures to reduce them.

Water and Water Pollution

Water pollution is any physical or chemical change in water that may adversely affect organisms. It is global in scope, but the types of pollution vary according to a country's level of development. In the poorer nations water pollution is predominantly caused by human and animal wastes, pathogenic organisms from this waste, and sediment from unsound farming and timbering practices. The rich nations also suffer from these problems, but with their more extravagant life-styles and widespread industry they create an additional assortment of potentially hazardous pollutants: heat, toxic metals, acids, pesticides, and organic chemicals. Surely, as many Third World nations industrialize they too will begin to produce a broader assortment of harmful water pollutants.

Like air pollutants, water pollutants come from numerous natural and anthropogenic sources. Because water

(a)

(b)

Figure 16-1. Water pollution. (a) This boat dumping waste into the Cuyahoga River in Cleveland is an example of a point source. (b) Runoff from a feedlot is a nonpoint source.

respects no boundaries, pollutants produced in one country often end up in another's drinking or bathing water. The thoughtless dumping of wastes in rivers, unfortunate accidents, and uncontrolled growth without controls on pollution can have dire consequences for important commercial fisheries. For years the Mediterranean Sea has been viewed as an unlimited dump for domestic and industrial wastes. Extensive negotiations finally brought forth a regionwide plan that slows the increase in pollution.

Movement of pollutants from lakes and rivers to oceans is only half the problem. In recent years scientists have revealed **cross-media contamination**, that is, the movement of a pollutant from one medium (air) to another (water). Pesticides sprayed on crops can drift to nearby lakes and, from there, flow to the oceans. Toxic organics dumped in evaporating ponds ascend to the clouds, only to rain down on land and lakes. Hazardous wastes buried in the ground leak into aquifers, whose waters replenish streams. A joint committee of the US National Research Council and the Royal Society of Canada announced in 1986 that levels of DDT and PCBs in the water of the Great Lakes had declined only slightly since 1978 despite tight controls on industrial emissions in both countries. Controlling industrial releases has been quite successful in reducing emissions, they noted; however, significant

levels of these and other toxic chemicals persist in the lakes because of contaminated groundwater and atmospheric pollution (see Case Study 16-1). Unfortunately, very little has been done since then to eliminate airborne toxins.

Point and Nonpoint Sources

When we ponder the sources of water pollution, we generally think of factories, power plants, and sewage treatment plants that pour tons of sometimes toxic chemicals into sewers and lakes and rivers (Figure 16-1a). These **point sources**, so named because they are in discrete locations, are relatively easy to control. But they are only half the problem. The other half includes sources we rarely think about, the **nonpoint sources**—less discrete sites like farms, forests, lawns, and urban streets (Figure 16-1b). Rainwater carries oil from driveways and streets, and pesticides and fertilizers from urban lawns, into storm sewers and then into streams (Table 16-1). Nonpoint sources release half the pollutants that end up in our waters. Some of the substances include dust, sediment, pesticides, asbestos, fertilizers, heavy metals, salts, oil, grease, litter, and even air pollutants washed out of the sky by rain. Because the sources are many and spread out, control has proved difficult.

The Great Lakes: Alive but Not Well

The Great Lakes are among North America's most important natural resources. Carved by ancient glaciers, these five mammoth lakes hold one-fifth of the world's standing fresh water. Approximately 50 million people live within their drainage basin. The interconnected lakes flow slowly toward the St. Lawrence River and, beyond that, the ocean. Like so many of the continent's waters, they have suffered years of abuse from pollution and poor land management. Especially hard hit were the lower lakes, Ontario and Erie.

Cultural Stress: The Death of a Lake?

The story of Lake Erie serves as a reminder of the immense impact of human civilization and a lesson on ways to prevent future deterioration of the world's water bodies.

Lake Erie was once surrounded by dense forest land. Streams ran clean and free of sediment. Today, however, more than 13 million people live in the lake's watershed. The dense woodlands that once protected the soil were cut down to make room for farms, homes, industries, and roadways. As a result, large quantities of topsoil were washed into the rivers and the lake, clogging navigable channels and destroying spawning areas essential to the lake's once-rich fish life. In the early 1900s many of the swamps along the lake's shore were drained, and hundreds of small dams were constructed to provide power to mills, thus blocking the upstream migration of fish such as the walleye and sturgeon.

Before the widespread settlement of Lake Erie's shores, the water was clear most of the year. By the 1960s, however, the pristine waters had become polluted by organic and inorganic nutrients. Raw sewage floated on the water's surface, and algal blooms were common. Dissolved oxygen levels frequently dropped to dangerously low levels, especially in the profundal zone that occupies the large central basin of the lake. Blue-green algae proliferated in the warm summer months in the shallow western end of the lake, creating a foul-smelling, murky, green water. Lead, zinc, nickel, mercury, and other toxins from industry polluted harbors and built up in near-shore sediments. In 1970 and 1971 mercury levels in fish taken from Lake Erie often exceeded safe levels set by the US Food and Drug Administration. Reductions in mercury discharges in 1975 brought about a sharp drop in mercury in fish, but violations of health standards continue to be reported.

In Colonial days numerous fish species inhabited the lake and its tributaries. Largemouth and smallmouth bass, muskellunge, northern pike, and channel catfish were common in the tributaries of Lake Erie. Lake herring, blue pike, lake white-fish, lake sturgeon, and others made the open waters their home. By the 1940s, however, blue pike and native lake trout had vanished. Sturgeon, lake herring, whitefish, and muskellunge managed to hold on in reduced numbers.

Lake Erie suffered from severe cultural stress caused by overfishing, introduction of alien species, pollution, and destruction of shorelines and spawning grounds. Algal blooms, beach closings, thick deposits of sludge, oxygen depletion, taste and odor problems, and contaminated fish were the legacy of years of disregard and mismanagement.

By the late 1950s large areas of Lake Erie's central basin were anoxic (without oxygen) for weeks on end during the summer. Until the late 1970s anoxia spread cancerously. The lake was pronounced dead; many ecologists feared that the other lakes would follow suit.

The Joint Cleanup Program— Not Enough

In 1972, alarmed by the condition of Lake Erie and other lakes, the United States and Canada agreed to restrict pollutant discharge into the Great Lakes. The Great Lakes Water Quality Agreement, updated in 1978, called for controls of point and nonpoint pollution sources. It demanded that release into the lakes of "any or all persistent toxic substances" be "virtually eliminated." With widespread cooperation from industrial and municipal polluters, the lakes, among them Lake Erie, began to show signs of recovery. Lake Erie is the shallowest and fastest-flowing of the Great Lakes, which helped it to make a quick recovery from years of insult. Today the lake is teeming with fish. Gone are the raw sewage discharges that once discolored the waters. Controls on phosphorus have eliminated the massive algal blooms. The other lakes followed suit.

Despite these efforts about 50 areas still fail to meet the standards set out in the US–Canadian accord (see figure). In 1986, a Buffalo-based environmental group, Great Lakes United, held hearings involving hundreds of residents, environmental activists, government officials, and water pollution experts. Their collective testimony, published in 1987, showed that while visible, smelly pollutants such as sewage have been significantly reduced, many toxic substances not visible to the naked eye continue to pour into the Lakes. For instance, PCBs and pesticides still persist at unacceptable levels. Officials administering the 1978 Agreement conceded that the US and Canadian governments had not fully lived up to the terms of the Agreement.

Studies of toxic residues in fish and wildlife reaffirm that the lakes' pollution problems are far from solved. The basin acts as a huge sink for the 50 million people who live and work nearby. Toxins enter the lakes from factories and sewage treatment plants; nonpoint pollution, including farmland and

Case Study 16-1 (continued)

urban runoff; toxic fallout, i.e., pollutants deposited from the atmosphere; and, finally, resuspension of substances contained in the bottom sediments caused, in part, when harbors and river mouths are dredged.

One of the most significant avenues is atmospheric deposition. Trace metals, pesticides, phosphorus, nitric acid, nitrates, sulfates, sulfuric acid, and organic compounds are all deposited from the air. Available data show that 60% to 90% of all PCBs entering Lakes Superior and Michigan come from the atmosphere. An estimated 14,000 metric tons of aluminum is deposited in Lake Superior from the atmosphere, and nearly 29,000 metric tons annually falls from the skies into Lake Michigan. The atmosphere is also a major source of phosphorus, providing an estimated 59% of the total input to Lake Superior. The 1987 report released by Great Lakes United charged that the United States was not adhering to the provisions of the Clean Air Act that require control of airborne toxic pollutants. Without such controls, it will be impossible to achieve the Agreement's stated goal of no toxic discharges into the Great Lakes. Canada has come under similar criticisms.

What does all this mean? For one, commercial fishing, which was once an economic mainstay in the Great Lakes, continues to be banned in all of the lakes except for Lake Superior. Even there, the commercial fishermen operate under continual uncertainty, never knowing when their catch will exceed safe limits and be declared unsafe by the Food and Drug Administration. Second, introduced salmon and lake trout survive in the lakes, but the salmon population must be restocked each year because the fish do not reproduce successfully. Restocking costs millions of dollars a year. Third, continued pollution

has forced some states to issue warnings advising women who are pregnant, lactating, or of childbearing age not to eat certain fish caught in the lakes. Parents are also advised to keep children away from lake-caught fish. This warning came after a scientific study on chronic low-level exposure, which showed that infants of women who had eaten PCB-contaminated fish two to three times a month were smaller, more sluggish, and had weaker reflexes than infants of women who had not eaten contaminated fish. Acceptable levels for most of the 800 pollutants found in the lakes simply have not been established, primarily because of a lack of information on health effects.

On first glance the Great Lakes appear to have been brought back to life. Gone are the smelly sewage sludge and thick tangles of algae from days past. In their place, however, is a host of toxins that make the fish inedible, leaving the lake a sick patient in need of renewed efforts to cut back on nonpoint pollution and atmospheric deposition. Reducing these problems is complicated. Seven states, two provinces, numerous tribal councils, and two nations share an interest in managing the waters of the Great Lakes. The result is often conflict that holds up the important steps needed to bring the Great Lakes back to a full, productive, and healthy life.

In recent talks Canada and the United States agreed on new ways to control airborne pollutants that settle in the Great Lakes. They also agreed to remove contaminated sediments from the lakes and to better control groundwater pollution that contributes to the degradation of the Great Lakes. At least 15 to 30 years of rehabilitation are needed, but improvements are already beginning to be seen.

Some Features of Surface Waters

This chapter is concerned primarily with pollution of surface water. Therefore, it is important to examine some features of surface waters that affect pollution. Information on groundwater can be found in Chapters 7 and 10.

Freshwater ecosystems fall into two categories. **Standing systems**, such as lakes and ponds, are usually more susceptible to pollution, because water is replaced at a slow rate. A complete replacement of a lake's water may take 10 to 100 years or more; thus, pollutants can build up to hazardous levels. **Flowing systems** include rivers and streams. Because water flows more quickly in them, they tend to purge their pollutants. However, this purging effect is useless if the supply of pollutants is constant or is spread evenly along its banks, as is common along many rivers.

As shown in Figure 16-2, lakes may consist of three zones: (1) the **littoral zone**, the shallow waters along the

shore where rooted vegetation (such as cattails and arrowheads) can grow; (2) the **limnetic zone**, the open water that sunlight penetrates and where phytoplankton such as algae live; and (3) the **profundal**, or deep, zone, into which sunlight does not penetrate. The fish population in the profundal zone is denser in shallow lakes than in deep lakes.

In the warmer months lakes can also be divided into three temperature zones (Figure 16-3). The upper, warm water is called the **epilimnion**; the deeper, cold water forms the **hypolimnion**; and the transition between the two is called the **thermocline**, or **metalimnion**.

In temperate regions lakes go through an important mixing process twice a year, which allows upper and lower waters to be exchanged. In the fall the air temperature begins to drop, and the surface waters begin to cool. When the surface water reaches 4° C (39° F), it becomes cooler and heavier than the water below. The

Table 16-1 Major Nonpoint Pollution Sources in the US

Activity	Explanation
Silviculture	Growing and harvesting trees for lumber and paper production can produce large quantities of sediment.
Agriculture	Disruption of natural vegetation leads to increased erosion; pesticide and fertilizer use, coupled with poor land management, can pollute neighboring surface water and groundwater.
Mining	Leaching from mine wastes and drainage from mines themselves can pollute surface and groundwater with metals and acids; disruption of natural vegetation accelerates sediment erosion.
Construction	Road and building construction disrupts vegetation and increases sediment erosion.
Salt use and groundwater overuse	Salt from roads and storage piles can pollute groundwater and surface water; saltwater intrusion from groundwater overdraft pollutes ground and surface water.
Drilling and waste disposal	Injection wells for waste disposal, septic tanks, hazardous waste dumps, and landfills for municipal garbage can contaminate groundwater.
Hydrological modification	Dam construction and diversion of water both can pollute surface waters.
Urban runoff	Pesticides, herbicides, and fertilizers applied to lawns and residues from roads can be washed into surface waters by rain.

denser surface water sinks to the bottom. The thermal stratification disappears during this **fall overturn**. Winds help churn the waters.

In the spring the lake turns over again. Water expands when it freezes, which is why ice floats. As ice melts it warms from 0° C; when the meltwater is at 4° C, it becomes denser than the slightly warmer water below and then sinks to the bottom, causing the **spring overturn**. Winds may again participate by churning the waters. The seasonal turnover of lakes is important because it helps circulate oxygen from surface waters to deeper waters in the fall, which allows organisms to survive in the profundal zone. It also carries important nutrients from the lower levels of a lake to its upper levels, where they can be used by plants and algae.

With these few basics in mind, let's look more closely at water pollution.

Types of Water Pollution

Nutrient Pollution and Eutrophication

Rivers, streams, and lakes contain many organic and inorganic nutrients needed by the plants and animals that live in them. In higher-than-normal concentrations they become pollutants.

Organic Nutrients Feedlots, sewage treatment plants, and some industries such as paper mills and meat-packing plants may release large quantities of organic pollutants. These substances stimulate bacterial growth. Bacteria, in turn, consume the organics, helping to purify the waters. But there's a catch. During the degradation of organic pollutants, bacteria consume dissolved oxygen (Figure 16-4). As oxygen levels drop, fish and other aquatic organisms perish. When oxygen levels become very low, **anaerobic** (nonoxygen-requiring) bacteria take over, breaking down what's left but producing foul-smelling and toxic gases (methane and hydrogen sulfide) in the process. Oxygen depletion in rivers and streams occurs more readily in the hot summer months, because stream flow is generally lower and organic pollutant concentrations are higher. In addition, increased water temperatures speed up bacterial decay.

As the organic matter is depleted, oxygen levels return to normal. When numerous sources of organic pollutants are found along the course of a river, recovery may be impossible. Dissolved oxygen levels are replenished by oxygen from the air and from photosynthesis in aquatic plants, but oxygen replacement is generally quite slow unless the water is turbulent. Lakes recover from organic pollutants, but usually much more slowly than rivers.

The organic nutrient concentration in streams is measured by determining the rate at which oxygen is depleted from a test sample. Polluted water is saturated with oxy-

Figure 16-2. The three ecological zones of a lake. Note the difference between shallow entropic lakes, which tend to have high levels of plant nutrients, and deep oligotrophic lakes, which support fewer fish because they lack nutrients that stimulate plant growth.

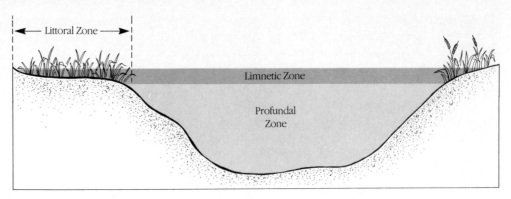

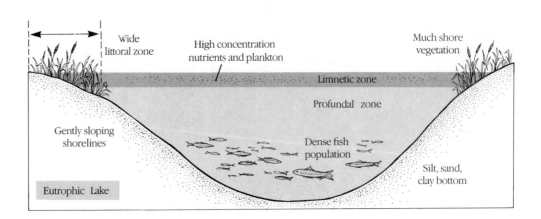

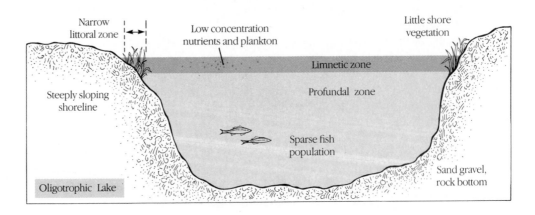

gen and kept in a closed bottle for five days; during this period bacteria degrade the organic matter and consume the oxygen. The amount of oxygen remaining after five days gives a measurement of the organic matter present; the more polluted a sample, the less oxygen left. This standard measurement is called the **biochemical oxygen demand**, or **BOD**.

Inorganic Plant Nutrients   Whereas organic nutrients nourish bacteria, certain inorganic nutrients stimulate the growth of aquatic plants. These plant foods include nitrogen, phosphorus, iron, sulfur, sodium, and potassium.

Nitrogen, in the form of ammonia and nitrates, and phosphorus, in the form of phosphates, are often limiting factors for populations of algae and other plants. Consequently, if levels become high, plant growth can go wild, choking lakes and rivers with thick mats of algae or dense growths of aquatic plants. In freshwater lakes and reservoirs phosphate is usually the limiting nutrient for plant growth; marine waters are usually nitrate-limited.

Excessive plant growth impairs fishing, swimming, navigation, and recreational boating. In the fall most of these

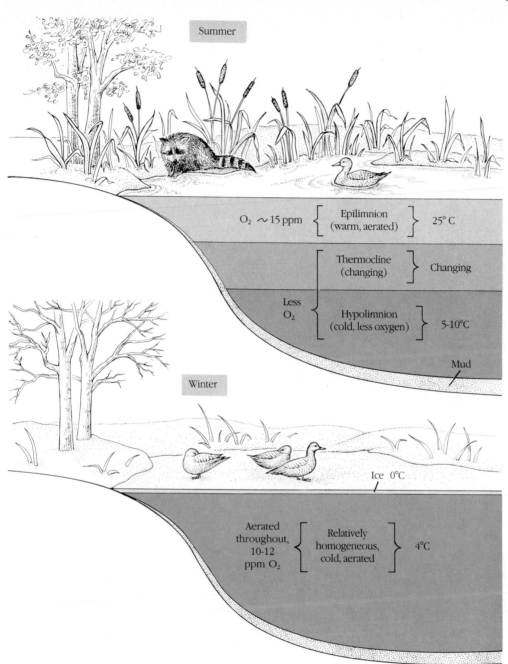

Figure 16-3. The three thermal zones of a lake.

Summer

$O_2 \sim 15$ ppm $\left\{\begin{array}{c} \text{Epilimnion} \\ \text{(warm, aerated)} \end{array}\right\}$ 25° C

$\left\{\begin{array}{c} \text{Thermocline} \\ \text{(changing)} \end{array}\right\}$ Changing

Less O_2 $\left\{\begin{array}{c} \text{Hypolimnion} \\ \text{(cold, less oxygen)} \end{array}\right\}$ 5-10°C

Mud

Winter

Ice 0°C

Aerated throughout, 10-12 ppm O_2 $\left\{\begin{array}{c} \text{Relatively} \\ \text{homogeneous,} \\ \text{cold, aerated} \end{array}\right\}$ 4°C

plants die and are degraded by aerobic bacteria, which can deplete dissolved oxygen, killing aquatic organisms. As oxygen levels drop, anaerobic bacteria resume the breakdown and produce noxious products. Thus, inorganic nutrients ultimately create many of the same problems that organic nutrients do.

Inorganic fertilizer from croplands is the major anthropogenic source of plant nutrients in fresh waters. When highly soluble fertilizers are used in excess, as much as 25% may be washed into streams and lakes by the rain. More careful use could greatly reduce this problem.

Laundry detergents are the second most important

anthropogenic source of inorganic nutrient pollution in this country. Many detergents contain synthetic phosphates, called tripolyphosphates (TPPs). These chemicals cling to dirt particles and grease, keeping them in suspension until the wash water is flushed out of the washing machine. Unfortunately, the phosphates stimulate the growth of aquatic algae, causing sudden spurts in growth called **blooms**.

Nearly 60% of the US population lives in soft-water regions where soap-based cleansing agents work as well as detergents. In the hard-water regions, harmless substitutes for TPPs can be used. For example, lime soap-

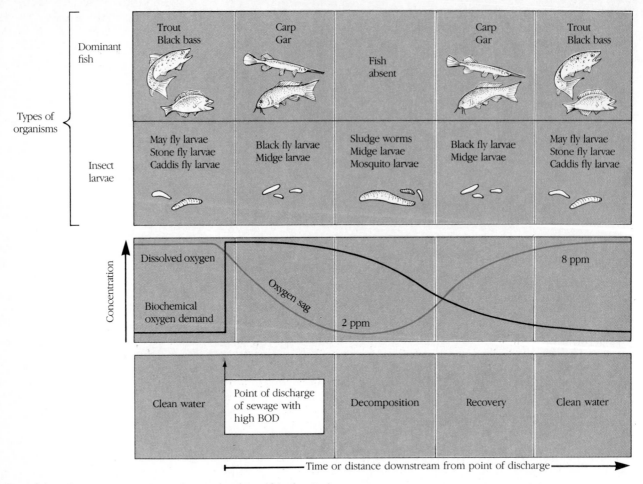

Figure 16-4. The oxygen sag curve. Oxygen levels and biochemical oxygen demand are shown below a point source of organic nutrients.

dispersing agents have been used in bar soaps for years and could easily be used for laundry detergents.

Nitrates and nitric acid can also enter surface waters from the atmosphere, as described in Chapter Supplement 15-2. According to the Environmental Defense Fund about 25% of the nitrogen polluting the Chesapeake Bay comes from wet and dry deposition.

Eutrophication and Natural Succession Lakes naturally pick up nutrients from surface runoff and rainfall. The amount becomes excessive and lakes suffer when watersheds are disturbed by farming or timber cutting. Major pollutant sources such as sewage treatment plants can also disturb the chemical balance in lakes.

The natural accumulation of nutrients in lakes is called **natural eutrophication**. Nutrient accumulation and natural erosion can, given sufficient time, transform lakes into swampland and then into dry land, a process called natural succession (discussed in Chapter 3). In this process inorganic nutrients stimulate plant growth; plants eventually die and contribute organic sediment to the

lake's bottom (Figure 16-5). This sediment combines with silt from erosion and may gradually fill in a lake.

Accelerated erosion, caused by human activities, and **cultural eutrophication**, resulting from inorganic nutrients released from farms, feedlots, and sewage treatment plants speed up the process. Good, productive lakes become choked with vegetation that rots in the fall, depleting oxygen and giving off an offensive odor. The fate of lakes overfed with nutrients from sewage treatment plants and farms, however, is not as dim as we once believed; if nutrient inflow is greatly reduced or stopped, a lake may make a comeback. For example, Lake Washington near Seattle became a foul-smelling, eutrophic eyesore after decades during which millions of gallons of sewage were dumped into its waters. In 1968 local communities began to divert their wastes to Puget Sound, an arm of the sea with a greater capacity to assimilate the wastes. Lake Washington began a slow recovery.

Of course, the diversion of sewage to Puget Sound has had some negative effects. Although the open sound can cleanse itself more easily than the lake, certain toxins in

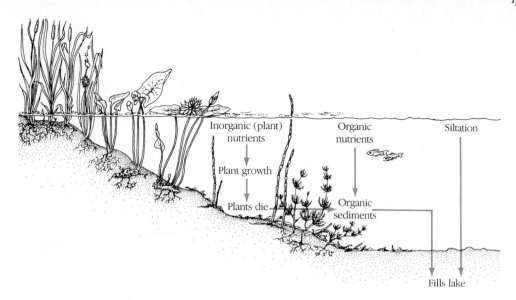

Figure 16-5. Contributions of inorganic and organic nutrients and sediment to succession of a lake into swampland.

the waste and in surface runoff (nonpoint pollution) are having a harmful effect on marine life in the sound. The effects are especially noticeable in areas where industries have discharged their wastes for decades. Shellfish beds have been closed to prevent contamination. Sediments in the sound contain high levels of toxic organic wastes and heavy metals, which may be responsible for the high incidence of tumors in sole, a fish that lives on the bottom in Puget Sound. Efforts are now under way to alleviate the harm done by sewage and industrial discharges in the sound.

Eutrophication is the most widespread problem in US lakes. Today, seven out of eight lakes are experiencing accelerated eutrophication. Nearly all receive wastes from industry and municipalities, but even if these sources were eliminated, only half of the lakes would improve because of continued pollution from nonpoint sources.

Infectious Agents

Water may be polluted by pathogenic (disease-causing) bacteria, viruses, and protozoans. Waterborne infectious diseases are a problem of immense proportions in the less developed nations of Africa, Asia, and Latin America. They were once the major water pollutants of now-developed nations before sewage treatment plants and disinfection of drinking water became commonplace.

The major sources of infectious agents are (1) untreated or improperly treated sewage; (2) animal wastes in fields and feedlots beside waterways; (3) meat-packing and tanning plants that release untreated animal wastes into water; and (4) some wildlife species, which transmit waterborne diseases. The major infectious diseases include viral hepatitis, polio (viral), typhoid fever (bac-

terial), amoebic dysentery (protozoan), cholera (bacterial), schistosomiasis (parasitic worm), and salmonellosis (bacterial). These diseases are especially harmful to the young, old, and already ill.

Measuring the level of each pathogenic organism would be costly and time-consuming. By measuring levels of a naturally occurring intestinal bacterium, the *coliform bacterium*, water-quality personnel can determine how much fecal contamination has occurred. The higher the coliform count, the more likely the water is to contain some pathogenic agent from fecal contamination. About one-third of our rivers now violate standards for coliform bacteria.

Toxic Organic Water Pollutants

About 10,000 synthetic organic compounds are in use today. Many of these find their way into our water, creating what may be our most important water pollution problem.

The reasons for concern over these pollutants are several: (1) Many toxic organic compounds are nonbiodegradable or are degraded slowly, so they persist in the ecosystem. (2) Some are magnified in the food web (Chapter 14). (3) Some may cause cancer in humans; others are converted into carcinogens when they react with the chlorine used to disinfect water. (4) Some kill fish and other aquatic organisms. (5) Some are nuisances, giving water and fish an offensive taste or odor.

Unfortunately, our knowledge of the effects of synthetic organics, which are often found in low concentrations, is rudimentary. Reports of diseases traceable to a single chemical are few; but many experts worry that cancer and genetic damage may result from long-term exposure.

Toxic Inorganic Water Pollutants

Inorganic water pollutants encompass a wide range of chemicals, including metals, acids, and salts. Most states report that toxic metals, such as mercury and lead, are a major water pollution problem. Metals come from industrial discharge, urban runoff, mining, sewage effluents, air pollution fallout, and some natural sources. Recent surveys of US drinking water show contamination from pipes and groundwater supplies.

Mercury   One of the more common and potentially most harmful toxic metals is mercury. In the 1950s mercury was thought to be an innocuous water pollutant, although it was known to have been hazardous to miners and to nineteenth-century hat makers, who frequently developed tremors, or "hatter's shakes," and lost hair and teeth.

In the 1950s an outbreak of mercury poisonings in Japan raised awareness of the hazard. Residents who ate seafood from Minamata Bay, which was contaminated with methyl mercury, developed numbness of the limbs, lips, and tongue. Muscle control was lost. Deafness, blurring of vision, clumsiness, apathy, and mental derangement also occurred. Of 52 reported cases, 17 people died and 23 were permanently disabled.

Mercury is a by-product of manufacturing the plastic vinyl chloride. It is also emitted in aqueous wastes of the chemical industry and incinerators, power plants, laboratories, and even hospitals. Worldwide, about 10,000 metric tons of mercury are released into the air and water each year.

In streams and lakes inorganic mercury is converted by bacteria into two organic forms. One of these, dimethyl mercury, evaporates quickly from the water. But the other, methyl mercury, remains in the bottom sediments and is slowly released into the water, where it enters organisms in the food chain and is biologically magnified.

Nitrates and Nitrites   Nitrates and nitrites are common inorganic pollutants of water. **Nitrates** come from septic tanks, barnyards, heavily fertilized crops, and sewage treatment plants; they are converted to toxic **nitrites** in the intestines of humans.

Nitrites combine with the hemoglobin in red blood corpuscles and form methemoglobin, which has a reduced oxygen-carrying capacity. Nitrites can be fatal to infants. Over 2000 cases of infant nitrite poisoning, about 160 of which resulted in death, have been reported in Europe and North America in the last 40 years. Most poisonings occurred in rural areas, where drinking water is contaminated by septic tanks and farmyards. Today, cases are extremely rare in the US and Canada.

Salts   Sodium chloride and calcium chloride are used on winter roads to melt snow. Melting snow carries these salts into streams and groundwater. Salts kill sensitive plants, such as the sugar maple. In surface waters they may kill salt-intolerant organisms, allowing salt-tolerant species to thrive. However, the fluctuations in the flow of salt lead to varying concentrations, so that neither salt-tolerant species that thrive in high salt concentrations nor salt-intolerant organisms can survive.

Chlorine   Chlorine is a highly reactive inorganic chemical commonly used (1) to kill bacteria in drinking water, (2) to destroy potentially harmful organisms in treated wastewater released from sewage treatment plants into streams, and (3) to kill algae, bacteria, fungi, and other organisms that grow inside and clog the pipes of the cooling systems of power plants. Chlorine and some of the products it forms in water are highly toxic to fish and other organisms.

Chlorine reacts with organic compounds to form chlorinated organics. These chemicals may show up in drinking water downstream from sewage treatment plants and other sources. Many of them are known carcinogens and teratogens. However, medical studies indicate that the rates of certain cancers (liver, intestinal tract) are only slightly elevated in populations consuming water contaminated by these compounds.

Sediment

Sediment, the leading water pollutant in the United States in terms of volume, is a by-product of timber cutting (see Chapter 9), agriculture (Chapter 7), mining (Chapters 11–13), and construction of roads and buildings. Agriculture increases erosion rates four to eight times above normal. Poor construction and mining may increase the rate of erosion by 10 to 200 times. Sediment destroys spawning and feeding grounds for fish, reduces fish and shellfish populations, smothers eggs and fry, fills in lakes and streams, and decreases light penetration, which destroys aquatic plants.

The deposition of sediment in lakes speeds up natural succession. The filling in of streambeds, or **streambed aggradation**, results in a gradual widening of the channel, and as streams become wider, they become shallower. Water temperature may rise, lowering the amount of dissolved oxygen and making streams more vulnerable to organic pollutants that deplete oxygen. Streambed aggradation also makes streams more susceptible to flooding. Sediment can fill shipping channels, which must then be dredged. Hydroelectric equipment may be worn out by sediments. Finally, some pollutants—such as pesticides, nitrates, phosphates from agricultural fertilizers, and pathogenic organisms—bind to sediment. This extends their lifetime and impacts.

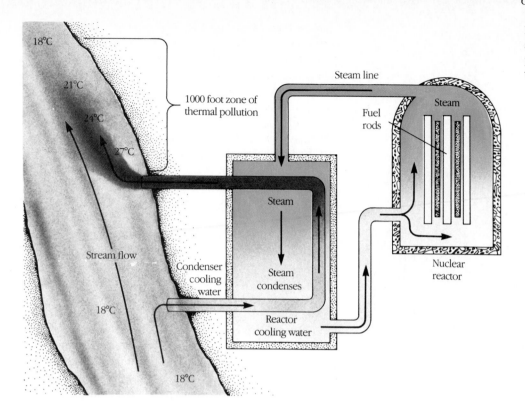

Figure 16-6. The cooling system of an electric power plant and its effect on surface waters and organisms.

Sediment pollution can be checked, and even eliminated, by good land management (described in Chapters 7 and 9).

Thermal Pollution

Rapid or even gradual changes in water temperature can disrupt aquatic ecosystems. Industries frequently bring about such change by using water to cool various industrial processes. The US electric power industry is a major contributor (Figure 16-6). It uses about 86% of all cooling water in the United States, or about 730 billion liters (190 billion gallons) per day. Steel mills, oil refineries, and paper mills also use large amounts of water for cooling.

Small amounts of heat have no serious effect on the aquatic ecosystem; but large quantities can kill heat-intolerant plants and animals outright, disrupting the web of life dependent on the aquatic food chain. Elimination of heat-intolerant species may allow heat-tolerant species to take over. These are usually less desirable species.

Thermal pollution lowers the dissolved oxygen content of water, at the same time increasing the metabolic rate of aquatic organisms. Since metabolism requires oxygen, some species may be eliminated entirely if the water temperature rises 10° C (18° F). At the Savannah River nuclear power plant the number of rooted plant species and turtles was at least 75% lower in ponds receiving hot water than in ponds at normal temperature. The number of fish species was reduced by one-third.

Sharp changes in water temperature cause **thermal shock**, a sudden death of fish and other organisms that cannot escape. Thermal shock is frequently experienced when power plants begin operation or when they temporarily shut down for repair. The latter can devastate heat-tolerant species that inhabit artificially warmed waters. Aquatic life cannot adapt to sudden, unpredictable temperature changes.

Fish spawn and migrate in response to changes in water temperature, so heated water may interfere with these processes. Water temperature influences survival and early development of aquatic organisms. For instance, trout eggs may not hatch if water is too warm. Thermal pollution can also increase the susceptibility of aquatic organisms to parasites, certain toxins, and pathogens.

Thermal pollution can be controlled by constructing ponds for collecting and cooling water before its release into nearby lakes and streams. Cooling towers are another way to dissipate heat (see Figure 10-12).

Groundwater Pollution

Aquifers supply drinking water for about 120 million Americans. That water, scientists are now reporting, is increasingly threatened by pollution. In fact, many pollutants are present at much higher concentrations in groundwater than they are in most contaminated surface supplies. And many contaminants are tasteless and odor-

Figure 16-7. Air stripper at a site in Massachusetts removes dissolved volatile organic compounds from water. Although the removal rate depends on the contaminant, most air strippers remove 98% of the contaminants.

less at concentrations thought to threaten human health.

About 4500 billion liters (1185 billion gallons) of contaminated water seeps into the ground in the United States every day from septic tanks, cesspools, oil wells, landfills, agriculture, and ponds holding hazardous wastes. Unfortunately, very little is known about the extent of groundwater contamination. Some experts believe that groundwater pollution is a minor problem. They estimate that 1% to 2% of US groundwater is polluted. However, an EPA study completed in 1981 showed groundwater contamination in 28% of 954 cities with populations over 10,000. By more recent estimates, at least 8000 private, public, and industrial wells are contaminated. In 1989, the EPA launched a program to assess the extent of groundwater contamination.

Thousands of chemicals, many of them potentially harmful to health, turn up in water samples from polluted wells. The most common chemical pollutants are chlorides, nitrates, heavy metals, and various toxic organics like pesticides and degreasing agents. The low-molecular-weight organic compounds are particularly worrisome, since many of them are carcinogenic. Concern among medical experts is great because some fear that there is no threshold level for these compounds—that is, there is no level free from risk of cancer or other problems.

Others fear that many chemicals may act synergistically, turning a potentially difficult problem into a health nightmare (Chapter 14). Beverly Paigen, a researcher in Oakland, California, renowned for her studies at Love Canal in Niagara Falls, New York, published a summary of health studies of Americans exposed to groundwater pollutants. The most common problems include miscarriage, low birth weight, birth defects, and premature infant death. Adults and children suffer skin rashes, eye irritation, and a whole host of neurological problems, including dizziness, headaches, seizures, and fainting spells. In a now widely publicized case in San Jose, California, pollutants from a leaky underground storage tank owned by the Fairchild Camera and Instrument Company are thought to have doubled the rate of miscarriage in pregnant women and tripled the rate of heart defects in newborns. In Woburn, Massachusetts, contaminated groundwater is blamed for a doubling in the childhood leukemia rate.

Many people think of groundwater as fast-flowing underground rivers. Nothing could be further from the truth. Groundwater typically moves from 5 centimeters (2 inches) to 64 centimeters (2 feet) a day. Since groundwater moves so slowly, it may take years for water polluted in one location to appear in another. Additionally, once an aquifer is contaminated, it may take several hundred years for it to cleanse itself.

Detecting groundwater pollution is expensive and time-consuming. Numerous test wells must be drilled to sample water and determine the rate and direction of flow. Despite intensive drilling health officials can easily miss a tiny stream of pollutants that flows through one portion of a large aquifer. For example, liquids that do not readily mix with water may travel along the top or bottom of the aquifer in thin layers and are often difficult to detect.

Groundwater supplies one-fourth of the annual water demand in the United States. Preventing groundwater pollution is generally the cheapest way to protect this vital resource. Cutting down the production of hazardous wastes would be an important first step. Improvements in the ways we dispose of wastes would also help. Ways to achieve this goal are discussed in Chapter 18.

To reclaim polluted aquifers it may be necessary to pump contaminated water to the surface, purify it, and then return it to the aquifer (Figure 16-7). New techniques are also being developed to use naturally occurring bacteria found in soil and groundwater to clean up some contamination. For instance, hydrocarbons (such as crude oil, gasoline, and creosote) that have leaked from storage tanks or are spilled from vehicles have polluted more groundwater used for drinking than any other class of chemicals in the United States. Microbiologists have known since the late 1970s that some bacteria can digest or break down hydrocarbons in the soil and groundwater, converting them into carbon dioxide and methane gases.

Case Study 16-2

The Case of the Dying Seals

In the Spring of 1988, the harbor seals in the North Sea began to die. Adult seals floated aimlessly in the water, too weak to eat or play (see Figure 10-20 in Chapter 10). Pregnant females aborted their fetuses. The mysterious disease began off the coast of Denmark and spread quickly to seal colonies throughout the North and Baltic seas. By the middle of the summer, seals were dying along hundreds of miles of North Sea coastline. By September, the disease had spread to the Atlantic coast of Ireland.

Some people called this scourge the "black death of the sea," for it recalled the epidemics of bubonic plague, or black death, that devastated Europe in the 1300s. In this tragic turn of events, a population of harbor seals once containing 18,000 animals has been cut to only 6000. This, the largest die-off of seals in recorded history, may be caused by the canine distemper virus. Biologists believe, however, that the virus is not working alone. Pollution in the seas, they say, may have greatly weakened the immune systems of the seals, making them vulnerable.

The North and Baltic seas have been polluted for years. The North Sea alone annually receives 60 billion liters (15 billion gallons) of waste water from factories and waste-treatment facilities in bordering industrial nations.

Germany's environmental minister argued that the industrial pollution is a principal cause of seal deaths and many biologists who are studying the die-off agree. The pollution problem in the North and Baltic seas, however, is compounded by the nature of the seas themselves. Both are shallow and cleanse themselves very slowly. The North Sea, in fact, renews itself only twice every ten years. The Baltic Sea turns over once every 20 to 30 years. Because of this, pollution levels can increase locally, causing adverse impacts on fish and wildlife, and, possibly, people.

Seals living in the waters off the coast of West Germany and the Netherlands are heavily contaminated with a toxic chemical called PCB (polychlorinated biphenyl), a substance once used as an insulator in electrical devices. The PCBs and possibly other chemical contaminants are believed responsible for the reproductive problems and the suppression of the seals' immune systems.

The seal plague may be the latest manifestation of a chronic pollution problem in the North and Baltic seas. Northern Europe and Scandinavia have taken steps to clean up the seas, but many key nations seem uninterested in helping. Great Britain, for instance, has been remarkably complacent about the seal deaths. East Germany and Czechoslovakia, two principal polluters, failed to attend the conference held in 1987 in which most North Sea states agreed to cut industrial emissions to the rivers of Europe by half.

Seal deaths off the coast of Europe are a symptom of a global problem. Similar events are occurring elsewhere. Since June of 1987, as many as four out of every ten dolphins off the Atlantic coast of the United States have perished. Studies of gulls in the Great Lakes have shown an alarming reproductive failure due to PCBs and other organic pollutants.

Despite an outpouring of laws to control pollution, America has hardly come to grips with the problem. Tens of thousands of hazardous waste sites litter the American landscape. Pollution control laws passed in the 1970s initially decreased water pollution nationwide, but since the early 1980s, pollution levels have remained more or less constant. Making matters worse, regulation and enforcement of hazardous waste laws has been lax.

William K. Reilly, Jr., former President of the World Wildlife Fund and the Conservation Foundation, notes that despite the successes of pollution control laws, America "faces an array of environmental problems even more daunting than [the] pollution crises of the past generation." Global climate change, acid precipitation, worldwide deforestation, and ozone depletion, he says, are all unanswered by current policies.

Solutions to global problems require new laws and tighter controls. Critical thinking demands a search for additional solutions. New technologies, for example, can help us reduce waste and use the earth's resources more judiciously. Individuals can also chip in. A personal commitment to conserve, to recycle, to use renewable resources (for example, paper rather than plastic), and to limit family size can go a long way in helping to solve the environmental problems facing this nation. Individual actions, multiplied many times, must be a part of the solution. All of these efforts must be brought to bear on the global environmental crisis, making the world a healthier place for all life.

Bacteria in the soil, however, as a rule can only degrade about 1% of the hydrocarbon pollution flowing past them. Why? Quite simply, they lack key chemical nutrients needed for metabolism. By supplying these nutrients researchers may be able to accelerate the bacterial decomposition of hydrocarbons.

Ocean Pollution

"When we go down to the low-tide line," Rachel Carson wrote, "we enter a world that is as old as the earth itself—the primeval meeting place of the elements of earth and water, a place of compromise and conflict and eternal

Figure 16-8. The anatomy of an oil spill from an improperly drilled hole.

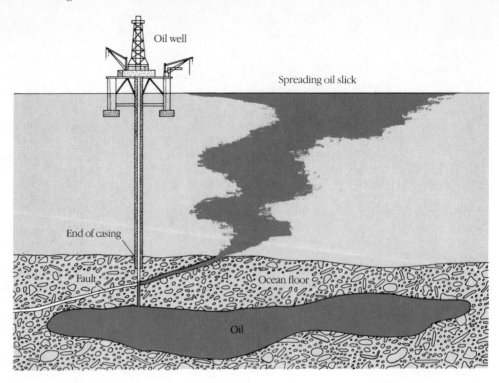

change." Today this compromise, conflict, and eternal change has taken on a new meaning as humankind forges out into the oceans in search of food, fuel, and minerals. Even the Kansas farmer and Minnesota factory have an impact on this vast body containing more than 1.3 billion cubic kilometers of water, for many inland pollutants make their way to the ocean.

The hazards of pollution in the biologically rich coastal zones were discussed in Chapter 10. This section deals with two problems: oil and plastic pollution.

Oil in the Seas

About 3.2 million metric tons of oil enters the world's seas every year. About half of the oil that contaminates the ocean comes from natural seepage from offshore deposits. One-fifth comes from well blowouts, breaks in pipelines, and tanker spills (Figure 16-8). The rest, quite surprisingly, comes from oil disposed of inland and carried to the ocean in rivers.

Leakage from offshore wells occurs during the transfer of oil to shore and also during normal operations. Contamination from this source has not captured the public attention, even though its effect on marine life and birds can be quite significant. What captures the headlines are the large spills—well blowouts or wrecked tankers that spill out tons of black, viscous oil. From 1980 to 1986 there were 77,000 oil and chemical spills of various sizes

in US waters, an average of 11,000 incidents per year (over 90% of these were oil spills).

The harmful effects of oil spills are many. Oil kills plants and animals in the estuarine zone. Especially hard hit are the barnacles, mussels, crabs, and rock weed (a type of algae). Recovery after a major spill may take two to ten years. Oil settles on beaches and kills organisms that live there. It also settles to the ocean floor and kills benthic (bottom-dwelling) organisms such as crabs. Those that survive may accumulate oil in their tissue, making them inedible. Oil poisons algae and may disrupt major food chains and decrease the yield of edible fish. It also coats birds, impairing flight or reducing the insulative property of feathers, thus making the birds more vulnerable to the cold (Figure 16-9). Oil endangers fish hatcheries in coastal waters and can contaminate the flesh of commercially valuable fish, as it did in Prince William Sound in Alaska after the *Valdez* spill.

About 25% of the crude oil in a spill is volatile and evaporates within three months. Other compounds, relatively nonvolatile but lighter than water, float on the surface, where they are broken down by bacteria over the next few months. Nearly 60% of the oil spill is destroyed in this way. The remaining 15% consists of heavier compounds that stick together and sink to the bottom in huge globs. In cold polar waters oil decomposes very slowly. In some cases it may become incorporated in sea ice and be released for years afterward.

The amount of damage caused by oil pollution depends partly on the direction in which it is carried by the wind and ocean currents. If slicks reach land, they damage beaches and shorelines, recreational areas, and marine organisms. Oil may be driven over portions of the continental shelf, a highly productive marine zone, and there it can poison clams, scallops, flounder, haddock, and other important food species. Oil driven out to sea has fewer environmental consequences, because there is little life in the deep ocean. The damage resulting from an oil spill also depends on when it occurs. For example, the devastating spill off the coast of Alaska in 1989 occurred only two weeks before hundreds of thousands of ducks and other waterfowl migrated through the sound or came to nest. Had the spill occurred after the migration the damage might have been much less.

Oil pollution of the oceans poses less of a threat to the overall marine environment than was once feared, a National Research Council committee has concluded. Oil can have serious local effects, the group noted. And these can persist for decades. But overall, the marine environment has not suffered irrevocable damage from oil. The committee was quick to point out that marine scientists have only limited knowledge about the potential damage of oil in tropical and arctic regions, where much of the current oil development is occurring. Studies of the 1989 oil spill in Alaska showed that the hydrocarbons in the oil are more stable and thus more persistent in the cold waters than in warmer seas.

Thanks to public outcry and stricter controls, the number of oil spills began to fall after 1980 and has remained relatively stable in recent years. Tougher governmental standards for new oil tankers went into effect in 1979. New safety standards for older tankers were phased in between 1981 and 1985. Dual radar systems, backup steering controls, collision avoidance aids, and improved inspection and certification were instrumental in reducing spills. Under the new regulations crude oil must be cleaned to eliminate sludge buildup in tanks. This sludge was once rinsed out at sea. New regulations also require tankers to have separate ballast tanks. These tanks are filled with salt water to help ships keep their balance when returning after discharging their cargo. In older ships emptied oil tanks were filled with water for ballast. When the ship arrived at port, the oil-contaminated water was dumped into the sea. The serious effects of oil washing and combined ballast–oil tanks must not be underestimated. About 1.3 million metric tons of oil was released each year during tank purging and ballast tank discharge, or over six times the amount released by tanker spills.

Although the number of oil spills has decreased, the quantity of oil released fluctuates wildly. In 1984, for example, approximately twice as much oil was released as in 1983 in about the same number of spills. Much room for improvement remains.

At this writing both the US House and Senate were

Figure 16-9. This bird is beyond the help of volunteers, who, after a 1981 oil spill in San Francisco, attempted to save the oil-covered animals.

working on legislation that may help prevent tanker accidents. One law passed by the Senate in August 1989 would establish a $1-billion fund to be used to clean up oil spills. The money would come from a 3-cents-per-barrel tax on domestic and imported oil and would be used for immediate costs of cleanup, passing the cost to customers. Under the bill, oil companies would ultimately be responsible for cleanup costs, but only to a point, because the law sets strict limits on their financial liability. A company responsible for a spill, for instance, would pay only $10 million for cleanup. If the spill occurred near an onshore facility where damage is much greater, the company would pay $350 million for each spill. Presumably, the fund would be used to pay the additional costs. Many critics are disappointed with the liability limits because the costs of cleaning oil are likely to be far greater. Cleanup in Prince William Sound stood at nearly $1 billion in August 1989, only four months after the spill. By then only a fraction of the coastline had been cleaned.

Several key provisions were defeated by the Senate. One of them would have required oil companies to fit *existing* tankers with double steel hulls. The law requires the Secretary of Transportation to order double hulls on all *new* tankers unless safer methods of prevention are found. Unfortunately, industry and the Coast Guard, say critics, have always found a way to avoid double hulls each time Congress has proposed them in the past.

On the bright side, the law would establish regional oil-spill response teams that would be deployed immediately after a spill to coordinate cleanup efforts. It also allows states to set stricter guidelines. A state, for instance, could mandate double hulls for use in its harbors.

Plastic Pollution

A young seal swims playfully in the coastal waters of San Diego Bay. Floating in its watery domain is a piece of a plastic fishing net that has drifted with the currents for months. The seal swims around and around curiously, checking out its new plaything, and then plunges through an opening in the net only to be entrapped.

At first the net is just a mild nuisance, but as the seal grows, the filament begins to tighten around its neck. Eventually, it cuts into the seal's skin, leaving an open ring of raw flesh exposed to bacteria. Unless it is helped, the seal will perish along with an estimated 100,000 other sea mammals each year that tangle with the 160,000 metric tons of plastics discarded into the ocean annually by American fishermen and sailors. Tens of thousands of tons may also come from private boats and factories.

According to some estimates, a million seabirds die each year from plastics. No one knows the number of fish that perish from this growing problem, caused by the careless discarding of nylon fishing nets, plastic bags, six-pack yokes, plastic straps, and a myriad of other objects made from nonbiodegradable plastic.

Plastic nets entangle fish, birds, and sea mammals. They may strangle, starve to death, or drown their victims. Plastic bags, looking like jellyfish, are eaten by sea turtles. One scientist pulled enough plastic out of a leatherback turtle's stomach to make a ball several feet in diameter. Starvation generally results, because the animal's stomach is packed with undigestible plastic that cannot pass through its digestive tract. Birds and fish gobble plastic beads resembling the tiny crustaceans that are a normal part of the food chain. They may become poisoned and die. Discarded plastic eating utensils, when swallowed, may cut into an animal's stomach lining, causing it to bleed to death.

Growing awareness has created a groundswell of activity. The Oregon Fish and Wildlife Department organizes annual beach cleanups to remove plastic, netting many tons of plastic garbage. Italy recently placed a ban on all nonbiodegradable plastics, which becomes effective in 1991. Oregon and Alaska have passed laws requiring that all six-pack yokes be biodegradable. In 1988, the US Congress passed legislation making it unlawful for any US vessel to discard plastic garbage in the ocean. Congress is also considering passing legislation requiring all manufacturers of six-pack yokes to use biodegradable plastic. In 1988 the Senate unanimously approved an international treaty banning the disposal of plastics in the ocean. The treaty also prohibits the dumping of other garbage within 20 to 40 kilometers (12 to 25 miles) of the world's coasts. The agreement was signed by 28 nations and went into effect in December 1988. Although the treaty may be difficult to enforce given the enormous size of the ocean, it could help reduce plastic pollution.

On another front, in what may prove to be a precedent-setting move, New Jersey legislators introduced legislation to ban the release of nonbiodegradable balloons. Why? Many balloons released to celebrate national holidays and other occasions eventually end up in the ocean. Here they are often mistaken for food by marine organisms. They accumulate in their digestive tracts and cause starvation.

Despite these steps millions of fish, birds, and sea mammals will perish in coming years. Without stricter controls, biodegradable plastic, and widespread public cooperation, rising use of plastic will bring the unnecessary death of innumerable sea creatures.

Medical Wastes and Sewage Sludge

In the summer of 1988, many Americans were shocked to learn that medical wastes were being illegally dumped into the ocean. Bloody bandages, sutures, vials of AIDS-infected blood, and used syringes washed up onto the eastern shores of the United States as well as the shores of Lake Erie.

Because there was no way to track the wastes to their source, the US Congress passed the **Medical Waste Tracking Act** (1988). It is a two-year program that covers ten states. It requires that people generating, storing, treating, and disposing of medical wastes keep records that are accessible to the public. If wastes show up on shorelines, they can be traced back to their source so that responsible parties can be brought to justice.

In 1988 Congress also passed the **Ocean Dumping Ban Act**, which prohibited the dumping of **sewage sludge** (organic material from sewage treatment plants, discussed later) in the ocean. Dumping will end by December 31, 1991. Cities still discarding their sludge at sea must produce plans to phase out the practice as soon as possible.

According to the Natural Resources Defense Council, 8.9 trillion liters (2.3 trillion gallons) of liquid waste generated from sewage treatment plants is also dumped directly into the ocean. Some of the waste receives little or no treatment before it is discharged. Much of it is industrial waste containing toxic organic chemicals and toxic metals.

Water Pollution Control

Legal Controls

Most efforts to reduce water pollution in the United States and abroad have been aimed at point sources. Building sewage treatment plants to handle municipal wastes and reducing waste discharges by industry have been the key strategies of the US **Clean Water Act** (Table 16-2). Since

1981, 10,000 sewage treatment plants have been constructed (Figure 16-10). In 1987, a program of federal assistance to the states for construction of these plants was reauthorized in an amendment to the Clean Water Act, providing a total of $9.6 billion in grants through 1990. From 1991 until 1994, states will be able to get loans only for treatment facility construction. After 1994, all federal loans for this purpose will stop.

Unfortunately, in many cases nonpoint pollution from expanding cities offsets the gains of sewage treatment plants. Thus, water quality has not improved or has improved only slightly in many areas experiencing rapid growth. Many experts argue that tighter controls are needed to clean up the nation's waters. The 1987 amendment to the Clean Water Act addressed this necessity. Under a major provision of the amendment, states are required to identify the problem areas and nonpoint sources in question within 18 months. The states must also draft a management program, outlining the ways in which they will implement needed controls on nonpoint sources. On the basis of the reports and programs, the EPA will distribute $400 million over four years to the states' programs. Critics point out that this amount of funding will hardly put a dent in nonpoint pollution. Some local governments have already taken the initiative, however, through zoning ordinances to reduce agricultural and urban runoff. Local soil conservation districts identify trouble spots and work with farmers. Much work is needed in the years ahead.

New laws at all levels of government can help reduce pollution. Laws requiring terracing of steep roadbanks, revegetation of denuded land, use of mulches to hold soil in place while grasses are growing on newly constructed highways and housing sites, sediment ponds to collect runoff before it can reach streams, and porous pavements that soak up rainwater could help.

Some experts believe that a groundwater pollution control act is needed, although various provisions of the hazardous waste laws already address much of the problem (see Chapter 18). In recent years the EPA has begun to establish standards for acceptable levels of 22 groundwater contaminants. Progress in this area has been slow. In 1984 the EPA issued a formal groundwater protection plan that, among many things, established an Office of Groundwater Protection. Under the plan the EPA provides technical assistance in analyzing problems and advice needed by states to establish their own groundwater-protection programs. According to the plan, the states are expected to play the chief role in groundwater protection, a point with which many critics take issue, mainly because states often lack necessary funds, experience, and expertise. Many states have adopted their own groundwater standards. Some have taken preventive measures by mapping out aquifers and banning industrial development over them. The EPA approves groundwater protection

---

Table 16-2 Major Provisions of the US Clean Water Act

Planning

The states receive planning grants from the EPA to review their water pollution problems and to determine ways to solve them by reducing or eliminating point and nonpoint water pollutants.

Standards development

The states adopt water quality standards for their streams. These standards define a use for each stream and prescribe the water quality needed to achieve that use.

Effluent standards

The EPA develops limits on how much pollution may be released by industries and municipalities. These limits are developed at the national level based on engineering and economic judgments. The EPA or the states are required to make the discharge limits for individual plants more stringent if necessary to meet the state water quality standards.

Grants and loans

The EPA provides financial assistance to state water programs for the construction of sewage treatment plants, permit applications, water quality monitoring, and enforcement. Federal grants for treatment facilities will be phased out by 1990.

Dredge and fill program

The EPA develops environmental guidelines to protect wetlands from dredge and fill activities. These guidelines are used to assess whether permits should be issued.

Permits and enforcement

All industries and municipal dischargers receive permits from either the EPA or the states. The EPA and the states regularly inspect these dischargers to determine whether they are in compliance with the permit and take appropriate enforcement actions if necessary.

Source: US Environmental Protection Agency.

---

plans, but as of June 1989 only 30 states had submitted plans for approval.

In 1986 Congress toughened the US drinking water law by adding regulations that protect groundwater. Congress also recently passed legislation that requires the EPA to set standards of drinking water quality for 85 additional chemical substances, including pesticides and various industrial chemicals. The new law requires drinking water suppliers to test for these chemicals and maintain the drinking water standards. It also requires them to

Figure 16-10. A vast sewage treatment plant in Des Moines, Iowa.

monitor drinking water supplies for other substances that might pose a threat to human health. Even though the EPA is now monitoring these chemicals, hundreds more could be monitored.

States have become more involved in groundwater protection in recent years, but many of the poorer states have no programs or only modest ones. The federal government provides only modest funding ($7–10 million a year) to support groundwater protection in all 50 states. Without much better funding, groundwater protection is bound to remain inadequate, potentially exposing millions of people to harmful chemicals.

Control Technologies

The first sewage treatment plant in the United States was built in Memphis, Tennessee, in 1880; today, there are over 13,000 of them (Figure 16-10). Sewage entering a treatment plant contains pollutants from homes, hospitals, schools, and industries. It contains human wastes, paper, soap, detergent, dirt, cloth, food residues, microorganisms, and a variety of household chemicals. In some cases water from storm drainage systems is mixed with municipal waste drainage systems to save the cost of building separate pipes for each, which can be exorbitant. Combined systems generally work well, but during storms inflow may exceed plant capacity. Consequently, some untreated storm runoff and sewage passes directly into waterways, raising the coliform count in downstream waters and rendering them unfit for swimming.

Sewage treatment can take place in three stages: primary, secondary, and tertiary. **Primary treatment** physically removes large objects by first passing the sewage through a series of grates and screens. Sand, dirt, and other solids settle out in grit chambers (Figure 16-11a). The solid organic matter, or sludge, settles out in a settling tank.

Secondary Treatment The secondary stage destroys biodegradable organic matter through biological decay (Figure 16-11b). Sludge from primary treatment enters a large tank, where bacteria and other organisms decompose the agitated waste. Another common way is to pass the liquid sludge through a **trickling filter** (Figure 16-12). Here, long pipes rotate slowly over a bed of stones (and sometimes bark), dripping wastes on an artificial detritus food chain consisting of bacteria, protozoa, fungi, snails, worms, and insects. The bacteria and fungi consume the organics and, in turn, are consumed by protozoans. Snails and insects feed on the protozoans. Some inorganic nutrients are also removed. This step is followed by a secondary settling basin or clarifier to remove residual organic matter.

In most municipalities liquid remaining after secondary treatment is chlorinated to kill potentially pathogenic bacteria and protozoans and is then released into receiving streams, lakes, or bays. The efficiency of primary and secondary treatment is shown in Table 16-3.

Tertiary Treatment Many methods exist for removing the chemicals that remain after secondary treatment. Most of these **tertiary treatments** are costly and therefore are rarely used unless water is being released into bodies of water that require a high level of purity. Fortunately, some cheaper options are gaining recognition. For example, effluents can be transferred to holding ponds after secondary treatment. Algae and water hyacinths growing in the water consume the remaining nitrates and phos-

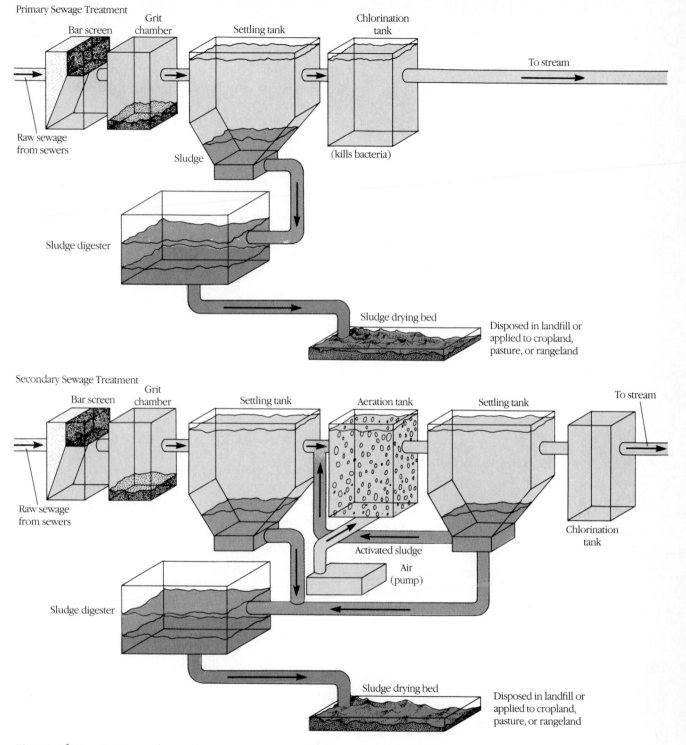

Figure 16-11. Primary and secondary sewage treatment facilities.

phates. Certain aquatic plants such as duckweed absorb dissolved organic materials directly from the water.

Aquatic plants grown in sewage ponds can be harvested and converted into food for humans or livestock. In Burma, Laos, and Thailand duckweed has been consumed by farmers for years. The protein yield of a duckweed pond is six times greater than that of an equivalent field of soybeans. One of the problems with this approach, however, is that water hyacinths and duckweed also absorb toxic metals from the water. Therefore, consumption by humans and livestock must be carefully monitored.

Figure 16-12. Trickling filter. A detritus food chain consisting of bacteria and other microorganisms in the rock or bark bed of the system consumes organic matter, nitrates, and phosphates in the liquid sewage.

Another suggestion is the age-old process of **land disposal**. In ancient times land disposal of human sewage was commonplace; it is still practiced in many developing nations, such as China and India. As the populations of developing countries grew and became more urbanized, this natural method of recycling wastes was gradually eliminated. Land disposal, discussed in Chapter 10 as a means of recharging groundwater, uses the surface vegetation, soil, and soil microorganisms as a natural filter for many potentially harmful chemicals. Sewage can be piped to pastures, fields, and forests (Figure 16-13). Organic matter in the effluent enriches the soil and improves its ability to retain water. Nitrates and phos-

phates are taken up by plants. The water supports plant growth and helps recharge aquifers. Crops nourished by effluents from treated sewage show a remarkable increase in yield.

Land disposal of sludge from sewage treatment plants has some problems. First, treated sewage may contain harmful bacteria, protozoans, and viruses that could adhere to plants consumed by humans or livestock or become airborne after the effluent dries. To get around this problem the Swiss and West Germans heat their sludge to destroy such organisms before applying it to pastures and cropland. Alternatively, sewage sludge can be decayed in compost piles before application. The heat given off during composting kills virtually all of the viruses, bacteria, and parasite eggs.

The second problem is that toxic metals found in some sewage may accumulate in soils and be taken up by plants and livestock. Metal-contaminated sewage usually comes from industries. By removing metals from their waste stream, an option called pretreatment, industries can eliminate the problem. Third, transporting sludge to fields increases the cost of sewage treatment, limiting some of the incentive to use this method. Experts are quick to point out, however, that land disposal is ten times cheaper than building and operating a tertiary treatment plant.

Scientists at the University of Maryland recently developed an unusual way to put sludge to good use. By combining it with clay and slate they formed odorless "biobricks," which look like ordinary bricks. Washington's Suburban Sanitary Commission recently put the idea to the test by building a 750-square-meter (8300-square-foot) maintenance building in Maryland with 20,000 biobricks. If successful, biobricks could help reduce land disturbance from mining materials for brick making and could help cut sewage disposal costs and environmental contamination.

Personal Solutions

Personal efforts can improve our waterways. Limiting family size and reducing the consumption of unnecessary goods can help cut down on water pollutants. Reducing the use of goods eliminates hazardous waste that sometimes seeps into our waterways. Another effective way to cut your personal contribution is to install a composting toilet. The composted wastes can be added safely to gardens, eliminating the need for synthetic fertilizer. When you become a homeowner, restrict your use of synthetic fertilizers, insecticides, herbicides, bleaches, detergents, disinfectants, and other household chemicals. Select low-phosphate or no-phosphate detergents for washing your clothes. You may also contact local and federal officials in support of further cleanup efforts.

Table 16-3	Removal of Pollutants by Sewage Treatment Plants	
	Percentage Removed by Treatment	
Substance	**Primary**	**Primary and Secondary**
Solids	60%	90%
Organic wastes	30	90
Phosphorus		30
Nitrates		50
Salts		5
Radioisotopes		0
Pesticides		0

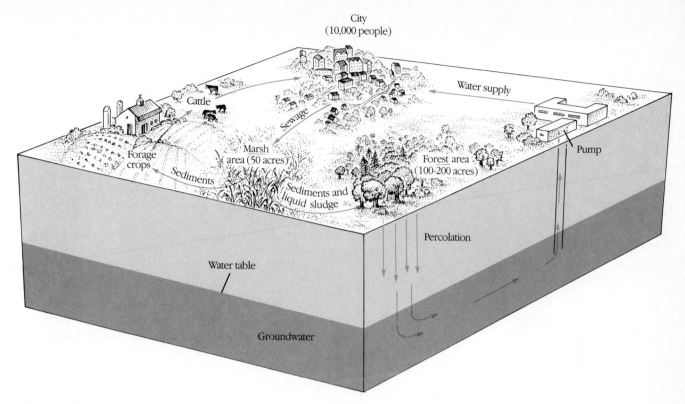

Figure 16-13. Land disposal of sewage helps fertilize farmlands and forests, reduce surface water pollution, and replenish groundwater supplies.

It is astonishing with how little wisdom mankind can be governed, when that little wisdom is its own.

W. R. INGE

Summary

Water pollutants come from **nonpoint** and **point** sources. The effects on aquatic systems depend, in large part, on whether polluted waters are standing or flowing. Standing systems (lakes and ponds) are generally more susceptible because of slow turnover.

The major water pollutants are organic nutrients, infectious agents, toxic organics, toxic inorganics, sediment, and heat. **Organic nutrients** come from feedlots, municipal sewage treatment plants, and industry. They serve as foodstuffs for aerobic bacteria and cause proliferation of natural populations of aquatic bacteria. Bacterial decomposition of these materials results in a drop in dissolved oxygen, with dire effects on other oxygen-requiring organisms.

Two **inorganic plant nutrients**, nitrogen and phosphorus, are also major anthropogenic pollutants. Inorganic fertilizer and laundry detergents are the leading sources in the United States. Both cause excessive plant growth and clogging of navigable waterways. Bacterial decay of plants in the fall results in a drop in dissolved oxygen, which may suffocate fish and other organisms.

Water may contain pathogenic bacteria, viruses, protozoans, and parasites. Waterborne infectious diseases are a problem in developing nations. Untreated or improperly treated sewage, animal wastes, meat-packing wastes, and some wild species are the major sources. While uncommon in the United States and other developed countries, infectious agents are not altogether absent.

Toxic organic pollutants include a large number of chemicals, many of which are nonbiodegradable or only slowly degraded, biologically magnified, and carcinogenic in humans.

Toxic inorganic pollutants include a wide range of chemicals such as metals and salts from an array of sources. Most states report that toxic metals, such as mercury and lead, are major pollutants. Mercury can be converted into methyl and dimethyl mercury by aerobic bacteria. These forms are more toxic than inorganic mercury. Methyl mercury is biologically magnified in the food chain.

Nitrates and nitrites are common inorganic water pollutants coming from septic tanks, barnyards, heavily fertilized crops, and sewage treatment plants. They are converted to nitrites in the intestines of humans. In the blood they combine with hemoglobin, reducing its oxygen-carrying capacity. At high levels they

Why We Have Failed

Barry Commoner

Barry Commoner is director of the Center for the Biology of Natural Systems, Queens College, City University of New York.

In 1970, in response to growing concern for the environment, the US Congress began a massive effort to control pollution (discussed in Chapters 14–18). The National Environmental Policy Act and the Environmental Protection Agency form the cornerstone of what is indisputably the world's most vigorous pollution control effort. Now, nearly 20 years later, it is time to ask an important and perhaps embarrassing question: how far have we progressed toward the goal of restoring environmental quality?

Apart from a few exceptions, environmental quality has improved only slightly, and in some cases worsened. Since 1975, for example, emissions of sulfur dioxide and carbon monoxide are down by about 19%, but nitrogen oxides are up about 4%. Overall improvement in major pollutants amounts to only about 15% to 20% and the rate of improvement has actually slowed considerably in recent years.

There are, of course, some notable exceptions. Pollution levels of a few chemicals, such as DDT and PCBs in wildlife and people, mercury levels in fish of the Great Lakes, and phosphate pollution in some local rivers, have been reduced by 70% or more. Levels of airborne lead have declined more than 90% since 1975.

The successes explain what works and what does not. Every success on the very short list of significant improvements reflects the same action: production of the pollutant has been stopped. DDT and PCBs levels, for example, have fallen because their production and use have been banned.

The lesson is plain: pollution prevention works; pollution control does not. Only where production technology has been changed to eliminate the pollutant has the environment been

substantially improved. Where the technology remains unchanged and where attempts are made to trap pollutants in an appended control device, such as the automobile's catalytic converter or a power plant's smokestack scrubber, environmental improvements are modest or nil.

Most of our environmental problems are the result of the sweeping technological changes that took place after World War II (see Chapter 2): the shift from fuel-efficient railroads to gas-guzzling automobiles and trucks, the substitution of fertilizers for manure and crop rotation, and the use of toxic synthetic pesticides to control weeds and insects.

These technological changes were the root cause of environmental pollution. Because environmental legislation ignored the origin of the assault on environmental quality, it has dealt only with subsequent symptoms. As a result, all environmental legislation mandates only palliative measures. The notion of preventing pollution has yet to be given any administrative force.

The goal established by the Clean Air Act in 1970 could have been met if the EPA had confronted the auto industry with a demand for fundamental changes in engine design, changes that were practical and possible. And had American farmers been required to reduce the high rate of nitrogen fertilizer they used, nitrate water pollution would be falling rather than increasing. If the railroads and mass transit were expanded, if the electric power system were decentralized and increasingly based on cogenerators and solar sources, if American homes were weatherized, fuel consumption and air pollution would be sharply reduced.

Of course, all of this is easier said than done. I am fully aware that what I am proposing requires sweeping changes in major systems of production undertaken for a social purpose: effective environmental improvement. This represents social (as contrasted with private) governance of the means of production—an idea that is so foreign to what passes as our national economic ideology that even to mention it violates a deep-seated taboo. A major consequence of this powerful taboo is our failure to reach the goals in environmental quality that motivated the legislation of the 1970s.

In the absence of a prevention policy, the EPA adopted a convoluted pollution control process. The EPA first must estimate the degree of harm caused by different levels of numerous pollutants. Next, they determine some level of "acceptable" harm. Polluters are then expected to respond by introducing control measures to bring emissions to the required levels. The net result is that an "acceptable" level of pollution is frozen in place.

This process, however, is the opposite of the preventative approach to public health, which seeks to reduce the risk to

◀ *Viewpoint*

health as much as possible—or eliminate it entirely. Infectious disease is a good example. The medical professions did not decide that the smallpox prevention program could stop when the risk reached one case in a million—the EPA's "acceptable" cancer risk for many pollutants. Instead, smallpox has been eliminated, worldwide.

The current process puts much of the burden of environmental risk on the poor (described in Chapter 14). One reason is that it relies on cost–benefit analysis and costs are often assessed by the number of lives lost and their value. Value, in turn, is determined by earning power. In this comparison the poor lose out. On such a risk–benefit scale, the poor can be exposed to more pollution than the rich. In fact, this is happening in the United States. Municipal incinerators and haz-ardous waste landfills are disproportionately sited near the poor who lack the political and financial clout to deter the risk. In this way, risk–benefit analysis conceals a profound, unresolved moral question: should poor people be subjected to a more severe environmental burden than rich people? Risk–benefit masquerades as science and relieves society of the duty to confront this moral question.

Instituting the practice of prevention rather than control will require the courage to challenge the taboo against questioning the dominance of private interests over the public interest. But I suggest that we begin with an open public discussion of what has gone wrong and why. This is the necessary first step on the road toward realizing the nation's unswerving goal: restoring the quality of the environment.

may be fatal to children under the age of three.

Salts used on roadways to melt snow and ice enter streams and lakes and can upset the ecological balance. Salt-sensitive trees along salted roads may also be killed as the salt water percolates down to the roots. Chlorine is used to kill bacteria in drinking water and treated sewage and to kill organisms that might clog pipes used to cool water in power plants and industry. Chlorine and chlorine by-products can severely affect aquatic organisms.

Sediment, the leading water pollutant in the United States, is a by-product of timber cutting, agriculture, ranching, mining, and construction of roads and buildings. Sediment destroys spawning and feeding grounds for fish, reduces fish and shellfish populations, destroys deep pools used for resting, smothers eggs and fry, fills in lakes and streams, and decreases light penetration, thus endangering aquatic plants.

Thermal pollution refers to the heating or cooling of water, both of which drastically alter biota in a body of water. Large quantities of heat can kill heat-sensitive organisms and harm organisms dependent on the aquatic ecosystem. Heat lowers the dissolved oxygen levels but increases the metabolic rate of organisms. Sudden changes in temperature bring about rapid death by **thermal shock**. Temperature changes can also interfere with migrations, spawning, and early development.

The concentration of many pollutants in groundwater is often higher than that in most contaminated surface water supplies. Many of these chemicals are tasteless and odorless at concentrations believed to pose a threat to human health. The major groundwater pollutants are chlorides, nitrates, heavy metals, and organics. Since groundwater usually moves slowly through an aquifer, it make take years for pollution to show up in nearby locations. Additionally, once an aquifer is contaminated, the pollutants may remain for centuries.

Oil pollution is another major problem. About half of the oil that contaminates the ocean comes from human sources: oilwell blowouts, tanker spills, and inland disposal of oil. Oil harms many organisms, especially if a spill occurs near an estuarine zone. After a spill it may take two to ten years for aquatic life to recover. Thanks to public outcry and stricter controls, the number of oil spills has decreased substantially, although the problem is far from solved.

Plastic pollution has also become a major problem throughout the world. Plastic nets, plastic garbage, and plastic medical wastes are killing countless marine mammals, turtles, and fish. Animals may become tangled in the plastic debris or may eat it and die. Because of public outcry, many governments have banned the dumping of plastics in oceans.

Most control efforts in the United States and abroad have sought to reduce point-source pollution. Sewage treatment plants and limitations on factory discharges have been key strategies of the US **Clean Water Act**. Sewage treatment can take place in three stages. **Primary treatment** removes large objects by passing the sewage through a series of grates and screens. Solid organic matter, or sludge, settles out in a primary settling tank. **Secondary treatment** removes biodegradable organic material and some inorganic substances. Sludge may be decomposed by bacteria in large aerated tanks or may be passed through a trickling filter, where it is dripped over a bed of rocks or bark housing a detritus food chain. In most municipalities liquid remaining after secondary treatment is chlorinated to kill potentially pathogenic organisms and released into receiving streams. Many methods exist for **tertiary treatment**, the final cleanup stage, but most are expensive and therefore rarely used. Fortunately, there are some cheaper options, such as algae and water hyacinth ponds and land disposal.

Progress in cleaning up US waters has been slow. Continued population growth, industrial growth, and agricultural expansion have negated many gains. Nonpoint sources often offset gains at point sources, making it imperative that further controls be directed at them.

Discussion Questions

1. List the major types of water pollutants found in developing and developed nations.

2. Define the terms *point source* and *nonpoint source*. Give some examples of each, and explain why nonpoint sources of water pollution are often more difficult to control than point sources.

3. What are the three major ecological zones of a lake? Describe each one.

4. Explain why a lake "turns over" in the spring and fall and how this natural turnover benefits aquatic organisms.

5. Describe where organic nutrients come from, what effects they have on aquatic ecosystems, and how they can be controlled.

6. What are the inorganic plant nutrients, and how do they affect the aquatic environment?

7. Define the term *eutrophication*. Describe how inorganic and organic nutrients accelerate natural succession of a lake.

8. What are the major sources of infectious agents in polluted water? How can they be controlled?

9. What dangers do synthetic organic water pollutants pose?

10. What are some of the major inorganic water pollutants? How do they affect the aquatic environment and human populations?

11. Why is chlorine used in the treatment of human sewage and drinking water? What dangers does its use pose? How would you determine if the risks of chlorine use outweigh the benefits?

12. What are the major sources of sediment, and how can they be controlled? What are the costs and benefits of sediment control?

13. Describe ways to eliminate or reduce thermal pollution.

14. If you were going to survey your state for groundwater pollution, how would you go about locating sources and taking samples of groundwater?

15. What are the major sources of groundwater pollution? What factors determine whether a toxic waste disposal site or some other source will pollute groundwater? How can groundwater pollution be reduced or eliminated? What can you do?

16. Discuss the sources of oil in the ocean and ways to reduce oil contamination.

17. Even though thousands of new wastewater treatment plants have been constructed in the United States, water quality in many areas has not improved. Why?

18. Describe primary and secondary wastewater treatment. What happens at each stage, and what pollutants are removed?

19. Debate the pros and cons of land disposal of wastewater.

Suggested Readings

Belton, T., Roundy, R., and Weinstein, N. (1986). Urban Fisherman: Managing the Risk of Toxic Exposure. *Environment* 28 (9): 18–20, 30–37. Detailed look at ways to communicate the risk to the public.

Borelli, P. (1989). Troubled Waters. Alaska's Rude Awakening to the Price of Oil Development. *The Amicus Journal* 11 (3): 10–20. Excellent overview of the Alaskan oil spill.

Bowker, M. (1986). Caught in a Plastic Trap. *International Wildlife* 16 (3): 22–23. Excellent discussion of the impacts on wildlife of plastics discarded in the ocean.

Maurits la Rivière, J. W. (1989). Threats to the World's Water. *Scientific American* 261 (3): 80–94. Good up-to-date survey of global pollution problems.

Peskin, H. M. (1986). Cropland Sources of Water Pollution. *Environment* 28 (4): 30–34, 44. Important article on nonpoint water pollution control.

Robertson, G. P. (1986). Nitrogen: Regional Contributions to the Global Cycle. *Environment* 28 (10): 16–20, 29. Detailed report on nitrogen pollution.

Schmidt, S. (1989). The Oil Doctors. *The Amicus Journal* 11 (3): 21–23. Interesting view of ways to combat oil pollution.

Weber, M., Bierce, R., Atkins, N., McManus, R. E., Roet, E., and Yolen, N. (1985). *The 1985 Citizen's Guide to the Ocean*. Washington, DC: Potomac Publishing. Wonderful overview of the major environmental issues pertaining to the ocean.

World Resources Institute and International Institute for Environment and Development. (1988). *World Resources 1988–89*. New York: Basic Books. Chapter 9 contains an excellent survey of problems facing the oceans.

Pesticides:
A Double-Edged Sword

*What we do for ourselves dies with us. What we do for
others and the world remains and is immortal.*

ALBERT PINE

"What is a weed?" Ralph Waldo Emerson asked. "A plant
whose virtues have not yet been discovered." Most farmers would scoff at such a remark. To them, weeds—and
other pests—are an impediment to efficient farming.
They hold down production and cut into profits. To them,
pests are pests.

Each year weeds, insects, bacteria, fungi, viruses, birds,
rodents, mammals, and other pests consume or destroy
an estimated 48% of the world's food production (this
includes both pre- and post-harvest losses). The highest
rate of destruction is in the tropics and subtropics, where
as many as three crops are grown each year on the same
field. Pest destruction is high even in the developed
nations despite elaborate control strategies costing many
billions of dollars a year. In the United States, for instance,
preharvest losses are estimated to be about 37%, and
postharvest losses amount to 9% of what is left (Figure
17-1). Together, about two-fifths of annual production is
lost to various pests, or about $64 billion worth of food
for humans and livestock.

Similar losses are reported in other developed countries. For example, wheat production in Saskatchewan has
been reduced in some years by three-fourths as a result
of wireworm infestation. In a world where about 20% of
the people are hungry, such losses are tragic. Reducing
losses through pest control can help increase world food
supplies. This is, however, a task that must be embarked
on judiciously, bearing in mind the ecological lessons we
have learned.

Figure 17-1. Prominent pests in the United States and the areas where they are found.

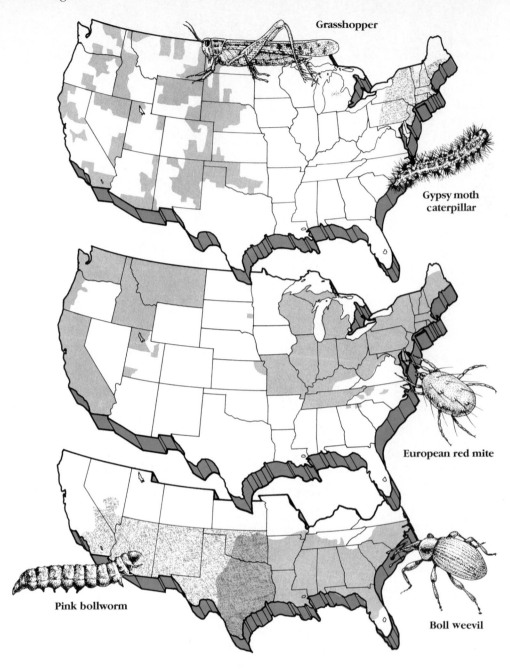

A Historical Overview

Pest control measures have been used throughout the centuries. In China over 3000 years ago, for example, farmers controlled locusts by burning infested fields. In the ancient Middle East, open ditches were used to trap immature locusts. Arsenic poisons were used as pesticides in AD 900, and in 1182 Chinese citizens were required to collect and kill locusts in an effort to control an outbreak. Techniques similar to these have been in use since the beginning of agriculture in many other parts of the world.

Development of Chemical Pesticides

In recent years chemical pesticides (biocides) have come to form the cornerstone of pest management. **Pesticides** are chemicals used to kill troublesome pests, including weeds, rodents, and insects. Early pesticides, known as **first generation pesticides**, were simple preparations made of ashes, sulfur, arsenic compounds, ground tobacco, or hydrogen cyanide. Lead, zinc, and mercury compounds were also used. Today few of these remain in use, because they are toxic or relatively ineffective. Nevertheless, many of the compounds persist in the soil 50 years later.

In 1939 the Swiss chemist Paul Müller discovered the insecticidal properties of a synthetic organic compound called DDT (*d*ichloro*d*iphenyl*t*richloroethane). This was the first in a long line of **second-generation pesticides**, synthetic organic compounds that kill a variety of pests. For the 25 years following its development DDT was viewed by many as the savior of humankind. It quickly removed insects and increased crop yield. Being relatively inexpensive to produce, it spread to all corners of the world. In 1944 Müller was awarded a Nobel Prize.

Over the years thousands of new chemicals have been synthesized and tested. Today 1500 different substances are used in over 33,000 commercial formulations of herbicides, insecticides, fungicides, miticides, and rodenticides, with one goal in mind: to reduce pests to tolerable levels (Figure 17-2). Some pesticides, like DDT, are **broad-spectrum** biocides, which attack a wide variety of organisms; others are **narrow-spectrum** pesticides, used in controlling a few pests.

Most chemical insecticides fall into three chemical families: chlorinated hydrocarbons (organochlorines), organic phosphates (organophosphates), and carbamates. The **chlorinated hydrocarbons** are a high-risk group including DDT, aldrin, kepone, dieldrin, chlordane, heptachlor, endrin, mirex, toxaphene, and lindane. All of these have been banned or drastically restricted or are being considered for such actions because of their ability to cause cancer, birth defects, neurological disorders, and damage to wildlife and the environment. Generally, the chlorinated hydrocarbons are extremely resistant to breakdown, persist in the environment, are passed up the food chain, and remain for long periods in body fat.

The second group, the **organic phosphates**, consists of chemicals such as malathion and parathion. These break down more rapidly than chlorinated hydrocarbons, are water soluble, and can be excreted in the urine. Because they are water soluble they are less likely to bioaccumulate. However, most are highly toxic. Humans exposed to even low levels may suffer from drowsiness, confusion, cramps, diarrhea, vomiting, headaches, and breathing difficulty. Higher levels cause severe convulsions, paralysis, tremors, coma, and death.

The third group, the **carbamates**, are widely used today as insecticides, herbicides, and fungicides. One of the most common is carbaryl (commonly known as Sevin). As a group the carbamates are also less persistent than the chlorinated hydrocarbons, remaining a few days to two weeks after application. But like the organochlorines and organophosphates, they are nerve poisons that have been shown to cause birth defects and genetic damage.

Approximately 2.5 million metric tons of chemical pesticides are used annually throughout the world, approximately 34% in North America, and about 45% in Europe and other developed countries. The remaining 21% is used primarily in developing countries. About one-fifth

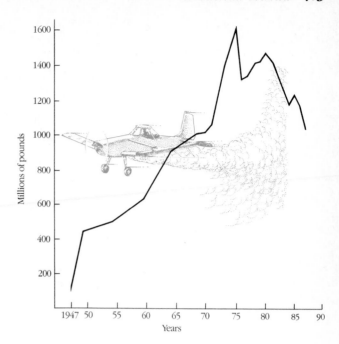

Figure 17-2. Increasing pesticide production in the United States.

of all pesticides are applied to US crops each year. Of these pesticides 25% are insecticides for insect control, 60% are herbicides for weed control, and the rest are mostly fungicides used to reduce fungal growth. According to the University of Illinois entomologist Robert Metcalf, author of a standard college textbook on pest management, farmers apply more than twice the pesticide they need. Adding to the unnecessary environmental contamination are a cadre of misinformed homeowners who apply about one-tenth of all US chemical pesticides on gardens, lawns, and trees in higher amounts per acre than farmers do.

Pesticides constitute only about 3% of the commonly used commercial chemicals in the United States each year. Nonetheless, because they are released into the environment in large quantities and have the potential to alter ecosystem balance and threaten human health, their use has created widespread and often heated controversy.

Exploration, Exploitation, and Reflection

Pesticide use has progressed through three developmental stages: (1) exploratory, (2) exploitive, and (3) reflective. These periods illustrate a common progression of human behavior and attitude that follows the introduction of many chemicals and technologies.

Exploration During the exploratory stage, starting in the 1940s, DDT and other new pesticides were applied to a variety of crops, with astonishing results. DDT was even used to delouse soldiers and civilians in World War II.

Figure 17-3. Pesticide resistance. A tobacco budworm crawls through deadly DDT unaffected.

DDT and other new chemical preparations proved to be fast and efficient in controlling insects, weeds, and other pests. In India, for example, before the use of DDT in the 1950s to control malaria-carrying mosquitoes, there were over 100 million cases of malaria a year. By 1961 the annual incidence had been reduced to 50,000. Pesticides also allowed farmers to respond quickly to outbreaks, often avoiding economic disaster. Second-generation pesticides were cheap and relatively easy to apply, and their use resulted in substantial financial gains as yields increased. Some insecticides, such as DDT and dieldrin, persisted long after application, giving extended protection.

Exploitation The successes of the explorative phase led to the **exploitive phase**, during which pesticide production and use expanded considerably. Researchers developed many new pesticides. Although some steps were taken to protect farm workers from exposure and consumers from toxic residues remaining on vegetables and grains, they often proved inadequate.

In the rush to expand pesticide use, however, many problems arose. First, many pesticides, especially broad-spectrum chemicals, often killed predatory and parasitic insects that help control pests and potential pests. Thus, insects that had not been pests suddenly increased in number, creating a need for additional pesticides. Agronomists call this population explosion of new pests an **upset**.

A classic example of such an upset has taken place in California. Spider mites, once only a minor crop pest, have become a major pest because pesticides killed off many of their natural enemies, which were more sensitive to the sprays. Today the mites cause twice as much dam-

age as any other insect pest in California and cost farmers (in damage and control) five times what they cost 20 years ago. Two of modern farming's most costly pests, the cotton bollworm and corn-root worm, were minor problems 50 years ago before widespread pesticide use reduced their natural predators.

Pesticides also destroy beneficial insects, such as honeybees, which play an important role in pollination. Honeybees pollinate crops annually worth about $20 billion. Apple orchards are particularly hard hit. Over 400,000 bee colonies are destroyed or severely damaged in the United States each year by pesticides.

Another unanticipated effect of pesticide use was a dramatic increase in genetically resistant insects. Because of genetic diversity (see Chapter 2), a small portion of any insect population (roughly 5%) is genetically resistant to the pesticides and is not killed by a normal application (Figure 17-3). Therefore, even though pesticides initially reduce pests to a level at which they do little damage, the surviving resistant insects reproduce and eventually form a new population that can be killed only by large doses or new pesticides.

Genetic resistance to DDT was first reported in 1947 by Italian researchers. Today, according to some estimates, well over 440 insect pests are resistant to DDT and other insecticides. Twenty of the worst pests are now resistant to most types of insecticide. This unexpected development called for new strategies. The first was to increase the amount of pesticide. When DDT and other insecticides were introduced in Central America, cotton fields were sprayed eight times each growing season; today, 30 to 40 applications are necessary. The second strategy was the development of new pesticides. However, scientists found that it was expensive to create pes-

Case Study 17-1

Are We Poisoning Our Children with Pesticides?

In the Spring of 1989, the Natural Resources Defense Council (NRDC), one of the nation's leading environmental groups, announced the results of a two-year health study on children and pesticides. NRDC's researchers concluded that children are exposed to dangerous levels of pesticides in fruits and vegetables. According to the study, some 5500 to 6200 children alive today will develop cancer in their lifetimes from just eight of the many pesticides they are exposed to in their preschool years.

Moreover, the NRDC charged the EPA, the federal agency responsible for establishing safe pesticide exposure levels, with routinely neglecting children in setting their standards. The EPA attempted to refute the charges, but the facts show otherwise.

Children face a higher risk from pesticides because they eat more fruit than adults. Several years ago, the EPA began taking into account the higher level of fruit consumption in children. Unfortunately, say critics, the agency used 1977 data, which were out of date and led to erroneous estimates of risk. A 1985 survey by the US Department of Agriculture, for example, shows that fruit consumption in preschoolers has increased 30% since 1977. Clearly, say the NRDC researchers, the EPA should have used the more recent data.

The EPA has also failed to ensure the public health on another count. Safety levels for 300 pesticides now in use were set in the 1950s and 1960s, before the EPA was established. The agency should have gone back and lowered existing tolerance levels based on the new consumption data, but it didn't. Although the EPA is in the process of resetting standards now, the task will not be complete until 1998.

Approximately one-third of a preschooler's diet consists of fruit. Fruit constitutes about one-fifth of a preschooler's mother's diet. Safety standards must reflect this difference to protect our children. But children are at risk for other reasons. Some research suggests that children are more sensitive to chemical exposure than adults. Lead studies clearly show this to be true (see Chapter Supplement 14-1). Rapidly dividing cells are prime targets for carcinogens (see Chapter 14). Furthermore, enzymes needed to detoxify chemicals may not have fully developed in children. Exposure early on also carries a greater cancer risk than exposure later in life because early exposure ensures a longer lead time to develop cancer.

The main culprit in pesticide-caused cancer, the NRDC charges, is a chemical called Alar. Alar is not a pesticide, as such, but rather a growth regulator that delays ripening so apples do not fall off the tree prematurely. Alar also promotes the reddening of apples, and delays overripening so that apples stay fresher in storage. Alar penetrates the flesh of apples and cannot be washed off. By NRDC estimates, 86% to 96% of the total pesticide cancer risk comes from Alar and its breakdown product, UDMH.

The EPA's safety standards seek to limit cancer risk to one in one million. But UDMH poses a cancer risk of 1 in 4200— 240 times greater than the routinely accepted standard, says the NRDC. The EPA's own calculations, made after NRDC published its report, show that Alar poses a higher than acceptable level of risk and, as a result, the agency has moved to ban it from domestic use.

Several researchers who reviewed the NRDC study think that the NRDC report actually underestimates the cancer risk from pesticide residues, for several reasons. First, there are at least 500 pesticides in common use that can leave residues on fruit and vegetables. The NRDC only looked at 27 pesticides and only calculated the risk for eight of them. Second, the NRDC research team omitted several foods, such as milk, which are an important component of children's diets. Third, the NRDC study only looked at the effects of childhood exposure, ignoring the continued exposure to pesticide residues in the adult diet.

Critical thinking suggests that the issue is not as simple as it may seem. Dr. Bruce Ames of the University of California at Berkeley thinks that Alar poses a much lower threat than some naturally occurring chemicals found in some foods. The ban on Alar, he argues, may require orchard owners to increase pesticide use. An insect called the leafminer, for instance, causes apples to fall prematurely. To control them, apple growers may turn to more pesticides, increasing human exposure to yet another potentially carcinogenic substance. Molds may increase in apples on trees and in storage, says Ames, because apples are less firm and more susceptible. Naturally occurring mold toxins could increase, exposing people to a greater danger than Alar itself. Finally, Ames points out that regulatory agencies choose safety limits for low-level human exposure based on high-level animal exposure, but the former are really quite speculative.

One of the chief problems in this debate is that the study of human cancers based on animal studies is an imperfect science. But risk management is a political, not a scientific process. Science only provides the raw material. Given their druthers, most people would err on the conservative side, especially when it comes to their children's health. Pesticides and other additives are in our foods without our choice, and that angers many people. Most people do not know whether Alar causes cancer, but they know that a few experts think that it may. That's enough for them. They would prefer not to take a gamble on their children's future.

Edward Groth III of the Consumer's Union summed it up best: "We must teach people to see risks in perspective. At the same time, we (scientists and public policy makers) must listen to what people say about risks. It is not the size of the risk but its moral offensiveness that makes the public respond so strongly."

Rain 0.1-0.3 ppb

Tradewinds 0.1-0.3 ppb

Fat of man
6-12 ppm

Rivers and lakes
0.001-0.2 ppb

Groundwater
0.001-0.2 ppb

Fat of cows 0.5 ppm

Figure 17-4. Pesticide sprayed from planes contaminates the ecosystem because much of the pesticide drifts away. Various avenues for the dispersal of pesticides are shown. Average values for DDT concentrations are indicated in parts per million and billion.

ticides and that the insects developed genetic resistance to these new chemicals almost as quickly as they were developed. Even weeds and plant pathogens can develop a resistance to pesticides, although not so quickly as insects.

Another problem was that many of the second-generation pesticides proved to be hazardous environmental contaminants. These synthetic pesticides persist in the ecosystem because microconsumers (bacteria) lack the enzymes to degrade them. Increased spraying resulted in widespread contamination of the ecosystem (Figure 17-4). Trout in upstate New York showed elevated levels of DDT and its chief breakdown product, DDE (dichlorodiphenyldichloroethylene), because of spraying in nearby forests. These toxins did not affect the adults but reduced survival among newly hatched fry. Insect- and worm-eating birds also perished in areas where aerial spraying of insecticide had occurred. As a result of widespread pesticide use, populations of many birds plummeted.

The manufacturers of pesticides argued that the chemicals were found in only minute concentrations in the environment and could not be the cause of declining populations of fish and wildlife. But numerous experimental studies showed that certain persistent pesticides—even when present in small amounts in the environment—could drastically affect the reproduction and survival rate of birds and other animals.

One of the most important findings was that although DDT and DDE levels in aquatic ecosystems were quite low, concentrations were higher in producers and still higher in consumers, a phenomenon called biomagnification (Chapter 14). Fish-eating birds, the consumers at the top trophic level, had the highest concentrations of DDT and DDE. Although these levels were not lethal to adults, they impaired reproduction. In birds that feed on fish and other birds, such as peregrine falcons, brown pelicans, cormorants, bald eagles, gulls, and ospreys, DDE and DDT reduced the deposition of calcium in eggshells. Of the two compounds, biologists discovered that DDE posed the greater physiological threat to birds and was persistent in the environment. Reduced eggshell calcium levels create a thinner, more fragile shell that cracks easily during incubation. As a result of widespread DDT contamination many predatory populations were nearly wiped out. (For a more detailed discussion see Chapter

Figure 17-5. Crop duster spraying potato fields. One-half to three-fourths of the pesticide is carried off by wind, landing on nearby fields and homes.

8.) In one study of bald eagle reproduction James Grier, a zoologist at North Dakota State University, showed that the number of young per nest in northwestern Ontario declined by about 70% between 1966 and 1974.

Other work soon showed the presence of DDT in fish, beef, and other foods. DDT also appeared in the fatty tissues of seals and Eskimos in the Arctic, far from its point of use, indicating that it was traveling in the atmosphere to remote parts of the globe, being washed from the sky by rain, and passing through the food chain. DDT has also been detected in human breast milk, a discovery that caused considerable alarm, although the long-term effects of low levels on humans remain unknown.

Farm and chemical workers suffered from direct exposure to pesticides on the job as well, indicating widespread misuse. Workers pick up pesticide on their clothing and skin through accidents, negligence, or prematurely entering sprayed fields. Symptoms of poisoning include insomnia, nausea, loss of sex drive, reduced powers of concentration, irritability, nervous disorders, and, in severe cases, death. At least 45,000 US workers are seriously poisoned each year, but many experts believe that this figure grossly underestimates the number of serious poisonings. Surveys in California, for instance, show that three-fourths of all serious poisonings go unreported. Each year 200 to 1000 people die from pesticide poisoning in the United States. Worldwide, there are at least 500,000 (perhaps as many as two million) pesticide poisonings annually, according to some experts. These result in at least 5000 to 10,000 deaths and numerous chronic and fatal illnesses.

Although farm and chemical workers are the groups most heavily exposed to pesticides, residents of rural and even suburban areas are often exposed to high levels if they live near agricultural lands. Families living near fields sprayed with herbicides and pesticides outside Scottsdale, Arizona, for example, suffered from persistent headaches, cramps, skin rashes, dizziness, high blood pressure, chest pains, persistent coughs, internal bleeding, and leukemia. People living near sprayed fields are so heavily exposed because one-half to three-fourths of the sprayed material never reaches the ground but is carried away by light winds (Figure 17-5). Health officials are worried the most by possible long-term problems from exposure.

In their zeal to control mosquitoes, many cities in the southern United States routinely spray insecticides in neighborhoods and nearby breeding areas. In Florida various mosquito control agencies argue that controlling the insect is vital to real estate interests and tourism, because it minimizes the risk of encephalitis, a disease carried by mosquitoes. The price citizens pay to have trucks and aircraft spewing out mists of pesticides while they sleep is unknown. Florida wildlife officials, however, think that pesticide use is largely responsible for a 70% decline in the population of snook, a popular sport fish. Adding to the problem, city officials use a variety of pesticides in city parks, and lawn-care companies and individuals douse lawns and trees with a variety of toxic substances, often incorrectly and without warning neighbors.

Consumers may also become the victims of pesticide poisonings. In the summer of 1985, for instance, 1400 people on the West Coast were stricken with nausea, diarrhea, vomiting, and blurred vision after eating watermelons contaminated with the pesticide aldicarb. The EPA permits aldicarb for use on cotton and food crops that

◀ *Viewpoint*

The Myth of the "Banned" Pesticides

Lewis Regenstein

The author wrote How To Survive in America the Poisoned, *published by Acropolis Books. A writer and conservationist in Atlanta, Georgia, he is past president of the Monitor Consortium, a coalition of 35 national environmental and animal protection groups. He is also the author of* The Politics of Extinction: The Story of the World's Endangered Wildlife *(Macmillan, 1975).*

Today, almost three decades after the publication in 1962 of Rachel Carson's epic work, *Silent Spring*, almost all of the toxic pesticides she discussed are still in widespread use. The few that have been restricted have often been replaced by equally or more hazardous compounds. And a thousand new and largely untested chemicals are introduced in America each year. Our society is being overwhelmed by deadly chemicals, which are affecting the lives and health of millions of Americans, and future generations as well.

Over 100 of the pesticides in general use today are thought to cause cancer. The EPA estimates that about one-third of the active ingredients used in pesticides are toxic, and that one-fourth are carcinogenic (capable of causing cancer).

Since its beginning the EPA has effectively banned the domestic use of only a handful of these by canceling their registration. (In addition, an unknown number of pesticides have either not been granted registration or been withdrawn voluntarily by the manufacturer.)

One often comes across erroneous references to certain highly toxic pesticides, such as DDT, as having been "banned." In fact, the widely-publicized 1972 "ban" on DDT only outlawed *most* uses of the chemical, but permitted several other uses to continue. And unknown to the public, the pesticide dicofol, which contained up to 15% DDT, was still in wide-

spread use a decade and a half after the supposed ban on DDT.

The EPA has refused to ban dicofol (sold under the names Kelthane, Acarin, and Mitigan), instead requiring only that DDT levels be substantially reduced. In 1984, DDT was the most commonly detected pesticide on fresh produce on supermarket shelves in California, where dicofol is widely used; and DDT levels have remained high in wildlife there and in Texas and other states where dicofol is used on cotton and citrus crops.

The sad fact is that after many years of efforts by scientists, conservationists, and some government officials, very few restrictions have been placed on pesticides. Despite the overwhelming evidence that many pesticides cause cancer and are extremely damaging to humans and the environment, almost none of these chemicals has ever been "banned" by the government in the true sense of that word. In the very few cases where pesticides have been the subject of suspensions, cancellation proceedings, or court actions, the results have usually been restrictions or bans placed on some or most uses while other applications are allowed to continue.

Even in the handful of cases where all domestic use has been prohibited (such as with kepone), manufacture for export can still be (and often is) legally undertaken. When production for export continues, it inevitably results in exposure of workers and the public through pollution, dumping of wastes, and other accidental and intentional releases of the substance, as well as through foodstuffs imported from foreign countries to which the chemical is exported.

The way the EPA has carried out its policy on "banning" toxic pesticides is contradictory and nonsensical. In numerous instances the agency has proclaimed a chemical hazardous to humans and the environment and prohibited major uses of the compound. At the same time, it has allowed other uses to continue as before, often depending on label instructions to ensure adherence to the prohibitions. Several cancer-causing chemicals that have been restricted but not banned are aldrin, dieldrin, chlordane, heptachlor, DDT, 2,4,5-T, mirex, and DBCP. In these and other instances, the EPA action has often been presented to the public by the news media as another pesticide "ban."

In canceling the registrations for major uses of these most obviously dangerous pesticides, the EPA in many cases allowed them to be slowly phased out or permitted the sale and distribution of existing stocks. This made it possible for large quantities to continue to be sold and used long after they were "banned" and even for users to stockpile them for future use.

The implications of this EPA policy should be obvious. It is impossible to know what use will be made of these products. Indeed, EPA end-use restrictions are considered a joke within the agricultural community, and many farmers use certain chemicals as they always have, regardless of whether such applications have been banned.

Moreover, there is no way to determine how and where such carefully restricted and regulated chemicals are being disposed of. Common sense would dictate that if a chemical is so dangerous that it has been declared a threat to humans and the environment, it should simply be banned: its manufacture, sale, transport, use, or possession prohibited.

Pesticides that have been restricted are listed in the EPA publication "Suspended and Cancelled Pesticides," which gives the conditions, if any, under which these substances may be used. The rules and regulations outlined in this largely incomprehensible document are so complex and unfathomable that their only real use and value are to lawyers and bureaucrats interested in the byzantine-world of federal regulations. It is difficult to believe that farmers and agricultural workers, particularly illiterate or Spanish-speaking migrants, could understand the meaning of these rules or the labels on pesticides.

What happens in real life is that a migrant worker is given a container of pesticide and told to spray it on the crops, and that's what he does.

Label requirements for many potentially toxic chemicals, even when clearly written, are equally useless. For example,

the brochure requires that metaldehyde, used to kill slugs and snails, "must have the following statement on the front panel of the product label: 'This pesticide may be fatal to children and dogs or other pets if eaten. Keep children and pets out of treated areas.' " But such a warning in no way prevents this extremely toxic chemical from being widely used throughout the nation in precisely the way the label cautions against. US government agencies, including those oriented toward the environment, pay little attention to the labels. In recent years, the National Park Service has used metaldehyde on the White House lawn and the public grounds around and between the White House and the Capitol. Such use in these areas potentially endangers not only the president and his family but also countless thousands of tourists who flock to the Ellipse, the Mall, and other areas around the White House.

The EPA's labeling and restricted-use policies, unintelligible and contradictory as they may seem, do serve several purposes. They keep the public reassured that it is being protected from "banned" and "restricted" pesticides while allowing chemical and agribusiness interests to carry on business as usual with many of these products. It is, in fact, an ideal arrangement: it keeps the bureaucrats, the politicians, and industry happy, even if the public gets poisoned in the process.

are cooked before consumption, such as beans and potatoes. This incident points out how difficult it is for government officials to regulate the way farmers use pesticides.

Herbicides have also drawn a considerable amount of attention in recent years. Their uses in times of peace and war are discussed later in this chapter.

In addition to the ecological and health concerns discussed above, many experts have begun to question the efficacy of pesticide use. Despite the increased application of chemical pesticides, annual losses due to pests have continued to climb. In the first 30 years after the introduction of second-generation insecticides, use increased tenfold, and insect damage doubled.

Reflection As a result of growing concern over the biological and ecological effects of pesticides and growing skepticism regarding their effectiveness, industrial societies have entered the reflective stage, a period of caution. Gone are the days of unbridled enthusiasm for pesticide use.

Caution regarding pesticide use has grown enormously since the publication of the late Rachel Carson's book *Silent Spring*, which pointed out many of the real and potential impacts of pesticide use. Our understanding of

pesticide impacts continues to grow, casting further doubt on conventional pest-control practices. Quite recently researchers found that herbicides penetrate the soil up to ten times deeper than laboratory tests had indicated. Such findings explain why groundwater is often polluted with herbicides. Researchers are finding that bans on pesticides have benefited wildlife. The endangered bald eagle appears to be on the upswing. Recent studies show that DDE and DDT levels have dropped in wild populations and that normal reproductive rates have returned. Researchers caution, however, that domestic pesticide bans are only part of the answer. Continued use of pesticides outside of the United States poisons migratory species, such as songbirds. Much of the produce imported into the United States has recently been found to be contaminated with pesticides, many of which were banned in the US. (The reasons for this are explained in Lewis Regenstein's Viewpoint on page 408.) One-fourth of the fruit and vegetables sold in this country come from foreign soil. Global bans of some harmful pesticides are needed to protect the biosphere.

Pesticides are indispensable for agriculture and will remain in use, despite their many drawbacks. However, many experts believe that pesticides will play a much smaller role in agriculture in years to come as other, less harmful, controls are developed and integrated into pest management strategies. For example, the red spider mite can be kept under control in apple orchards by insecticides applied early in the season, well before the mite's natural predators emerge. Throughout the rest of the season farmers hold back on pesticide, letting the natural predators do their work. Using a similar approach on cotton, researchers at Texas A&M have cut pesticide use by 70% while maintaining normal cotton crop production.

Integrated Pest Management

Integrated pest management calls for the combined use of four basic means of pest control: environmental, genetic, chemical, and cultural. But these methods require adequate education for practitioners, and increased monitoring of pests. When combined, these techniques can provide an effective, long-lasting and environmentally sound way of controlling pest damage.

Education and Monitoring

An insect pest appears in a farmer's field. Without really thinking much, the farmer hooks up the sprayer and heads to the field to douse it with insecticide. That's the typical scenario. With a little imagination and thought, however, farmers could find ways to reduce pest damage without chemicals or with much less. But some think that training farmers to try other approaches would require a massive educational effort. High schools, universities, university extension services, and even agricultural magazines could all play a part in retraining farmers to think about pest management in a new light. As it is now, farmers are bombarded with advice on pest control from the chemical manufacturers and their sales representatives. In some universities, much of the research on pest control is sponsored by pesticide manufacturers, and that biases the system toward this approach.

Farmers need education in other areas as well to move to integrated pest management. A better understanding of insect biology, better skills in recognizing insects, and improvements in monitoring insect populations can help farmers become better stewards of the land. Farmers especially need training in methods to monitor insect populations, so they know when they really have a problem and don't overreact to a stray pest or two.

Education and monitoring are prerequisites to integrated pest management. Without them, the heavy dependency on chemical pesticides is bound to continue.

Environmental Controls

Environmental control methods are designed to alter the biotic and abiotic environment, making it inhospitable to the pest. Because they generally rely on knowledge more than technology, these practices are especially suitable for poor nations. Still, they can be equally effective if used properly in modern agricultural societies.

Increasing Crop Diversity In Chapter 7 we saw that monocultures generally promote the proliferation of insects and disease organisms. Crop diversity, on the other hand, reduces the amount of food available to any one pest and helps prevent such rapid population growth. Two basic techniques increase crop diversity: heteroculture and crop rotation.

The planting of several crops side by side is called **heteroculture**. It works because it provides environmental resistance so the biotic potential of pests cannot be reached. Pest populations are often much smaller in heterocultures than in monocultures. For example, intercropping corn and peanuts can reduce corn borers by as much as 80%. Part of the reason for this success may be that peanuts harbor predatory insects that feed on the corn borer.

Crop rotation increases soil fertility and can help reduce erosion, as discussed in Chapter Supplement 7-1, but it also helps hold down pest populations. For instance, wireworms feed on potatoes but not alfalfa. Therefore, if potatoes and alfalfa are alternated from year to year, food becomes a limiting factor that holds the wireworm population in check.

Altering the Time of Planting Some plants naturally escape insect pests by sprouting early or late in the growing season. A good example of this adaptation is the wild radish, which sprouts early in the season before the emergence of the troublesome cabbage maggot fly.

Agriculturalists can use their knowledge of an insect's life cycle to their advantage by coordinating plantings with the expected date of hatching. Delayed planting of wheat, for example, helps protect this crop against the destructive Hessian fly. In general, if a pest emerges early in the spring, planting can be delayed to avoid that pest. Without food the pest will perish. If the pest emerges late in the growing season, an early planting may prove effective.

Altering Plant and Soil Nutrients The levels of certain nutrients in soil and plants can also affect pest population size. Thus, by regulating soil nutrients a farmer may be able to control pests.

Nitrogen is one of the important nutrients that insects and parasites derive from plants. Too much or too little of this key element can alter the population size of various pests. For example, grain aphids reproduce better on grain high in nitrogen. Other insects, such as the greenhouse thrip and mites, do poorly on high-nitrogen spinach and tomatoes, respectively. Therefore, knowledge of pest nutrient requirements, soil nutrient levels, and plant nutrient levels can be helpful in controlling pests. Plants rich or poor in nitrogen can be selected to control pests as long as the level of nitrogen is adequate for human consumption.

Controlling Adjacent Crops and Weeds

Adjacent crops and weeds may provide food and habitat for pests, especially insects. In some cases plants adjacent to valuable food crops harbor viruses that can infest pest species and later be transmitted to crops. Thus, elimination of adjacent crops and weeds can prove helpful in the control of insects and other pests.

Sometimes adjacent low-value crops (trap crops) attract pests away from more valuable crops. Alfalfa is a good example. When planted adjacent to cotton, it lures the harmful lygus bug and thus prevents serious damage to the cotton.

Introducing Predators, Parasites, and Disease Organisms

In nature thousands of potential insect pests never become real pests because of natural controls exerted by predators, diseases, and parasites (that is, biotic components of environmental resistance). Farmers can capitalize on this knowledge through **biological control** or **food chain control** to manage weeds, insects, rodents, and other pests.

There are over 300 examples of partial or complete control of pests through predators and parasites. For example, the prickly pear cactus was introduced in Australia from its native Mexico. By 1925 over 24 million hectares (60 million acres) of land had been badly infested; half of this land was abandoned because of the thick carpet of cactus. Farmers introduced a cactus-eating insect to Australia to eradicate the pest, and seven years later much of the land had been cleared and could once again be used.

The predatory lady beetle was introduced into California from Australia in the 1880s to control an insect that destroyed citrus trees. Parasitic insects from Iran, Iraq, and Pakistan have been introduced to control the olive scale, an insect that once threatened the state's olive trees. Both lady beetles and the predatory insects now exert complete control on their prey, keeping their populations at manageable levels without the use of pesticides.

Entomologists in the United States are currently experimenting with a new method of controlling mosquitoes using *Toxorhynchites rutilis* (Big Tox, for short), a large, nonbiting mosquito whose larvae feed on the larvae of other mosquitoes. Bred in captivity, this predatory mosquito will be released in infested regions to control biting mosquitoes.

A few insect pests can also be controlled by birds, a natural control organism whose potential has been overlooked. Brown thrashers can eat over 6000 insects in one day. A swallow consumes 1000 leafhoppers in 12 hours,

Figure 17-6. Gypsy moths kill large sections of trees by eating leaves near their hatching site. Efforts are now under way to control the moth using biological control techniques, such as the bacterium BT.

and a pair of flickers can snack on 500 ants and go away hungry. In China thousands of ducklings are driven through rice fields; in some places they reduce the populations of insects by 60% to 75%, allowing farmers to reduce insecticide use considerably.

Bacteria and other microorganisms can be brought to bear on pests. One common example is the bacterium *Bacillus thuringiensis* (BT), used to control many leaf-eating caterpillars. Cultivated in the lab and sold commercially, it can be applied as a powder or mixed with water and then sprayed on plants. Caterpillars that eat the bacteria die because BT produces a toxic protein that paralyzes the digestive system of insects. Humans and other organisms are usually unaffected.

BT is used by organic gardeners with considerable success. It has been sprayed in China to control pine caterpillars and cabbage army worms. In California it has been used for more than 20 years to control various troublesome caterpillars, and it is currently applied in the northeastern United States to help control gypsy moths, which devastate forests (Figure 17-6). Another strain of BT has also been employed in the battle against mosquitoes. The use of BT and other techniques has resulted in a measurable reduction in insecticide use in certain crops, especially almonds and tomatoes.

Researchers have also successfully inoculated corn plants with genetically altered bacteria containing the BT gene. The bacteria multiply in the corn plant as they grow and kill European corn borers that feed on the stalks. Studies show that the bacteria do not migrate into the kernels of the corn plant.

Viruses and fungi may be used similarly. In Australia, after years of fruitless efforts to control rabbits, scientists introduced a pathogenic *myxoma* virus, which eliminated almost all of the rabbits within one year. Unfortunately, the virus has evolved to an avirulent form and the rabbit has evolved resistance. Control is no longer as effective

Figure 17-7. The USDA prevents buildup of screwworms by rearing and releasing sterile male flies.

as it once was. Cabbage loopers can be controlled with .5 gram of an experimentally produced virus applied to a hectare of cropland. Other viruses are being used to control pests such as the pink bollworm, which damages cotton, and the gypsy moth. US scientists are now testing a fungus (*Tolypocladium cylindrosporum*) to manage mosquitoes. Special traps are set out to attract adult females. Females enter the traps and are contaminated with spores, but they are allowed to escape, thus carrying the fungal spores back into the environment, where they infect eggs and larvae.

These biological control agents must be developed with caution to ensure that they are not harmful to humans, livestock, and other members of the ecosystem. Pest control agents must not become pests themselves.

Organisms can develop genetic resistance to biological controls, as did the Australian rabbits mentioned earlier. Researchers in Kansas also recently found that larvae that eat stored grain (Indian meal moth larvae) develop genetic resistance to BT. In such cases new controls could be introduced, but in some instances biological control agents themselves may undergo genetic changes that off-set the newly acquired resistance of the pest. This process is called **co-evolution**. As yet, there is no record of such changes in biological control agents.

Genetic Controls

We will consider two major genetic control strategies, the sterile male technique and the breeding of genetically resistant plants and animals. Both are important components of integrated pest management.

Sterile Male Technique The sterile male technique has been effective against several species of insect pests, including the screwworm fly in Mexico and the United States, the Mediterranean fruit fly in Capri, the melon fly on the island of Rota, near Guam, and the Oriental fruit fly in Guam.

In this technique males of the pest species are raised in captivity and sterilized by irradiation or special chemicals. Then they are released in large numbers in infested areas, where they mate with wild females. Since many insect species mate only once, eggs produced by such a union are infertile. If the population of sterilized males greatly exceeds that of the wild males, most of the matings will be with sterile males. Populations can be brought under control swiftly.

In the United States screwworms have been controlled by this method, saving millions of dollars each year (Figure 17-7). The screwworm fly lays eggs in open wounds of cattle and other warm-blooded animals. The eggs hatch within a few hours, and the larvae feed off blood and tissue fluid. This keeps the wound open, allowing a bacterial infection to set in, which eventually kills the host.

In 1976 there were over 29,000 cases of screwworm infestation in the United States. Because of inclement weather and an extensive release of sterile males, costing $6 million, the following year there were only 457 reported cases. The screwworm has practically been eliminated from the southeastern United States, but it still remains in the Southwest, where new flies migrate in from Mexico. With continued cooperation between Mexico and the United States, though, the future of the screwworm as a major pest may be a short one. Even if the fly is never eradicated, controlling it provides substantial economic benefits at a relatively low cost. The program is saving the cattle industry about $120 million annually.

Sterile males have been introduced in other instances with much less success. For example, California imported sterilized Mediterranean fruit flies (medflies) from Hawaii in efforts to control this pest in 1980 and 1981. The medfly lays its eggs in 235 different fruits, nuts, and vegetables; its larvae develop in the ripening fruits and eventually destroy them. Agricultural interests argued that if the medfly proliferated, it would cause $2.6 billion in damage per year in California alone.

At first the state tried a combined approach, using sterile males and baited traps laced with malathion to avoid widespread aerial spraying, which, scientists had argued, would cause undue health risk. Combined programs had worked well in two areas in the state in 1975 and 1980. In the 1980–1981 incident, however, poorly funded control efforts were inadequate and applied too late. Fertile medflies kept appearing both inside and outside of areas believed to be infested. Farmers became nervous, and the US Department of Agriculture threatened to quarantine all California produce, forcing the governor to order a massive aerial spraying program that eradicated the medfly. Follow-up studies showed that the spraying, although successful in controlling the medfly, had also reduced the populations of many predatory and parasitic insects that control populations of other insect pests. Researchers found, for instance, that aphids and whitefly populations in home gardens had increased dramatically because of the loss of natural controls. Insects that destroy olive trees increased substantially in sprayed agricultural regions, for similar reasons.

The sterile male technique has also proved unsuccessful in mosquito control. Scientists believe that the chief reason for these failures is the lower sexual activity of sterilized males compared with wild males. Other reasons may include an inadequate number of sterile males, ignorance of the insect pest's breeding cycle, and the inmigration of additional pests. Some researchers also suggest that through natural selection a new race of insects may evolve that recognizes and avoids sterile males.

Despite these problems the sterile male technique is an important tool in pest control. It is species-specific, can be used with environmental controls, and can be effective in eliminating pests in low-density infestations.

Developing Resistant Crops and Animals Genetically resistant plants and animals can be developed through genetic engineering and artificial selection. Genetic engineering may be especially useful in this regard, because it allows for a much more rapid development of new strains than artificial selection.

In a recent example of artificial selection, scientists have found that certain oils in the skins of oranges, grapefruits, and lemons are highly toxic to the eggs and larvae of the Caribbean fruit fly, which lays its eggs in the skins of these fruits. The flies' larvae destroy the fruit, but scientists may now be able to selectively breed citrus to increase the amount of toxic oils in their peels.

Cornell University scientists are developing a new type of potato plant whose leaves, stems, and sprouts are covered with tiny, sticky hairs, which trap insects and immobilize their legs and mouth parts. Field tests show that this plant can reduce green peach aphids by half. The new variety was developed by crossing cultivated potatoes with a wild species with sticky hairs, which grows as a weed in Bolivia.

Other genetic research has led to Hessian fly–resistant wheat and leafhopper-resistant soybeans, alfalfa, cotton, and potatoes. Work on chemical factors that attract insects to plants may help scientists selectively remove them to make plants unappealing.

Monsanto Company recently announced another promising weapon in the fight against pests. Robert Kaufman and his colleagues have isolated the gene that gives BT its pesticidal action. The scientists have transferred that gene to another bacterium, *Pseudomonas fluorescens*, that lives on the roots of corn and several other plants. The transplanted gene renders the host bacterium lethal to insects and other organisms, such as the black cutworm, that feed on the roots of important commercial plants. Simply by planting seeds that have been pretreated with *P. fluorescens* bearing the toxic gene, farmers may be able to provide long-term protection without many of the dangers of pesticides. The world renowned entomologist, David Pimentel, however, believes that widespread use of BT in corn and other crops will probably result in the rapid development of resistance and loss of the BT control farmers now enjoy.

Monsanto hopes that more insecticidal genes can be added to *P. fluorescens* in the years to come, giving corn a wider range of protection, reducing chemical pesticide use and in the process protecting wildlife from the harmful toxic pesticides that have been the mainstay of agriculture for decades.

Root-zone protection is not the only strategy that geneticists are developing. Numerous bacteria colonize aboveground plant parts; fitted with insecticidal genes from BT and other naturally occurring biological agents, these bacteria could create a protective barrier to ward off dozens of insect pests.

Genetic resistance is a necessary element of effective pest management. The major problem is the time, money, and labor involved in producing resistant varieties. Furthermore, genetic resistance can be overcome when pests adapt. In this case, scientists must be ready with new varieties.

Chemical Controls

Second-Generation Pesticides Even with wider use of natural control agents and other nonchemical strategies, second-generation pesticides will remain a part of our pest control strategy. However, several principles should guide their use: (1) they should be applied sparingly; (2) they should be applied at the most effective time to reduce the number of applications; (3) they should destroy as few natural predators, nonpest species, and biological control agents as possible; (4) they should not be applied near drinking water supplies; (5) they should be carefully tested for toxic effects; (6) persistent pesticides and pesticides that bioaccumulate should be avoided; (7) exposure to workers should be as low as possible; and (8) pesticides should be used to reduce populations to low levels, and environmental, genetic, and cultural control measures should then be used to keep populations low.

Chemical pesticides will continue to play an important role in pest management. Today, however, several new chemical agents have been developed, such as phero-

Figure 17-8. Pheromone trap, containing a sticky substance to immobilize male gypsy moths in search of mates.

mones, hormones, and natural chemical repellants. These are **third-generation pesticides**.

Pheromones Pheromones are externally released chemicals produced by insects and other animals. One well-known group of pheromones is the **sex attractants**, which are emitted by female insects to attract males at the time of breeding. Effective in extraordinarily small concentrations, pheromones draw males to females, an evolutionary adaptation that ensures a high rate of reproductive success.

Some sex attractants are now synthesized in the laboratory and are commercially available for pest control. They are used in three ways. First, **pheromone traps** can be used to lure males. Traps may contain a pesticide-laden bait or a sticky substance that immobilizes insects (Figure 17-8). Second, pheromones can be sprayed widely at breeding time. This is known as the **confusion technique**, as the males are drawn by the pheromone from all directions and may never find a partner. One modification of this technique involves the release of wood chips treated with sex attractants. Males are drawn to the wood chips and may attempt to breed with them. Third, pheromone traps can be used to pinpoint the time when insect eggs hatch. Males emerging at this time are attracted to the traps. By knowing precisely when insects emerge farmers can time their pesticide applications for maximum effectiveness. This technique helps reduce the amounts of pesticide applied. Pheromones can also be

used to lure beneficial insects from fields so that pesticide sprays can be applied.

Pheromone traps of various sorts have been used to control at least 25 insect species. Early work on clearwing moths, cotton bollworms, and other species has been quite successful. Pheromones offer many advantages: they can be used with other methods, they are nontoxic and biodegradable, they can be used at low concentrations, they are highly species-specific, and they are not expected to present any environmental hazards. The major disadvantage is the high cost of developing new pheromones.

Insect Hormones The life cycle of many insects is shown in Figure 17-9. Insects pass through the larval and pupal stages before reaching adulthood. The entire life cycle is regulated by two hormones, **juvenile hormone** and **molting hormone**. **Hormones** are chemical substances that are produced by specific cells in the body and travel through the bloodstream to distant sites, where they exert some effect. Altering the levels of juvenile and molting hormones disrupts an insect's life cycle, resulting in death. For example, larvae treated with juvenile hormone are prevented from maturing and eventually die. If given molting hormone, they will enter the pupal stage too early and die. Experiments are under way to assess the effectiveness of spraying crops with insect hormones.

Insect hormones offer many of the same advantages that pheromones offer, including biodegradability, lack of toxicity, and low persistence in the environment. Like pheromones, however, they have a high cost and long production time. In addition, insect hormones act rather slowly, sometimes taking a week or two to eliminate a pest, by which time extensive damage may have been done. Insect hormones are not as species-specific as pheromones and therefore may affect natural predators and other nonpest species. The timing of application is also critical, for hormones are effective only at certain times in an insect's life cycle.

Researchers recently discovered that a plant from Malaysia produces juvenile hormone. Feeding the plant to juvenile insects caused considerable abnormalities. Some researchers hope that the genes responsible for the production of this hormone can be transferred to commercially important crops, offering another avenue of protection.

Natural Chemical Pesticides Natives of the South American tropics have used the seeds and leaves of the neem tree for many years to control pests. Researchers have found that this tree produces chemicals that kill or repel a variety of insects. This extract may become useful in the control of larvae that feed on vegetables and ornamental crops.

Egyptian researchers noticed that flies ignored a species of brown algae left out on a counter to dry. Curious,

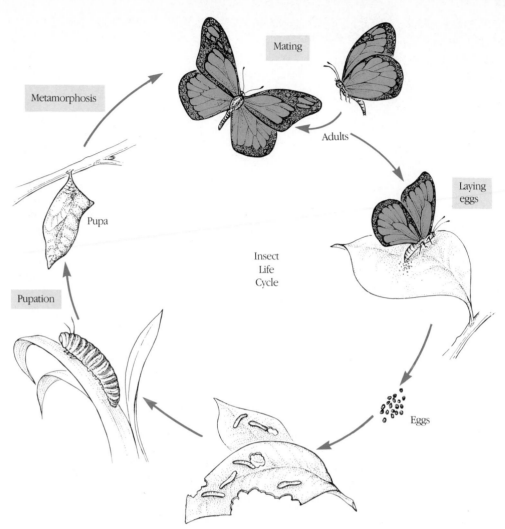

Figure 17-9. The insect life cycle.

Mating

Metamorphosis

Adults

Laying eggs

Pupa

Insect Life Cycle

Pupation

Eggs

they extracted a mixture of chemicals from the algae and found that they, too, repel a variety of insects that attack cotton and rice.

Natural chemicals like these may prove useful in years to come. Like other third-generation pesticides, they are biodegradable and nonpersistent.

Researchers have also found that some plants produce chemicals that alter insect metabolism. Petunias, for example, synthesize a chemical that dramatically stunts the growth of corn earworms. Scientists hope that they can transfer the genes to crop plants either through genetic engineering or more conventional means to provide a natural protection against pests, thus providing an on-site means of control. Nicotine, caffeine, and citrus oil are all natural insecticides and are under investigation today.

Cultural Controls

Cultural control refers to many other simple techniques to control pest populations. These methods include cul-

tivation to control weeds, noisemakers to frighten birds, manual removal of insects from crops (especially suitable for smaller gardens), destruction of insect breeding grounds, improved forecasting of insect emergence, quarantines to regulate the spread of pests, and water and fertilizer management to ensure optimum crop health and resistance to pests. Computers may also help farmers in the battle to protect crops. A computer model developed by researchers at the University of California at Riverside, for example, may help farmers predict when citrus trees will bloom far in advance of the event. State law prohibits the use of pesticides during the bloom to protect honeybees. Inaccurate predictions in the past have cost growers millions of dollars.

As a final note, a researcher from the State University of New York in Syracuse has suggested microwaving all books and magazines in libraries to control insect damage. A number of insects, such as cockroaches, silverfish, and termites, feed on crumbs left by readers (one reason why libraries forbid eating on the premises) but also dine on the pages and binding glue of books. To save their

Indonesia Turns to Biological Pest Control

Indonesia is a country of islands—nearly 14,000 of them—in Southeast Asia. In 1983, this rural nation, once the world's leading importer of rice, succeeded in growing enough rice to feed its own people. New strains of rice, fertilizers, and an intricate irrigation system deserve credit for the success.

In 1985, however, the notorious brown planthopper threatened the progress of the previous years. This insect causes rice to dry out, rot, and fall in the field. Infestations of the insect can cause enormous damage. From 1975 to 1979, in fact, the brown planthopper destroyed 4 million hectares (10 million acres) of rice crop.

To combat the planthopper, the government decided in 1985 to try integrated pest management. They were advised to do so by an Indonesian entomologist, Dr. Ida Oka, who had received his Ph.D. from Cornell University under David Pimentel, an expert on natural insect control. Oka had been in charge of pest management in the late 1970s and early 1980s and had implemented sound integrated pest management for rice and other crops. During that time, pesticide use had been greatly reduced and rice yields soared. But Oka had resigned when a new Minister of Agriculture was appointed. This official's pro-pesticide policies resulted in widespread use of chemical pesticides and were largely responsible for the outbreak of the brown planthopper in 1985.

Indonesian scientists had found that the use of pesticides to control the planthopper and other insects killed many beneficial insects that preyed on pests. One of the beneficial insects destroyed in the spraying is the wolf spider, which can devour 5 to 20 brown planthoppers a day. If left alone, the beneficial insects can often control the harmful ones. Researchers also found that farmers sprayed fields regularly, whether they needed it or not. The overuse of pesticides actually increased the severity of infestations. It was time to try something different.

The Indonesian government asked the United Nations Food and Agricultural Organization (FAO) to help them promote an integrated pest management program (IPM), which the FAO had begun in the Philippines. In IPM, indigenous predatory insects are used to control pests. Farmers can reduce—sometimes eliminate—spraying altogether.

The government had its work cut out for it; convincing Indonesia's farmers to try an alternative method of pest control became a major undertaking. Farmers see pesticides as a kind of insurance against insect pests. Since the government also subsidized 75% of the cost of pesticides, most farmers saw little reason to switch.

Nonetheless, Indonesian officials were convinced of the danger of continuing pesticide use. In 1986, with the help of the FAO, they embarked on a crash program to educate farmers on IPM and the dangers of pesticides. Experts went out into the rice paddies and showed farmers how to diagnose problems, how to calculate the ratio of good bugs to bad ones, and how to decide how much damage the crop could stand without harming the yield.

Early results showed that IPM worked. The average yield on farms using pesticides was 2.47 tons per acre compared with 2.55 tons per acre on fields controlled by IPM. Despite the high subsidies for insecticides, the farms using IPM proved more profitable than those sprayed more frequently.

The success of the pilot program in Indonesia convinced the government to adopt IPM as a national pest control strategy. The government, in fact, banned 56 of the 57 pesticides previously approved for farming in Indonesia to help protect predatory insects. The government then launched a massive campaign to educate Indonesia's farmers in IPM. By 1994, they hope to have reached all of them.

IPM helped reduce pesticide use substantially. Trained farmers, for example, apply one-ninth as much pesticide as they did before training, with no decrease—sometimes even an increase—in crop yield. Farmers have learned to discern more carefully insect damage from fungal damage, and this helps to reduce pesticide use. The government saves an estimated $120 million a year on pesticide subsidies. Indonesia's streams and wildlife are also showing signs of recovery.

Integrated pest management is spreading to Thailand, Bangladesh, Sri Lanka, Malaysia, India, and China. Farmers in Indonesia who have not yet been introduced to the technique are calling for action. They want to be included in this colossal experiment, which, if successful, could help the world move to a more sustainable, environmentally safe form of farming. Agricultural experts believe that IPM could be used on fields that provide 45% of the rice for people living in southern and southeast Asia and could save millions of dollars, preserve wildlife, and protect human health without endangering high crop yields.

precious books, libraries typically fumigate books, set out poisoned bait, and spray insecticides. But microwaving may prove to be a cheaper and safer alternative to destroy eggs and larvae.

Economics, Risk, and Pest Control

The central goal of pest control is to reduce pest populations to levels that do not cause economic damage. The cost of pest control, including internal and external costs, should not exceed the economic benefits of increased yield (Chapter 14). To be certain that benefits do not outstrip the actual economic, environmental, and health costs, thorough and fair studies must be made comparing each of the pest control strategies.

Today, most of the money for pest control research goes for conventional chemical methods. Many critics agree that we must be willing to spend money today on newer methods, such as natural predators, that can be part of a long-term, sustainable pest management strategy. Though costs may be high initially, over the long term they may become insignificant.

Society might also rethink some of its demands for unblemished fruits and vegetables. Pesticide use just to prevent harmless blemishes is probably a waste of energy, time, and money. A few more spots on our oranges could mean many more birds overhead, cleaner waterways, improved health for workers and the general public, and cheaper oranges.

As for poor nations, pest management programs can (1) make use of people to destroy bugs and weeds by hand; (2) employ environmentally sound measures, such as crop rotation, heteroculture, timed plantings, natural predators, and genetically resistant plants and animals; and (3) minimize the use of first- and second-generation pesticides. Developed nations must keep companies from exporting pesticides that have been banned at home. (For more on banned chemicals see the Viewpoint in this chapter.) The rich can also help the poor develop a sustainable pest management program through technical and financial assistance.

Herbicides in Peace and War

Herbicides are a form of pesticide used to control weeds. Before World War II few herbicides were used in the United States. Today, however, the herbicide industry is large and prosperous, with sales of over $1.3 billion a year.

In the United States herbicides account for over 60% of the chemical pesticides applied each year. Although there are over 180 types of synthetic herbicides on the market, the leading ones are atrazine and alachlor, which collectively account for half of the total sales. In addition, butylate and 2,4-D (2,4-Dichlorophenoxyacetic acid) are used in large quantities.

Despite the number of herbicides in use, two particular chemicals have received most of the attention; these are 2,4-D and 2,4,5-T, nonpersistent synthetic organic compounds similar in function to plant hormones, called **auxins**. When sprayed on plants, 2,4-D and 2,4,5-T increase the metabolic rate of cells so much that plants cannot keep up with increased nutrient demands and literally "grow to death."

Peacetime Uses: Pros and Cons

In peacetime 2,4-D and 2,4,5-T have been sprayed on brush and plants along roadways, power lines, and pipelines and have been used to control poison ivy and ragweed, to eliminate unwanted trees in commercial tree farms and in national forests, to control aquatic weeds, and to rid rangelands of brush and poisonous plants. Overall, three-fourths of these chemicals are used for weed control on farms.

The benefits of herbicides are many:

1. They decrease the amount of mechanical cultivation needed to control weeds and thus reduce labor costs for weed control.

2. They reduce weed damage when soils are too wet to cultivate, because crops can be sprayed by plane.

3. They help farmers reduce water usage, since water escapes more rapidly from ground that has been tilled to control weeds.

However beneficial their use may be, herbicides have many drawbacks:

1. Some weeds that are not normally pests are resistant to herbicides and tend to proliferate after spraying. Elimination of newly created pests may require the use of an additional herbicide.

2. Weeds can become resistant to herbicides. Many agronomists believe that the problem of plant resistance has been underestimated.

3. Herbicide use may increase the need for insecticides. When herbicides are used, farmers often reduce tillage; weeds killed by herbicides remain on the ground and provide food and habitat for insect pests. In addition, herbicides that are used to control weeds around the periphery of fields may destroy the habitat of predatory insects. Herbicides may also decrease the farmer's incentive to rotate crops, an effective way of reducing insect pests.

4. Herbicides may make some crops more susceptible to insects and disease. For example, some herbicides reduce the waxy coating on plants, change their metabolic rates, and retard or stimulate plant growth; all of these may make plants more susceptible to disease.

5. There is great concern that some herbicides are toxic and may cause birth defects, cancer, and other illnesses.

Critics argue that **integrated weed management** could reduce the use of herbicides. Such management would employ special equipment such as wick applicators that apply herbicide only on weeds between rows. Wick applicators use much less herbicide than aerial spraying and create less environmental contamination. The reasonable use of herbicides could be complemented by mechanical cultivation, proper spacing of rows for healthy crops, biological weed controls, and crop rotation.

Controversy over Wartime Use of 2,4-D and 2,4,5-T

Herbicides were used extensively in the Vietnam War as defoliants to prevent guerrilla ambushes along roads and waterways; to deter the movement of soldiers through demilitarized zones and across the border of Laos; to destroy crops, which might be eaten by the enemy; and to clear areas around camps. Three herbicide preparations were sprayed from planes, helicopters, boats, trucks, and portable units from 1962 to 1970.

Ecological Effects of Agent Orange   The most effective and most controversial of the herbicides was Agent Orange, a 50-50 mixture of 2,4-D and 2,4,5-T. During the war over 42 million kilograms (93 million pounds) of Agent Orange was sprayed on the swamps and forests of Vietnam, decimating 1.8 million hectares (4.5 million acres) of countryside and at least 190,000 hectares (470,000 acres) of farmland. Over half of the mangrove vegetation of South Vietnam (1930 square kilometers, or 744 square miles) and about 5% of the hardwood forests were destroyed. The forests alone represent about $500 million worth of wood, a supply that would last the country 30 years.

The prospect for these forests seems dim, especially since hardy weeds such as cogon grass and bamboo have invaded the deforested zones. Ecologists fear that these species may greatly delay recovery or prevent it altogether. The US National Academy of Sciences estimates that defoliated mangroves may take 100 years to recover, because the destruction was so great that few seed sources remain.

Defoliants and numerous insecticides sprayed to control mosquitoes created an unprecedented ecological disaster in Vietnam, resulting in the death of numerous fish and animals. In one survey of a heavily sprayed forest visited years after the war ended, a Harvard University biologist, Peter Ashton, found 24 species of birds and 5 species of mammals, compared with 145 and 170 bird species and 30 and 55 mammal species in two nearby forests that had not been sprayed. The total impact of such actions will never be known.

Health Effects of Agent Orange   Agent Orange may have been the cause of serious medical problems that developed in soldiers and villagers throughout Vietnam. In 1969 the first indication of such health effects came from Saigon newspapers, which reported an increase in miscarriages and birth defects in babies born in local hospitals. Agent Orange sprayed in Vietnam was contaminated with dioxin, a chemical that causes birth defects and cancer in mice and rats. Dioxin, discussed in more detail in Chapter 18, is believed to be 100,000 times more potent than the tranquilizer thalidomide, which has caused many birth defects in the United States (for a discussion of birth defects see Chapter 14).

Soldiers from the United States and Australia who fought in herbicide-defoliated areas drank water and bathed in bomb craters believed to have been contaminated by Agent Orange, ate sweet potatoes and other vegetables from contaminated fields, and in some intances were doused with herbicides while in the field. Soldiers developed severe headaches, nausea, diarrhea, internal bleeding, chloracne (a severe skin rash similar to acne), and depression. In 1970, as a result of the public outcry in the United States, the government banned the use of Agent Orange in Vietnam, although 2,4-D and 2,4,5-T are still used today in the US to reduce the competition from hardwood trees in commercial evergreen forests.

Starting in the late 1970s Vietnam veterans who had returned from the war began to suffer an unusual number of medical disorders. A high proportion of the men have fathered fetuses that were born dead or aborted prior to term and infants with multiple birth defects. Experts estimate that over 2000 defective babies will be born to Vietnam veterans. Other veterans have developed cancers such as lymphoma, leukemia, and testicular cancer.

Perhaps the most compelling evidence linking adverse health effects to Agent Orange came from a study carried out by Vietnamese doctors. They examined the rate of birth defects in the offspring of 40,000 Vietnamese couples. The researchers found that women whose husbands had fought in South Vietnam, where Agent Orange was sprayed, were 3.5 times more likely to miscarry or give birth to defective children than women whose husbands had remained in the north during the war. American scientists who have studied the results, although expressing caution, find the study convincing.

Data recently released from an independent study of military veterans indicates an increased risk of the following problems: elevated blood pressure, benign fatty tumors, wife's miscarriage, visual and skin sensitivity to light, and depression. Other evidence has implicated dioxin as the cause of soft-tissue sarcomas (cancers); veterans exposed to dioxin had a rate of soft-tissue sarcomas seven times higher than normal.

Such findings have shifted the Veterans Administration's position on Agent Orange. Over 1200 disability claims have been filed with the Veterans Administration by US soldiers. In May 1984 manufacturers and Vietnam veterans reached an out-of-court settlement that established a $180-million fund for veterans and their families who claim injury from Agent Orange. In 1989, seven companies who were being sued by veterans for damages caused by Agent Orange agreed to pay them $240 million.

Unfinished Business

Pesticides are currently regulated by two laws: the Federal Insecticide, Fungicide, and Rodenticide Act (FIFRA) and the Federal Food, Drug, and Cosmetic Act (FFDCA).

FIFRA is administered by the Environmental Protection Agency. When passed in 1947, the **Federal Insecticide, Fungicide, and Rodenticide Act** had narrow intentions. Its chief purpose was to protect farmers from dangerous (and also ineffective) chemicals. FIFRA required manufacturers to register pesticides transported across state boundaries. Those produced and used within state boundaries were free from federal scrutiny.

In 1972, the law was broadened in an attempt to protect public health and the environment from new chemical pesticides. From that point on, manufacturers were required to test their new products for health effects *before* the pesticides could be registered for use. Registration is a process by which the EPA attempts to limit harmful effects. When a company comes up with a new pesticide that it wants to market, it first performs routine tests on plants and animals to determine its toxicity. Test plots are sprayed with the pesticide to determine residues. These data are then handed over to the EPA. Taking into account the average American diet, residue levels, and toxic effects, the EPA then determines which crops, if any, the pesticide can be used on. Critical thinking suggests a problem right away. How does the EPA define the average diet?

The average American diet includes a great deal of red meat, chicken, and other meat products. But that automatically excludes two large groups: vegetarians and children. Fruits and vegetables make up the bulk of their diets. Ironically, many vegetarians who select a special diet for health reasons are inadvertently exposed to the most pesticide.

Pesticides can be registered for general or restricted use. General use means that anyone can purchase and use them. Technically, restricted-use pesticides are to be used by licensed applicators. The EPA has restricted their use to prevent unacceptable risk. Restrictions can be lifted when special needs arise.

Pesticide use is largely an honor system. Labels on restricted-use products describe their legal uses, and only licensed applicators can buy them, but other than those limitations there's very little, if anything, to stop people from using pesticides for whatever purposes they desire.

To help protect public health, FIFRA also authorizes the EPA to set tolerance levels for pesticides on foods. **Tolerance levels** are concentrations in or on foods that are believed to pose an acceptable health risk. For cancer, the EPA (theoretically) sets the concentration at the level that will cause no more than one additional cancer death in one million people. This determination is fraught with difficulty, as explained in Case Study 16-1 and Chapter 14.

In 1988, after several years of struggle, FIFRA was amended to correct some weaknesses. One of the most significant gains was a plan to reregister many pesticides introduced before the EPA took over the pesticide registration process. (The EPA was founded in 1972.) Pesticide registrations in the 1950s and 1960s were made with very little, if any, sound toxicological data, say critics. Reregistering the over 300 chemical pesticides will take nine years to complete and will be financed by fees paid by chemical companies. Some experts see this as a weeding-out process. They believe that many pesticides will be cancelled. Companies afraid that their product won't be approved (for health and environmental reasons) won't invest the money to reregister, especially if it is a low-profit item.

Despite recent improvements in FIFRA, pesticide management needs considerable reform. Currently, pesticide registration does not require manufacturers to test for neurotoxicity—toxic effects on the brain, spinal cord, and nerves. To many critics, this is a glaring omission since most insecticides are insect neurotoxins that have profound effects on the human nervous system. New research shows that some pesticides damage the immune system, the body's defense against bacteria, viruses, and even cancer. Pesticide registration, however, requires no test of immune system effects, another glaring omission, say critics. Last, but not least, pesticide registration does not take into account possible synergism, the superadditive effect described in Chapter 14.

In addition, FIFRA provides virtually no monitoring of end use. The end users are free to do more or less as they please. They can apply as much pesticide as they want and can apply it wherever they want. In the field, there is no one to watch over pesticide application. There is no one to see that the solution is mixed correctly or applied in proper amounts. There is no one to ensure that equipment is working properly. As a result, farmers sometimes apply much more than is needed.

Most farmers are licensed applicators; however, there is virtually no training offered to them. Pesticide sales people become a primary source of information. As noted earlier in the chapter, programs that teach farmers to become better at identifying pests and monitoring pest populations in the field could decrease insecticide use. Crop scouts could be trained to monitor pest populations and to determine when they have reached the threshold level, at which time spraying is beneficial and cost effective. Better instruction on mixing and applying pesticides could help as well. As it is, farmers often jump the gun, spraying their fields when pest populations are below the level at which they cause harm. Under FIFRA, states are required to certify and license those who apply restricted pesticides. But the process is far from optimum. Critics argue that more rigorous testing and education are needed.

Under FIFRA, the EPA sets tolerance levels for pesticide residues on fruit and vegetables and other foods. But it is up to the Food and Drug Administration and state agencies to monitor our food supply and to enforce tolerance levels. FDA can, for example, seize and condemn foods containing residues which exceed EPA levels or which contain illegal pesticides, but only when they are shipped from one state to another. Inside the states, the responsibility for monitoring food lies with state agencies.

Both the FDA and state agencies suffer from a lack of funds and a shortage of inspectors. Given the shortages of inspectors and money and the massive amount of food consumed by the American public, it can be no surprise that only a small portion of our food is actually tested. Furthermore, examiners only look for a handful of the pesticides that could be on our food. Unless you grow your own food, you are probably eating more pesticide than you would like. Why? Because more than 50% of all food in supermarkets contains detectable levels of pesticides.

In 1987, the EPA reported that at least 55 of the pesticides that could leave residues on food are carcinogenic. In 1987, the National Academy of Sciences issued a report concluding that one million Americans alive today will develop cancer as a result of pesticide contamination of their food. Add to that possible birth defects, miscarriages, mutations, neurological effects, and other milder symptoms, and it is little wonder that the EPA ranks pesticides in food as one of the nation's most serious health prob-

lems. Despite their recognition of a potential problem, the EPA has a long way to go. Most critics think that the tolerance levels set for pesticide residues are inadequate, failing to take into account the special diets of children and vegetarians. The EPA rarely revises tolerance levels when new scientific data about risks become available, say its critics.

Pesticide residues may or may not cause cancer. We may never know. For some people, the mere presence of a potential carcinogen on their food is reason enough for concern, indeed, outrage. Some scientists, manufacturers, and regulators argue that low levels of pesticide are not worth worrying about. A slightly elevated risk of cancer, a poisoned bird or fish are the price we pay for progress. Pesticides, they argue, are one of those necessary evils.

But as we've seen in this chapter, pesticides are not indispensable. Pesticide use can be greatly reduced through integrated pest management. For now, pesticides are a part of our lives. Individuals can avoid them by growing some of their own fruits and vegetables or by purchasing organically grown produce. Washing fruits and vegetables can help, but won't eliminate all pesticide residues. Beyond that, you can exercise your democratic prerogatives by writing local, state, and federal officials and asking for tougher laws to regulate pesticide registration and use. Requirements that store owners label produce indicating the pesticides that have been used on them could inspire farmers to find alternatives to these potentially harmful substances. Many farmers have already switched. In 1987, the US Department of Agriculture announced that 30,000 American farmers have forsaken pesticides and artificial fertilizers altogether. At least 100,000 more have made substantial reductions in chemical use. That's good news to some, but it is only a small fraction of the two million farmers now supplying us with food.

Our doubts are traitors and make us lose the good we oft might win by fearing to attempt.

SHAKESPEARE

Summary

Each year, insects, weeds, bacteria, fungi, viruses, parasites, birds, rodents, and mammals consume or destroy an estimated 48% of the world's food crop. Reducing losses through pest control represents an enormous opportunity to increase world food supplies.

In recent years chemical pesticides have come to form the cornerstone of pest management. **Pesticides** are chemicals used

to kill troublesome pests. The **first-generation pesticides** were simple preparations made of ashes, sulfur, and other chemicals. **Second-generation pesticides** are synthetic organic compounds like DDT. Some are **broad-spectrum** chemicals able to kill a wide variety of organisms. Others are **narrow-spectrum** pesticides used to control a few pests.

Second-generation pesticides fall into three classes: (1) chlorinated hydrocarbons, a high-risk group that is largely banned today; (2) organic phosphates, less risky, but still quite toxic;

and (3) carbamates, widely used today because they tend to break down quickly and do not bioconcentrate as chlorinated hydrocarbons do.

Pesticide use has helped control pests but not without many problems. For example, some broad-spectrum pesticides kill predatory and parasitic insects that help control pests and potential pests. Pesticides also destroy beneficial insects such as honeybees. Perhaps the most striking problem is genetic resistance among pest populations, which has been met with increasing doses and new chemical preparations, resulting in considerable environmental pollution. DDT proved especially harmful to predatory bird populations, because it impaired eggshell deposition. Growing evidence shows that many pesticides are harmful to people, especially workers carelessly exposed on farms.

The problems with pesticide use have led many to suggest the need for an **integrated pest management** strategy that relies on environmental, genetic, chemical, and cultural controls. Education and monitoring are essential to make the transition.

Environmental controls are designed to alter a pest's immediate environment, making it lethal or inhospitable. Some options include increasing crop diversity, varying the time of planting, altering plant and soil nutrients, controlling adjacent crops and weeds, and introducing predators, parasites, and disease organisms. **Genetic controls** are alterations in the genetic composition of pests, crops, and livestock. In the sterile male technique, for example, captive-raised males of the pest species are sterilized and then released into fields, where they greatly outnumber wild, fertile males. This technique ensures a high percentage of infertile matings. Genetic resistance can be bred into plants and livestock or introduced through genetic engineering. A new arsenal of naturally occurring chemicals may also play a big role in pest control. For example, **sex attractant pheromones**, chemicals released by females to attract males, have been synthetically produced and marketed. Synthetic sex attractants can be used to draw males into traps laced with poison or sticky materials. Insect hormones that control vital life processes can also be applied to crops, throwing off the normal developmental cycle and killing the pests. Naturally occurring pesticides may also provide a partial answer to low-risk pest control.

The central goal of pest control is to reduce pest populations to levels that do not cause economic damage. Logic dictates that the cost of pest control should not exceed the economic benefits. External costs must be factored into the cost equation.

Herbicides are a type of pesticide used to control weeds. Despite the large number of herbicides in use today, 2,4-D and 2,4,5-T have received the most attention. They are generally nonpersistent synthetic organic compounds that, when sprayed on plants, increase the metabolic rate in cells so much that plants "grow to death."

Herbicides reduce the mechanical cultivation needed to control weeds, cut energy demand, and cut costs, but also have some drawbacks. For instance, some weeds that are normally no trouble are resistant to herbicides and grow wildly after others are killed. Some weeds also develop resistance to herbicides, requiring farmers to apply more or different varieties. Herbicides may also increase the need for insecticides and fungicides and may make some crops more susceptible to disease and insects.

Integrated weed management could reduce the use of herbicides and their adverse effects. Special wick applicators could be used to apply herbicide only on the weeds between rows. Proper spacing of rows for healthier crops, biological weed control, and crop rotation could all be used, too.

Herbicides were used extensively in the Vietnam War as defoliants to prevent guerrilla ambushes along roads and waterways and to deter the movement of enemy soldiers. The most effective, and most controversial, was Agent Orange, a 50-50 mixture of 2,4-D and 2,4,5-T. Sprayed on nearly 2 million hectares of countryside and farmland, Agent Orange created an unprecedented ecological disaster. It may also have been responsible for a number of medical problems in civilians and soldiers, because it was found to be contaminated with dioxin, a chemical that causes birth defects and cancer in mice and rats. Mounting evidence from the United States and abroad supports the view that Agent Orange and dioxin are responsible for cancer and illness in US veterans and birth defects in their offspring. In 1970, the US government banned its further use in Vietnam. The use of the herbicides in Agent Orange has continued virtually unchecked in the United States.

The EPA ranks pesticides in foods as one of the nation's leading health problems. Pesticides in foods are currently regulated by the EPA and FDA through the **Federal Insecticide, Fungicide, and Rodenticide Act** and the Federal Food, Drug, and Cosmetic Act, respectively. FIFRA creates a system of registration that is supposed to protect humans and the environment from adverse effects. Under powers granted by FIFRA, the EPA sets **tolerance levels** for pesticides in food, but many critics think that the tolerance levels do not accurately reflect health risks and need to be lowered. Better regulation of pesticide end use is also sorely needed.

The FDA monitors our food supply for pesticide residues and has the power to seize shipments containing levels that exceed those permitted by the EPA. Unfortunately, the FDA does not test for all pesticides and tests only a fraction of our food.

Discussion Questions

1. List and discuss reasons why pest damage is high in developed nations despite elaborate pest control strategies.

2. Describe some of the problems encountered with the use of pesticides. How can they be avoided?

3. Why do DDT and other chlorinated hydrocarbons persist in the environment?

4. What is integrated pest management? What advantages does it offer over current management techniques?

5. Why do crop rotation and increasing crop diversity cut back on pests?

6. Describe some of the biological control methods. Give examples.

7. Describe why the sterile-male technique works.

8. Discuss some ways in which genetic engineering may be used to help cut down on pest damage.

9. Outline how you would go about determining the costs and benefits of modern pest control techniques.

10. What is Agent Orange, why was it used in Vietnam, and why were some soldiers exposed to high levels of it?

11. List and describe the pros and cons of herbicide use.

Suggested Readings

Ames, B. N. and Gold, L. S. (1989). Pesticides, Risk, and Applesauce. *Science* 244 (4906): 755–767. Excellent look at natural carcinogens with important implications for controlling human-produced pesticides.

Bosch, R. van der. (1978). *The Pesticide Conspiracy.* New York: Doubleday. A study of the influence of pesticide manufacturers on integrated pest management.

Carson, R. (1962). *Silent Spring.* Boston: Houghton Mifflin. The book that raised worldwide alarm over the use of pesticides.

Dreistadt, S. H. and Dahlsten, D. L. (1986). California's Medfly Campaign: Lessons from the Field. *Environment* 28 (6): 18–20, 40–44. Sober look at pest control.

Goldberg, E. D. (1986). TBT: An Environmental Dilemma. *Environment* 28 (8): 17–20, 42–44. Worthwhile article.

Groth, E. (1989). Alar in Apples. *Science* 244 (4906): 755. Excellent overview of the Alar controversy from a consumer's standpoint.

Mott, L. and Snyder, K. (1988). Pesticide Alert. *The Amicus Journal* 10 (2): 20–29. Nice overview of the pesticide-in-food controversy.

Postel, S. (1987). *Defusing the Toxics Threat: Controlling Pesticides and Industrial Waste.* Worldwatch Paper 79. Washington, DC: Worldwatch Institute. Excellent overview of toxics control.

Roberts, L. (1989). Pesticides and Kids. *Science* 243 (4896): 1280–1281. Good summary of the NRDC report on the exposure of children to pesticide residues in fruits and vegetables.

Sewell, B. H., Whyatt, R. M., Hathaway, J., and Mott, L. (1989). *Intolerable Risk: Pesticides in Our Children's Food.* Washington, DC: Natural Resources Defense Council. Controversial paper worth reading.

Wilcox, F. A. (1983). *Waiting for an Army to Die: The Tragedy of Agent Orange.* New York: Vintage Books. Moving account of the herbicide's effects on Vietnam veterans.

Hazardous Wastes: Progress and Pollution

There is nothing more frightful than ignorance in action.

GOETHE

Shenandoah Stables was a large and successful quarter-horse ranch in Moscow Hills, Missouri, northwest of St. Louis. But in the last week of May 1971 a routine procedure triggered a disastrous chain of events. To control dust, the owners had hired a man to spread 3800 liters (1000 gallons) of waste automobile oil on the arena. Shortly after the oil had been applied, one of the owners noticed strong chemical odors in the air in the barn and on the grounds. A day later she discovered dozens of dead sparrows on the floor of the barn. Then the dogs and cats at the ranch began to lose their fur and to dehydrate. By mid-June 11 cats and 5 dogs had died, each with the same symptoms.

Of 85 horses routinely exercised in the arena, 43 died within a year. Autopsies revealed that their internal organs were swollen and bloody. In 1971, 41 horses on the ranch had been bred; most pregnancies ended in spontaneous abortion. Of those born alive, all but one died within a few months.

Soon afterward, both of the owner's daughters, who lived at the stables, became ill, complaining of severe headaches; one developed sores on her hands. The other began bleeding internally and was hospitalized. The owner herself suffered chest pains, headaches, and diarrhea.

What caused the poisonings and deaths? Tests on the oil applied to the arena showed dioxin, PCBs, and other highly toxic contaminants. Investigators found that the oil had been sold to a company that was supposed to remove the contaminants before reuse. Instead, the company

(a)

(b)

Figure 18-1. (a) In May 1984, workers from the Neville Chemical Plant in Santa Fe Springs, California, began covering soil believed to be contaminated with highly toxic chemical wastes. The covering prevented the toxic soil from blowing from the site. Company officials were recently found guilty of illegally disposing of hazardous wastes. (b) Pollution Abatement Services chemical dump at Oswego, New York. The dump is now inactive and awaiting cleanup.

spread the oil on at least four sites in Missouri, leaving behind "a trail of sickness and death," in the words of the state assistant attorney general.

The tragedy at Shenandoah Stables is representative of a much larger problem: illegal and irresponsible hazardous waste disposal that continues even today (Figure 18-1). What we can do to clean up the mess and what we can do to prevent further disasters are the topics of this chapter.

Hazardous Wastes: Coming to Terms with the Problem

Love Canal: The Awakening

Hazardous wastes are waste products of industry that, if not disposed of properly or destroyed, pose a threat to the environment. The problem of hazardous waste caught the attention of the American public in the 1970s when toxic chemicals began to ooze out of a dump known as Love Canal, in New York near Niagara Falls. The incident

has forever changed the way Americans view hazardous wastes.

The story of Love Canal began in the 1880s when William T. Love began digging a canal that would run from the Niagara River just above Niagara Falls to a point on the river below the falls. For one reason or another, the canal was never completed. Only a small remnant of the canal remained in the early 1900s. In 1942 the Hooker Chemical Company signed an agreement with the canal's new owner, the Niagara Power and Development Corporation, to dump wastes there. In 1946 Hooker bought the site, and from 1947 to 1952 it dumped over 20,000 metric tons of highly toxic and carcinogenic wastes, including the deadly poisonous dioxin.

In 1952 the story took an ironic twist. In that year the city of Niagara Falls began condemnation proceedings on the property that would allow it to use the land for an elementary school and residential community. With no other choice, Hooker sold the land for $1 in exchange for a release from any future liability. Hooker insists that it warned against any construction on the dump site itself, but it allegedly never disclosed the real danger of building on the abandoned dump. Before turning the land over to the city, Hooker sealed the dump with a clay cap and topsoil, once thought sufficient to protect hazardous waste dumps.

Troubles began in January 1954, however, when workers removed the clay cap during the construction of the school. In the late 1950s rusting and leaking barrels of toxic waste began to surface. Children playing near them suffered chemical burns; some became ill and died. Hooker said that it warned the school board not to let children play in contaminated areas, but made no effort to warn local residents of the potential problems.

The problem continued for years. Chemical fumes took the bark off trees and wiped out grass and garden vegetables. Smelly pools of toxins welled up on the surface. In the early 1970s after a period of heavy rainfall, the water table rose, and basements in homes near the dump began to flood with a thick, black sludge of toxic chemicals. The chemical smells in homes around the dump site became intolerable.

Tests in 1978 on water, air, and soil in the area detected 82 different chemical contaminants, a dozen of which were known or suspected carcinogens. The State Health Department found in 1978 that nearly one of every three pregnant women in the area had miscarried, a rate much higher than expected. Birth defects were observed in 5 of 24 children. Another study, released in 1979 by Dr. Beverly Paigen of the Roswell Cancer Institute, showed that over half of the children born between 1974 and 1978 to families living in areas where groundwater was leaching toxic chemicals from the dump had birth defects. In this study the overall incidence of birth defects in the Love Canal area was one in five, compared with a normal rate of less than one in ten (see Chapter 14). The miscarriage rate was 25 in 100, compared with 8 in 100 women moving into the area. Asthma was four times as prevalent in wet areas as dry areas in the region; the incidence of urinary and convulsive disorders was almost three times higher than expected. The incidence of nasal and sinus infections, respiratory diseases, rashes, and headaches was also elevated.

As a result of public outcry the school was soon closed. The state fenced off the canal and evacuated several hundred families (Figure 18-2). President Carter declared the dump a disaster area. In May 1980 a new study revealed high levels of genetic damage among residents living near the canal. An additional 780 families were evacuated from outlying areas.

As of 1987 Love Canal had cost the state of New York and the federal government about $200 million for cleanup, research, and relocation of residents. In 1987 the EPA announced plans to dredge the sewers and creeks in the Love Canal area to remove sediments that have been contaminated with toxic materials. Dredged material will be incinerated, then buried in a landfill. All told,

Figure 18-2. House is bulldozed in the Love Canal area of Niagara Falls, New York.

about 35,000 cubic meters of sediment will be burned, making this the largest single application of thermal destruction in modern history. The EPA estimates that the dredging and incineration will take about five years to complete and will cost $26 to $31 million.

A 1980 study by the EPA showed that chemical contamination was pretty much limited to the canal area (the actual dump), an area immediately south of it, and two rows of houses on either side of the canal (Figure 18-3). The last group of residents to be evacuated, the report said, were probably moved out unnecessarily. The EPA study also showed that the dump had contaminated shallow groundwater, but not the deeper aquifers. The EPA concluded that further migration of toxic chemicals was highly unlikely. Based on this study and other work, the EPA and the state of New York declared two-thirds of the evacuated Love Canal site "habitable" and proposed to move families back in. Lois Gibbs, the Love Canal resident largely responsible for drawing public attention to the disaster and getting the state and federal governments to take action, argues that the decision to resettle the area has been made without adequate risk assessment. The New York State Health Department, she says, spent $14 million to compare the Love Canal site to two other sites in the city. It found that the site was *as* contaminated as two other sites, both contaminated by industrial wastes, and thus deemed it suitable for resettlement. Gibbs warns that resettling Love Canal will put more people at risk.

The Dimensions of a Toxic Nightmare

Love Canal began the frenetic search for hazardous waste dumps and illegal waste-disposal practices that is still

Figure 18-3. Love Canal. Area within the colored rectangle was closed off, and citizens were evacuated. Citizens were also evacuated from the declaration area, but tests have shown that hazardous wastes have not migrated into this area.

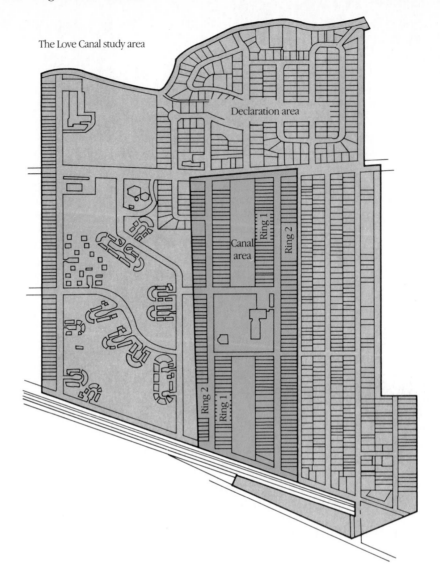

The Love Canal study area

Declaration area

Ring 1

Ring 2

Canal area

Ring 2

Ring 1

going on. Many people, previously unable to explain bizarre diseases in their family, soon found the answer in nearby waste dumps or factories that leaked hazardous wastes into groundwater, nearby streams, or the air.

In the years following the Love Canal incident the American public has been barraged by a list of startling statistics showing that what appeared to be an isolated incident was in fact just the tip of the iceberg. The EPA, for instance, estimated that there were 14 other sites in Niagara Falls alone that it considered an "imminent hazard." Nationwide, the EPA announced, Love Canal was one of a thousand or more sites in need of a cleanup. Today, some experts put the number of dangerous hazardous waste facilities that need to be cleaned up at well over 10,000.

Making matters worse, each year factories create an estimated 54 to 72 million metric tons of waste considered hazardous by federal standards and about 200 million metric tons of hazardous waste covered by state regula-

tions. The total is well over a ton per person. But the United States is not alone. European countries also produce millions of tons of hazardous waste each year.

Even more startling than the sheer amount of waste produced is its fate. Until quite recently 90% of the hazardous wastes in the United States were improperly disposed of, ending up in abandoned warehouses; in rivers, streams, and lakes; in leaky landfills that contaminate groundwater; in fields and forests; and along highways. Because of improvements in hazardous waste management, smaller amounts of toxic waste are being improperly discharged. But don't be misled; the nation needs to do a great deal more to curb this problem. In 1987 in California alone businesses produced 2.5 million metric tons of pollution. Here's where it ended up:

1. 65% was discharged into the ocean, lakes, and rivers.

2. 27% was injected into deep wells.

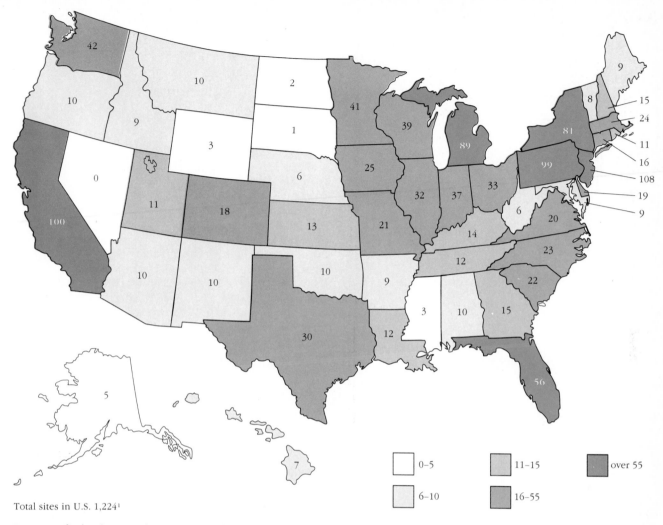

Total sites in U.S. 1,224[1]

Represents final and proposed sites on National Priority List.
[1]Includes nine in Puerto Rico; and one in Guam.
Source: U.S. Environmental Protection Agency, National Priorities List Fact Book.

Figure 18-4. Proposed and potential hazardous waste sites on the EPA's 1989 priority list for cleanup. All told, 889 sites are slated for cleanup; 335 additional sites are proposed for cleanup.

3. 4.3% went to public sewers and sewage treatment plants.

4. 1.8% went to treatment, storage, and disposal facilities.

5. 1.3% escaped into the air.

Surprisingly little is known about hazardous waste production and disposal. This we do know: Ill-conceived and irresponsible waste disposal creates a legacy of polluted groundwater and contaminated land. The US Office of Technology Assessment believes that it will cost $100 billion to clean up the 10,000 sites in the United States that pose a serious threat to health (Figure 18-4).

The ecologist Barry Commoner recently remarked about hazardous wastes, "We are poisoning ourselves and our posterity." In truth, improper waste disposal has left behind a long list of costly effects: (1) groundwater contamination, (2) well closures, (3) habitat destruction, (4) human disease, (5) soil contamination, (6) fish kills, (7) livestock disease, (8) sewage treatment plant damage, (9) town closures, and (10) difficult or impossible cleanups.

A decade after the United States first awakened to the hazardous waste issue, experts believe that we are in for a much longer, more difficult battle than once anticipated. Lee Thomas, former head of the EPA, argued, "There are far more sites that are far more difficult to deal with than anybody ever anticipated."

>< *Point/Counterpoint*

Are We Facing an Epidemic of Cancer?
America's Epidemic of Chemicals and Cancer

Lewis G. Regenstein

Lewis Regenstein, an Atlanta writer and conservationist, is author of How to Survive in America the Poisoned (*Acropolis Books*).

America is in the throes of an unprecedented cancer epidemic, caused in large part by the pervasive presence in our environment and food chain of deadly, cancer-causing pesticides and industrial chemicals.

Today, significant levels of hundreds of toxic chemicals known to cause cancer, miscarriages, birth defects, and other health effects, are found regularly in our food, our air, our water—and our own bodies. Accompanying this widespread pollution has been a dramatic and alarming rise in the cancer rate in recent decades.

Each year, almost a million Americans (about 985,000 in 1988, not including skin cancers) are diagnosed as having cancer—almost 3000 people a day! The disease now strikes almost one American in three, and kills over a thousand of us *every day*! This means that of the Americans now alive, some 70 to 80 million people can expect to contract cancer in their lifetimes. More Americans die of cancer *every year* (an estimated 494,000 in 1988) than were killed in combat in World War II, Korea, and Vietnam *combined*.

Man-made chemicals are also depleting the Earth's protective ozone layer, which makes life on the planet possible by shielding us from most of the sun's ultraviolet rays. The US Environmental Protection Agency (EPA) projects that the increase in radiation hitting the Earth will cause Americans to suffer 40 million cases of skin cancer, 800,000 deaths in the next 88 years, and 12 million incidences of eye cataracts.

Cancer is the leading cause of death for women between the ages of 30 and 40, and for children aged 1 to 10. Thus, the elevated cancer rate is not caused solely by people living longer, as the chemical industry often claims; it has now become a common disease of the young as well as the old.

In 1978, the President's Council on Environmental Quality (CEQ) reported unequivocally that "most researchers agree that 70% to 90% of all cancers are caused by environmental influences and are hence theoretically preventable." Only about 30% of the nation's annual cancer deaths (mainly lung cancers) are thought to be caused by cigarette smoking, and many of these may be related to carcinogenic pesticides and chemicals in tobacco. The remaining 70% of cancer deaths are caused by a variety of other factors, with toxic chemicals thought to play a major role. For instance, in 1978 the Department of Health, Education, and Welfare warned that 20% to 38% of all cancers could be attributed to occupational exposure to just six industrial chemicals.

In 1980, the federal inter-agency Toxic Substances Strategy Committee issued its report clearly demonstrating that synthetic chemicals are a threat to the lives and health of untold millions of Americans, and confirming the alarming rise in the cancer rate. In announcing the findings, CEQ emphasized that "man-made toxic chemicals are a significant source of death and disease in the United States today." The agency stressed the link between the increase in production of such chemicals between 1950 and 1960 and the large increase in the cancer rate showing up 20 to 25 years later, "the lag time one might expect."

Evidence continues to mount demonstrating that toxic chemicals are heavily contributing to the cancer epidemic. In general, the most polluted areas of the country have the highest cancer rates, with heavily industrialized New Jersey having the greatest concentration of chemical and petroleum facilities. In July, 1982, the University of Medicine and Dentistry of New Jersey released a study showing a correlation between the presence of toxic waste dumps and elevated cancer death rates (up to 50% above average) in areas of the state.

In February, 1984, a report by researchers at the Harvard School of Public Health demonstrated a link between the consumption of chemically contaminated well water near Woburn, Massachusetts and the extraordinary incidence of childhood leukemia, stillbirths, birth defects, and childhood disorders of the kidneys, lungs, and skin among local residents.

An October 1984 report by the Council on Economic Priorities found an "alarming" increase in cancer death rates in rural counties with high employment in chemical and petroleum industries that generate large amounts of toxic waste.

And a July 1985 Congressional study found an apparent link between high respiratory cancer rates and the presence of petrochemical plants in 133 counties east of the Rockies.

Some scientists and government officials, pointing to the decrease in rates of some cancers, continue to insist that the cancer scare has gotten out of hand, that there is no "cancer epidemic," and that toxic chemicals play a small role in causing cancer. But Dr. Samuel Epstein of the University of Illinois Medical Center, perhaps the foremost authority on the subject, points out that apart from AIDS, "cancer is the only major killing disease which is on the increase," with incidence rising by at least 2% a year, and death rates at 1% annually over the last decade. He concludes that "the facts show very clearly that we are in a cancer epidemic now," in large part because of "the carcinogenizing of our environment, the increasing contamination of our air and our water and our food and the workplace."

Today, every American is regularly and unavoidably exposed to a variety of dangerous, health-destroying chemicals. Dozens of pesticides used on our food are known or thought to cause cancer and birth defects in animals. By the time restrictions were placed on some of the deadliest chemicals, such as DDT, dieldrin, BHC, and PCBs, these carcinogenic poisons were being found in the flesh tissues of literally 99% of all Americans tested, as well as in the food chain and even mother's milk. In fact, breast milk is heavily contaminated with high levels of banned, cancer-causing chemicals.

Virtually all Americans carry in their bodies traces of dioxin (TCDD), the most deadly manmade chemical known. The sources of this dioxin to which we are regularly exposed include food (such as fish and beef, contaminated from waste dumping and herbicide spraying), emissions from municipal and industrial incinerators (which produce dioxins when plastics are burned), milk from white cardboard cartons, and exposure to other white paper products, in which the bleaching process creates dioxins.

The response of the US government has been largely weak or nonexistent enforcement of the nation's health and environmental protection laws. For example, with few exceptions, the EPA has refused to carry out its legal duty to ban or restrict pesticides known to cause cancer. Nor has the government adequately implemented or enforced the laws regulating hazardous waste, which is being generated at a rate of up to 275 million metric tons a year—over a ton for every man, woman, and child in the nation. Much of this is disposed of in a manner that will ultimately threaten the health of nearby residents.

Thus we are even now sowing the seeds for cancer epidemics of the future. Only time will tell what will be the effect on this generation, and future ones, of Americans—the chemical industry's ultimate guinea pigs. By the time we know the answers, it may be too late to do anything effective about the problem.

"America's Epidemic of Chemicals and Cancer"—Myth or Fact?

David L. Eaton

David L. Eaton is Associate Professor of Environmental Health and Environmental Studies, and Director of Toxicology at the University of Washington. He has an active research program on the mechanisms by which chemicals cause cancer.

There is no debate that cancer is a devastating and deadly disease. One in three people living in the US today will contract some form of cancer in his or her lifetime, and one in four will die from it, if current rates continue. However, the contention that we are in the throes of an unprecedented cancer epidemic, and that this "epidemic" is caused in large part by cancer-causing pesticides and industrial chemicals is simply not supported by the scientific data available on cancer incidence and etiology (causes).

Are cancer rates increasing in epidemic proportions? The total number of people and the fraction of all deaths attributable to cancer have increased dramatically in the past 50 years. However, cancer is largely a disease of old age, and thus it is necessary to adjust such statistics for changes in the age distribution of our population. A 1988 report from the National Cancer Institute states that "the age adjusted mortality rates for all [types of] cancers combined, except lung cancer, have been declining since 1950 for all individual age groups except 85 and above." Statistics from the American Cancer Society yield the same conclusion.

Excluding lung cancer in this evaluation allows us to examine cancer trends without the strong influence that the dramatic increase in lung cancer incidence (and deaths) have on *total* cancer rates. We *are* in an "epidemic" of lung cancer. For most of the first half of this century, lung cancer mortality was not even in the "top five" types of cancer-related deaths. Lung cancer is now the leading cause of cancer-related deaths in both men and women. About 85–90% of all lung cancers in men, and perhaps 70% in women, is directly attributable to smoking. *Per capita* consumption of cigarettes increased 5-fold in men from 1900 to 1960, and with it a concomitant increase in lung cancer. This pattern was repeated, 20 years later, in women. The risks of several other types of common cancers are also increased by smoking (e.g., cancers of the bladder and esophagus). Approximately one-third of *all* cancer deaths could be eliminated by eliminating smoking from our society. The supposition that many of these smoking-related cancers may be related to carcinogenic pesticides and chemicals in tobacco is grossly misleading, as the *source* of these carcinogenic pesticides and chemicals is *mother nature*, not the chemical and agricultural industry.

What proportion of cancers can be related to environmental pollution from synthetic pesticides and industrial chemicals? The often-cited statistic that "70 to 90% of all cancers are caused by environmental influences and are hence theoretically preventable" is frequently incorrectly interpreted as meaning that *chemical pollution* is responsible for 70–90% of cancers. Studies in the 1960s suggested that most cancers could not be directly traced to genetic or hereditary factors; therefore, it was concluded that the majority of cancers must have an "environmental" cause. In this context, the term "environmental" includes not only chemicals, but lifestyle factors such as smoking and alcohol, dietary factors such as the proportion of fat and fiber in the diet, "natural" carcinogens that occur in nearly all foods, cancer-causing viruses such as hepatitis B, and occupational exposures to substances such as asbestos.

Of the variety of environmental factors other than smoking, dietary factors are now generally thought to represent the largest source of cancer risk, perhaps related to 30–40% of all cancers. Although synthetic chemicals such as industrial pollutants and pesticides present in trace amounts in our food supply may contribute to dietary risk, recent studies have suggested that this contribution is trivial relative to other "non-pollutant" factors. For example, the risk of breast cancer in women (second only to lung cancer in incidence and mortality) is significantly increased by high fat diets, and the amount of fiber in the diet substantially influences the risk of

colon cancer, a major site of cancer in both men and women.

The largest source of exposure to cancer-causing chemicals may not be industrial pollution, but chemicals that occur naturally in our diet. All plants produce toxic chemicals as a means of protection against insects, fungi and animal predators. It has been estimated that we ingest in our diet about 10,000 times more of "nature's pesticides" than man-made chemical residues. Many of these chemicals are potent mutagens and carcinogens, and are frequently present at levels thousands of times higher than the trace levels of synthetic pesticide residues and industrial chemicals sometimes found in food crops. For example, natural chemicals which are potentially carcinogenic are found in mushrooms, parsley, basil, fennel, pepper, celery, figs, and mustard, to name a few. The vast majority of natural chemicals present in foods (tens of thousands) have never been tested to determine if they might be carcinogenic. Also, cooking of foods, especially meats, forms highly mutagenic and perhaps carcinogenic chemicals from natural precursors such as certain protein components. Taken together, the dietary risk factors from natural sources, often present in relatively high amounts, are far more important than the pesticide residues and industrial chemicals that can often be detected at exceedingly small concentrations in our diets. Unfortunately, because of the relatively high exposure to carcinogens from natural sources, the complete elimination of synthetic industrial chemicals from our diet, if it were possible, would not likely have any significant beneficial effect on cancer incidence and mortality.

There is no question that extensive exposure to some industrial chemicals can increase the risk of certain types of cancer. Every effort should be made to identify and reduce workplace exposure to these chemicals. However, the reliance upon a 1978 report of the Department of Health, Education and Welfare which stated that "20 to 38% of all cancers could be attributed to occupational exposure to just 6 industrial chemicals" is no longer valid. The authors of this report acknowledge that this was a substantial overestimation of the contribution of occupational exposures to total cancer incidence. Most authoritative sources now estimate that occupational exposures account for no more than 5 to 10% of all cancers, and many of these cancers are a result of extensive exposures to asbestos and a few other industrial carcinogens that were commonplace in the 1950s and 1960s, but have since been greatly reduced (although certainly not eliminated). Of course, 5% is too high, and every effort should be made to reduce this contribution further.

Finally, recent advances in the understanding of the biology of cancer suggest that "spontaneous" or "background" alterations in DNA may explain much of the cause of cancer. The use of modern techniques in molecular biology has revealed that DNA is inherently unstable and can be altered by normal errors in DNA replication. DNA is subject to extensive damage from processes associated with normal cellular metabolism. Within our life span, our cells undergo about 10 million billion

⋈ *Point/Counterpoint*

cell divisions. Spontaneous errors in this process, which lead to mutations and cancer, accumulate with age. It is not surprising then that cancer seems to be a frequent outcome of old age. It has been estimated that the metabolism of oxygen in cells, essential to survival, results in approximately 10,000 oxygen-induced alterations in each of our ten trillion cells each day. Although the ability of our bodies to repair such damage is remarkable, it is not perfect, and some of this daily DNA damage will result in mutations which could give rise to cancer.

The views that we are in an overall cancer epidemic and that these cancers are largely a result of industrial chemicals are not supported by the vast majority of cancer researchers throughout the world. One has only to read the modern scientific literature on the causes of cancer to arrive at that conclusion. Unfortunately, it will take some time for the political arena, influenced greatly by public fears, to come to grips with the fact that further reduction in public exposure (i.e. non-occupational) to synthetic chemicals will not have much impact on cancer incidence, and that such results will come only at great social and economic expense. The United States is currently spending about $80 billion per year on pollution reduction, about nine times the total budget for all basic scientific research. I believe that much of this is justified to enhance the quality of our environment, and ensure the habitability of our planet for our children, their children's children, and other species. However, there is also little question

that huge sums of money are spent each year to reduce what is in all likelihood a trivial cancer risk, with few other environmental benefits. If our society is truly concerned about reducing the human tragedy from cancer, more efforts should be focused on eliminating smoking and alcohol abuse, better research and education on dietary risk factors, more research into the biochemical and molecular events that lead to cancer (which in turn will lead to more effective preventive and curative measures), and continued identification and reduction in occupational exposures to those chemicals which pose a significant cancer risk.

Critical Thinking

1. Summarize each author's major points and supporting data. In your view, how well have each of the authors supported their contentions? Point out potential weaknesses in arguments.

2. Review the critical thinking skills presented in Chapter Supplement 1-1. How can they help you in sorting out the facts and fallacies in the arguments presented here?

3. Do you agree with Dr. Eaton that to determine whether we are in a cancer epidemic caused by synthetic chemicals, such as pesticides, we should eliminate lung cancer from the calculation? Why or why not?

LUST—It's Not What You Think

You feel dizzy. Your head spins. Your insides ache. You haven't been yourself for weeks. What may be ailing you is LUST—but not the usual kind. Your symptoms may be caused by the latest in a long list of hazardous chemical problems, groundwater pollution from a leaking underground storage tank, which EPA's top acronymists have dubbed LUST.

Underground tanks containing petroleum by-products, toxic chemicals, and hazardous wastes deteriorate over time, springing leaks and dripping out their toxic substances. Moisture and soil acidity are primarily responsible for the leaks. The main concern is the potential effect on groundwater and human health. Even a small leak can contaminate large quantities of groundwater. For example, a leak of only 1.5 cups of hazardous liquid per hour can contaminate nearly 3.8 million liters (1 million gallons) of groundwater in a single day. Contaminated groundwater is very difficult and expensive to clean up (Chapter 16). In some cases, it may be impossible to clean.

Many people in affected areas have switched to bottled water, only a temporary solution at best. Contaminated water used for baths and showers can also be dangerous. Benzene, a component of gasoline that can cause cancer, is absorbed through the skin when bathing. Showering generates dangerous vapors that can cause skin and eye irritation.

A report by the New York Department of Environmental Conservation suggests that at least half of the state's underground steel tanks containing petroleum products over 15 years old may now be leaking. Nationwide, 3 to 5 million underground storage tanks containing hazardous materials dot the United States. The EPA estimates that 200,000 to 400,000 (maybe more) of these tanks are leaking.

Major oil companies have already spent millions to clean up polluted groundwater and soil. Many other companies have installed new tanks at a quicker pace to avoid further contamination, but the cost of such actions can be exorbitant. Chevron alone estimates its replacement costs at about $100 million. Unfortunately, half of US service stations are owned by independent dealers, who generally are not financially able to replace the leaking tanks. Many tanks that could be leaking are under schools, police stations, and private homes.

About 90% of the cleanup and replacement of leaking tanks is being financed and performed by private industry. The rest is being done by the states themselves. The EPA sets guidelines for cleanup and replacement and also provides financial assistance to help states. Today, more than 30 states have their own funds to pay for part of the cleanup cost. State and federal funds are derived from taxes on gasoline.

Attacking Hazardous Wastes on Two Fronts

Two hazardous waste problems face society today: (1) what to do with the enormous amounts of hazardous waste produced each year, and (2) what to do with the leaking waste disposal sites or with areas like Times Beach, Missouri, which was contaminated by road oil containing dioxin.

What to Do with Today's Waste

The Legal Approach In 1976 congressional representatives proudly announced a tough, new law—the **Resource Conservation and Recovery Act (RCRA)**—designed to cut back on illegal and improper waste disposal. What are its major provisions? First, RCRA named the Environmental Protection Agency as the hazardous waste watchdog. The EPA's first role was to determine which wastes were hazardous. Second, RCRA called on the agency to establish a nationwide reporting system for all companies handling hazardous chemicals. This requirement created a trail of paperwork that would follow hazardous wastes from the moment they were generated to the moment they were disposed of—from cradle to grave. This stipulation, Congress believed, would make it difficult for waste generators to dump wastes improperly. Third, and perhaps most important, RCRA directed the EPA to set industrywide standards for packaging, shipping, and disposal of wastes. Only licensed facilities could receive wastes.

Unfortunately, RCRA's implementation has been slow. It was not until four years after Congress passed the act that the EPA came up with its first hazardous waste regulations. To the dismay of many, the regulations were full of loopholes, so much so that about 40 million metric tons of pollutants escaped control each year!

Public pressure mounted, and in 1984 Congress passed a set of tough amendments to RCRA to eliminate loopholes and ensure proper waste disposal. For example, under the original law if a company produced under 1000 kilograms (2200 pounds) of hazardous wastes per month, it could dump them in a local garbage dump if it wanted. Today any generator of waste over 100 kilograms (220 pounds) must follow the same guidelines imposed on large waste producers. In a bold move Congress's 1984 amendments declared that it was the national policy to reduce or eliminate land disposal of hazardous waste. Congress made it clear that land-disposal technologies must be considered a last resort. Preference is given to recycling, destruction, and other processes, discussed below. Unfortunately, these recommendations are often ignored. The newest additions to RCRA also address the problem of leaking underground storage tanks. Beginning in May 1985 all newly installed underground tanks must be protected from corrosion for the life of the tank. The lining of the tank must be compatible with stored substances. Owners and operators must have methods for detecting leaks, must take corrective action when a leak occurs, and must report such action.

RCRA, while an important first step in helping us solve the hazardous waste problem, still has many loopholes that critics want eliminated. The most important change would be broadening the definition of hazardous waste to include many wastes that now escape very stringent controls. Michael Picker of the National Toxics Campaign, for example, thinks that municipal waste itself should be classified as hazardous waste because it contains toxic chemicals, such as pesticides, that are routinely discarded in municipal trash. Leachates from some municipal landfills are as toxic as those coming from regulated hazardous waste facilities. Picker also thinks that sewage and untreated wastewater handled by publicly owned sewage treatment plants should be considered a hazardous waste. Toxic chemicals in the sewer system, which are released by factories and by homeowners, can escape into the air and into waterways. Agricultural wastes, mostly pesticides, are also not regulated. In California, for example, rules require that leftover pesticides be diluted and sprayed into the environment. Mill and mine tailings are also excluded from most regulatory control. By expanding the definition of what is toxic and instituting better controls, the government could greatly cut back on the constant flow of hazardous materials into the increasingly poisoned environment.

One lesson we have learned in the past 20 years is that the passing of a law is not a guarantee of protection. Why? For one, agencies whose responsibility it is to administer and enforce new programs under new environmental laws don't always do as they are instructed. Some drag their feet because they don't approve of the law. More commonly, agencies are so badly underfunded and so overworked that they can't take on new responsibilities or, if they do, they do a shoddy job. The EPA is a case in point. Understaffed and underfunded, the EPA today struggles to implement RCRA and other laws aimed at pro-

Case Study 18-1

Exporting Toxic Troubles

Tough regulations and rising costs for hazardous waste disposal in the United States have led many industrial hazardous waste producers to look to foreign countries as a place to dump their hazardous materials. Today, most US waste, including municipal solid waste, is dealt with domestically. Some goes to Canada and Mexico, but an increasing amount is finding its way to Third World nations.

Developing nations in need of cash to pay foreign debt have fallen victim to unscrupulous waste traders. Many of the traders have no qualms about characterizing a product as less hazardous than it really is. Cities with tight budgets, like Philadelphia, have found that environmental laws have made costs prohibitive and are looking elsewhere for waste disposal sites. They are finding that it is much cheaper for them to ship their waste to Africa or Central America where it can be disposed of at a fraction of the cost in the United States.

The problem with exporting waste is that many of the countries that receive the waste don't know what is in it, don't know how toxic the materials really are, and don't have facilities to store it or dispose of it properly.

In March 1988 a Norwegian freighter arrived on Africa's west coast to deliver a cargo listed as "raw materials for bricks." A Guinea concrete manufacturer had purchased the material to build roads and cinder blocks. Unfortunately the material proved to be terribly inadequate. Bricks crumbled in the hands of laborers. Moreover, trees near the piles where the material was stored before use began to die. On closer examination, the brick material turned out to be ash from a Philadelphia garbage incinerator that contained heavy metals and dioxins.

The ash had originally been destined for Panama, where it was to be used to build a road through a wetlands, but the environmental group Greenpeace warned the government that the material would cause incredible damage. The government halted the shipment and the seller found a new home in Africa.

In Tecate, Mexico, Mexican officials found that 400,000 liters (100,000 gallons) of hazardous materials had been carelessly dumped on the ground. The waste came from the United States and had been shipped to a bogus, unlicensed Mexican recycling company and dumped near Tecate.

These incidents stirred debate on what the United States can do to regulate the export of hazardous materials. In 1986, Congress amended the Resource Conservation and Recovery Act (RCRA) by establishing procedures to notify importing countries and also to require a prior written consent. These regulations, however, are already under fire. EPA officials think that hundreds of tons of hazardous wastes are being exported illegally.

The practice of exporting hazardous waste, besides polluting Third World nations, may also prove to be bad politics. The relationship between Nigeria and Italy, for example, was severely strained after nearly 3000 metric tons of highly toxic waste were dumped in the African country in 1988 by an Italian businessman. Citizens expressed outrage over the incident. The military government of Nigeria, in fact, threatened executions, arrested 15 alleged dumpers, and seized two Italian ships. If the US continues to allow hazardous waste to be shipped to developing nations, similar political nightmares await.

Exporting hazardous waste to a nation without its full consent goes against important principles of international law. Many African nations have agreed to a resolution condemning the disposal of toxic wastes on African soil. Other nations in the developing world are following the African lead. The Organization of Eastern Caribbean States in 22 Latin American countries, for instance, has joined forces to stop the dumping of hazardous wastes on their soils. In May 1988 the European parliament passed a resolution calling for an end to large-scale hazardous waste export to developing countries. Currently Europe annually exports approximately 450,000 metric tons of toxic waste. The twelve-member European Community adopted initiatives that require prior notification as well as assurance that the recipient is capable of properly handling the waste. In the United Nations talks are under way for the development of global standards to regulate transboundary movement of hazardous waste. This action, say some environmentalists, is a step in the wrong direction, for it could end up promoting export and not encouraging waste reduction. Furthermore, regulating such a system and enforcing such a program could prove to be a nightmare.

On the legal front, there are several major options open to the United States. We could maintain the existing regulatory structure under RCRA, which is ineffective and full of loopholes, or amend RCRA so that the government can more effectively monitor the hazardous waste trading.

Others believe that a complete ban on the international movement of waste may be the best answer. This would help protect the environment from inadequate disposal methods and would help nations develop long-term solutions that reduce hazardous waste production.

Ultimately the waste trade is not a problem in itself, but a symptom of the failure of the developed world to intelligently deal with its overproduction of waste, says Greenpeace author Judy Christrup. Banning the international waste trade would help stop the contamination of the environment. Providing waste makers with an escape valve such as export moves us in the wrong direction. The only real solution is to reduce toxic waste at its source, that is, to stop it before it is produced.

Adapted from: Christrup, J. (1988). Return to Sender: Clamping down on the International Waste Trade. *Greenpeace* Nov./Dec. 1988; 13 (6): 8–11.

Figure 18-5. A three-tier hierarchy of options for handling hazardous wastes. The top tier reduces the hazardous waste stream; it contains the most desirable options. The middle tier converts hazardous materials into nonhazardous or less hazardous substances. Perpetual storage, the lowest tier, is the least desirable, but often the cheapest, alternative.

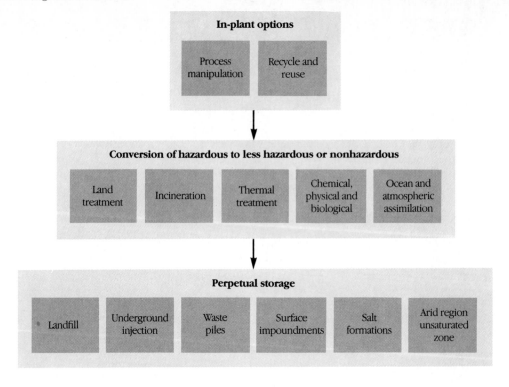

tecting public health and the environment. Since the agency was formed its workload has more than doubled, but funding has remained at more or less the same level (when we adjust for inflation) until very recently. Even now, increases only partly offset the remarkable increase in potential workload.

Another lesson that time has taught us is that tougher legislation often has unanticipated effects. Because disposing of wastes properly invariably costs more than dumping wastes in sewers and in lakes and streams, tougher laws can lead to increased illegal dumping by unscrupulous business people. The high cost of hazardous waste management has also led some companies to seek foreign countries that are willing to take their waste. Hungry for money, these countries will take the waste and dump it in ways that would be illegal here. (For more on this practice, see Case Study 18-1.)

Technological Answers In 1983 the prestigious National Academy of Sciences issued a lengthy report outlining, in order of desirability, many of the United States' options for handling hazardous wastes (Figure 18-5). At the top of the list are **in-plant** options, changes that can be made to reduce hazardous waste production. Changes in the way products are manufactured can drastically cut waste. In a report called *Cutting Chemical Wastes*, a group called Inform outlines numerous options that can help cut toxics use. Industry, for example, can change the raw materials it uses, substituting a nontoxic or less toxic substance for highly toxic materials. Cleo Wrap, the

world's largest producer of gift wrapping, for example, switched inks and cut the production of hazardous waste by 140,000 kilograms (300,000 pounds) a year. Industries can also change the components of a product, eliminating ones that might be harmful or produce harmful by-products during manufacturing. Companies can also better monitor their processes to locate leaks that may be emitting toxic wastes into the air, ground, or water. Exxon Chemical, for example, installed floating lids over some vats containing volatile organic compounds, thus greatly reducing losses from evaporation. The Office of Technology Assessment argues that American industries could reduce or prevent more than 50% of the nation's hazardous waste generation by applying such measures. In some instances wastes can be separated and purified, yielding salable or reusable chemicals. **Recycling and reuse** strategies help cut waste and may save companies millions of dollars a year in operating costs. The recycling-and-reuse option also eliminates the cost of waste disposal, cuts down on potential environmental and health damage, and saves valuable raw materials, including energy.

Not every waste product can be used, however. Therefore, some waste will always be produced—the less the better. The academy recommended that these remaining wastes be destroyed or detoxified, converted to less hazardous materials. **Detoxification** can be accomplished for certain types of waste by applying them to land and mixing them with the top layer of soil, where they are broken down by chemical reactions, oxidation by sunlight, or by

Figure 18-6. A mobile hazardous waste incinerator owned and operated by the EPA avoids the problem of transporting exceptionally dangerous materials.

bacteria or other organisms in the soil. Some nondegradable wastes may be absorbed onto soil particles and held there indefinitely. Others may migrate into deeper layers. Land treatment is an expensive option, requiring care to avoid polluting ecosystems, poisoning cattle and other animals, and contaminating groundwater.

Destroying hazardous wastes by **incineration** is a questionable option. High-temperature furnaces at stationary waste disposal sites, on ships that can burn their wastes at sea, and on mobile trailers can all burn organic wastes (Figure 18-6). Oil or natural gas is used as a fuel. Hazardous substances are injected into the furnace or mixed with the fuel before combustion. In 1985 the EPA announced that its new mobile incinerator destroyed 99.9999% of the dioxin wastes in soil and liquids. Officials at the EPA are optimistic that the incinerator will be useful in cleaning up dioxin wastes at sites targeted for cleanup by Congress. Critics point out, however, that these results come from tests run under optimal conditions. In real life, incinerators must handle a wide variety of wastes that may not burn so efficiently.

Incineration can provide energy for plant operations, can eliminate toxic organic wastes such as dioxin and PCBs, and can reduce perpetual storage. However, communities often object to incinerators, fearing environmental contamination because of possible spills during transport or emissions and leaks from the plants. Incinerators may not always perform adequately and operating personnel may bypass regulations. Communities have real cause to be concerned about the placement of a toxic incinerator.

Low-temperature decomposition of cyanide and toxic organics such as pesticides offers some promise. In this technique wastes are mixed with air and maintained under high pressure while being heated to 450° to 600° C (840° to 1100° F); during the process organic compounds are broken into smaller biodegradable molecules. Valuable materials can be extracted and recycled. This process uses less energy than incineration.

Chemical, physical, and biological agents can be used to detoxify or neutralize hazardous wastes. For example, lime can neutralize sulfuric acid. Ozone can be used to break up small organic molecules, nitrogen compounds, and cyanides. Toxic wastes can be encapsulated with a plastic waterproof seal, lowering the risk of land disposal. Many bacteria can degrade or detoxify organic wastes and may prove helpful in the future. New strains capable of destroying a wide variety of organic wastes may be developed through genetic engineering.

In-plant modifications and conversion technologies that destroy or detoxify wastes cannot rid us of all of our waste. By various estimates, 25% to 40% of the waste stream will remain even after the best efforts to recycle, reuse, and destroy it.

Wastes that cannot be destroyed entirely and detoxified wastes as well must be stored. Residual waste can be dumped in **secured landfills**, excavated pits lined by synthetic liners and thick, theoretically impermeable layers of clay. To lower the risk of leakage, landfills should be placed in arid regions—neither over aquifers nor near major water supplies. Special drains must be installed to catch any liquids that leak out of the site. Groundwater and air should be monitored to detect leaks.

Growing public opposition to hazardous wastes makes it more difficult for companies to find dump sites. Observers have labeled this the NIMBY syndrome: get rid of the stuff, but *not in my* backyard. It seems that most people want the products that make waste or give little thought to their purchase, but no one wants the wastes dumped (or even burned) nearby. New studies show that landfill liners are not as reliable as once thought, further supporting community doubts.

Even though the EPA has issued tough new regulations for landfills, critics argue that landfills are only a temporary solution. No matter how well constructed they are, they will eventually leak. Landfills are one of the cheapest waste disposal practices in use today and are therefore highly favored by industry. But savings today, critics warn, are inevitably charged to future generations. In an attempt to avoid making problems for future generations, the EPA has drawn up a list of chemicals that cannot be disposed of in landfills.

Other methods of **perpetual storage** include (1) use of surface impoundments and specially built warehouses that hold wastes in ideal conditions and prevent any material from leaking into the environment, (2) deposition in geologically stable salt formations, and (3) deposition deep in the ground in arid regions where groundwater is absent.

Barriers to Waste Reduction The most effective way of reducing hazardous waste problems is to cut back on the generation of wastes in the first place. Unfortunately, when the EPA began its hazardous waste control program in the United States it spawned a large and lucrative treatment and disposal industry. This politically powerful industry with its close ties to government may be thwarting the development of waste minimization strategies.

The EPA today spends less than 5% of its hazardous waste budget on preventing pollution. Decreasing hazardous waste production would undercut the hazardous waste management industry and discourage investment in an industry spawned by EPA action. "Hazardous waste regulators," says Picker, "tied to the needs of the industry they foster have a powerful incentive to encourage waste production."

This bias toward waste disposal and treatment, the riches that can be made in the industry, and some Americans' blind faith in technologies such as incinerators and secured landfills, are three barriers to a system of hazardous waste management. (For more on barriers to sustainability, see Chapter 20.)

Disposing of High-Level Radioactive Wastes High-level radioactive wastes are some of the United States' most hazardous wastes, but they have long been ignored. Even today, despite the buildup of 18,000 metric tons of spent uranium from commercial reactors and 8000 to 9000 metric tons of Department of Defense high-level waste from weapons production, the country is without a solution. The seriousness of continued delay is underscored by three facts: many radioactive wastes have a long lifetime, some materials can concentrate in animal tissues, and radiation poses a serious threat to human health (Chapter 12).

High-level radioactive waste is not a problem that will go away, even though the American nuclear power industry faces bad times. By 2000, experts predict, 40,000 more tons of radioactive wastes will have been generated by existing commercial plants. Expanded nuclear capacity (possible in the future), continued operation of the existing facilities, and construction of nuclear weapons necessitate long-term, low-risk storage of nuclear wastes, and soon. But progress in developing storage sites has been painfully slow.

The 1982 **Nuclear Waste Policy Act** established a strict timetable for the Department of Energy to choose sites for disposal of high-level radioactive wastes. According to the act, the Department of Energy (DOE) had to select a site by 1987, and the site must be in operation by 1998. States have the right to veto sites unless overridden by both houses of Congress. If a veto stands, a new site must be chosen. Sites in Nevada, Washington, and Texas have been selected, among many others. In 1986 the Washington site was dropped. In 1987, the Texas site was also dropped by Congress. The Department of Energy must focus all of its resources on studying the suitability of the Yucca Mountain site in Nevada. The decision to limit the search to one site was made to save money and to quiet public protest over the two other sites. Studying a single site will cost over $1 billion. The Yucca Mountain site seemed the most suitable based on preliminary data. This decision, however, has not pleased many Nevadans. In order for federal decision makers to approve the site, they must be relatively sure that future earthquakes, volcanic eruptions, and climate change will not threaten the stability of the repository. State geologists question the accuracy of some of the federal research on the geological stability of the Yucca Mountain site, 100 miles northwest of Las Vegas. Federal scientists may be analyzing their data in a way that underestimates the hazards of the Yucca Mountain site.

If the Nevada site passes a set of criteria, and the state doesn't oppose its construction, workers will dig a repository 600 meters (2000 feet) below the surface of the earth to store high-level wastes. Nuclear wastes from power plants and defense facilities all over the United States (and possibily from other countries) will be shipped in casks to their (hopefully) permanent home. However, if the Nevada site proves unsatisfactory, the process begins again, delaying construction even more.

Many authorities believe that deep geological disposal

is the best option for dealing with dangerous radioactive wastes. However, little is known about the interaction between heat generated from radioactive wastes and the rock and nearby groundwater. Still, deep rock and salt formations 600 to 1200 meters (2000 to 4000 feet) below the surface in geologically stable regions are believed to be the best option for keeping wastes from entering groundwater and contaminating the environment.

Some people have suggested transporting wastes into space. Cost, energy requirements, and material requirements would be major problems. Disposal of radioactive wastes from a single 1000-megawatt nuclear plant would cost over $1 million a year. Furthermore, radioactive capsules shot into space might someday return to earth or, in a replay of the Challenger tragedy, never make it out of our atmosphere.

Others have suggested dumping radioactive wastes on uninhabited lands in the Arctic and Antarctica. Too little is known about the effects of this disposal technique for experts to assess its safety and effectiveness.

Radioactive waste can be bombarded with neutrons in special reactors to **transmute**, or convert, some of it into less harmful substances. However, existing reactors do a poor job of altering cesium-137 and strontium-90, two of the more dangerous by-products of nuclear fission.

Seabed disposal has been used in the past by the United States and European countries but is now forbidden. Still, some scientists suggest that the seabed may provide a site for radioactive wastes; the effects are difficult to predict.

A final suggestion has been to build special tanks on the ground: individual canisters would be placed in enormous 35-ton steel casks surrounded by a thick concrete covering. Canisters might also be stored in cooled and guarded warehouses.

Ironically, the United States has spent billions of dollars of private and public money on nuclear reactors but very little on research on radioactive wastes. The disposal issue is independent of the future of nuclear weapons and nuclear power; a complete ban on nuclear weapons and nuclear power would not solve the problems of accumulated waste. Therefore, wisdom dictates establishing a cost-effective and low-risk disposal method, keeping in mind the costs to future societies.

Disposing of Low- and Medium-Level Radioactive Wastes

Low-level waste from hospitals and research laboratories is packaged and shipped to three sites—in Nevada, Washington, and South Carolina—for disposal. The Nuclear Regulatory Commission is currently trying to reclassify low-level wastes so they can be discarded in ordinary sanitary landfills, the dumps where your garbage ends up. This proposal has some health officials and environmentalists up in arms.

Medium-level waste from nuclear power plants and weapons facilities is another matter altogether. The United States also has, as yet, no permanent repository for these wastes.

The Department of Energy built what could be the first medium-level radioactive waste depository in New Mexico, called WIPP (Waste Isolation Pilot Plant). WIPP is an experimental project, which, if successful, could be expanded to full operation. In August, 1988, however, two months before the site was to begin receiving wastes, the Department of Energy announced that it would postpone the opening. Researchers were concerned about the safety of the $700-million facility, which has been carved out of salt deposits 630 meters (2100 feet) below the ground near Carlsbad, New Mexico. Early in 1988, water had leaked into the facility, leading a New Mexico State scientific advisory committee to question the safety of the site. Water might corrode the steel canisters, allowing radiation to leak. According to the EPA, the Energy Department has not yet demonstrated that the facility will meet EPA standards for radioactive waste storage. At this writing, while dangerous medium-level wastes build up, the Department of Energy has offered no revised date for opening.

Some Obstacles Hazardous waste production has dropped steadily between 1980 and the present in the United States. This encouraging trend resulted partly because of a slump in the chemical industry but also because of in-plant efforts to reduce hazardous waste production by process modification, recycling, and reuse. The Chemical Manufacturers Association's sixth annual hazardous waste survey of 221 plants studied from 1981 to 1986 showed that they had reduced solid hazardous waste production by 56%. The plants had reduced hazardous waste in water by nearly 10%. The survey indicated a trend away from the disposal of wastes in landfills in favor of more permanently effective techniques, such as recycling and process redesign. Critical thinking, however, suggests a closer look. Industry surveys may be biased and overstate progress in reducing hazardous wastes. A downturn in chemical production during this period may be partly responsible for the apparent reduction in solid hazardous wastes. The nation is still not out of the woods. Tens of thousands of tons of waste are cranked out every day. Some companies continue to dump their wastes illegally whenever they can get away with it.

Another problem is that much of the United States' hazardous waste (85%) is highly diluted in water that is released by various industrial processes. This dilute waste stream is typically pumped into deep wells, from which hazardous substances may leak into groundwater. Besides contaminating drinking water, industrial wastes pumped into deep wells can increase the incidence of earthquakes. Two Ohio geologists believe that industrial waste injected into the ground by a chemical company in Ohio may have

Redefining National Security: Waste from the Nuclear Weapons Industry

It is a strange irony. Each year, the federal government spends billions of dollars to build atomic weapons at federal sites all over the country. The weapons presumably protect us from foreign aggression. They are part of a $300-billion system of national security. However, the weapons plants themselves may be some of the most hazardous facilities in the entire country. They are poisoning our people in the process of protecting the country.

The Department of Energy builds nuclear weapons at 45 sites, covering a land area equal to Delaware and Rhode Island combined. Recent disclosures of ineptitude and careless waste management at the nuclear facilities have created a storm of controversy. In July of 1988, the General Accounting Office estimated that cleaning up 17 of the Department of Energy's defense facilities would cost taxpayers about $20 billion. More recent estimates put the cost of cleanup at $100 billion. Most cleanup funds would go to the three most heavily contaminated sites: the Rocky Flats plant near Golden, Colorado, the Hanford plant in Richland, Washington, and the Fernald materials production center in Ohio. Cleanup could take 20 to 30 years to complete.

For years, the nation's nuclear weapons industry has been supervised by lax government officials and even, some say, ordered by government officials to stifle concerns over health and safety. Private contractors run the weapons facilities under the control of the federal government. For whatever reasons, the weapons plants have been run poorly. Studies show that some of the facilities have released large quantities of radioactive material into the air and have dumped tons of potentially carcinogenic waste into nearby water bodies and leaking pits. This has contaminated groundwater supplies. One of the worst problems occurred at the Hanford facility in Washington. Documents secured by a Spokane environmental group under the Freedom of Information Act showed that between 1944 and 1956, 530,000 curies of radioactive iodine were released into the air by the Hanford facility. Radioactive iodine was deposited on the land around the facility for as far as 15 miles.

Local residents were disturbed by the revelations and some have noted an unusually high number of cancer deaths in that area. Residents were never informed of the releases and some suffer thyroid problems possibly related to the radioactive iodine releases.

Tom Bailie was born near Hanford in 1947. His father had surgery for colon cancer at age 39. His mother had skin cancer. His two sisters have had their lower colons removed. Bailie himself is sterile and has only 90% of his lung capacity. He lays the blame on the Hanford facility.

The Centers for Disease Control in Atlanta estimates that 20,000 children in eastern Washington may have been exposed

to potentially harmful levels of radioactive iodine. The iodine came from milk produced by cows that graze on contaminated grasslands.

The Feed Materials Production Center in Fernald, Ohio, is a uranium processing plant. Its name and the red and white checkerboard design on a nearby water tower led some local residents to think that the facility produced cattle feed or pet food. To their dismay they learned that the company did not produce animal feed, but uranium fuel rods for nuclear reactors and components for nuclear warheads. The plant also released substantial amounts of radiation.

Richard Shank, the director of Ohio's Environmental Protection Agency, estimates that the Fernald plant has released nearly 300,000 pounds of uranium into the air since the plant began production. The plant has also deliberately discharged nearly 170,000 pounds of waste into a nearby river over the past 37 years. In addition, 12.7 million pounds of waste were dumped in pits, all of which are thought to be leaking.

The Fernald plant began operation in 1953. The contractor, National Lead of Ohio, was told by the Atomic Energy Commission to dump radioactive waste into pits in the ground. This was a standard practice for disposing of radioactive material at the time. But rainwater caused the pits to overflow; National Lead suggested ways to fix the problem, but their efforts were thwarted by the Atomic Energy Commission. In 1958, the company warned the Atomic Energy Commission that concrete storage tanks that held radioactive waste were leaking. The Atomic Energy Commission suggested that they merely not fill the tanks above the leaks. The flawed tanks are still in use today.

Ohio's governor, Richard Celeste, has demanded that the Feed Materials Production Center be closed permanently. Richard Shank says, "The US government is the single biggest polluter in Ohio and probably the nation."

Problems also abound at the Department of Energy's Savannah River operation. Studies show that many shallow aquifers in the region are contaminated with radioactive wastes. A deeper aquifer has been contaminated with chemical toxins.

The facility's manager, DuPont, has been accused of numerous wrongdoings over the past 30 years. On May 10, 1965, for instance, operators ignored an alarm for 15 minutes. The alarm was indicating that radioactive water was spilling out of a reactor onto the floor. About 8000 liters (2000 gallons) of fluid leaked out of the reactor. The coolant level dropped dangerously and the reactor shut down automatically.

In 1982, a technician at the Savannah River plant left a valve open for 12 hours. Radioactive liquid flooded a plutonium processing room, filling it two feet deep. Radioactive liquid waste has also been stored in tanks with corroded bottoms,

Case Study 18-2

which leak into underlying groundwater.

Trouble surfaced at the Rocky Flats facility downwind from Denver, Colorado, on October 8, 1988. At that time the Energy Department ordered work stopped in one of the buildings where vital operations took place. This severely cut back activity at the plant. The shutdown occurred after three people walked into a room containing contaminated equipment. The warning sign that should have indicated that the room was off-limits was covered by an electrical panel. The workers were not seriously exposed, but a sloppy attitude toward safety seemed to be prevalent at the plant.

This is the least of Rocky Flat's troubles. The plant has a record of bad management. For many years, barrels containing oil contaminated with plutonium were buried on site, where

they leaked into the soil. Cleanup could cost as much as $750 million. The Rocky Flats plant also releases small but significant amounts of plutonium into the air every day. Several fires at the plant have released incredible amounts of plutonium as well. Studies of nearby residents show a higher than average rate of leukemia and other cancers. Damage from a fire in May of 1969 cost $21 million. Toxic wastes and radioactivity have been detected in ponds on the property, worrying health officials about nearby water supplies.

The Department of Energy took over the responsibility of the nuclear weapons network in 1977 from the now-defunct Atomic Energy Commission. At least on the surface, the Department of Energy seems intent on reforming the nuclear weapons industry.

---

created underground pressure that triggered the 4.9-richter-scale earthquake that shook northeastern Ohio and a nearby nuclear power plant in 1986. They think that pumping waste into the sandstone increased fluid pressure in the pores of a bed of rock that lies directly above the crystalline bedrock, causing the earthquake along a fault in the bedrock. The sheer volume of polluted water has made it hard to regulate. Removing the hazardous substances from the water is extremely costly, so most plant owners are unwilling to make the investment in waste water treatment. To cut down on deep-well discharges, inexpensive techniques must be developed to separate the hazardous wastes.

The push to reduce hazardous waste disposal on land and in water has stirred considerable interest in incinerators. On the surface, incineration seems like an attractive option. Incinerators are equipped with pollution-control devices and can remove many of the chemical pollutants. They can also be used to generate energy and reduce landfilling of hazardous materials. In rural areas, they provide employment. Companies that want to locate them in rural areas are often willing to pay huge sums of cash to support schools, roads, water supply systems, and other needed projects as incentives.

One person's solution, however, often becomes another person's problem. For example, in Ritzville, Washington, a farming town in the eastern part of the state, a hazardous waste incinerator is being proposed by an incinerator company. It promises jobs and more taxes for this depressed region. But not all residents think it is such a good idea. One of the most significant problems is cross-media contamination. The incinerator will require a landfill to dispose of the highly toxic residue produced by the incinerator. Residents are wary of leaks

that will pollute their groundwater, which is used for irrigation and drinking. Residents are also concerned about toxic emissions. Although the plant would have state-of-the-art pollution control equipment, small amounts of toxins, including dioxin, would be emitted and could accumulate in the soil and crops downwind. Residents are also concerned about hazardous waste shipments, which will increase dramatically if the incinerator is approved.

The story in Ritzville is being repeated all over the United States. Because urban residents don't want incinerators in their areas, companies are eyeing rural regions where there is less political opposition and where the promises of economic benefits may be better received.

Opponents of incineration think that the rural siting strategy may be shrewd business, but it keeps us from finding permanent solutions, like reduction, recycling, and reuse, which are environmentally more acceptable and, in the long run, are the best strategy for dealing with hazardous wastes.

Cleaning Up Past Mistakes

In June 1983 the 2400 residents of Times Beach, Missouri, agreed to sell nearly all of their 800 homes and 30 businesses to the federal government for $35 million. The roads in Times Beach, like the arena at Shenandoah Stables, had been sprayed with oil containing hazardous wastes, including dioxin. Studies showed that dioxin levels in the soil were 100 to 300 times higher than levels considered to be harmful during long-term exposure (Figure 18-7). The town had to go. The federal government bought it.

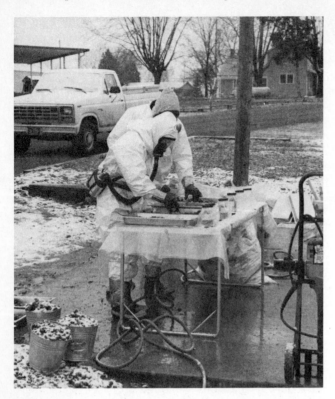

Figure 18-7. Times Beach, Missouri. Two EPA workers testing the soil for dioxin.

Today, Times Beach is a ghost town bordered by a tall chain-link fence. Its only occupants are occasional EPA officials and scientists from a handful of companies who are there to test ways of detoxifying the soil.

The $35-million purchase price for this contaminated piece of real estate came from a special fund, called the **Superfund**. Created in 1980, the Superfund was part of an extensive piece of legislation with an equally unwieldy title, the **Comprehensive Environmental Response, Compensation and Liability Act**, CERCLA for short. Commonly called the Superfund Act, the law makes owners and operators of hazardous waste dump sites and contaminated areas, as well as their customers, responsible for cleanup costs and property damage. More than anything else, it has forced the hazardous waste community to begin to take serious measures to prevent further contamination. As one industry representative put it, "You are liable for your waste forever."

The Superfund Act created a $1.6 billion fund, derived mostly from taxes on oil and chemical industries, with a small portion from the taxpayers. The money was earmarked for two purposes: cleaning up contaminated sites and paying for damage to property.

Legislators hoped the Superfund Act would permit the EPA to begin cleaning up contaminated areas and leaking dump sites throughout the United States. Unfortunately,

it didn't work out that way. The law required states to pick up the tab for part of the cleanup, but only eight states had funds for this purpose. Without money, the other states were disqualified; their dump sites could not be touched.

A second problem arose from the unexpectedly high cost of cleanup. Stabilizing a leaking pond designed to hold hazardous wastes cost the EPA $500,000. A study to determine what chemicals were leaking from another site cost the agency $800,000. Five years after the fund was set up, more than $1 billion had been spent and only six sites had been cleaned up; 1800 sites remained on the EPA's cleanup list. Experts cautiously predict that full cleanup will require 50 years and at least $100 billion.

A third problem with the Superfund is that the money cannot be used to compensate victims of illegal dumping of hazardous wastes for personal injury or death. The act provides only for the cleanup of contaminated areas and compensation for damage to property. According to Senator George Mitchell of Maine, "Under the legislation it's all right to hurt people but not trees." Many people believe that a fund similar to worker's compensation is needed to provide victims with some immediate assistance.

Initial mismanagement by top EPA officials delayed serious action by the agency. Officials negotiated with owners of hazardous waste dumps to begin private cleanups, partly to eliminate the need for state money. Critics argued that the EPA let some companies off too easily, that cleanups were superficial, and that future liability was waived in some agreements. Thus, if problems developed in the future, companies would bear no responsibility. Investigations conducted in 1983 led the EPA's top leadership to resign or be fired because of the issue. One EPA official went to jail for perjury.

Reauthorized in October of 1986, the Superfund Act will provide $9 billion over a five-year period for cleanup. Lee Thomas, then the EPA's administrator, pledged that 650 of the nation's 2000 worst sites would be cleaned up by 1990. A 1988 report by the US Congressional Office of Technology Assessment (OTA), which examined the effectiveness of the EPA's Superfund project, found that the program has been largely ineffective and inefficient. Major problems, says the OTA, make the program a failure. Inconsistencies in decision making, decentralization, which has led to minimal direction and oversight from the EPA headquarters to regional offices, and, most importantly, the agency's decision to opt for short-term quick fixes have all impaired the effectiveness of the Superfund program.

The last point is particularly troublesome to many observers. The EPA often opted for quick-fix solutions to clean up an area. In Love Canal, for instance, they put a clay cap over the dump site and dug a ditch around it with monitoring wells to determine whether hazardous wastes were leaking out. In other sites, contaminated soil

Resources Misuse

Throughout much of history, humans have been able to ignore the impact of their activities on natural resources. However, in recent years, we have been forced to recognize the error of our ways; air and water pollution, unsafe disposal of toxic wastes, the near catastrophe at Three Mile Island, depletion of precious minerals, extinction of endangered species, and more. These misuses threaten the resources needed to sustain not only our civilization, but our planet as well. Today, the earth's future lies in a combined and difficult task; not only must we correct our past mistakes, we must *understand* how and why they happened, and take steps to ensure that they don't happen again.

1 There goes the neighborhood... A wet cooling tower stands in bold contrast to a farmhouse in rural northern Ohio.

2 A denuded hill in Alaska has been clear-cut to remove timber for domestic and foreign markets.

3 Controlled fires burn the grassy vegetation of cleared rain forest. Some experts feel that removal of tropical rain forests will have a direct effect on climatic conditions. The release of great amounts of carbon dioxide into the atmosphere is said to contribute to the greenhouse effect, a phenomenon which may produce an unfavorable warming trend in the earth's atmosphere.

4 Land in a tropical rain forest is cleared for farming or grazing. After removal of vegetation, the nutrient-poor soil is often washed away or baked into a brick-like consistency, rendering it useless.

5 Overgrazing turned a once-rich grassland into a lifeless desert. Without the plant-root network that once bound it, the soil is unable to retain moisture and is easily blown away.

6 An infrared aerial view of Love Canal, Niagara Falls, New York. The dump, home to dozens of chemicals including PCBs and dioxin, was sold to the city in the 1950s as a school and playground site. Healthy vegetation appears red; brown vegetation indicates contamination. The toxic wastes were spread throughout the neighborhood via underground streams; 237 families were evacuated from the area.

7 Toxic wastes are stored in steel drums, then buried or stored in toxic waste dumps like this one. The steel drums, which are supposed to separate the dangerous wastes from you and me, rust due to their corrosive contents.

8 Even in the wilderness of Antarctica massive amounts of solid waste, unhidden by the snow, are a constant ecological threat.

9 Industrial society produces massive quantities of refuse, some of which gets buried, burned, or recycled. The rest ends up along highways, in empty lots, or, as in this case, in someone's backyard.

10 Boats at dock on the Amazon sit immersed in a sea of floating garbage.

2

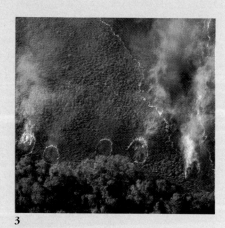

3

4

5

6

7

8

9

10

11 Waste from an iron ore processing plant colors the water of this artificial pond a bright orange. Such wastes seep into the earth, contaminating groundwater supplies.

12 Nitrogen fertilizer added to irrigation water to boost crop production in the Oro Valley, Arizona, may contaminate groundwater in this area.

13 New York City lies enshrouded in a layer of filthy air.

14 At McMurdo Station (U.S.A.) solid refuse covers the embankment, while raw sewage flows into the Antarctic Ocean.

13

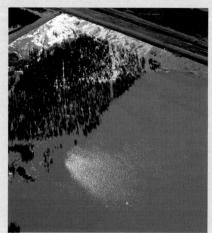

11

12

14

was simply excavated and hauled off to another landfill. Incineration and biological destruction of the wastes might have been more long-lasting solutions. The conclusions of the OTA's report were substantiated by a report from the Environmental Defense Fund, the trade association for the hazardous waste treatment industry and four other groups. Since Love Canal, the EPA has been looking more carefully at options and trying more permanent solutions, but the jury is still out on how well they will do.

Rectifying the improper disposal of hazardous wastes and improving waste disposal practices is as gargantuan a problem as the waste heap itself. With 60,000 generators of hazardous materials and 15,000 haulers in the United States, complicated legal and technological solutions are necessary. We are on our way, but there's much to be done.

But what about our personal actions? Is there anything we can do individually to reduce hazardous wastes? Each of us can contribute by properly disposing of oil and other potentially toxic wastes, avoiding toxic chemicals such as pesticides and herbicides, and reducing our consumption of nonessential goods, whose production creates the problem that poisons our land and water. By one estimate, each kilogram (2.2 pounds) of trash a person generates results in 20 kilograms (44 pounds) of waste upstream. Much of that waste is hazardous. You can cut back on potentially hazardous materials by purchasing environmentally safe cleaning products and insecticides. You can learn about waste sites in your area and become active in grass-roots organizations working on reduction, recycling, and safer disposal methods. Together, millions of Americans using resources wisely can make significant inroads into the hazardous waste problem.

It is not enough for a nation to have a handful of heroes. What we need are generations of responsible people.

RICHARD D. LAMM

Summary

The problem of hazardous waste caught the attention of the American public in the 1970s when toxic chemicals began to ooze out of a dump known as Love Canal, near Niagara Falls. That incident forever changed the way Americans view hazardous wastes. It began a frenetic search for hazardous dump sites and ways to halt illegal waste disposal practices that continues today.

Shortly after Love Canal the true proportions of the hazardous waste problem became evident. Ten thousand waste dumps need cleaning up, and millions of tons of waste are produced every year. Ill-conceived and irresponsible waste disposal has left a legacy of polluted groundwater and contaminated land. Cleanup could take 50 years and cost around $100 billion.

The latest in a long list of hazardous chemical problems stems from LUST—leaking underground storage tanks. Millions of tanks holding gasoline and other petroleum by-products and hazardous wastes of all kinds may be leaking. Health officials' main concern is the potential effect on groundwater and human health.

Two hazardous waste problems face society today: (1) what to do with the enormous amounts of waste produced each year and (2) how to deal with leaking and contaminated waste disposal sites.

In 1976 Congress passed the **Resource Conservation and Recovery Act**, aimed at cutting back on illegal and improper waste disposal. RCRA called on the EPA to determine which wastes were hazardous and how they should be handled and disposed of. A reporting network was established to trace hazardous wastes from cradle to grave. The EPA also began issuing permits for waste disposal. Unfortunately, implementation has been slow. Too many loopholes existed in the legislation, allowing much hazardous waste to escape control. New amendments closed many of the loopholes and tightened the net on hazardous wastes.

There are many technological options for controlling hazardous wastes. The first line of defense is to reduce hazardous waste production through **process manipulation**, **recycling**, and **reuse** of waste products. **Detoxification** and **stabilization** are the second line of attack. Detoxification can take place through land treatment, which breaks down wastes by chemical reactions, bacteria, or even sunlight. Incineration is also an alternative for organic wastes. High-temperature furnaces can operate at stationary waste disposal sites, on ships, or on mobile trailers.

Hazardous wastes that cannot be broken down can be stored permanently in **secured landfills** lined by synthetic liners and thick, theoretically impermeable layers of clay. Although landfills are the cheapest option, they have many disadvantages, the most noteworthy being inevitable leaks.

Disposal of radioactive wastes creates a special problem for society. The 1982 Nuclear Waste Policy Act establishes a strict timetable for the Department of Energy to choose appropriate sites for disposing of high-level radioactive wastes. The government is now concentrating its efforts on a site in Nevada. Many experts believe that deep geological disposal is the best technical option for getting rid of most dangerous radioactive wastes. Medium- and low-level radioactive wastes could go to a new facility in New Mexico.

To deal with leaking dumps and contaminated areas, Congress passed the **Comprehensive Environmental Response, Compensation and Liability Act** in 1980. It makes owners and operators of hazardous waste disposal sites, as well as their customers, liable for cleanup and property damage. This legislation established a $1.6-billion Superfund, to be used to clean up imminent hazards. The Superfund had some major drawbacks, notably that the law required states to pay part of the money for cleanup costs. Few states have funds for this purpose. One of the most critical weaknesses in the Superfund legislation is that it fails to address damage to human health. Unfortunately, the cost of cleanup has proved far greater than anticipated in the original Superfund Act. In the first five years of the program the agency had managed to clean up only six sites. As a result, when Congress reauthorized the Superfund Act in 1986, it increased the fund to $9 billion.

Solving the problems of hazardous waste is an enormous task that could be made easier by individual contribution. Reduced consumption, especially of unnecessary items, could cut back on currently generated waste.

Discussion Questions

1. Summarize the major events occurring at Love Canal. Who was to blame for this problem? What might have been done to avoid it?

2. You are appointed to head a state agency on hazardous waste disposal. You and your staff are to make recommendations for a statewide plan to handle hazardous wastes. Draw up a plan for eliminating dumping. Which techniques would have the highest priority? How would you bring your plan into effect?

3. Discuss the major provisions of the Resource Conservation and Recovery Act (1976) and the Comprehensive Environmental Response, Compensation and Liability Act (1980), the so-called Superfund Act. What are the weaknesses of each?

4. Describe the pros and cons of the major technological controls on hazardous wastes, including process modification, recycling and reuse, conversion to nonhazardous or less hazardous materials, and perpetual disposal.

5. Debate the statement "All hazardous wastes should be recycled and reused to eliminate disposal."

6. A hazardous waste site is going to be placed in your community. What information would you want to know about the site? How would you go about getting the information you need? Would you oppose it? Why or why not?

7. List personal ways in which we can each contribute to lessening the hazardous waste problem.

8. Discuss some of the options we have for getting rid of radioactive wastes. Which ones seem the most intelligent to you? Why?

9. Debate the statement "Victims of improper hazardous waste disposal practices should be compensated by a victim compensation fund developed by taxing the producers of toxic waste."

Suggested Readings

Burns, M. E. (ed.) (1987). *Low-Level Radioactive Waste Regulation: Science, Politics and Fear.* New York: Lewis Publications, Inc. A comprehensive analysis of radiation.

Epstein, S. S., Brown, L. O., and Pope, C. (1982). *Hazardous Waste in America.* San Francisco: Sierra Club Books. Thorough and dramatic coverage of hazardous wastes.

Friedlander, S. K. (1989). Pollution Prevention: Implications for Engineering Design, Research, and Education. *Environment* 31 (4): 10–15, 36–38. Important source of information on the pollution prevention strategy.

League of Women Voters. (1985). *The Nuclear Waste Primer: A Handbook for Citizens.* New York: Nick Lyons Books. Excellent overview of the problem, still relevant today.

Martin, L. (1986). The Case for Stopping Wastes at Their Source. *Environment* 28 (3): 35–37. An interesting case study.

Piasecki, B. W. and Davis, G. A. (1987). *America's Future in Toxic Waste Management. Lessons From Europe.* New York: Quorum Books, Greenwood Press. Comprehensive survey of European successes and failures to reduce toxic waste dumping.

Popkin, R. (1986). Hazardous Waste Cleanup and Disaster Management. *Environment* 28 (3): 2–5. Good overview of the US hazardous waste problem.

Postel, S. (1987). *Defusing the Toxics Threat: Controlling Pesticides and Industrial Waste.* Worldwatch Paper 29. Washington, DC: Worldwatch Institute. Excellent reading.

Regenstein, L. (1982). *America the Poisoned.* Washington, DC: Acropolis Books. Chapter 3 is a good overview of the hazardous waste problem.

Steinhart, P. (1990). Innocent Victims of a Toxic World. *National Wildlife* 28 (2): 20–27. Discusses effects of toxins on wildlife.

Stranahan, S. Q. (1985). Putting the Heat on Polluters. *National Wildlife* 23 (5): 30–33. A look at how citizens are fighting to rid their communities of hazardous waste dumps.

Tschinkel, V. J. (1986). The Transition toward Long-Term Management. *Environment* 28 (3): 19–20, 25–31. An informative look at an important topic.

Tucker, S. P. and Carson, G. A. (1985). Deactivation of Hazardous Chemical Wastes. *Environ. Sci. and Technol.* 19 (3): 215–220. Concise but technical coverage of hazardous waste deactivation.

Wexler, M. (1985). Strike Force. *National Wildlife* 23 (4): 38–41. A look at ways in which Los Angeles is fighting corporate polluters.

Solid Wastes: Solving a Growing Problem

A society in which consumption has to be artificially stimulated to keep production going is a society founded on trash and waste.

DOROTHY L. SAYERS

All organisms produce wastes, but none produces as many wastes of such diverse composition as humans. Society's wastes arise from many different activities. This section concentrates on municipal solid wastes: discarded materials from businesses, hospitals, airports, schools, stores, and homes (Figure S18-1).

Each year over 145 million metric tons of municipal solid waste is generated in the United States, about 580 kilograms (1300 pounds) per person. On average each man, woman, and child produces 1.6 kilograms (3.5 pounds) per person per day. A city of one million people produces enough solid waste each year to fill a football stadium. Municipal solid wastes make up only about 4% of the total solid waste discarded in the United States each year. However, municipal waste is growing from 2% to 4% per year. This growth and the sheer volume of waste create major problems in cities, where land for disposal is in short supply.

Arthur C. Clarke wrote that "solid wastes are only raw materials we're too stupid to use." Currently, about 14.5 million metric tons of American trash is recycled, about 10% of the total. Another 10% is burned in incinerators. The remainder—millions of tons of paper, glass, metals, tires, and plastics—is dumped into the ground and buried.

Unknown to many, the landfill disposal option costs communities millions of dollars each year. Landfilling the disposable diapers used in the United States alone costs the nation an estimated $300 million a year. For most local governments, the cost of trash disposal is usually exceeded only by the cost of education and highway construction and maintenance.

Waste dumped in the ground produces water and air pollution, squanders energy, and consumes large quantities of land in and around urban centers—land that is growing more scarce and more costly by the day. The incinerator option also wastes valuable resources and produces toxic air pollutants. Ash

remaining after the trash is burned may also contain toxic materials that can leak out of landfills where it is deposited.

Like so many other problems, municipal solid waste is the end product of many interacting factors: (1) large populations, (2) high per capita consumption, especially of products of marginal to low necessity, (3) low product durability, (4) a rash of disposable products, (5) low reuse and recycling rates nationwide, (6) a lack of personal and governmental commitment to reduce waste, (7) widely dispersed populations where producers of recyclable and reusable items are separated from those willing to purchase these materials, (8) traditionally cheap energy and abundant land for disposal, and (9) powerful business interests that make large profits from waste disposal and disposable products.

Reducing the problem requires action on three fronts (Figure S18-2). First is the **input approach**, attempts to reduce the amount of materials traveling through the production–consumption cycle. This is the most important but also the most often overlooked strategy. The input approach includes ways to reduce excessive consumption, to increase product durability, to reduce overpackaging, and others. Second is the **throughput**

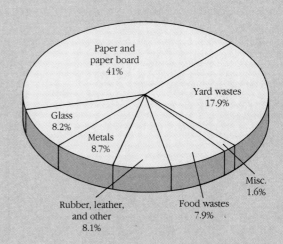

Figure S18-1. Composition of typical municipal solid waste by weight.

Figure S18-2. Three strategies for reducing solid waste. A combination of all three must be applied to alleviate the solid waste problem.

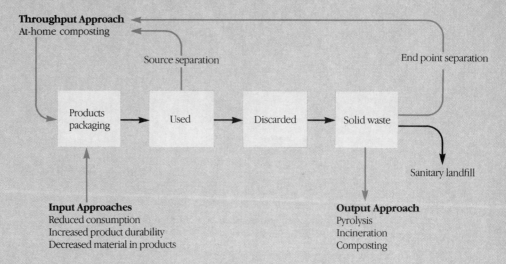

Throughput Approach
At-home composting

Source separation

End point separation

Products packaging → Used → Discarded → Solid waste

Sanitary landfill

Input Approaches
Reduced consumption
Increased product durability
Decreased material in products

Output Approach
Pyrolysis
Incineration
Composting

approach, reducing solid waste production by reusing and recycling materials before they enter the waste stream. This is one of the most popular approaches today. Third is the **output approach**, finding better ways of dealing with trash that pours out of cities and towns throughout the United States. Most often this means incinerating trash or finding ways to build safer landfills. Unfortunately, these approaches have received most of the attention over the years and are still looked on favorably.

Before we look at these strategies in detail, it is instructive to examine the goals of modern waste management. This will help you think more critically about the options and help you decide which strategies are best.

Solid waste management has, for years, simply meant finding ways to get rid of garbage to protect people from disease, odors, and unsightly dumps. Today, the goals have begun to shift. In many cases, solid waste managers are looking for ways to (1) reduce landfill requirements, (2) save energy, (3) save resources, such as trees and aluminum, (4) protect groundwater, (5) reduce other forms of air pollution that contribute to acid deposition and global warming, and (6) protect wildlife habitat, wilderness, and farmland. This broad set of goals may not be on the mind of every city manager, but it is becoming increasingly important as people begin to understand that problems require broad, comprehensive solutions.

The Input Approach

Reductions in solid waste and in its attendant problems can also be achieved by reducing the amount of material entering the consumption phase of the production–consumption cycle. The three ways of doing this are (1) increasing product lifespan, (2) reducing the amount of materials in goods, and (3) reducing consumption (demand for goods).

More durable toys, garden tools, cars, and clothing require less frequent replacement and thus decrease resource use. The trouble is that, as John Ruskin noted, "There is hardly anything in the world that some man cannot make a little worse and sell a little cheaper." In the long run, however, cheaply made goods end up costing consumers more than well-made and slightly

more expensive goods. The rapid turnover may be profitable to businesses but is suicidal from an ecological viewpoint. Planned obsolescence destroys the air, water, and land.

High-waste production is also an attitude problem. The attitude that "new is better" leads many consumers to purchase new goods when old ones still work. The fashion industry thrives on its ability to convince the public that new fashions are in—and that anyone wearing the old is "out of fashion." Advertisers capitalize on this strategy as well and many consumers fall into the trap.

A great deal of resource waste and solid waste production comes from packaging. Each year, for example, packaging requires 90% of the glass consumed in the United States, 50% of the paper, 11% of the aluminum, and 8% of the steel, according to Concern, Inc., a citizen's group interested in reducing the solid waste problem in the United States. All told, about 30% of the municipal solid waste (by weight) is discarded packaging.

Of course, packaging is necessary, but much of it is superfluous and wasteful. The Campbell Soup Company realized this and redesigned its soup cans; today, they use 30% less material. Some beverage companies now package drinks in aseptic containers, boxes constructed of several thin layers of polyethylene, foil, and paper. The containers hold milk, a variety of juices, and wine and keep them fresh for several months without refrigeration. Beverages in aseptic containers cost the consumer about 10% less than those in cans. The reason is that canned drinks must be pasteurized for 45 minutes, whereas the contents of aseptic packages are sterilized out of the package for only one minute. This reduces energy demand and preserves flavor. Milk and juices in aseptics do not require refrigeration during transportation and storage, which also lowers the energy demand. Being lighter than cans also helps cut down on transportation costs. Unfortunately, the aseptic container is completely unrecyclable. Therefore, this strategy to reduce waste has grave disadvantages.

Some manufacturers are experimenting with biodegradable packages. Most notable are plastics and food containers that dissolve in water when cooked, thus reducing solid waste and also adding nutrients to the food. Biodegradable plastic shopping bags began to appear in grocery stores throughout the United States in 1989. For the most part, these bags are a blend of corn starch (which is biodegradable) and polyethylene (a plastic that is not). Bacteria in landfills can break down the starch, leaving behind tiny particles of plastic. Initial work shows that the breakdown occurs very slowly in the oxygen-free environment of the landfill. Thus, bags that may "decompose" in a few weeks in lab tests may take more than 100 years to decay in a landfill. Environmentalists are concerned that the plastic particles left after biodegradation may end up polluting groundwater. A far better approach might be reusable shopping bags—sturdy nylon or canvas bags that individuals can use over and over again.

Virtually all products can be redesigned to reduce waste. Many large newspapers, for example, have gone to a more economical design that has cut the use of newsprint by 5%. Smaller cars and trucks have emerged, too. Smaller computers and calculators have also helped save valuable materials. Smaller houses could help as well.

H. W. Shaw once wrote, "Our necessities are few, but our wants are endless." Our ceaseless efforts to satisfy our endless wants are a big part of the solid waste dilemma the United States, Canada, and other countries face. By one estimate, for each kilogram of waste that gets dumped in a landfill, 20 kilograms of waste were produced "upstream" during manufacturing and other stages of production. By cutting back on consumption, individuals can help reduce solid wastes and many other problems described in this book.

Some cities have taken steps to reduce waste by banning objectionable materials—notably plastics. In 1988, for example, Suffolk County, New York, passed an ordinance that bans plastic grocery bags and many plastic food containers in the county starting in July 1989. Fast food restaurants will have to switch to biodegradable products. Minneapolis–St. Paul, Minnesota; Berkeley and Palo Alto, California; Newark, New Jersey; and other cities have passed similar ordinances, stirring the plastics industry to step up measures to increase recycling.

Each of us can make a personal effort to reduce consumption with little noticeable change in life-style. There are limitless possibilities for reducing consumption; all we have to do is try them. (See the Action Pack for suggestions.)

The Throughput Approach

Throughput approaches—reuse and recycling—take out useful materials before they enter the waste stream (Table S18-1).

Reuse

Reuse is the return of operable or reparable goods into the market system for someone to use. In most cities organizations such as Goodwill and the Disabled American Veterans pick up usable discards or slightly damaged products, including clothes, furniture, books, and appliances. Many of them provide drop-off stations as well. These products are cleaned, repaired, then resold at a low price to needy and frugal people. So before you throw out a still-usable product, why not give one of these organizations a call? You will be helping increase the product's useful lifespan and helping to reduce resource demand.

Table S18-1 Reuse and Recycling of Solid Wastes

Material	Reuse and Recycling
Paper	Repulped and made into cardboard, paper, and a number of paper products. Incinerated to generate heat. Shredded and used as mulch or insulation.
Organic matter	Composted and added to gardens and farms to enrich the soil. Incinerated to generate heat.
Clothing and textiles	Shredded and reused for new fiber products, or burned to generate energy. Donated to charities or sold at garage sales.
Glass	Returned and refilled. Crushed and used to make new glass. Crushed and mixed with asphalt. Crushed and added to bricks and cinderblocks.
Metals	Remelted and used to manufacture new metal for containers, building, and other uses.

Source: Modified from Nebel, B. J. (1981). *Environmental Science.* Englewood Cliffs, NJ: Prentice-Hall, p. 297.

Packaging materials, such as cardboard boxes, bottles, and grocery bags, can also be reused, saving both energy and materials. In some states, shopping bags can be reused by consumers for their own groceries. Reusable beverage containers can be sterilized, refilled, and returned to the shelf, sometimes completing the cycle as many as 50 times. Unfortunately, disposable and recyclable bottles and cans have nearly eliminated the reusable container from the market.

The advantages of reuse are many: (1) it saves energy, (2) it reduces the land area needed for solid waste disposal, (3) it provides jobs, (4) it provides inexpensive products for the poor and the thrifty, (5) it reduces litter, (6) it decreases the amount of materials consumed by society, and (7) it helps reduce pollution and environmental degradation.

Recycling

Recycling is the return of materials to manufacturers, where they can be broken down and reincorporated into new products. Recycling alleviates future resource shortages, reduces energy demand, cuts pollution, saves water, and decreases solid waste disposal and incineration (Table S18-1).

Consider some important recycling facts: Each 1.2-meter (4-foot) stack of newspaper you recycle saves a 12-meter (40-foot) Douglas fir tree. Recycling a ton of newspaper saves 17 trees. Paper recycling uses one-third to one-half as much energy as the conventional process of making paper from wood pulp. Small savings can add up to make incredible changes. If the

United States, for example, increased paper recycling by 30%, we would save an estimated 350 million trees each year. Paper recycling has an added benefit of reducing air pollution by 95%.

Aluminum recycling offers great benefits as well. To begin with, aluminum recycling is 95% energy efficient. That means it requires 95% less energy to recycle aluminum than to make it from raw ore (bauxite). Translated, a manufacturer can make 20 aluminum cans from recycled metal with the same energy it takes to make one can from bauxite ore. Aluminum recycling produces 95% less air pollution as well. Similar environmental benefits are available from recycling other metals and plastics.

In the United States and many other countries, however, recycling efforts have fallen short of their full potential. One exception is the automobile. In the United States, approximately 90% of all American cars are recycled. Recycling rates for other important materials are extremely low. Only about 1% of the plastic, 8.5% of the glass, 25% of the paper and cardboard, and 25% of the aluminum is recycled. Recycling rates for metals could easily double or triple. Glass and plastic recycling could increase even more.

Obstacles to Recycling From an energy and resource standpoint, recycling is generally not as good an option as reuse, but it is far better than burning materials and infinitely better than throwing them away. If recycling is such a good idea, why don't Americans and Canadians do more of it? The reasons are complex.

First, industrial societies grew up with abundant resources. They saw little need for recycling, except perhaps in times of war when recycling became commonplace. Factories were set up to handle virgin material. The entire production–consumption system was built without recycling in mind. Corporate empires were built on the profits of extractive industries and those companies today wield enormous political power. Changing this ingrained and wasteful system will not be easy.

Second, the nation's tax laws have grown up around extraction. The laws work against recycling. Even today, for example, mining companies receive generous tax breaks (depletion allowances) that give them an unfair advantage over recyclers (Chapter 13). These tax breaks often make virgin materials cheaper than recycled ones. Logging companies that supply wood for paper mills (and other uses) are also heavily subsidized by the federal government—and thus by the taxpayer. The Forest Service, for example, charges logging companies $2 for a giant old-growth tree, which they sell at enormous profit. Logging roads in national forests are built with public money and timber sales on public land are frequently made below cost (Chapter 9), further benefiting virgin paper industry over recycling. Federal subsidies create unfair economics.

The traditional extractive industries receive another hidden subsidy that's far more difficult to quantify. It is called an **economic externality**—a cost that is passed on to the public from pollution and other harmful effects of these activities. Because the traditional ways of making paper, tin cans, and other products produce more pollution, they have a bigger impact on our health and our environment than recycling, which uses less energy and produces far less pollution. But the higher cost of virgin materials is not reflected in the price of the product. It is, instead, paid in federal taxes that go to clean up our air and water. It is paid in higher health bills and dozens of other ways

that few of us are aware of. Were economic externalities part of the price of products made from raw materials, recycled materials would probably outcompete them.

Another difficulty is the built-in transportation price differential, mandated by law, for scrap metal and ore. For example, ore travels more cheaply than metal bound for a recycling mill. Fourth, many products contain a mixture of materials, making them more difficult to recycle.

Recycling sometimes suffers from other economic disadvantages. One of those is the shifting market demand. For recycling to be feasible, dealers must have a steady market for recyclables; if not, they must store them until a market is opened up. Storage costs money and cuts into profits. Fluctuating markets for recyclables, therefore, make it difficult for marginally profitable operations to survive.

In much of the West and Midwest, cheap landfill costs are proving to be a difficult barrier to overcome. In Colorado, for example, landfill tipping fees—the cost to dump a ton of trash in a landfill—range from $4 to $7 per ton. On the East and West Coasts, the costs may be $40 to $100, making recycling far more profitable. Cheap landfilling is bound to end in the next decade as landfills are closed and new, more expensive sites are developed. Colorado, like other interior states, is also a long way from the nation's paper recycling mills, making it more expensive and less profitable to recycle.

Recycling suffers from an image problem as well. In the 1970s, many recycled paper products were admittedly of inferior quality. Many people who used them were dissatisfied and soon returned to virgin materials. Since that time, however, recycled paper products have improved immensely. Recycled office paper, stationery, and computer paper are indistinguishable from virgin stock. Still, the notion that recycled paper is inferior persists.

Plastics pose a special problem for recycling. Most plastic is perfectly recyclable. The problem is that there are more than 45 different types of plastic commonly used for packaging. Making matters worse, many packages contain two or more types of plastic, making it difficult to recycle. A plastic ketchup bottle, for example, has five layers of plastic, each one different, each one providing a special feature needed to make a perfectly squeezable bottle to deliver our ketchup on time.

One way to promote plastics recycling is to place codes on plastic packaging so that individuals and recyclers can tell exactly what type of plastic they are made of. You may have noticed some codes on plastic bags and containers already, but the industry is moving very slowly in this regard.

Recycling on the Rise Despite these barriers, recycling is on the rise and is bound to increase substantially in the 1990s. In 1988, the US Environmental Protection Agency announced a nationwide goal of reducing the solid waste stream by 25% through waste reduction and recycling by 1994. But lofty government goals are not the only forces that will increase recycling. Rising energy prices, decreasing landfill space, and depletion of high-grade ores will surely spur and increase this vital activity. Changes in transportation rates, subsidies for recycling that put it on even ground with the use of virgin materials, and widespread citizen interest could greatly help.

Many major US cities are facing a crisis in waste management:

too much garbage and not enough land to bury it in. As a result, many cities have developed recycling programs to reduce landfilling. In 1985, for instance, Philadelphia topped out its last landfill. The next closest dump is a 335-kilometer (210-mile) round trip. Because transportation costs would be prohibitive, the city has begun an active recycling program. Faced with similar problems, Seattle began a **curbside recycling** program in 1988. Recyclables are picked up on the curb once a month and sent to a recycling facility. City officials had hoped for a 30% participation rate at the end of the first year and were surprised when they found out that 60% of the residents had joined the program. Most of the northeastern states are planning to recycle large amounts of their waste streams, between 20% and 50% in the not-too-distant future. Through a variety of measures, New York City plans to recycle half of its solid waste by 1997.

The trash crisis has struck many American cities, including New York, Chicago, Miami, Los Angeles, Minneapolis, and Berkeley, California. Shrinking land and public opposition to landfills in their neighborhoods have forced city officials to look for alternatives; not surprisingly, recycling often comes out on top.

Cities and towns are trying several options. Some are using drop-off sites, where residents can deposit their recyclables on the way to work or to the grocery store. San Francisco now has six redemption centers in operation. In California, hundreds of igloo-shaped glass banks are situated near busy intersections. Dozens of paper redemption centers are also scattered throughout Los Angeles and San Diego. Drop-off centers can be successful, but only if containers are conveniently placed, for example, at train stations, near parks, or near heavy-use intersections. Compared to curbside recycling, however, drop-off programs return only a small fraction of the potentially recyclable material.

Curbside recycling, in which recyclables are periodically picked up by trash haulers at the curb, is by far the most successful type of recycling program. Participation rates as high as 60% to 80% can be expected if recyclables and trash are picked up on the same day. Curbside may actually save energy. One study in Canada showed that curbside recycling requires about 10% less energy than a drop-off program.

Because of declining landfill space and rising costs, recycling is catching on like wildfire in the United States. Today, over 600 cities and towns have some kind of recycling program. Hundreds more are considering them.

Recycling is not *the* answer to the world's problems, but it is one of them. It is but one of the key strategies in building a sustainable society. Recycling must be convenient and cost effective. It will, however, require individual action. Separating trash before collection requires a little effort, but not much more than the effort we expend to separate our white clothes from our colored clothes when we do our laundry. Source separation is the small rent we have to pay for the riches the earth gives us. It is a small sacrifice that pays off in a better world.

Raw trash can be shredded at a resource recovery center, then separated using machines, but this technique (end-point separation) is generally more costly. End-point separation, in fact, is sometimes more costly than landfilling because it requires extensive equipment which requires maintenance and runs on energy. Complete separation may not be possible at these facilities, thus lowering the value of recyclable materials.

Procuring Recycled Materials The EPA estimates that the United States could reduce urban solid waste by 25% by 1994, in large part by recycling. Although impressive, a goal of 60% to 80% would be far better and is achievable, but several changes must occur. One of those is a change in attitude from the exploitive one so prevalent today to a nurturing one in which we are willing to make small sacrifices for the good of the earth and future generations. The other is a change in the materials we use. That means a dramatic shift to recycled products. Recycling is not enough. People must be willing to purchase recycled paper and other materials to help build stronger and larger markets for them. If recyclers cannot find a market for their materials, all the recycling in the world is of no use. Individuals can help build markets by insisting on recycled paper.

Perhaps one of the most effective changes would be to require governmental agencies—at all levels of government—to purchase recycled paper and other materials. All told, local, state, and federal governments account for about 20% of the US GNP (all the money in our economy that changes hands for goods and services). Government is the single most important purchaser in the US economy. Think of the impact that local, state, and federal governments could have if they started buying recycled paper. The enormous demand might inspire many businesses to switch from conventional practices. It would surely increase markets and create supplies for the rest of us.

Several states have already started procurement programs. Government agencies in Maryland, for example, have purchased well over a million reams of recycled office paper. New York allows state agencies to purchase recycled paper if it comes within 5% of the cost of virgin stock; California allows purchase if it comes within 10%. These are called **price preference policies**. In these two states, about one-fourth of all the office paper, tissue, paper towels, and cardboard purchased by state agencies is made from recycled stock. On average, the states pay only about 2% more for office paper. Who knows what they saved in reduced pollution and health bills.

The federal government could also become a major purchaser of recycled materials. Unknown to many, the 1976 Resource Conservation and Recovery Act (RCRA) allows the purchase of recycled materials by federal agencies and departments, if the products are reasonably priced. Unfortunately, two problems have arisen. First, the law called on the EPA to publish a list of guidelines for a dozen or so recycled products, outlining what would be acceptable. The EPA, however, did not begin drawing up the guidelines for over a decade—and only then after it was sued by environmentalists. The job was not completed until 1989. The second problem is that the EPA interprets the "reasonable cost" provisions to mean the "lowest cost." Given federal transportation policy, the smaller markets, and the failure of current pricing systems to reflect external costs, recycled products are often slightly more expensive.

Recycling programs are hindered by small markets. Many times when new recycling programs start up, officials find that the "waste" that has been diverted from the landfills creates an apparent glut or surplus of recyclable material. Prices often fall and markets dry up. When New York State began recycling plastic soft drink bottles it couldn't find a buyer; the recycling program became an expensive trip to the landfill. But over time markets developed, and the bottles are now ground up and used to make filler for pillows and jackets, among other products. New Jersey couldn't find buyers for newspapers after it began a statewide mandatory recycling program. Prices for newspaper on the East and West Coasts both fell precipitously as recycling programs got under way. The chief problem, says Leon Swartzendruber, cofounder of Friends of Curbside, is not a glut of recycled material, but rather a glut of virgin materials on the market, leaving no markets for recycled paper.

Researchers at the Environmental Defense Fund note that temporary "gluts" of recyclable materials always seem to end as markets develop. As recycling becomes more in vogue, more and more markets are bound to shift to recycled materials. Recycling facilities are bound to open in many states, creating jobs and economic opportunities. By the time most of you graduate and start your families, recycling could well be the chief source of materials.

Increased recycling requires our personal attention. Each of us can make a commitment to recycle cans, bottles, and paper. We can encourage our friends to follow our lead. Tax breaks for recycling companies may prove helpful. Some states are already trying them.

It may be surprising to many, but recycling is much more prevalent in the developing countries than in the industrialized nations. The poor raid the dumps for food, clothing, and materials for shelter and also seek out discarded metals and other goods they can sell. The developed nations can help promote recycling as they assist developing nations in their efforts to raise their economic well-being. Information on the energy and material savings as well as the technologies for recycling could be built into global industrialization at the start. To be credible, however, the developed nations will need to increase their own recycling efforts.

The Output Approach

The final and often least desirable option is the output approach. It looks on solid waste as something to be discarded as cheaply and as safely as possible.

Dumps and Landfills

The garbage dump. By any other name it would smell as bad. Until the 1960s garbage dumps were prevalent features of the American landscape. But public objection to wafting odors, rat- and insect-infested midden heaps, and dark plumes of smoke that billowed out of burning dumps forced cities to look for other ways to deal with their growing trash problem. The federal government also contributed to the demise of the dump by passing RCRA. It required all open dumps to be closed or upgraded by 1983.

The open dump has been replaced by a second cousin from a better part of town, the sanitary landfill. A sanitary landfill is a natural or manmade depression into which solid wastes are dumped, compressed, and daily covered with a layer of dirt. Because solid wastes are no longer burned, as they were in many open dumps, air pollution is greatly reduced. Because trash is covered each day with a layer of dirt, odors, flies, insects, rodents, and potential health problems are eliminated or sharply reduced.

Despite their immediate benefits, landfills have some notable problems. First, and most important, landfills require land. The trash from 10,000 people in a year will cover 1 acre 10 feet deep. Around many cities usable land is in short supply. Second, landfills, like dumps, require a lot of energy for excavation, filling, and hauling trash. Third, they can pollute groundwater. Toxic household wastes (paint thinner, pesticides, and other poisons) and feces (from disposable diapers) are discarded in municipal landfills where they can leak into groundwater. Fourth, they produce methane gas from the decomposition of organic materials. Methane can seep through the ground into buildings built above and around reclaimed sites, and it is explosive at relatively low concentrations. Fifth, they sink or subside as the organic trash decays, requiring additional regrading and filling. Buildings constructed on top of reclaimed landfills may suffer serious structural damage. Sixth, they have low social acceptability. Quite understandably, most people don't want the noise, traffic, and blowing debris.

Fortunately, there are many solutions for reducing the energy and land requirements of landfills. Energy requirements can be cut by new methods of waste collection. For example, packer trucks now reduce waste volume by 60%; thus, fewer trucks are needed to haul garbage to landfills. Vacuum collection systems can also be used to save energy. In these systems solid waste is dumped into pipes that carry it to a central collection point. Recyclable and burnable materials can be separated there before the trash is trucked off to landfills. Recycling reduces the volume of garbage, helps cut back on energy demand, and reduces the land required to dispose of waste. Vacuum collection systems are feasible in urban areas where population density is high. Apartment complexes are extraordinarily well suited to this method of trash disposal. One such system is in operation in Sundbyberg, Sweden. Garbage is whisked away from wall chutes to a central collection facility, where the glass and metals are removed by an automated process. The burnables are incinerated, providing heat for the 1100 apartments using the system. A similar system handles 50 tons of waste per day at Disney World, Florida. Today, over 400 such systems are in operation in Europe in hospitals, apartment buildings, and housing tracts.

Water pollution problems may be reduced or eliminated by locating landfills away from streams, lakes, and wells. Test wells around the site can be used to monitor the movement of pollutants, if any, away from the site. Special drainage systems and careful landscaping can reduce the flow of water over the surface of a landfill, thus reducing the amount of water penetrating it. Impermeable clay caps and liners can reduce water infiltration and the escape of pollutants. In addition, pollutants leaking from the site can be collected by specially built drainage systems and then detoxified. In some cases, the toxic seepage is merely shipped to a hazardous waste facility.

Methane gas produced in landfills can be drawn off by special pipes and sold as fuel, in many cases supplementing or replacing natural gas. Subsidence damage to buildings built on reclaimed sites can be reduced by removing organic wastes before disposal and by allowing organic decay to proceed for a number of years before construction.

Ocean Dumping

What do you think of when you think of the ocean? White beaches, gentle waves, and the sweet smell of coconut oil? For

Figure S18-3. World's largest composting facility, at Wijster in the Netherlands.

years many city officials have had a different view of the ocean, that of a limitless garbage dump for solid waste and human sewage. Until just a few years ago 126 offshore dump sites were being used in the United States: 33 in the Gulf of Mexico, 42 in the Pacific Ocean, and 51 in the Atlantic. Most of the solid waste dumped at sea was mud and sediment from dredging of harbors, estuaries, and rivers. The rest consisted of industrial wastes (acids, alkaline substances, PCBs, and arsenic compounds) and sludge left over from sewage treatment plants (Chapter 16). Even today, city garbage specialists are trying to find ways to justify dumping their trash offshore. Just think of it: offshore "islands of trash," built from discarded automobiles, demolished buildings, and other solid wastes, supporting hotels and airports. Or even artificial reefs made of discarded stone, cement slabs, and automobiles.

The concerns over ocean dumping are many. The disposal of toxic wastes might contaminate commercially valuable fish and shellfish. Biomagnification of these chemicals might harm organisms higher on the food chain, including birds and humans (see Chapter 14). The long-term effects of ocean dumping are uncertain. Will ocean currents eventually wash the wastes ashore onto beaches, or will they be carried out to sea? Finally, aesthetic concerns have been raised by people like the explorer Thor Heyerdahl, who has witnessed oil and trash floating out at sea and deadened areas of the ocean that have long been the dumping sites for garbage and sludge from sewage treatment plants.

Ocean dumping began a steady decline after the Marine Protection, Research and Sanctuaries Act was passed by Congress in 1972. The long-term goal of this act was to phase out all ocean dumping. Thus, in January 1977 the EPA issued regulations calling for an end by 1981 to all dumping of industrial wastes and sludge in the Atlantic Ocean, where 90% of the waste disposal was occurring. At this writing, however, New York City and others continue to dump sewage sludge in the ocean.

Dredged materials can also be dumped at sea in some locations.

Appalled by the illegal dumping of medical wastes and other garbage, Congress passed two laws in 1988 to help clean up the oceans. The first was the Medical Waste Tracking Act (1988), discussed in Chapter 16. The second was the Ocean Dumping Ban Act (1988), putting an end to sewage sludge disposal at sea by December 21, 1991.

Composting

Composting is the most useful output strategy devised. Composting is a process in which organic wastes are allowed to undergo aerobic bacterial decay. The resulting product, compost, is used to build soil fertility (Chapter Supplement 7-1).

Where there is an abundance of organic matter, such as at slaughterhouses and vegetable- and fruit-packing plants, composting is a good strategy. Some successful composting operations can be found in the United States. In Seattle, for example, zoo officials now compost all of the manure produced by animals. When decomposed, the manure forms a rich organic soil supplement they call "zoodoo," which is sold to gardeners and homeowners for use around the yard. The program nets the zoo about $64,000 a year. Palo Alto, California, has had a successful composting program for yard wastes since 1979. Portland, Oregon and Berkeley, California are just beginning yard-waste composting programs. Most of the large-scale composting programs are in Europe—in the Netherlands, Belgium, England, and Italy—and in Israel (Figure S18-3).

Large-scale composting has a few drawbacks: (1) it requires large tracts of land and may produce odor and provide breeding sites for pests, (2) sorting out the noncompostable materials such as metals and glass is costly, (3) the demand for the organic compost is often low, and (4) sites are aesthetically unappealing.

For successful municipal composting, cities might use abandoned lots or outlying plots that could be cared for by convicts, welfare recipients, or the unemployed. Citizens could be required to sort out recyclable metals, plastics, and glass, thus eliminating the cost of separating the wastes later.

Composting can be practiced very successfully at home. When this is done, composting becomes a throughput approach to solid waste, helping reduce the waste stream by recycling perfectly reusable organic matter. Gardeners can make their own compost piles of leaves, grass clippings, and vegetable wastes from the kitchen. By mixing these materials with soil, which contains the bacteria that do the breakdown, individuals can help cut waste and pollution and help maintain healthy soils. A commercially available container or a simple wooden enclosure can help keep neighbors' dogs out of the pile. The product can be added to vegetable gardens and flower beds. Simply burying organic wastes in flower beds and gardens also may work well, reducing the need for artificial fertilizers.

Incineration

Incineration reduces waste volume by two-thirds, saving landfill space. It also produces heat and electricity, turning waste into energy. For that reason, incinerators are often called **waste-to-energy (WTE) plants**. One ton of garbage is equivalent to about one barrel of oil. But even with the energy gain, WTE plants are often more costly to build and operate than landfills. As a result, they can increase the cost of waste disposal. For the most part, WTE plants are also much more expensive than recycling programs. The Seattle curbside recycling program, for example, cost one-tenth as much as a proposed incinerator. Other recycling projects cost about one-third as much as incinerators.

Each incinerator must be individually designed to accommodate the particular local mixture of burnable and nonburnable refuse. Operating these incinerators is made more difficult because the mixture varies from season to season. In the spring and fall, for example, yard and garden waste increases dramatically. Wet leaves, however, do not burn well. To reduce this problem municipalities may require homeowners to separate combustible material from wet organic matter. Incinerators also emit toxic pollutants, especially when plastics are burned. The ash produced by an incinerator may also be hazardous to human health. In what may prove to be a landmark case, two environmental groups recently sued the city of Chicago and Westchester County, New York for supposedly violating federal and state hazardous waste laws. The municipalities routinely dump ash from WTE plants in landfills, which the environmentalists contend is a direct violation of RCRA. Test data compiled by the Environmental Defense Fund show that the ash from incinerators contains toxic metals, such as lead and cadmium, and dioxin, in concentrations considered hazardous. The WTE industry and municipalities hold that toxic ash from waste incinerators is exempt from RCRA disposal regulations. The Congress and EPA have been struggling for a number of years to determine whether wastes from municipal incinerators should be regulated by RCRA provisions. The EPA currently has a policy which states that if an ash tests hazardous then it must be dumped in a hazardous waste facility, which is much more costly than in an ordinary landfill. Today, many municipalities operating waste-to-energy plants are not required to dispose of ash in any special way.

If the court rules that ash from WTE plants is indeed a hazardous waste, special precautions will be required and the cost of operating these facilities will increase substantially. Americans may come to regret the disposal of ash in ordinary landfills in years to come if wastes begin to leak into surrounding groundwater, polluting public and private drinking water. Cleaning up these sites could cost many millions of dollars. Today, in fact, many ordinary landfills have become hazardous waste sites and require costly measures to close. If the EPA and Congress decide to include incinerator ash as hazardous, the cost of operating a WTE plant could rise substantially, making this option even more questionable.

Many municipalities are shifting from **mass burn** facilities, where the entire waste stream is burned without separation, to **refuse-derived fuel plants**, where nonburnable materials are first removed (and sometimes recycled). This increases the amount of combustible fuel, is much cleaner, and is more efficient than mass burn, but it still produces hazardous materials. Refuse-derived fuel plants are a step in the right direction, but are not as cost-effective as programs that rely on waste reduction, composting, and recycling.

Denmark burns over 60% of its solid wastes; the Netherlands and Sweden both burn about one-third of theirs. The United States, on the other hand, burns about 10% of its solid waste, up considerably in the past few years. In a move that disappointed many, McDonald's Corporation recently announced that it would be experimenting with on-site incinerators to burn styrofoam waste. The incinerators are now operating in three locations in the Chicago area. Public pressure to eliminate styrofoam first mounted because the plastic was produced with CFCs, gases that deplete the ozone layer. McDonald's agreed to eliminate the CFCs but continued using plastic. Then pressure mounted to recycle the styrofoam to help reduce waste in landfills. McDonald's agreed to try recycling, but soon gave up because of the low cost of virgin styrofoam. That is when they announced plans to burn the waste at each facility in incinerators they call Archie McPuffs. The company is applying for permits in other states as well, but may face considerable public pressure, because of fears that they will emit toxic pollutants.

Incinerators may seem like a good way to solve the growing trash problem, but critical thinking suggests a broader look at the original question. The question is not just how do we reduce landfilling, but how do we cut our waste; save valuable resources; protect our air, water, and land; and ensure vital wildlife habitat? Incinerators reduce trash but don't solve the other problems. Incinerators are riddled with problems, not the least of which is a lack of community support. Residents of Lowell, Massachusetts, defeated an incinerator that its city council was planning on building because they found, among other things, that Lowell only produced 225 metric tons of waste per day, but the plant would require 1350 metric tons to operate. That meant that the city council would have had to enter into agreement with neighboring towns to accept their trash to meet the needs of the plant. Lowell would become a waste magnet. Citizens in Spokane, Washington, were not so lucky. Their city council approved a massive incinerator that will require a considerable amount of outside trash to keep it running.

Suggested Readings

Chandler, W. U. (1983). *Materials Recycling: The Virtue of Necessity.* Worldwatch Paper 56. Washington, DC: Worldwatch Institute. Detailed study of recycling.

Concern, Inc. (1988). *Waste: Choice for Communities.* Washington, DC: Concern, Inc. Excellent booklet on waste management, including recycling.

Environmental Defense Fund. (1985). *To Burn or Not to Burn: The Economic Advantages of Recycling Over Garbage Incineration for New York City.* Washington, DC: Environmental Defense Fund. Excellent comparison of WTE and recycling.

Environmental Defense Fund. (1988). *Coming Full Circle: Successful Recycling Today.* Washington, DC: Environmental Defense Fund. Superb study of successful recycling programs. A must for anyone interested in recycling.

Iker, S. (1986). New Life from Old Junk. *National Wildlife* 24 (2): 12–17. Look at artificial reefs built from discarded junk.

Pollock, C. (1987). *Mining Urban Wastes: The Potential for Recycling.* Worldwatch Paper 76. Washington, DC: Worldwatch Institute. Fact-filled study of recycling. Excellent reading.

Pollock, C. (1988). Building a Market for Recyclables. *Worldwatch* 1 (3): 12–18. Good overview of ways to promote the recycling business.

Stump, K. and Doiron, K. (1989). The System Works: Seattle's Recycling Success. *Greenpeace* 14 (1): 16–17. Great overview of Seattle's recycling program.

V

ENVIRONMENT AND SOCIETY

Environmental Ethics: The Foundation of a Sustainable Society

The Frontier Mentality

Sustainable-Earth Ethics

Making the Transition

Viewpoint: Why We Should Feel Responsible for Future Generations / Robert Mellert

Modern man is the victim of the very instruments he values most. Every gain in power, every mastery of natural forces, every scientific addition to knowledge, has proved potentially dangerous, because it has not been accompanied by equal gains in self-understanding and self-discipline.

LEWIS MUMFORD

This book began with an outline of today's environmental crisis. Each part has explored a facet of it. Part 2 covered the population question, Part 3 surveyed resource problems, and Part 4 described the many faces of pollution. Each chapter suggested remedies for many of our problems, solutions that were technical, legal, and personal. All were aimed at developing a sustainable society.

The diverse solutions presented in earlier chapters amount to a hodgepodge without another ingredient: a set of informed values, an appropriate ethical system that serves as a foundation for the new society. This chapter outlines the old attitudes and sets forth a new set of attitudes called "sustainable ethics."

This new ethical system is already being adopted by many people in the United States and abroad. For sustainable ethics to achieve widespread acceptance will require that many more people replace long-standing beliefs that have deep roots and far-reaching effects. These ecologically out-of-date beliefs constitute the "frontier mentality" and underlie many of our environmental problems.

The Frontier Mentality

Today's industrial world operates largely with a frontier mentality, an attitude described by economist Kenneth

Table 19-1 Frontier and Sustainable Ethics Compared

Frontier Mentality	Sustainable Ethics
The earth is an unlimited bank of resources.	The earth has a limited supply of resources.
When the supply runs out, move elsewhere.	Recycling and the use of renewable resources will prevent depletion.
Life will be made better if we just continue to add to our material wealth.	Life's value is not simply the sum total of our banking accounts.
The cost of any project is determined by the cost of materials, energy, and labor. Economics is all that matters.	The cost is more than the sum of the energy, labor, and materials. External costs such as damage to health and the environment must be calculated.
Nature is to be overcome.	We must understand and cooperate with nature.
New laws and technologies will solve our environmental problems.	Individual efforts to solve the pressing problems must be combined with tough laws and new technologies.
We are above nature, somehow separated from it and superior to it.	We are a part of nature, ruled by its rules and respectful of its components. We are not superior to nature.
Waste is to be expected in all human endeavor.	Waste is intolerable; every wasted object should have a use.

Boulding, historian Roderick Nash, and others. The **frontier mentality** is characterized by three precepts: First, the world has an unlimited supply of resources for human use, not necessarily to be shared by all life forms; in other words, "There is always more, and it's all ours." Part of this belief is the notion that the earth has an unlimited capacity to assimilate pollution. Second, humans are apart from nature rather than a part of it. Third, nature is something to conquer. Technology has become the tool by which humans subdue nature, the answer to many of the conflicts between human society and nature. See Table 19-1 for a summary of the frontier mentality. (For two examples see Julian Simon's essay on page 110 and Ben Bova's essay on page 308.)

The frontier mentality has been a part of human thinking for many tens of thousands of years, perhaps all of human history. It was present in both the hunting and gathering and agricultural societies, but more so in the latter. The European settlers in North America, for example, cut down the forests and grew crops on the soil until the nutrients had been depleted or the soil had been eroded away by rain. Then they moved into new territory to start the cycle over again. One of the most serious aspects of the notion that there is always more is the disregard for the consequences of our actions. Why worry about soil erosion or water pollution, because there is always more soil and plenty of clean water.

The frontier mentality sees humans as separate from nature and superior to all other life forms; many societies have continually sought to dominate nature. This view, so prevalent today, was expressed by the nineteenth-century English poet Matthew Arnold: "Nature and man can never be fast friends. Fool, if thou canst not pass her, rest her slave!" The attitude of domination has spawned a frenzy of dam building to tame rivers, highways to tame wilderness, breakwalls to hold back ocean tides, levees to hold back floods, streambed channelizations to reduce flooding, chemical pesticides to control insects, and so on. Much of this book has looked at the ecological backlashes of attempts at domination. In many instances experience has shown that cooperation with nature would have been better for humans as well as nature. Not building homes on barrier islands or in river floodplains, not grazing cattle above the carrying capacity of grasslands, and letting natural forest fires burn are three examples of cooperation that pay huge dividends in the long term (Figure 19-1). (Case Studies 3-1, 10-1, and 17-2 all illustrate further ways of cooperating with nature.)

The frontier mentality is the dominant belief structure, or paradigm, of modern society (paradigms are discussed in Chapter Supplement 1-1). If you listen to politicians or newscasters you will hear the rhetoric of unlimited possibilities—for continued growth, bold new frontiers, and vast, virtually limitless resources. In fact, much of the excitement over cold fusion (Chapter 12) probably stemmed from people's desire to find an unlimited source of energy. So deeply imbedded is the frontier mentality that it affects not only how we view our problems but also how we go about solving them. For instance, fossil fuel shortages are "solved" by increased exploration and drilling or by developing alternatives such as oil shale and tar sands. Our single-minded dedication to the frontier mentality—for instance, the idea that there is always more—often makes us overlook the simple answers, such as conservation or reducing our energy consumption to make supplies last longer.

Figure 19-1. This Santa Cruz, California, beachside home was damaged by coastal storms in the winter of 1983. Homes placed farther from the beach were undamaged, suggesting the wisdom of playing by nature's rules: avoiding floodplains and coastal areas where storm swells frequently rush on shore.

The frontier mentality also influences individual goals and expectations. It is used to justify our actions. Moreover, nearly all of the social and political institutions function to maintain it. This makes it difficult to dislodge. Rediscovering that we *are* part of nature, however, could change our outlook entirely.

Roots of Our Attitudes toward Nature

Where did our attitudes toward nature come from? Most historians exploring that question have been interested primarily in the origin of the view that humans are "apart from and above nature" and "here to dominate it." But equally important are the roots of the attitude that "there is always more." Let's look at the first set of attitudes.

"Be fruitful and multiply, and replenish the earth, and subdue it; and have dominion over the fish of the sea, and over the fowl of the air, and over every living thing that moves upon the earth" (Gen. 1:28). This quotation from Genesis proclaims human supremacy over nature. A University of California historian, Lynn White, argues that it is the root of Western attitudes toward nature. In a widely read paper, "The Historical Roots of Our Ecologic Crisis," he writes that the Judeo-Christian ethic holds that nature has only one purpose: to serve humans. Furthermore, White argues, Christians believe it to be God's will that we exploit nature for our own purposes. Critics point out that the Hebrew word that has been translated as "dominion" connotes supremacy but along with it love, concern, and responsibility. White's reasoning, his critics

contend, is not entirely accurate. Moreover, religious scholars point out that some passages in the Bible call on humankind to protect the earth. Revelations, for example, says "hurt not the earth, neither the sea, nor the trees" (7:3). In Genesis, Adam and Eve are placed in the garden of Eden and told to "till it and keep [*shomer*, or guard, in Hebrew] it" (2:15). They are instructed not to destroy, but to supervise and maintain it—to protect the environment. White, say some scholars, has wrenched one verse out of the Bible and built an argument on it, ignoring others. Critical thinking skills warn against such partial study.

The biologist René Dubos presents an alternative viewpoint. He notes that exploitation of the natural world is a universal human trait regardless of religion. Erosion of the land, destruction of animal and plant species, excessive exploitation of natural resources, and ecological disasters are not unique to the Judeo-Christian peoples. At all times and all over the world, thoughtless interventions into nature have had disastrous consequences. Two examples were the widespread extinction of many large animals, such as the wooly mammoth and saber-toothed tiger, caused by early North American hunters and gatherers, and overgrazing and general mismanagement of the once-fertile Tigris-Euphrates region, which has led to widespread desertification.

The debate over our belief that we are above nature, outlined briefly above, illustrates a problem common to debates in general: the search for single causes when, in fact, several may be involved. Chapter Supplement 1-1 discusses critical thinking and presents a simplified model aimed at getting to the real roots of environmental problems. In the debate in question, the narrow views are somewhat misleading; they present only part of the truth.

Taking a broader view, we see that environmental damage stems in part from **biological imperialism**, the tendency for every organism to convert as much of the environment as possible into itself and its offspring. Humans are no different from the rest of the biological world, except that we have technologies to achieve our goals. And we have exceeded the carrying capacity in many ways, thus wreaking considerable havoc. Other species have little ability to exceed their carrying capacity.

Dubos is correct in pointing out that humans have always damaged the environment through exploitation. Judeo-Christian teachings, however, may have reinforced this tendency. To say that these teachings are responsible for our feelings of superiority and control over nature is only a half truth; it would perhaps be more accurate to say they are a fine veneer over a deeply rooted biological trait.

Some insights from psychology shed more light on the attitude that we are apart from nature. Psychologists assert that humans create mental models that determine what

Why We Should Feel Responsible for Future Generations

Robert Mellert

The author teaches philosophy and future studies at Brookdale Community College in Lincroft, New Jersey. He has published What Is Process Theology? *and numerous articles on process philosophy, ethics, and future studies.*

"Why should I feel obligated to future generations? We're inevitably separated by time and space. My presence here on earth now will have no influence on someone living 200 years from now."

You may have heard this opinion expressed by your friends—perhaps you even hold it yourself. But if you have ever explored a wilderness preserve, used a library, or visited a historical monument, you already have some reasons for being responsible. Much of what we value in our family, our society, and our world has been provided by our predecessors, sometimes at considerable cost and effort on their part.

In today's world we face a number of issues that will affect future generations even more profoundly than they affect us now. Exploding world populations, shrinking nonrenewable resources, and plant and animal organisms threatened with extinction all add up to one thing—an ailing environment.

These are not isolated problems. Each stems from a common perception of our relationship to the world and our future. This perception can be characterized by a description of people and things as unique, immediate, individual, and separate from everything else. Any solutions that we might use to resolve our problems would have to start by challenging this perception.

Let me propose four basic considerations that may suggest a new paradigm for understanding our relationship to the world and our future. I believe our moral responsibility to coming generations will follow directly from these.

1. Future generations will be essentially the same as we are. They may have different wants and priorities, but they will manifest the same basic needs for food, water, air, and space. In addition, they will have the same basic physical and mental capacities with which to interact with their environment. Once born, they too will claim a right to life and protection from life-threatening conditions, such as extreme temperatures, toxins, famine, and disease. To give them life without also providing the basic means to sustain and enhance life would be cruel. If we expect the species to continue, we are obliged to leave a hospitable environment for those still to come.

2. One is born into a given generation by historical accident. None of us chose when to be born, or to whom. Because we have no special claim to the time and place of our birth, justice would require that we have no more rights over the world and its resources than anyone else.

3. Our survival as a species is more important than our individual survival. This is confirmed in nature every day; parents, whether they be rabbits, wolves, whales, or humans, spend their energies to reproduce and care for their young before they themselves die. Many will even risk their own lives for their offspring. This is because life is not ours to keep, but to share with others.

4. Even after we die, the effects of our life continue. We will be present in the memories of others and in the habits and traditions we shared with them. Our ideas will continue to enlarge the range of options for others; and even when these memories and ideas are no longer consciously a part of the future, they will ripple onwards, actively influencing the course of future events and people. What has been can never die. We are what we have been given and what we have chosen to make of these "gifts." In short, we are the product of our ancestors—all that they died for and believed in—and we are the product of our decisions. Future generations will be the result of what we are now and how they use what we leave them.

If we accept these four simple ideas, it is easy to see why we have an obligation to future generations. Our obligation is based on the truth that we are more than unique and separate individuals, living only the immediacy of the now. We are, rather, parts of a much larger whole, one that transcends space and time.

As John Locke, the great English philosopher, once said, we owe the future "enough and as good" as we received from the past.

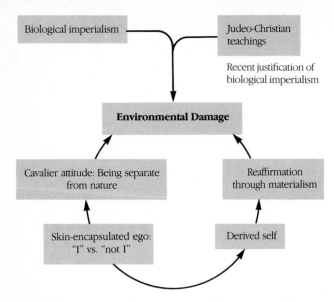

Figure 19-2. Roots of environmental damage caused by humans.

The Technological Fix

Many of today's problems are the result of the unbridled development and careless application of technology. In pursuing what some perceive as the "limitless" resources of our planet, newer and often more damaging technologies continue to be developed (see Chapter 13). Much of this book describes the impacts of such technologies on the environment.

Many of our problems also stem from an undying optimism in the power of technology to solve our problems. This is called **technological optimism**. Signs of this confidence are everywhere. For example, in efforts to reduce stratospheric ozone depletion, DuPont and other companies are looking for chemical substitutes for chlorofluorocarbons (CFCs). One substitute that is widely touted as a "technological solution" to the problem is a chemical called HCFC-22, described in Chapter Supplement 16-1. It is 20 times less destructive than the chemical it could replace. That sounds good. But on closer examination it turns out that the chemical is also a greenhouse gas. Making matters worse, HCFC traps much more heat than carbon dioxide. In attempting to solve one problem, we could be worsening another.

Here's another example: To solve the greenhouse warming, some experts have proposed replacing oil, natural gas, and coal-fired power plants with nuclear power. On the surface, this may seem like a logical solution. But nuclear power is riddled with problems, as you learned in Chapter 12. Moreover, to replace all of the fossil-fuel power plants would require utilities to build a new nuclear plant every 2.5 days for 38 years! The cost would be prohibitive. In addition, full replacement would only reduce global carbon dioxide by about 4%.

In our zeal to find high-tech solutions, logical and inexpensive low-tech solutions, such as conservation, are often overlooked. Conservation, for example, is a far more effective way of reducing global carbon dioxide and is far cheaper than the proposed nuclear solution. A dollar invested in conservation, in fact, reduces carbon dioxide emissions seven times more than each dollar invested in nuclear power.

Recycling, a topic discussed in Chapter Supplement 18-1, is a far cheaper solution than high-tech trash incinerators. But many municipalities are building incinerators. Unfortunately, our unqualified optimism in the powers of technology blinds us to the obvious solutions at our feet. Spurred by thoughts of limitless resources, many applied scientists, technologists, politicians, and business people promise a bright and unlimited future. Faithful that these experts can't be wrong, many people place their faith in solutions only to find that they backlash on us.

As time passes, though, more and more people are beginning to see the fallacy of technological optimism

reality is. These models dictate how we view the world. The psychologist Alan Watts coined the term *skin-encapsulated ego* to define the prevalent model we use: what is inside the skin is "I," and what is outside is "not I." Thus, many people operate with this view of nature as "not I." Knowing how dependent humans are on natural processes, you can see the problem with this narrow but prevalent view.

Besides supporting the notion that humans are apart from nature, a skin-encapsulated ego causes some direct damage from attempts at ego-building. According to psychologists, as we humans pass from infancy to adulthood, we develop a sense of the self, our own separate identity. Psychologists call this the **derived self**. Once the derived self becomes established, it also needs reaffirmation. New clothes, fast cars, and extravagant homes are some of the ways in which some of us build our egos, or reaffirm who we are. Reaffirmation can lead to the accumulation of materials, which depletes essential resources, causes pollution, and reduces wildlife habitat (Figure 19-2).

Having viewed some possible roots of our attitudes toward nature, let us turn to the dangerous notion that "there is always more." Tracing this viewpoint is relatively easy. For most of human history population size has been small in comparison with the earth's resource supply and capacity to assimilate wastes. We know from history that civilization has been constantly on the move in search of new resources, and it has almost always found what it wanted. In fact, there *was* always more. Now that the human population is five billion, however, many people have questioned this attitude. Extinction of species, depletion of natural resources, and worldwide pollution are blatant reminders that the world we live in has limits.

and are calling for simpler, more effective solutions that depend more on individual responsibility and on actions that benefit others alive today, future generations, and the great many species that make this planet their home.

A More Personal Look

Many modern societies operate under a frontier mentality. Many of our political leaders, who subscribe to frontierism, talk about continued economic growth and increased material wealth despite the limits to the world's resources. Many people, untrained in the environmental sciences, are caught up in their rhetoric. Why worry, if there is always more? Certainly technology will come to the rescue.

This book points out the fallacies of the frontier mentality. It shows that fossil fuels and mineral supplies are limited: there is not always more. It also shows that renewable resources can be used up. Our study of extinction, for instance, tells us that once a species is wiped out by habitat destruction or overhunting, it is gone forever. No amount of technology will bring back the species that we are currently destroying. Surely, the attitudes passed down by some of our leaders are maladaptive; they are disadvantageous to our species to the point that they could threaten our survival. This section looks at the personal actions and attitudes of many people and points out the problems they create.

Apathy Many people, while understanding that the earth is finite, remain apathetic about the course of modern society. Involved in their own lives, they see resource limitations and pollution as problems for which someone else must take responsibility. Apathy is effortless, noncontroversial, and cheap.

Where does apathy come from? In part, many of us are taught or conditioned to be apathetic—not to rock the boat or make waves. Someone bulldozes a favorite forest of ours to put up a shopping mall and our parents calm our indignation by saying, "That's the price of progress."

Government and history teachers must also shoulder some of the blame, for they often fail to teach students about participating in government. For many people, democracy means freedom. But democracy is much more. First and foremost, it is an invitation to participate: an invitation to vote, to write congressional representatives, to sit on citizen advisory committees. But many people don't even vote. Why bother?

Not voting leads to another cause of apathy, powerlessness. Why vote? It doesn't seem to matter. People are paralyzed by a sense of powerlessness. If you can't do anything, why worry? Figure 19-3 shows a hypothetical curve that plots apathy and its close cousin, despair, against empowerment—taking control of your life and taking action. It suggests, and with good reason, that as

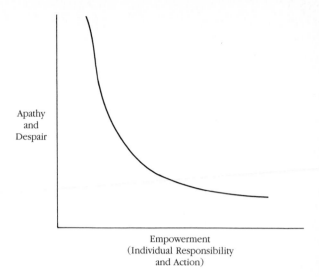

Figure 19-3. Hypothetical curve showing that apathy and despair are at their peak when people are powerless or feel powerless to make changes. When individuals take action, apathy and despair often fall, opening the doors to positive change.

people take control of their lives, get active, and work toward change, apathy and despair tend to fall.

Apathy also stems from our explicit faith in technology, technological optimism. Experts warn that blind optimism in our technology may be the fatal flaw of modern society. As previous chapters have noted, technological solutions are only part of the answer to our problems.

The Self-Centered View The average man and woman on the street often take a self-centered view. Their economic and noneconomic welfare governs their actions: what kinds of homes they buy, what size cars they buy, how much luxury they surround themselves with, and so on. Author and social critic Tom Wolfe coined the phrase the me-generation to describe the people of the 1980s who seem intent on self-indulgence. Many people of the me-generation buy what they can afford, giving little thought to the effects of consumption. Replacing the self-centered approach to life with a global environmental perspective is the main thrust of environmental education.

Ironically, many environmental groups have failed to promote individual responsibility and action among their members and the general public. Much of their educational outreach is through direct mail, appeals that generally outline the problems we face and propose to solve them through legal and legislative action. Individuals are asked to help, but individual involvement is generally limited to writing a check. Wendell Berry, philosopher and author of *The Unsettling of America*, writes that "The giving of money has . . . become our characteristic virtue.

But to give is not to do. The money is given *in lieu* of action, thought, care, time." I call it the cash conscience. Environmental groups also spend much of their time pointing accusatory fingers at lawmakers, inept government regulators, and avaricious business executives.

This lopsided approach nurtures the idea that the blame for the environmental crisis lies almost entirely in someone else's hands and can only be solved by regulating someone else's activities or by applying new and more efficient technologies to control someone else's pollution. It nurtures the idea that money alone can solve our problems. The legislative and legal work of environmental organizations such as the Environmental Defense Fund, the Natural Resources Defense Council, Greenpeace, Sierra Club, and others is vital to our efforts to improve the environment, but individual action is equally important. The me-generation will never get the message unless someone starts saying it.

Feelings of Insignificance Feelings of insignificance also grip many of us. This pervasive feeling may have a powerful effect on our own lives and the world around us. In large part, feelings of insignificance create the problems we have and keep us from solving them. Let me explain this paradox. Being just one of billions of people on earth is an excuse many of us use for our wasteful ways. What difference, we ask, does it make if I drive 10 miles per hour over the speed limit and waste a little gasoline? I'm just one of 250 million Americans. I'm just a small part of the problem, and an insignificant one at that. The trouble with this thinking is that millions of people think the same thing. Together, their actions add up to a lot of waste and pollution.

Feelings of insignificance also keep us from solving many problems. If I drive the speed limit, recycle aluminum and paper, and keep my thermostat at a reasonable temperature, what difference will it make? I'm only one of 250 million, and my contribution to the resource depletion and pollution is insignificant, so why do it?

There you have it: feelings of insignificance create many of our problems and keep us from solving them. We need to develop an understanding that there are billions of people and that their actions, taken together, can make or break us. Education can go a long way in this regard.

Restricted Space–Time Values Chapter 14 discussed the concerns of people related to space and time. As we saw, most people's concerns are restricted to the self, family, and community. With regard to time, most people concern themselves with the present or the near future. This restricted view, though very natural, can be a detriment to modern society in light of the rapid rate of change in population, resources, and pollution (Figure 19-4a). Resource demand, pollution, and population growth have all rounded the bend of the exponential curve (Figure 19-4b). Each doubling brings an enormous increase in the resources we use, the pollution we produce, and the people we must feed and house. Humanity's future hinges largely on how well we expand our space–time values to come to grips with the realities of exponential growth.

A Low-Synergy Society

The frontier mentality, the economic-growth orientation of modern society, and the personal attitudes described above combine to create a low-synergy society. Toxicologists use the term synergy to describe a phenomenon in which two chemicals act together to produce an effect larger than one might expect by adding their individual effects together (see Chapter 14). In common usage, synergy simply means "working together." Low-synergy societies are those in which the individual parts work against the benefit of the whole.

The earliest human societies were high-synergy societies. The early agriculturalists in tropical climates, for instance, practiced slash-and-burn farming that did little damage to the jungle. But back then human population size was small, the technology was simple. What damage humans caused was repaired naturally.

Today, however, synergy has been eroded. Large numbers of people, advanced technologies, and a seemingly infinite variety of pollutants have created worldwide ecological disruption. Our air, water, and land are suffering as society stretches well beyond the earth's carrying capacity.

To rebuild a high-synergy, or sustainable, society under the new conditions of human civilization will require some fundamental changes in economics and government (Chapters 20 and 21) in addition to the changes in ethics discussed here. Further alterations in housing, urban development, industry, transportation, and technology are discussed in the epilogue.

Sustainable-Earth Ethics

Few writers have had as much influence on environmental ethics as Aldo Leopold, a wildlife ecologist, best known for his book, *A Sand County Almanac.* Leopold carried on the battle of John Muir, the founder of Sierra Club and a long-time crusader for wilderness. Leopold wrote about the need to include nature in our ethical concerns, that is, to extend our concerns beyond people. Leopold called his ethic a **land ethic**. It proposed that humans were a part of a larger community that included soil, water, plants, animals—in short, the land. Leopold suggested caution and deferred rewards in our use of natural resources.

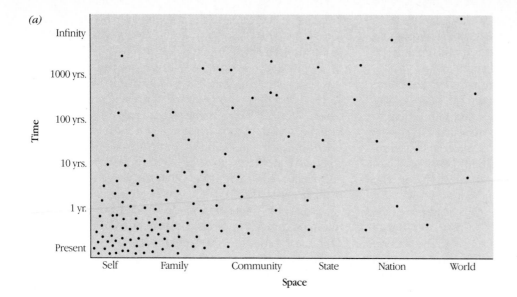

(a)

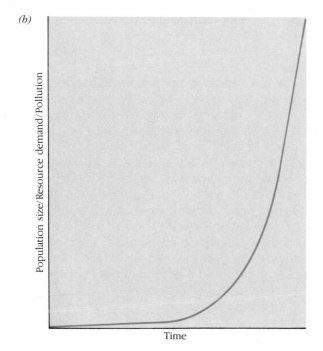

(b)

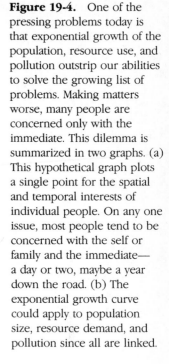

Figure 19-4. One of the pressing problems today is that exponential growth of the population, resource use, and pollution outstrip our abilities to solve the growing list of problems. Making matters worse, many people are concerned only with the immediate. This dilemma is summarized in two graphs. (a) This hypothetical graph plots a single point for the spatial and temporal interests of individual people. On any one issue, most people tend to be concerned with the self or family and the immediate—a day or two, maybe a year down the road. (b) The exponential growth curve could apply to population size, resource demand, and pollution since all are linked.

Leopold first suggested the need for a land ethic in 1933. *A Sand County Almanac*, first published in 1949, took the message further. Charles E. Little, author and founder of the American Land Forum (now the American Land Resource Association), calls the land ethic "one of the most important ideas of the century."

Leopold's view was considerably more encompassing than the self-serving attitudes of Theodore Roosevelt-style conservationists, who protected resources principally because they were of value to humans. It has helped to change the thinking of many people the world over. Still, the land ethic is a minority opinion. Some think that it doesn't go far enough. Why?

The Norwegian ecological philosopher Arne Naess says that "A philosophy, as articulated wisdom, has to be a synthesis of theory and practice." The land ethic teaches us to respect the land and ecosystems. It instructs us, in Leopold's own words, to enlarge "the boundaries of the community to include soils, water, plants and animals, or collectively: the land." In so doing, Leopold thought that the role of *Homo sapiens* would shift from "conqueror . . . to plain member and citizen of it." What the land ethic fails to offer is a concrete way to make the transition.

Enter the **sustainable-earth ethic**, or sustainable ethic for short. The sustainable ethic holds that the earth has a limited supply of resources and that humans are a part of nature and not in any way superior to it.

The main concept of the sustainable ethic is that "there is not always more." As the resource chapters of this book have pointed out, the earth has a limited supply of non-renewable resources such as metals and oil. Even renewable resources can be depleted by improper management and should be harvested only at a fixed rate (equal to the replacement rate). To develop a sustainable society we must learn that infinite growth of material consumption in a finite world is an impossibility. We must learn that ever-increasing production and consumption can only damage the life-giving environment.

These important realizations will lead to new resource–consumption strategies. The first is conservation, curtailing excessive use of resources. The second strategy is reuse and recycling of all materials. Third is the use of more renewable resources (sunlight) and fewer nonrenewable resources (coal and oil). The fourth, and most important, is the control of population growth.

Sustainable ethics also holds that humans are not apart from but rather a part of nature. This simple message is articulated in a quote attributed to the Indian chief Seattle, who watched in despair as white people usurped the Indian lands of the Pacific Northwest:

You must teach your children that the ground beneath their feet is the ashes of our grandfathers. So they will respect the land, tell your children that the earth is rich with the lives of our kin. Teach your children what we have taught our children—that the earth is our mother. Whatever befalls the earth, befalls the sons of the earth. If men spit upon the ground, they spit upon themselves.

This we know. The earth does not belong to man; man belongs to the earth. This we know. All things are connected like the blood which unites one family. All things are connected. . . .

What befalls the earth befalls the sons of the earth. Man did not weave the web of life; he is merely a strand in it. Whatever he does to the web, he does to himself.

Sustainable ethics embraces a respect for the land, air, water, and all living things, including other people in other lands and in future generations. It nurtures a reverence for life that would inevitably result in a curtailment of some of our activities, would diminish our view of self-importance, and would result in a decrease in the destructive, narrow (human-centered) thinking so prevalent in frontier societies.

Sustainable ethics outlines four operating principles by which society must work, thus putting ethical guidelines into action. They are: conservation, recycling, renewable resources, and population control. Thus, the sustainable ethic goes beyond the land ethic, giving shape to ideals. It is, therefore, a pragmatic philosophy, rooted in the science of ecology.

The ethical and practical aspects of this new philosophy could have many benefits. Consider some of the following benefits as you think about the possibilities of this new idea.

First, sustainable ethics would teach us to examine our economic and resource decisions more carefully to see how they affect the integrity, stability, and beauty of the world. Short-term exploitive approaches would be frowned on. Citizens might take a critical look at traditional economics, too, insisting that not everything has a price in dollars and cents. We might take a dim view of those activities that rob us of security, happiness, beauty, and health. To strive for a quality existence means controlling materialism.

Second, we might become more and more aware of the interconnection of all components of the earth and of the fact that our actions often have many unforeseen effects. Our growing knowledge of the global interconnection might create a more thoughtful approach to all human activities.

Third, as an outgrowth of our changes in view, we might exercise more restraint in all facets of our lives. In regard to technology and development, the ability to say "I can" would not inevitably be followed by "I will." Instead, new questions would be asked: Should we build this dam? Should we introduce this product? Should we build more nuclear weapons? Should we have another child? Instead, "I can" would be followed by two important questions: "What are the environmental consequences?" and "Should I?"

The notion of restraint seems foreign now. As resources become scarce, however, restraint will become more natural and will not carry the negative connotations that it does for many people today. As the futurist John Naisbitt has written, "Change occurs where there is a confluence of both changing values and economic necessity." In essence, necessity can promote restraint. Restraint might cause us to focus on our responsibility to the earth, future generations, and other organisms.

A growing number of environmentalists believe that animals, plants, and other organisms have rights similar to those we claim for ourselves. A tree, they say, has a right to exist, as does the praying mantis. Animal and plant rights, say supporters of this controversial and far-reaching view, exist *irrespective of the organism's usefulness to humans*. Given our population size and our technological prowess, we have a responsibility to protect other organisms. Sustainable ethics holds that humans are a part of nature and thus extends our boundaries and may help us fit more lastingly into the web of life.

Sustainable-earth ethics are a long way from happening and many who would like to see them become a reality fear that human society may never achieve them. If we cannot treat our human neighbors with respect and kindness, and cannot refrain from exploiting residents of Third World nations by treating them like colonies, how can we learn to treat nonhuman species more equitably?

More and more sustainable ethics could help us turn away from self-centered thinking, favoring what is good for the whole of society and the whole of the earth. Restraint is exercised because it benefits the whole of society, future generations, and the earth. Restraint could help create a high-synergy society in which the individual parts function for the good of the whole.

The sustainable ethical system is a new paradigm that lays the foundation for a sustainable society. However different it may be, it does not necessarily renounce all technology, all growth, or all material goods. Instead, it advocates a thoughtful look at the long-term health of the planet and an evaluation of the consequences of technology, population growth, and materialism.

Value Judgments and Decision Making

Each day people are faced with dozens of decisions, many of which have an influence on the environment. For example, should you ride your bicycle or drive your car to school or work? Should you turn up the heat or throw out an old pair of pants? The sum of the decisions by many millions of individuals like yourself add up, profoundly influencing the world we live in. Corporations and governments face many decisions as well. Single decisions at this level have impacts as profound as many millions of individual decisions. For example, should the government develop nuclear power or push for energy conservation? Should the government use recycled paper or continue buying virgin stock? Understanding how decisions are made and learning alternative ways of making decisions can help us, individually and collectively, learn to be more responsible.

Ultimately, most of our decisions are influenced by our values—what we view as right or wrong, desirable or undesirable. Values are learned from our parents, relatives, peers, teachers, religious leaders, politicians, writers, and even news commentators and reporters. Values may shift over the years, sometimes drastically.

Many people today base their judgments on **utilitarianism**, a doctrine by which the worth of anything is determined by its usefulness. This view puts human needs above most, if not all, others. Economics, discussed in the next chapter, is the yardstick of many utilitarian decisions. Utilitarian resource management, for example, finds the fastest, cheapest ways of acquiring resources. It seeks to protect the environment, but mostly to protect human needs. Forests may be reseeded, but not so much to provide habitat for animals as to provide wood for future generations.

At the opposite end of the spectrum is a new and controversial view of **natural rights**, described earlier in this chapter and also in Chapter 8. It says that all living things have rights irrespective of their value to human society. Wilderness should be set aside, not just so people can use it, but to protect the species that have lived on the land for thousands of years. Animals, plants, and insects should be preserved, not because they are of use to humans, but because they have a right to live.

Another value system is based on **divine law**, the word of God as written in the Bible. For the most part, divine law dictates personal behavior—interactions between people. The Ten Commandments instruct us on ways to treat one another: Thou shalt not kill. Thou shalt not commit adultery. One religious group has recently added an eleventh commandment: protect the earth.

These are some of the principal value systems that affect our decisions. As you think about them, you may realize that no one system is at work at all times. In some cases, you may act out of utility. In others, you may make a sacrifice for the good of the whole.

Sustainable ethics could be considered another system of values. It calls on people to consider a new set of parameters in making decisions. It reminds us of our place in the world and implores us to act in cooperation and consideration, rather than in isolation and strict self-interest.

Making the Transition

As solid as entrenched belief systems seem, they do change. Colonialism and slavery and the beliefs that supported them, for instance, have fallen by the wayside in the United States. New paradigms have replaced these outmoded systems, but not without considerable turmoil. The same may hold true for the frontier mentality. But how can the changes be made?

Three Approaches

The shift to a sustainable society is a tall order for a society steeped in frontierism. A profound shift in attitude, many experts argue, is needed to begin the process. As people adopt sustainable ethics, they will make the shifts in lifestyle. This change in values leading to a new societal order can be called the **bottom-up approach**. A **top-down** approach calls for new laws and regulations to regulate behavior. The thinking behind this approach is that forced behavioral changes will lead to belief changes. The National Environmental Policy Act, discussed in Chapter Supplement 21-1, attempts to change beliefs by forcing developers to study their potential impacts before construction. Some people believe that disaster is necessary to bring about changes in attitude and life-style. We can call this the **crisis approach**. In reality, crisis brings about policy change (top-down) and shifts in beliefs (bottom-up). Critics argue that crises are too disruptive and that this approach must be avoided.

The bottom-up approach would cause the least hardship of the three options. However, it is probably the most difficult one. It requires some insight into our problems and their solutions and a willingness to make changes before troubles begin. Sustainability on a global scale calls for insight, commitment to change, and action among the peoples of the world.

Some Attitudinal Changes Are Already Evident

Without a doubt, attitudes are changing. The frontier mentality of "doing more with more," prevalent for so long, is now being replaced by a new attitude, an attitude of "doing more with less." This new attitude, thrust on developed countries by the high cost of energy, dwindling resource supplies, and economic problems, is an important change, but it will not suffice to bring about a sustainable society. The principal element in the strategy of achieving "more with less" is reforming inefficient ways. According to this view, we can have our fancy cars, extravagant homes, wasteful appliances, and more, but we will do it with a little less energy and fewer resources.

To achieve a truly sustainable society, we would have to shift away from the "more with less" attitude. This may require a stepwise progression. The first step would be to acquire the attitude of "doing the same with less." In other words, we will maintain our standard of living but reduce our resource demands. There are signs today that some people have already adopted this attitude. The next, and perhaps most difficult, step toward a sustainable society is the development of the attitude of "doing less with less." This requires us to examine the ways we do business and run our daily lives. It means having only one small, energy-efficient car per family, using mass transit and walking; sharing tools with neighbors or renting them instead of buying a tool used only once a year; digging gardens and shoveling sidewalks by hand; forgoing the latest fashion or the newest gadget designed to make life easier; taking a train home from college or for business trips; and building a small, energy-efficient, solar-powered, earth-sheltered home that is warm and comfortable instead of an oversized, electric-powered home with inadequate insulation. It means breaking our wasteful habits.

Already, there are many signs that the change from "doing more with more" to "doing less with less" may be taking place. Many people have simplified their lifestyles. Some have moved to rural areas where they can grow their own food, and many are turning to nonmotorized recreation, such as cross-country skiing and canoeing.

With the changes in attitude and life-style there is room for optimism, but there is a lot more to be done. Clearly, any optimism we may have should be cautious optimism.

Avoiding Pitfalls

To make the transition to a sustainable society is the challenge and the opportunity of a lifetime. But to do so, we must avoid numerous pitfalls.

First, we must avoid the paralysis that comes with an attitude that we are doomed to perish by overpopulation, food shortage, war, or pollution. This paralysis will do no good. It will keep us from thinking creatively and acting with determination to make necessary changes in our own lives.

We must also avoid excessive optimism in technological solutions to our population, resource, and pollution problems. Such a view can be dangerous in the long run, for it blunts personal responsibility and action. By avoiding blind faith in technology and by carefully developing those technologies that are most useful and least harmful to the environment and human health, we can go a long way toward solving our problems. Technology is not *the* answer, but it can be a part of the answer to our problems.

We must also avoid outdated solutions, for example, building large, expensive power plants to heat our homes when insulation and solar energy will provide the warmth we need, more cheaply and with fewer impacts on the environment. The problems of today are caused by the conditions of today. New solutions to new problems are what the times call for, but we must be careful not to overlook obvious answers.

We must also avoid narrow thinking, and restricted imagination. All our creativity, cooperation, and patience will be called on to achieve a sustainable society. Even though we don't know the exact form of the future society we're working for, we know the principles that will work. That gives us enough to start thinking now.

We must also avoid apathy and the attitude that someone else will solve the problem. The sustainable society comes closer to being a reality with each person's effort, no matter how small or large. We must become active and make the changes we need to achieve a less resource-intensive life-style. We must start with our immediate surroundings. Which changes we make are not as important as the choice to change.

Once we've achieved our personal goals, we might take a more active role in our communities by organizing recycling on campus or in our apartment buildings or dormitories. Working with schools or employers to reduce energy consumption is another option. Writing our congressional representatives can also help.

Remember the saying, "If you're not part of the solution, you're part of the problem." We live in an age of cooperation; we succeed as we work together. We succeed as each of us makes the small changes that eventually transform the globe.

A human being is part of the whole, called by us "Universe." . . . He experiences himself, his thoughts and feelings as something separated from the rest—a kind of optical delusion of his consciousness. This delusion is a kind of prison for us, restricting us to our personal desires and to affection for a few persons nearest to us. Our task must be to free ourselves from this prison by widening our circle of compassion to embrace all living creatures and the whole of nature in its beauty.

ALBERT EINSTEIN

Summary

Building a sustainable society requires a new set of beliefs—an appropriate ethical system—because our attitudes toward nature determine how we interact with the environment. Today, many people operate with a frontier mentality based on three main ideas: (1) the world has an unlimited supply of resources for human use, (2) humans are apart from nature, not a part of nature, and (3) nature is something to overcome. These attitudes toward nature stem in part from Judeo-Christian teachings and in part from biological imperialism—the tendency for all organisms to convert as much of the environment into themselves and their offspring as possible. The psychological roots of our behavior are found in the common perception of the self as "I" versus nature as "not I." As a result, we think of ourselves as apart from nature. This psychology also causes damage from the ego-building that leads many to seek material wealth.

The roots of the attitude that there is always more are easily seen. For most of human history, population size has been small in comparison with the earth's resource supplies. There always *has* been more, until recently.

The frontier mentality is the underpinning of contemporary human thought. Made aware of the limits of resources, many people simply express apathy. Resource limitations and pollution seem like problems for which someone else must take responsibility. Many people take a self-centered view of things: what they can afford determines what they buy. Feelings of insignificance also grip the average citizen. They greatly contribute to the problems we suffer and also keep us from solving them. Finally, restricted space–time values create narrow thinking that is dangerous in these times of rapid change.

The frontier mentality creates a low-synergy society whose parts work at cross purposes. A high-synergy society might be created by adopting sustainable-earth ethics. This system of ethics holds that the earth has a limited supply of resources and that humans are a part of nature and not superior to it. It advises us to pursue conservation, recycling, renewable resources, and population control. The new ethical system will help us learn to fully examine our economic and resource decisions in terms of what creates a sustainable society.

Human decisions about the environment are affected by our values. Many people base their decisions on the value of things by considering only their usefulness. This is called **utilitarianism**. Natural rights hold that all things have value and rights, irrespective of their usefulness to humans. Divine law is contained in the Bible. It dictates how some people act. Sustainable-earth ethics is an additional system with great importance.

There are encouraging signs that we are shifting toward a sustainable society. Still, we are long way from realizing this goal. To make the changes we must avoid the paralysis that comes with the attitude that humans are doomed. We must also avoid blind optimism about technology and the "good-old-days" trap, which seeks to answer today's problems with yesterday's answers. And we must avoid narrow thought and restricted imagination. We must avoid apathy and the attitude that someone else will solve the problem.

Discussion Questions

1. Describe the three major concepts of the frontier mentality. Using your critical thinking skills discuss these new attitudes. Will they help build a sustainable society? Are they attainable? Are they practical?

2. How is your personal ethic similar to and how is it different from the frontier mentality?

3. Debate the statement "The frontier mentality comes from Judeo-Christian teachings that give humans dominion over the world."

4. Describe the terms *biological imperialism, skin-encapsulated ego,* and *derived self*. How are these related to environmental damage?

5. Describe the role apathy, self-centeredness, feelings of insignificance, and technological optimism play in creating environmental problems.

6. What is a low-synergy society? Give some examples.

7. Discuss the tenets of sustainable ethics. Indicate which of these tenets coincide with your personal beliefs. Which ones don't? Why?

8. In what ways would society change if it adopted sustainable ethics?

9. Debate the statement "We are a long way from achieving a sustainable society."

10. Which of the following attitudes best fits yours? "More with more," "more with less," "same with less," or "less with less."

11. What is the attitude of most of your friends?

12. For one week, when you listen to the news, watch TV, read articles, and listen to people talk, note examples of the frontier mentality. Make a list of them. What attitudes could replace them? Would new attitudes be more effective?

13. "Animals and plants have rights," says a leading philosopher, "irrespective of their value to humans." Do you agree? Why or why not?

Suggested Readings

Berry, T. (1988). *The Dream of the Earth*. San Francisco: Sierra Club Books. Important book on a theology based on ecology.

Berry, W. (1987). *Home Economics*. Berkeley: North Point Press. Thoughtful collection of essays on wise stewardship.

Brown, L. R. (1981). *Building a Sustainable Society*. New York: Norton. Highly readable discussion of the need for a sustainable society and ways to build it.

Chiras, D. D. (1990). *Beyond the Fray: Reshaping the American Environmental Response*. Boulder, CO: Johnson Books. Author's description of important changes needed in the environmental movement.

DeVall, B. (1988). *Simple in Means, Rich in Ends. Practicing Deep Ecology*. Salt Lake City: Peregrine Smith. Important discussion of deep ecology and ways to put reverence for nature into action.

Hardin, G. (1985). *Filters against Folly*. New York: Viking. Thoughtful treatise on ethics. Important reading.

Leopold, A. (1966). *A Sand County Almanac*. New York: Ballantine. Collection of essays on nature and conservation.

Milbrath, L. W. (1984). *Environmentalists: Vanguard for a New Society*. Albany: State University of New York Press. Contains an excellent attitude survey.

Milbrath, L. W. (1989). *Envisioning a Sustainable Society. Learning Our Way Out*. Albany: State University of New York Press. Early chapters discuss values of the dominant paradigm (frontierism) and need for a new value system.

Naisbitt, J. (1982). *Megatrends: Ten New Directions Transforming Our Lives*. New York: Warner Books. Superb analysis of social trends.

Nash, R. F. (1989). *The Rights of Nature. A History of Environmental Ethics*. Madison: University of Wisconsin Press. Important reading on the evolution of environmental ethics.

Partridge, D. (ed.). (1980). *Responsibilities to Future Generations*. Buffalo, NY: Prometheus Books. Good collection of essays on ethics.

Rolston, H. (1987). *Environmental Ethics. Duties to and Values in the Natural World*. Philadelphia: Temple University Press. Explains the rights of other creatures.

Ruckelshaus, W. D. (1989). Toward a Sustainable World. *Scientific American* 261 (3): 166–175. Worthwhile reading on sustainability.

Russell, P. (1982). *The Global Brain*. Los Angeles: Tarcher. A thought-provoking book on achieving a unified society.

Schumacher, E. F. (1973). *Small Is Beautiful: Economics as if People Mattered*. New York: Harper and Row. One of the best books ever written on the subject of a sustainable society and new ethical systems necessary for survival on our finite planet.

Van Matre, S. and Weiler, B. (1983). *The Earth Speaks*. Warrenville, IL: The Institute for Earth Education. Superb collection of writings on nature.

Economics and the Environment

A penny will hide the biggest star in the Universe if you hold it close enough to your eye.

SAMUEL GRAFTON

In southern India people once made traps for monkeys by drilling small holes in coconuts, filling the shells with rice, and chaining them to the ground. The success of this trap was based on a simple principle: the hole was large enough for a monkey to insert its empty hand, but too small for it to pull a handful of rice out. Monkeys were trapped by their own refusal to let go of the rice.

Some economists believe that the same plight grips humankind: we're caught in a spiraling cycle of greed, trapped by our own refusal to let go of our desire for continual economic growth. This chapter looks at economics, pointing out some of its major laws. It examines how the prevalent economic thinking contributes to environmental problems, and it evaluates an economic system for a sustainable society.

Economics and the Environment

Economic activity began with the advent of towns and villages, which, in turn, owe their origin to the agricultural surpluses that allowed men and women to turn to crafts and trades. But the science of economics, many say, really began in earnest only two centuries ago, with the publication of the Scotsman Adam Smith's monumental book, *The Wealth of Nations*. Economics is the study of the production, distribution, and consumption of goods and services in a society. It concerns itself with two key variables: inputs and outputs. Inputs include labor, land, and commodities, such as energy or minerals, which com-

467

panies require to turn out their products. Outputs are the goods and services that companies produce for consumption or for further production.

Economics, like environmental science, is concerned with relationships. It also employs scientific tools to discover the laws that regulate economies. The description of economic facts and relationships falls within the purview of **descriptive economics**, so named because it is supposed to be free of judgment. Descriptive economics is, relatively speaking, a pure science. Its questions can be answered only by facts. Economics melds with political science and sociology when it attempts to answer value-laden questions. For example, should companies pay for pollution controls? How rapidly should the economy grow? Such questions cannot be answered by empirical facts and figures. There are no right or wrong answers to them, for they are value judgments and are left to the political process. This realm of economics is called **normative economics**.

Economic Systems

Economics is a tool that can help society solve three fundamental problems: (1) *what* commodities it should produce and in what quantity, (2) *how* it should produce its goods, and (3) *for whom* it should produce them. Of course, there are many ways for a society to solve the three basic problems. In a **command economy** such as that existing in the Soviet Union for many years (but now apparently changing), the government dictates production and distribution goals. Decisions are left to bureaucrats. In a **market economy** the government takes a back seat to the marketplace. In such an economy, companies generally produce the goods and services that yield the highest profit, thus answering the first question, what goods and services and in what quantity? Profit dictates how goods are produced. Generally, the least costly method of production yields the greatest profit. In a market economy the question "for whom?" is determined by money. As a general rule, whoever can afford a good or service will get it.

One of the key principles of economics is the **law of scarcity**. It states that most things that people want are limited. As a result their sale is rationed. In a market economy price is the key tool that rations output. For instance, few of us drive Porsches because the price greatly exceeds our ability to pay for them. In a command economy governments ration output, although prices can also play an important role.

In truth, most nations' economies are *mixed*, that is, they contain elements of market and command economies. The mixed economies of Great Britain, Canada, and the United States, for example, all promote free enterprise, a system driven by supply and demand, described later. But these and other countries with market econ-

omies also influence and even regulate their economies to one degree or another.

One way governments regulate the economy is through laws and regulations requiring companies to control pollution. Laws that stipulate how much pollution a company can emit are a kind of government imposition on the production process. Outright government bans on dangerous products also limit product availability. Through bans governments dictate production and consumption, interfering in a free economy to protect health or the environment. Another way of influencing the market is through federally mandated freight rates. As noted in Chapter 13, freight rates are mandated by the federal government. Virgin material travels cheaper than scrap for recycling, a great benefit to the mining industry and a deterrent to recycling. State and federal governments also subsidize various activities by special tax breaks, or by sponsoring or funding research. These and other subsidies create an uneven field of competition that benefits some and harms others. Last, but far from least, governments also impose tariffs, taxes on imports, to regulate the flow of goods into a country, thus stifling free international competition.

Regulations, bans, subsidies, and other policy instruments are the ways governments influence free market economies. Command economies are also a mixture of free and not-so-free practices. China and the Soviet Union, for example, allow some free market enterprise within their borders.

The Law of Supply and Demand

In market economies the three essential questions—what to produce, how to produce it, and for whom—are solved by price, which is theoretically dependent on supply and demand. Figure 20-1a shows a demand curve for rice. On the vertical axis of the graph is the price (P) of rice in dollars per bushel. On the horizontal axis is the quantity (Q) that people will buy at each price. This graph shows that a rise in price decreases the demand. Conversely, a lowering of price will generally increase demand. P and Q are inversely related. Most of you are familiar with this phenomenon. Consider a special sale on jeans offered by a local store. The day before the sale business is slow, but the day the price drops, people flock to get a bargain.

The supply curve is of interest chiefly to producers (Figure 20-1b). It plots the relationship between price and the quantity that suppliers will produce. It illustrates an intuitively simple concept: the higher the price the more producers are willing and able to produce. At lower prices, producers reduce production. The fall in oil prices in the early 1980s, for example, put many oil drillers out of business and caused incredible economic hardship in Denver and Houston. What happened? As oil prices fell due to conservation (which reduced demand) and

(a) Demand Curve

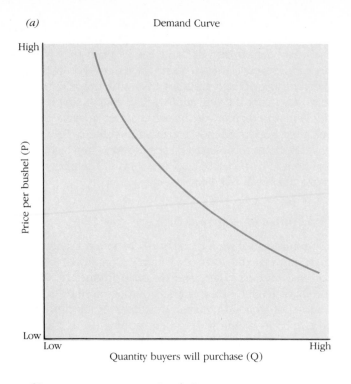

(b) Supply Curve

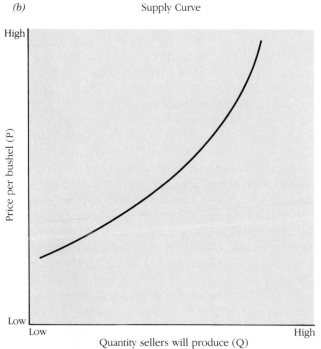

(c)

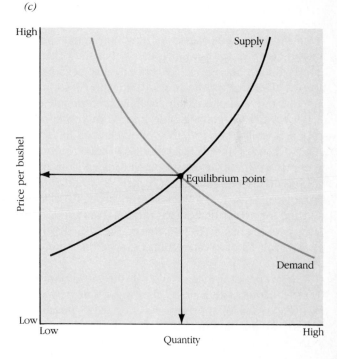

Figure 20-1. Supply and demand curves for rice. (a) The demand curve shows the relationship between the price (P) on the vertical axis and demand (Q) on the horizontal. This graph shows that a rise in prices reduces the demand. Falling prices increase demand. (b) The supply curve shows the relationship between the price and supply, or amount produced. The higher the price, the more farmers will produce. (c) The equilibrium point is the intersection of the supply and demand curves. It's the price people will pay for rice and the amount farmers will produce.

cheaper foreign oil, many companies went out of business or shut down wells because they couldn't produce oil profitably at the lower prices.

The economy is a balancing act (offset by some government tampering as mentioned earlier) with two principal players—supply and demand. The balance of these two results in the price. Graphically, this is represented by the intersection of the supply and demand curves and is called the **market price equilibrium** (Figure 20-1c). The market price equilibrium is a compromise point. It represents the price at which consumers can afford to buy a product and the price at which producers can afford to produce it.

Imagine an economy in which literally thousands of prices are set by this kind of interaction. That is the free market system of the Western world—of course, with a little government interference here and there. The system may also be tampered with by business itself. Although professing to like competition and the free market system, businesses sometimes undermine the process by finding illegal ways to control the market to their advantage. They can, for example, buy up competitors, creating a monopoly. The monopoly eliminates its competition and can set prices at any level it wants (provided the public will pay). Antitrust laws in the United States are aimed at protecting individuals from monopolies. Companies may try to drive others out of business. For example, Browning Ferris Industries (BFI), a major US trash hauler, was found guilty of trying to illegally drive a smaller Vermont trash hauler out of business. BFI was fined $51,000 in damages and $6 million in punitive damages, a punishment for illegal activities.

The law of supply and demand has some very real implications for the environment. Consider ivory trade (discussed in more detail in Chapter 8). Elephants have been slaughtered by the thousands in Africa to support the profitable ivory trade. Even after African countries made it illegal to shoot elephants, poachers continued. Why? The dramatic decline in elephants has raised the price of ivory. That gives poachers considerable economic incentive to continue illegal hunting, even at the risk of being killed by game wardens. The supply graph predicts such activity, for it shows that the higher the price, the more willing someone is to produce. People become rich and the elephant is pushed toward extinction. It's a simple line of cause and effect with devastating consequences.

Price can have considerable impact on conservation efforts as well. When the price of oil climbed in the late 1970s because of the artificial shortages created by embargoes on foreign oil, Americans, Canadians, and Europeans got busy conserving energy (Chapter 12). The fall of oil prices in the early 1980s, however, brought back the demand for powerful, fast (gas-wasting) cars and eroded many people's resolve to save energy. The US government lost virtually all of its interest in conservation and energy efficiency.

To many people, price should rule the world economy. "Let the market system alone," they warn. This view, however, has resource specialists worried. Why? Because it fails to take into account the limited supply of many natural resources: oil, natural gas, and minerals.

As described in Chapter 13, optimists point out that supply and demand can solve the dilemma. If supplies fall, they reason, prices will rise. That will stimulate exploration and more discovery, thus creating new supplies. At some point, however, nonrenewable resources will become economically depleted, that is, in such short supply that they will no longer be economically affordable (Chapter 13). If substitutes are not available, and there are many important minerals for which there are no substitutes, the economy could suffer. Long before that point,

though, the rising prices of declining resources could stimulate global inflation with crippling effects.

Clearly, supply and demand has its problems, which impair progress toward sustainability. To overcome these barriers, many observers argue that modern society must learn to accept the limitations of the law of supply and demand, in other words, realize that our nonrenewable resource supply cannot expand indefinitely. The next chapter, on government, looks at the ways in which governments can affect supply and demand and help promote more responsible resource management.

Economic Measures: Beyond the GNP

Economists need ways to measure economic activity. The most important measure of a nation's economy is its gross national product. The **gross national product** (GNP) is the market value of the nation's output, in other words, the cost of all goods and services that a nation produces and sells, including government purchases, in a given year. **Real GNP** is the GNP adjusted for inflation. **Per capita GNP** is defined as follows:

$$\text{Per capita GNP} = \frac{\text{Real GNP}}{\text{Total population}}$$

GNP is widely used to track economies. It gives a general picture of the relative wealth of nations and the living standards of their people. However, MIT's Paul Samuelson and Yale's William Nordhaus note that the "GNP is a flawed index of a nation's true economic welfare." Why? The GNP includes many goods and services that make no contribution to the welfare of the people. It fails to differentiate between good and bad expenditures. For example, the GNP includes all expenditures on homes, books, concerts, and food—deemed good because they improve the standard of living; at the same time, however, the GNP includes expenditures on cancer treatment, air pollution damage, hospitals and health services, and water pollution projects—all necessary evils brought on by pollution. Therefore, a country with filthy air and polluted water that is faced with rising cancer rates and other disorders might register a high GNP. Politicians looking in from the sidelines might mistake the high GNP for an enviable condition of living, while residents would decry the putrid mess they were living in. By carefully examining measures and defining terms, critical thinking shows us the fallacy of overdependence on a flawed measure of success.

Nordhaus and another Yale economist, James Tobin, devised a measure that adjusts the GNP, making it a more accurate representation of the good that people receive from their nation's output. This measure, called the **net economic welfare** (NEW), subtracts the "disamenities" of an economy—the cost of pollution and other activities that do not improve the quality of life—from the GNP. It also adds the cost of certain activities, such as household services provided by men and women, that are not part of the traditional GNP calculations but that improve well-being.

Figure 20-2 shows that the United States' NEW is lower than its real GNP. That is to be expected. Advocates of a clean environment and a healthy economy believe that nations should strive to reduce the difference between the two measures. Figure 20-2 also shows a disturbing trend: the NEW is increasing more slowly than the GNP. In other words, as the nation's output grows, the economic benefits fall behind, largely due to rising pollution. Some economists believe that this trend may be inevitable as the world becomes more congested and more dependent on fossil fuels, nuclear power, large-scale technologies, synthetic chemicals, disposable goods, and excessive consumption.

Growth of GNP has become one of the central measures of a nation's success. In fact, countries frequently compare themselves with one another on the basis of their growth in GNP. This criterion, however, fails to take into account the accumulated wealth of a country. A newly developed nation may have a 5% annual growth in its GNP, compared with a 2% growth rate in a developed country such as the United States. Does this mean the newly developed nation is doing better? Not really. It only means that its economy is expanding more rapidly. Most likely, it has more room to expand but its people are far less wealthy than those of the United States. Keeping in mind their accumulated wealth can help nations set aside a blind dedication to growth.

Besides accumulated wealth, a nation's well-being can be measured by the health of its people. Most often that is done by comparing life expectancy and infant mortality. But a nation's health might be better measured by more qualitative measures. It's not enough to say Americans are living well into their seventies. What's more important is how well they are living. How healthy are we? Are our later years fraught with disease? The literacy rate and educational level are also measures that go beyond the per capita GNP. A measure of recreational opportunities—miles of raftable, fishable rivers per capita, for example—and the number of people availing themselves of such opportunities tells much more than the stark economic per capita GNP figure that is the usual measure of a country's standard of living.

Economics and Pollution Control

Economics plays an important role in how pollution is controlled. This section looks at some of the factors that influence our choices and discusses some practical issues.

The free market economic system in the United States

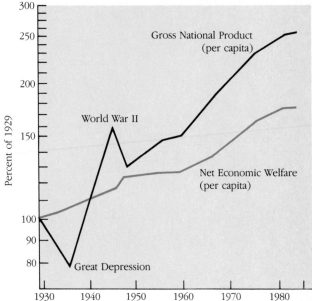

Net Economic Welfare Versus Gross National Product

Figure 20-2. The GNP is the market value of all goods and services produced by a country. US per capita GNP has risen continuously for many years. The NEW, or net economic welfare, is a measure of beneficial goods and services. It is derived by subtracting the negative aspects of the GNP which do nothing to improve the quality of life, such as damage from air pollution. The per capita NEW is lower than the GNP and rises at a slower rate, suggesting diminishing returns from economic growth.

and other countries, as well as the command economies of socialist and communist nations, treated pollution with almost uniform disregard until the late 1960s and early 1970s. Pollution was something that issued from the smokestacks and symbolized economic progress. Where it landed no one seemed to know or care. Under such a system pollution and its impacts are considered an **economic externality**, a cost to society not paid by the manufacturer or its customers.

In many instances businesses were simply unaware of the external costs of pollution. Gradually, though, citizens throughout the world began suing polluters (if their system of government allowed it). Governments established pollution standards to protect the public as well as industry, and businesses began to curb pollution by redesigning processes and, most often, installing pollution control devices. Economic externalities came home to roost. Rather than foisting the costs onto society, companies began to control pollution. In a way, the costs were **internalized**. Preventing external costs became part of the cost of doing business.

Chapters 14–18, on pollution, described laws that have forced cost internalization. In the United States the major

(a)

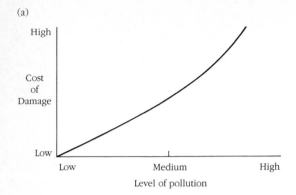

(b)

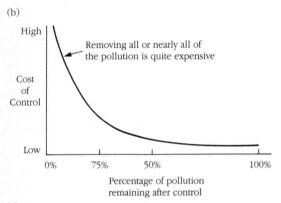

(c)

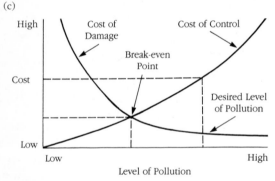

Figure 20-3. (a) As pollution levels rise, the amount of damage increases. (b) Pollution control can cut damage, but it costs money. Removing 100% of the pollution is quite expensive. Removing 75% costs less per unit removed. (c) This graph shows the point at which pollution control costs equal damage. This is the break-even point. Unfortunately, determining the full costs of pollution is not very easy and estimates may be in serious error. Society may wish to lower pollution levels past the break-even point. The desired level of pollution may technically cost society more than it benefits in reducing damage.

laws are the Clean Air Act, the Clean Water Act, the Resource Conservation and Recovery Act, and the Surface Mine Control and Reclamation Act.

More and more, governments are tightening regulations to protect the public from the tens of thousands of chemical substances and the physical agents, such as noise and radiation, we are exposed to at work and at home.

But what factors determine the level of control, that is, just how much money should we spend on controlling potentially harmful substances?

Cost–Benefit Analysis and Pollution Control

Economics plays a key role in determining how much pollution control businesses can afford. As we saw in Chapter 14, in our study of risk assessment, the chief goal of pollution control is to achieve a maximum reduction in pollution, yielding the maximum benefit at the lowest cost. This goal is made possible by **cost–benefit analysis**.

The cost of pollution control and the cost of the harmful effects of pollution can be plotted on a graph (Figure 20-3). Figure 20-3a illustrates an intuitive concept: the higher the level of pollution, the higher the cost. Thus, the more acid deposition, the more damage one can expect. Removing pollutants decreases the cost to society. Figure 20-3b shows that a little bit of effort can remove large quantities of pollutants cheaply. As more and more pollutants are removed, however, the cost to remove each additional increment rises. In the initial phase, for example, removing 100 units of pollution may cost only $1. Later on, the cost of removing pollution increases rapidly, for example, to $1 per unit of pollution. The phenomenon is called the **law of diminishing returns**. The law states that for each dollar invested, we get a smaller return; in this case, return is measured as units of pollution removed from industrial smokestacks.

Through risk analysis scientists determine how harmful a pollutant is—the level of risk (see Chapter 14). Once the risk is calculated, it is up to regulators to determine what risk is acceptable. This level is called risk acceptability. In Figure 20-3c, risk acceptability is measured as the economic (or other) cost a society will bear, or accept. The cost of reducing pollution to that particular level can be determined on the graph. Ideally, the benefits of controlling pollution should be equal to the costs of control.

Determining the cost of pollution is exceedingly difficult. The first problem arises in determining the amount of damage caused. How many people will die from air-pollution-induced cancer? How many fish will be poisoned? The second problem comes in assigning a value to lost lives, lost wilderness, polluted air, extinct species, or obstructed views.

As noted in Chapter 19, some individuals argue that other species have rights. They have a right to live, irrespective of their usefulness to human beings. They possess **intrinsic value** because they have no price tag. This is a wrench in the works of cost–benefit analysis that frustrates regulators and business people. How do you calculate a cost–benefit ratio if something has no price tag? For example, how much are the many free services (watershed protection, oxygen production, and carbon dioxide trapping, for example) provided by nature and discussed in Chapter 8 and elsewhere really worth? Chapter 15 points out that air pollution also destroys priceless statuary and ancient buildings. These have intrinsic value. They are nonreproducible. They are priceless because they cannot be replaced. One way around the dilemma of the unpriceable good is to calculate **mitigation costs**, the cost of offsetting damage. How much would it cost to restore an eroded statue? How much would it cost to move an endangered species to a new habitat?

Perhaps one of the thorniest issues in cost–benefit analysis is determining acceptable risk to humans. In an effort to reduce red tape, President Reagan ordered decision makers in the government to use cost–benefit analysis when drafting regulations. At first glance that may seem like a good idea, but how do you put a value on human life lost because of technology? As explained in Barry Commoner's essay in Point/Counterpoint in Chapter 16, the value of human life is generally calculated by determining lost earning power. Some critics argue that this is a narrow way of assessing a person's worth. How can we be sure, for example, that a janitor earning $10,000 a year might not spend his later years writing Pulitzer Prize-winning plays describing prejudice in America, which stir the American people to press harder to eliminate discrimination? Or how can we be sure that a businessman worth millions of dollars might not become a drug trafficker who illegally imports heroin into Canada and ends up spending his life in prison, where he becomes an economic liability to society?

Reducing human life to dollars and cents bothers many people for other reasons as well. It strikes many as morally wrong to consider sacrificing people's lives so that society can have its endless supply of disposable pens, diapers, and razors. It strikes many as a cruel kind of business logic that people can be sacrificed willingly for profit. Clearly, it is not an easy issue.

Chapter 17 notes that the EPA calculates tolerance levels for pesticides on the foods we eat. One additional cancer in one million people is deemed acceptable. To many that sounds fair; it is not much of a sacrifice in order to control insect pests on crops that feed thousands of people, but what if that person is you or your child? Are you willing to be the one who dies so that thousands of others can have unblemished fruit?

Because of the difficulties in assessing damage and the elusive nature of determining value, modern society is bound to find itself torn by two factions. On the one side are those who would control pollution, allowing some acceptable damage as a tradeoff for the benefits of modern technology. On the other are those who argue for eliminating harmful pollutants altogether. Bans are the only effective way of eliminating a harmful pollutant and protecting people, says Barry Commoner in his Viewpoint. You may want to read it to consider his ideas more

carefully. The next section introduces another troublesome economic question that is answered almost every day in the courts and legislative bodies of the free countries: who should pay for pollution control?

Who Should Pay for Pollution Control?

Should the corporation pay to clean up the environment, in which case the costs will be passed on to the consumers of its products? This is the **consumer-pays option**. It places the obligation to pay for damage from pollutants on the users of the products. Another option, the **taxpayer-pays option**, places the responsibility on the general public, the noncorporate taxpayers. Government payments to coal miners who suffer from black lung is an example of this option. Government programs to lime lakes to neutralize acids is another.

Individuals who favor the consumer-pays option argue that the people who use the products that create pollution should bear the cost. Nonusers should not have to subsidize the cost of users. Passing the cost directly to the consumer, they add, could create more frugal buying habits. Those who support the taxpayer-pays option argue that taxpayers have allowed industry to pollute for years with impunity. Today, new standards are imposed on industry that place costly burdens on companies that have been operating under the law for long periods. Shouldn't society, which has suddenly changed the rules of the game, pay for the controls? Advocates of the taxpayer-pays option also argue that society has elected and continues to elect officials who make deals with polluters to entice them and their polluting businesses into the community. In other cases, elected officials have overlooked flagrant violations of environmental laws. If society is responsible for elected officials who permitted pollution and other forms of environmental destruction, then society must bear at least some of the cost.

As in most controversies there is validity to both arguments. In the case of old industries, taxpayers should probably bear much of the burden; but in industries that started up after the laws had been passed, pollution control costs should be borne by the corporation and consumers.

This discussion leaves the impression that pollution prevention *always* costs inordinate sums of money. Far from it. In some cases, as discussed in Chapter 18, redesigning chemical and industrial processes can sharply reduce energy use and waste. Pollution control devices in many industries can be used to capture useful products that otherwise might be dispersed into the air. Some projects can generate profits by the sale or reuse of these materials. Savings in raw materials or revenues from such sales can pay for the cost of installing and maintaining pollution control equipment and may also generate a profit. Preventing pollution, in many cases, pays. Nowhere

Case Study 20-1

Washington's Historic Timber/Fish/Wildlife Agreement

In the Pacific Northwest, conflict and litigation are perhaps the most common adjectives used to describe the management of public forests. Since the early 1970s, a succession of battles has been waged between timber companies, environmentalists, anglers, Indian tribes, wildlife managers, and whitewater enthusiasts. The battles are complex and bitter; rivals fight over ways to harvest timber while protecting recreation, wilderness, fisheries, and other important environmental values.

In 1986, however, the warring factions paused to explore another way to manage forests. The process began when representatives from the Indian tribes and the timber industry decided to meet to find a way to put differences behind them and to find creative ways to solve several thorny resource management problems.

Two of the thorniest issues were timber-cutting near lakes and streams—riparian zones—and construction of logging roads, both practices that can lead to excessive erosion. The Forest Practices Board of the state of Washington's Department of Natural Resources had been struggling for four years to draft rules for timber-cutting in riparian zones. The board had developed technical reports, sponsored workshops, and was beginning to arrive at some management recommendations when it was approached by a few individuals who were embarking on a new way to negotiate their own solutions to the problems the board was considering. Brian Boyle, Washington's Commissioner of Public Lands, liked what he heard and temporarily suspended the Forest Practices Board's activities to give the new proponents some time to hammer out an agreement.

The first step was for the technical people—the researchers—from the state of Washington, the Indian tribes, environmental groups, and timber companies to sit down and come up with some resolutions. This process made it more clear what individual interests needed.

Within six months, the Indian and timber representatives, working with environmentalists, government agencies, and other individuals, hammered out the Timber, Fish, and Wildlife (TFW) agreement. The TFW agreement established a new kind of management process called adaptive management, which creates a procedure to monitor and evaluate forestry practices. Using the information from these evaluations, researchers will find out what works and what doesn't. This information will then be reviewed by a TFW policy group. If

the forestry practices are found wanting, the policy group can suggest changes in the regulations, which, if adopted, could improve forest management and better protect the environment.

To help the process along, certain ground rules were set up. First, there would be no votes. All decisions would be achieved by consensus. Second, each group would have all the time it needed to present its side. Spokespersons for fish, timber, water, wildlife, archeological, and cultural resources were allowed time to outline their interests. The combined group focused on ways to achieve win/win answers and solutions to a broad range of issues. Furthermore, posturing and advocacy were set aside and old enemies found that they actually could agree.

The Forest Practices Board conducted public hearings and adopted the regulations recommended by the TFW participants, who then convinced the state legislature to appropriate $4.5 million, an unprecedented amount of money, to implement the agreement. The rules adopted by the state became effective January 1, 1988, and seem to be working well. When the program started, only 1.5 million acres of commercial forest land were placed under TFW management. Today, however, at least 4 million acres are under the TFW program.

One of the benefits of this process is that it has given the environmental community unprecedented involvement in timber harvesting on both state and private lands. Overall, adaptive management will help improve water quality in streams and lakes, could boost fish populations, and could improve wildlife habitat. Decreased litigation and widespread public involvement are additional benefits.

The TFW agreement was developed by the affected parties themselves, not by a government regulator. During the process, new working relationships were formed which should carry over in time, uniting traditional foes for the common good. Those who participated in the project think that it worked because of a willingness on the part of the participants to go beyond confrontation to cooperation, so vitally needed to build a sustainable society.

The TFW agreement may provide a model for other conflicting groups throughout the United States to come together around issues and reach creative solutions that allow for change and wide participation.

is this more obvious than in the recent oil spill in Alaska. Double-hulled ships, closer scrutiny of the drinking habits of the ships' captains, and cleanup personnel and equipment on site are three preventive measures that could have saved Exxon, the federal government, and the

state government hundreds of millions of dollars. Richard Dunford, an environmental economist, points out that accidents have three costs: direct costs, indirect costs, and repercussion costs.

Direct costs from an oil spill are those that oil companies incur in the weeks and months following an accident. These include the costs of lost oil, cleanup, waste disposal, and ship repairs. The oil lost from the *Valdez* was worth an estimated $4.8 million. Cleanup costs could easily run to $1 billion the first year. Lawsuits may cost the company $1 million. Clearly, a little prevention would have been far cheaper in the long run.

Indirect costs are those costs that are picked up by state and federal agencies. Oil companies usually are required to reimburse the government for some or all of these costs. Another indirect cost results from damage to the local economy: reductions in tourism, fishing revenues, and others. Profits from the salmon fishery in Prince William Sound amount to $100 million a year. Losing some of that revenue because of contamination seems inevitable. This is a bill that Exxon will pay. Another indirect loss is damage to wildlife. The Clean Water Act holds companies responsible for oil spills liable for the cost of damage to natural resources. The law permits federal agencies to collect money for lost sea otters, waterfowl, and eagles. All told, indirect costs could come to several hundred million dollars.

Repercussion costs, the image problems arising from the spill, cause people to boycott the company or reduce their patronage. Adverse publicity may also result in more costly restrictions on oil tankers and has certainly cost the oil companies their plans to explore for oil in the Arctic National Wildlife Refuge (see Case Study 11-1).

Companies invariably balance the costs of prevention against the possible costs of an accident. But, as Dunford notes, if they don't incorporate the full costs, their cost–benefit analysis is likely to be flawed and may end up costing them hundreds of millions of dollars. Clearly, pollution prevention pays—not just in dollars and cents, but in a cleaner, more healthy environment for all living things.

Harnessing Market Forces to Protect the Environment

Democratic Senator Tim Wirth (Colorado) and his Republican colleague John Heinz (Pennsylvania) sponsored a recent study called Project 88 to assess ways to harness market forces and entrepreneurial ingenuity to protect the environment. A 50-member team from businesses, colleges and universities, the environmental community, and government spent months looking for ways to employ economic forces to increase environmental protection and sustainable management—hopefully at a lower cost than previous methods. Such measures, they argue, could help supplement traditional laws aimed at regulating pollution and resource use.

The main reason for embarking on the study was the growing cost of regulation. Moreover, companies argued that if they were left on their own, they could probably find ways to reduce pollution at a much lower cost than governmentally mandated controls.

Project 88 researchers outlined a number of "marketplace solutions," including: (1) pollution charges, or economic disincentives, (2) economic incentives, (3) tradable and marketable permits, (4) laws that eliminate market barriers that promote inefficient resource use, and (5) laws that remove unwarranted subsidies of environmentally destructive activities.

Economic Disincentives Governments have many tools to control pollution. Rules and regulations described in the previous chapters are the key weapon when backed by legal recourse. To enforce their laws, governments have typically relied on fines levied against violators. These are a kind of **economic disincentive**, a tax, charge, or fine aimed at discouraging a certain activity. More and more, corporate executives of companies that violate pollution laws are finding themselves behind bars as well.

Speeding tickets are an economic disincentive. They cost you money for violating the law. There are other economic disincentives as well. In Chapter 15 you learned about gas-guzzler taxes, taxes on cars that get poor gas mileage. By imposing a fee or by charging companies that produce environmentally undesirable products, governments provide economic disincentives. It may be to the polluter's advantage to cease the activity. To help create a market for recycled paper, proposed legislation introduced to Congress in 1989 would put a tax on paper mills and publishers that do not manufacture or use recycled paper.

Economic Incentives In recent years, government officials have also been exploring a variety of **economic incentives**, far gentler options to induce companies to comply with pollution laws. Billions of dollars of grants, for example, were provided by the Clean Water Act to assist cities and towns to improve their municipal sewage treatment plants (Chapter 16). Tax credits can also be effective incentives. A tax credit works this way: A government gives a company a tax credit—say 10%—for investing in recycling equipment or buying recycled material. This lowers the cost of business and encourages "responsible" business. The 10% credit for a $10,000 purchase would amount to a $1000 savings on taxes. Tax credits can be given to individuals who invest in environmentally responsible products, for example, solar energy, wind energy, and conservation. Wisconsin offers a 5% sales tax credit for companies that invest in recycling equipment. Governments are often wary of tax credits because they can lower tax receipts. Careful analysis must be made before offering tax credits to avoid several problems: losing revenues, investing in businesses that could make it without support, and investing in businesses that cannot survive even with tax credits. Smart community leaders, however, can find ways to help environmentally responsible businesses get started that *save* the city money. For example, a city that is responsible for its trash pickup might find it advantageous to subsidize private recycling businesses. In the process, the city can save enormous amounts of money. How? Recycling reduces the amount of garbage that needs to be landfilled. This reduces landfill tipping fees, labor costs, wear and tear on trucks, and fuel needed by the city to transport the rubbish to the landfill—all savings that accrue to the city for a small initial investment.

The Permit System One incentive under consideration is a permit system. Permits could be traded or sold. A **marketable permit** works this way: The EPA grants a permit to Company A, allowing it to emit 100,000 metric tons of sulfur dioxide each year, but the company finds a cheap way to cut its emissions to 50,000 metric tons. It can then sell its permit for 50,000 metric tons to Company B. For Company B, purchasing the permit may be cheaper than installing pollution control devices. Under this system, the total emission of pollution in the region remains the same. However, permit systems can also be designed to reduce pollution levels. In this example, the EPA would simply lower permitted levels of emissions.

In Colorado, a **tradable permit system** was used to control water pollution entering a large reservoir in the heart of the ski country. Nitrogen and phosphorus entering Dillon reservoir from farms, sewage treatment plants, and other sources had begun to cause problems, making the reservoir eutrophic. A study of various options to reduce phosphorus pollution showed that additional controls in treatment plants would cost the towns about $1800 per kilogram of phosphorus removed. Controlling nonpoint pollution would cost only about $150 per kilogram—over ten times less. The legislature and the EPA approved a tradable permit plan allowing the publicly owned sewage treatment plants to pay for nonpoint pollution controls on farmland, saving an estimated $1 million per year. They could continue emitting the same amount of pollution, but had to offset it by paying for controls elsewhere.

New Laws Numerous laws and regulations create market barriers and provide unwarranted subsidies for environmentally destructive activities. **Depletion allowances**, for example, provide tax breaks for fuel and mineral companies. As these companies deplete their reserves, they are allowed a tax credit. The money is supposed to be used to invest in more exploration, but many companies use it to diversify, that is, to buy other companies unrelated to fuel and mineral production. In Chapter Supplement 18-1 and in this chapter you learned about preferential freight rates—mandated by federal regulations—that make it cheaper to haul virgin materials than

scrap bound for recycling. In Chapter 12 you learned about subsidies granted to oil, gas, coal, and nuclear industries that give them a tremendous advantage over renewable fuels. By removing the subsidies and shipping regulations, economically inefficient and environmentally destructive practices can be eliminated. Removing these barriers can help protect the environment, ensure sustainable practices, and reduce government budget deficits.

In truth, the solutions proposed by Project 88 are not new. They have been discussed and some have been tried successfully for well over a decade. The report puts a new and appealing label (harnessing market forces—what could be more American than that?) on some radical ideas, making them appear perhaps a bit more palatable to business people. If its recommendations are implemented, we will all be the better for it.

Individuals can also become part of the market force that promotes environmentally and socially responsible businesses. By purchasing products made from recycled or recyclable material, by avoiding discardables, by purchasing products from companies with good environmental products, and by other measures individuals can help shape the economy to their liking. The Council on Economic Priorities, a nonprofit consumer/environmental group, published a handbook in 1988 called *Shopping for a Better World*. It lists hundreds of products that people routinely buy and rates the companies in ten categories, including environmental policy, charitable giving, minority advancement, and women's advancement. You may find it useful in your shopping.

One additional way of helping shape the economy is investing in socially and environmentally responsible companies and mutual funds. The New Alternatives Fund, for example, is a mutual fund that invests in companies that are involved in cogeneration, insulation, efficient light bulbs, and other forms of energy conservation. Over the past five years, it has earned investors about 16% per year, proving that good business and environmental responsibility can go hand in hand.

Environmental Regulations: Do They Impede Business?

Enormous federal and private expenditures on pollution control are seen by many people in the business community as an impediment. Environmental regulations and permits, business leaders argue, can delay projects. However, studies by the Conservation Foundation and other groups show that delays are often the result of uncertainty and poor planning by companies and agencies, not of compliance with environmental laws. Environmental protests, which do block some projects, and often for good reason, could be greatly reduced if corporations and governmental agencies were less reluctant to invite the public

Environmental Protection: Job Maker or Job Taker?
Job Blackmail and the Environment

Richard Grossman and Richard Kazis

The authors are staff members of Environmentalists for Full Employment, a policy research organization. They are the authors of the book Fear at Work: Job Blackmail, Labor and the Environment.

About 100,000 of the Americans fortunate enough to have jobs die each year from exposure to toxins at work; additionally, some 400,000 contract occupationally caused diseases. Environmental destruction from air and water pollution and diseases caused by industrial practices and waste disposal cost the United States tens of billions of dollars annually.

Yet employers and politicians keep saying that we must put up with high levels of occupational disease and extensive environmental damage. This, they say, is the price of economic growth, jobs, and the good life. The price of progress.

When workers and communities have demanded and organized for health and environmental protections along with their jobs, employers and politicians have risen to the occasion. Now that major national health and environmental laws have been in place for over a decade, we can see overwhelming evidence that protecting the public and the natural environment from industrial and government pollution is technically possible. We can also see that sound environmental policies make economic and employment sense. We don't have to be engineers or economists to understand that in a land ravaged by toxic dumps, chemical abuse, soil erosion, poisoned wells, polluted work places and choking cities, there can't be much "life" worth living.

Then what was going on when President Ronald Reagan announced that a strong Clean Air Act was a barrier to economic recovery? Why has it seemed that his appointees were doing their best to undermine health, environmental, and consumer protection laws in the name of jobs and economic growth? Why has the nuclear power industry been countering nuclear opponents over the past decade with threats of "freezing and starving in the dark"? Why did Interior Secretary James Watt sell publicly owned coal for bargain-basement prices because he "loves jobs"? Why did the *Wall Street Journal* editorialize that increased lead in the air would create "more wealth and more jobs," which "will most likely do more for ghetto children" than stringent health standards?

Call it job blackmail. Reagan, Watt, the nuclear industry, the *Wall Street Journal*—all of them have used jobs, the promise of new jobs, or the threat of unemployment in order to control decisions over investment, resource and labor use, and profit. Business and government leaders hope they can use job blackmail to keep a citizens' challenge in check, to "persuade" the public to support certain investments that are often harmful to public health and the environment.

Job blackmail is not new. Employers have long used it to fight off child labor laws, worker's compensation, Social Security, and the minimum wage. Here's an example: In 1982, the fat-melting industry of New York City threatened to shut down and throw all its employees out of work if it was forced to change production processes to eliminate the stench of slaughtered animals. When the city stood firm, however, the melters developed a closed-tank process that solved the problem, saved money, and kept workers on the job.

Now, as then, environmental protection is good for public health, good for the environment, good for jobs, and good for the economy. Since 1970 many pollution trends have been reversed; air is getting cleaner, and water quality is improving. More than 100,000 new jobs have been created because of clean-air laws, and more than 200,000 have resulted from clean-water regulations.

By contrast, according to the Environmental Protection Agency, fewer than 3000 workers a year since 1971—out of a work force of more than 100 million—are even alleged to have lost their jobs because of environmental regulations. The Industrial Union Department of the AFL-CIO has concluded that "environmental regulations have not been the primary cause of even one plant shutdown."

Despite evidence that sound environmentalism is sound economics, employers, polluters, and their political supporters in government can use job blackmail effectively as long as they control where and how investments are made and how many and what kinds of jobs are created. Today, they are eyeing trillions of taxpayer dollars and vast public resources that they say they want to use to "rebuild" the nation. But don't expect extensive public discussions on (1) the values on which investment policies will be based, (2) the kinds of jobs to be created, (3) the rights of workers and the distribution of the jobs, (4) health and environmental protection, (5) how the profits will be shared, or (6) how much control communities across the nation will have over their futures. Rather, they will try to convince us that there is no way but their way, that suggested alternatives are "impractical" or "extremist." We can anticipate that they will dangle the job claims and roll out the threats of lost jobs.

The public's fight against job blackmail is a fight for control over our lives and work. This is fundamentally a struggle for an equitable share of the nation's wealth and democratic participation in decision making.

Fortunately, there is a history of successful opposition to corporate and government job blackmail. The key has been the joining of workers, communities, environmentalists, and other constituencies to challenge "business as usual." Together, such coalitions have been able to make progress toward jobs and health, jobs and environment, jobs and civil rights, jobs and worker rights, jobs and women's rights, and jobs and democratic participation in decision making.

Today, with unemployment, pollution, and environmental mismanagement all around us, people in communities across the United States must lead the struggle to integrate employment and environmental policies. There is no other option. If we are forced to choose between our jobs and our health, we'll end up with neither.

The Impact of Environmental Laws on Jobs

Catherine England

The author is a policy analyst at the Heritage Foundation. Previously, she was on the faculty of the economics department at American University in Washington, DC. She specializes in regulatory issues.

Statutes as complex and far-reaching as the US environmental laws certainly have an impact on the number of jobs generated by the economy. Since the major pieces of legislation were passed in the early 1970s, however, there has been some debate about the exact nature of that impact. Many have argued that the environmental statutes and their accompanying regulations have increased employment opportunities in the United States, and there are some econometric studies that would seem to support such a claim. There is certainly no doubt that job opportunities have increased in at least one industry over the past decade—the manufacturing of pollution control equipment.

However, there are strong reasons to believe that environmental laws, as currently written and administered, have done more than merely shift jobs from regulated sectors to control equipment industries. Rather, the current enforcement of environmental regulations has almost certainly reduced the employment-creating potential of the US economy, thus costing a substantial number of jobs.

From the beginning it should be understood that the criticism here is not directed at the broad goals of the environmental statutes. There is a clear need to ensure that US citizens enjoy clean air, safe water, and freedom from toxic substances and hazardous wastes. But the means used to achieve these ends have been needlessly inefficient, posing higher-than-necessary costs on society.

Let us agree that businesses, like individuals, families, or even governments, have limited financial resources. There is only so much money a company can borrow or generate through selling its product or issuing stock. Therefore, any change that increases the costs of doing business in one area requires cutbacks, however minor, somewhere else. That is not to say that a regulation may not be worth the cost, but to

ensure continued economic growth, the full costs of government regulations should be understood and minimized.

In the case of environmental regulations, it is generally argued that firms cut back on investments in new plants and equipment to make the sometimes substantial expenditures needed to meet environmental requirements. But it is exactly this investment in new plants and equipment that is necessary to maintain worker productivity at a level enabling US firms to remain competitive in world markets. Growing concerns about trade deficits and the relative expense of US goods in international markets must, then, be laid at least partially at the door of the EPA and Congress.

The important question is, could we meet the fundamental goals of environmental statutes at a lower cost? Yes. This response means that corporate funds could be freed to invest in new equipment, to lower costs to consumers, or for a myriad of other uses that would lead to more jobs in the country. But how?

First, the focus of environmental regulations should be changed. Currently, most EPA regulations are extremely detailed, describing the specific technology a firm must use to comply with environmental standards. This general policy has substantially reduced the potential payoff to independent research in pollution control technologies. One author has described this focus on specific technologies as "chok[ing] off the imagination and innovation of American industry." Redirecting attention to performance standards, that is, specifying acceptable behavior and leaving businesses to decide how to comply, should encourage the more rapid introduction of less expensive means of accomplishing stated goals.

Secondly, the centralized nature of decision making in the environmental area has needlessly raised compliance costs. The United States is a country marked by geographic diversity. Accordingly, the environmental impact of a particular plant in the Northeast may be very different from its impact on the desert Southwest. Not only does population density vary, but also climate, elevation, water tables, prevailing winds, and so on. It shouldn't be surprising, then, that a single national policy has satisfied almost no one.

State officials are in a much better position to recognize and react to the existing regional diversity. Furthermore, decisions made at the state level are closer to the citizens who are actually affected by changes in environmental quality. As a result, some states may wish to impose standards tougher than those contained in national guidelines. State experimentation with mechanisms for setting standards and achieving compliance could result in creative ways of reaching environmental goals.

Finally, and perhaps most important for the job-creation question, current environmental laws and regulations are biased against emerging industries and new technologies. Environmental standards for new plants are much more stringent than those applied to existing sources of pollution. Yet the newer technologies employed in modern factories often cause less pollution than older ways of doing things. Despite this, federal regulators are only beginning to explore market-oriented means of encouraging economic growth while reducing pollution levels.

To summarize, existing environmental laws and the regulations they have spawned have cost the US economy an unknown number of jobs by inhibiting the job-creating potential of American business. The costly way in which current environmental statutes are enforced has sapped the financial resources of affected firms, placed US products at a cost disadvantage in world markets, and inhibited growth of new technologies and industries. Providing more flexibility through the use of plant-wide performance standards would encourage businesses to search for less expensive means of attaining environmental goals. Shifting more decision-making responsibilities to the state level would allow recognition of and response to regional diversity. Encouraging the trading of

emissions rights would lead to economic growth without sacrifices in environmental quality.

The underlying aims of the environmental laws are important, but just as important is the nation's economic health. Luckily, the two goals need not be mutually exclusive.

Critical Thinking

1. Summarize each of the author's main points and the data they used to support them. In your view, how well have the authors supported their contentions? Taking an objective stand, point out strengths and weaknesses in both sides of the argument.

2. Kazis and Grossman believe that jobs are often used to blackmail communities into accepting environmentally unsound practices or to support new projects that would destroy natural resources. Have you witnessed this phenomenon? How real was the threat? When looking at the big picture are job losses offset by job gains?

3. England contends that existing environmental laws and regulations have cost the US economy an unknown number of jobs by inhibiting the job-creating potential of American business. These regulations and laws have sapped the financial resources of companies, placed US products at a cost disadvantage in world markets, and inhibited the growth of new technologies and industries. Japan has environmental regulations every bit as tough as ours but remains highly competitive. How can you account for this difference? Could it be that the assertions made by England and industry are not entirely accurate? Are there other interpretations?

4. Which viewpoint do you align with most closely? Why? Is your leaning affected by certain biases?

or the government to participate in the early planning stages of projects.

Business leaders argue that environmental regulations also cut productivity. **Productivity** is the dollar value of goods per hour of paid employment. This figure has been used to show how healthy an economy is. Business people argue that environmental regulations divert workers from productive jobs (miner) to nonproductive ones (mine safety inspector). The Point/Counterpoint in this chapter shows, however, that far more jobs have been created by environmental controls than have been lost. Careful analysis shows that environmental regulations do indeed diminish the output of industry. However, they decrease output much less than other factors, such as high energy prices and the general shift to a service economy. On balance, environmental regulations decrease productivity by an estimated 5% to 15%. This is the price for safer

working conditions and a cleaner environment. Environmental laws can, however, help save companies money by preventing problems. Tighter regulations on hazardous waste disposal in the 1950s, 1960s, and 1970s, for example, could have saved millions on cleanup (Chapter 18). And requirements for double-hulled ships would undoubtedly have cost Exxon far less than the cost of the Alaskan oil spill.

The Economics of Resource Management

The previous section looked at the science of economics, concentrating on the GNP and the law of supply and demand. This section looks at our economic behavior, that is, the influence of economics on our actions. Two key areas are presented to introduce you to this topic: resource management and pollution control. By examining economic behavior and how it affects our decisions about natural resources and pollution, you can gain some insight into the problems of modern society and can find some new solutions to them as well.

Many private decisions about natural resources are influenced by economic considerations. One of the most important of these factors is called time preference.

Time Preference

Time Preference is a measure of one's willingness to postpone some current income for greater returns in the future. For example, suppose a friend offers you $100 today or $108 a year from now. If you are short on cash and need to pay for books for school, you may take the money now. Your decision to accept the money now is based on your current needs, which in this case outweigh the benefits of waiting the year, even though you would be $8 ahead. Economists would say that your need for current income outweighs greater returns in the future.

As this example shows, your preference for immediate income is determined by your *current needs*. It may also be influenced by *uncertainty*. How certain are you that your friend will be around in a year to give you the $108 she promised? A third, and very important, factor affecting time preference is the *rate of return*. The higher the rate of return, the more likely you will wait for the income. For instance, if your friend had offered you $150 a year from now, it might be advisable to wait. You could borrow $100 from your parents at 10% interest and pay them back at the end of the year with the $150 your friend gave you and still have money left over. *Inflation* also affects time preference. In times of inflation, people are apt to invest now to avoid higher costs later. But inflation can also drive interest rates up, making savings more appealing.

Time preference applies equally well to the ways in which we manage many of our natural resources, such as water, farmland, and forests. Take agriculture as an example. Farmers have two basic choices when it comes to managing their land. They can choose a **depletion strategy** to acquire an immediate high rate of return for a short period. This might involve the use of multiple cropping, artificial fertilizers, herbicides, and pesticides to maximize production (Chapter 7). Soil erosion control and other techniques might be ignored. Alternatively, farmers may choose a **conservation strategy**, techniques to conserve topsoil and maintain soil fertility; these actions require immediate monetary investments. They may cost the farmers a little more in the short run and cut into their immediate profits. Over the long run, however, the conservation strategy is the more profitable.

The choice of strategy in this example depends on the time preference. Will farmers choose the cheapest route of production, which gives the highest profit in the short term? Or will they choose a more expensive route, forgoing immediate high profits in favor of much higher gains in the future? The economic needs of farmers determine, to a large extent, their time preference. For example, a young farmer looking forward to a productive career may opt for the conservation strategy. His immediate needs may be small. He may have no family and few debts. He can sacrifice income now for larger returns in the long run. However, an established farmer may have a family to support and excessive debt. He may, therefore, want to maximize his profits through a depletion strategy.

Farmers' willingness to give up potentially higher income from conservation may also result from uncertainty about future prices, the long-range prospects for farming, and interest rates. If the price of corn is high this year but could drop significantly in coming years, farmers may choose to make their money now. If the bottom falls out of the market in the next few years, they will have made the most of this short-term opportunity. If interest rates are likely to rise, short-term profit making may be the preferable choice. High interest rates on land and machinery that farmers purchase tend to encourage the depletion strategy.

Opportunity Cost

Another factor that greatly influences economic decisions regarding resource management is the opportunity cost. **Opportunity cost** is the cost of lost money-making opportunities. For instance, the conservation strategy requires a monetary investment. The money put into conservation could have been invested in the stock market or a new business venture, possibly yielding more profit with less work than the conservation strategy. As a result, when opportunity costs are high, farmers are likely to choose options other than conservation.

Opportunity costs are also incurred when resources are wantonly destroyed. Chapter 8 on wildlife extinction, for example, describes the economic benefits of medicines derived from plants from the tropical rain forest. Losing them through disregard creates a significant opportunity cost—a loss of profit and potentially life-saving drugs. As another example, many of the world's ocean fisheries have been badly depleted. Salmon runs have also been ruined by dams and water diversion projects, pollution, and outright habitat destruction. The economic loss to commercial fishing interests and the lost recreational opportunities are enormous.

These losses suggest that a broader view of opportunities be considered during resource-management decisions and when making new laws and regulations.

Ethics

Several noneconomic factors greatly affect our economic decisions. One of the key factors, not to be overlooked, is the ethics of the farmer. Ethics can be as powerful as—or even more powerful than—economic factors. A long-term view, seeking to maximize yield for future generations and ensure the survival and well-being of all life, may foster wiser management of our natural resources. Sustainable ethics, discussed in Chapter 19, contrasts sharply with the frontier mentality, which seeks to maximize personal gain with little concern for the future, in effect leaving future generations to fend for themselves.

With this brief overview of the factors that influence the way we manage our farms and forests in mind, let us look at the topic of pollution control and the underlying economic issues.

Differing Perspectives on Growth and the Future

The Growth Issue

As we saw in Chapter Supplement 1-1, bias is difficult to eliminate from science. It creeps into the interpretations of scientific findings no matter how objective the researcher is. Descriptive economics, like other "objective" sciences, suffers from biases, the key one being economic growth. In the 1967 edition of a popular textbook Paul Samuelson, a Nobel Prize-winning MIT economist, calls economics "the science of growth." Subsequent editions have dropped the wording, but the bias remains. In his 1984 edition, for example, Samuelson writes: "Today, the ultimate measure of economic success is a country's ability to generate a high level of and rapid growth in the output of economic goods and services. Greater output of food and clothing, cars and education, radios and concerts—what else is an economy for if not to produce an appropriate mix of these in high quantity and fidelity?" While not all economists subscribe to this view, many do, especially those in the business world or in government. To them growth has become something of a god, sacred. Economic growth is a measure of progress.

Critics call this undying dedication to economic growth "growthmania." Its roots are firmly embedded in the frontier notion that "there is always more." For many decades human civilization has been caught up in it and has been willing to pay almost any price for it. Today, the dominant social view (or paradigm) that economic growth is desirable is in question. More and more Americans express skepticism over unlimited growth. More importantly, many express a strong preference for environmental pro-

tection over economic growth. In fact, a recent survey by Lester Milbrath at the State University of New York at Buffalo showed that three-fourths of the American public favored environmental protection over economic growth. Several other surveys support his findings.

Economic growth means increased consumption of goods and services. Such an increase may arise from (1) an increase in population size and (2) an increase in the amount each of us buys—per capita consumption. Because it means greater production and, presumably, greater economic wealth, population growth has traditionally been viewed as an asset to society. Each new baby is seen as a new consumer.

In a world of limited resources (such as minerals and oil) and greatly exceeded carrying capacity (witnessed in global pollution and widespread extinction of plants and animals), ecologists argue that continued population growth and rising per capita consumption, particularly among the wealthy, is unwise. But what of the billions of poor people in the world who have little food, who suffer with inadequate housing and education, and who lack sufficient clean drinking water and sanitation facilities? These value-laden questions have no easy answers. Chapter 6 pointed out that most observers believe that a combination of population control and reasonable economic development are the foundations of an effective strategy to improve the welfare of these people (Figure 20-1).

Differing Perspectives on the Future

Business people and environmentalists often battle over issues of environmental protection and resource management. One of the reasons for the conflict is that the groups see the future differently. Perceptions of the dividing line between the short-term and long-term future can be quite different.

For a business economist the dividing line between the near term and the far term is typically five years. Political leaders, tied to upcoming elections, also often subscribe to the short-term outlook (see Chapter 21). The environmentalist, on the other hand, is likely to set the dividing line at a century or more. "To say that economics is near-sighted is not necessarily to condemn it," Garrett Hardin argues. "The near-sightedness is understandable. Most of the time economics is a handmaiden of business, and experience has shown that it is very difficult to predict what business will be like more than five years in the future."

In a world of rapid change, however, a five-year outlook can be dangerous. Our large population base and resource use, which grow exponentially, render the five-year outlook obsolete. Setting a new course for humanity almost certainly means aiming our sights much further into the future.

Table 20-1 Characteristics of a Sustainable Economy
Strives for constant GNP (growth in some areas of the economy, shrinkage in others).
Emphasizes essential goods and services.
Stresses product durability.
Avoids throwaway (disposable) products.
Reduces resource use to ensure long-term supplies.
Minimizes waste and pollution.
Relies on recycling and conservation.
Maximizes use of renewable resources.
Decentralizes certain businesses.
Uses appropriate technology.
May strive for equitable distribution of wealth.

Sustainable Economies

Kenneth Boulding was one of the first to write about changes needed in the economies of developed countries, such as the United States. In 1966 he coined the phrase "cowboy economy" to describe the present economic system, characterized by maximum flow of money, maximum production, maximum consumption, maximum resource use, and maximum profit. He suggested that the cowboy, or frontier, economy be replaced by a "spaceship economy," recognizing that the earth, much like the spaceship, is a closed system wholly dependent on a fragile life-support system. Individuals within such an economy would value recycling, conservation, the use of renewable resources, product durability, and a clean and healthy environment. Throwaways would be eliminated. Repair of broken-down goods would be encouraged to minimize the need for replacements. Waste would be intolerable. In short, people would live within the limits posed by the closed system, the earth.

An economic system of this nature is also called a **sustainable economy** (Table 20-1). It can succeed only with widespread population control (Chapter 6), new environmental ethics (Chapter 19), and new political directions (Chapter 21). The spaceship economy is founded on the important concept of living comfortably and well, but perhaps not extravagantly and especially not wastefully. The key feature of the economy is a constant, rather than a continually growing, stock of physical wealth. The tradeoff is between more material wealth and a better environment with more healthful lives (Table 20-1).

The notion of a sustainable economy was not widely accepted when Boulding first proposed it in 1966 and still remains out of vogue. Today, however, more and more people recognize the inevitability of the concept.

Dissatisfaction with the frontier economy stems from its widely recognized inadequacies in managing natural resources and controlling environmental pollution. Critics recognize, as some economists and many business people do not, that the production and consumption of goods and services hinge on the availability of resources. Resource availability, in turn, is governed by the finite nature of nonrenewable resources. Supporters of a sustainable economy recognize the implications of exponential growth of population, resource consumption, and pollution. They seek to avoid the catastrophe that will result from our current dedication to unlimited growth.

The Steady-State Economy

Herman Daly, a professor of economics at Louisiana State University, has long argued in favor of a sustainable economic system, which he calls a **steady-state economy**. He writes, "A growth economy and a steady-state economy are as different as an airplane and a helicopter. An airplane is designed for forward motion—if it cannot keep moving it will crash." What we need is a more maneuverable, steady-state helicopter. "Throughout most of its tenure on earth," Daly points out, "humanity has existed in near steady-state conditions. Only in the past two centuries has growth become the norm." The sustainable economy is deeply rooted in history as well as in sound ecological thinking.

Eventually our economic system must conform to the design principles of the ecosystem. But that does not mean that the economic future is bleak. The sustainable economy need not mean dull living and no growth. What it means is that growth will occur in some sectors, such as solar energy and conservation, while stagnation strikes others, such as oil and steel production. A sustainable economy, based largely on renewable resources, need not mean a retreat into the Dark Ages. Nonrenewable resources would remain in use, recycling through the system many times. Renewable fuels such as ethanol and perhaps even hydrogen could power our transportation system. Solar energy could heat our well-insulated homes. Solar voltaics along with wind energy might provide the electricity we need.

Despite the abundant political rhetoric to the contrary, America's economy has begun to slow down. Critical thinking helps us understand why. For well over two decades the *growth in per capita GNP has been offset by inflation*. This means that the increase in economic growth that you hear about on the evening news and read about in the paper is illusory. A 4% inflation rate, for example, negates a 4% growth in per capita GNP—the dollar value of goods and services per person. Put another way, people may be earning more, but they are spending more—they're not getting ahead, they're treading water.

A closer look at inflation and growth in per capita GNP shows that inflation has actually outstripped per capita GNP. Translated, that means that the average American has less buying power today than he or she did 20 years ago. Population also negates gains in GNP. As the population grows, it reduces the per capita gain.

Statistics have an odd way of confusing the picture. This offsetting of growth in the GNP is a case in point. Just because per capita GNP has remained the same does not mean that our economy is not growing. Each year, more cars, computers, textbooks, houses, and just about anything else you can think about are sold. Gross production continues to climb. In Daly's steady-state economy the GNP would stabilize. American factories would turn out the same number of cars as the year before. This is a tall order for a society based on growth. Is it desirable? Will it create economic turmoil? No one knows the answer to these questions, but studies of the European nations whose population growth rates have severely slowed, stopped, or even reversed could provide some answers in the years to come.

Today's economy and that proposed by economists such as Daly are worlds apart. One of the principal differences is the continued high level of consumption and waste, which will deplete our nonrenewable (oil and aluminum) and renewable (wood from the tropical rain forest) resources. A truly sustainable economy achieves a level of production and consumption that can be sustained forever. The key elements of a sustainable society are a dedication to population control to help us live within the carrying capacity of the earth, a reliance on renewable resources, an unflagging dedication to conservation, and an efficient system of recycling nonrenewable materials. Despite the key differences between today's steady-state economy and the one Daly proposes, we can learn from our current economic state that no growth is not a prescription for doom.

A vision of the future is one thing; a plan to get there is another. How do we make the dramatic shift from the frontier system to the sustainable system of economics?

We may, of course, attain zero growth by eco-catastrophe, that is, by overshooting the earth's carrying capacity. Daly, however, argues that "the idea is to avoid that—to make a soft landing in the steady state rather than a crash landing." Our government can play a key role by studying the economies of European nations, helping us to prepare for the day when US population begins to level off, even shrink. Governments everywhere can help by promoting conservation, recycling, renewable resource use, and population control. (For more on the role of government, see Chapter 21.)

One step toward sustainability is the **decentralization** of some industries. Proponents of this action believe that decentralization is necessary because large-scale industries may often be less efficient than smaller industries

turning out the same product. Communication failures, inflexibility, and alienation of workers create the inefficiency. Furthermore, rising energy prices will increase the cost of shipping the raw materials needed for production to distant large industries and also increase the cost of shipping goods to consumers.

A more **decentralized economic system**, proponents such as the late E. F. Schumacher argue, would rely on local materials to produce products for local consumption. Small decentralized industries might be appropriate for the manufacture of soap, toothpaste, and towels, but surely not for the production of ships or cars. Thus, the primary goal would be to achieve a proper mixture of large and small industry that used all resources efficiently. (See the author's Case Study 20-2 on economic health.)

Ethical Changes

The most fundamental change is the ethical change. It is hard to bring about, but concerted efforts on the part of teachers, parents, and even governmental leaders could make inroads into the outdated frontier mentality. To be convinced that such changes are needed, we need to better understand the future and the dangers implicit in the frontier mentality (for more on this, see Chapter 21). Research is needed now more than ever to guide us wisely into the future.

Population Control

Controlling population is necessary to achieve a sustainable economy. For some countries this may mean capping population growth; for others it may mean achieving replacement-level fertility and then decreasing the size of the population by subreplacement-level fertility. Chapter 6 discusses these goals.

Global Economic Challenges

Building a sustainable economy will require a new way of thinking about our relationship to the earth, to other species, and to one another. It will also require a fundamental ethical change, described in part in the last chapter.

Table 20-1 lists the characteristics of a sustainable economy. You may want to take a moment to study them, then think about the differences between this economy and our current system.

Challenges for the Developed Countries

Achieving a sustainable economy will require enormous efforts on the part of the more developed nations. Their economies are, for the most part, based on numerous

Economic Health: Plugging Up the Leaks/ Finding Hidden Opportunities

The economy is on the minds of nearly everyone these days. Unfortunately, most economic thinking consists of finding ways to do more of what we've been doing all along. Current economic experts continue to prescribe more growth to cure ailing state economies.

In Colorado, for example, business and government leaders are looking for new businesses, mostly from other states, to give the economy a boost. Texans, Pennsylvanians, and others are engaged in similar tactics. To entice new business, states will often give away land and offer generous tax breaks. New business, however, often means new people, which results in more pollution and more crowding on our highways and in our wilderness areas.

It's time that citizens and their representatives start looking at economic development more broadly. First and foremost, we must begin to think of economic development and environmental protection as parts of one whole. We need to balance economic aspirations with environmental needs: the need for clean air, clean water, and uncrowded wilderness. Economic concerns must also take into account the needs of the thousands of species that share this planet with us. Next, we need to reexamine the goals of economic development and, if possible, find alternative ways to meet our goals without destroying, and preferably with improving, the environment.

What are the goals of economic development? In the broadest sense, economic development is a way to increase wealth. To a business owner, additional wealth means more spending money and, possibly, business expansion. To the jobless, business expansion means employment. To the government official, a thriving business means more tax revenues to meet citizen needs.

In the traditional equation, economic development makes almost everyone happy—business owners, government officials, and job seekers. But some environmentalists are bothered by continued economic expansion. It results in the gradual destruction of wilderness, wildlife habitat, air quality, and more. Continued economic development is a path to oblivion. It is unsustainable in a world of limits.

It is time Americans stopped thinking about economic growth and began to strive for economic health. Our economy has matured. Does it need to continue growing? How do we build a healthy economy and ensure a healthy environment?

Perhaps the most important step is to set aside our growth notions and begin healing the earth, cleaning our rivers, restoring salmon runs and rangeland and forests. Massive efforts are needed to repair the damage of the past two centuries. By making these efforts, we can restore productive fisheries and rangelands. We can ensure long-lasting supplies of renewable resources and a clean and healthy environment.

There is plenty to be done in our cities and towns as well. Consider two of the most important steps needed to end waste and build economic strength: The first is plugging up the leaks. The second is finding hidden opportunities. We will consider them one at a time.

Plugging up the leaks means making existing businesses more efficient in how they use energy, materials, water, and other resources. It's a term I borrowed from the people at the Rocky Mountain Institute. By reducing waste, business owners can improve their profit. They can then afford to expand their businesses. Increased profits ultimately mean more jobs and increased tax revenue, which help states solve unemployment problems. It is a way of tapping local labor resources.

Plugging up the leaks should be the first line of defense in any economic "development" strategy. It's a way of stopping the outflow of dollars from a business. But making a business more efficient is only part of the strategy. Entire communities can join in, increasing efficiency in transportation, housing, and government. All these efforts help retain dollars within a local economy to circulate among the citizens. Before the 1973 oil embargo, a dollar used to circulate 26 times in the US before leaving the country. Now it circulates fewer than ten times, largely as a result of increased dependence on foreign oil and foreign imports. Plugging the leaks stops this unnecessary outflow of dollars.

The benefits of a more efficient community are not just more wealth, but a cleaner environment: less air pollution, less water pollution, reduced hazardous waste, and so on. Clearly, economic health and environmental protection can go hand in hand.

Japanese business owners probably understand the benefits of plugging up the leaks better than anyone in the world today. Because their factories use energy more efficiently than US manufacturers, their products require, on average, about 5% less energy to produce than ours. This gives the Japanese a slight edge in pricing. The relatively inefficient use of energy costs the US economy about $220 billion a year, according to the Worldwatch Institute.

The next economic development strategy is to find hidden opportunities. Here's an example of what I mean: Residents of Denver and the surrounding suburbs annually discard an estimated 4.5 million plastic milk jugs each year. Milk jugs can be melted down and reused for a variety of useful products. Denverites discard nearly 700 tons of trash a day loaded with recyclable materials. A recent study showed that recycling 10,000 tons of trash through curbside programs produces, on average, 36 new jobs. If Denver recycled only half of its waste stream almost 400 new jobs would be created. Add the suburbs

Case Study 20-2

to the equation and the employment potential would increase substantially.

There are tons of jobs in our trash and lots of profit to be made by the materials we ordinarily throw away. We are trashing our environment and overlooking an enormous business opportunity. Reducing the flow of milk jugs, tires, and other reusable and recyclable materials would reduce landfilling and save taxpayers money, and that means a healthier personal economy. Recycling would reduce pollution and save energy. Government could help by pointing out the hidden oppor-

tunities for business and directing the unemployed to new lines of work.

My message is simple. Economic development is viewed too narrowly. We need to develop strategies to plug up the leaks and find hidden opportunities. The environmental and economic dividends would be enormous. We can put people to work without a net decrease in environmental quality—in fact, with an improvement.

Adapted from Chiras, D. D. (1990). *Beyond the Fray: Reshaping America's Environmental Response.* Boulder, CO: Johnson Books.

---

unsustainable practices: overfishing, waste and depletion, and others described in this book. Below is a list of ways developed countries can make the transition to a sustainable economy:

1. Reduce population growth, then gradually reduce population size through attrition.

2. Reduce resource consumption and waste by reducing demand, increasing product durability, increasing efficiency, recycling, and reuse.

3. Increase national self-sufficiency by using renewable resources whenever possible.

4. Protect and conserve renewable resources, such as farmland, fisheries, forests, grasslands, air, and water.

5. Improve renewable resource management to ensure sustainability.

6. Repair past damage to natural resources by replanting forests and grasslands, reseeding roadsides, restoring streams, cleaning groundwater, and reducing overgrazing.

7. Cut pollution by 50% to 75%.

8. Support sustainable development projects in Third World nations through the use of appropriate technology, appropriate policy, and local knowledge.

9. Strive for global peace and cooperation.

10. Develop sustainable-earth ethics and promote individual responsibility and action.

Many of these changes have already begun to take place in the industrial nations, but don't be lulled into com-

placency by the early signs of progress; there is much much more to do.

Challenges for the Less Developed Nations

The less developed nations face serious challenges as well. First and foremost, they must slow the growth of population. Considerable growth will occur in the coming years. Therefore, like developed countries, Third World nations may find it advantageous to reduce population size, bringing human numbers in line with the earth's carrying capacity (Chapter 6). This could help ensure a better life-style and a more livable environment, not just for people but for all living things.

The less developed countries must build sustainable agricultural systems. For Third World nations to achieve sustainable agriculture, wealthy nations may need to relieve the burden of international debt—money lent to support many unsustainable development projects. As described in Chapter 7, servicing the debt has forced Third World nations to convert food crops for domestic production to cash crops (tea and citrus fruit, for instance) for export.

Third World countries must concentrate their financial resources on education, housing, sanitation, clean water, food, and other basic needs (Chapter 7). Unless nations rid themselves of their debt and reduce their expenditure on weapons, there is little hope for them. (See Viewpoint 15-1 for more on the cost of arms and how it hinders Third World nations from improving their conditions.)

Many individuals think that a more equitable distribution of world resources could help the poorer nations feed, clothe, and house their people. How that would be done and how successful it would be is debatable. Food handouts, while helping alleviate suffering, don't get at the root causes of the problem: rapid population growth, poverty, unsustainable farming, and environmental decay.

Table 20-2   Characteristics of Appropriate Technology

Machines are small to medium-sized.

Human labor is favored over automation.

Machines are easy to understand and repair.

Production is decentralized.

Factories use local resources.

Factories use renewable resources whenever possible.

Equipment uses energy and materials efficiently.

Production facilities are relatively free of pollution.

Production is less capital intensive than conventional technology.

Management stresses meaningful work, allowing workers to perform a variety of tasks.

Products are generally for local consumption.

Products are durable.

The means of production are compatible with local culture.

What is needed, argue others, is assistance to help Third World nations become more self-reliant.

Self-reliance can become a reality through **sustainable development**. The 1987 World Commission on Environment and Development defined sustainable development as that which meets the needs and aspirations of the present *without* compromising the ability of future generations to meet their own needs. It includes projects to create jobs, use local resources, conserve energy, protect topsoil, and produce adequate food and fuel.

At the very least, wealthy nations must stop exploiting Third World nations, robbing them of their natural resources at dirt-cheap prices. Rich multinational corporations, wealthy landowners, and industrialists in Third World nations are growing richer as the natural resource bases of the poor countries are dwindling and as people fall deeper into poverty. The people who are displaced from the land or who work for these exploiters will be no better off, perhaps worse off, when it is all over. For what is often left is a landscape in ruin, a soil too depleted to produce enough food for people to survive.

Appropriate Technology and Sustainable Economic Development

Sustainable economic development is needed in the poor (as well as the rich) nations of the world. In Third World nations, economic development plans should rely on **appropriate technology**, technology scaled to the local needs and local resources (Table 20-2). It should also rely on **appropriate policy**, laws and regulations that promote sustainable practices.

Appropriate technology is a concept popularized by the late E. F. Schumacher in his classic book, *Small Is Beautiful*. Schumacher emphasizes that appropriate technology fits well in developing countries as well as in highly industrialized countries. Its chief advantages are that it puts people to work in a meaningful way, requires less money to construct and operate, is efficient on a small scale, uses locally available resources, and is more compatible with the environment because of its low energy requirements and minimal pollution.

In developing nations without the capital and energy resources needed to support modern industries, appropriate technology may be the key to economic development. Countries such as India and China have an abundance of people who can be employed by factories using appropriate technology. Highly automated factories that use fewer people to produce their goods would waste the human labor potential and would require large amounts of capital.

Appropriate technology can find a niche on farms throughout the world to replace conventional technology, which is costly and wasteful of indigenous labor. For example, a tractor is an inappropriate use of technology in rural India. Although it increases production, it puts people out of work. These people then move to already overpopulated cities in search of employment, creating serious problems in burgeoning cities. In addition, the costs of tractors, repair, and energy are prohibitive in poor, rural regions of developing countries. Gandhi put it best when he said that the poor of the world cannot be helped by mass production, only by production of the masses. What will work in India, Latin America, and Africa are slightly improved methods of agriculture that keep farm workers employed while using a minimum amount of energy. A well-designed metal plow produced from local resources and easily repaired by farmers, for instance, would utilize abundant human labor (Figure 20-4).

It is doubtful that the world could support economic development in the Third World on a scale similar to that in the developed nations. (It is doubtful that the developed nations can support this level of economic activity in the long term as well.) But without some improvement in life-styles, hunger, disease, environmental destruction, and death are bound to continue.

Appropriate technology can also help the more developed countries reduce their resource demands, freeing up resources for other people and making a better life possible for a larger number of people. Putting people to better use and using resources more efficiently through appropriate technology can help reduce the enormous impact of modern industrial society. Passive solar heating for homes and business, wind generators and photovoltaic panels for making electricity, bicycles for commuting to work when the weather permits, and compost piles to decompose yard wastes are all examples of simple technologies geared to a less exploitive life-style.

In developed countries the shift to appropriate technology will eliminate some jobs that have been a part of the economy for decades. Auto and steel workers and miners, for example, may be hard hit by a large-scale shift to a more efficient way of life. But many new jobs will emerge as priorities shift. Energy auditors, for example, will find ample opportunity as they help people improve energy conservation in homes and factories. Jobs for solar and wind energy experts will open up as well. Road builders will become road repairers. Construction workers will find employment refurbishing run-down buildings, bridges, and water systems.

Appropriate technology is not a panacea, the one solution to the world's problems, but it is an essential part of a sustainable future. Combined with changes in attitude, shifts in politics and economics, and changes in education, appropriate technology will be part of a new world order, a global economic system that seeks to meet basic human needs.

Making Sustainable Development Work

"Developing countries are littered with the rusting good intentions of projects that did not achieve social or economic success," writes Walter V. C. Reid, a leading authority on Third World development. Making matters worse, he notes, ill-conceived projects have wrought considerable damage to the Third World. Erosion, desertified landscapes, pesticide poisoning, pollution, and deforestation are some of the results of a multibillion dollar annual budget for Third World development. Why?

The reasons are many. One of the first is the good intentions of the international lending agencies, the **multilateral development banks** (MDBs) and development agencies. There are four MDBs: the World Bank, the Inter-American Development Bank, the Asian Development Bank, and the African Development Bank. They are funded by groups of nations. The World Bank, headquartered in New York City, for example, is supported by the United States, France, West Germany, the United Kingdom, and others. MDBs lend billions of dollars a year to Third World nations to finance economic development projects, from road construction to farming to dam building.

The MDBs are joined by private commercial banks and international development agencies. The US Agency for International Development, for example, is a key player. It provides outright grants for development projects.

All this is well and good. Problems arise, in part, because of the *types* of projects the MDBs fund or support. Many projects are unsustainable. Thus, the MDBs have

Figure 20-4. Young agricultural students in Burkina Faso (West Africa) learn to use and maintain plows—the technology is appropriate to the region's agricultural needs and capabilities. Limited economic growth may help individuals of Third World nations eradicate hunger, improve the standard of living, and control population growth.

not only failed to stop environmental deterioration, they have, in many cases, worsened it. Large dam projects, for example, flood productive farmland, displace people, increase the prevalence of waterborne disease, and reduce sediment that nourishes estuarine life. The costs of these projects often exceed the economic benefits gained from hydroelectric power and irrigation water. Irrigation water in arid regions leads to salinization and waterlogging, both of which decrease productivity and can render land useless (Chapter 7). Over half of the Third World's irrigated cropland, for instance, suffers from salinization.

MDBs have helped finance deforestation and colonization projects as well. In western Brazil, forests were cut to make way for farmers and ranchers, but 80% of the people soon left because the soils quickly washed away or lost their fertility (Chapter 9).

Pesticide use to support large agricultural projects funded by MDBs has also proved costly economically and environmentally. In Central America, in fact, chronic and acute pesticide poisoning is among the region's most serious problems.

Volumes of horror stories could be written about well-intended projects gone awry. Fortunately, the MDBs and others are beginning to see the need for new practices—for sustainable development. Fortunately for all parties concerned, sustainable development does not require new technologies or new knowledge; they already exist.

Considerable gains can be made in the Third World by improving energy efficiency. Simple changes in wood cookstoves, for instance, could cut energy demand drastically and help reduce deforestation in the Third World. Energy efficiency, combined with projects to replant trees and better manage forests, could go a long way toward helping Third World nations meet their needs sustainably. Today, however, energy efficiency projects constitute less than 1% of the international aid.

Agriculture, forestry, and animal husbandry can be made sustainable, but important changes are needed. First, funding agencies must rely more on local knowledge than agricultural experts from the West who often attempt to transfer costly technology to Third World nations. Second, MDBs and other agencies must begin to study the environmental, social, and political effects of proposed projects. Until very recently, the environment and the long-term sustainability of projects have received little or no attention at all. A more careful analysis will help MDBs and others avoid projects that are doomed to fail or destined to create widespread environmental damage. Third, developers must design with nature, rather than continue to redesign nature. "Modification of the environment to fit the needs of a production system," writes Walter Reid, "is much less likely to be sustainable than modifications of the production system to fit the environment." That's because most environmental modifications have serious repercussions. Fourth, development should preserve genetic diversity whenever possible. Extractive reserves in Brazil and other countries are a case in point (Chapter 9). They can provide sustainable income at a far lower cost than traditional development practices like clearing and cutting forests—while preserving species. Fifth, inappropriate laws and policies must be removed. Subsidies for pesticide use and for ranching in cleared tropical rain forests are two examples of ruinous policy. Sixth, widespread participation among locals should be encouraged. Planners have long ignored the input of people who will be affected by their projects. They have also ignored knowledgeable local experts, who better understand the people's needs, cultures, beliefs, and the environmental constraints of areas being affected. "Planners should not assume that they know people's needs," writes Walter Reid. Development is unlikely to be sustained unless the needs of people are identified and unless local residents support the project. Seventh, MDBs and other agencies must be more flexible, allowing projects to shift as problems arise. Inflexible bureaucracies often impair projects by refusing to allow for changes, even in the face of serious problems. By funneling money through smaller, nongovernmental organizations, the large bureaucracies can be separated from the management of projects, allowing greater flexibility and ensuring greater success.

These changes can help the Third World develop along a sustainable path, operating within the limits posed by the earth. They are guidelines that could be applied to our own future development as well.

Civilization is a slow process of adopting the ideas of minorities.

ANONYMOUS

Summary

Economics is a science that concerns itself with the production, distribution, and consumption of goods and services. It is in many ways an applied science, for it helps society solve three fundamental problems: (1) what commodities it should produce and in what quantity, (2) how it should produce its goods, and (3) for whom it should produce them. In a **command economy**, in which government dictates production and distribution goals, decisions are left to bureaucrats. In a **market economy**, where prices, supply, and demand prevail, the government usually takes a back seat to the marketplace. Most economies are mixed, having features of both.

One of the key principles of economics is the **law of scarcity**, which states that most things that people want are limited. As a result, they must be rationed by price or some other means. Price is set by supply and demand. Supply and demand interact to establish the market price of goods and services. The **law of supply and demand** operates well in most instances, but resource scientists note that rising prices cannot expand the supply of nonrenewables indefinitely and that a point will be reached when a resource becomes economically depleted. One of the major challenges facing market economies is realizing that our supply of nonrenewable resources cannot expand indefinitely.

Economists use a variety of measures to study a nation's output. The most widely used is the **gross national product** (GNP). The GNP is the value of the national output of goods and services. It gives a general picture of the relative wealth of nations and the living standards of their people. However, it fails to take into account some of the ill effects of the economy, such as damage from pollution. Economists have thus devised a new measure, the **net economic welfare** (NEW), which subtracts the negatives. The NEW in the United States is lower than the GNP and is increasing at a much slower rate. Advocates of a clean environment and a healthy economy believe that nations should strive to reduce the difference between GNP and NEW.

Growth of the GNP is central to capitalistic economies, but a single-minded concern for growth fails to take into account the accumulated wealth of nations. Keeping in mind accumulated wealth can help nations set their economic priorities. Using other measures such as life expectancy, overall health of the population, and recreational opportunities can also help nations stay off the economic growth treadmill.

Economics plays an important role in pollution control, one form of resource management. One of the chief goals of society is to achieve the maximum reduction of pollution, which yields the maximum benefit, at the lowest cost. This can be done by cost–benefit analysis, but determining the true cost of pollution, including all externalities, can be quite difficult. In addition, cost determinations are impossible for many free services such as recreation, watershed protection, and oxygen production.

Who pays the cost of pollution control is an issue of heated debate. Some argue that consumers should bear the burden, because pollution is a by-product of the goods they buy. Others argue that taxpayers should pay, because they have long allowed many industries to pollute and, now that the rules have changed, should bear some if not all of the responsibility for cleaning up the environment. In reality, there are times when consumers should pay; when new plants are being built the cost of pollution control should be incorporated into prices. In older industries now under new laws and regulations, however, the taxpayer should at least help bear the burden.

Historically, pollution control has been enforced by punitive measures, mostly fines. Efforts are under way to harness market forces to control pollution. Pollution charges, economic incentives, pollution permits, laws that eliminate market barriers that promote inefficient resource use, and laws that remove harmful subsidies could create incentives within the business community to control pollution more effectively and at a lower cost than traditional legal avenues.

Conventional wisdom suggests that environmental regulations and laws delay projects, increase the cost of doing business, decrease productivity, and ultimately cost society jobs. Careful studies show that these claims are true but blown out of proportion. For instance, pollution controls have created far more jobs than they have destroyed. Pollution control can also save companies money. Accidents such as oil spills, for example, cost companies far more than preventive measures.

Resource management is significantly affected by economics in many ways. One of the most important economic variables that governs resource use is **time preference**, a measure of one's willingness to give up current income for greater returns in the future. For example, farmers may choose depletion strategies, which favor immediate income, or conservation strategies, which favor long-term gains. The choice of a strategy hinges on time preference for accruing income. For someone who is close to starving there is no choice.

Another factor that affects resource management decisions is **opportunity cost**, the economic cost of lost money-making opportunities. Money put into conservation represents a short-term opportunity cost. However, resource mismanagement creates long-term opportunity costs that are rarely considered in such decisions.

Economics, like other sciences, suffers from biases, the key one being economic growth. Many economists, especially those in business and government, see economic growth as a true measure of success. Since economic growth hinges on rising population as well as increases in per capita consumption, ecologists and environmentalists often take issue with these economists. In a world of limited resources and widespread pollution, ecologists argue that continued population growth and rising per capita consumption, particularly among the wealthy, is unwise.

Economics and ecology have developed sharp differences, few as clear as their view of the future. For an economist the dividing line between the near term and the far term is typically five years. Ecologists, on the other hand, are likely to set the dividing line at a century or more.

Some economists suggest that a new economic system is necessary for the long-term survival of the human race. A spaceship economy, or sustainable economy, would promote recycling, conservation, use of renewable resources, product durability, and a clean and healthy environment. People would live within the limits posed by the earth. Such a system can succeed only with widespread population control and new political directions. The most fundamental change needed is an ethical shift, promoted by teachers, government officials, and parents. Government can take important steps by developing laws that encourage resource conservation, recycling, reduction in wastes, uses of renewable resources, and product durability, to mention a few. Some selective decentralization of industries may be required. Some redistribution of wealth and ways to ration resource use may also be important. Overall, the shift to a sustainable society can come from a combination of economic, governmental, and personal actions.

Sustainable economic development, a strategy that meets current needs without destroying the potential for future generations, could help build a sustainable world economy. Such plans must call on ways to use **appropriate technology**, technology geared to local resources, local needs, and local conditions.

Discussion Questions

1. Define the following terms: *command economy, market economy, law of scarcity, descriptive economics,* and *normative economics.*

2. Draw and describe the supply and demand curves. Describe how an increase in demand affects price. How does a decrease in supply affect price? Why do price increases generally result in an increase in supply?

3. What is the market price equilibrium?

4. What is the major weakness of the law of supply and demand?

5. Using your critical thinking skills, debate the statement "Economic growth has become a dangerous preoccupation of

our times, so much so that we have become singularly attached to making money and blind to the quality of the environment."

6. Define your own economic goals. Would you classify them as consistent with a frontier economy or a sustainable economy?

7. Using your critical thinking skills, debate the statement "Population growth is economically beneficial."

8. In your view, is continued economic growth possible? If so, why and for how long?

9. Describe the gross national product and its strengths and weaknesses.

10. Define the term *net economic welfare*. How does it differ from the GNP? Is it greater than the GNP or less? Does it grow as quickly or more slowly than the GNP? Is this good or bad?

11. Discuss how time preference and opportunity costs affect the ways in which people manage natural resources.

12. Define the term *economic externality* and describe why economic externalities are increasingly becoming internalized.

13. What is the economically optimal level of pollution control? Why is it impractical to consider reducing pollution from factories and other sources to zero?

14. Describe the law of diminishing returns. How does it apply to pollution control? Can you think of any other examples where the law applies?

15. There are three factories in your community. A new factory wants to move in. How could it be allowed to start operation without increasing existing levels of air pollution?

16. Describe a sustainable economy. What are its main goals? In your view, is it a practical alternative to the current economic system? What are its strengths and weaknesses?

Suggested Readings

Brown, L. R., et al. (1987). *State of the World, 1987*. New York: Norton. See Chapter 10 for a discussion of ways to achieve a sustainable economy.

Daly, H. E. (1984). Economics and Sustainability: In Defense of a Steady-State Economy. In *Deep Ecology*, ed. M. Tobias. San Diego: Avant Books. Good synopsis of Daly's thoughts on a steady-state economy.

Daly, H. E. (1988). Moving to a Steady-State Economy. In *The Cassandra Conference: Resources and the Human Predicament*. P. R. Ehrlich and J. P. Holdren, eds. College Station, TX: Texas A & M University Press. Good coverage of ways to achieve a steady-state economy.

Hardin, G. (1985). *Filters against Folly*. New York: Viking. Superb reading on economics and ethics.

Pope, C. (1985). An Immodest Proposal. *Sierra* 70 (5): 43–48. An important look at the role of economic incentives in promoting pollution control.

Reid, W. V. C. (1989). Sustainable Development: Lessons from Success. *Environment* 31 (4): 6–9, 29–35. Detailed but highly readable account of the problems and challenges of Third World development.

Reid, W. V. C., Barnes, J. N., and Blackwelder, B. (1988). *Bankrolling Success: A Portfolio of Sustainable Development Projects*. Washington, DC: Environmental Policy Institute and National Wildlife Federation. Study of development projects that have succeeded.

Samuelson, P. A. and Nordhaus, W. D. (1985). *Economics* (12th ed.). New York: McGraw-Hill. Chapters 1–6 provide an excellent introduction to economics.

World Commission on Environment and Development (1987). *Our Common Future*. Oxford: Oxford University Press. See Chapter 2 on sustainable development.

Government and the Environment

The loftier the building the deeper must the foundation be laid.

THOMAS À KEMPIS

We need a future we can believe in, one that is neither so optimistic as to be unrealistic nor so grim as to invite apathy or despair. In short, we need a future that is not only hopeful but also attainable. This book has outlined one of many possible futures, that is, a sustainable society that lives within the limits of nature. Creating such a system requires changes in people's attitudes (Chapter 19) and changes in the economic system as well (Chapter 20). Such changes may come about in many ways. Perhaps the most effective means of social and economic change is through government.

This section looks at government: the roles of government, the principal participants in governmental policy, how decisions are made, and the governmental barriers to creating a fully sustainable society. Like the previous chapter, it discusses some ideas beyond the mainstream of current thinking in hopes of stimulating discussion on ways to build a sustainable society.

Government: An Overview

Chapter 20 described two basic economic systems: the market economy, in which the marketplace determines the availability and price of goods and services, and the command economy, in which the government largely assumes this task. Chapter 20 also noted that most countries have features of both command and market economies but lean significantly in one direction or the other. But what about governments? Do they correspond at all with the economic systems, and, if so, how?

491

Forms of Government

For the most part governments and economic systems go hand in hand. Countries with market economies are, in general, **democratic nations**, that is, governments in which officials are elected by the citizenry and are supposed to be responsive to the people's desires. Representative governments operate by rules agreed on by the majority. The means of production and distribution of goods and services are, for the most part, privately held.

Countries with command economies are, as a rule, communist or socialist nations in which society owns and operates the production and distribution network. Such nations emphasize the requirements of the state, rather than individual liberties. **Communist** nations believe that goods should be distributed equally, although in practice this is rare. **Socialism**, at least in Marxist theory, is a stage of government between capitalism and communism in which private ownership of the means of production and distribution has been eliminated.

Just as economies are mixed, so are governments. New Zealand, West Germany, and Sweden are all democratic. However, in all three nations the state provides health care and other services, which are financed by taxes. In China and the Soviet Union, two communist nations, private sale of products for individual profit is now allowed, although in limited quantity. More and more, these countries seem to be edging toward the democratic, capitalistic system present in the United States, Canada, and other countries. Democratic rule seems to be emerging in many communist nations.

With this highly simplified overview of governments in mind, let us take a brief look at the ways governments conduct resource management and pollution control.

The Role of Government in Environmental Protection

Governments employ three major tools to regulate the economy and perform other activities, such as environmental protection: (1) taxes, (2) expenditures, and (3) regulations.

Taxes reduce private expenditure (putting a lid on consumption) and thus free money for public expenditures—rail systems, pollution control, wildlife programs, and the like. Taxes may also be used to promote or discourage certain activities. For instance, a heavy tax on gasoline in Great Britain discourages driving and promotes energy-efficient automobiles. Tax breaks, special rules that favor some activity, can also work to the benefit of society. For example, US tax credits promoted the development of solar energy. Tax policies that favor undesirable practices can also have significant impacts on the environment, as noted in previous chapters.

Government expenditures may have a profound effect on the economy and the environment. For instance, government grants for water pollution control projects created over 200,000 jobs in the United States and helped reduce the rise in water pollution that might otherwise have occurred. Additionally, government-funded research programs in energy conservation, waste disposal, acid rain, and solar energy have yielded a wealth of new knowledge that will prove helpful in controlling pollution and better managing resources. Government **procurement programs** (preferential purchase programs) for photovoltaics (solar cells that produce electricity using sunlight energy) and recycled products could help create larger markets for these environmentally desirable products, as discussed in Chapter 12 and Chapter Supplement 18-1. Larger markets, in turn, could result in substantial price cuts. This could make the products affordable, thus helping desirable technologies and desirable products squeeze out less environmentally suitable ones. On still another front, government expenditures on less-than-desirable products and activities divert money from other important tasks, like conservation and renewable energy development. Cuts in expenditures on military spending, nuclear energy research, and other activities could free up considerable sums of money. Lester Brown, president of the Worldwatch Institute, argues that sharp reductions in global military spending could help finance a massive campaign to restore the earth and build a sustainable society (see Brown's Viewpoint in Chapter 15).

Government regulation also affects resource policy. Regulation takes two forms. First are the laws (see Chapter Supplement 21-1) that emanate from legislative bodies. Laws affect some activities directly. For example, the **Corporate Average Fuel Economy Act** (1975) set gas mileage standards for automobile manufacturers in the United States, calling for a new fleet average of 27.5 miles per gallon by 1985. Further increases were scheduled for subsequent years. But laws are only as good as their enforcement and funding. If an agency lacks funds, human power, or the will to carry out a law, it is just so many words on a piece of paper. The mileage law is a good case in point. The Department of Transportation was responsible for implementing and enforcing the standards. Under heavy pressure from Ford and General Motors, however, mileage standards were rolled back twice. Thus, by 1988 the average fleet mileage was far below Congress's goals. When President Bush took office, however, he pledged to raise standards: By 1989 the average new car got 27.5 miles per gallon. Further improvements were called for in the coming years.

This example also illustrates the role of a president in setting the agenda and determining governmental resolve. According to many of his critics, President Reagan cared very little about environmental laws. Under his leadership and with his appointees in office, very little progress was made in cleaning up the environment. Important gains were threatened.

Table 21-1 Some Federal Agencies and Their Responsibilities for the Environment and Health

Agency	Responsibility
Environmental Protection Agency	Research, demonstration programs, and enforcement of most environmental laws
Occupational Safety and Health Administration	Research and enforcement of worker safety and health laws
Food and Drug Administration	Research and enforcement of laws to protect consumers from harmful foods, drugs, and cosmetics
Health Services Administration	Family planning programs and community health programs
Health Resources Administration	Research, planning, and training; collection of statistics on health in the US
National Institutes of Health	Study of cancer, through the National Cancer Institute, and of radiation and other environmentally related diseases, through the National Institute of Environmental Health Services
Centers for Disease Control	Epidemiological studies on disease
National Oceanic and Atmospheric Administration	Research into and monitoring of the oceans and atmosphere; ecological baseline information and models to better predict the impacts of air and water pollution

Environmental regulations also emanate from governmental agencies, such as the Environmental Protection Agency (EPA), the Occupational Safety and Health Administration (OSHA), and the Food and Drug Administration (FDA). Since Congress lacks the expertise and time to set specific standards for pollution control, it usually leaves the task to federal agencies. They draft rules and regulations, open them to public comment, then, barring significant protest, put them into effect. In Chapter 17, for example, you learned that the EPA sets tolerance limits for pesticides on foods. It was directed to do so under the Federal Insecticide, Fungicide and Rodenticide Act.

The EPA is currently empowered by nine federal laws, such as the Clean Water Act, the Clean Air Act, the Superfund Act, and the Resource Conservation and Recovery Act. Each law gives it authority to put into effect regulations governing a great many activities, from hazardous waste to emissions standards for new factories. (Table 21-1 lists the federal agencies most directly involved in protecting public health and the environment.)

Through taxes, expenditures, and regulations, governments throughout the world have made tremendous strides in protecting the environment, managing resources, and controlling population. Many chapters in this book have pointed out the most significant advances. For a quick review of US progress, see Table 14-2. It lists nearly two dozen environmental laws and amendments enacted since 1958 to regulate toxic chemicals. Similar progress has been made in most other developed nations. Have the new laws resulted in substantial environmental progress? Are we better off than we were ten or twenty years ago? For one answer see the Point/Counterpoint in Chapter 20.

But what about the developing world, where over four billion of the earth's residents now live? Many developing nations have adopted policies to control population growth, save wildlife, and reduce pollution. However, progress has been hindered in many cases by a lack of money, rapid growth of population, and governmental ineptitude, apathy, or mismanagement. Interference by foreign interests and projects funded by international aid organizations and multilateral development banks have also served to worsen environmental quality (see Chapter 20). In many cases hunger and poverty are so severe that environmental matters assume a lesser significance. Without population stabilization and programs to raise the standard of living and the availability of food, little progress can be expected in environmental protection. But, as pointed out in several sections, environmental quality is not a luxury but a necessity.

Political Decision Making: Who Contributes?

Government isn't the only answer to our environmental problems, but it is a major determinant and has grown in power consistently for 200 years. How much should a government do to protect the environment? How much

Figure 21-1. Diagram showing the many lines of connection between government, individuals, and corporations. Laws are passed by lawmakers. These, in turn, affect individual behavior and business behaviors. Individuals, environmental groups, and businesses also influence the laws through lobbyists and the court system.

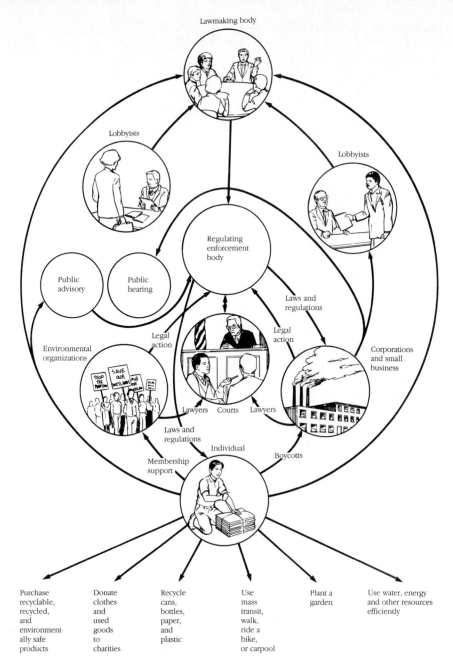

money should it spend to conserve resources or develop renewable resources?

Government Officials

In communist governments the answers to these questions come from the officials in charge, the ruling elite. The people have little or no public input, although recent changes in Poland, the Soviet Union, and other communist nations suggest that things may be changing for the better. Given the profound economic problems facing these nations, however, it is unlikely that their rampant environmental problems will be solved soon.

The Soviet Union, for example, has some of the most stringent environmental standards in the world, but enforcement is virtually nonexistent. The Soviet Union suffers from serious air and water pollution. Soot and sewage are the most obvious signs, but a wide number of toxic pollutants, such as toxic metals, are also let loose on the environment. Progress in correcting the problems is going slowly.

Why aren't the problems being solved more quickly? The reasons are many. First, the country lacks the technical expertise. Second, it lacks the financial resources to control pollution and other problems. Third, the economic and social problems are so acute that they are given priority by the government. Fourth, the nation suffers from the same kind of frontier mentality (there is always more) that keeps Americans and Canadians from practicing better resource management.

In democratic nations, officials at all levels of government participate in important decisions that affect the environment, sometimes favorably, sometimes not (Table 21-1). The 1989 nuclear weapons scandal, described in Chapter 18, showed the American public that government officials can be some of our worst enemies. Under the Department of Energy's control, dozens of nuclear weapons plants have, over the past 45 years, released large amounts of radiation into the air, water, and soil. Government regulators either turned a deaf ear to reports or openly endorsed illegal acts. An executive order by President Reagan made matters worse, because it rendered these facilities immune from US environmental laws. Officials in the EPA and state health departments have also been known to overlook violations of environmental and health laws at these facilities. But by the same token, dedicated, hard-working men and women at all levels of government have also made important contributions to bettering the environment. Individuals who push for new laws and regulations or who risk their lives to put criminal polluters in jail must not be overlooked.

The Public

In a democratic society the voters can have a say in policymaking (Figure 21-1). Voters select their representatives and can influence how their elected officials vote on important legislation. Voters' letters, phone calls, and responses on surveys are direct lines of communication that allow citizens to keep in touch with their governmental representatives. Since democracies are designed to serve the public and voter preference is the guiding force, politicians are public servants. Their job is to interpret public preferences and find ways of supplying the public policies that satisfy them. Unfortunately, many people have grown cynical about their influence and individual power to make changes. In the long run, this view is counterproductive.

Citizens can make a difference. A letter to a congressional representative is an indicator of public interest, say legislative assistants. In fact, for each letter they receive, legislators generally estimate that there are 5000 to 10,000 people who think and feel the same way. One thousand letters means there are five million people who think similarly. That's pretty compelling to an elected official.

A letter to a state legislator goes even further, because these officials rarely hear from the public. Twenty or thirty letters are cause for concern. Forty or fifty indicate, in their minds, a disaster requiring action. So don't be dismayed. Write and vote. Let your voice be heard. (For more

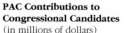

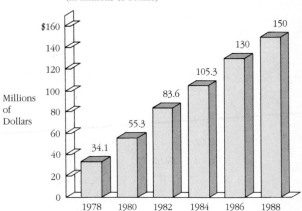

Figure 21-2. Political action committee contributions to congressional candidates from 1978 to 1988.

information on becoming politically active and writing your representatives, see the Action Guide.)

Special Interest Groups

Making policy that satisfies the public is not always as easy as it may sound. The **theory of public choice** tells us why. It states that politicians act in such a way as to maximize their chances of reelection. Accordingly, they must appease the general public—the voters—and the special interest groups, which apply enormous pressure on elected officials to adopt policies sympathetic to their cause (Figure 21-1). Special interest groups, including environmental organizations, often work through lobbyists. When Truman was President, there were about 450 lobbyists in Washington, DC; today there are an estimated 23,000 of them, swarming the halls of Congress, writing laws, convincing legislators and legislative aides of the merits of their views. As a further inducement the special interests often reward politicians with monetary support, which is badly needed to run political campaigns. To make an even larger impact, many special interests have banded together to form **political action committees**, or PACs. These are consortia that pool their individual financial resources, allowing them to make substantial donations to political campaigns. In the 1988 elections PACs donated more than $150 million to congressional candidates, up from $34 million a decade earlier (Figure 21-2).

Common Cause, a nonprofit citizen's lobbying group, argues that PAC contributions pay off "in billions of dollars worth of government favors for the corporations and other special interest groups that make them." Contributions from the corporations help defeat important environmental legislation. In recent years, for instance, PAC

Figure 21-3. Greenpeace activists position themselves between Soviet whaling ships and whales to thwart the slaughter of whales.

contributions have helped stall controls on acid deposition, toxic air pollution, and other important problems. In 1984 auto manufacturers successfully lobbied the US Congress to pass import quotas on energy-efficient Japanese automobiles, which yielded the companies $300 million in profit and cost the consumer an estimated $2 billion or more.

The import quota represents a dangerous political maneuver, called the **double-C/double-P game**. Double-C/double-P stands for "commonize the costs and privatize the profits." In other words, let the taxpayer or general public pay the costs of something that will make profits for the private sector. Many US water projects were built with large federal subsidies, thus providing farmers with cheap, publicly subsidized water to irrigate their crops. The taxpayer paid the bill, and the farmer benefited.

Democracy is a government of the people, by the people, and for the people, but the influence of special interest groups, especially the wealthy business PACs, may distort the process. But money does not always reign supreme in American politics. Powerful antipollution laws, auto safety standards, hazardous waste laws, and other important environmental statutes have been enacted in the last three decades despite strong opposition. A good measure of this success can be attributed to the efforts of environmental groups, such as the Environmental Defense Fund, the National Wildlife Federation, the Wilderness Society, the Natural Resource Defense Council, Greenpeace, the Sierra Club, the Population Institute, the Fund for Animals, the Animal Protection Institute, and hundreds of others. (For a listing of some of the major environmental groups in the United States see the Action Guide.)

Environmental groups affect public policy in several ways. Some, such as Greenpeace, take an active role in the outdoors, meeting face to face with whalers, seal hunters, and polluters, endangering their own lives to protest actions they oppose (Figure 21-3). Such public displays have proved highly successful in raising awareness on various issues. In 1980, an even more radical group of environmentalists began to take action. They called themselves Earth First! Under the leadership of Dave Foreman, they took more active, frequently illegal steps to protect the environment. They put their bodies on the line to stop bulldozers. They embedded spikes in old growth trees, which thwarted chainsaws. Their tactics were called *ecotage*—sabotage in the name of the environment.

Some years later, Paul Watson, one of the original founders of Greenpeace, began an oceanic equivalent of Earth First! Called the Seashepherd Conservation Society, they roam the high seas, ramming ships that are illegally whaling, flying miniature airplanes to disrupt sonar, and generally disrupting illegal activities. They have never been arrested or charged with a crime because they stop people engaged in illegal activities who don't want public attention drawn to them.

Most environmental groups operate on other levels. Some groups are involved in education, preparing educational materials for schools. A few groups spend their time and energy buying up land to be set aside for wildlife, plants, and future generations. Other groups are involved in lobbying efforts, writing new laws, getting them passed, or strengthening existing laws. Still others serve as watchdogs, making sure that polluters obey the rules and that government agencies do their jobs. As Frederic Krupp pointed out in his viewpoint in Chapter 10,

Case Study 21-1

Two Groups at Work

One of the great strengths of the environmental movement is its diversity of approach. Admittedly, some areas are weaker than others, but the environmental movement encompasses a broad spectrum of styles and strategies. This case study illustrates two divergent approaches to environmental protection.

Consider the Nature Conservancy first: Called the Donald Trump of environmental organizations, the Nature Conservancy has also been called Mother Nature's real-estate broker or a real estate "undeveloper." Why?

The Conservancy is in the business of buying habitat—land for animals and plants—all over the world. Working within the system, the Conservancy often buys critical plant and wildlife habitat that state agencies want but can't purchase immediately because of budgetary constraints. The organization purchases land for such programs through an $85-million revolving fund—a fund that has been growing steadily for years. When the state agencies get the money they need, the Nature Conservancy sells them the land—with interest, of course.

The Nature Conservancy also buys land outright, setting it aside for wildlife, plants, and future generations. Since it began in 1951, the organization has purchased 3.5 million acres of land and is currently setting aside 400 hectares (1000 acres) a day.

The Nature Conservancy has also played a major role in the debt-for-nature swaps. For example, it purchased $5.6 million of Costa Rican debt for $750,000 from the American Express Bank. Rather than have the country fail to pay the debt, the bank was willing to sell it at a discounted rate. In turn, the bank can write off the loss on their taxes and help preserve nature. The Costa Rican government put up $1.7 million (some of the money it would have paid the bank) and promised to use it for conservation education, land acquisition, and park protection.

Other groups are more politically oriented. Take the Natural Resources Council of Maine (NRCM): The NRCM is a membership organization that has been in business for about 30 years. It has a 15-member staff with attorneys, scientists, and resource specialists. They study local issues, publicize their findings, and even lobby for new laws.

In 1988, the Natural Resources Council of Maine succeeded in tackling an issue that has baffled many environmentalists throughout the nation: local growth control. Through skillful lobbying and tireless efforts to educate the public on unregulated growth and development, they overcame the objections of the governor, developers, and local government to a statewide land-use planning act. In 1988 the Maine legislature passed the law, which required each town to develop a land-

use plan to put an end to the haphazard development that is tearing the state into pieces. The law provides $4 million in state money to assist towns in this process.

Brownie Carson, executive director of the NRCM, says the Natural Resources Council of Maine took a leadership role, outlined a strategy, then systematically worked to see their plan through. "We started two years before," says Carson. "We started by researching what was being done in other states to address this kind of problem. We looked at the North Carolina coastal programs, the Florida program, the Oregon programs. . . . We liked what we saw in Oregon the best."

With one staffer, Carson spent ten days in Oregon, talking with leaders of the conservation community and the political leadership, town planners, business people, bankers, ranchers. They crisscrossed the state to get a feel for how Oregon's land-use planning works.

When they got back to Maine, they began talking with the governor's cabinet, business people, and newspaper editors to garner support and hear their concerns. Next, they convinced a major Maine newspaper to do a piece on the growth issue, showing how developers were ruining vital open space and critical wildlife habitat, building condominiums and golf courses on swamps they filled. NRCM members started writing editorials about the issue. "We went to the homes of major political leaders . . . sponsored a conference. Just kept building a constituency," says Carson.

"Essentially what we tried to do was to frame the problem and then propose a solution," says Carson. "We got a legislative study commission created in the end of the 1987 legislative session." NRCM worked with the legislators on the commission, which sponsored a series of hearings around the state. NRCM packed the hearings with people who supported land-use planning. By sending alerts and calling friends and allies, they ignited a groundswell of public support for their land-use planning proposal. People who were angry with developers showed up en masse. NRCM worked with legislators to get these hearings held in places either where there was tremendous growth pressure or where developers had been particularly destructive.

NRCM drafted a proposal for the legislation, then worked with state representatives to refine the language. Much to their dismay, the governor threatened to veto the bill at the eleventh hour, but NRCM managed to get newspaper editorials within 24 hours of the announcement, arguing that a veto would be ill advised.

This well-orchestrated effort represents the kind of political work that can be done to help build a sustainable society. It's a far cry from the Nature Conservancy's real-estate ventures, but it's part of a vital strategy.

other groups are involved in proactive planning, devising sustainable alternatives to harmful projects. Some groups hire economists, public policy experts and scientists who research issues and offer factual reports that help shape public opinion and public policy. Environmental groups often spend much of their time in court, suing governmental agencies when they are not doing their job or are doing it poorly, or suing businesses that violate environmental laws. They are a force to be contended with. Legal expertise has grown substantially since Earthday 1970, the year many believe the environmental movement officially began.

One of the most influential environmental groups is the Environmental Defense Fund. Like the Sierra Club, the National Wildlife Federation, and the Audubon Society, it operates on many levels. It fights for new legislation, sues governmental agencies to make them enforce the law, studies issues, draws up alternative plans, and publishes books that influence public opinion and public policy.

Environmental groups can offset the disproportionate influence of business, but they are usually at a disadvantage. Corporations like Exxon, which boasts a $5 billion annual profit, wield considerably more influence than environmentalists. Environmental groups have the potential, however, to be much more effective and could exert much wider influence on society. Here are ten guidelines environmental groups can follow to become more effective agents of change:

1. Define the concept of sustainability and promote it more widely in articles, speeches, and television specials.

2. Promote individual responsibility and action, not just among group members, but among the public at large.

3. Reach minorities, the religious community, children, and senior citizens, engaging them in the battle to save the environment.

4. Take a more proactive approach, that is, help devise sustainable alternatives.

5. Build stronger coalitions among environmentalists, business, and government.

6. Reduce duplication of efforts.

7. Promote restoration.

8. Act more consistently with environmental values (for instance, by reducing environmental junk mail and using recycled paper for newsletters, direct-mail appeals, and magazines).

9. Support state and local efforts to control pollution.

10. Become better at plotting long-term strategies to address issues on many different levels.

Case Study 21-1 features two environmental groups at work, the Natural Resources Defense Council of Maine and the Nature Conservancy. It may help you better understand how environmental organizations work.

Some Barriers to Sustainability and Some Suggestions

From an environmentalist's viewpoint government must be made responsive to the needs of future generations and the needs of other species that share this planet with us. But this is no easy task. Many barriers lie in the way. Two of the chief barriers are apathy and cynicism among the members of democratic societies. Apathy was discussed in Chapter 19. Cynicism was touched on earlier in this chapter when describing why people don't choose to participate in democratic governments. In a nutshell, people are cynical about governmental change. Why fight? You can't do anything about it anyway. Knowing how powerful industry's influence is and how unmoveable and unresponsive governmental bureaucracies can be, many people shrink behind a wall of cynicism. Americans also suffer from a rather short attention span. Today's crises are tomorrow's forgotten memories. Political will to support a pollution control project dies when issues fall out of fashion. There are other barriers as well.

Lack of Consensus

One of the key barriers to building a sustainable future is a lack of consensus about the future. The United States and Canada are amalgamations of people with markedly different ideas and philosophies. Even within a single political party, opinions may vary widely on individual issues. These differences present a colossal obstacle to the long-range planning needed to achieve a sustainable society. As Herbert Prochnow once asked, "How can a government know what the people want when the people don't know?" How can we agree on a course for the future when few agree on which future is best?

Research One solution is to promote more thorough scientific study of resource depletion, the long-term effects of pollution, and the effects present generations may have on the future. A major goal of this research would be to identify the forces that affect the future and explore alternative futures. Social scientists can play a big role in this research by identifying values and studying the ways current values can be changed to achieve a sustainable society. We must also learn more about the need for long-range planning, and we must better understand our obligations to future generations and other species. Support for this work could come from private and government grants.

Education The information we gain from our intensified study of the future must be made available to all citizens—to schoolchildren and adults. Long-term problems of resource shortages, overpopulation, and species extinction must become more immediate concerns of everyday citizens and their elected officials.

Teachers and religious leaders can help by promoting a better understanding of the long-term threats posed by modern society. Environmental groups can do more to educate children on the need for environmental protection, conservation, population control, renewable resource use, and recycling (Figure 21-4).

Crisis Politics

Henry Kissinger once said that in government the "urgent often displaces the important." Immediate problems, such as strikes, emergency aid, and oil embargoes, often reduce the amount of time spent working on long-range problems. This way of operating is called **crisis politics** and is a second major barrier that hinders our building of a sustainable future. With immediate issues taking precedence over important long-term problems, governments may lumber along from crisis to crisis, applying temporary remedies to cover up the symptoms of deeper, more complex problems. One solution is to develop governments that seek a balance between proactive measures and reactive measures, as discussed below.

Proactive and Reactive Government A government that "lives and acts for today" is by definition a **reactive government**. Its laws and regulations are sometimes ill-conceived and ineffective, but they satisfy a basic need of politicians: they give the impression that something is being done. Such laws are often valuable vote-getters; in the long run, though, they can further complicate an already serious problem, making a truly effective solution harder to reach and closing down options of future generations.

Many of the laws passed by a reactive government are retrospective, that is, they attempt to regulate something that has gotten out of hand. For example, the **Superfund Act** technically provides money to clean up thousands of existing toxic waste dumps in the United States. (As noted in Chapter 18, though, Superfund has failed to have much effect.) Retrospective laws are part of a "patch-it-up-and-move-to-the-next-crisis" syndrome.

The long-term outlook requires **proactive laws**, which seek to prevent potentially hazardous events from occurring or help build a sustainable society. One example is

Figure 21-4. Outdoor classroom discussion on wildlife preservation. Education of children of all ages is a key to building a sustainable society.

the **Toxic Substances Control Act**, with its provisions for screening new chemicals before they are introduced into the marketplace (Chapter 14). The **Resource Conservation and Recovery Act** is another example. It established a system to track hazardous substances in the United States from their production to disposal, allowing for a better accounting of the nation's wastes. Laws that promoted solar energy, conservation, recycling, and population control would be fine examples of legislation that could help today's democracies deal with environmental constraints. Ultimately, they would help bridge the gap between the largely wasteful frontier economy and the sustainable society.

Many examples of future-oriented acts by Congress exist:

1. Establishment in 1946 of the National Science Foundation to promote research

2. Passage in 1969 of the **National Environmental Policy Act**, which requires environmental impact statements for all projects sponsored or supported by the federal government

3. Creation in 1972 of the Office of Technology Assessment, to examine the costs and benefits of new technologies

4. Authorization of the Congressional Research Service to create a Futures Research Group

5. Passage in 1974 of the **Resources Planning Act** and in 1977 of the **Resources Conservation Act**, both of which require the Department of Agriculture to make periodic assessments of the supply of and demand for wildlife, farmland, water, rangeland, and recreation

6. Passage of the National Appliance Energy Conservation Act (1987), which establishes efficiency standards for appliances

American government today is a mix of proaction and reaction, but current economic policy favors a reactive approach by letting the marketplace control resource allocation (the limitations of which were discussed in Chapter 20). One of the casualties of the market approach is long-range energy policy. In 1985, for instance, tax benefits for solar and wind energy were terminated. The economic boost both industries had received was removed, and today they have been severely crippled. The Solar Energy Research Institute's budget was also cut, further impeding the transition to a sustainable society. Federal funding for conservation was slashed by 75%. Money for population control in developing countries has also been reduced. Ironically, the government still promotes oil use through special tax incentives. Raw materials still enjoy cheaper freight costs than materials destined for recycling plants. Tobacco farmers still receive federal subsidies.

As we move into the last years of the 1900s and as the world population continues to grow well past the bend in the exponential growth curve, proactive government becomes more necessary than ever before, as discussed in Chapter 19. But how can we make our governments more oriented to the long term?

Long-Range Planning and Follow-Up Better research on the future of our resources and a better dissemination of that knowledge will make citizens more aware of the long-range problems. In representative governments that awareness should result in more long-range planning.

Special commissions on the future could be appointed. In 1977 President Carter directed the Council on Environmental Quality, the State Department, and several federal agencies to study population, resources, and pollution through the end of the century. The report was to serve as a "foundation for longer-term planning." The resulting study, **The Global 2000 Report to the President**, is a gold mine of information about the future on which many decisions can be based. This work should not stop here. Estimates and projections should be periodically updated and revised as necessary. Great lessons can be learned in the coming decades from the projections. Were they too pessimistic? Were the methods of projecting resource and environmental trends faulty, and how can they be improved?

Special study sections on the future in various governmental agencies such as the Departments of Agriculture, Energy, and Interior should be better funded and better equipped to project into the future. Reports from futures groups should be widely distributed to political leaders, teachers, and the public.

Some observers have suggested that the United States needs a Department of Long-Range Planning and Coordination, similar to one in Sweden. The department would research future issues and make recommendations to the President and Congress regarding population, pollution, and resources. Its main purpose would be to assess long-term trends and help US political leaders develop a long-term strategy. It might serve as a central clearinghouse for new ideas. Those in favor of keeping the federal government from growing argue that a central interagency committee might better coordinate the future activities of all federal agencies, be less costly, and require less expansion of an already swollen bureaucracy (Figure 21-5).

Long-range vision and planning can also be provided by the United Nations. The World Commission on Environment and Development, for instance, was asked to produce a global agenda for change. Its report, *Our Common Future*, called on the General Assembly of the United Nations to find ways to achieve sustainable development, improve cooperation, and improve environmental protection. The UN's role is described more fully at the end of this chapter.

Few would deny that long-range planning is badly needed. But planning is a task fraught with difficulties, which are largely due to the uncertainty of resource supplies and future demands. American utilities, for instance, planned for a regular 7% increase in electrical demand in the United States. To accommodate the expected demand they began a massive buildup of nuclear power plants, only to find in the late 1970s that there was an excess of electricity. When demand began to increase again, it climbed at a much slower 3%. The lesson we can learn from this example is that we must plan, but we must do so carefully. And we must realize that planning will be imperfect. Bias and self-interest also distort planning. The frontier notion that there is always more, for instance, leads planners to erroneous conclusions or potentially harmful projects.

The 1980s taught environmentalists in the United States a valuable lesson—that environmental laws are useless without regulation and enforcement. Many people had overlooked this simple fact. Cynthia Wilson, former executive director of Friends of the Earth, suggests why. Wilson argues that the regulatory process that follows a law's

Figure 21-5. President Bush addresses the US Congress in the Capitol Building. As environmental problems increase in number and complexity, advocates argue for more regulations and spending, leading to an ever-burgeoning central government. Some think the job should be left to the states, but they often lack the expertise and the human resources. Besides, most pollution issues are of national concern—pollution knows no boundaries.

enactment is much less glamorous and considerably more demanding than the work required to get a law passed. Because of these factors and because new problems are constantly unfolding, environmentalists quickly moved on to new legislative battles, naively trusting that the government would do as it had been mandated by law. The 1980s showed how wrong these assumptions were.

Follow-up is needed to ensure that (1) funds that are earmarked by new laws are made available to the regulatory agencies, (2) business interests don't seriously weaken regulations being drafted by federal agencies, and (3) agencies follow their mandate. Follow-up is needed in other areas as well. Monumental reports often create a storm of interest, but people quickly forget today's pressing issues. Environmental groups and educators can help end this short attention span by reminding people that environmental protection is like preventive medicine. It requires a lifetime of attention and care. Educating children is one of the most important tasks we have to accomplish.

Limited Planning Horizons

Another barrier to sustainability is the limited planning horizon of political decision makers. Concerns for re-election can restrict the vision of many politicians. Politicians are often weather vanes, reflecting the current thinking of their people. Since their constituents are mired in the present, so are politicians. Start a parade and soon enough a politician will arrive to lead it. But woe to the politician who gets too far in front of the parade of contemporary concern.

Budgetary constraints also limit the planning horizons. Politicians and government officials want immediate returns on investments. That can eliminate some long-term conservation measures, whose benefits accrue to future generations.

New Leaders Overcoming the barrier of limited planning horizons will require more far-sighted leaders who are willing to implement policies that have implications well past the time of their next election—policies that may not favor the immediate electorate as much as the future electorate. But citizens must be willing to elect such leaders—men and women with a vision of the future and the skills to articulate and popularize that vision. If we expect visionaries, though, we must become more visionary ourselves.

Environmental groups at the national and state level are helping support responsible leadership. You can help by supporting candidates who promote recycling, renewable resource use, pollution control, intelligent management of resources, protection of wildlife and plants, and population control. Citizens can write or visit their elected officials. (For some tips on writing congressional representatives see your Action Guide.) One effective way for citizens to make their voice heard is to join advocacy groups.

Election Reforms Another possible solution that is tossed around from time to time is to push for longer terms of office for House and Senate members and perhaps even the president. A six-year nonrenewable term for the president, for example, might reduce the reelection pressures that lead to shortsighted political decisions and take valuable time away from the office. Longer terms would reduce the amount of time spent on the campaign trail and increase the amount of work time. Longer terms of office, of course, could have damaging effects as well. Shorter campaigns could help, too. In England, campaigning begins only six weeks before general elections. In the United States campaigns run a full year and can cost millions of dollars.

Inadequate Land-Use Planning

A new dam is built in a Third World country, promising much-needed water for irrigation. Upstream, however, the hillsides are stripped of trees or heavily overgrazed. When the rains come, soil washes from the hillsides, filling the reservoir with sediment and rendering it useless in no time. This is an example of poor management and faulty land-use planning—putting land to uses that are unsustainable and often ruinous.

Countless examples of near-sighted land-use planning exist, in the developed as well as the less developed nations. You have read about many of them in earlier chapters. Building suburbs, cities, and airports on prime agricultural land, constructing homes in floodplains, farming steep hillsides, filling marshes to build homes, and placing factories upwind from cities and towns are all examples of improper land-use planning.

Sustainability through Land-Use Planning Ideally, **land-use planning** is a process by which governments can put the land to the best and most sustainable uses. It can be used as a tool to preserve farmland, recreational areas, wetlands, scenic views, watersheds, aquifer recharge zones, and wildlife habitat. Land-use regulations determine where people can live and do business, where water pipes, electrical lines, roads, and shopping malls can go with the least amount of damage. Done properly, land-use plans take into account the slope of the land, soil quality, water drainage, location of wildlife habitat, and many other features. Land-use planners then set out to use the land optimally—with a minimum of disturbance. Designing with nature, rather than redesigning nature, is the chief goal of proper land-use planning. Case Study 10-1 illustrates an extraordinary land-use planning success in Woodland, Texas. When properly done, land-use planning can help us achieve a more sustainable relationship with the earth.

In need of special protection are the renewable resources, which are the foundation of tomorrow's civilization. Farmland, pastures, forests, fisheries, and wild species are human insurance for sustainability, but, as noted earlier, they can be irretrievably lost by careless actions and overexploitation. Overfishing, deforestation, and intensive farming can all destroy the renewable resource base needed for the future of humankind.

Historically, the marketplace has determined use of our resources. Chapter 20, however, explains that the market often seeks immediate gains at the expense of future generations. Agricultural land conversion provides a telling example. On the outskirts of many growing cities, bulldozers now rip up prime farmland to clear the way for tract housing, shopping centers, and highways. The short-term value of suburban development outweighs the long-term value of agriculture. Such is the market's way of prioritizing resources. For the long-term future, however, such actions can be dangerous.

The Japanese provide a model that many countries could adopt to protect their land. In 1968 the entire country was placed under a nationwide land-use planning program. Lands were divided into three classes: urban, agricultural, and "other." Several years later, the zoning classifications were expanded to include forests, natural parks, and nature reserves. The success of the Japanese plan lies in protecting land from the market system, which, left on its own, appropriates land irrespective of its intrinsic value.

Many European nations have adopted similar programs. In Belgium, France, the Netherlands, and West Germany, for instance, national guidelines for land-use planning were established in the 1960s. Administered by local governments, these guidelines protect farmland, prevent urban sprawl, and help to establish greenbelts, undeveloped areas in or around cities and towns. The Netherlands has perhaps the best program of all. Its national planning program affects not only land but also water and energy.

Land-use planning in the United States is rudimentary at best. So far, except for establishing national parks, wilderness areas, national forests, and wildlife preserves, the federal government has done little to systematically protect its land. Most zoning occurs on the community level, and much of that is inadequate. On the local level, planners primarily concern themselves with restrictions on land use for commercial purposes—housing developments and industrial development. As a result of community-level planning, states become a patchwork quilt of conflicting rules and regulations. One result is that companies often deliberately locate in areas with lenient rules and regulations. Areas in the same state that try to protect their air and water may suffer.

Statewide land-use planning could bring some order to the chaos. But statewide land-use planning is an idea that is slow in coming. Oregon recently passed such a program. A number of companies have chosen to locate in Oregon despite its tough environmental standards. Even though its rules may be stricter than those of other states, many companies prefer Oregon because they know where they stand. They have confidence that laws will not tighten and force drastic changes in the way they do business. (Case Study 21-1 describes a recent land-use planning program in Maine.)

Zoning and Other Approaches The main tool of land-use planning for years has been zoning, classifying land according to use. In a city, zoning helps separate potentially noisy, odoriferous, or hazardous activities from residential areas. When used properly, zoning can also protect farmland and other lands from urban development. In rural Black Hawk County in Iowa, for instance, the zoning laws provide permits for well-conceived housing developments on farmland with lower productivity. Prime farmland cannot be used for housing.

New approaches are also being adopted, especially to protect farmland. One of those is **differential tax assessment laws**, under which land is assessed for taxes according to its value for farming, regardless of potential value for housing development or other uses. In past years farmers often sold their land when city planners began taxing it as if it were residential. High taxes make farming unprofitable and force farmers off the land. Currently, most states have laws that permit differential taxation.

Another technique to keep farmers from selling their land to developers is for states to buy up the **development rights** for important lands. To do this, two assessments of the land are made, one of its value as farmland and one of its value for development. The difference between the two is the development right. States may buy the development rights from the farmer and hold them in perpetuity. The land must then be used for farming, no matter how many times it changes hands.

Land-use planning is essential in the developing nations as well. Urbanization in these countries is a major problem, and millions of hectares of farmland fall to the bulldozer each year as cities expand. In some areas land reform is badly needed. Wealthy landowners in many Latin American countries, for example, graze their cattle in rich valleys while peasants scratch out a living on the erodible hillsides. Hilly terrain that should be protected from erosion is being torn up by plows and is eroding away by rainfall. Lester Brown, an expert on world resource management, argues that "sensible land use hinges on reform of these feudal land-holding systems."

Agroecological Zones Land-use planning is generally designed around the needs and wishes of urban and suburban residents. It solves the question of where to put people, factories, roads, and malls. But land-use planning can also be applied to agriculture and could, some experts argue, help make agriculture more sustainable.

The first step requires mapping the earth's arable and potentially arable land. Maps of climate, soil types, and landform (rolling hills, flat plains, etc.) would provide the basis for determining **agroecological zones**, regions characterized by similar soil, landforms, and climate. Peter Oram, a research fellow at the International Food Policy Research Institute, argues that mapping the world's agricultural zones could provide data needed to tailor present and future agricultural development to the local environment. A failure to understand the agroecological zone often leads to the demise of agricultural development projects. Crops unsuited to a particular soil or climate invariably fail, often at considerable cost.

More detailed studies of the various zones could help agricultural experts determine which crops are best suited and most likely to succeed in a given zone. Fit the crop to the zone; don't try to force a crop to grow in an unsuitable zone.

Mapping the millions of hectares of arable and potentially arable cropland is no easy task. It will require an enormous number of researchers and huge sums of money. Some steps have already been made. In the long run, though, it could help humanity learn to farm within the constraints of nature.

A Sustainable World Community

A sustainable system is needed in both the rich and poor nations of the world. Some world leaders are sitting back and letting things happen with the hopes that the market will steer their nations in the right direction. Their plan is not to plan. This **passive approach** could work, but more likely than not it will only create a long string of crises, many of which could cripple the world economy and create widespread environmental destruction.

An **active approach**, many argue, would be much better. The active approach entails concerted efforts to reshape civilization. Governments and their people must be part of that movement. Some suggested changes include (1) a shift away from the growing war economies, with their enormous expenditures, to a sustainable economy, promoting restoration, recycling, conservation, renewable resources, and population control; (2) more sharing of knowledge between the rich and the poor, especially in areas of population, food, and energy; (3) investments in recycling, conservation, and renewable resources; and (4) strict population control policies.

West Germany's Green Party: An Ecological Approach to Politics

The nations of the world may build a sustainable society in a piecemeal fashion, one country at a time. Already, signs are evident. One example is West Germany's political party, the Greens. Rallying under the banner "We are neither right nor left, but in the front," the Greens are actively pushing for a sustainable society and winning seats in the national legislature (Figure 21-6).

Largely maligned and misunderstood by the press, write the futurists Fritjof Capra and Charlene Spretnak in

Figure 21-6. One member of Germany's Green Party shows up at parliament in a solar-powered vehicle to demonstrate this alternative, relatively clean form of transportation.

their book *Green Politics*, the Greens advocate measures to protect the environment, including reductions in pollution and hazardous wastes, an end to nuclear power, conservation, use of renewable resources, wise use of energy, and an end to nuclear weapons. If they get their way, West Germany will slowly be transformed into a nation of ecologically benign (nonpolluting) industries that concentrate mainly on socially necessary products. In Western Europe a dozen similar parties are now seeking office.

Despite their sometimes disruptive ways of pressing their points and the fringe elements that have adhered to them, and despite conflicts that have divided their leadership on many occasions, the Greens represent a major shift in political thinking. They take a long-term view of the future, calling for redirection of policy consistent with the sustainable ethics discussed in Chapter 19.

The Green Party is alive and well in the United States. Colorado, Connecticut, Washington, and dozens of other states now have small but active Green Parties. They are promoting individual action—through conservation and recycling—and social justice. They stand for a clean and healthy environment. They are writing letters and making phone calls to political leaders, staging peaceful demonstrations, and even working on getting some of their

own members elected to state legislatures and other offices.

Conventional political wisdom holds that in the United States popular ideas are absorbed by major political parties. They enter the mainstream of political thought and could, conceivably, become the cornerstone of important public policy. Thus, no matter whether the Greens survive or are simply absorbed into the mainstream, they may have begun a new movement that could reshape American politics.

Achieving a Global Sustainable Society

Not too long ago the world seemed unlimited and our problems few. Today, though, the globe has begun to shrink with better transportation and communication and rapid depletion of resources. Our problems now seem as unlimited as the world once appeared. One step in solving these problems—and by no means the only one—is achieving sustainability on a global scale. But how?

As more and more countries recognize the limits of resources and the need to protect the life-giving biosphere, they will undoubtedly begin to adopt sustainable resource policies. Thus, sustainability may well evolve over time in much the same way that organisms evolve to meet changing environments.

Building a global sustainable society could be facilitated by the United Nations (Figure 21-7). Already, its population and agricultural programs have helped Third World nations in many ways. Conferences on population and its impact, for instance, have convinced virtually the entire Third World that something must be done to control rampant growth before it is too late. In 1987, the UN facilitated negotiations for an agreement among 31 countries to reduce the current rate of ozone depletion (Chapter Supplement 15-1). In 1988, the UN Commission for Europe negotiated an agreement that would freeze nitrogen oxide emissions at their 1987 levels. The agreement requires the use of the best available control technology on new vehicles and power plants in the US and Europe. Further gains could be made in solving problems of global importance, including the greenhouse effect, marine fisheries, seabed minerals, and others. Unfortunately, conflicting national policies often keep the United Nations from realizing its potential.

Many people agree that some sort of international cooperation is needed. The United Nations is part of the answer, but other avenues are also available. The International Whaling Commission (IWC), described in Chapter 8, is a good example. Composed of members from all of the whaling nations, the IWC sets quotas on whale kills and enacts outright bans. The IWC, however, has no enforcement power; it relies principally on cooperation. As noted in Chapter 8, not all nations comply with the regulations. To help give the IWC a little muscle, the United States and other nations prohibit trade with countries that violate the agreements.

The European Economic Community—a consortium of western European nations and the United Kingdom—also has the potential of providing a framework for cooperation. Europe is a major source of air and water pollution and a major producer of hazardous waste. Agreements between member nations could go a long way toward cleaning up the environment and helping to build a more sustainable economy.

Global Resource Sharing: Is It a Good Idea?

Reports of millions of starving children in Sudan and elsewhere have resulted in an outpouring of aid from countries throughout the world. Global resource sharing, such as this, is seen by some individuals—many environmentalists and textbook writers—as a way of benefiting those less fortunate than the citizens of wealthy countries.

On the surface, global resource sharing sounds like an excellent solution to the shortages now plaguing many countries. The environmentalist Garrett Hardin takes strong exception to this view, arguing that moving wealth from the rich to the poor will create universal poverty and instability. The rich will become poorer, and the poor will have no incentive to better their lot in life. Population growth and good resource management will be ignored if necessities are guaranteed under a world government that seeks to distribute the earth's wealth according to the Marxian principle "To each according to his need."

Hardin argues that redistribution of food and other resources creates, in effect, a "zero-sum" society in which one person's material gain is another's loss. Universal sharing of the earth's goods could lead to widespread deterioration of world resources—the air, water, and land (see Chapter 9). For example, US grain destined for Third World nations could contribute to the impoverishment of American soils and could contribute to sediment and nitrate pollution (see Chapter 7), because farmers would probably ignore conservation measures to lower production costs.

Hardin proposes, instead, that nations share knowledge, which is a renewable resource not subject to the zero-sum principle. Sharing of knowledge would create a "plus-sum" society in which one person's gain is also another's gain. This means that knowledge shared with other countries often comes back augmented, or in a richer form. It can then be passed on and refined once

Figure 21-7. The UN has helped increase global cooperation in pollution control and other important issues. It could provide a forum for more progress on building a sustainable society.

again, benefiting everyone. Chapter 20 described efforts to promote sustainable economic development. Shared expertise and financial assistance are necessary ingredients of such plans. A more equitable sharing of resources may help, too. But this poses an incredible challenge to market economies based on growth and free enterprise.

To be successful in the long run, both the rich and poor nations of the world must develop sustainable systems of agriculture and industry. A global sustainable society and government may seem like utopia, an unrealistic dream. As Rolf Edberg reminds us, "The utopia of one generation may be recognized as a practical necessity by the next." He adds that such goals depend on our ability to free ourselves from "ideas and emotions that once had a function in our battle for survival but have since become useless." Starving masses and the universal fear of nuclear war may be the psychological forces that set the stage for a new world society governed by sustainable principles.

> *The great thing in this world is not so much where we stand as in what direction we are moving.*
>
> OLIVER WENDELL HOLMES

Summary

We need a future that is optimistic, but attainable as well. Now, more than ever, we must take some decisive actions to reshape society to fit within the constraints of the biosphere.

For the most part governments and their economic systems go hand in hand. Countries with market economies are, as a rule, democratic nations, which are ruled by elected officials responsive to the needs and desires of the people. The means of production and distribution of goods and services are, for the most part, privately held.

Countries with command economies are generally communist or socialist nations, in which the government owns the production and distribution network. Communist nations are unrepresentative and put the requirements of the state above those of individual liberty.

In democratic nations, political decisions crucial to the environment can be affected by a great many people—especially if people choose to participate. Government officials, the general public, special interest groups (including business, labor, and environmentalists), all play a role in determining policy. Because of their sometimes vast financial resources businesses often wield considerable power. But environmental groups have learned how to counter that power and become effective agents of social change. Environmental groups operate in many ways: by purchasing land for protection, educating the public, pushing for new and tighter laws and regulations, acting as watchdogs over corporate and government organizations to ensure compliance, suing recalcitrant agencies and corporations that violate the laws, and staging public protests to draw attention to important issues.

Governments employ three major tools to regulate the economy and other activities, such as environmental protection. They are (1) taxes, (2) expenditures, and (3) laws and regulations. Through these tools governments throughout the world—especially in developed countries—have made tremendous strides to protect the environment, manage resources, and control population. Still, new problems seem to be cropping up as fast as old ones are addressed. Many laws are not funded adequately or are not enforced. Unfair regulations and taxes favor environmentally harmful activities. Developing nations have been less successful because of lack of money, rapid growth of population, and governmental ineptitude, apathy, or mismanagement. In the poor nations hunger and poverty are so severe that environmental matters assume a lesser significance.

How much should a government do to protect the environment? In socialist and communist systems that answer comes from the ruling party, generally with little input from the populace. In democratic nations voters have a much larger say in policymaking. But even in representative governments the balance of power may shift away from the people because of the influence of special interest groups, especially industry. Unfortunately, the desires of business or other lobbies do not always coincide with those of the public. Nevertheless, money does not always reign supreme in politics. Powerful antipollution laws, auto safety standards, hazardous waste laws, and other important environmental legislation have been enacted in the last three decades despite active lobbying on the part of their opponents. A good measure of this success can be attributed to the efforts of environmental groups. They affect policy in many ways: by raising public awareness of issues, by researching issues, by lobbying, by campaign contributions, and by lawsuits.

From an environmentalist's viewpoint government must be made responsive to the needs of future generations and the needs of other species that share this planet with us. Four major barriers to this goal are (1) a lack of consensus, (2) crisis politics and ignoring of long-range issues, (3) a limited planning horizon of elected officials, and (4) poor land-use planning.

Research on the future can help solve these problems. Better educating children and adults about our obligations to future generations can also go a long way toward overcoming these barriers. Changes in government may result from fundamental shifts in our perceptions about the future. For instance, a wider acceptance of our ability and obligation to make the future habitable and healthful may produce a broader outlook among governmental representatives. The net effect would be a better balance between reactive government, which primarily concerns itself with immediate problems, and proactive government, which takes a look at long-range issues as well. Land-use planning is needed to protect renewable resources, but historically the marketplace has determined how land is used. To stop the unnecessary loss of resources, statewide—perhaps even nationwide—land-use planning must be adopted. Japan and many European nations have national plans that could be emulated.

The chief tool of land-use planning has been zoning, classifying land according to use. New approaches are being adopted, especially to protect farmland. One of those is differential tax assessment laws, which allow city officials to tax farmland according to its agricultural value, regardless of its potential for development. Some states purchase the development rights, that is, the difference between the land's assessed value for farming and its assessed value for development. States hold the development rights in perpetuity, ensuring that the land must be farmed.

Land-use planning is also important in developing nations, because rapidly growing populations, largely found in urban centers, are threatening wildlife and good farmland. All the changes outlined here can help build a sustainable society.

Revisions in the present economic system might also help us build a sustainable state. New leaders and election reforms that promote long-range planning to tackle tough issues could help considerably. More self-reliance in communities could bring resource and pollution questions to the doorstep of the average citizen.

A global sustainable society seems imperative if we are to survive in the long run. Such a society may evolve in response to rising pollution and upcoming shortages. New political parties, such as Germany's Green party, or revisions of the ideology of existing parties could help bring the new sustainable ethic into practice.

Discussion Questions

1. Describe the sustainable society. What changes in our current economic and political systems would be required to make the transition to a sustainable society?

2. How do governments affect resource use and pollution? In other words, what are the major tools by which governments control economic activity and environmental protection?

3. How do citizens in democracies influence their elected officials?

4. In what ways are democracies unrepresentative? Can you give some examples?

5. Look around your community. What aspects of it could have been planned better? Why?

6. Give some examples of ways new cities and towns could design with nature.

7. How can we create a global sustainable society? Outline major goals for agriculture, energy, wildlife, population, economics, and politics.

Suggested Readings

Bardes, B. A., Shelley, M. C., and Schmidt, S. W. (1988). *American Government and Politics Today: The Essentials* (2nd ed.). St. Paul, MN: West Publishing. Excellent survey of American government and politics.

Brown, L. R., et al. (1987). *State of the World, 1987.* New York: Norton. See Chapters 1, 10, and 11.

Burke, T. (1984). The Politics of Ecology. In *Ecology 2000: The Changing Face of Earth,* ed. E. Hillary. New York: Beaufort Books. Interesting historical look at environmental law and politics.

Chiras, D. D. (1990). *Beyond the Fray: Reshaping America's Environmental Response.* Boulder, CO: Johnson Books. Describes alternative governmental solutions to help achieve sustainability.

Hardin, G. (1981). An Ecolate View of the Human Predicament. *Alternatives* 7: 241–262. Controversial and thought-provoking. As important as his classic paper, "The Tragedy of the Commons."

Hawken, P., Ogilvy, J., and Schwartz, P. (1982). *Seven Tomorrows.* New York: Bantam Books. Interesting book promoting a thoughtful consideration of our possible futures.

Little, D. L., Dils, R. E., and Gray, J. eds. (1982). *Renewable Natural Resources. A Management Handbook for the 1980s.* Boulder, CO: Westview Press. Collection of thoughtful and thought-provoking essays.

Morgenstern, R. and Sessions, S. (1989). Weighing Environmental Risks: EPA's Unfinished Business. *Environment* 30 (6): 14–17, 34–39. Overview of EPA's constraints and upcoming challenges.

Oram, P. A. (1988). Moving Toward Sustainability: Building the Agroecological Framework. *Environment* 30 (9): 14–17, 30–36. More detailed description of the agroecological zones discussed in this chapter.

Perry, J. S. (1986). Managing the World Environment. *Environment* 28 (1): 10–15, 37–40. Detailed look at the complex problem of global environmental management.

Renner, M. (1988). *Rethinking the Role of the Automobile.* Worldwatch Paper 84. Washington, DC: Worldwatch Institute. Detailed coverage of the growth in automobile use and alternative transportation systems.

Van der Ryn, S. and Calthorpe, P. (1986). *Sustainable Communities: A New Design Synthesis for Cities, Suburbs and Towns.* San Francisco: Sierra Club Books. Full of interesting information and case studies on sustainable community development.

Welch, S., Gruhl, J., Steinman, M., and Comer, J. (1988). *American Government* (2nd ed.). St. Paul, MN: West Publishing. Excellent coverage of interest groups, public participation in government, and other topics.

A Primer on Environmental Law

Our environmental laws are not ordinary laws, they are laws of survival.

EDMUND MUSKIE

An ancient Roman legal axiom proclaims, "The people's safety is the highest law." Today, many environmental laws embrace this principle, but it was not until the late 1960s that this idea took hold in the United States.

During the late 1960s and 1970s environmental protection became increasingly important to US citizens and their congressional representatives. Numerous laws were passed during those years, so that today the United States has the world's most comprehensive and tough set of environmental laws and regulations. The US campaign has had a positive effect on the rest of the world. Many nations have patterned their environmental protection laws after US statutes, benefiting from some of Americans' errors, resulting from our eagerness to make important inroads into resource and pollution issues.

National Environmental Policy Act

One of the most significant advances in environmental protection was the **National Environmental Policy Act** (1969). NEPA is a brief, rather general statute with several major goals. First, it declares a national policy calling on the federal government to "use all practicable means" to minimize environmental impact in its actions. It requires decisions regarding federally controlled or subsidized projects such as dams, highways, and airports to describe possible adverse impacts in an **environmental impact statement** (EIS). Among other things, an EIS must describe (1) what the project is; (2) the need for it; (3) its environmental impact, both in the short term and the long term; and (4) proposals to minimize the impact, including alternatives to the project. Drafts of the EIS are available for public comment and review by federal agencies at least 90 days before commencement of the project. Written comments from the public are included and must be addressed in the final EIS, which is issued at least 30 days before undertaking the proposed action.

The EIS has been an effective way of getting businesses and governmental agencies to focus on the environmental impacts of their projects. The underlying idea is that individuals who become aware of their potential impact will act responsibly to avert it as much as possible. In this sense the EIS is a political carrot (a gentle inducement) rather than a political stick (a punishment). The EIS is a superb legal tool in a sustainable society. Available early in the planning stage, it can help decision makers determine whether they are directing their policies, programs, and plans in compliance with the national environmental goals expressed in NEPA and other legislation.

Between 1970 and 1983 approximately 24,000 EIS's were prepared. NEPA also established the **Council on Environmental Quality** in the executive branch. The Council publishes an annual report (*Environmental Quality*) on the environment and on environmental protection efforts of the federal government. It also develops and recommends to the president new environmental policies.

NEPA has been one of the few statutes to significantly affect federal decision making. It has led to hundreds of lawsuits filed by environmental groups against the government, perhaps more than any other environmental statute. In addition, several states and nations throughout the world have passed laws or issued executive or administrative orders patterned after NEPA. France, Canada, Australia, New Zealand, and Sweden all require EIS's. California passed an Environmental Quality Act in 1970 that requires EIS's for all projects—private and public—that will affect the environment in a significant way.

The success of NEPA has been great, but the law is not without flaws. One of the most frequent criticisms is that the EIS's are too lengthy and deal with too many peripheral subjects. Reports are often ignored; they may show serious adverse impacts, but the project will be carried out regardless, often without ameliorative actions. A common complaint from environmentalists is that reports are often based on inadequate information. Projections of environmental impact are difficult to make and often too subjective. Practically no work has been done to see if the projected impacts actually materialize; thus Americans continue to be unprepared to make sound projections about impact. EIS's may be "doctored" by agencies or private consulting firms that write them for federal agencies to hide the real impacts. Some agencies can avoid writing EIS's by simply stating that there will be no adverse environmental impact; it is then up to others to prove the need in court. Other critics of the EIS contend that it is too costly and often leads to delays in important projects. The paperwork and time involved seem excessive.

To answer some of the complaints, the Council on Environmental Quality issued streamlined procedures for preparing EIS's in 1979. They call for (1) a maximum length of 150 pages, except for more involved projects; (2) a summary of no more than 15 pages that describes major findings and conclusions; (3) documentation and referencing of projected impacts; and (4) the use of clear, concise, and plain language.

Some environmentalists believe that NEPA should require agencies to select the most environmentally benign and cost-effective approach, both in the short term and in the long run. Currently, environmental groups can sue an agency that they believe should have filed an EIS or one that has filed an inadequate EIS, but they cannot recover attorney's fees from such suits. If they could, that would relieve the costly burden of forcing governmental agencies to heed the law; however, it might also open the door to numerous costly lawsuits, which would be a burden on the taxpayer.

Although it has its critics and still stands in need of improvement, NEPA is the cornerstone of US governmental policy. Numerous federal agencies have reported important environmental benefits from it and economic savings from recently revised rules. NEPA is a landmark law of fundamental importance to a sustainable society.

Environmental Protection Agency

Another major environmental accomplishment is the establishment of the **Environmental Protection Agency** in 1970. The EPA was founded by a presidential executive order calling for a major reorganization of 15 existing federal agencies working on important environmental issues.

The EPA was directed to carry out the Federal Water Pollution Control Act and the Clean Air Act. Having grown in size, it now manages many of the environmental protection laws that issue from Congress. Current responsibilities of the EPA include research on the health and environmental impacts of a wide range of pollutants as well as the development and enforcement of health and environmental standards for pollutants outside of the workplace. The EPA is concerned with a variety of areas, including pesticides, hazardous wastes, toxic substances, water pollution, air pollution, radiation, and noise pollution.

The EPA can provide incentives to state and local communities through grants for substantial portions of water pollution control projects. Grants to universities have helped expand the research capability of the agency. The EPA carries on much of its own research at four National Environmental Research Centers, located at Cincinnati, Ohio; Research Triangle Park, North Carolina; Las Vegas, Nevada; and Corvallis, Oregon.

The EPA is often caught in crossfire between opposing groups, for example, between environmentalists who seek tighter controls and the businesses the EPA regulates, which commonly complain that regulations are too stringent and costly.

In the late 1970s widespread interest in the environment was overridden by economic hardship brought on by the oil embargoes and inflation. A powerful political movement arose in the early 1980s to dismantle or weaken the EPA. The most common argument from the business sector was that the cost of protec-

tion was too excessive and that stiff environmental protection laws were preventing a healthy economy. But Japan has tough laws, continues to fight pollution, and has done well economically. Furthermore, as Russell Peterson, former administrator of the EPA and president of the National Audubon Society, noted, "We cannot have a thriving economy without a thriving ecosystem."

The American people continue to express a strong concern for a healthy environment; they support maintaining existing laws or even strengthening them. New problems such as acid precipitation and hazardous wastes constantly crop up, demanding the attention of the EPA and other federal bureaucracies. A growing population and an expanding economy create an ever-increasing burden on the environment and on the agencies that regulate environmental issues. Thus, many argue that Americans must continue to expand their support of the EPA and other agencies involved in environmental protection.

Evolution of US Environmental Law

Environmental laws and regulations have become an integral part of the complex US legal system. How did this happen?

State and federal environmental laws evolved gradually over the years from scattered ordinances imposed by local governments. Interested in protecting health and environmental quality, officials of cities and towns passed local ordinances in the 19th century limiting activities of private citizens for the good of the whole. For example, municipal ordinances regulated burning of trash within city limits to reduce air pollution.

By the end of the 1800s, however, it was clear that local control of many problems such as water pollution was inadequate in densely populated regions. Disputes arose between neighboring municipalities with different laws. Regions with strict laws were hampered in cleaning up their rivers by upstream cities with lax pollution laws. Thus, states began drafting legislation to regulate water pollution.

Soon state laws proved inadequate, too, because air and rivers flow freely across state borders. Thus, interstate conflicts over pollution replaced the conflicts between neighboring municipalities. State programs were inadequately funded and lacked the technical expertise to set pollution standards. State agencies also found themselves powerless against large corporate polluters with political influence in the courts and legislatures. Because of these problems and the growing effectiveness of special interest groups, environmental controls shifted to the federal government, gradually in the 1940s and 1950s and then more rapidly.

Initially, the federal government restricted itself to research on health and development of pollution control technologies. This approach met little opposition from state and local governments. Next, the federal government stepped in with grants to fund pollution control projects and the formation of state pollution enforcement agencies. But with increasing pressure from environmental groups and citizens, it began to take a larger role in enforcement. Today it sets ambient pollution standards and standards for emissions from factories, automobiles, and other sources and can take strict enforcement actions if needed.

The shift to federal control is based on at least two important principles of American federalism: (1) When it is important to maintain uniform standards, the federal government provides the best route. Uniform policies help minimize interstate conflicts and help create an economically fairer system for businesses. (2) The power of the federal government to tax is much stronger than a state's. To control pollution effectively requires much expensive research, which the federal government can more easily afford. Furthermore, it would be costly, time-consuming, and redundant for each state to carry out this extensive research.

The shift to the federal level has some disadvantages, however. First, the federal government may not always understand the problems within the regions it regulates as well as local officials. Another criticism is that states should have the right to do as they please with their own resources, in other words, federal control diminishes self-determination and self-governance. But without central controls, states impinge on one another's quality of life. For example, poor watershed management in the Rocky Mountain states leading to erosion could have long-term adverse impacts on downstream users to the east and west.

One way of addressing these problems is to develop federal standards but allow the states to manage their own programs. Thus, the Clean Air Act, the Surface Mining Control and Reclamation Act, and the Resource Conservation and Recovery Act all permit the states to run their own programs as long as they are at least as stringent as the one set up by federal law. These acts also provide money to assist the states in setting up their own programs.

Principles of Environmental Law

In the United States, government's power to protect the environment is conferred by the US Constitution, state constitutions, common law, federal and state statutes and local ordinances, and regulations promulgated by state and federal agencies. Statutory law and common law are the mainstays of environmental protection, and the concepts they share enter into every law and lawsuit that involves the environment.

Statutory Law

Throughout this book have appeared many examples of state and federal laws for environmental protection and resource management. These **statutory laws** generally state broad principles, such as the protection of health and environment by reducing air pollution, or the judicious use of natural resources. However, Congress and the state legislatures lack the time and expertise to determine specifically how these goals can be met. Thus, Congress assigns the setting of standards, pollution control requirements, and resource management programs to executive agencies such as the EPA. (See the discussions of the Clean Air Act in Chapter 15 and the Toxic Substance Control Act in Chapter 14.)

Common Law

Many environmental cases are tried on the basis of **common law**, a body of unwritten rules and principles derived from thousands of years of legal decisions. It is based on proper or reasonable behavior and has been replaced in many states by statutes.

Common law is a rather flexible form of law that attempts to balance competing societal interests. As an example, a company that generates noise may be brought to court by a nearby landowner who argues that the factory is a nuisance. The landowner may sue to have the action stopped through an injunction. In deciding the case, the court relies on common-law principles. It weighs the legitimate interests of the company in doing business (and thus making noise) and the interests of society, which wants its citizens employed and wants to collect taxes from the company, against the interests of the landowner, who is trying to protect the family's rest, health, and enjoyment of property.

The court may favor the plaintiff (the one(s) who files the lawsuit) if the damage (loss of sleep, health effects, and inconvenience) is greater than the cost of preventing the risk (costs of noise abatement, loss of jobs, and loss of tax revenues). But the court may not issue an injunction causing the factory to shut down; instead, it may simply require the defendant (the one defending the case or whose actions are being contested) to reduce noise levels within a certain period. This way, a balance is struck between competing interests.

Cases such as this one illustrate the balance principle. But on what legal principles are cases involving common law decided? Basically, there are two: nuisance and negligence.

Nuisance   **Nuisance** is the most common ground for action in the field of environmental common law. A nuisance is a class of wrongs that arise from the unreasonable, unwarranted, or unlawful use of a person's own property that obstructs or injures the right of the public or another individual, producing annoyance, inconvenience, discomfort, or hurt. What this means is that one can use one's personal property or land in any way one sees fit, but only in a reasonable manner and as long as that use of the property does not cause material injury or annoyance to the public or another individual.

Generally, two types of remedy are available in a nuisance suit: compensation and injunction. **Compensation** is a monetary award for damage caused. **Injunctions** are court orders requiring the nuisance to be stopped.

Nuisances are often characterized as either public or private. Until recently the two were distinctly different concepts. A public nuisance is an activity that harms or interferes with the rights of the general public. Typically, public nuisance suits are brought to court by public officials. A private nuisance is one that affects only a few people. For example, the pollution of a well affecting only one or two families is considered a private nuisance. A public nuisance would be pollution that affects hundreds, perhaps thousands, of landowners along a river's shores. The most common environmental nuisance is noise (see Chapter Supplement 15-4). Water pollutants and air pollutants such as smoke, dust, odors, and other chemicals are other major nuisances.

Historically, the distinction between private and public nuisance has hampered pollution abatement, because the courts have traditionally held that an individual's nuisance suit could be brought against a public nuisance only when the individual had suffered a unique injury, or one different from that suffered by others. An individual would be unable to sue a company for polluting a river shared by many others. Relief was possible only through a public nuisance suit brought by an official (the local health department, for instance). Public officials may be unwilling to file suit against local businesses that provide important tax dollars for the community and campaign support as well.

Increasingly, the distinction between private and public nuisance is fading; private persons can bring suit to stop a public nuisance. As a result, the private individual is gradually getting more power to stop polluters.

Several common defenses are used to fight nuisance suits. Since most nuisance suits are decided by balancing the rights and interests of the opposing parties, **good-faith efforts** of the polluter may influence the decision. For example, if a small company had installed pollution control devices and had attempted to keep them operating properly but was still creating a nuisance, the court would hold it liable but might be more lenient in damages or conditions of abatement. If, on the other hand, the company had made no attempt to eliminate pollution and had created a public or private nuisance, the court would generally be more severe.

The availability of pollution control must also be considered. If a company is using state-of-the-art pollution control and still creates a nuisance, the court may not impose damages or an injunction. In contrast, if the company has failed to keep pace with pollution control equipment, the court may order it to install such controls.

Class-action suits can be used in states that still distinguish between public and private nuisance. Class-action suits are brought by a group on behalf of many people. They emphasize the composite damage caused by a nuisance. In order for a federal class-action suit for compensation to be allowable, however, each person named in the suit must have suffered at least $10,000 in damage. If not, the suit can be dismissed. This requirement, then, provides an opportunity for a defense.

Another defense is that the plaintiff has "come to the nuisance." Coming to the nuisance occurs when an individual moves into an area where a nuisance—such as an airport, animal feedlot, or factory—already exists and then complains. An old common-law principle holds that if you voluntarily place yourself in a situation in which you suffer injury, you have no legal right to sue either for damages or an injunction. In most courts, however, even though you purchase property and know of the existence of a nuisance, you still have the right to file suit to abate it or recover for damages. This is based on another common-law principle that clean air and the enjoyment of property are rights that go along with owning the property. Thus, if population expands toward a nuisance, it may be the responsibility of the party creating the nuisance to put an end to it.

According to the environmental attorney Thomas Sullivan, "The courts are moving to strict liability for environmental nuisances, so that practically speaking, there are no good defenses. The solution is: do not create nuisances."

Negligence A second major principle of common law is negligence. From a legal viewpoint a person is negligent if he or she acts in an unreasonable manner and if these actions cause personal or property damage.

Negligence provides a basis for liability, just as nuisance does, but negligence is generally more difficult to prove. What is reasonable action in one instance may not be reasonable in another. Statutory laws and regulations help the courts determine whether behavior is reasonable. For example, regulations drawn up by the EPA specify how certain hazardous wastes should be treated. Failure to comply with those standards may be evidence of negligence.

Negligence may be shown in instances where a company fails to use common practices in the industry. For example, a company may be found negligent if it fails to transport hazardous wastes in containers like those used by other companies.

In a much broader sense, the courts may decide that a company is negligent simply if it fails to do something that a reasonable person would have done. For example, negligence might be demonstrated if a company failed to test its wastes for the presence of harmful chemicals when a reasonable person would have done so. Likewise, negligence may stem from action that a reasonable person would not have taken.

In summary, negligence can result from either inaction or action that may be deemed unreasonable considering the circumstances. The concept of knowing also plays a role. Briefly, negligence can be determined on the basis of what a defendant knew or should have known about a particular risk. The standard of comparison is what a reasonable person should have known under similar circumstances. For example, a man on trial may argue that he is not negligent because he did not know that a harmful chemical was in the materials that he dumped into a municipal waste dump and that subsequently polluted nearby groundwater. His argument will be valid if a reasonable person in his position could not have known about the wastes.

The **standard of reasonableness** applies also to past mistakes. In other words, even though an operator of a hazardous waste dump did not know of the hazards it created 20 years ago, he or she may be ruled negligent for having failed to eliminate the risk when he or she learned of it or should have learned of it.

Liability for damage or harm need not be based on negligence in cases where the risk is extraordinarily large (in legal terms, an "abnormal risk"). Practically speaking, in proving liability for an activity of abnormally high risk, such as the housing of hazardous materials, one need prove only that injury or damage occurred, not that the operator was negligent or acted unreasonably.

Business interests often try to lessen their liability by getting Congress to pass laws that put ceilings on liability. The Price-Anderson Act, for example, frees utility companies from all financial liability incurred by a nuclear power plant accident. The act requires the government to pay for all damages outside the plant, but only up to $560 million. The airlines have a similar law; at this writing, the oil companies are pressing for similar limits on liability resulting from tanker spills.

Problems in Environmental Lawsuits

Legal actions to stop a nuisance or collect damages from a nuisance or act of negligence carry with them a burden of proof. Plaintiffs must prove that they have been harmed in some significant way and that the defendant is responsible for that harm. This is not always easy, for several reasons. First, the cause-and-effect connection between a pollutant and disease may not have been definitely established by the medical community. If doubt exists, the case is weakened. Second, diseases such as cancer may occur decades after the exposure, making it extremely difficult to prove causation. It is generally easier to link cause and effect with acute diseases. Third, it is often difficult or impossible to identify the party responsible for damage, especially in areas where there are many industries or where illegal acts, such as midnight dumping of hazardous wastes, have occurred. Any reasonable doubt about the party responsible for personal or property damage may severely cripple a lawsuit.

The **statute of limitations**, which limits the length of time within which a person can sue after a particular event, also creates problems in cases of delayed diseases. Statutes of limitations help reduce old lawsuits where evidence is unavailable or memories of potential witnesses have faded and become unreliable. In latent-disease cases, though, they create a major obstacle to individuals seeking compensation for damage in states that apply the time limitations to the onset of exposure. This essentially makes it impossible for cancer victims to file suit. Other states start the judicial clocks from the time the victim learns of the disease. This makes it a little easier to collect compensation for diseases such as cancer, black lung, and emphysema.

Out-of-court settlements present another legal problem. Such settlements have hindered environmental law. Eager to avoid a costly settlement, a company may pay victims if they agree to dismiss the company from further liability. Out-of-court settlements may also benefit plaintiffs, saving them the time, headaches, and costs of environmental litigation.

While advantageous to both parties, these settlements provide no precedents for environmental laws. In short, the fewer cases that make it to court, the fewer precedents courts have to settle cases. This lack of precedents may discourage attorneys and citizens from filing court cases. Without clear examples from the past, they may simply be unwilling to face costly, time-consuming legal battles.

Resolving Environmental Disputes Out of Court

An increasing number of disputes between environmentalists and businesses are being settled out of court by **dispute resolution**, or **mediation**. This innovative approach often employs a neutral party who mediates the discussion between opposing parties. The mediator keeps the proceedings on track, encourages rivals to work together, and tries to resolve the dispute in a way that is satisfactory to both groups.

The benefits of mediation over litigation are many. Mediation is much less costly and time-consuming. Mediation also tends to create better feelings among disputants, whereas court settlements create winners and losers and often leave bitter feelings. In addition, and perhaps most important of all, mediation may bring about a more satisfactory resolution. For instance, environmental lawsuits often hinge on specific points of law rather than substantive issues and thus may have little to do with what the plaintiff really wants. An environmental group might bring a suit over the adequacy of an EIS but, in reality, might want the government to ensure protection of a valuable species that would be affected by the project. In mediation, this will be the central issue. Mediation may therefore result in more appropriate solutions.

Mediation also tends to promote a more accurate view of problems. For instance, in lawsuits each party tends to bring up evidence that favors its goal and ignore or dismiss unfavorable information or ambiguous information that might weaken its stand. In mediation, both parties are encouraged to openly discuss the uncertainties of their positions, discovering many points of agreement and building a better understanding of opposing positions.

Mediation has its drawbacks, too. First, funding is inadequate. In the past, mediation has been financed largely by foundations. Federal, state, and local governments need to develop programs to fund mediation. Second, some groups fear they will lose their constituency if they enter into negotiations on certain issues, because they will be giving the impression that they are failing at their stated goals and compromising with the "enemy." A third drawback is a lack of faith in the outcome of mediation. Unlike court orders, resolutions drawn up in mediation are not legally enforceable. Thus, months of discussion may produce nothing but a piece of paper that polluters will ignore.

Effective mediation requires the following: (1) A truly neutral party must serve as mediator. (2) A formally agreed-on agenda for discussion and a point of focus are also important. Resource and pollution issues should be the focus of discussion; disputes over values should not dominate the proceedings, because they cannot be solved by mediation. Although values will surely come out in the debate, they should not be the central point. (3) There must be a willingness to explore new ideas and possibilities on both sides. (4) Disputants must deal honestly with each other. (5) An adequate representation of all interested parties must also be achieved. If someone is not represented, a solution that is unsatisfactory may result. This could lead to a lawsuit. (6) Strict rules should be imposed regarding news releases. The media should not be employed as a lever by either group.

Dispute resolution is growing in the United States, but it will not replace litigation. Still, it can play a valuable role in the future. The first environmental mediations began in 1975, and as of 1990 over 100 disputes had been settled using this approach. Eight states have organizations that offer mediation services, and a growing number of private organizations have been formed to provide professional mediators with experience and knowledge in environmental issues. Thus, more and more disputes may be settled by this noncombative approach.

Suggested Readings

Arbuckle, J. G., et al. (1985). *Environmental Law Handbook* (8th ed.). Washington, DC: Government Institutes. Superb overview of environmental law.

Epstein, S. S., Brown, L. O., and Pope, C. (1982). *Hazardous Waste in America*. San Francisco: Sierra Club Books. See Chapter 10 for a good overview of principles of environmental law.

Findley, R. W. and Farber, D. A. (1981). *Environmental Law: Cases and Materials*. St. Paul, MN: West Publishing. In-depth presentation of important cases in environmental law.

Morgenstern, R. and Sessions, S. (1988). Unfinished Business. *Environment* 30 (6): 14–17, 34–39. Excellent survey of EPA's constraints and of work needing to be done.

Turner, T. (1988). The Legal Eagles. *The Amicus Journal* 10 (1): 24–37. Excellent history of advances in environmental law.

Wenner, L. M. (1976). *One Environment Under Law: A Public-Policy Dilemma*. Pacific Palisades, CA: Goodyear. Superb reading.

Rethinking the Past/ Remaking the Future

What we do for ourselves dies with us. What we do for others and the world remains and is immortal.

ALBERT PINE

The transition from a hunting and gathering society to an agricultural society and the transition from an agricultural to an industrial society are two of the most profound changes that have occurred in the history of human civilization. Each of these transitions has been marked by an increase in efforts to gain control over the environment, a decrease in environmental quality, and a growing conflict between humans and the natural order.

The benefits of these two major transitions have been extraordinary. But, as this book has pointed out, the consequences have often been earth shattering. Acid deposition, global warming, crowding, hunger, erosion, deforestation, salinization, species extinction, groundwater contamination, and a host of other serious threats call into question many of our basic assumptions and actions. The growing list of environmental problems that have worsened social problems reinforces the often-ignored truth that planet care is self-care.

Over hundreds of years humans have experimented with a great many ways of eking out a living on earth. Many of those experiments have ended in disaster. It is time to examine the results of these experiments honestly and with an eye to the future. It is time to draw up a set of operating plans that will ensure sustainability. It is time to begin a third transition, the shift to sustainability.

The Third Transition

Conservation, recycling, renewable resources, and population control are the essential elements of this transition. These are the operational principles—the guidelines for sustainability. They must be accompanied by changes in attitude. In order to begin, we must abandon the reckless notion that there is always more. And we must come to accept the idea that humans are not apart from but rather a part of nature. Cooperation rather than dominance and control must become our guiding principle.

Sustainability does not mean sustaining what we are doing now, because in the long run that's impossible. It has been said many times but merits repeating: Our society is spending nature's principal—depleting our bank account of natural resources—when it should be living off the interest.

Sustainability requires that we use only those resources, including energy, that a region can continuously supply *through natural processes*. It requires that we live in balance with our environment. The immediate implication of such an agenda is a greatly reduced resource budget. A sustainable society is one that meets its needs and aspirations without compromising the ability of future generations to meet theirs.

Sustainable management systems are incompatible with systems of management with single-minded values. A stream is not a source of hydroelectric power, but rather a source of fish, recreation, irrigation water, wildlife habitat, and power. A sustainable system attempts to find ways that optimize all of these worthy ends. Ultimately, though, a sustainable system is one in which human economic objectives are balanced against biological constraints, and woven into ecological cycles. It is a system in which human activities respect natural laws.

Sustainability requires a better understanding of the subtle long-term needs of the land. It requires critical thinking. Using artificial fertilizer to replace nutrients lost from the soil, for example, meets only part of the soil's needs. Artificial fertilizer does not replace organic matter that holds water and nourishes valuable soil microbes.

Sustainability requires flexibility: applying different solutions to different situations. Appropriate technology and appropriate policy are good examples. They show the intelligence of designing with nature rather than redesigning nature.

Sustainability requires adaptive management: treading lightly and learning from our mistakes. Human understanding of nature is imperfect and thus suggests that our interactions with nature should be guided by caution. Our interactions can become opportunities to learn from experience and find a sustainable equilibrium.

A sustainable society replaces short-term profit motives with considerations of long-term stability. It requires that we forgo potential income now for the long-term benefits of productive land.

Sustainability requires a lifetime of commitment, in much the same way that good health requires a lifetime of care and attention. It requires individual action, too. A sustainable society may seem utopian. It is not. It is a strategy for survival.

Signs of the Transition

The transition to a sustainable society has already begun. Many people, for example, are simplifying their life-styles, either because materialism has lost its appeal or because economic forces leave them no other choice. Many Americans have also turned to smaller, more fuel-efficient cars and public transportation not just to save money but to cut air pollution. More and more builders are installing better insulation and domestic solar collectors for heating water. The National Association of Home Builders reports that home buyers list energy efficiency as a priority more than any other factor. And new legislation will result in sharp declines in energy consumption by appliances over the next decade.

People throughout the world have made some dramatic improvements in energy efficiency. In fact, in many areas Americans lag behind the Europeans and the Japanese. Brazil, more than most countries, has set the pace for the transition to a sustainable energy system. Brazil hopes to eliminate most of its oil imports. Energy will come from hydroelectric power, wood, and alcohol. Brazil is building a society whose homes, industries, and automobiles will be fueled by renewable energy, showing the whole world that it can be done.

Environmental protection has also increased in the last two decades. A general awareness of environmental issues can be found in rich as well as poor nations.

Each year more and more companies are turning from the old attitude that it pays to pollute to a new philosophy that pollution prevention pays. The Dow Chemical Company, for example, invested $2.7 million to recover hydrogen and harmful chlorine gas once released into the atmosphere from its chemical plant in Hemlock, Michigan. By so doing, the plant now saves approximately $900,000 a year.

On a broader scale, world population growth has slowed in the last two decades, dropping from 2.5% 20 years ago to 1.7% today. Even though population growth continues at a fast pace in many Third World nations, it has all but come to a stop in many developed nations. Some Third World countries are recognizing the need for population control and have taken measures to stop the rampant growth.

What's Needed: More of the Same Positive Changes

Despite these gains and many others too numerous to list here, much work is needed to build a sustainable society. Foremost on the list are attitudinal changes. The frontier mentality must be replaced by a sustainable mentality. Feelings of insignificance must be replaced by the positive attitude that what we do makes a difference.

As a future parent, teacher, scientist, journalist, business executive, accountant, or political leader, you can play a big role in the transition. You may even find a career in environmental law, research, teaching, or writing. But even if you choose another route you can live consistently with the values of sustainability—treading lightly on the planet as you go through life.

Some see things as they are and say why? I dream of things that never were and say why not?

GEORGE BERNARD SHAW

Glossary

Abiotic factors Nonliving components of the ecosystem, including chemical and physical factors such as availability of nitrogen, temperature, and rainfall.

Accelerated erosion Loss of soil due to wind or water in land disturbed by human activities.

Accelerated extinction Elimination of species due to human activities such as habitat destruction, commercial hunting, sport hunting, and pollution.

Acid deposition Rain or snow that has a lower pH than precipitation from unpolluted skies, also includes dry forms of deposition such as nitrate and sulfate particles.

Acid mine drainage Sulfuric acid that drains from mines, especially abandoned underground coal mines in the East (Appalachia). Created by the chemical reaction between oxygen, water, and iron sulfides found in coal and surrounding rocks.

Active solar Capturing and storage of the sun's energy through special collection devices (solar panels) that absorb heat and transfer it to air, water, or some other medium, which is then pumped to a storage site (usually a water tank) for later use. Contrast with *passive solar*.

Actual risk An accurate measure of the hazard posed by a certain technology or action.

Acute effects In general, effects that occur shortly after exposure to toxic agents. Contrast with *chronic effects*.

Acute toxicity Poisoning generally caused by short-term exposure to high levels of one or more agents. Symptoms appear soon after exposure.

Adaptation A genetically determined structural or functional characteristic of an organism that enhances its chances of reproducing and passing on its genes.

Adaptive radiation Evolution of several life forms from a common ancestor.

Advanced industrial society Post-World War II *industrial society* characterized by great rises in production and consumption, increased energy demand, and a shift toward synthetics and nonrenewable resources.

Age-specific fertility rate Number of live births per 1000 women of a specific age group.

Agricultural land conversion Transformation of farmland to other purposes, primarily cities, highways, airports, and the like.

Agricultural society A group of people living in villages or towns and relying on domestic animals and crops grown in nearby fields. Characterized by specialization of work roles.

Algal bloom Rapid growth of algae in surface waters due to increase in inorganic nutrients, usually either nitrogen or phosphorus.

Alien species (or foreign species) Any species introduced into or living in a new habitat. Also known as an exotic.

Alpha particles Positively charged particles consisting of two protons and two neutrons, emitted from radioactive nuclei.

Alveoli Small sacs in the lungs where exchange of oxygen and carbon dioxide between air and blood occurs.

Ambient air quality standard Maximum permissible concentration of a pollutant in the air around us. Contrast with *emissions standard*.

Annuals Plants that grow from seeds, for example, domestic corn and radishes.

Antagonism In toxicology, when two chemical or physical agents (often toxins) counteract each other to produce a lesser response than would be expected if individual effects were added together.

Anthropogenic hazard A danger created by humans.

Appropriate technology A term coined by the late E. F. Schumacher to refer to technology that is "appropriate" for the economy, resources, and culture of a region. It is characterized by small- to medium-sized machines, maximum human labor, ease of understanding, meaningful employment, use of local resources, decentralized production, production of durable products, emphasis on renewable resources, especially energy, and compatibility with the environment and culture.

Aquaculture Cultivation of fish and other aquatic organisms in freshwater ponds, lakes, irrigation ditches, and other bodies of water.

Aquifer Underground stratum of porous material (sandstone) containing water (groundwater), which may be withdrawn from wells for human use.

Aquifer recharge zone Region in which water from rain or snow percolates into an aquifer, replenishing the supply of groundwater.

Artificial selection Selective breeding to create new plant and animal breeds to bring out desirable characteristics.

Asbestos One of several naturally occurring silicate fibers. Useful in society as an insulator but deadly to breathe even in small amounts. Causes mesothelioma, asbestosis, and lung cancer.

Asbestosis Lung disease characterized by buildup of scar tissue in the lungs. Caused by inhalation of asbestos.

Asthma Lung disorder characterized by constriction and excessive mucus production in the bronchioles, resulting in periodic difficulty in breathing, shortness of breath, coughing. Usually caused by allergy and often aggravated by air pollution.

Atmosphere Layer of air surrounding the earth.

Atom A basic unit of matter consisting of a nucleus of positively charged protons and uncharged neutrons, and an outer cloud of electrons orbiting the nucleus.

Auxins Plant hormones responsible for stimulating growth.

Bacteria A group of single-celled organisms, each surrounded by a cell wall and containing circular DNA. Responsible for some diseases and many beneficial functions, such as the decay of organic materials and nutrient recycling.

Barrier islands Small, sandy islands off a coast, separated from the mainland by lagoons or bays.

Beach drift Wave-caused movement of sand along a beach.

Beta particles Negatively charged particles emitted from nuclei of radioactive elements when a neutron is converted to a proton.

Big bang Theory of the universe's formation. States that all matter in the universe was infinitely compressed 15 to 20 billion years ago and then exploded, sending energy and matter out into space. The matter was in the form of subatomic particles which formed atoms as the universe cooled over millions of years.

Biochemical oxygen demand (BOD) Measure of oxygen depletion of water (largely from bacterial decay) due to presence of biodegradable organic pollutants. Gives scientists an indication of how much organic matter is in water.

Bioconcentration Ability of an organism to selectively accumulate certain chemicals, elements, or substances within its body or within certain cells.

Biogas A gas containing methane and carbon dioxide. Produced by anaerobic decay of organic matter, especially manure and crop residues.

Biogeochemical cycle Complex cyclical transfer of nutrients from the environment to organisms and back to the environment. Examples include the carbon, nitrogen, and phosphorus cycles.

Biological control Use of naturally occurring predators, parasites, bacteria, and viruses to control pests. Also called food chain control.

Biological extinction Disappearance of a species from part or all of its range.

Biological magnification Buildup of chemical elements or substances in organisms in successively higher trophic levels. Also called biomagnification.

Biomass As measured by ecologists, the dried weight of all organic matter in the ecosystem. In the energy field, any form of organic material (from both plants and animals) from which energy can be derived by burning or bioconversion such as fermentation. Includes wood, cow dung, agricultural crop residues, forestry residues, scrap paper.

Biomass pyramid See *pyramid of biomass*.

Biome One of several immense terrestrial regions, each characterized throughout its extent by similar plants, animals, climate, and soil type.

Biosphere All the life-supporting regions (ecosystems) of the earth and all the interactions that occur between organisms and between organisms and the environment.

Biotic factor The biological component of the ecosystem, consisting of populations of plants, animals, and microorganisms in complex communities.

Biotic (reproductive) potential Maximum reproductive potential of a species.

Birth control Any measure designed to reduce births, including contraception and abortion.

Birth defect An anatomical (structural) or physiological (functional) defect in a newborn.

Bloom See *algal bloom*.

Breeder reactor Fission reactor that produces electricity and also converts abundant but nonfissile uranium-238 into fissile plutonium-239, which can be used in other fission reactors.

Broad-spectrum pesticide (or biocide) Chemical agent effective in controlling a large number of pests.

Bronchitis Persistent inflammation of the bronchi caused by smoking and air pollutants. Symptoms include mucus buildup, chronic cough, and throat irritation.

Brown-air cities Newer, relatively nonindustrialized cities whose polluted skies contain photochemical oxidants (especially ozone) and nitrogen oxides, largely from automobiles and power plants. Tend to have dry, sunny climates. Contrast with *gray-air cities*.

Cancer Uncontrolled proliferation of cells in humans and other living organisms. In humans, includes more than 100 different types afflicting individuals of all races and ages.

Carbon cycle The cycling of carbon between organisms and the environment.

Carcinogen A chemical or physical agent that causes cancer to develop, often decades after the original exposure.

Carrying capacity Maximum population size that a given ecosystem can support for an indefinite period or on a sustainable basis.

Catalyst Substance that accelerates chemical reactions but is not used up in the process. Enzymes are biological catalysts. Also see *catalytic converter*.

Catalytic converter Device attached to the exhaust system of automobiles and trucks to rid the exhaust gases of harmful pollutants.

Cation Any one of many kinds of positively charged ions.

Cellular respiration Process by which a cell breaks down glucose and other organic molecules to acquire energy. Also called oxidative metabolism.

Chlorofluorocarbons Organic molecules consisting of chlorine and fluorine covalently bonded to carbon. CCl_3F (Freon-11) and CCl_2F_2 (Freon-12). Used as spray-can propellants and coolants. Previously thought to be inert, but now known to destroy the stratospheric ozone layer. Also called chlorofluoromethanes and freon gases.

Chlorophyll Pigment of plant cells that absorbs sunlight, thus allowing plants to capture solar energy.

Chromosomes Genetic material of organisms, containing DNA and protein. Carries the genetic information that controls all cellular activity.

Chronic bronchitis Persistent inflammation of the bronchi due to pollutants in ambient air and tobacco smoke. Characterized by persistent cough.

Chronic effects In general, the delayed health results of toxic agents, for example, emphysema, bronchitis, and cancer. Contrast with *acute effects*.

Chronic obstructive lung disease Any one of several lung diseases characterized by obstruction of breathing. Includes emphysema, bronchitis, and diseases with symptoms of both of these.

Chronic toxicity Poisoning generally caused by long-term exposures to low levels of one or more toxic agents. Symptoms appear long after exposure. Examples: emphysema and cancer.

Clear-cutting Removal of all trees from a forested area.

Climate The average weather conditions: temperature, solar radiation, precipitation, and humidity.

Climax community or ecosystem See *mature community*.

Closed system A system that can exchange energy, but does not exchange matter, with the surrounding environment. Example: the earth. Contrast with *open system*.

Coal gasification Production of combustible organic gases (mostly methane) by applying heat and steam to coal in an oxygen-enriched environment. Carried out in surface vessels or *in situ*.

Coal liquefaction Production of synthetic oil from coal.

Coastal wetlands Wet or flooded regions along coastlines, including mangrove swamps, salt marshes, bays, and lagoons. Contrast with *inland wetlands*.

Coevolution Process whereby two species evolve adaptations as a result of extensive interactions with each other.

Cogeneration Production of two or more forms of useful energy from one process. For example, production of electricity and steam heat from combustion of coal. Increases energy efficiency.

Coliform bacterium Common bacterium found in the intestinal tracts of humans and other species. Used in water quality analysis to determine the extent of fecal contamination.

Commensalism Relationship between two organisms that is beneficial to one and neither harmful nor helpful to the other.

Common law Body of rules and principles based on judicial precedent rather than legislative enactments. Founded on an innate sense of justice, good conscience, and reason. Flexible and adaptive. Contrast with *statutory law*.

Commons Any resource used in common by many people, such as air, water, and grazing land.

Community Also called a biological community. The populations of plants, animals, and microorganisms living and interacting in a given locality.

Competition Vying for resources between members of the same or different species.

Composting Aerobic decay of organic matter to generate a humus-like substance used to supplement soil.

Confusion technique (of pest control) Release of insect sex-attractant pheromones identical to pheromones released by normal breeding females to attract males for mating. Release in large quantities confuses males as to the location of the females, thus minimizing the chances of males finding females and helping to control pest populations.

Conservation A strategy to reduce the use of resources, especially through increased efficiency, reuse, recycling, and decreased demand.

Conservation biology See *restoration ecology*.

Consumer (or consumer organism) An organism in the ecosystem that feeds on autotrophs and/or heterotrophs. Synonym: heterotroph.

Continental drift Movement of the earth's tectonic plates on a semiliquid layer of mantle, forcing continents to shift position over hundreds of thousands of years.

Contour farming Soil erosion control technique in which row crops (corn) are planted along the contour lines in sloping or hilly fields rather than up and down the hills.

Contraceptive Any device or chemical substance used to prevent conception.

Control group In scientific experimentation, a group that is untreated and compared with a treated, or experimental, group.

Control rods Special rods containing neutron-absorbing materials. Inserted into a reactor core to control the rate of fission or to shut down fission reactions.

Convergent evolution The independent evolution of similar traits among unrelated organisms resulting from similar selective pressures.

Cosmic radiation High-energy electromagnetic radiation similar to cosmic rays but originating from periodic solar flare-ups. Possesses extraordinary ability to penetrate materials, including cement walls.

Cost–benefit analysis Way of determining the economic, social, and environmental costs and benefits of a proposed action such as construction of a dam or highway. Still a crude analytical tool because of the difficulty of measuring environmental costs.

Critical population size Population level below which a species cannot successfully reproduce.

Crop rotation Alternating crops in fields to help restore soil fertility and also control pests.

Crossing over Transfer of genetic material from one chromosome to another during the formation of gametes.

Cross-media contamination The movement of pollution from one medium, such as air, to another, such as water.

Crude birth rate Number of births per 1000 people in a population at the midpoint of the year.

Crude death rate Number of deaths per 1000 people in a population at midyear.

Cultural control (of pests) Techniques to control pest populations not involving chemical pesticides, environmental controls, or genetic controls. Examples: cultivation to control weeds and manual removal of insects from crops.

Cultural eutrophication *Eutrophication* (see definition) due largely to human activities.

Daughter nuclei Atomic nuclei that are produced during fission of uranium.

DDT Dichlorodiphenyltrichloroethane. An organochlorine insecticide used first to control malaria-carrying mosquitoes and lice and later to control a variety of insect pests, but now banned in the United States because of its persistence in the environment and its ability to bioaccumulate.

Decibel (dB) A unit to measure the loudness of sound.

Decomposer Also microconsumer. An organism that breaks down nonliving organic material. Examples: bacteria and fungi.

Decomposer food chain A specific nutrient and energy pathway in an ecosystem in which decomposer organisms (bacteria and fungi) consume dead plants and animals as well as animal wastes. Essential for the return of nutrients to soil and carbon dioxide to the atmosphere. Also called detritus food chain.

Deforestation Destruction of forests by clear-cutting.

Demographic transition A phenomenon witnessed in populations of industrializing nations. As industrialization proceeds and wealth accumulates, crude birth rate and crude death rate decline, resulting in zero or low population growth. Decline in death rate usually precedes the decline in birth rate, producing a period of rapid growth before stabilization.

Demography The science of population.

Depletion allowance Tax relief given to extractive industries as they deplete reserves. Intended to allow the companies to invest more in exploration. Gives extraction industries unfair advantage over recycling companies.

Desert Biome located throughout the world. Often found on the downwind side of mountain ranges. Characterized by low humidity, high summertime temperatures, and plants and animals especially adapted to lack of water.

Desertification The formation of desert in arid and semiarid regions from overgrazing, deforestation, poor agricultural practices, and climate change. Found today in Africa, the Middle East, and the southwestern United States.

Detoxification Rendering a substance harmless by reacting it with another chemical, chemically modifying it, or destroying the molecule through combustion or thermal decomposition.

Detritus Any organic waste from plants and animals.

Detritus feeders Organisms in the decomposer food chain that feed primarily on organic waste (detritus), such as fallen leaves.

Detritus food chain See *decomposer food chain*.

Deuterium An isotope of hydrogen whose nucleus contains one proton and one neutron (a hydrogen atom has only one proton).

Developed country A convenient term that describes industrialized nations, generally characterized by high standard of living, low population growth rate, low infant mortality, excessive material consumption, high per capita energy consumption, high per capita income, urban population, and low illiteracy.

Developing country Same as *less developed country*.

Dioxin A large group of highly toxic, carcinogenic compounds containing some herbicides (2,4-D and 2,4,5-T) and Agent Orange. Once disposed of by mixing with waste crankcase oil that was spread on dirt roads to control dust.

Diversity A measure of the number of different species in an ecosystem.

DNA (deoxyribonucleic acid) A long-chained organic molecule that is found in chromosomes and carries the genetic information that controls cellular function and is the basis of heredity.

Dose–response curve Graphical representation of the effects of varying doses of chemical or physical agents.

Doubling time The length of time it takes some measured entity (population) to double in size at a given growth rate.

Ecological backlashes Ecological effects of seemingly harmless activities, for example, the *greenhouse effect*.

Ecological equivalents Organisms that occupy similar ecological niches in different regions of the world.

Ecological niche See *niche*.

Ecological system See *ecosystem*.

Ecology Study of living organisms and their relationships to one another and the environment.

Economic depletion Reduction in the supply of a resource to the point at which it is no longer economically feasible to continue mining, extracting, or harvesting it.

Economic externality A cost (environmental damage, illness) of manufacturing, road building, or other action that is not taken into account when determining the total cost of production or construction. A cost generally passed on to the general public and taxpayers; external cost.

Ecosphere See *biosphere*.

Ecosystem Short for ecological system. A community of organisms occupying a given region within a biome. Also, the physical and chemical environment of that community and all the interactions between organisms and between organisms and their environment.

Ecosystem stability Dynamic equilibrium of the ecosystem. Also a characteristic of ecosystems causing them to return to their previous state (resilience) and their resistance to change (inertia).

Ecotone Transition zone between adjacent ecosystems.

Element A substance, such as oxygen, gold, or carbon, that is distinguished from all other elements by the number of protons in its atomic nucleus. The atoms of an element cannot be decomposed by chemical means.

Emigration Movement of people out of a country to establish residence elsewhere.

Emissions offset policy Strategy to control air pollution in areas meeting federal ambient air quality standards, whereby new factories must secure emissions reductions from existing factories to begin operation; thus the overall pollution level does not increase.

Emissions standard The maximum amount of a pollutant permitted to be released from a *point source* (see definition).

Emphysema A progressive, debilitating lung disease caused by smoking and pollution at work and in the environment. Characterized by gradual breakdown of the *alveoli* (see definition) and difficulty in catching one's breath.

Endangered species A plant, animal, or microorganism that is in immediate danger of biological extinction. See *threatened* and *rare species*.

Energy The capacity to do work. Found in many forms, including heat, light, sound, electricity, coal, oil, and gasoline.

Energy pyramid See *pyramid of energy*.

Energy quality The amount of useful work acquired from a given form of energy. High-quality energy forms are concentrated (e.g., oil and coal); low-quality energy forms are less concentrated (e.g., solar heat).

Energy system The complete production–consumption process for energy resources, including exploration, mining, refining, transportation, and waste disposal.

Entropy A measure of disorder. The second law of thermodynamics applied to matter says that all systems proceed to maximum disorder (maximum entropy).

Environment All the biological and nonbiological factors that affect an organism's life.

Environmental control (of pests) Methods designed to alter the abiotic and biotic environments of pests, making them inhospitable or intolerable. Examples include increasing crop diversity, altering time of planting, and altering soil nutrient levels.

Environmental impact statement (EIS or ES) Document prepared primarily to outline potential impacts of projects supported in part or in their entirety by federal funds.

Environmental phase (of the nutrient cycle) Part of the nutrient or biogeochemical cycle in which the nutrient is deposited or cycles through the environment (air, water, and soil).

Environmental resistance Abiotic and biotic factors that can potentially reduce population size.

Environmental science The interdisciplinary study of the complex and interconnected issues of population, resources, and pollution.

Epidemiology Study of disease and death in human populations, which attempts to find links between causes and effects through statistical methods.

Epilimnion Upper, warm waters of a lake. Contrast with *hypolimnion*.

Estuarine zones *Coastal wetlands* and *estuaries*.

Estuary Coastal regions such as inlets or mouths of rivers where salt and fresh water mix.

Ethanol Grain alcohol, or ethyl alcohol, produced by fermentation of organic matter. Can be used as a fuel for a variety of vehicles and as a chemical feedstock.

Eukaryotes The first aerobic cells complete with nuclei and energy-releasing organelles.

Eutrophication Accumulation of nutrients in a lake or pond due to human intervention (cultural eutrophication) or natural causes (natural eutrophication). Contributes to process of *succession* (see definition).

Evapotranspiration Evaporation of water from soil and transpiration of water from plants.

Evolution A long-term process of change in organisms caused by random genetic changes that favor the survival and reproduction of the organism possessing the genetic change. Through evolution, organisms become better adapted to their environment.

Exclusion principle Ecological law holding that no two species can occupy the exact same *niche*.

Experimental group In scientific experimentation, a group that is treated and compared with an untreated, or control, group.

Exponential curve See *J curve*.

Exponential growth Increase in any measurable thing by a fixed percentage. When plotted on graph paper, it forms a J-shaped curve.

Externality A spillover effect that benefits or harms others. The source of the effect (say, pollution) does not pay for the effect.

Extinction See *biological extinction*.

Fallout Radioactive materials produced during an atomic detonation and later deposited from the air.

Fall overturn Annual cycle in deep lakes in temperate climates, in which the warm surface water and cool subsurface water mix.

Family planning Process by which couples determine the number and spacing of children.

Feedlot Fenced area where cattle are raised in close confinement to minimize energy loss and maximize weight gain.

First-generation pesticides Earliest known chemical pesticides such as ashes, sulfur, ground tobacco, and hydrogen cyanide. Contrast with *second-* and *third-generation pesticides*.

First-law efficiency A measure of the efficiency of energy use. Total amount of useful work derived from a system divided by the total amount of energy put into a system.

First law of thermodynamics Also called the law of conservation of energy. States that energy is neither created nor destroyed; it can only be transformed from one form to another.

Fission Splitting of atomic nuclei when they are struck by neutrons or other subatomic particles.

Fission fragments See *daughter nuclei*.

Floodplain Low-lying region along river or stream, periodically subject to natural flooding. Common site for human habitation and farming.

Fly ash Mineral matter escaping with smokestack gases from combustion of coal.

Food chain A specific nutrient and energy pathway in ecosystems proceeding from producer to consumer. Part of a bigger network called the food web. See *decomposer food chain* and *grazer food chain*.

Food web Complex intermeshing of individual food chains in an ecosystem.

Foreign species See *alien species*.

Fossil fuel Any one of the organic fuels (coal, natural gas, oil, tar sands, and oil shale) derived from once-living plants or animals.

Freons See *chlorofluorocarbons*.

Frontier mentality A mind-set that views humans as "above" all other forms of life rather than as an integral part of nature and sees the world as an unlimited supply of resources for human use regardless of the impacts on other species. Implicit in this view are the notions that bigger is better, continued material wealth will improve life, and nature must be subdued.

Fuel rods Rods packed with small pellets of radioactive fuel (usually a mixture of fissionable uranium-235 and uranium-238) for use in fission reactors.

Gaia hypothesis Term coined by James Lovelock to describe the earth's capacity to maintain the physical and chemical conditions necessary for life.

Galaxy Grouping of billions of stars, gas, and dust, such as the Milky Way galaxy.

Gamma rays A high-energy form of radiation given off by certain *radionuclides*. Can easily penetrate the skin and damage cells.

Gasohol Liquid fuel for vehicles, containing nine parts gasoline and one part ethanol.

Gene Segment of the DNA that either codes for proteins produced by the cell (structural gene) or regulates structural genes.

Gene pool Sum total of all the genes and their alternate forms in a population.

Generalists Organisms that have a broad *niche*, usually feeding on a variety of food materials and sometimes adapted to a large number of habitats.

Genetic control (of pests) Development of plants and animals genetically resistant to pests through breeding programs and genetic engineering. Also, introduction of sterilized males of pest species (see *sterile-male technique*).

Genetic engineering Isolation and production of genes that are then inserted in bacteria or other organisms. Can be used to produce insulin and other hormones. May someday also be used to treat genetic diseases.

Geopressurized zone Aquifer containing superheated, pressurized water and steam trapped by impermeable rock strata and heated by underlying magma.

Geothermal energy Energy derived from the earth's heat that comes from decay of naturally occurring radioactive materials in the earth's crust, magma, and friction caused by movement of tectonic plates.

GNP See *gross national product*.

Gradualism Theory of evolution holding that species evolve over long periods. Contrast with *punctuated equilibrium*.

Grasslands Biome found in both temperate and tropical regions and characterized by periodic drought, flat or slightly rolling terrain, and large grazers that feed off the lush grasses.

Gray-air cities Older industrial cities characterized by predominantly sulfur dioxide and particulate pollution. Contrast with *brown-air cities*.

Grazer food chain A specific nutrient and energy pathway starting with plants that are consumed by grazers (herbivores).

Greenhouse effect Mechanism that explains atmospheric heating caused by increasing carbon dioxide. Carbon dioxide is believed to act like the glass in a greenhouse, permitting visible light to penetrate but impeding the escape of infrared radiation, or heat.

Green Revolution Developments in plant genetics in the late '50s and early '60s resulting in high-yield varieties producing three to five times more grain than previous plants but requiring intensive irrigation and fertilizer use.

Gross national product (GNP) Total national output of goods and services valued at market prices, including net exports and private investment.

Gross primary productivity The total amount of sunlight converted into chemical-bond energy by a plant. This measure does not take into account how much energy a plant uses for normal cellular functions. See *net primary productivity*.

Groundwater Water below the earth's surface in the saturated zone.

Habitat The specific region in which an organism lives.

Half-life Time required for one-half of a given amount of radioactive material to decay, producing one-half the original mass. Can also be used to describe the length of residence of chemicals in tissues. Biological half-life refers to the time it takes for one-half of a given amount of a substance to be excreted or catabolized.

Hard path A term coined by Amory Lovins to describe large, centralized energy systems such as coal, oil, or nuclear power, characterized by extensive power distribution, central control, and lack of renewability.

Hazardous waste Any potentially harmful solid, liquid, or gaseous waste product of manufacturing or other human activities.

Herbicide Chemical agent used to control weeds.

Herbivore Heterotrophic organism that feeds exclusively on plants.

Heteroculture Agriculture in which several plant species are grown simultaneously to reduce insect infestation and disease.

Heterotroph An organism that feeds on other organisms such as plants and animals. It cannot make its own foodstuffs.

Hot-rock zones Most widespread geothermal resource. Regions where bedrock is heated by underlying magma.

Humus Mixture of decaying organic matter and inorganic matter that increases soil fertility, aeration, and water retention.

Hunting and gathering society People who lived as nomads or in semipermanent sites from the beginning of human evolution until approximately 5000 BC. Some remnant populations still survive. They gathered seeds, fruits, roots, and other plant products and hunted indigenous species for food.

Hybrid Offspring produced by cross-mating of two different strains or varieties of plants or animals.

Hydrocarbons Organic molecules containing hydrogen and carbon. Released during the incomplete combustion of organic fuels. React with nitrogen oxides and sunlight to form photochemical oxidants in photochemical smog.

Hydroelectric power Electricity produced in turbines powered by running water.

Hydrological cycle The movement of water through the environment from atmosphere to earth and back again. Major events include evaporation and precipitation. Also called the water cycle.

Hydrosphere The watery portion of the planet. Contrast with *atmosphere* and *lithosphere*.

Hydrothermal convection zone Rock strata containing large amounts of water heated by underlying magma and driven to the surface through cracks and fissures in overlying rock layers. Forms hot springs and geysers.

Hypolimnion Deep, cold waters of a lake. Contrast with *epilimnion*.

Hypothesis Tentative explanation for a natural phenomenon.

Immature ecosystem An early successional community characterized by low species diversity and low stability. Contrast with *mature ecosystem*.

Immigration Movement of people into a country to set up residence there.

Indoor air pollution Generally refers to air pollutants in homes from internal sources such as smokers, fireplaces, wood stoves, carpets, paneling, furniture, foam insulation, and cooking stoves.

Induced abortion Surgical procedure to interrupt pregnancy by removing the embryo or fetus from the uterus. In the first trimester, generally carried out by vacuum aspiration. Contrast with *spontaneous abortion*.

Industrial smog Air pollution from industrial cities (gray-air cities), consisting mostly of particulates and sulfur oxides. Contrast with *photochemical smog*.

Industrial society Group of people living in urban or rural environments that are characterized by mechanization of industrial production and agriculture. Widespread machine labor causes high energy demands and pollution. Increasing control over natural processes leads to feelings that humans are apart from nature and superior to it.

Inertia Tendency of an ecosystem to resist change.

Infant mortality rate Number of infants under 1 year of age dying per 1000 births in any given year.

Infectious disease Generally, a disease caused by a virus, bacterium, or parasite that can be transmitted from one organism to another (example: viral hepatitis).

Infrared radiation Heat, an electromagnetic radiation of wavelength outside the red end of the visible spectrum.

Inland wetlands Wet or flooded regions along inland surface waters. Includes marshes, bogs, and river outflow lands. Contrast with *coastal wetlands*.

In-migration Movement of people into a state or region within a country to set up residence.

Inorganic fertilizer Synthetic plant nutrient added to the soil to replace lost nutrients. Major components include nitrogen, phosphorus, and potassium. Also called artificial fertilizer or synthetic fertilizer.

Input approach A method of solving an environmental problem by reducing the inputs. For example, reducing consumption and increasing product durability can cut production of solid wastes, pollution, or hazardous wastes.

Insecticide One form of pesticide used specifically to control insect populations.

Integrated pest management Pest control with minimum risk to humans and the environment through use of a variety of control techniques (including pesticides and biological controls).

Integrated wildlife or species management Control of populations through the use of many techniques, including the reintroduction of natural predators, habitat improvement, reduction in habitat destruction, establishment of preserves, reduced pollution, and captive breeding.

Interspecific competition Competition between members of different species.

Ion A particle formed when an atom loses or gains an electron.

Ionizing radiation Electromagnetic radiation with the capacity to form ions in body tissues and other substances.

Isotopes Atoms of the same element that differ in their atomic weight because of variations in the number of neutrons in their nuclei.

J curve A graphical representation of exponential growth.

Juvenile hormone Chemical substance in insects that stimulates growth through early life stages. Used with some success as an insecticide. When applied to infested fields, JH alters normal growth and development of insect pests, resulting in their death.

Kerogen Solid, insoluble organic material found in oil shale.

Keystone species Critical species in an ecosystem whose loss profoundly affects several or many others.

Kilowatt One thousand watts. See *watt*.

Kinetic energy The energy of objects in motion.

Kwashiorkor Dietary deficiency caused by insufficient protein intake; common in children one to three years of age in less developed countries. Characterized by growth retardation, wasting of muscles in limbs, and accumulation of fluids in the body, especially in feet, legs, hands, and face.

Lag effect The tendency for a population to continue growing even after it has reached replacement-level fertility. Caused by an expanding number of women reaching reproductive age.

Land ethic View that extends ethical concerns beyond humans to the ecosystem.

Land-use planning Process whereby land uses are matched with the needs of the community and environmental considerations, for example, need for open space and agricultural land and for control of water and air pollution.

Laterite Soil found in some tropical rain forests. Rich in iron and aluminum but generally of poor fertility. Turns bricklike if exposed to sunlight.

Less developed country Term describing the nonindustrialized nations, generally characterized by low standard of living, high population growth rate, high infant mortality, low material consumption, low per capita energy consumption, low per capita income, rural population, and high illiteracy.

Light water reactor Most common fission reactor for generating electricity. Water bathes the core of the reactor and is used to generate steam, which turns the turbines that generate electricity. Contrast with *liquid metal fast breeder*.

Light year Astronomical unit that measures the distance that light can travel in a year.

Limiting factor A chemical or physical factor that determines whether an organism can survive in a given ecosystem. In most ecosystems, rainfall is the limiting factor.

Limnetic zone Open water zone of lakes through which sunlight penetrates; contains algae and other microscopic organisms that feed on dissolved nutrients.

Liquefaction Production of liquid fuel from coal.

Liquid metal fast breeder Fission reactor that uses liquid metals such as sodium as a coolant.

Lithosphere Crust of the earth. Contrast with *hydrosphere* and *atmosphere*.

Littoral drift Movement of beach sand parallel to the shoreline. Caused by waves and longshore currents parallel to the beach.

Littoral zone Shallow waters along a lakeshore where rooted vegetation often grows.

Macronutrient A chemical substance needed by living organisms in large quantities (for example, carbon, oxygen, hydrogen, and nitrogen). Contrast with *micronutrient*.

Magma Molten rock beneath the earth's crust.

Malnourishment A dietary deficiency caused by lack of vital nutrients and vitamins.

Manganese nodules Nodular accumulations of manganese and other minerals such as iron and copper found on the ocean floor at depths of 300 to 6000 meters. Particularly abundant in the Pacific Ocean.

Marasmus A dietary deficiency caused by insufficient intake of protein and calories and occurring primarily in infants under the age of 1, usually as the result of discontinuation of breast feeding.

Mariculture Cultivation of fish and other aquatic organisms in salt water (estuaries and bays).

Mature community A community that remains more or less the same over a long period of time. Climax stage of succession. Also called a climax community.

Mature ecosystem An ecosystem in the climax stage of succession, characterized by high species diversity and high stability. Contrast with *immature ecosystem*.

Measure of economic welfare Proposed standard that takes into account the accumulated wealth of a nation.

Megawatt Measure of electrical power equal to a million watts, or 1000 kilowatts. See *watt*.

Mesothelioma A tumor of the lining of the lung (pleura). Caused by asbestos.

Metalimnion See *thermocline*.

Metastasis Movement of cancer cells to another location where new tumors are formed.

Methyl mercury Water-soluble organic form of mercury formed by bacteria in aquatic ecosystems from inorganic (insoluble) mercury pollution. Able to undergo biological magnification.

Microconsumers Bacteria and single-celled fungi that are part of the decomposer food chain.

Micronutrient An element needed by organisms, but only in small quantities, such as copper, iron, and zinc. Contrast with *macronutrient*.

Migration Movement of people across state and national boundaries to set up new residence. See *immigration, emigration, in-migration,* and *out-migration*.

Mill tailings Residue from uranium processing plants. Spent uranium ore that is contaminated with radioactivity.

Mineral A chemical element (e.g., gold) or inorganic compound (e.g., iron ore) existing naturally.

Minimum tillage Reduced plowing and cultivating of cropland between and during growing seasons to help reduce soil erosion and save energy. Also called conservation tillage.

Molecule Particle consisting of two or more atoms bonded together. The atoms in a molecule can be of the same element but are usually of different elements.

Monoculture Cultivation of one plant species (such as wheat) over a large area. Highly susceptible to disease and insects.

Mutagen A chemical or physical agent capable of damaging the genetic material (DNA and chromosomes) of living organisms in both germ cells and somatic cells.

Mutation In general, any damage to the DNA and chromosomes.

Mutualism Relationship between two organisms that is beneficial to both.

Narrow-spectrum pesticide A chemical agent effective in controlling a small number of pests.

Natural erosion Loss of soil occurring at a slow rate but not caused by human activities. A natural event in all terrestrial ecosystems.

Natural eutrophication See *eutrophication*.

Natural gas Gaseous fuel containing 50%–90% methane and lesser amounts of other burnable organic gases such as propane and butane.

Natural hazards Dangers that result from normal meteorologic, atmospheric, oceanic, biological, and geological phenomena.

Natural resource See *resource*.

Natural selection Process in which slight variations in organisms (adaptations) are preserved if they are useful and help the organism to better respond to its environment.

Negative feedback Control mechanism present in the ecosystem and in all organisms. Information in the form of chemical, physical, and biological agents influences processes, causing them to shut down or reduce their activity.

Net energy See *net useful energy production*.

Net migration Number of immigrants minus the number of emigrants. Can be expressed as a rate by determining immigration and emigration rates.

Net primary productivity Gross primary productivity (the total amount of energy that plants produce) minus the energy plants use during cellular respiration.

Net useful energy production Amount of useful energy extracted from an *energy system*.

Neutralism A relationship without ties.

Niche Also called an ecological niche. An organism's place in the ecosystem: where it lives, what it consumes, what consumes it, and how it interacts with all biotic and abiotic factors.

Nitrate (NO_3^-) Inorganic anion containing three oxygen atoms and one nitrogen atom linked by covalent bonds.

Nitrite (NO_2^-) Inorganic anion containing two oxygen atoms and one nitrogen atom. Combines with hemoglobin and may cause serious health impairment and death in children.

Nitrogen cycle The cycling of nitrogen between organisms and the environment.

Nitrogen fixation Conversion of atmospheric nitrogen (a gas) into nitrate and ammonium ions (inorganic form), which can be used by plants.

Nitrogen oxides Nitric oxide (NO) and nitrogen dioxide (NO_2), produced during combustion when atmospheric nitrogen (N_2) combines with oxygen. Can be converted into nitric acid (HNO_3). All are harmful to humans and other organisms.

Noise An unwanted or unpleasant sound.

Nonattainment area Region that violates EPA air pollution standards.

Nonpoint source (of pollution) Diffuse source of pollution such as an eroding field, urban and suburban lands, and forests. Contrast with *point source*.

Nonrenewable resource Resource that is not replaced or regenerated naturally within a reasonable period (fossil fuel, mineral).

Nuclear fall Hypothesis suggesting that the effects on the earth's climate of dust and smoke released in nuclear explosions would be more temporary and less severe than predicted by the nuclear winter hypothesis. Contrast with *nuclear winter*.

Nuclear fission Splitting of an atomic nucleus when neutrons strike the nucleus. Products are two or more smaller nuclei, neutrons (which can cause further fission reactions), and an enormous amount of heat and radiation energy.

Nuclear fusion Joining of two small atomic nuclei (such as hydrogen and deuterium) to form a new and larger nucleus (such as helium) accompanied by an enormous release of energy. Source of light and heat from the sun.

Nuclear power (or energy) Energy from the fission or fusion of atomic nuclei.

Nuclear winter Hypothesis suggesting that dust from nuclear explosions and smoke from burning cities would reduce solar radiation, resulting in a dramatic decrease in global temperature. Contrast with *nuclear fall*.

Nutrient cycle Same as *biogeochemical cycle*.

Off-road vehicle (ORV) Any vehicle used cross-country, especially in a recreational capacity (four-wheel-drive vehicles, dune buggies, all-terrain vehicles, snowmobiles, and trail bikes).

Oil See *petroleum*.

Oil shale A fine-grained sedimentary rock called marlstone and containing an organic substance known as kerogen. When heated, it gives off shale oil, which is much like crude oil.

Old growth forest Ancient forests with trees often 150 to 1000 or more years old.

Omnivore An organism that eats both plants and animals.

Open system A system that freely exchanges energy and matter with the environment. Example: any living organism. Contrast with *closed system*.

Opportunity costs Costs of lost money-making opportunities (and potentially higher income) incurred when we make a decision to invest our money in a particular way.

Ore Rock bearing important minerals, for example, uranium ore.

Ore deposit A valuable mineral located in high concentration in a given region.

Organic farming Agricultural system in which natural fertilizers (manure and crop residues), crop rotation, contour planting, biological insect control measures, and other techniques are used to ensure soil fertility, erosion control, and pest control.

Organic fertilizer Material such as plant and animal wastes added to cropland and pastures to improve soil. Provides valuable soil nutrients and increases the organic content of soil (thus increasing moisture content).

Organismic phase The part of the nutrient cycle in which nutrients are located in organisms: plants, animals, bacteria, fungi, or others.

Out-migration Movement of people out of a state or region within a country to set up residence elsewhere in that country.

Output approach A method of solving an environmental problem by controlling the outputs. For example, composting or burning trash reduces the land requirements for solid waste disposal. Control devices reduce air and water pollution.

Overgrazing Excessive consumption of producer organisms (plants) by grazers such as deer, rabbits, and domestic livestock. Indication that the ecosystem is out of balance.

Overpopulation A condition resulting when the number of organisms in an ecosystem exceeds its ability to assimilate wastes and provide resources. Creates physical and mental stress on a species as a result of competition for limited resources and deterioration of the environment.

Oxidants Oxidizing chemicals (for example, ozone) found in the atmosphere.

Oxygen-demanding wastes Organic wastes that are broken down in water by aerobic bacteria. Aerobic breakdown causes the oxygen levels to drop.

Ozone (O_3) Inorganic molecule found in the atmosphere, where it is a pollutant because of its harmful effects on living tissue and rubber. Also found in the stratosphere, where it helps screen ultraviolet light. Used in some advanced sewage treatment plants.

Ozone layer Thin layer of ozone molecules in the stratosphere. Absorbs ultraviolet light and converts it to infrared radiation. Effectively screens out 99% of the ultraviolet light.

Paradigm A major theoretical construct that is central to a field of study. For example, the theory of evolution and the structure of DNA are two paradigms that are central to biological science.

Parasitism Relationship in which one species lives in or on another, its host.

Particulates Solid particles (dust, pollen, soot) or water droplets in the atmosphere.

Passive solar Capture and retention of the sun's energy within a building through south-facing windows and some form of heat storage in the building (brick or cement floors and walls). Contrast with *active solar*.

PCBs See *polychlorinated biphenyls*.

Perennial A plant that grows from the same root structure year after year (for example, rose bushes).

Permafrost Permanently frozen ground found in the tundra.

Permanent threshold shift Loss of hearing after continued exposure to noise. Contrast with *temporary threshold shift*.

Pesticide A general term referring to a chemical, physical, or biological agent that kills organisms we classify as pests, such as insects and rodents. Also called biocide.

Petroleum A viscous liquid containing numerous burnable hydrocarbons. Distilled into a variety of useful fuels (fuel oil, gasoline, and diesel) and petrochemicals (chemicals that can be used as a chemical feedstock for the production of drugs, plastics, and other substances).

pH Measure of acidity on a scale from 0 to 14, with pH 7 being neutral, numbers greater than 7 being basic, and numbers less than 7 being acidic.

Pheromone Chemical substance given off by insects and other species. Sex-attractant pheromones released into the atmosphere in small quantity by female insects attract males at breeding time. Can be used in pest control. See *pheromone traps* and *confusion technique*.

Pheromone traps Traps containing pheromones to attract insect pests and pesticide to kill pests or a sticky substance to immobilize them. These traps may be used to pinpoint the emergence of insects, allowing conventional pesticides to be used in moderation.

Photochemical oxidants Ozone and a variety of oxygenated organic compounds produced when sunlight, hydrocarbons, and nitrogen oxides react in the atmosphere.

Photochemical reaction A chemical reaction that occurs in the atmosphere involving sunlight or heat, pollutants, and sometimes natural atmospheric chemicals.

Photochemical smog A complex mixture of *photochemical oxidants* and nitrogen oxides. Usually has a brownish-orange color.

Photosynthesis A two-part process involving (1) the capture of sunlight and its conversion into cellular energy and (2) the production of organic molecules such as glucose and amino acids from carbon dioxide, water, and energy from the sun.

Photovoltaic cell Thin wafer of silicon or other material that emits electrons when struck by sunlight, thus generating an electrical current. Also solar cell.

Pioneer community The first community to become established in a once-lifeless environment during *primary succession*.

Pitch (or frequency) Measure of the frequency of a sound in cycles per second (cps) (hertz, Hz)—compressional sound waves passing a given point per second. The higher the cps, the higher the pitch.

Pneumoconiosis (black lung) A debilitating lung disease caused by prolonged inhalation of coal and other mineral dusts. Results in a decreased elasticity and gradual breakdown of alveoli in the lungs. Eventually leads to death.

Point source (of pollution) Easily discernible source of pollution such as a factory. Contrast with *nonpoint source*.

Pollution Any physical, chemical, or biological alteration of air, water, or land that is harmful to living organisms.

Polychlorinated biphenyls (PCBs) Group of at least 50 organic compounds, used for many years as insulation in electrical equipment. Capable of biological magnification. Disrupts reproduction in gulls and possibly other organisms high on the food chain.

Population A group of organisms of the same species living within a specified region.

Population control In human populations, all methods of reducing birth rate, primarily through pregnancy prevention and abortion. In an ecological sense, regulation of population size by a myriad of abiotic and biotic factors.

Population crash (dieback) Sudden decrease in population that results when an organism exceeds the carrying capacity of its environment.

Population growth rate Rate at which a population increases on a yearly basis, expressed as a percentage. For world population: GR = (crude birth rate − crude death rate) × 100. For a given country, population growth rate must also take into account the net migration rate.

Population histogram Graphical representation of population by age and sex.

Positive feedback Control mechanism in ecosystems and organisms in which information influences some process, causing it to increase.

Potential energy Stored energy.

Predator An organism that actively hunts its prey.

Presbycusis Loss of hearing with age through natural deterioration of the organ of Corti, the sound receptor in the ear. Contrast with *sociocusis*.

Prey Organism (e.g., deer) attacked and killed by predator.

Primary air pollutant A pollutant that has not undergone any chemical transformation; emitted by either a natural or an anthropogenic source.

Primary consumer First consuming organism in a given food chain. A grazer in grazer food chains or a decomposer organism or insect in decomposer food chains. Belongs to the second trophic level.

Primary succession The sequential development of biotic communities where none previously existed.

Primary treatment (of sewage) First step in sewage treatment to remove large solid objects by screens (filters) and sediment and organic matter in settling chambers. See *secondary* and *tertiary treatment*.

Proactive government One that is concerned with long-range problems and lasting solutions. Contrast with *reactive government*.

Producer (autotroph or producer organism) One of the organisms that produces the organic matter cycling through the ecosystem. Producers include plants and photosynthetic algae.

Productivity The rate of conversion of sunlight by plants into chemical-bond energy (covalent bonds in organic molecules). See *gross* and *net primary productivity*.

Profundal zone Deeper lake water, into which sunlight does not penetrate. Below the *limnetic zone*.

Prospective law One designed to address future problems and generate long-lasting solutions. Contrast with *retrospective law*.

Punctuated equilibrium A theory of evolution stating that species are fairly stable for long periods and that new species evolve rapidly over short periods of thousands of years that punctuate the equilibrium. Contrast with *gradualism*.

Pyramid of biomass Graphical representation of the amount of biomass (organic matter) at each trophic level in an ecosystem.

Pyramid of numbers Graphical representation of the number of organisms of different species at each trophic level in an ecosystem.

Quad One quadrillion (10^{15}) BTUs of heat.

Rad (radiation absorbed dose) Measure of the amount of energy deposited in a tissue or some other medium struck by radiation. One rad = 100 ergs of energy deposited in one gram of tissue.

Radioactive waste Any solid or liquid waste material containing radioactivity. Produced by research labs, hospitals, nuclear weapons factories, and fission reactors.

Radioactivity Radiation released from unstable nuclei. See *alpha* and *beta particles* and *gamma rays*.

Radionuclides Radioactive forms (isotopes) of elements.

Rain shadow Arid downwind (leeward) side of mountain range.

Rangeland Grazing land for cattle, sheep, and other domestic livestock.

Range of tolerance Range of physical and chemical factors in which an organism can survive. When the upper or lower limits of this range are exceeded, growth, reproduction, and survival are threatened.

Reactive government A government that lives and acts for today, addressing present-day problems as they arise. Shows little or no concern for long-term issues and solutions. Contrast with *proactive government*.

Reactor core Assemblage of fuel rods and control rods inside a reactor vessel. Bathed by water to help control the rate of fission and absorb the heat.

Real price (or cost) The price of a commodity or service in fixed dollars, that is, the value of a dollar at an earlier time. Helpful way to determine whether a resource has experienced a real increase in cost or whether higher costs are simply due to inflation.

Reclamation As used here, the process of returning land to its prior use. Common usage: to convert deserts and other areas into habitable, productive land.

Recycling A strategy to reduce resource use by returning used or waste materials from the consumption phase to the production phase of the economy.

Reduction factors Abiotic and biotic factors that tend to decrease population growth and help balance populations and ecosystems, offsetting growth factors.

Relative humidity The amount of moisture in a given quantity of air divided by the amount the air could hold at that temperature. Expressed as a percentage.

Rem (roentgen equivalent man) Measure that accounts for the damage done by a given type of radiation. One *rad* = one rem for X rays, gamma rays, and beta particles, but one rad = 10 to 20 rems for alpha particles, because they do more damage.

Renewable resource A resource replaced by natural ecological cycles (water, plants, animals) or natural chemical or physical processes (sunlight, wind).

Replacement-level fertility Number of children a couple must have to replace themselves in the population.

Reproductive age Age during which most women bear their offspring (ages 14–44).

Reproductive isolation Any of many mechanisms that prevent species from interbreeding or producing viable offspring.

Reserve Deposit of energy or minerals that is economically and geologically feasible to remove with current and foreseeable technology.

Residence time Length of time a chemical spends in the environment.

Resilience Ability of an ecosystem to return to normal after a disturbance.

Resource (in general) Anything used by organisms to meet their needs, including air, water, minerals, plants, fuels, and animals.

Resource (as a measurement of a mineral or fuel) Total amount of a mineral or fuel on earth. Generally, only a small fraction can be recovered. Compare with *reserve*.

Restoration ecology Study of restoring ecosystems to their natural state after human interference. Also called conservation biology.

Retorting Process of removing kerogen from oil shale, usually by burning or heating the shale. Can be carried out in surface vessels (surface retorting) or underground in fractured shale (*in situ* retorting).

Retrospective law One that attempts to solve a problem without giving much attention to potential future problems. Contrast with *prospective law*.

Reverse osmosis Means of purifying water for pollution control and desalination. Water is forced through porous membranes; pores allow passage of water molecules but not impurities.

Risk acceptability A measure of how acceptable a hazard is to a population.

Risk assessment The science of determining what hazards a society is exposed to from natural and human causes and the probability and severity of those risks.

Risk probability The likelihood a hazardous event will occur.

Risk severity A measure of the total damage a hazardous event would cause.

Salinization Deposition of salts in irrigated soils, making soil unfit for most crops. Caused by rising water table due to inadequate drainage of irrigated soils.

Saltwater intrusion Movement of saltwater from oceans or saltwater aquifers into freshwater aquifers, caused by depletion of the freshwater aquifers or low precipitation or both.

Sanitary landfill Solid waste disposal site where garbage is dumped and covered daily with a layer of dirt to reduce odors, insects, and rats.

Scrubber Pollution control device that removes particulates and sulfur oxides from smokestacks by passing exhaust gases through a fine spray of water containing lime.

Secondary consumer Second consuming organism in food chain. Belongs to the third trophic level.

Secondary pollutant A chemical pollutant from a natural or anthropogenic source that undergoes chemical change as a result of reacting with another pollutant, sunlight, atmospheric moisture, or some other environmental agent.

Secondary succession The sequential development of biotic communities occurring after the complete or partial destruction of an existing community by natural or anthropogenic forces.

Secondary treatment (of sewage) After *primary treatment*, removal of biodegradable organic matter from sewage using bacteria and other microconsumers in activated sludge or trickle filters. Also removes some of the phosphorus (30%) and nitrate (50%). See also *tertiary treatment*.

Second-generation pesticides Synthetic organic chemicals such as DDT that replaced older pesticides such as sulfur, ground tobacco, and ashes. Generally resistant to bacterial breakdown.

Second-law efficiency Measure of the efficiency of energy use, taking into account the unavoidable loss (described by the second law of thermodynamics) of energy during energy conversions. Calculated by dividing the minimum amount of energy required to perform a task by the actual amount used.

Second law of thermodynamics States that when energy is converted from one form to another, it is degraded; that is, it is converted from a concentrated to a less concentrated form. The amount of useful energy decreases during such conversions.

Secured landfill One lined by clay and synthetic liners in an effort to prevent leakage.

Sediment Soil particles, sand, and other mineral matter eroded from land and carried in surface waters.

Selective advantage An advantage one member of a species has over others by virtue of some adaptation it has acquired.

Selective cutting Restricted removal of trees. Especially useful for mixed hardwood stands. Contrast with *clear-cutting* and *shelter-wood cutting*.

Sewage treatment plant Facility where human solid and liquid wastes from homes, hospitals, and industries are treated, primarily to remove organic matter, nitrates, and phosphates.

Shale oil Thick, heavy oil formed when shale is heated (retorted). Can be refined to produce fuel oil, kerosene, diesel fuel, and other petroleum products and petrochemicals.

Shelterbelts Rows of trees and shrubs planted alongside fields to reduce wind erosion and retain snow to increase soil moisture. May also be used to reduce heat loss from wind and thus conserve energy around homes and farms.

Shelter-wood cutting Three-step process spread out over years: (1) removal of poor-quality trees to improve growth of commercially valuable trees and allow new seedlings to become established, (2) removal of commercially valuable trees once seedlings are established, and (3) cutting remaining mature trees grown from seedlings.

Sigmoidal curve An S-shaped curve.

Simplified ecosystem One with lowered species' diversity, usually as a result of human intervention.

Sinkhole Hole created by sudden collapse of the earth's surface due to groundwater overdraft. A form of subsidence.

Slash-and-burn agriculture Farming practice in which small plots are cleared of vegetation by cutting and burning. Crops are grown until the soil is depleted; then the land is abandoned. This allows the natural vegetation and soil to recover. Common practice of early agricultural societies living in the tropics.

Sludge Solid organic material produced during sewage treatment.

Smelter A factory where ores are melted to separate impurities from the valuable minerals.

Smog Originally referred to a grayish haze (combination of smoke and fog) found in industrial cities. Also pertains to pollution called photochemical smog, found in newer cities. See *industrial smog*.

Social Darwinism The application (or misapplication) of the theory of evolution to social behavior.

Sociocusis Hearing loss from human activities. Contrast with *presbycusis*.

Soft path A term coined by Amory Lovins to describe such practices as conservation, efficient use of energy, and renewable energy systems such as solar and wind. Characterized by high labor intensity, decentralized energy production, and small-scale technology. Contrast with *hard path*.

Soil horizons Layers found in most soils.

Solar collector Device to absorb sunlight and convert it into heat.

Solar energy Energy derived from the sun (heat) and natural phenomena driven by the sun (wind, biomass, running water).

Solar system Group of planets revolving around a star.

Sonic boom A high-energy wake creating an explosive boom that trails after jets traveling faster than the speed of sound.

Spaceship earth Metaphor introduced in the 1960s to foster a greater appreciation of the finite nature of earth's resources and the ecological cycles that replenish oxygen and other important nutrients.

Specialist Organism that has a narrow niche, usually feeding on one or a few food materials and adapted to a particular habitat.

Speciation Formation of new species.

Species A group of plants, animals, or microorganisms that have a high degree of similarity and generally can interbreed only among themselves.

Species diversity Measure of the number of different species in a biological community.

Spontaneous abortion Loss of an embryo or fetus from the uterus not caused by surgery. Generally the result of chromosomal abnormalities. Contrast with *induced abortion*.

Spring overturn Annual cycle in deep lakes in temperate climates in which surface and subsurface waters mix.

SST (supersonic transport) Jet that travels faster than the speed of sound.

Stable runoff Amount of *surface runoff* that can be counted on from year to year.

Star Spherical cloud of hot gas, such as the sun, fueled by *nuclear fusion* reactions in its core.

Statutory law Law enacted by Congress or a state legislature.

Contrast with *common law*.

Steady-state economy Economic system characterized by relatively constant GNP, dedication to essential goods and services, and maximum reliance on recycling, conservation, and use of renewable resources. Also spaceship, or sustainable, economy.

Sterile-male technique Pest control strategy whereby males of the pest species are grown in captivity, sterilized, then released en masse in infested areas at breeding time. Sterile males far exceed normal wild males and mate with normal females, resulting in infertile matings and control of the pest.

Sterilization A highly successful procedure in males and females to prevent pregnancy. In males the ducts (vas deferens) that carry sperm from the testicles are cut and tied (vasectomy); in females the Fallopian tubes, or oviducts, which transport ova from the ovary to the uterus, are cut and tied (tubal ligation). Sterilization is not to be confused with castration in males (complete removal of the gonads).

Stratosphere Outer region of the earth's atmosphere, found outside the troposphere, extending 7 to 25 miles above the earth's surface. Outermost layer of the stratosphere contains the ozone layer.

Streambed aggradation Deposition of sediment in streams or rivers, thereby reducing their water-carrying capacity.

Streambed channelization An ecologically unsound way of reducing flooding by deepening and straightening of streams, accompanied by removal of trees and other vegetation along the banks.

Strip cropping Soil conservation technique in which alternating crop varieties are planted in huge strips across fields to reduce wind and water erosion of soil.

Subsidence Sinking of land caused by collapse of underground mines or depletion of groundwater.

Succession The natural replacement of one biotic community by another. See *primary* and *secondary succession*.

Sulfur dioxide (SO₂) Colorless gas produced during combustion of fossil fuels contaminated with organic and inorganic sulfur compounds. Can be converted into sulfuric acid in the atmosphere.

Sulfur oxides (SOₓ) Sulfur dioxide and sulfur trioxide, common air pollutants arising from combustion of coal, oil, gasoline, and diesel fuel. Also produced by natural sources such as bacterial decay and hot springs. Sulfur dioxide reacts with oxygen to form sulfur trioxide, which may react with water to form sulfuric acid.

Supply and demand theory Economic theory explaining the price of goods and services. The supply of and demand for goods and services are primary price determinants. High demand diminishes supply, creating competition for existing goods and services, thus driving up prices.

Surface mining Any of several mining techniques in which all the dirt and rock overlying a desirable mineral (coal, for example) are first removed, exposing the mineral.

Surface runoff Water flowing in streams and over the ground during rainstorm or snowmelt.

Sustainable development Economic development that meets current needs without compromising ability of future generations to meet their needs. Relies on appropriate technology.

Sustainable ethics (mentality) A mind-set that views humans as a part of nature and earth as a limited supply of resources, which must be carefully managed to prevent irreparable damage. Obligations to future generations require us to exercise restraint to ensure adequate resources and a clean and healthy environment.

Sustainable society One based on sustainable ethics. Lives within the limits imposed by nature. Based on maximum use of renewable resources, recycling, conservation, and population control.

Sustained yield concept Use of renewable natural resources, such as forests and grassland, that will not cause their destruction and will ensure continued use.

Sympatric speciation Formation of new species without geographical isolation. Common in plants.

Synergism The acting of two or more agents (often toxins) together to produce an effect larger than expected based on knowledge of the effect of each alone.

Synfuel See *synthetic fuel*.

Synthetic fertilizer Same as *inorganic fertilizer*.

Synthetic fuel Gaseous or liquid organic fuel derived from coal, oil shale, or tar sands.

Taiga Biome found south of the tundra across North America, Europe, and Asia, characterized by coniferous forests, soil that thaws during the summer months, abundant precipitation, and high species diversity.

Tar sands Also known as oil sands or bituminous sands. Sand impregnated with a viscous, petroleumlike substance, bitumen, which can be driven off by heat, producing a synthetic oil.

Technological fix A purely technological answer to a problem. Also called a technical fix.

Technological optimism Undying faith in technological fixes.

Tectonic plates Huge segments of the earth's crust that often contain entire continents or parts of them and that float on an underlying semiliquid layer.

Temperate deciduous forest Biome located in the eastern United States, Europe, and northeastern China below the taiga. Characterized by deciduous and nondeciduous trees, warm growing season, abundant rainfall, and a rich species diversity.

Temperature inversion Alteration in the normal atmospheric temperature profile so that air temperature increases with altitude rather than decreases.

Temporary threshold shift Momentary dulling of the sense of hearing after exposure to loud sounds. Can lead to *permanent threshold shift*.

Teratogen A chemical or physical agent capable of causing birth defects.

Terracing Construction of small earthen embankments on hilly or mountainous terrain to reduce the velocity of water flowing across the soil and thus reduce soil erosion.

Tertiary treatment (of sewage) Removal of nitrates, phosphates, chlorinated compounds, salts, acids, metals, and toxic organics after *secondary treatment*.

Thermal pollution Heat added to air or water that adversely affects living organisms and may alter climate.

Thermocline Sharp transition between upper, warm waters (epilimnion) and deeper, cold waters (hypolimnion) of a lake. Also called metalimnion.

Thermodynamics The study of energy conversions. See the *first* and *second laws of thermodynamics.*

Third-generation pesticides Newer chemical agents to control pests, such as pheromones and insect hormones.

Threshold level A level of exposure below which no effect is observed or measured.

Throughput approach A method of solving an environmental problem by recycling and reuse. For example, recycling or reusing hazardous wastes reduces their output.

Time preference A measure of the value of an immediate gain in comparison with a long-term gain.

Tolerance level (for pesticides) Level of residue on fruits and vegetables permitted by EPA because it is considered "safe."

Total fertility rate Average number of children that would be born alive to a woman if she were to pass through all her childbearing years conforming to the age-specific fertility rates.

Toxin A chemical, physical, or biological agent that causes disease or some alteration of the normal structure and function of an organism. Impairments may be slight or severe. Onset of effects may be immediate or delayed.

Transpiration Escape of water from plants through pores (stomata) in the leaves.

Tree farms Private forests devoted to maximum timber growth and relying heavily on herbicides, insecticides, and fertilizers.

Tritium (hydrogen-3) Radioactive isotope of hydrogen whose nucleus contains two neutrons and one proton. Can be used in fusion reactors.

Trophic level Describes the position of the organism in the food chain.

Tropical rain forest Lush forests near the equator with high annual rainfall, high average temperature, and notoriously nutrient-poor soil. Possibly the richest ecosystem on earth.

Tundra (alpine) Life zone found on mountaintops. Closely resembles the *arctic tundra* in terms of precipitation, temperature, growing season, plants, and animals. Extraordinarily fragile.

Tundra (arctic) First major life zone or biome south of the North Pole. Vast region on far northern borders of North America, Europe, and Asia. Characterized by lack of trees, low precipitation, and low temperatures.

Ultimate production Total amount of a nonrenewable resource that could ultimately be extracted at a reasonable price.

Ultraviolet (UV) light or radiation Electromagnetic radiation from sun and special lamps. Causes sunburn and mutations in bacteria and other living cells.

Undernourishment A lack of calories in the diet. Contrast with *malnourishment.*

Variation Genetically based differences in behavior, structure, or function in a population.

Waste-to-energy plant Incinerator for rubbish that produces small amounts of electricity from heat given off by combustion.

Water cycle See *hydrological cycle.*

Waterlogging High water table causing saturation of soils due to poor soil drainage and irrigation. Decreases soil oxygen and kills plants.

Watershed Land area drained by a given stream or river.

Water table Top of the zone of saturation.

Watt Unit of power indicating rate at which electrical work is being performed.

Wave power Energy derived from sea waves.

Wet cooling tower Device used for cooling water from power plants. Hot water flows through rising air, which draws off heat. Cool water is then returned to the system.

Wetlands Land areas along fresh water (*inland wetlands*) and salt water (*coastal wetlands*) that are flooded all or part of the time.

Wilderness An area where the biological community is relatively undisturbed by humans. Seen by developers as an untapped supply of resources such as timber and minerals, seen by environmentalists as a haven from hectic urban life, an area for reflection and solitude.

Wilderness area An area established by the US Congress under the Wilderness Act (1964) where timber cutting and use of motorized vehicles are prohibited. Most are located in national forests.

Wind energy Energy captured from the wind to generate electricity or pump water. An indirect form of solar energy.

Wind generators Windmills that produce electrical energy.

Zero population growth A condition in which population is not increasing; the *population growth rate* is zero.

Zone of intolerance Range of environmental conditions that an organism cannot survive in.

Zone of physiological stress Upper and lower limits of range of tolerance where organisms have difficulty surviving.

Photo and Text Credits

ARE; WHERE WE'RE GOING. Washington D.C., Population Reference Bureau.

Tables
5-2 World Population Data Sheet
5-4 1989. Population Reference
5-5 Bureau

Chapter 6 Figures

6-1 Modified and redrawn with
6-2 permission from THE LIMITS
6-3 TO GROWTH: A REPORT FOR THE CLUB OF ROME'S PROJECT ON THE PREDICAMENT OF MANKIND, by Donella H. Meadows, Dennis L. Meadows, Jorgen Randers, William W. Behrens, III. A Potomac Associates book published by Universe Books, N.Y., 1972. Graphics by Potomac Associates.
6-4 AP/Wide World Photos
6-5 UNICEF/Abigail Heyman
6-8 AP/Wide World Photos

Table
6-1 Modified from A. Haupt and T. Kane. (1978). POPULATION HANDBOOK. Washington, D.C.: Population Reference Bureau.

Part III Opener Photograph by Ansel Adams. Courtesy of the Ansel Adams Publishing Rights Trust. All rights reserved.

Chapter 7 Figures

7-1 FAO
7-2 UNICEF/Horst Max Cerni (81)
7-3 USDA photo/Tim McCabe
7-5 UN photo/John Isaac
7-8 Wayne Miller/Magnum Photos, Inc.
7-9 Modified and redrawn with permission from L. R. Brown (1982). "Fuel Farms: Croplands of the Future?" THE FUTURIST 14 (3): 16–28. Published by the World Future Society, 4916 St. Elmo Avenue, Washington, D.C. 20014
7-10 USDA
7-11 USDA—Soil Conservation Service
7-12a United Nations/Bill Graham
7-12b USDA—Soil Conservation Service
7-13a USDA—Soil Conservation Service
7-13b FAO photo/F. Mattioli

Table
7-1 Reprinted with permission from W. Reichert (1982). "Agriculture's Diminishing Diversity." ENVIRONMENT 24 (6):

6–11; 33–38. A publication of the Helen Dwight Reid Educational Foundation.

Chapter Supplement Figures

S7-4 USDA—Soil Conservation Service
S7-5 © Grant Heilman
S7-6 USDA—Soil Conservation Service
S7-7 USDA—Soil Conservation Service
S7-8 USDA—Soil Conservation Service

Chapter 8 Figures

8-4 John Launois/Black Star
8-8 USFWS Photo/Gary R. Falirer
8-9 S. W. Woo, U.C. Davis, courtesy USFWS
8-10 Carl Koford/Photo Researchers, Inc.
8-11 Courtesy of Multimedia Publications, Ltd.
8-12 USDA/Agricultural Research Services
8-13 Courtesy of Tennessee Wildlife Resources Agency
8-14 © Janet Robertson

Table
8-1 Reprinted with permission of the Center for Environmental Education, Washington, D.C. 20001.

Chapter 9 Figures

9-1 © Ric Ergenbright
9-2 Bureau of Land Management (USDI)
9-4 USDA—Soil Conservation Service
9-6a U.S. Forest Service
9-6b Hank Lebo/Jeroboam, Inc.
9-7 From CONSERVATION OF NATURAL RESOURCES. Guy-Harold Smilth, © 1965. Reprinted with permission of John Wiley and Sons, Inc.
9-8a U.S. Forest Service
9-8b U.S. Forest Service
9-8c Steve Botti, National Park Service
9-9 Photograph by Ansel Adams. Courtesy of the Ansel Adams Publishing Rights Trust. All rights reserved.
Case Study 9-1 Rainforest Action Network

Chapter 10 Figures

10-5 Modified and redrawn from A. N. Strahler and A. H. Strahler (1973). ENVIRONMENTAL GEOSCIENCE. Hamilton Pub-

lishing Company, Santa Barbara, California. © 1973 by John Wiley and Sons.
10-6 U.S. Geological Survey
10-7 USDA—Soil Conservation Service
10-10 Doug Lee Photography
10-11 Water Services of America
Case Study 10-1 Woodlands Development Office
Case Study 10-2 Florida State Archives

Chapter 11 Figures

11-8 U.S. Geological Survey 187
Case Study 11-1 Steve McCutcheon/Alaska Pictorial Service

Chapter 12 Figures

12-1 Washington Public Power Supply System
12-4 © Doug Lee
12-11 Courtesy Solar Station
12-13 Emilio A. Mercado/Jeroboam, Inc.
12-14 U.S. Forest Service
12-17 From Kendall and Nadis' ENERGY STRATEGIES: TOWARD A SOLAR FUTURE, © 1980, Union of Concerned Scientists. Reprinted with permission from Ballinger Publishing Company.
12-18 The League of American Wheelmen
Case Study 12-1 TASS from Sovfoto

Tables
12-2 From Kendall and Nadis'
12-4 ENERGY STRATEGIES: TOWARD A SOLAR FUTURE, © 1980, Union of Concerned Scientists. Reprinted with permission from Ballinger Publishing Company.
12-5 Worldwatch Institute

Chapter Supplement Figure

S12-3 Gilles Peress/Magnum Photos, Inc.

Chapter 13 Figure

13-9 NOAA

Part IV Opener Dan Morrill

Chapter 14 Figures

14-1 © Gamma-Liaison (Pablo Bartholomew)

14-2 From "Cancer and Industrial Chemical Production," D. L. Davis and B. H. Magee, SCIENCE, Vol. 206, pp. 1356–1358, 21 December 1979. © 1979 by AAAS. Modified and redrawn with permission of the American Association for the Advancement of Science and the Authors.

14-4
14-5 Modified and redrawn with permission of W. G. Thilly and H. L. Lieber (1980). "Genetic Toxicology," in TOXICOLOGY, Doull, Klaassen, and Amdur, eds. Macmillan: New York.

14-6 Modified and redrawn with permission from R. D. Harbison (1980). "Teratogens," in TOXICOLOGY, Doull, Klaassen, and Amdur, eds. Macmillan: New York.

14-7 C. D. Klaassen and J. Doull (1980). "Evaluation of Safety: Toxicologic Evidence," in TOXICOLOGY, Doull, Klaassen, and Amdur, eds. Macmillan: New York.

14-12 Bruce Larson/Tacoma News Tribune

Chapter 15 Figures

15-3a David C. McElroy/FPG International
15-3b Dave Baird/Tom Stack & Associates
15-8 Courtesy of William W. Kellogg. CLIMATE CHANGE AND SOCIETY. (1980). Westview Press: Boulder, Colorado.

Chapter Supplement Figures

Acid Rain
S15-5 From "Acid Precipitation and Sulfate Deposition in Florida," Brezonik et al., SCIENCE, Vol. 208, 30 May 1980, © by AAAS. Reprinted with permission of the American Association for the Advancement of Science.
S15-7 Ted Spiegel/Black Star
Indoor Pollution
S15-1
S15-2 From T. D. Sterling and E. Sterling (1979). "Carbon Monoxide Levels in Kitchens and Homes with Gas Cookers." J. AMER. POLLUTION CONTROL ASSOC. 29(3): 238–241. Reprinted with permission.

Chapter 16 Figures

16-1a Dave Baird/Tom Stack & Associates
16-1b USDA—Soil Conservation Service
16-7 Courtesy of Groundwater Technology, Inc.
16-9 Hank Lebo/Jeroboam, Inc.
16-10 The Des Moines Register
16-12 Bureau of Reclamation

Chapter 17 Figures

17-1 Modified and redrawn with permission from A. A. Boraiko and F. Ward (1980). "The Pesticide Dilemma." NATIONAL GEOGRAPHIC 157(2), p. 149.
17-3 Fred Ward/Black Star
17-5 Grant Heilman
17-6 AP/Wide World Photos
17-7 USDA
17-8 USDA photo/Larry Lana

Chapter 18 Figures

18-1a Michael A. Leonard
18-1b FPG/Dennie Cody
18-2a Joe Traver/Gamma-Liaison
18-6 Courtesy of the U.S. EPA, MSB-MERL, Edison, N.J.
18-8 Richard Nichols/Gamma-Liaison

Chapter Supplement Figure

S18-3 © Louie Psihoyos/Contact

Part V Opener Bruce Davidson/Magnum Photos

Chapter 19 Figures

19-1 Dennis A. Noonan

Chapter 20 Figures

20-4 FAO photo/J. Van Acker

Chapter 21 Figures

21-3 © Rex Weyler/Greenpeace
21-4 © Scott Reuman/Keystone Science School
21-5 Carol T. Powers/The White House

21-6 Reuters/Bettmann Newsphotos
21-7 © Forrest Anderson/Gamma-Liaison

Color Photo Credits
Front End Rainforest Action Network
Papers

Gallery 1

1 NASA

Gallery 2   Biomes

1 Stock Imagery
2 Tom Till
3 Tom Till
4 G. T. Bernard/Animals Animals
5 Tom Till
6 W. Perry Conway
7 E. R. Dagginger/Animals Animals
8 David Muench
9 W. Perry Conway
10 Tom Till
11 S. Belyavoi/Tass Sovfoto
12 David Muench
13 C. W. Perkins/Animals Animals

Gallery 3   Endangered Species

1 Frans Lanting
2 C. C. Lockwood/Animals Animals
3 Tom McHugh/Photo Researchers, Inc.
4 Michael Fogden/Animals Animals
5 Tom Stack/Tom Stack & Associates
6 Jeff Fotte/Bruce Coleman, Inc.
7 Art Wolfe/Aperture
8 W. Perry Conway
9 Zoological Society of San Diego
10 Zoological Society of San Diego
11 Zoological Society of San Diego
12 Ron Garrison/Zoological Society of San Diego

Gallery 4

1 Gary Randall/Tom Stack & Associates
2 Joel W. Rogers/Aperture
3 Gesig Gerster/Photo Researchers, Inc.
4 Fiona Funquist/Tom Stack & Associates
5 Tom Stack/Tom Stack & Associates
6 New York State Department of Health Division Laboratories
7 Gary Milburn/Tom Stack & Associates
8 Keith-Nels Swenson/Greenpeace
9 Dan Morrill
10 Walt Anderson/Tom Stack & Associates
11 Russ Kinne/Photo Researchers, Inc.
12 Suzi Barnes/Tom Stack & Associates
13 Porterfield-Chickering/Photo Researchers, Inc.
14 Keith-Nels Swenson/Greenpeace

Index

Metalimnion, 380, 383*i*
Metals
  air pollution affecting, 345*t*
  changes in reserves of, 302*i*
  consumption of, 299*t*
Metamorphic rocks, 297
Methyl isocyanate, release of in Bhopal, 6,
    314–315
MHD. *See* Magnetohydrodynamics
Microbes, emergence of, 25–26
Microconsumers, in food chain, 55
Micronutrients, 67
Migration in population growth, 116–117, 122
  net, 116, 122
Mileage, gasoline, improving, 285
Mill tailings, radioactive, 265, 266*i*
Miller, Stanley, 24–25
Mine tailings, pollution caused by, 244
Mineral resources, 296–311
  changes in reserves of, 302*i*
  conservation of, 309–310, 311
  consumption of, 289, 310
  continental movement and, 297–298, 310
  earth's crust and, 296–298, 310
  economic depletion of, 302
  environmental costs of expansion of, 204
  expanding reserves of, 301–306
  future needs for, 300–310
  global interdependence and, 299–300
  outer space and, 302–306
  recycling and, 306, 310–311
  reduction of supplies of, 301–302
  and rising energy costs, 302, 305*i*
  rock cycle and, 298
  from the sea, 302–306, 310
  society and, 298–300
  substitutes for, 306
  United States imports of, 301*i*
Minerals, definition of, 297
Ming, L., 147
Minimum tillage, in erosion control, 164,
    165*i*, 166*i*
Mitchell, George, 440
Mitigation costs, 473
Molecules, 42
Molting hormone, in integrated pest
    management, 414
Molybdenum consumption of, 299*t*
Monocultures, 92, 97
  and loss of genetic diversity, 152
Montaigne, 52
Montreal Protocol, 361–362
Mosquito control, pesticides for, 407
Mountains, air pollution affected by, 342
Müller, Paul, 403
Multilateral development banks, 487–488
Multiple Use-Sustained Yield Act, 196
Multiple-cause-and-effect model, 18–20
Mumford, Lewis, 454
Murphy, Dennis, 349
Mutagens, 317, 331
Mutations
  in evolution, 27–28, 42
  and exposure to toxic substances,
    317–318, 331
    cancer incidence and, 317
Mutualism, 50
Myers, Norman, 181, 182
Myxoma virus, 411–412

Naess, Arne, 461
National Environmental Policy Act, 499,
    508–509

National Institutes of Health, 493*t*
National Oceanic and Atmospheric
    Administration, 493*t*
National Wilderness Preservation System, 205
National Wildlife Refuge System, 231
Natural chemical pesticides, 414–415
Natural gas
  electrical generation costs and, 276*t*
  future supplies of, 248
  impacts of use of, 245*t*
  as nonrenewable energy source, 270
Natural hazards, 327
Natural Resources Council of Maine, 497
Natural rights, 472
  environmental ethics and, 463, 465
Natural selection, in evolution, 27, 42
Natural succession, in water pollution,
    384–385
Nature, roots of attitude toward, 456–458
Nature Conservancy, 497
Neanderthals, evolution of, 33
Negative-feedback loop, 11
Neglect, benign, 52
Negligence, 511
Nelson, Norton, 374
Net economic welfare, 470–471, 489
Net migration, population growth and, 116,
    122
Net primary productivity, 65
Neutralism, 50
Neutrons, 22, 42, 291
NEW. *See* Net economic welfare
Newhall, Nancy, 296
Niche, 52–53, 71
Nickel consumption, 299*t*
  imports as percentage of, 301*i*
NIMBY syndrome, hazardous waste disposal
    and, 436
Nitrates, water pollution caused by, 386,
    397–399
Nitrites, water pollution caused by, 386,
    397–399
Nitrogen cycle, 67–69
Nitrogen fixation, 68
Nitrogen oxides
  health effects of, 344*t*
  sources of, 339*i*, 344*t*
Nonbiodegradable pollutants, 10
Nonpoint sources, for water pollution, 378,
    397
Nonrenewable energy resources, 259–271
  coal, 269–270
  coal gasification and liquefaction, 271,
    287
  natural gas, 270
  nuclear fission, 259–267, 287
  nuclear fusion, 267–269, 287
  oil shale, 270, 287
  synthetic fuels, 270–271, 287
  tar sands, 271, 287
Nonrenewable resources, 4
NPP. *See* Net primary productivity
NRCM. *See* Natural Resources Council of
    Maine
Nuclear fallout, 100
  waste disposal and, 264–265
Nuclear fission
  breeder reactors for, 266–267, 287
  cost of, 265–266
  electrical generation costs and, 276*t*
  as nonrenewable energy source, 259–267
  and proliferation of nuclear weapons, 266
  pros and cons of, 259–262

reactor safety and, 261–264
  social acceptability of, 265–266
  water disposal and, 264–265
Nuclear fusion
  in formation of stars, 23
  as nonrenewable energy source,
    267–269, 287
Nuclear radiation, 100, 101*t*
Nuclear reactors
  breeder reactors, 266–267, 287
    fast liquid metal, 267, 287
  radionuclides from, 294*t*
  safety of, 261–264
Nuclear war
  environmental effects of, 99–102
  predicting effects of, 102
Nuclear Waste Policy Act, 436, 441
Nuclear weapons
  proliferation of, 266
  radionuclides from, 294*t*
  waste from production of, 438–439
Nuclear winter, 101–102
Nucleus, 22, 42, 291
Nuisance suits, 510–511
Numbers, pyramid of, 63–64, 65*i*, 71
Nutrient cycles, 66–69, 70*i*, 71
  carbon, 67, 68*i*
  environmental phase of, 66, 67*i*
  nitrogen, 67–69
  organismic phase of, 66, 67*i*
  phases of, 66, 67*i*
  phosphorus, 69, 70*i*
Nutrient water pollution, 381–385
Nutrients
  chemical, 55
  inorganic plant, water pollution caused by,
    382–384, 397
  organic, water pollution caused by,
    381–382, 384*i*, 397
  plant, alteration of, in integrated pest
    management, 410–411
  soil
    alteration of, in integrated pest
      management, 410–411
    depletion of, 146–147
    preventing, 165–168

O₃. *See* Ozone
O horizon, 164
Occupational Safety and Health Act, 326*t*
Occupational Safety and Health
    Administration, 493*t*
  federal laws regulated by, 326*t*
  responsibilities of, 493*t*
Ocean. *See also* Sea
  mineral resources from, 304–306, 310
  pollution of, 389–392, 449
    by medical wastes, 392
    by oil, 178–179, 390–391, 399
    by plastic, 392, 399
    by sewage sludge, 392
Ocean dumping, for solid waste disposal, 449
Ocean Dumping Ban Act, 392, 449
Oil. *See also* Fossil fuels
  electrical generation costs and, 276*t*
  exploration for in Arctic National Wildlife
    Refuge, 247–248
  future supplies of, 246–248
  impacts of use of, 245*t*
  ocean pollution caused by, 178–179,
    390–391, 399
  proven reserves of, 238*i*, 246
  ultimate production of, 246

Principles of Sustainability

What is a Sustainable Society?

A sustainable society is one in which all human activity takes place and is maintained over time within the limits set by the environment—the capacity of the environment to assimilate waste, provide food, and supply other resources. It is a society that meets its needs without compromising the needs of other species and all future generations. A sustainable society is based on two sets of principles: (a) ethical and (b) operational.

Ethical Principles

1. The world is a limited supply of resources ("there's not always more"), which must be shared with all living things ("and it's not all for us").
2. Humans are a part of nature and subject to its rules. There are no exceptions. We violate the laws of nature at our own risk.
3. Humans must not dominate nature, but learn to cooperate with its forces.

Operating Principles

A sustainable society is built on four pillars: (1) conservation (efficiency), (2) recycling, (3) renewable resources, and (4) population control.

1. Conservation. Conservation means cutting back on unnecessary consumption. It means becoming a more conscientious consumer. Buy only what you need. Buy durable goods. Reuse products whenever possible.

Conservation also means using resources more efficiently. Currently we waste about 50% of the energy we consume. Every day, huge amounts of water and other resources are wasted. Although this wastefulness is often viewed as one of society's greatest faults, it is also one of our greatest opportunities for improvement. By becoming more efficient we can cut waste, reduce environmental damage, and ensure a steady supply of resources for future generations.

2. Recycling. To recycle means to use our materials again and again. Recycling saves energy and reduces all forms of pollution. It conserves resources, helps ensure supplies for future generations, protects wildlife habitats, and creates employment and business opportunities.

Currently, we recycle only about 11% of our municipal garbage. We could, however, recycle and compost as much as 50% to 90%. There's a gold mine in our garbage—glass, aluminum, newspaper, office paper, computer paper, plastic milk jugs, plastic bags, plastic soda bottles, plastic shampoo bottles, and paper bags could all be returned to recycling outlets.

3. Renewable Resources. Renewable resources, like trees or wind energy, are those that are generated by natural processes. We could conceivably acquire 10 times as much energy—each year—from our renewable resources as we could from all of the remaining fossil fuel.

Whenever possible, choose renewable or abundant resources (wood, paper, glass, cotton, or wool) over non-renewable resources (plastics and synthetic cloth).

4. Population Control. In order for humans to achieve a sustainable future, we must find ways to stabilize (and eventually reduce) the size of the human population. Limiting family size to two children and giving support to international organizations that promote family planning in less developed nations are important ways of controlling population growth.